AF355968

# Lithosphere-Atmosphere-Ionosphere Coupling during Earthquake Preparation: Recent Advances and Future Perspectives

# Lithosphere-Atmosphere-Ionosphere Coupling during Earthquake Preparation: Recent Advances and Future Perspectives

Editor

**Masashi Hayakawa**

Basel • Beijing • Wuhan • Barcelona • Belgrade • Novi Sad • Cluj • Manchester

*Editor*
Masashi Hayakawa
Hayakawa Institute of Seismo
Electromagnetics, Co. Ltd.
(Hi-SEM)
Chofu
Japan

*Editorial Office*
MDPI AG
Grosspeteranlage 5
4052 Basel, Switzerland

This is a reprint of articles from the Special Issue published online in the open access journal *Geosciences* (ISSN 2076-3263) (available at: https://www.mdpi.com/journal/geosciences/special_issues/IO16NZ733D).

For citation purposes, cite each article independently as indicated on the article page online and as indicated below:

Lastname, A.A.; Lastname, B.B. Article Title. *Journal Name* **Year**, *Volume Number*, Page Range.

ISBN 978-3-7258-1903-4 (Hbk)
ISBN 978-3-7258-1904-1 (PDF)
doi.org/10.3390/books978-3-7258-1904-1

# Contents

*Article*

# Earthquake Precursors: The Physics, Identification, and Application

Sergey Pulinets [1,*] and Victor Manuel Velasco Herrera [2]

[1] Space Research Institute (IKI), Russian Academy of Sciences, Moscow 117997, Russia
[2] Instituto de Geofísica, Universidad Nacional Autónoma De México, Mexico City 04510, Mexico; vmv@igeofisica.unam.mx
* Correspondence: pulse@cosmos.ru

**Abstract:** The paper presents the author's vision of the problem of earthquake hazards from the physical point of view. The first part is concerned with the processes of precursor's generation. These processes are a part of the complex system of the lithosphere–atmosphere–ionosphere–magnetosphere coupling, which is characteristic of many other natural phenomena, where air ionization, atmospheric thermodynamic instability, and the Global Electric Circuit are involved in the processes of the geosphere's interaction. The second part of the paper is concentrated on the reliable precursor's identification. The specific features helping to identify precursors are separated into two groups: the absolute signatures such as the precursor's locality or equatorial anomaly crests generation in conditions of absence of natural east-directed electric field and the conditional signatures due to the physical uniqueness mechanism of their generation, or necessity of the presence of additional precursors as multiple consequences of air ionization demonstrating the precursor's synergy. The last part of the paper is devoted to the possible practical applications of the described precursors for purposes of the short-term earthquake forecast. A change in the paradigm of the earthquake forecast is proposed. The problem should be placed into the same category as weather forecasting or space weather forecasting.

**Keywords:** earthquake precursors; air ionization; ionospheric precursors of earthquakes; thermal precursors of earthquakes; Global Electric Circuit; geospheres interaction

**Citation:** Pulinets, S.; Herrera, V.M.V. Earthquake Precursors: The Physics, Identification, and Application. *Geosciences* **2024**, *14*, 209. https://doi.org/10.3390/geosciences14080209

Academic Editors: Jesus Martinez-Frias and Enrico Priolo

Received: 11 June 2024
Revised: 5 July 2024
Accepted: 30 July 2024
Published: 5 August 2024

## 1. Introduction

The problem of earthquake hazards still remains in the field of discussion and uncertainty. Several scientific communities are involved in these discussions and the field is far from consensus. One important reason is that the physical mechanism of seismogenesis has not yet been determined to the end in seismology, which leaves room for uncertainty. Thus, priority was given to statistical methods, instead of physical approaches [1]. However, the physical approach to earthquake prediction was practiced in seismology 50 years ago [2]. Unfortunately, harsh criticism of this approach led to the virtual cessation of physical precursor monitoring; instead, priority was given to statistical approaches to earthquake forecasting [3]. At the same time, measurements on artificial satellites, ground-based measurements of electromagnetic emissions and atmosphere parameters regularly demonstrated the appearance of unusual variations in the electron concentration in the ionosphere, the occurrence of electromagnetic radiation in the VLF range, anomalous propagation of radio waves, and variations in atmospheric parameters over the areas of strong earthquakes preparation [4–7]. One of the first systematizations of the ionospheric observations over earthquake-prone areas permitted the formulation of the main morphological features of the ionospheric precursors of earthquakes [8].

The first decade of the XXI century permitted us to gain the critical mass of experimental data collection concerning different types of earthquake precursors: GPS TEC [9], magnetic variations [10], atmospheric electric field variations [11], perturbations in VLF/LF electromagnetic waves propagation [12], ground surface thermal anomalies [13], OLR anomalies [14], ionospheric plasma parameters variations [15], ELF/VLF emissions in the magnetosphere [16], particle precipitations [17], and others.

Major contributions to these efforts were made by the French dedicated satellite DEME-TER (Detection of Electro-Magnetic Emissions Transmitted from Earthquake Regions). It was in orbit for almost 6 years (2004–2010) and dispelled all doubts about the existence of ionospheric precursors of earthquakes and their statistical significance [18].

The second decade was marked by a sharp increase in the number of publications on earthquake precursors but, more importantly, the rising number of publications on the physical models explaining how these precursors are generated [19–21] and the launch of the second dedicated satellite, China Seismo-Ionospheric Satellite (CSES1), in 2018 has continued the global monitoring of the seismo-ionospheric phenomena with the great success [22].

It seems that having the physical understanding of the precursor's generation and instruments of their registration at different levels from the ground surface to the magnetosphere in hand, we can start to use this knowledge in practical applications, including earthquake forecasts. But still, the results of the precursor's monitoring were met with criticism and the reason for this lies in the demonstration of the precursor's statistical reliability and detectability. That is why the recent publications on the different types of precursor registration are accompanied by the Molchan [23] and ROC (Receiver Operation Characteristic) [24] diagrams [25] and the majority of results of measurements of the cases demonstrate the high reliability of detected precursors.

Using the word "detection", we should keep in mind that the atmosphere and ionosphere are highly variable media and that earthquake precursors could be masked by the other kinds of environmental variability such as geomagnetic storms or active solar events. Different algorithms were developed to identify the precursors under the variable background impact [26], including technologies of the precursor's cognitive recognition [27].

Summarizing all that has been said above, one can conclude that we are ready for the real earthquake forecast but, unfortunately, this is very far from reality. In not one country is there a systemic approach to this problem. First of all, we do not have education in the field of earthquake forecasting using the physical precursors. All investigations conducted up to date are the results of the work financed from scientific grants or even performed by enthusiasts. The majority of them came from the Space Weather scientific direction. It means that we have no natural continuity of research; many results are connected with the names of individual scientists who do not have successors.

With regard to the perspective of professional seismologists, it seems that their claims on the impossibility of short-term earthquake forecasts are possibly not connected with science itself but with the fear of responsibility, the level of which is extremely high in comparison with the weather or Space Weather forecast. It seems that the paradigm of the approach to the earthquake forecast should be changed and this problem will be discussed in the last part of the paper.

## 2. Physics of the Earthquake Precursors in the Ionosphere

The possibility of registering from the satellite something that could be connected with earthquake preparation was initially met with hostility in the scientific community. The problem was shrouded in some mystery: how could something from underground reach cosmic space? But here, we will demonstrate that there is no mystery and that the processes involved in the precursor's generation are characteristic of many other well-known phenomena of our environment, which should be considered as an open nonlinear system with dissipation that is described within the framework of a synergetic approach [28]. We will return to the thesis later; for now, let us consider the main physical processes responsible for the precursor's generation.

## 2.1. Air Ionization

Air ionization is one of the main processes in the whole near-Earth environment because, from most distant areas of the magnetosphere up to the lower boundary of the atmosphere, we encounter ions which are the result of ionization. We can divide the sources of ionization of the atmosphere into three groups: solar electromagnetic radiation and solar wind energetic particle fluxes, galactic cosmic rays, and natural Earth radioactivity. Different altitude levels of the atmosphere are subjected to the action of different sources of solar activity and galactic cosmic rays, see Figure 1 [29].

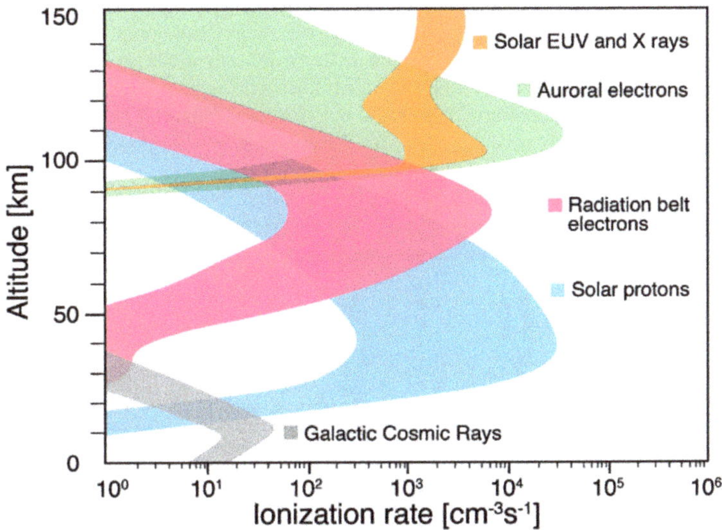


**Figure 1.** Vertical profiles of ionization of the atmosphere by solar electromagnetic radiation, energetic particle fluxes, and galactic cosmic rays [29].

We can see that Galactic Cosmic Rays (GCR) having energy larger than the most energetic solar protons can penetrate up to the ground surface. A more detailed picture of the ionization of the near-ground layer of the atmosphere is presented in Figure 2 [30].

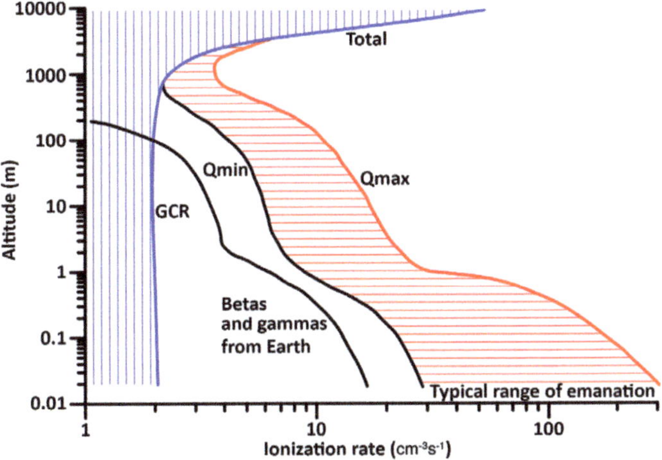


**Figure 2.** Vertical profiles of air ionization from the ground surface up to 10 km altitude. Blue hatching—galactic cosmic rays; red hatching—ground radioactivity (modified from [30]).

Looking at the figure, we can conclude that the main source of air ionization near the ground surface is the natural radioactivity to which the main contribution is provided by radon [31] up to the nearly 2 km altitude where ionization effects from GCR start to prevail. To understand what effect produces the ionization in lower layers of the atmosphere, we should introduce one more that is strongly responsible for the atmosphere–ionosphere coupling: it is the Global Electric Circuit (GEC) [32]. It controls the atmospheric electricity including the potential difference between the ground surface and ionosphere and, consequently, the vertical electric current. The potential difference is created by the convective air movements and global thunderstorm activity and is of the order 250–400 kV, where the ionosphere is a positive plate and the ground surface is a negative one. In fair weather areas, the electric current is directed down, while in the area of thunderstorm activity, it is in the opposite direction. The authors of [33] analyzed the variations in the ionospheric potential (IP) due to large-scale radioactivity enhancement during nuclear tests in the atmosphere. Their model calculations demonstrated that the IP 40% increase can be explained by the 46% decrease in thunderstorm cloud conductivity. These numbers will be necessary in our following discussions. The most simplified model of GEC (but quite enough for our considerations) is presented in Figure 3.

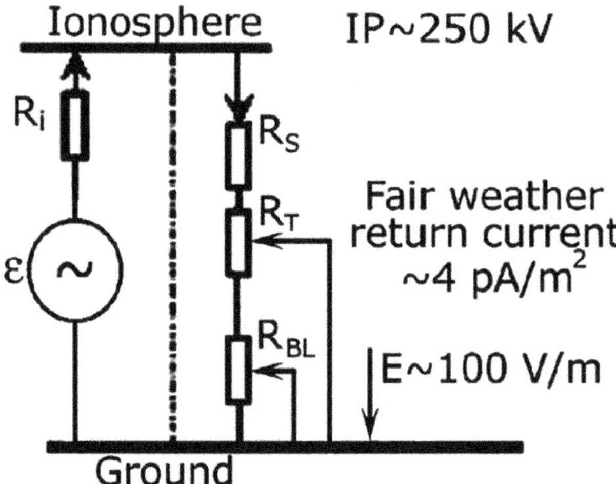


**Figure 3.** Simplified model of the Global Electric Circuit.

Now, looking at this model, we can conclude that under the condition of constant current in the GEC due to current continuity, by changing the boundary layer conductivity/resistance $R_{BL}$ or troposphere $R_T$, we can modulate the IP, which will lead to consequences within the ionosphere in the form of large-scale irregularities formation (negative or positive, depending on the sign of the air conductivity change) [20].

Let us return to the thesis that the mechanism of precursor generation (in our case, pre-seismic variations in the ionosphere) does not differ from other phenomena (natural or anthropogenic) where we deal with processes of air ionization. Following the model calculations for nuclear tests in the atmosphere [33], let us compare such events with the pre-seismic variations in the ionosphere. We do not have experimental data for the nuclear tests but we can use the case of the Chernobyl nuclear power plant explosion in 1986. The authors of [34] observed the anomaly of VLF signal propagation (16 kHz) registered on radio pass Rugby (Great Britain)—Kharkov (Ukraine) during the Chernobyl atomic plant catastrophe. The anomalies were registered both in amplitude and phase of the signal and they were very similar to those that are observed before strong earthquakes when the radio pass of the VLF signal is located over the earthquake preparation zone [35]. Usually, such variations are interpreted as a lowering of the upper boundary of the subionospheric resonator, which is equivalent to the increase in electron concentration in the D-layer [36]. The anomalous variations in the data around the time of the Chernobyl accident were registered even before the reactor explosion, which means that the emanation of radioactive materials started earlier. Both results are supported by the data of vertical ionospheric sounding. The main effect was observed during night-time hours, which corresponds to the formation of an earthquake precursory mask [37]. Figure 4a shows the deviation of critical frequency *foF2* from the 15-day median (color coded) registered by Kiev ionosonde in coordinates Days (horizontal axis)—UTC (vertical axis). Figure 4b shows the ionospheric precursor (so-called precursor mask) obtained by the epoch overlay method for 9 earthquakes of M $\geq$ 5.4 in Italy at a distance of less than 300 km from Rome by the data of the Rome ionosonde. On the horizontal axis, the day's numeration means days in relation to the earthquake day (minus—before the earthquake; plus—after it).

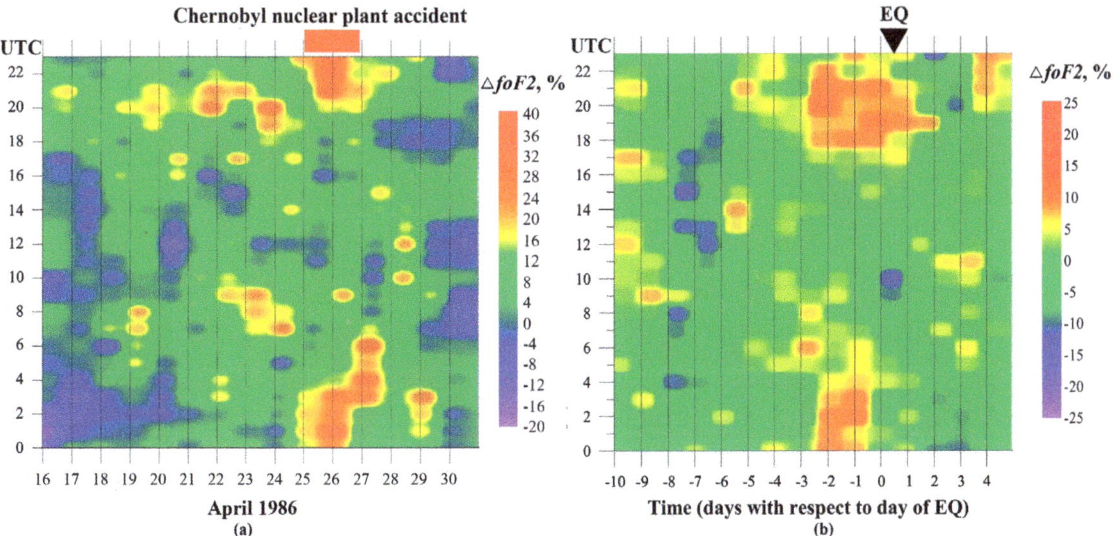


**Figure 4.** (**a**) Deviation of critical frequency *foF2* from 15-days median (color coded) registered by Kiev ionosonde in coordinates Days (horizontal axis)—UTC (vertical axis); (**b**) Deviation of critical frequency *foF2* obtained by epoch overlay method for 9 earthquakes of M $\geq$ 5.4 in Italy at the distance less than 300 km from Rome by the data of the Rome ionosonde.

The similarity of two different effects, subionospheric VLF wave propagation and formation of night-time positive irregularity in the ionosphere, both for nuclear power plant accidents and for earthquakes, suggests that they have a similar or identical physical mechanism to be discussed later. We selected the anthropogenic effect for comparison but the same result can be obtained for natural phenomena such as volcano eruptions or dust storms [38,39]. Figure 5a demonstrates the ionospheric effect created by the eruption of the Kirishima volcano in Japan on 26–27 January 2011. Figure 5b demonstrates the positive variations in the ionosphere on 8 March 2011, 3 days before the M9 Tohoku earthquake.

Again, the similarity of distributions demonstrated in differential TEC maps of Figure 5 is easily noticeable: in both cases, the positive deviation is concentrated at both crests of the equatorial anomaly and shifted south from the source of disturbance marked by a white cross. This effect is explained in [40] and is conditioned by the fact that, due to anisotropy of air conductivity, the vertical electric field at the ground surface is transformed in the electric field perpendicular to the geomagnetic field lines and contains components directed east. This field increases the development of equatorial anomaly and we observe the positive deviation in the differential maps. In the case of a volcano eruption, the increased electric field appears due to a voltage drop in the layer of volcano ash with very low conductivity. In case of an earthquake, the low conductivity is provided by the huge clouds of aerosol-sized cluster ions. The mechanism of their formation will be discussed in the next section.

Nevertheless, in both cases, we deal with the same mechanism of electric field increase due to the low conductivity of the boundary layer of the atmosphere and troposphere.

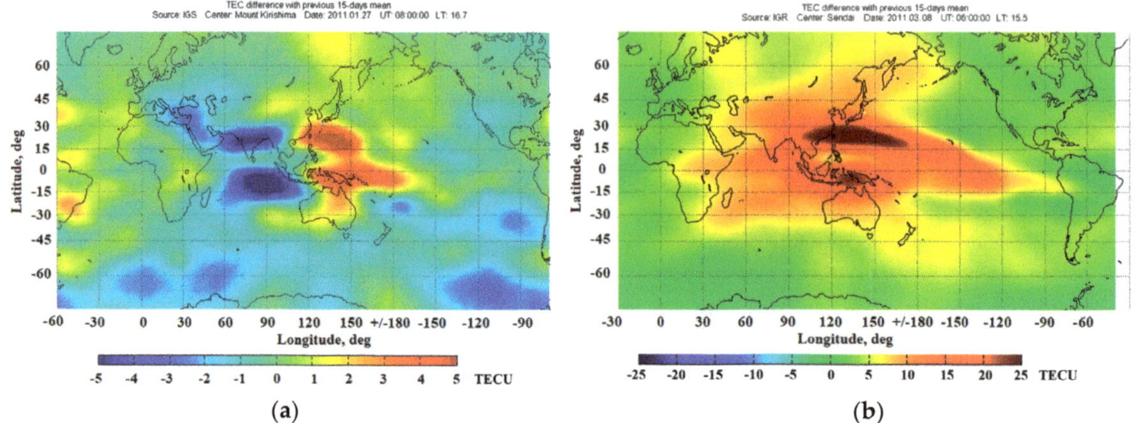

**Figure 5.** (**a**) Differential TEC map on 27 January 2011; (**b**) Differential TEC map on 8 March 2011, 3 days before the M9 Tohoku earthquake. Volcano position and earthquake epicenter are shown by a white cross.

### 2.2. Ion-Induced Nucleation (IIN)

In many publications estimating the effect of pre-earthquake radon emanation on the generation of earthquake precursors, only the first stage of the process is considered: air ionization by $\alpha$-particles emitted by radon during its decay and consequent increase in air conductivity [41]. But the increase in conductivity of the near-ground layer of the atmosphere has a minor effect [42]. It is quite natural if we look at the Figure 3. By decreasing $R_{BL}$, we leave the effect near the ground level. We have enough resistors over the boundary layer and even a strong increase in boundary layer conductivity will be limited by these resistors. In electrical engineering, they are called limiting resistors that prevent shortcuts.

Publications such as [41] do not take into account what happens after the appearance of new ions. This is described in many publications [43,44]. The initial stage of the process called ion hydration is described in [45]. The ions, immediately after their appearance, are subjected to hydration: the process of attachment of water molecules to newborn ions, Figure 6. We can see at the axis below that the process takes only 1 s.

According to [46,47], the process becomes explosive and, for a short period of time, the lite ions become replaced by heavy cluster ions having extremely low mobility; hence, the conductivity of the near-ground layer of the atmosphere dramatically drops. This is confirmed by the drop in conductivity of thunderstorm clouds described in [33]. According to the same publication [33], this process increases the IP leading to the formation of large-scale increases in electron concentration or a TEC value similar to those forming over the clouds of sandstorms or volcanic ash [39].

The total vertical current $I$ can be expressed as

$$I = e(n^{+}\mu^{+} + n^{-}\mu^{-})E = \sigma E \tag{1}$$

where $\sigma$ is air conductivity, $E$ is the electric field, and $\mu$ is the ion's mobility. From the discussion above, we should take into account that we have several types of ions with different sizes and, consequently, mobilities, so the integral conductivity can be expressed according to (1) as

$$\sigma = e\sum_{i=1}^{n}\left(n_{i}^{+}\mu_{i}^{+} + n_{i}^{-}\mu_{i}^{-}\right) \tag{2}$$

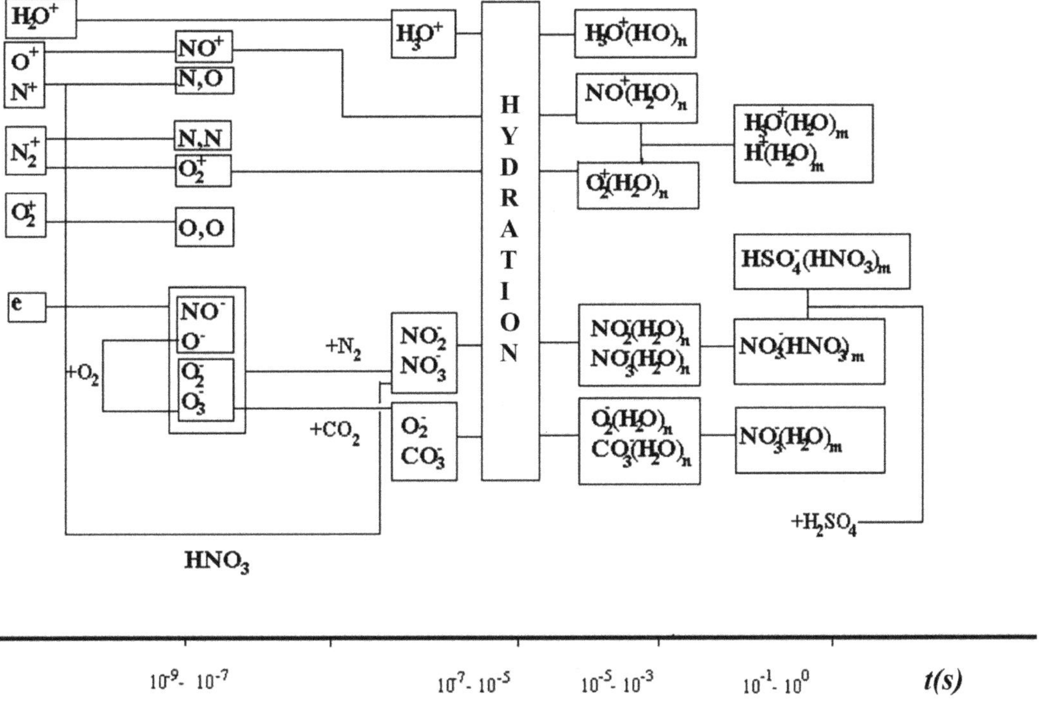


**Figure 6.** Schematic presentation of the near-ground kinetics with the basic ion's formation.

From the Table 1, we can conclude that the ion's mobility is also proportional to the square of the ion's size. Two orders of magnitude ion's size increase leads to the four orders of magnitude of the ion's mobility drop. According to (1), we should expect a drop in air conductivity, which is proportional to mobility. But, with regard to conductivity, the amount it diminishes will depend on the balance of concentration between the light and heavy ions. In any case, an increase in heavy ions concentration in air leads to a drop in air conductivity [47], see Figure 7.

**Table 1.** Variations in an ion's mobility as dependent on the ion's size [47].

| Ana-Lyzer | Fraction | Mobility $cm^2\ V^{-1}\ s^{-1}$ | Diameter nm |
|---|---|---|---|
| | | *Small Cluster Ions* | |
| $IS_1$ | $N_1/P_1$ | 2.51–3.14 | 0.36–0.45 |
| $IS_1$ | $N_2/P_2$ | 2.01–2.51 | 0.45–0.56 |
| $IS_1$ | $N_3/P_3$ | 1.60–2.01 | 0.56–0.70 |
| $IS_1$ | $N_4/P_4$ | 1.28–1.60 | 0.70–0.85 |
| | | *Big Cluster Ions* | |
| $IS_1$ | $N_5/P_5$ | 1.02–1.28 | 0.85–1.03 |
| $IS_1$ | $N_6/P_6$ | 0.79–1.02 | 1.03–1.24 |
| $IS_1$ | $N_7/P_7$ | 0.63–0.79 | 1.24–1.42 |
| $IS_1$ | $N_8/P_8$ | 0.50–0.63 | 1.42–1.60 |

**Table 1.** *Cont.*

| Ana-Lyzer | Fraction | Mobility cm$^2$ V$^{-1}$ s$^{-1}$ | Diameter nm |
|---|---|---|---|
| | | *Intermediate Ions* | |
| IS$_1$ | $N_9/P_9$ | 0.40–0.50 | 1.6–1.8 |
| IS$_1$ | $N_{10}/P_{10}$ | 0.32–0.40 | 1.8–2.0 |
| IS$_1$ | $N_{11}/P_{11}$ | 0.25–0.32 | 2.0–2.3 |
| IS$_2$ | $N_{12}/P_{12}$ | 0.150–0.293 | 2.1–3.2 |
| IS$_2$ | $N_{13}/P_{13}$ | 0.074–0.150 | 3.2–4.8 |
| IS$_2$ | $N_{14}/P_{14}$ | 0.034–0.074 | 4.8–7.4 |
| | | *Light Large Ions* | |
| IS$_2$ | *Ayp*$_{15}$ | 0.016–0.034 | 7.4–11.0 |
| IS$_3$ | *NJPu* | 0.0091–0.0205 | 9.7–14.8 |
| IS$_3$ | M *l*/P 17 | 0.0042–0.0091 | 15–22 |
| | | *Heavy Cluster ions* | |
| IS$_3$ | M8 *IP* 18 | 0.00192–0.00420 | 22–34 |
| IS$_3$ | $N_{l9}$/P 19 | 0.00087–0.00192 | 34–52 |
| IS$_3$ | Mo / P20 | 0.00041–0.00087 | 52–79 |

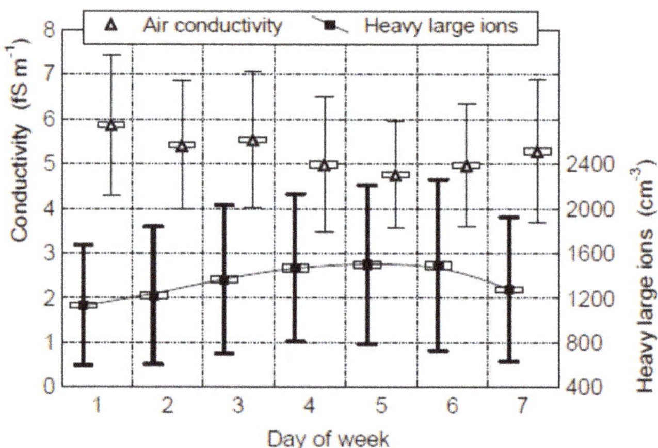


**Figure 7.** Variation in concentration of heavy ions and air conductivity during a week [47].

Let us again carefully look at the Figure 3. Resistors $R_{BL}$, $R_T$, and $R_S$ are connected consecutively and each of them can play a role of the limiting resistor, as determined in electrical engineering. What does this mean? It means that if even two of them are short-circuited, the current will not go to infinity because the third limiting resistor will still exist. But in the opposite case, if the resistivity of one of the resistors increases to infinity, even if both of the others are equal to zero, the total current will be equal to zero. What consequences does it have in the case of changing conductivity in one of the layers of the atmosphere? It means that increases and decreases in conductivity are not equivalent in their consequences for the total current and ionosphere potential. Taking into account the percentage of the total resistance of the troposphere, the drop in conductivity will for sure limit the total current or, more precisely, the effect will be divided by the drop in the total current and increase in the ionosphere potential. Taking into account that during daytime aerosols (or cluster ions) spread all over the troposphere, the effect of conductivity drop will be essential.

In case conductivity increases due to the first stage of radon emanation, the effect will not be so essential because the thin layer of the near-ground matter will be subjected to this effect and yet a large part of the troposphere will remain in the previous low conductivity condition. A significant effect could be reached during the sharp explosive conductivity growth as it happens during the nuclear explosion. Both cases could be validated even without resorting to the earthquake effects.

### 2.3. Additional Source of GEC Modification

The major conception of GEC implies that the air conductivity is much less than the conductivity of the ionosphere and the Earth's crust [32]; that is why only the vertical component of the electric field is registered in the atmosphere. Flowing from the ionosphere to the Earth's ground, the current spreads over the ground surface. But this conception works only when the crust conductivity is uniform. The situation can change if the current encounters ground conductivity irregularities. In this case, we should add the crust resistance to the system of resistors in Figure 3. This is not speculation but a real experimental fact. During the last decade, geoacoustic measurements of the specific resistivity of the Earth's crust in the deep wells (up to 3 km depth) at the Kamchatka peninsula have been provided [48]. In March 2020 when the strong M7.5 earthquake took place in northern Kurils on 25 March, synchronous measurements of the specific resistivity and GPS TEC using Petropavlovsk-Kamchatsky GPS receiver were conducted. Starting from the 16th until the 20th of March, the positive deviation of GPS TEC was registered synchronously with the increase in specific resistivity [49], which is shown in Figure 8.

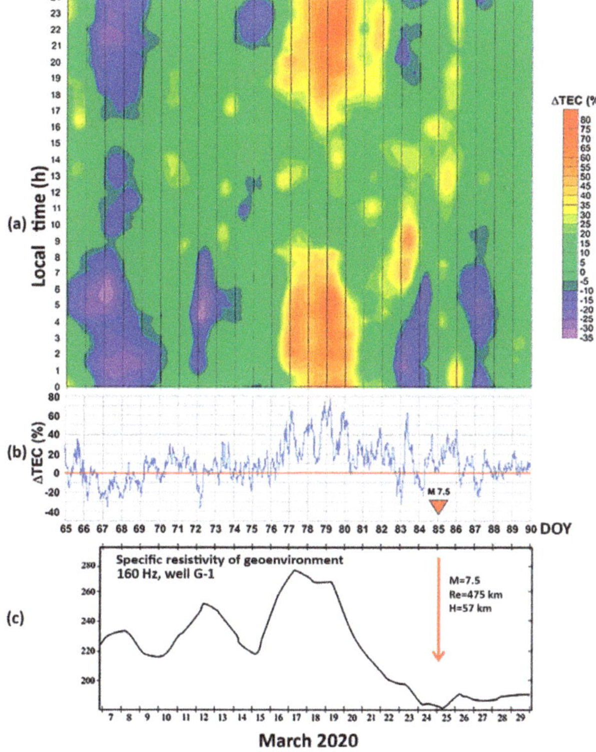


**Figure 8.** (**a**) Percental deviation of GPS TEC at Petropavlovsk-Kamchatsky (color coded) versus local time (vertical axis), DOY (horizontal axis); (**b**) Time series of the DPS TEC percental deviation, a red triangle indicates the moment of the M7.5 earthquake at northern Kurils; (**c**) Time series of specific resistivity measured by geoacoustic emission in the G-1 deep well on frequency 160 Hz at Petropavlovsk-Kamchatsky, a red arrow indicates the moment of the M7.5 earthquake at northern Kurils.

Another effect that is frequently observed in pre-earthquake variations in atmospheric electric fields is its negative bay-like variations [50,51]. A negative direction means that the field vector is directed up, opposite to the natural electric field direction, as is shown in Figure 9. This can be reached in several ways. The simplest option is the formation of the strong electrode layer [45] when the layer of negative ions is forming over the ground surface and the electric field sensor is located under it. The electrode layer forms in any condition even without earthquakes but ionization makes it more intensive and the higher mobility of negative ions in comparison with positive ones facilitates the formation of the negative volumetric charge over the ground surface. Again, speaking of the ionization by natural radioactivity, we find similarity with man-made nuclear test activity. A few minutes after the nuclear test in the atmosphere, the negative bay in the atmospheric electric field was registered at a distance of 7.8 km from the explosion [52], see Figure 10b.

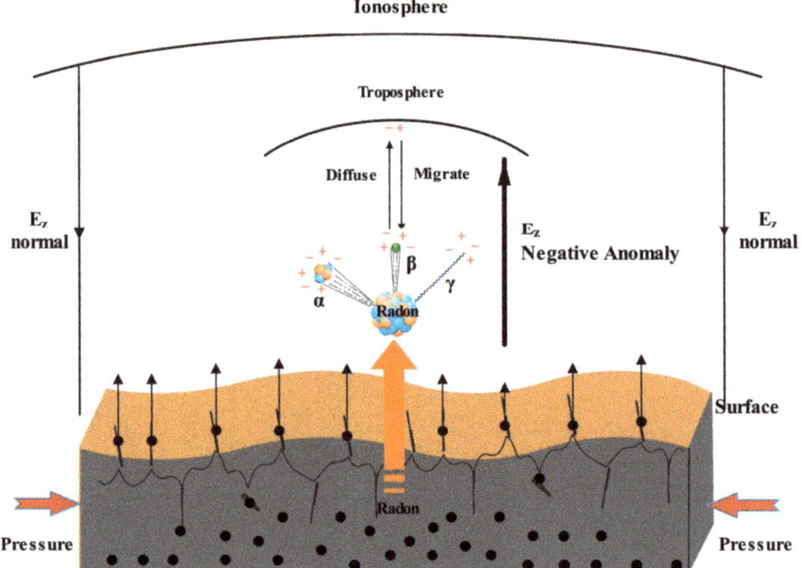


**Figure 9.** Schematic diagram of atmosphere ionization by radons (after [51]).

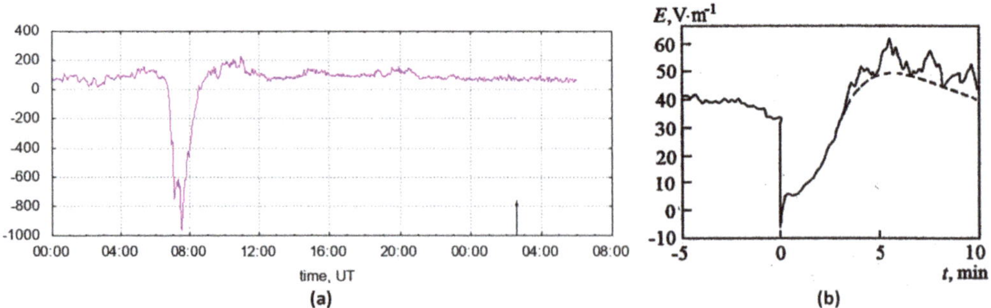


**Figure 10.** (**a**) Negative variation in atmospheric electric field 21 h before the M5 earthquake at Kamchatka peninsula on the 24th September 1997 (vertical black arrow indicates the moment of the earthquake [51]); (**b**) Variation in the vertical electric field around the time of the 20 kT nuclear test in the atmosphere in 1952 [52]. Zero indicates the moment of the explosion. Bold line indicates the electric field variations, dashed line shows the trend of natural value of atmospheric electric field recovering.

The second option is the formation of a positive charge at the ground surface proposed by the peroxy defect theory [53]. It explains the possibility of moving the positive holes to the ground surface, e.g., defect electrons in the $O_2^-$ sublattice, traveling via the O 2p-dominated valence band of the silicate minerals. Regardless of the large number of publications explaining the physics of the phenomena, there are practically no experimental proofs of such effects, not to mention the statistical results.

According to [51], the value of a negative (directed up) electric field may reach 1000 V/m (Figure 10a). The value of the negative electric field after a nuclear explosion was two orders of magnitude lower, only 10 V/m (Figure 10b).

*2.4. Problems of Modeling*

The problem of seismo-ionospheric coupling has a long and entangled history. First of all, it is because the majority of them have no idea of the real source leading to the variations in the ionosphere. The simplest approach is to suppose some electric field on the ground surface [53], external current [54], or surface positive charge or current [55] but nobody knows how these field, external current, or surface charge were generated. So, even if the calculations were made absolutely correctly (what never is correct to the end), we have the model without a real source directly related to the process of earthquake preparation. It looks like a castles in the air without a foundation. For the real modeling, we need the physical process generating the initial parameters used in the models mentioned above. Air ionization by radon is just a real process that is proved by many researchers, which is accepted as an earthquake precursor, and the effect of air ionization by radon is not only in atmospheric electricity but also in atmosphere thermodynamics, which will be demonstrated later.

In our LAIC model, some parts still remain on the conceptual or rush estimates level. The main problem is the correct calculation of ion kinetics in the boundary layer of the atmosphere, taking the boundary layer dynamics in local time into account. Nevertheless, the experimental results for many cases of major earthquakes convinced us that we were on the correct track. The most recent results both in theoretical and experimental proofs, which one can find in [56].

In the electromagnetic branch of our model, we use two conceptual approaches. The first one is the generation of near-ground electric field by ionization processes [45] and penetration of this field in the ionosphere [53,57]. In [57] our calculations the model value of a horizontal electric field in the ionosphere exactly corresponds to those measured experimentally onboard the FORMOSAT 5 satellite [58].

The main problem with Hegai's model is that it does not work at equatorial latitudes. For this case, we use the second approach: modification of the air conductivity and correspondent changes in IP [39,40]. According to [59], the fair-weather current $j$ can be estimated as

$$j = \frac{-\varphi_\infty}{\int_0^H \frac{dz}{\sigma(z)}} = \frac{-\varphi_\infty}{R_g}, \ R_g = \int_0^H \frac{dz}{\sigma(z)}, \tag{3}$$

where $\varphi_\infty$ is the ionospheric potential IP, $\sigma(z)$ is the air conductivity, and $R_g$ is the total column resistance.

If a certain area of fair weather is exposed to aerosol-size cluster ion particles and the area of action of thunderclouds as generators of the electric field is outside the area of earthquake preparation, then, using expression (3) for the ionospheric potential, we can write the expression for the changed value of the ionosphere potential as follows:

$$\varphi_\infty^a = \frac{R_g^a}{R_g} \varphi_\infty, \tag{4}$$

where $R_g^a$ is the global resistance where the areas filled with the cluster ions are taken into account.

But we do not need to calculate the global resistance; we need to estimate the local variation in the IP within the specific area where the cloud of cluster ions is located.

Let us determine the regional resistance over the area S filled by cluster ions as $R_g^a$ and $R_s^a$ and, from the global resistance $R_g$, calculate the regional resistance $R_s$ of the same area S but without heavy ions. We obtain

$$R_s = R_g \frac{S}{4\pi r^2},\tag{5}$$

where $r$ is the radius of the Earth.

Finally, we obtain that the regional IP will be equal to

$$\varphi_s^a = \frac{R_s^a}{R_s}\varphi_\infty,\tag{6}$$

We should keep in mind that the boundary layer and troposphere practically determine the total resistance over the area and this means that the ionospheric potential over the earthquake reparation zone will be proportional to the relation between the troposphere resistance and fair-weather resistance over the same area without heavy ions.

This part of the model still lacks an approach as to how to use the IP estimated at the D-layer altitude to obtain the electron concentration in the F-layer. We are still working on this problem. Nevertheless, the similarity of results for earthquakes and areas polluted by dust storms or volcano ash convinces us that we deal with the same physical mechanism of air conductivity variations in troposphere.

Taking into account the limited volume of the paper, I will not touch upon the modification of magnetospheric tubes and conjugated VLF emissions stimulating particle precipitation. I direct the readers to our previous publications [45,60].

### 2.5. Effects in the E-Region of the Ionosphere

From the point of view of electric field penetration into the ionosphere, the E-layer, which is 150–200 km lower in relation to the ground surface, is more susceptible to pre-earthquake electromagnetic impacts. That is why the effects of the Chernobyl emergency were registered both in the lower ionosphere [36] and in the F-region (Figure 4a) while the Fukushima emergency effects were registered only in E-layer [61], Figure 11. Because the GPS TEC measurements are not sensitive to this region of the ionosphere, the main source of information on the pre-earthquake effects in the E-layer is data of vertical ionospheric sounding. To reveal the Fukushima explosion effects, we used the parameter called the semi-transparence coefficient of the sporadic E-layer, which is shown by blue columns in the Figure 11.

According to [62], the coefficient $\Delta f_b E_s$, reflecting the effect of turbulization of the sporadic $E_s$ layer, can be expressed as

$$\Delta f_b E_s = (f_o E_s - f_b E_s)/F_b E_s\tag{7}$$

where $f_b E_s$ and $f_o E_s$ are the screening and critical frequencies of $E_s$.

In Figure 11, the daily mean values of $\Delta f_b E_s$ variations are demonstrated for the period from 1 February until the end of March 2011. During this period, we see three extremums of $\Delta f_b E_s$ marked by burgundy arrows. The first two coincide with the start of the period of thermal precursors of the Tohoku earthquake arising in the form of OLR emission, while the third one appears after the series of explosions at Fukushima NPP. This is one more confirmation that the observed effects are connected with air ionization from different sources but with the same physical mechanism and not any external currents proposed in the cited models.

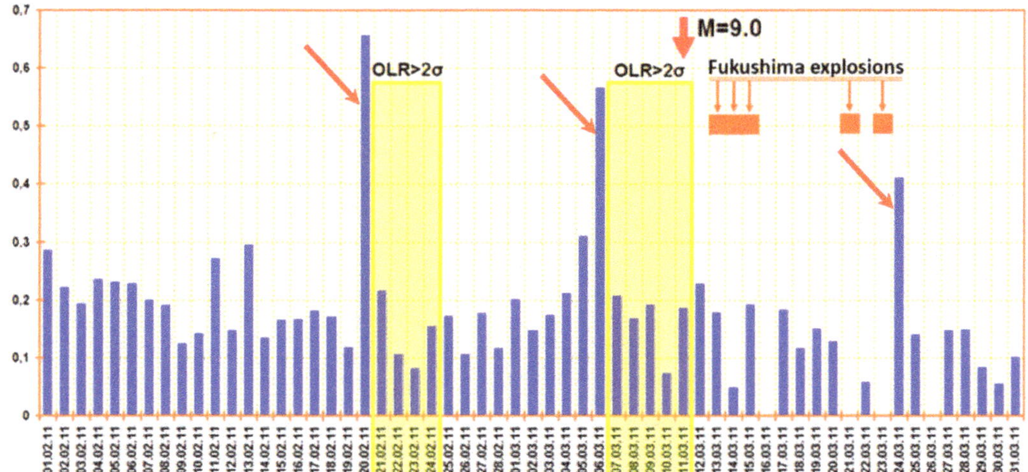



**Figure 11.** Variations in the semi-transparence coefficient of the sporadic *E*-layer $\Delta f_b E_s$ around the time of the Tohoku M9 earthquake on 11 March 2011 and after explosions at Fukushima NPP. The maximal peaks of $\Delta f_b E_s$ are marked by burgundy arrows.

One of the remarkable effects observed before earthquakes was described first in [63]. The authors proposed the physical mechanism of the formation of the layers of metallic aerosols at altitudes slightly higher than a normal altitude of $E_s$ due to the penetration of the seismogenic electric field. One of the recent examples of such an effect is demonstrated in Figure 12 [49] and was registered by the Petropavlovsk-Kamchatsky vertical ionosonde before the M7.5 earthquake on 25 April 2020 at northern Kurils.

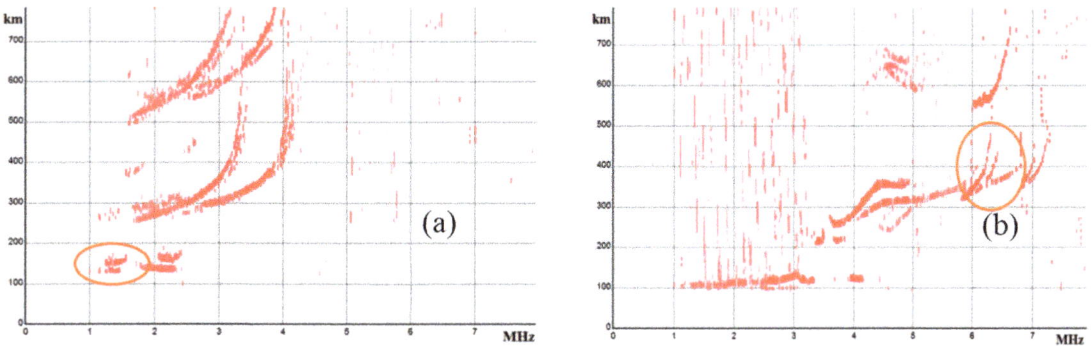



**Figure 12.** (**a**) Formation of elevated $E_s$ at 120 km altitude (orange oval) in comparison with (**b**) $E_s$ normal position at 100 km altitude [49].

The observed effect was confirmed by more extended and comprehensive model calculations [64].

### 2.6. Summary of the Electromagnetic Branch of the LAIC Model

The discussion presented in the previous sections aimed to demonstrate that only the effect of air ionization can provide the foundation to explain all of the complex phenomena observed in all layers of the ionosphere, which are interpreted as short-term earthquake precursors. These effects have the same nature as other natural and anthropogenic phenomena arising under the action of ionization. Different sources of ionization exist and among them, radon is responsible for the generation of the chain of processes leading to the formation of ionospheric precursors. The most important conclusions are the following:

- Ions created by radon ionization and consequent ion's hydration locally modify the GEC parameters, which leads to the generation of atmospheric electric field (positive and negative depending on atmospheric conditions) and air conductivity (increasing or decreasing with time due to ion's hydration);
- These processes change the conditions for the subionospheric propagation of VLF waves mainly by increasing the electron concentration in the $D$-layer of the ionosphere, which is detected by variations in amplitude and phase of VLF signal, mainly during night-time and sunrise-sunset conditions;
- Seismogenic electric field penetrating in the $E$-layer of the ionosphere creates additional sporadic layers of metallic ions at altitude ~120 km and provides turbulization of the ionosphere detected by the semi-transparency coefficient $\Delta f_b E_s$;
- By modification of atmospheric electric field and air conductivity, the large-scale irregularities of electron concentration (positive and negative) are formed in the $F$-layer of the ionosphere through changes in the IP over the earthquake preparation zone;
- Large-scale irregularities modify the geomagnetic tube loaned on this area. The stretched irregularities of electron concentration are formed along the tube where the VLF emissions are scattered and due to the increased level of VLF emission, they more effectively interact through the cyclotron resonance with energetic particles of radiation belts, which causes the stimulated particle precipitation;
- In the present moment, two approaches are used to model the F-layer effects: seismogenic electric field penetration (for high and mi-latitude ionosphere) and IP variations due to air conductivity changes for low latitudes. The model waits for future improvements;
- The model of near-ground ion kinetics still waiting for its development.

## 3. Physics of the Pre-Seismic Thermal Anomalies

In the previous paragraph, we demonstrated how the effects of ionization produce large-scale irregularities in the ionosphere. We proposed radon as the main agent of these irregularities' initiation. Regardless of existing publications neglecting the possibility of radon effects on the ionosphere [41,65], we have the most powerful proof of the importance of radon as the source of the earthquake precursors: it is the thermal effects that are also the consequences of the air ionization together with the ionospheric ones. The high correlation between radon variations and large-scale pre-seismic effects in the atmosphere and the direct cause–effect relationship will be demonstrated in this paragraph.

### 3.1. Air Ionization Latent Heat Release and Atmosphere Reaction

The first description in our LAIC model development of the processes produced by radon ionization and subsequent hydration was described in [45] but thermal effects were not yet clearly emphasized there. A more detailed description was made in [20,56] where the high energy effectiveness was demonstrated and we will take some information from there with some modifications.

The latent heat of water evaporation per molecule (chemical potential and work function of the droplet) can be estimated from the evaporation heat $Q = 40{,}683$ kJ/mol at boiling point (T = 100 °C) U0 = Q/NA = 0.422 eV where NA = 6.022 $10^{23}$/mol. So, during condensation, every molecule changing its phase state from free to bonded in a water droplet emits 0.422 eV of latent heat. Hydration is not equivalent to condensation because water molecules become attached to the ion through electrostatic attraction but still change their phase state and also emit latent heat. The theoretical consideration [56] confirmed by experimental measurements shows that in the case of hydration, the water molecule emits a larger amount of latent heat (later we will demonstrate this difference).

Radon (here we mean the main isotope $Rn^{222}$) decays in the following reaction:

$$_{86}Rn^{222} \rightarrow {}_{84}Po^{218} + {}_{2}He^{4} + 5.49 \text{ MeV} \tag{8}$$

In turn, polonium also emits the $\alpha$-particle with an energy of 5.99 MeV

$$_{84}Po^{218} \rightarrow {}_{82}Pb^{214} + {}_{2}He^{4} + 5.99 \text{ MeV} \tag{9}$$

It means that every atom of $Rn^{222}$ in the result of double decay produces 11.48 MeV, which can be spent for air ionization. Taking into account that the ionization potential of air molecules is between 10 and 30 eV, every radon decay produces $5 \div 6 \times 10^5$ ion–electron pairs.

If taking radon activity, which is usually measured in the seismically active areas of the order of ~2000 Bq/m$^3$, then, in view of the ionization capacity of radon $\alpha$-particles of ~$3 \times 10^5$ electron–ion pairs, the ion-formation rate is ~$6 \times 10^8$ m$^{-3}$ s$^{-1}$. A hydrated ion of a size of around 1–3 $\mu$m (particles of this size range are observed by the AERONET network a few days before the earthquake) contains around $0.4 \times 10^{12}$ water molecules. The latent heat constant $U_0$ is $U_0$ ~$40.68 \times 10^3$ J/mole (1 mole = $6.022 \times 10^{23}$).

For a given radon activity and the formation of hydrated ions of a size of ~1 $\mu$m, we obtain the release of latent s ~16 W/m$^2$, which is consistent with experimentally recorded fluxes of outgoing longwave infrared radiation [14]. Since 1 eV = $1.6 \times 10^{-19}$ J, a given radon activity of 2000 Bq/m$^3$ leads to an expenditure on ionization of $1.7 \times 10^{-9}$ J/m$^3$ s. The ratio of IIN-induced heat to the energy spent on the ionization of atmospheric gases is then $16/(1.7 \times 10^{-9}) \sim 10^{10}$, which is evidence that the energetic efficiency of the process is extremely high.

According to [46], the $H_3O^+(H_2O)_n$ complex ions named hydronium formed as a result of IIN playing the role of catalyzer; the process looks like an autocatalytic reaction producing more and more cluster ions capturing the water molecules from the air. We called this process thermodynamic instability [66] because we deal with the autocatalytic exothermic reaction with the transformation of energy and fast transition of water molecules from one phase state to another, which violates the thermodynamic equilibrium and leads to the formation of large-scale meteorologic anomalies in air temperature, relative humidity, and pressure.

Let us discuss what effects we may expect that accompany the latent heat release.

1. Increase in the air temperature;
2. Drop of relative humidity;
3. Drop of air pressure due to drop of partial pressure of the water vapor.

The last point was not considered in many publications. But according to Dalton's law, the total atmospheric pressure is the sum of the gas constituents' pressures. Here we can consider the sum of dry air pressure and the water vapor pressure. If we are able to register all three cases of air parameters variations, it means that we are on correct track. Figure 13 shows variations in all three parameters variations from 24 February to 20 March 2022 around the time of the M7.3 Fukushima earthquake on 16 March 2022 near the east coast of Honshu Island, Japan.

It should be noted that the negative variations in the air pressure are not too prominent because of the fact that we deal with the drop of partial pressure of water vapor, and the figure reflects variations in the total air pressure. Sometimes the pre-earthquake air pressure drop is so intensive that has a blocking character for regional air movements [67].

Here may arise a question: how can we prove that the observed variations have a relation to the impending earthquake? The answer is the following: all unusual variations in environment parameters participating in the earthquake preparation process are observed within the earthquake preparation zone [68] radius, of which $R$ is determined as

$$R \text{ (km)} = 10^{0.43M} \tag{10}$$

where $M$ is the earthquake magnitude [68].

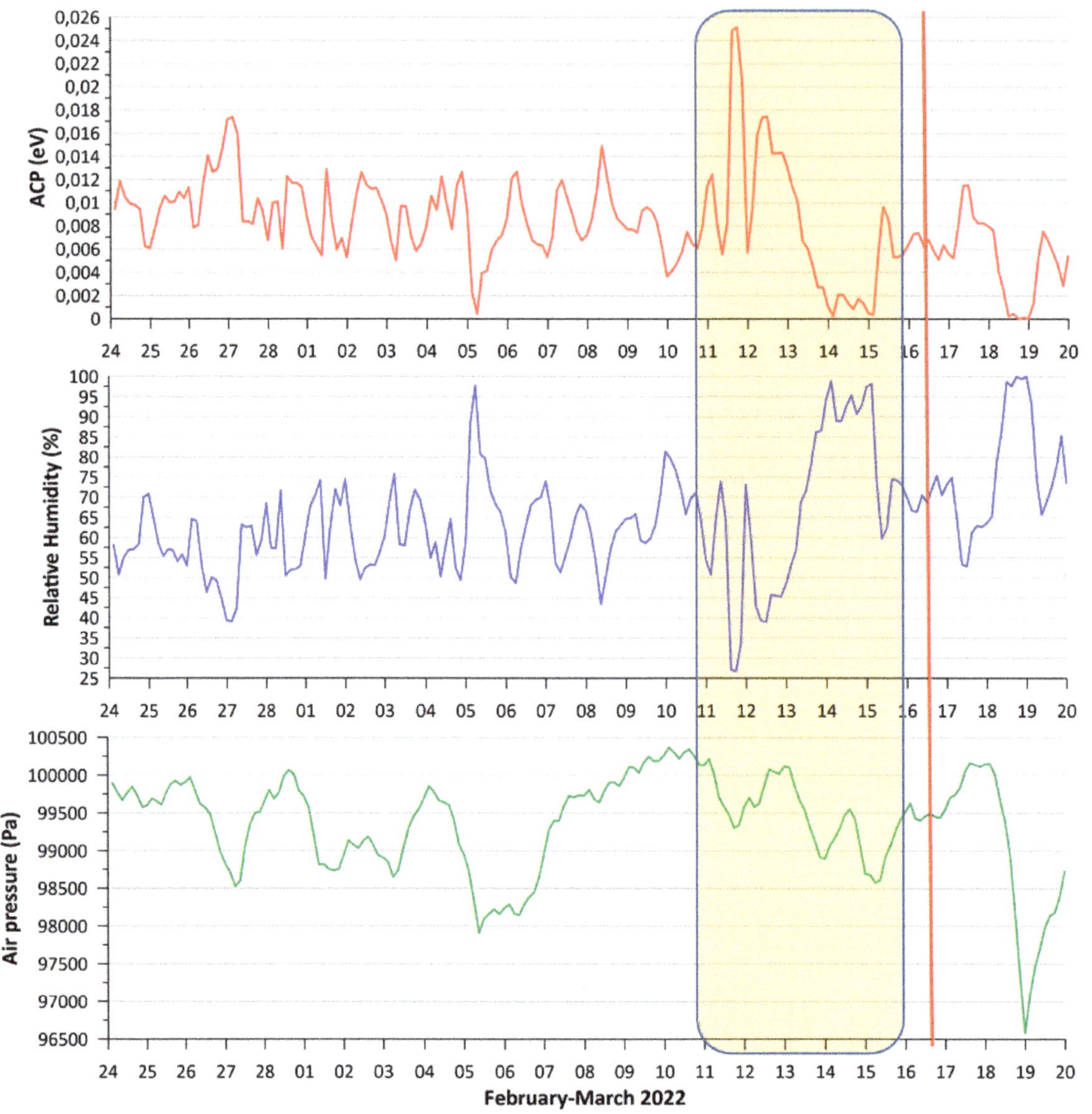

**Figure 13.** From **top** to **bottom**: air temperature, relative humidity, and air pressure over the epicenter of the M7.3 Fukushima earthquake. Coordinates of the epicenter: Lat 38.23° N and Lon 141.77° E. Shadowed rectangle is the precursory period; burgundy line marks the moment of the earthquake.

As proof, we demonstrate in the Figure 14 the map of the air relative humidity (a) and air pressure (b) for measurements taken within the precursory time interval for the same M7.3 earthquake in Japan on 16 March 2022.

The small size of the relative humidity drop area is probably a precursor of the Fukushima earthquake foreshock with a smaller magnitude of M6.0.

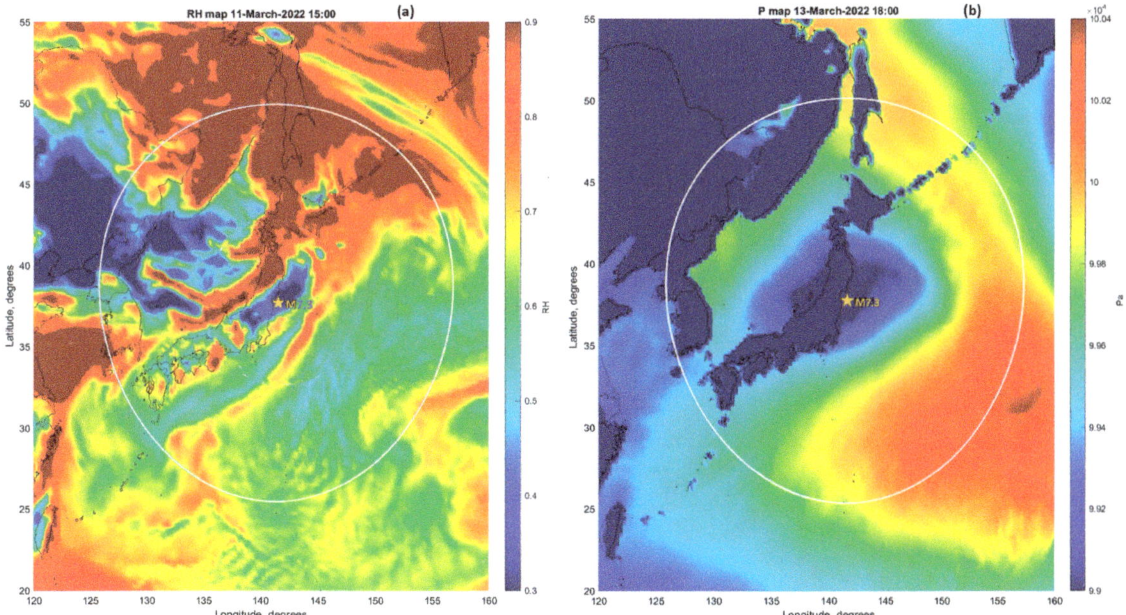



**Figure 14.** (**a**) Relative humidity map; (**b**) Air pressure map.

### 3.2. Atmosphere Chemical Potential (ACP) as an Integrated Diagnostic Parameter

The main advantage of understanding thermal precursors of earthquakes was the introduction of the conception of atmospheric chemical potential [20]. Regardless of the meteorologic parameters that demonstrate their reaction to the process of earthquake preparation, they belong to the system of global atmospheric circulation and weather anomalies. It was necessary to find parameters that, on the one hand, reflected the pre-seismic variations in atmospheric parameters and, on the other hand, were connected with some process in the Earth's crust. The studies of the ion kinetics under action of ionization [69] demonstrated that the bond energy of the water molecule connection with the ion (chemical potential) is larger than the bond energy of the water molecule with the water drop (latent heat constant 0.422 eV). Even more, they released the dependence of the bond energy of connection with ions on the ion production rate and on the concentration of ions [20]. It means that there appeared an opportunity to estimate the intensity of ionization through the chemical potential of the water molecule attached to the ion. One more advantage is that in a moment of condensation/evaporation or attachment/detachment in reaction with an ion, the chemical potential of the water molecule equals the latent heat, which can be expressed through the atmospheric parameters (air temperature and relative humidity). It appeared that the most informative parameter is the difference in the latent heat per water molecule (or chemical potential of a water molecule) between the chemical potential of a water molecule connected with an ion and the chemical potential of connection with a water drop using the possibility to express the chemical potentials through the latent heat [20]. This parameter was called the correction of the water vapor in the atmosphere's chemical potential. To shorten this term, in further publications, we started to use the term Atmospheric Chemical Potential (ACP). To not repeat the full derivation here (it is possible to find it in {20, 56]), we provide the final formula of the ACP as follows:

$$ACP = 5.8 \times 10^{-10}(20Tg + 5463)^2 \ln(100/H) \tag{11}$$

where ACP is in eV, $Tg$ (ground temperature) in °C, and $H$ (relative humidity)—in %.

Several years of exploring ACP in earthquake precursor monitoring demonstrated its high effectiveness and established its main advantages, as follows:

1.  It can be used as a proxy of radon activity [56];
2.  It has a high correlation with the shear traction, i.e., can indicate the level of tectonic activity [66];
3.  Instead of traditional point measurements of radon activity, we can now track its spatial distribution and estimate the size of the earthquake preparation zone.

Continuing the presentation of Fukushima M7.3 earthquake precursor monitoring, in Figure 15a, we present a correlation between the shear traction assimilative model (red curve) and ACP (clue curve). Figure 15b shows the long-term correlation between the shear traction (rose-shadowed contour) and ACP (blue curve).

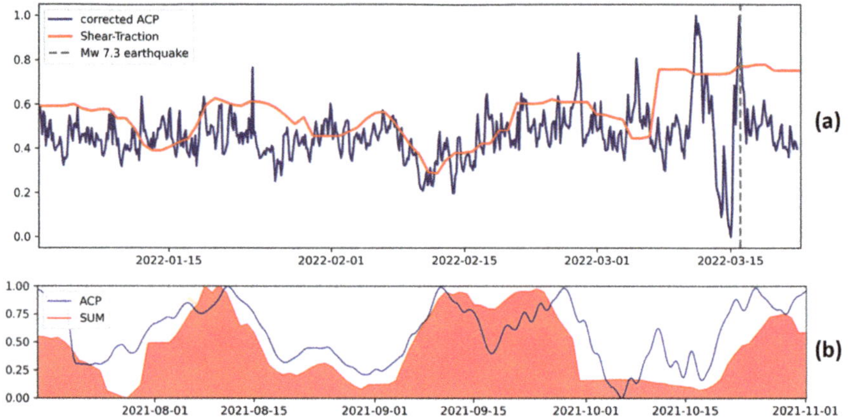


**Figure 15.** (**a**) Average of near- and intermediate-field of ACP (corrected for pressure changes—blue) and shear-traction field (red) in the epicentral area of the 16 March 2022, Fukushima, Japan, earthquake (time shown with grey vertical line). The ACP follows the temporal evolution of the shear-traction field before the earthquake, while the spike in ACP happens close to the increase in shear-traction; (**b**) Long-term correlation between shear-traction field assimilative model in rose and atmospheric chemical potential (ACP) in blue, close to the Andreanof Islands, Alaska, USA.

The ACP spatial distribution before the Fukushima earthquake at 0600 UTC on 11 March is shown in Figure 16. We can see that the increased level of ACP from burgundy to light green color with yellow splashes is inside the Dobrovolsky earthquake preparation zone shown by the white circle. A strong anomaly outside the preparation zone in the north-west is the thermal wave from China, which means that ACP should be calibrated for different geographic areas. The blue arc passing through the epicenter follows the Japan trench passing into the Kuril-Kamchatka trench in the north-east.

### 3.3. Effects of Ionization from Other Sources (Chernobyl NPP and Fukushima NPP Emergencies)

As was mentioned in the Introduction, the effects of ionization are not unique for earthquakes and radon as a source but are the same for any source of ionization. The simplest way to check is again to study thermal effects as results of emergencies at Chernobyl 1986 and Fukushima 2011 emergencies. Figure 17 shows atmospheric parameters measured over the Fukushima NPP in March 2011.

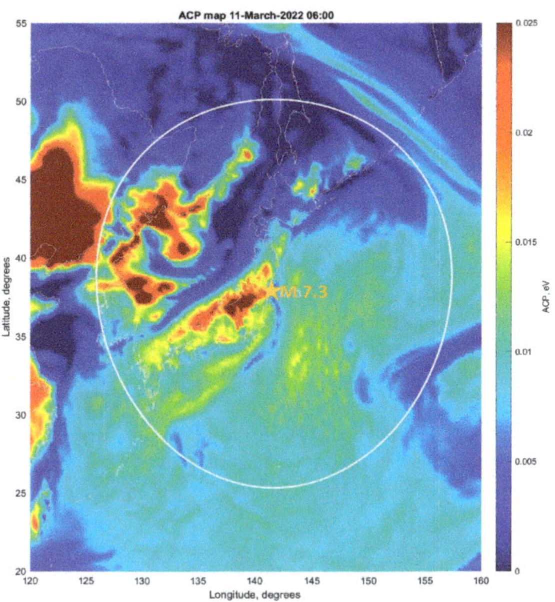



**Figure 16.** Map of ACP spatial distribution within the zone of Fukushima M7.3 earthquake preparation acquired on 11 March 2022, 5 days before the earthquake.

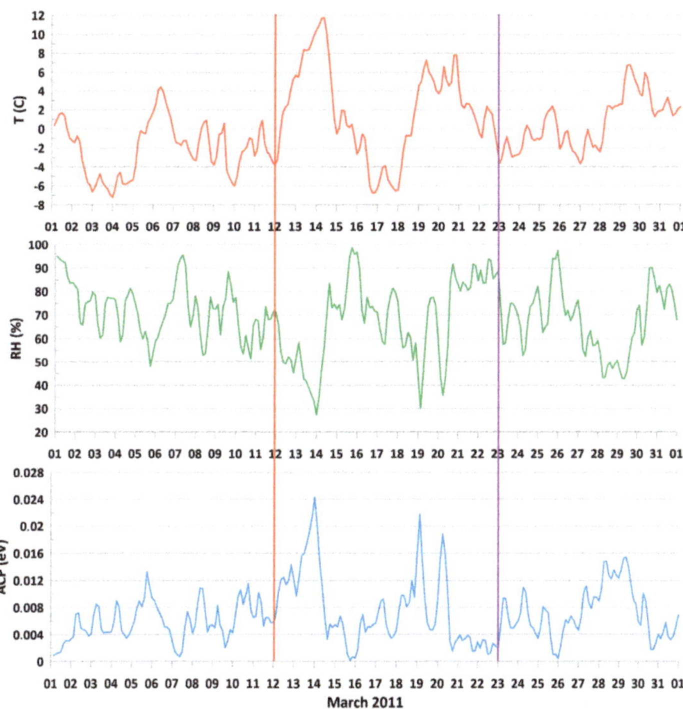

**Figure 17.** From **top** to **bottom**: air temperature, relative humidity, and ACP over Fukushima NPP in March 2011. Vertical lines indicate the period of explosions.

The initial release of radioactive substances into the atmosphere occurred from March 12 to 14 and was caused by the release of pressure from the containment and explosions at blocks No. 1 and 3 [70]. This is clearly seen in Figure 17 for all parameters. We see an increase in the air temperature, a drop in relative humidity, and an increase in ACP. Other variations relate to consequent explosions at different reactors.

The power of ACP was demonstrated in the Chernobyl case and was described in [71]. The French company Institute for Radiation Protection and Nuclear Safety made measurements and modeled the dynamics of propagation of the radioactive $Cs^{137}$ cloud in Europe in the form of a movie [72]. Taking into account the fact that we cannot insert the movie in the publication, we took one frame from this movie and compared it with the ACP distribution calculated for the same day. This comparison is presented in Figure 18. The position of Chernobyl NPP is shown by a yellow star. The high similarity is observed while comparing the distributions. Some details are connected by red arrows to underline this similarity. This figure confirms the power of ACP as an instrument of environment monitoring when, 36 years after the catastrophe, we are able to reconstruct the physical phenomena in detail and dynamics. This case confirms once more that the effects we observe before earthquakes have the same physical mechanism of radiation impact on the atmosphere.

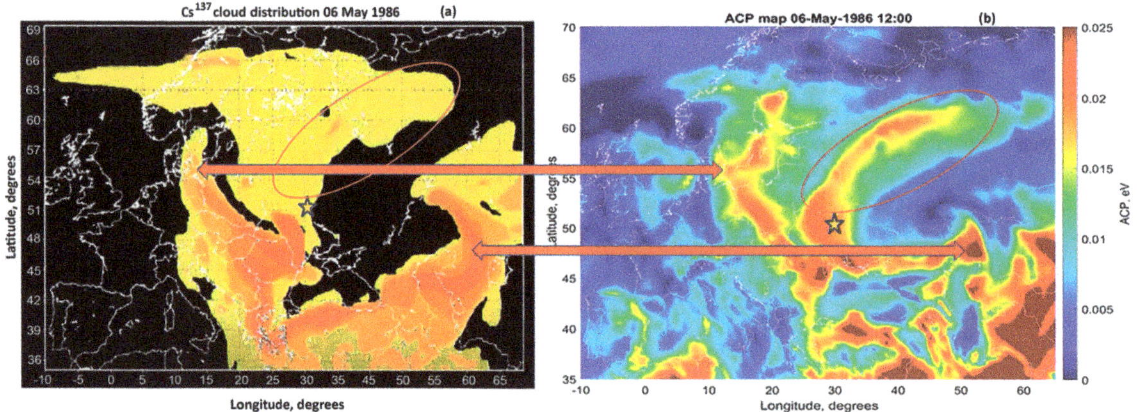


**Figure 18.** (**a**) Spatial distribution of $Cs^{137}$ on 6 May 1986 at 12:00 UTC; (**b**) Spatial distribution of ACP on 6 May 1986 at 12:00 UTC. Asterisks show the position of Chernobyl NPP and arrows and ovals show similarities in distributions.

### 3.4. Summary of the Thermal Branch of the LAIC Model

In paragraph 3, we presented the physical mechanism of the effects of ionization produced by radon emanation in the atmosphere. This is the process of energy extraction from the atmosphere due to launching the exothermic autocatalytic reaction. It should be underlined that regardless of the large individual energy of $\alpha$-particles emitted by radon, in general, it is negligible in comparison with energy, which is released after the launch of this reaction. This is the main point which (if not considered) leads many authors to neglect radon impacts as the main agent of earthquake precursors generation to the wrong conclusion. We do not describe other types of thermal effects before the earthquake here [56] because they are derived from the main source—radon emanation and primary latent heat release.

The LAIC model is a multi-branch composition of many processes/precursors that were described in our previous publications [20,21,56]. The task of this paper is to provide the main physical background for precursor generation initiation. The main consequences of ionization are the electromagnetic effects created by the local modification of GEC parameters through the modulation of near-ground air electric conductivity and generation of additional vertical electric field (sometimes leading to the overturn of the electric field in relation to its natural direction). These modifications induce the formation of large-scale irregularities in the *F*-layer of the ionosphere (positive and negative), the formation of additional sporadic layers in the ionosphere, and the modification of the *D*-layer, leading to anomalies in the propagation of VLF waves within the sub-ionospheric waveguide. The development of these irregularities leads to the modification of the magnetospheric tube in which the ELF/VLF emissions are scattered and the increased level of VLF emissions stimulates the precipitation of energetic particles from the radiation belt.

The thermal branch of the LAIC model describes the formation of large-scale atmospheric irregularities through the release of latent heat extracted from the atmosphere by the IIN process. Due to the large scale of the earthquake preparation zone, we observe not only observe small anomalies over the area of the epicenter but the large meteorological anomalies occupying thousands of square kilometers. The schematic diagram of the LAIC is presented in Figure 19.

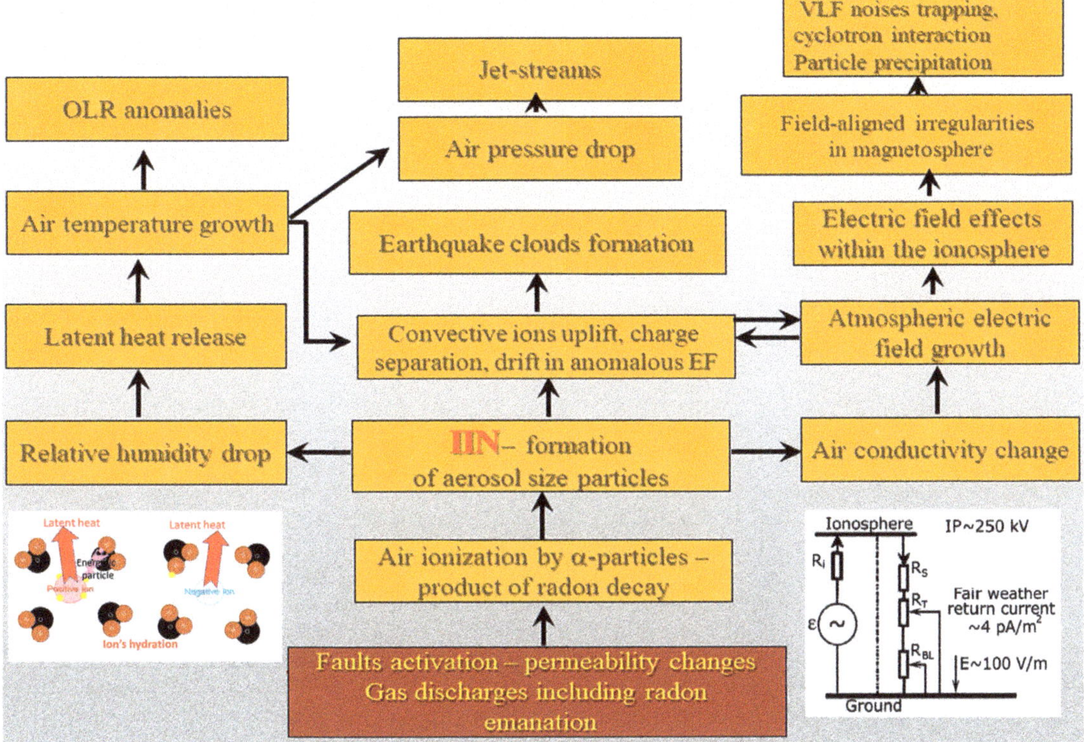


**Figure 19.** Schematic presentation of the LAIC model.

## 4. Precursor's Identification

In this paragraph, we will talk only about the physical precursors. Probably, the first publications determining the physical precursors were [2,73]. All of them were characterized by specific behavior during the seismic cycle and indicated the approaching of the final phase (period of short-term precursors) lasting two weeks/few days before the main event. We can mention the ground conductivity, seismic wave velocity, ground deformation, groundwater level, etc. Radon emanation also is a member of this family. Their main feature was that they were distributed within the area, which was called the "Earthquake preparation zone" by the authors [68]. This distribution has a logarithmic dependence on the magnitude of the impending earthquake (8). The dependence (8) needs explanation. The radius of the earthquake preparation zone indicates the maximal distance at which the precursors of a given earthquake can be registered but it does not mean that the whole zone should be homogeneously filled by precursors. They may appear in any place in this zone.

Radon perfectly meets these conditions. Even more, a special study of radon spatial distribution was conducted [74] and it turned out that it perfectly fits the Dobrovolsky magnitude–space relationship (Figure 20).

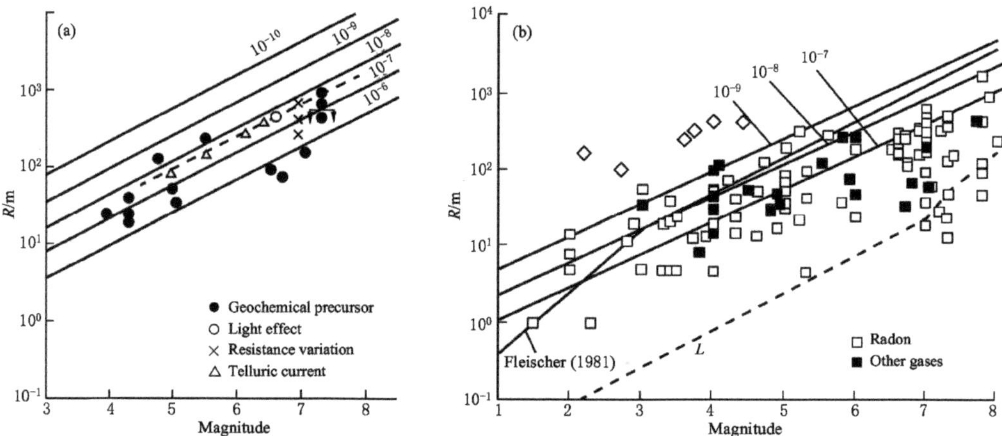


**Figure 20.** (**a**) Determination of the size of the earthquake preparation zone in relation to magnitude according to [71]. (**b**) Radon and geochemical precursors of earthquake distribution versus magnitude according to [74]. The single bold line characterizes the empirical relationship of [75] who calibrated the maximum distance of a radon anomaly for a given magnitude on the basis of a shear dislocation of an earthquake. Dashed line *L* characterizes the typical rupture length of active faults as a function of magnitude by using the empirical law of [76]. Modified from [45].

### 4.1. Absolute Precursory Signatures

Here, we should determine the precursor's features, which are absolutely necessary to be a precursor. From the introduction text of this paragraph, we already mentioned two of them: the locality of precursors (within the earthquake preparation zone) and the characteristic behavior within the seismic cycle.

One more signature follows from the physical nature of the precursor itself. It was described in [61]. For example, the equatorial anomaly in the ionosphere is formed due to the appearance of an east-directed electric field at the geomagnetic equator. This field appears in the afternoon hours of the local time and it follows that, for example, it cannot be formed during the downing hours of the local time. But before the M8.3 Illapel earthquake in Chile on 16 September 2015, the two-hump structure of the equatorial anomaly was detected by the Swarm-B satellite over the earthquake preparation zone two days before the earthquake. The same feature was regularly registered by the French DEMETER satellite at 10 LT, while the satellite passed over the preparation zone of the Wenchuan (China) M7.9 earthquake on 12 May 2008 earthquake [77].

Another property that may be classified in this category is the reaching of some environmental parameters to a magnitude that cannot be reached in natural conditions. Such an effect was registered around the time of the M6.3 earthquake close to the Chilean coast in the open ocean on 1 April 2023. The relative humidity, the drop of which is characteristic of a precursory phenomenon, reached a value 2% of what is an absolute impossible value over the open ocean.

The ionospheric precursor mask of the night-time increase in electron density in the *F*-layer a few days before the earthquake, demonstrated in Figures 4b and 8a, is also a signature of earthquake preparation. Just this behavior of the ionosphere before the Hector Mine M7.1 earthquake in California on 16 October 1999 permitted the revelation of the ionospheric precursor in a highly disturbed ionosphere and the distinguishment of this variation from variations during the geomagnetic storm that happened a few days later [78].

The formation of an additional sporadic E-layer at an altitude of 120 km [63] is also a unique event characteristic of pre-earthquake processes.

In conclusion, we can say that absolute precursors demonstrate themselves by the unique physical characteristics not observed without earthquakes.

### 4.2. Precursors' Identification

In the first decade of the XXI century when interest in ionospheric precursors grew sharply, the main technique of the precursor's identification was statistical analysis using the interquartile range [79]. This technology has been used up to now by many researchers [80]. In addition, many other statistical technologies were proposed. As interquartile analysis, these technologies are (regardless, some of them are very sophisticated) still on the floor of usual statistics not considering the specific features of parameters connected with the earthquake preparation process and their physical nature. Nevertheless, technologies providing absolute confidence in precursor identification appeared just using the main feature of the precursor—its locality [81], Figure 21.

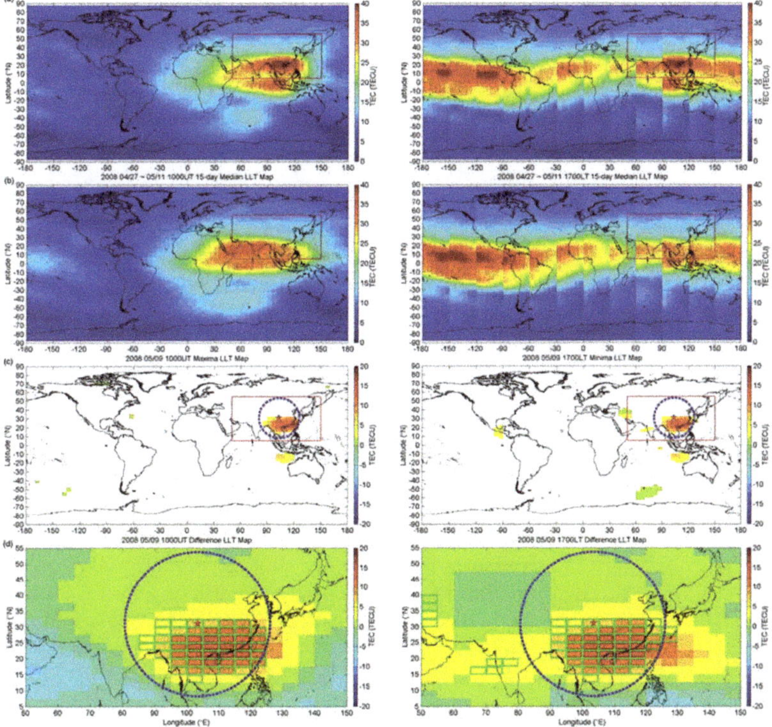

**Figure 21.** (**a**) Left panel—UT GIM map for 1000 UT on 9 May 2009, 3 days before the Wenchuan M7.9 earthquake in China; **right** panel—combined LT map for 1700 LT corresponding to 1000 UT in China; (**b**) **Left** panel—15-day median GIM map for 1000 UT; right panel—combined LT map for 1700 LT corresponding to 1000 UT in China; (**c**) Left panel and right panel—differences between the (**a**,**b**) corresponding maps; (**d**) Left panel and right panel—increased images outlined by red rectangles in the corresponding images (**c**). Blue circle—earthquake preparation zone. At all panels red star indicates the position of the Wenchuan earthquake epicenter.

This is a very powerful and important result. First of all, this is because it undoubtedly testifies that the detected positive large-scale formation is a precursor of the Wenchuan earthquake. It demonstrates that this is a long-lived formation. This conclusion is based on the fact that the LT map has the same shape and intensity as the UT map. For the construction of an LT map, it is necessary to take several GIM maps and select the longitudinal sector corresponding to 1700 LT from them (that is why at the right panels we see the vertical strips corresponding to different GIM maps taken every 2 h). The Wenchuan precursor occupies three strips, which means that it lasts at least 6 h. Moreover, this is because the UT map occupies two time zones by 15 degrees in longitude. It confirms the possibility of constructing the maps by interpolation of data from consecutive orbits of low-orbiting satellites as we conducted topside sounding monitoring of ionospheric precursors [45]. The result perfectly shows how the Dobrovolsky earthquake preparation zone works to estimate the earthquake magnitude. And finally, it perfectly confirms the proposed physical mechanism of ionospheric precursor generation on low latitudes proposed in [40]. One can see the excitation of the equatorial anomaly: the ionospheric irregularity is located south-east from the epicenter and occupies the northern crest of the equatorial anomaly; simultaneously, the south crest of it is seen as the yellow spot at the crest latitude on the southern hemisphere. Actually, the change in the sign of electron concentration deviation was also registered in [81]: the maps are similar as in Figure 21 but 2 days earlier a negative effect was also registered and this overturn is also described in [40].

In [81], one more powerful technology is proposed based on ionospheric tomography showing the formation of additional layers at different altitudes of the *F*-layer. But, it needs to have a dense network of GPS receivers, which limits its application to geographic areas with such networks.

Because of growing interest in earthquake precursor studies and the corresponding volume of available information, it is possible to bring more and more recipes for precursor detection and identification. But we should remember that we deal with the nonlinear system with multivariant behavior and that obtained information is always probabilistic. Here, it can help the synergetic approach describing the system behavior in critical situations [28]. In the case of the final stage of the earthquake cycle, it manifests itself as a concentration of different types of precursors in time and space, indicating the location and time of the future seismic event [56]. That is why the only way to "catch" this process is by providing the multiparameter measurements in real-time [25,82]. This approach becomes more and more popular but it leads to an erroneous approach: registering everything you can get your hands on regardless of the physical mechanism and synergetic coupling between the precursors. We consider that only precursors with a known physical mechanism and their correspondence to the LAIC model can give a positive result; otherwise, it can lead to chaos and mistakes.

## 5. Practical Applications and Problems of the Earthquake Forecast

For many years now, the short-term earthquake forecast has been a painful and pressing problem. We will not start a discussion about this again; it is enough to look through the introductions to our monographs [45,56,61]. We will start from what is necessary to do.

Taking into account that our approach is based on the use of physical precursors of earthquakes, except for fundamental knowledge of seismology, we should understand the physical mechanism of precursor generation, which was the subject of previous paragraphs. Except with regard to physics, usually, the questions are raised on the precursors' statistical reliability and credibility. For this purpose, the error diagram [23] or ROC criterion [24] are used, as was already mentioned in the Introduction. Frankly speaking, for every precursor, such a study should be performed only once to have proof of its statistical credibility because, for the real forecast, they are absolutely not necessary.

Real forecasting requires giving the answers here and now in real-time on the earthquake probability and its three parameters: time, location, and magnitude. For operative monitoring, we need to automate the process of the precursors' registration, processing, and purification in case other factors interfere with the precursor information.

### 5.1. Determination of the Earthquake Parameters

Precursors described in previous paragraphs give the possibility of determining all three parameters of earthquakes with reasonable accuracy. We prepared the synthetic figure demonstrating different possibilities to determine the earthquake location and magnitude using different types of precursors and the Dobrovolsky relationship (Figure 22).

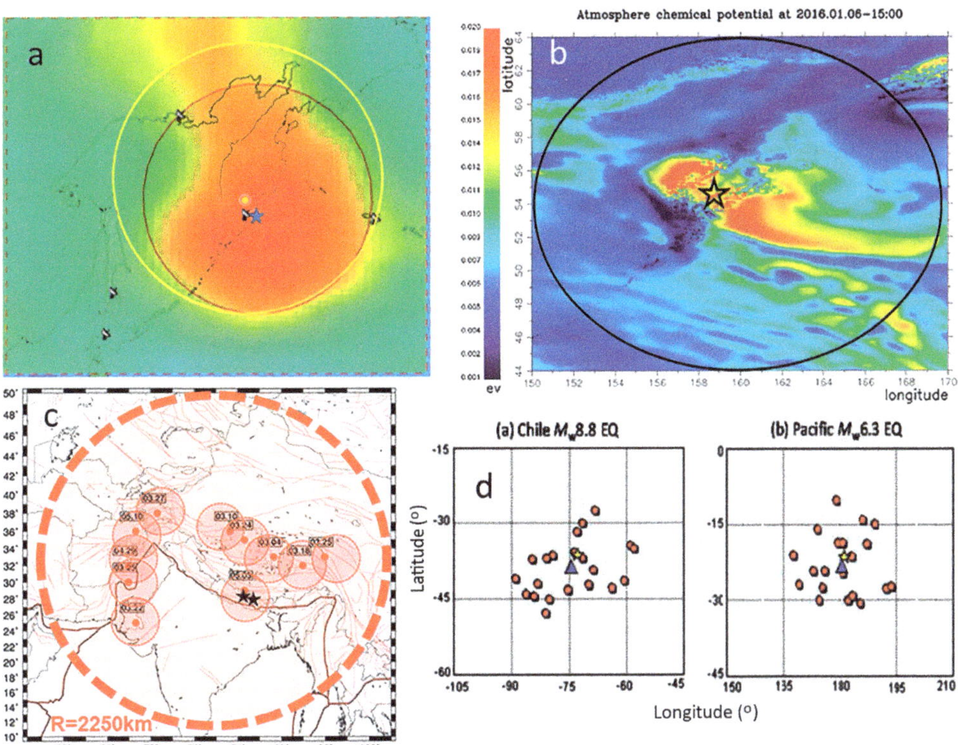


**Figure 22.** (**a**) Differential GIM TEC map registered 4 days before the Kamchatka M6.9 earthquake on 3 March 2023. Epicenter position is marked by a yellow spot, software detected epicenter position is marked by blue star; (**b**) ACP spatial distribution for Kamchatka M7.2 earthquake registered a few weeks before the main shock on 30 January 2016, Dobrovolsky zone is marked by black oval; (**c**) The series of OLR thermal spots positions (small circles) registered before the Nepal M7.8 earthquakes in 2015; black stars indicate positions of the Nepal M7.8 earthquakes on 25 April and 17 May 2015 [83], Dobrovolsky zone is marked by red circle; (**d**) Distribution of electron concentration large deviations along the DEMETER satellite orbits before the Chile M8.8 earthquake on 27 February 2010 and the Chile M6.3 earthquake on 19 November 2007; red circles indicate the ionospheric anomalies detected by DEMETER satellite while passing over the earthquake preparation zone, blue triangles indicate the position of epicenter determined automatically by the data processing software, yellow stars indicate the real epicenter position [84].

Figure 22 presents precursory spatial distributions acquired by different technologies and the different precursors. Figure 22a demonstrates the result of semi-automatic processing of differential GIM maps at the Kamchatka region. The software automatically detected the epicenter (blue star) and outlined the earthquake preparation zone (burgundy circle) and the real parameters of the earthquake are shown in yellow color. Figure 22b shows the results for the same Kamchatka region but with the use of the ACP parameter. The more complex picture is presented in the Figure 22c [83]. It shows the movements of OLR precursor spots with a time of approaching Nepal 2015 M7.8 earthquakes. The earthquake epicenters are marked by black asterisks. It can be associated with the movement of a strange attractor while approaching the critical point. As was mentioned in the discussion of Figure 21, the lifetime of ionospheric precursors is long enough to be registered by several consecutive passes of the low-orbiting satellite. It gives the opportunity to outline the area of earthquake preparation from several passes as it was conducted by the DEMETER satellite [84]. More detailed analysis of DEMETER data [85] confirmed our result that that the maximum affected area in the ionosphere does not coincide with the vertical projection of the epicenter of the impending earthquake and is shifted toward the equator in low and middle latitudes as one can see in Figure 21.

Several factors reduce the accuracy of the epicenter location determination:

4. The large size of the earthquake preparation zone and the fact that the precursor may appear at any point in this zone. If taking into account that for the M7 earthquake the Dobrovolsky radius is 1000 km, the error can be quite significant;
5. The equatorward shift of ionospheric precursors for the low and middle-low latitude events where the geomagnetic field lines inclination should be considered;
6. The low spatial resolution of satellite techniques of monitoring. It is a very rare occasion that the low-orbiting satellite passes exactly over the earthquake epicenter. For OLR measurements, the anomaly location is determined within the circle of 2.5°, which is too much for the desired accuracy of 50 km.

Here, the fusion of physical precursors with seismology may help to monitor the areas with the Gutenberg–Richter relationship b-value drop [21].

As concerns the time of the earthquake, this information can be extracted from data of statistical processing. But there are two different types of statistics: the first one is for fixed sites, for example, Taiwan [9], and the second one is global, which is possible only with the help of satellite measurements [84]. Of course, if taking into account the earthquake magnitude, the distance of the GPS receiver to the epicenter, or the longitudinal distance of the satellite orbit to the epicenter, the distributions of precursors become more complex and multidimensional [85]. Nevertheless, as a result of the more comprehensive analysis of the DEMETER satellite data, the statistically most probable leading time of ionospheric precursors for earthquakes M > 5 was determined as 5 days (Figure 23a) [86], which was established as early as 1998 [45]. The first years of Langmuir probe statistical data processing at the China Seismo-Electromagnetic Satellite CSES 1 launched in 2018 gave the same 5 days of leading time, which [87] makes us more than confident that this is the most probable value for the earthquake time forecast using the ionospheric precursors (Figure 23b).

It should be noted that the leading time of different types of precursors is different, which can be used for the purpose of multiparameter monitoring. For example, the leading time of thermal precursors, including ACP, is larger than ionospheric ones [14].

It is quite natural to put the following question forward: how accurate are the earthquake parameters obtained using the proposed technologies and how large is the percentage of false alarms? These questions, up to now, have no definite answer. It is necessary to organize the real-time continuous monitoring of precursors described above, which was never performed up to now. The only exception are the data of DEMETER and CSES 1 satellites but they also have a limitation: it was not real-time but post-event analysis. In addition, we should underline that both DEMETER and CSES-1 statistics are limited by the local time of the solar synchronized orbits of both satellites: 10 AM and 10 PM for DEMETER and 2 AM and 2 PM for CSES-1. What happened at other local times remains the secret. What we need to undertake to correct this situation will be discussed in the last paragraph.

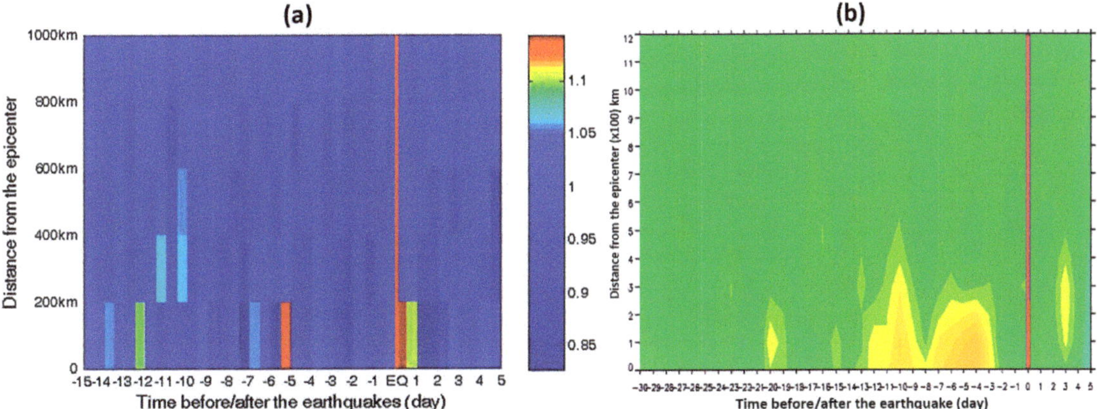

**Figure 23.** (**a**) Statistical distribution of the number of registered ionospheric signals as a function of time in relation to the day of earthquake and distance of satellite orbit from the epicenter for the DEMETER satellite (5426 earthquakes M > 5) [86]; (**b**) Statistical distribution of the number of registered ionospheric signals as a function of time in relation to the day of earthquake and distance of satellite orbit from the epicenter for the CSES 1 satellite [87].

*5.2. Data Purification from Other Types of Variations in Atmosphere and Ionosphere to Reveal the Earthquake Precursors*

In the first 4 paragraphs, the reader accepted information not only on the physical aspect of their generation but also on how these precursors look in different ways of their presentation. A lot of publications exist on the earthquake precursors, adding that they were registered and studied in quiet solar and geomagnetic conditions. It looks like parents protect their children from any complexities of real life and they will go out into independent life infantile and helpless. If we are talking about the real earthquake forecast, we should be ready to recognize precursors at any time and in any geophysical condition.

Let us consider the main factors that interfere with physical processes generating precursors. Let us consider the interfering factors from the ground surface to the upper layers of the ionosphere and magnetosphere. Usually, there is calm weather before earthquakes but in stormy areas, such as the Pacific near the Kamchatka peninsula, the series of cyclones can blow away the radon that happened before the M7.5 earthquake east of the Kuril Islands on 25 March 2020 and we were not able to register the ACP over the epicenter. Geomagnetic storms usually change conditions of propagation of electromagnetic waves and in data of VLF signals propagating in subionospheric waveguide appear anomalies, similar to the pre-earthquake effects. During geomagnetic storms, the additional sporadic layers in *E*-regions of the ionosphere could form, which could be confused with the pre-earthquake sporadic layers. Simultaneously geomagnetic storms create positive and negative anomalies in the *F*-layer depending on the phase of the geomagnetic storm, which also interferes with the pre-seismic effects. Sharp changes in the short-ultraviolet radiation monitored with the help of the F10.7 index of solar activity can also suddenly increase or decrease electron concentration in the *F*-layer. As a result of geomagnetic storms, the energetic particles precipitate from radiative belts similar to pre-earthquake precipitations [60].

We developed different technologies of data processing permitting to sift out interferences and identify precursors. They are published in several tens of papers, some summary ones can be found in [8,45]. Here, we will mention only the main principles.

1. Even a very strong earthquake is still a local event, so ionospheric effects will be observed only within the earthquake preparation zone, while the ionospheric effects of the geomagnetic storm are global;
2. A geomagnetic storm, as a compelling force, increases the correlation between the remote areas of the ionosphere while pre-seismic effects increase the small-scale variability of the ionosphere;
3. Strong variations in F10.7 can be filtered and the precursory variations may be identified due to their locality;
4. Pre-earthquake variations in the ionosphere strongly depend on the local time (precursor mask) and are shorter than the ionospheric effects of the geomagnetic storm;
5. The formation of sporadic layers at an altitude of 120 km in the $E$-region of the ionosphere is characteristic only to pre-earthquake effects;
6. Multiparameter monitoring strongly helps to reveal precursors. Geomagnetic storms do not create thermal anomalies or variations in the relative humidity.

### 5.3. Cognitive Recognition and Automation of the Precursors' Identification and Forecast

We developed several algorithms for the precursors' recognition, which use the main principles mentioned above and are named based on cognitive recognition, because we use our knowledge of the physical mechanism of precursor generation as well as pattern recognition technique and machine learning. Below, the algorithms used for the cognitive recognition of the ionospheric precursors are numbered [27]:

1. Analysis of $\Delta$TEC (or $\Delta foF2$) data sets with the pattern–recognition method for the correspondence of the ionospheric precursor mask to changes in the ionosphere current over the seismically active region [37];
2. Correlation analysis of arrays of the daily TEC values (or critical frequency $foF2$) between a pair of adjacent GPS/GLONASS receivers (or ground stations for vertical sounding of the ionosphere) [61];
3. Calculation of the coefficient of regional variability of the ionosphere in the presence of a dense local network of stationary GPS/GLONASS receivers [61];
4. Calculation and construction of differential maps of the global TEC $\Delta$TEC$_{\text{GIM}}$ to determine the position of the epicenter of the future earthquake and its magnitude [64]. If there is a dense local network of stationary GPS/GLONASS receivers, differential maps can be calculated with local data rather than GPS GIMs;
5. Comparison of variations in the global TEC with the local TEC with reference to the solar activity index F10.7 [61];
6. Calculation of the correction of the chemical potential of water vapor (ACP) according to the local temperature and relative humidity data to determine the time of the seismic event [20];
7. Construction of maps of the distribution of the correction of the chemical potential according to the data of local temperature and relative humidity to determine the position of the epicenter of the future earthquake and to estimate its magnitude [61];
8. Analysis of the dynamics of the equatorial anomaly (EA) for low-latitude earthquakes in order to detect the absolute anomaly and the longitudinal effect in the EA [40,61];
9. Multiparameter analysis using operational data on other physical precursors, if there are any (radon activity, crustal conductivity, OLR, and anomalous cloud structures) [82].

All these algorithms were compiled in the common software system presented in Figure 24.

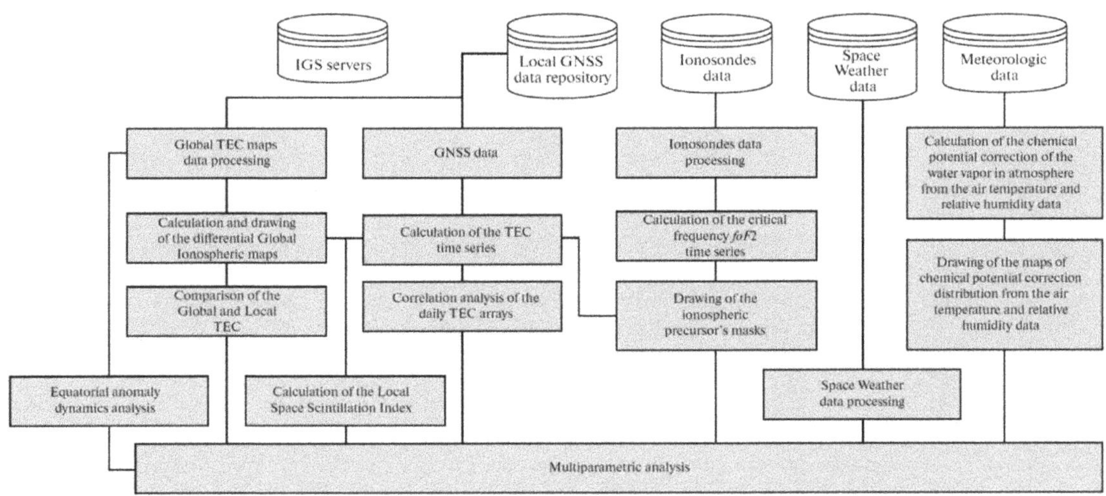


**Figure 24.** Schematic diagram of the machine processing of ionosphere monitoring and space weather data for the purpose of cognitive identification of earthquake precursors.

## 6. Discussion and Conclusions

This publication is an attempt to present the pathway from the physical idea to the practical applications using the technologies developed during the proposed idea studies. Not one of the parts presented here does not have a finished look but there is a wise saying: "The one who walks masters the road". But any road should have a clear goal. It can be to solve some complex problem or develop advanced technology. But some goals exist that are important to the whole of humanity and this is an earthquake forecast that we have not been able to resolve for almost two centuries. This road has bifurcation points and dead ends as in 1997 when seismologists decided that the short-term earthquake forecast was impossible [3].

Here, we try to demonstrate that the true way to the goal consists of three steps: firstly, understanding of physical background of the earthquake precursors' generation; secondly, developing technologies for their reliable identification, and, with their help, determining the main earthquake parameters, namely time, place, and magnitude; and thirdly, in a more complex step, organization of real-time monitoring with development of the short-term forecast.

The first step was very complex. Prof. Seiya Uyeda, who left us last year, is a full member of the Japan Emperor Academy of Sciences, foreign member of American and Russian Academy of Sciences, founder of the Inter Association Working Group on Electromagnetic Studies of Earthquakes and Volcanoes (EMSEV) of IAGA/IASPEI/IAVSEI, and told me many years ago that the real earthquake forecast will be possible only after we understand the physics of the precursor's generation. Almost 20 years passed that were spent on understanding the physics of the pre-earthquake atmospheric and ionospheric phenomena and the LAIC model [56] was created, part of which was presented in this paper.

Only after this, were we able to study the main morphological features of earthquake precursors and to develop the technologies of their definite identification.

But we cannot look at the future with optimism because the last step still looks like an impossible thing. The difficulty is not in technological problems and even not in financing: it is in human brains and politics. There is no consensus between the physicists and seismologists, the majority of whom stand on positions of the year 1997. It seems here that the problem is not only in a scientific position but also a fear of responsibility. Who will be responsible for the unfulfilled forecast and what will be the measure of responsibility? It seems that the problem can be resolved only in the way of changing the paradigm of earthquake forecast and responsibility. It should be put on the same level of responsibility as the weather forecast. It is necessary to put into the minds of the authorities that the forecast is a probabilistic value and that it can never reach 100%. We need to free seismologists from this fear and then perhaps the problem resolution will improve. The earthquake forecast should have the status of state services like medical or fire services. The earthquake forecast as a subject should be taught in universities and colleges in order to have enough specialists serving the forecast service.

**Author Contributions:** S.P.: Conceptualization, methodology, supervision, visualization; V.M.V.H.: Management activity, validation, review & editing, funding acquisition. All authors have read and agreed to the published version of the manuscript.

**Funding:** The work was carried out with the support of the Ministry of Science and Higher Education of the Russian Federation (theme/Monitoring, state registration No. 122042500031-8). Velasco Herrera thanks the UNAM Institute of Geophysics for its support.

**Data Availability Statement:** Data are available on request.

**Acknowledgments:** The authors want to acknowledge his many years of scientific collaborators: Dimitar Ouzounov, Alexander Karelin, and Valery Hegai, whose ideas and support helped to create this publication, and Pavel Budnikov who created powerful software for data processing.

**Conflicts of Interest:** The authors declares no conflicts of interest.

## References

1. Main, I. Statistical physics, seismogenesis, and seismic hazard. *Rev. Geophys.* **1996**, *34*, 433–462. [CrossRef]
2. Scholz, C.H.; Sykes, L.R.; Aggarwal, Y.P. Earthquake Prediction: A Physical Basis. *Science* **1973**, *181*, 803–810. [CrossRef] [PubMed]
3. Geller, R.J.; Jackson, D.D.; Kagan, Y.Y.; Mulargia, F. Earthquakes Cannot Be Predicted. *Science* **1997**, *275*, 1616–1617. [CrossRef]
4. Larkina, V.I.; Migulin, V.V.; Molchanov, O.A.; Khar'kov, I.P.; Inchin, A.S.; Schvetcova, V.B. Some statistical results on very low frequency radiowave emissions in the upper ionosphere over earthquake zones. *Phys. Earth Planet. Inter.* **1989**, *57*, 100–109. [CrossRef]
5. Akhmamedov, K. Interferometric measurements of the temperature of the F2 region of the ionosphere during the period of the Iranian Earthquake of June 20, 1990. *Geomagn. Aeron.* **1993**, *33*, 135–137.
6. Davies, K.; Baker, D.M. Ionospheric effects observed around the time of the Alaskan earthquake of March 28, 1964. *J. Geophys. Res.* **1965**, *70*, 2251–2253. [CrossRef]
7. Molchanov, O.A.; Hayakawa, M. VLF Monitoring of atmosphere-ionosphere boundary as a tool to study planetary waves evolution and seismic influence. *Phys. Chem. Earth* **2001**, *26*, 453–458.
8. Pulinets, S.A.; Legen'ka, A.D.; Gaivoronskaya, T.V.; Depuev, V.K. Main phenomenological features of ionospheric precursors of strong earthquakes. *J. Atmos. Sol. Terr. Phys.* **2003**, *65*, 1337–1347. [CrossRef]
9. Le, H.; Liu, J.-Y.; Liu, L. A statistical analysis of ionospheric anomalies before 736 M6.0+ earthquakes during 2002–2010. *J. Geophys. Res. Space Phys.* **2011**, *116*, A02303. [CrossRef]
10. Hattori, K. ULF Geomagnetic Changes Associated with Large Earthquakes. *Terr. Atmos. Ocean. Sci.* **2004**, *15*, 329–360. [CrossRef]
11. Smirnov, S. Association of the negative anomalies of the quasistatic electric field in atmosphere with Kamchatka seismicity. *Nat. Hazards Earth Syst. Sci.* **2008**, *8*, 745–749. [CrossRef]
12. Hayakawa, M.; Kasahara, Y.; Nakamura, T.; Muto, F.; Horie, T.; Maekawa, S.; Hobara, Y.; Rozhnoi, A.A.; Solovieva, M.; Molchanov, O.A. A statistical study on the correlation between lower ionospheric perturbations as seen by subionospheric VLF/LF propagation and earthquakes. *J. Geophys. Res.* **2010**, *115*, A09305. [CrossRef]
13. Genzano, N.; Aliano, C.; Filizzola, C.; Pergola, N.; Tramutoli, V. Robust satellite technique for monitoring seismically active areas: The case of Bhuj-Gujarat earthquake. *Tectonophysics* **2007**, *431*, 197–210. [CrossRef]
14. Ouzounov, D.; Liu, D.; Kang, C.; Cervone, G.; Kafatos, M.; Taylor, P. Outgoing long wave radiation variability from IR satellite data prior to major earthquakes. *Tectonophysics* **2007**, *431*, 211–220. [CrossRef]
15. Parrot, M.; Berthelier, J.J.; Lebreton, J.-P.; Sauvaud, J.A.; Santolík, O.; Blęcki, J. Examples of unusual ionospheric observations made by the DEMETER satellite over seismic regions. *Phys. Chem. Earth* **2006**, *31*, 486–495. [CrossRef]
16. Němec, F.; Santolík, O.; Parrot, M. Decrease of intensity of ELF/VLF waves observed in the upper ionosphere close to earthquakes: A statistical study. *J. Geophys. Res.* **2009**, *114*, A04303. [CrossRef]
17. Anagnostopoulos, G.C.; Vassiliadis, E.; Pulinets, S. Characteristics of flux-time profiles, temporal evolution, and spatial distribution of radiation-belt electron precipitation bursts in the upper ionosphere before great and giant earthquakes. *Ann. Geophys.* **2012**, *55*, 21–36.
18. Parrot, M. DEMETER Satellite and Detection of Earthquake Signals. In *Natural Hazards*, 1st ed.; Singh, R., Bartlett, D., Eds.; Taylor & Francis Group: Boca Raton, FL, USA, 2018; pp. 115–138.

19. Pulinets, S.; Ouzounov, D. Lithosphere-Atmosphere-Ionosphere Coupling (LAIC) model—An unified concept for earthquake precursors validation. *J. Asian Earth Sci.* **2011**, *41*, 371–382. [CrossRef]

20. Pulinets, S.A.; Ouzounov, D.P.; Karelin, A.V.; Davidenko, D.V. Physical Bases of the Generation of Short-Term Earthquake Precursors: A Complex Model of Ionization-Induced Geophysical Processes in the Lithosphere–Atmosphere–Ionosphere–Magnetosphere System. *Geomagn. Aeron.* **2015**, *55*, 540–558. [CrossRef]

21. Pulinets, S.; Ouzounov, D.; Karelin, A.; Davidenko, D. Lithosphere–Atmosphere–Ionosphere–Magnetosphere Coupling—A Concept for Pre-Earthquake Signals Generation. In *Pre-Earthquake Processes: A Multidisciplinary Approach to Earthquake Prediction Studies*; Ouzounov, D., Pulinets, S., Hattori, K., Taylor, P., Eds.; John Wiley & Sons, Inc.: Hoboken, NJ, USA, 2018; pp. 77–98.

22. Zhima, Z.; Yan, R.; Lin, J.; Wang, Q.; Yang, Y.; Lv, F.; Huang, J.; Cui, J.; Liu, Q.; Zhao, S.; et al. The Possible Seismo-Ionospheric Perturbations Recorded by the China-Seismo-Electromagnetic Satellite. *Remote Sens.* **2022**, *14*, 905. [CrossRef]

23. Molchan, G.M. Earthquake prediction as a decision-making problem. *Pure Appl. Geophys.* **1997**, *147*, 233–247. [CrossRef]

24. Junge, M.R.; Dettori, J.R. ROC Solid: Receiver Operator Characteristic (ROC) Curves as a Foundation for Better Diagnostic Tests. *Glob. Spine J.* **2018**, *8*, 424–429. [CrossRef] [PubMed]

25. Ouzounov, D.; Pulinets, S.; Liu, J.-Y.; Hattori, K.; Han, P. Multiparameter Assessment of Pre-Earthquake Atmospheric Signals. In *Pre-Earthquake Processes: A Multidisciplinary Approach to Earthquake Prediction Studies*; Ouzounov, D., Pulinets, S., Hattori, K., Taylor, P., Eds.; John Wiley & Sons, Inc.: Hoboken, NJ, USA, 2018; pp. 339–359. [CrossRef]

26. Akhoondzadeh, M. Genetic algorithm for TEC seismo-ionospheric anomalies detection around the time of the Solomon (Mw = 8.0) earthquake of 06 February 2013. *Adv. Space Res.* **2013**, *52*, 581–590. [CrossRef]

27. Pulinets, S.A.; Davidenko, D.V.; Budnikov, P.A. Method for Cognitive Identification of Ionospheric Precursors of Earthquakes. *Geomagn. Aeron.* **2021**, *61*, 14–24. [CrossRef]

28. Knyazeva, E.N.; Kurdiumov, S.P. Peculiar properties of nonequilibrium processes in open dissipative media. In *The Problems of XXI Century Geophysics*, 1st ed.; Nikolaev, A.A., Ed.; Nauka Publ.: Moscow, Russia, 2003; pp. 37–65.

29. Mironova, I.A.; Aplin, K.L.; Arnold, F.; Bazilevskaya, G.A.; Harrison, R.G.; Krivolutsky, A.A.; Nicoll, K.A.; Rozanov, E.V.; Turunen, E.; Usoskin, I.G. Energetic Particle Influence on the Earth's Atmosphere. *Space Sci. Rev.* **2015**, *194*, 1–96. [CrossRef]

30. Hoppel, W.A.; Anderson, R.V.; Willet, J.C. Atmospheric electricity in the planetary boundary layer. In *The Earth's Electrical Environment*; National Academic Press: Cambridge, MA, USA, 1986; pp. 149–165.

31. Tölgyessy, M.; Harangozó. RADIOCHEMICAL METHODS | Natural and Artificial Radioactivity. In *Encyclopedia of Analytical Science*, 2nd ed.; Worsfold, P., Townshend, A., Poole, C., Eds.; Elsevier: Amsterdam, The Netherlands, 2005; pp. 8–15.

32. Williams, E.R.; Mareev, E. Recent progress on the global electrical circuit. *Atmos. Res.* **2014**, *135–136*, 208–227. [CrossRef]

33. Slyunyaev, N.N.; Mareev, E.A.; Zhidkov, A.A. On the variation of the ionospheric potential due to large-scale radioactivity enhancement and solar activity. *J. Geophys. Res. Space Phys.* **2015**, *120*, 7060–7082. [CrossRef]

34. Fux, I.M.; Shubova, R.S. VLF signal anomalies as response on the processes within near-earth atmosphere. *Geomagn. Aeronom.* **1995**, *34*, 130–135.

35. Rozhnoi, A.; Solovieva, M.; Molchanov, O.; Schwingenschuh, K.; Boudjada, M.; Biagi, P.F.; Maggipinto, T.; Castellana, L.; Ermini, A.; Hayakawa, M. Anomalies in VLF radio signals prior the Abruzzo earthquake (M = 6.3) on 6 April 2009. *Nat. Hazards Earth Syst. Sci.* **2009**, *9*, 1727–1732. [CrossRef]

36. Martynenko, S.I.; Fuks, I.M.; Shubova, R.S. Ionospheric electric-field influence on the parameters of VLF signals connected with nuclear accidents and earthquakes. *J. Atmos. Electr.* **1996**, *16*, 259–269.

37. Pulinets, S.A.; Davidenko, D.V. The Nocturnal Positive Ionospheric Anomaly of Electron Density as a Short-Term Earthquake Precursor and the Possible Physical Mechanism of its Formation. *Geomagn. Aeron.* **2018**, *58*, 559–570. [CrossRef]

38. Tramutoli, V.; Marchese, F.; Falconieri, A.; Filizzola, C.; Genzano, N.; Hattori, K.; Lisi, M.; Liu, J.-Y.; Ouzounov, D.; Parrot, M.; et al. Tropospheric and Ionospheric Anomalies Induced by Volcanic and Saharan Dust Events as Part of Geosphere Interaction Phenomena. *Geosciences* **2019**, *9*, 177. [CrossRef]

39. Pulinets, S.A.; Davidenko, D.V.; Pulinets, M.S. Atmosphere-ionosphere coupling induced by volcanoes eruption and dust storms and role of GEC as the agent of geospheres interaction. *Adv. Space Res.* **2022**, *69*, 4319–4334. [CrossRef]

40. Pulinets, S. Low-Latitude Atmosphere-Ionosphere Effects Initiated by Strong Earthquakes Preparation Process. *Int. J. Geophys.* **2012**, *2012*, 131842. [CrossRef]

41. Denisenko, V.V.; Rozanov, E.V.; Belyuchenko, K.V.; Bessarab, F.S.; Golubenko, K.S.; Klimenko, M.V. The Ionospheric Electric Field Perturbation with an Increase in Radon Emanation. *Russ. J. Phys. Chem. B Focus. Phys.* **2024**, *18*, 837–843. [CrossRef]

42. Bliokh, P. Variations of Electric Fields and Currents in the Lower Ionosphere Produced by Conductivity Growth of the Air above the Future Earthquake Center. In *Atmospheric and Ionospheric Electromagnetic Phenomena Associated with Earthquakes*; Hayakawa, M., Ed.; TERRAPUB: Tokyo, Japan, 1999; pp. 829–838.

43. Kathmann, S.M.; Schenter, G.K.; Garrett, B.C. Ion-induced nucleation: The importance of chemistry. *Phys. Rev. Lett.* **2005**, *94*, 116104. [CrossRef] [PubMed]

44. Svensmark, H.; Enghoff, M.B.; Shaviv, N.J.; Svensmark, J. Increased ionization supports growth of aerosols into cloud condensation nuclei. *Nat. Commun.* **2017**, *8*, 2199. [CrossRef] [PubMed]

45. Pulinets, S.A.; Boyarchuk, K.A. *Ionospheric Precursors of Earthquakes*, 1st ed.; Springer: Berlin/Heidelberg, Germany, 2004; Volume 315. [CrossRef]

46. Timofeev, V.E.; Grigoryev, V.G.; Morozova, E.I.; Skryabin, N.G.; Samsonov, S.N. Action of cosmic rays on the latent energy of the atmosphere. *Geomag. Aeron.* **2003**, *43*, 636–640.

47. Hõrrak, U. Air Ion Mobility Spectrum at A Rural Area. Ph.D. Thesis, Tartu Ülikooli Kirjastuse, Trükikoda, Tartu City, Estonia, 2001.

48. Gavrilov, V.A.; Panteleev, I.A.; Deshcherevskii, A.V.; Lander, A.V.; Morozova, Y.V.; Buss, Y.Y.; Vlasov, Y.A. Stress–Strain State Monitoring of the Geological Medium Based on The Multi-instrumental Measurements in Boreholes: Experience of Research at the Petropavlovsk-Kamchatskii Geodynamic Testing Site (Kamchatka, Russia). *Pure Appl. Geophys.* **2020**, *177*, 397–419. [CrossRef]

49. Bogdanov, V.; Gavrilov, V.; Pulinets, S.; Ouzounov, D. Responses to the preparation of strong Kamchatka earthquakes in the lithosphere–atmosphere–ionosphere system, based on new data from integrated ground and ionospheric monitoring. *E3S Web Conf.* **2020**, *196*, 03005. [CrossRef]

50. Smirnov, S. Negative Anomalies of the Earth's Electric Field as Earthquake Precursors. *Geosciences* **2020**, *10*, 10. [CrossRef]

51. Han, R.; Cai, M.; Chen, T.; Yang, T.; Xu, L.; Xia, Q.; Jia, X.; Han, J. Preliminary Study on the Generating Mechanism of the Atmospheric Vertical Electric Field before Earthquakes. *Appl. Sci.* **2022**, *12*, 6896. [CrossRef]

52. Holzer, R.E. Atmospheric electrical effects of nuclear explosions. *J. Geophys. Res.* **1972**, *77*, 5845–5855. [CrossRef]

53. Khegai, V.V. Analytical model of a seismogenic electric field according to data of measurements in the surface layer of the midlatitude atmosphere and calculation of its magnitude at the ionospheric level. *Geomagn. Aeron.* **2020**, *60*, 507–520. [CrossRef]

54. Sorokin, V.; Novikov, V. A model of TEC perturbations possibly related to seismic activity. *Ann. Geophys.* **2023**, *66*, SE642. [CrossRef]

55. Kuo, C.L.; Lee, L.C.; Huba, J.D. An improved coupling model for the lithosphere-atmosphere-ionosphere system. *J. Geophys. Res. Space Phys.* **2014**, *119*, 3189–3205. [CrossRef]

56. Pulinets, S.; Ouzounov, D.; Karelin, A.; Boyarchuk, K. *Earthquake Precursors in the Atmosphere and Ionosphere*; Springer Nature: Dordrecht, The Netherlands, 2022; Available online: https://link.springer.com/book/10.1007/978-94-024-2172-9 (accessed on 30 July 2024).

57. Hegai, V.; Zeren, Z.; Pulinets, S. Seismogenic Field in the Ionosphere before Two Powerful Earthquakes: Possible Magnitude and Observed Ionospheric Effects (Case Study). *Atmosphere* **2023**, *14*, 819. [CrossRef]

58. Liu, J.-Y.; Chao, C.-K. An observing system simulation experiment for 1 FORMOSAT-5/AIP detecting seismoionospheric precursors. *Terr. Atmos. Ocean. Sci.* **2017**, *28*, 117–127. [CrossRef]

59. Morozov, V.N. *Mathematical Modeling of Atmospheric-Electrical Processes Taking into Account the Influence of Aerosol Particles and Radioactive Substances*; Russian State Hydrometeorological University Publishing: Saint-Petersburg, Russia, 2011.

60. Pulinets, S.; Krankowski, A.; Hernandez-Pajares, M.; Marra, S.; Cherniak, I.; Zakharenkova, I.; Rothkaehl, H.; Kotulak, K.; Davidenko, D.; Blaszkiewicz, L.; et al. Ionosphere Sounding for Pre-seismic Anomalies Identification (INSPIRE): Results of the Project and Perspectives for the Short-Term Earthquake Forecast. *Front. Earth Sci.* **2021**, *9*, 610193. [CrossRef]

61. Pulinets, S.; Ouzounov, D. *The Possibility of Earthquake Forecasting: Learning from Nature*; IOP Publishing: Bristol, UK, 2018; Available online: https://iopscience.iop.org/book/978-0-7503-1248-6 (accessed on 30 July 2024).

62. Liperovskaya, E.V.; Pokhotelov, O.A.; Hobara, Y.; Parrot, M. Variability of sporadic E-layer semi transparency ($f_o E_s - f_b E_s$) with magnitude and distance from earthquake epicenters to vertical sounding stations. *Nat. Hazards Earth Syst. Sci.* **2003**, *3*, 279–284. [CrossRef]

63. Kim, V.P.; Khegai, V.V.; Illich-Svitych, P.V. Probability of formation of a metallic ion layer in the nighttime midlatitude ionospheric E-region before strong earthquakes. *Geomagn. Aeron.* **1993**, *33*, 114–119.

64. Xu, T.; Hu, Y.; Deng, Z.; Zhang, Y.; Wu, J. Revisit to sporadic E layer response to presumably seismogenic electrostatic fields at middle latitudes by model simulation. *J. Geophys. Res. Space Phys.* **2020**, *125*, e2019JA026843. [CrossRef]

65. Surkov, V.V.; Pilipenko, V.A.; Silina, A.S. Can Radioactive Emanations in a Seismically Active Region Affect Atmospheric Electricity and the Ionosphere? *Izv. Phys. Solid Earth* **2022**, *58*, 297–305. [CrossRef]

66. Pulinets, S.; Budnikov, P.; Karelin, A.; Žalohar, J. Thermodynamic instability of the atmospheric boundary layer stimulated by tectonic and seismic activity. *J. Atmos. Sol. Terr. Phys.* **2023**, *246*, 106050. [CrossRef]

67. Daneshvar, M.R.M.; Tavousi, T.; Khosravi, M. Atmospheric blocking anomalies as the synoptic precursors prior to the induced earthquakes: A new climatic conceptual model. *Int. J. Environ. Sci. Technol.* **2015**, *12*, 1705–1718. [CrossRef]

68. Dobrovolsky, I.R.; Zubkov, S.I.; Myachkin, V.I. Estimation of the size of earthquake preparation zones. *Pure Appl. Geophys.* **1979**, *117*, 1025–1044. [CrossRef]

69. Boyarchuk, K.A.; Karelin, A.V.; Shirokov, R.V. Bazovaya model' kinetiki ionizirovannoi atmosfery. In *The Reference Model of Ionized Atmospheric Kinetics*; VNIIEM Publ.: Moscow, Russia, 2006.

70. Tsuruda, T. Nuclear power plant explosions at Fukushima-Daiichi. *Procedia Eng.* **2013**, *62*, 71–77. [CrossRef]

71. Pulinets, S.; Budnikov, P. Atmosphere Critical Processes Sensing with ACP. *Atmosphere* **2022**, *13*, 1920. [CrossRef]

72. Expansion of Radioactive Cloud after Chernobyl Disaster. Available online: https://www.youtube.com/watch?v=mqu_l29WioM (accessed on 6 June 2024).

73. Bolt, B.A. *Earthquakes: A Primer*; W.H. Freeman: San Francisco, CA, USA, 1978; p. 241.

74. Toutain, J.-P.; Baubron, J.-C. Gas geochemistry and seismotectonics: A review. *Tectonophysics* **1998**, *304*, 1–27. [CrossRef]

75. Fleischer, R.L. Dislocation model for radon response to distant earthquakes. *Geophys. Res. Lett.* **1981**, *8*, 477–480. [CrossRef]

76. Aki, K.; Richards, P.G. *Quantitative Seismology. Theory and Methods*; Freeman: New York, NY, USA, 1980.

77. Ryu, K.; Parrot, M.; Kim, S.G.; Jeong, K.S.; Chae, J.S.; Pulinets, S.; Oyama, K.-I. Suspected seismo-ionospheric coupling observed by satellite measurements and GPS TEC related to the M7.9 Wenchuan earthquake of 12 May 2008. *J. Geophys. Res. Space Phys.* **2014**, *119*, 305–323. [CrossRef]

78. Pulinets, S.; Tsidilina, M.; Ouzounov, D.; Davidenko, D. From Hector Mine M7.1 to Ridgecrest M7.1 Earthquake. A Look from a 20-Year Perspective. *Atmosphere* **2021**, *12*, 262. [CrossRef]

79. Liu, J.Y.; Chen, Y.I.; Chuo, Y.J.; Chen, C.S. A statistical investigation of preearthquake ionospheric anomaly. *J. Geophys. Res.* **2006**, *111*, A05304. [CrossRef]

80. Liu, J.-Y.; Hattori, K.; Chen, Y.-I. Application of Total Electron Content Derived from the Global Navigation Satellite System for Detecting Earthquake Precursors. In *Pre-Earthquake Processes: A Multidisciplinary Approach to Earthquake Prediction Studies*; Ouzounov, D., Pulinets, S., Hattori, K., Taylor, P., Eds.; John Wiley & Sons, Inc.: Washington DC, USA, 2018; pp. 305–317. [CrossRef]

81. Liu, J.Y.; Chen, Y.I.; Chen, C.H.; Liu, C.Y.; Chen, C.Y.; Nishihashi, M.; Li, J.Z.; Xia, Y.Q.; Oyama, K.I.; Hattori, K.; et al. Seismoionospheric GPS total electron content anomalies observed before the 12 May 2008 Mw7.9 Wenchuan earthquake. *J. Geophys. Res.* **2009**, *114*, A04320. [CrossRef]

82. Ouzounov, D.; Pulinets, S.; Hattori, K.; Kafatos, M.; Taylor, P. Atmospheric signals associated with major earthquakes. A multi-sensor approach. In *The Frontier of Earthquake Prediction Studies*; Hayakawa, M., Ed.; Nihon-Senmontosho-Shuppan: Tokyo, Japan, 2012; pp. 510–531.

83. Ouzounov, D.; Pulinets, S.; Davidenko, D.; Rozhnoi, A.; Solovieva, M.; Fedun, V.; Dwivedi, B.N.; Rybin, A.; Kafatos, M.; Taylor, P. Transient Effects in Atmosphere and Ionosphere Preceding the 2015 M7.8 and M7.3 Gorkha–Nepal Earthquakes. *Front. Earth Sci.* **2021**, *9*, 757358. [CrossRef]

84. Li, M.; Parrot, M. Statistical analysis of an ionospheric parameter as a base for earthquake prediction. *J. Geophys. Res.* **2013**, 3731–3739. [CrossRef]

85. He, Y.; Yang, D.; Qian, J.; Parrot, M. Response of the Ionospheric Electron Density to Different Types of Seismic Events. *Nat. Hazards Earth Syst. Sci.* **2011**, *11*, 2173–2180. [CrossRef]

86. Yan, R.; Parrot, M.; Pinçon, J.-L. Statistical study on variations of the ionospheric ion density observed by DEMETER and related to seismic activities. *J. Geophys. Res.* **2017**, *122*, 12,421–12,429. [CrossRef]

87. Shen, X. From CSES to IMCP. In Proceedings of the IMCP International Workshop, NSSC-Huairou Campus, Beijing, China, 14–17 September 2023.

*geosciences*

MDPI

*Article*

# The Preparation Phase of the 2022 $M_L$ 5.7 Offshore Fano (Italy) Earthquake: A Multiparametric–Multilayer Approach

**Martina Orlando** [1,2,*], **Angelo De Santis** [1,*], **Mariagrazia De Caro** [1], **Loredana Perrone** [1], **Saioa A. Campuzano** [1,3], **Gianfranco Cianchini** [1], **Alessandro Piscini** [1], **Serena D'Arcangelo** [1,3], **Massimo Calcara** [1], **Cristiano Fidani** [1], **Adriano Nardi** [1], **Dario Sabbagh** [1] and **Maurizio Soldani** [1]

[1]   Istituto Nazionale di Geofisica e Vulcanologia (INGV), 00143 Rome, Italy; mariagrazia.decaro@ingv.it (M.D.C.); loredana.perrone@ingv.it (L.P.); saioa.arquerocampuzano@ingv.it (S.A.C.); gianfranco.cianchini@ingv.it (G.C.); alessandro.piscini@ingv.it (A.P.); serena.darcangelo@ingv.it (S.D.); massimo.calcara@ingv.it (M.C.); cristiano.fidani@ingv.it (C.F.); adriano.nardi@ingv.it (A.N.); dario.sabbagh@ingv.it (D.S.); maurizio.soldani@ingv.it (M.S.)

[2]   Dipartimento di Scienze, Università Roma TRE, 00154 Rome, Italy

[3]   Department of Physics of the Earth and Astrophysics, Universidad Complutense de Madrid (UCM), 28040 Madrid, Spain

*   Correspondence: martina.orlando@ingv.it (M.O.); angelo.desantis@ingv.it (A.D.S.)

**Abstract:** This paper presents an analysis of anomalies detected during the preparatory phase of the 9 November 2022 $M_L$ = 5.7 earthquake, occurring approximately 30 km off the coast of the Marche region in the Adriatic Sea (Italy). It was the largest earthquake in Italy in the last 5 years. According to lithosphere–atmosphere–ionosphere coupling (LAIC) models, such earthquake could induce anomalies in various observable variables, from the Earth's surface to the ionosphere. Therefore, a multiparametric and multilayer approach based on ground and satellite data collected in each geolayer was adopted. This included the revised accelerated moment release method, the identification of anomalies in atmospheric parameters, such as Skin Temperature and Outgoing Longwave Radiation, and ionospheric signals, such as Es and F2 layer parameters from ionosonde measurements, magnetic field from Swarm satellites, and energetic electron precipitations from NOAA satellites. Several anomalies were detected in the days preceding the earthquake, revealing that their cumulative occurrence follows an exponential trend from the ground, progressing towards the upper atmosphere and the ionosphere. This progression of anomalies through different geolayers cannot simply be attributed to chance and is likely associated with the preparation phase of this earthquake, supporting the LAIC approach.

**Keywords:** earthquakes; preparation phase; LAIC models; R-AMR; Swarm satellites; NOAA satellites

**Citation:** Orlando, M.; De Santis, A.; De Caro, M.; Perrone, L.; A. Campuzano, S.; Cianchini, G.; Piscini, A.; D'Arcangelo, S.; Calcara, M.; Fidani, C.; et al. The Preparation Phase of the 2022 $M_L$ 5.7 Offshore Fano (Italy) Earthquake: A Multiparametric–Multilayer Approach. *Geosciences* **2024**, *14*, 191. https://doi.org/10.3390/geosciences 14070191

Academic Editors: Jesus Martinez-Frias and Dimitrios Nikolopoulos

Received: 31 May 2024
Revised: 6 July 2024
Accepted: 11 July 2024
Published: 16 July 2024

## 1. Introduction

The study of a possible coupling mechanism between the lithosphere, atmosphere, and ionosphere, also known as lithosphere–atmosphere–ionosphere coupling (LAIC), before an earthquake (EQ) or a volcanic eruption, is becoming increasingly relevant within the scientific community. Prior to the occurrence of such geophysical events, the Earth emits transient signals, sometimes strong but more often subtle and fleeting, which can manifest as local variations in the magnetic field, electromagnetic emissions across a wide range of frequencies (mostly 0.001 Hz–100 KHz, i.e., ULF-ELF-VLF), and a variety of atmospheric and ionospheric phenomena. This is a rather complex mechanism and there is considerable uncertainty about the nature of the processes that could produce these signals, both within the Earth's crust and on its surface. Over the years, various models have been developed to suggest a connection, i.e., coupling, among the geolayers following different channels.

Hayakawa [1] proposed three channels to establish connections between different observations in each layer: the chemical channel (also known as the electric field channel), the acoustic gravity wave (AGW) channel, and the electromagnetic (EM) channel. Freund [2] proposed an electrostatic channel in which stressed rocks release positive charge carriers known as "positive holes". Dahlgren et al. [3] contested this theory, in particular studies [4–6]. Scoville et al. [7] pointed out some pitfalls in the experiment.

Pulinets and Ouzounov [8] proposed a radon release from the crust in the earthquake preparation zone; radon is the radioactive gas produced in the decay chain of uranium or thorium emitted from the ground, affecting the electric field in the troposphere–ionosphere electric circuit. Surkov et al. [9] further investigated the potential impact of radon emissions on the atmosphere and ionosphere, finding that localized changes in atmospheric currents due to radon have minimal effects on the electron distribution in the ionosphere. Similarly, the study proposed by Schekotov et al. [10] does not find clear connections between electromagnetic variations and changes in pre-seismic temperatures, casting doubt on the hypothesis that radon emissions influence the ionosphere.

To explain seismo-ionospheric effects, other authors [11–13] suggested atmospheric processes producing acoustic and/or gravity waves in the seismic preparation region. Investigating LAIC processes, the multiparametric analysis around earthquakes has become a highly debated topic, involving various parameters across layers and utilizing different observation technologies within the same layer. A comprehensive study [14] on investigating the LAIC mechanism of the Mw = 7.2 Haiti earthquake on 14 August 2021, considered 52 precursors, including GPS Total Electron Content (TEC), 4 from CSES-01 satellite data, 7 lithospheric and atmospheric precursors from AIRS and OMI sensors, and 40 from the Swarm satellite constellation. This author observed a significant amount of anomalous values, suggesting a sequence possibly due to ion radiation from the Earth, i.e., a thin layer of particles transferring the electric field to the upper atmosphere and then to the ionosphere. Another study [15] examined two major channels of LAIC mechanisms, acoustic and electromagnetic, to analyze pre-seismic irregularities of the 2020 Samos (Greece) earthquake of M = 6.9. This study considered TEC, AGW, bursts of energetic particles in the radiation belt, magnetic field, electron density, and temperature, obtaining significant anomalies from 10 to 1 day before the seismic event. Pre-seismic low values of TEC were observed in regions with lower b-values in a study focused on the analysis [16] of the Mw = 7.7 Colima (Mexico) earthquake on 19 September 2022, indicating a higher probability of larger earthquakes.

Several studies focused on the correlation between the ionospheric TEC anomalies and the earthquakes [17–20]. Specifically, to explain the coupling of TEC anomalies and seismic events [18], they confirmed the existence of an anomalous electric field in seismogenic zones triggered by stressed rocks in earthquake regions associated with fault lineaments, while Tachema [20] proposed that the space between the lithosphere and ionosphere is occupied by a coherent structure of electrons and protons, transmitting electromagnetic waves generated during the seismic nucleation of rocks at depth.

De Santis et al. [21] analyzed several parameters from the lithosphere, atmosphere, and ionosphere on the occasion of the 2019 M = 7.1 Ridgecrest EQ, finding an accelerated progression of the cumulative number of all anomalies as the mainshock was approaching. Wang et al. [22] detected anomalous changes in the lithosphere, atmosphere, and ionosphere near the epicenter before the 2021 M = 7.4 Madoi EQ; meanwhile, Akhoondzadeh and Marchetti [23] analyzed the behavior of more than 50 different lithosphere–atmosphere–ionosphere anomalies during the preparation phase of the 2023 Turkey EQ, identifying a progressive increase in the number of anomalies starting about 10 days before, with the major peak the day before the mainshock.

On 9 November 2022, at 06:07:25 UTC, an $M_L$ = 5.7 (Mw = 5.5) EQ occurred approximately 30 km offshore the Marche coast in the Adriatic Sea, at latitude 43.984° N, longitude 13.324° E, and 5 km of depth. Just a minute after the mainshock, a strong aftershock of $M_L$ = 5.2 occurred about 8 km south of the main event, at latitude 43.913° N, longitude 13.345° E, and 8 km of depth. The two major shocks activated a seismic sequence of about 400 aftershocks lasting a week, thirteen of them with $M_L$ = 3.5. The seismic sequence occurred in correspondence with the frontal fault systems of the Northern Apennines, where ongoing convergence is accommodated on a series of buried faults, still poorly understood. The moment tensor solution of the mainshock indicates a reverse mechanism on a NW-SE trending fault plane, as well as the moment tensor solutions of the $M_L \geq 3.5$ events of the sequence. However, no moment tensor solution has been computed for the $M_L = 5.2$ event due to the overlap and interference of phases from the two events [24].

According to Pezzo et al. [24], the slip occurred along a thrust fault dipping approximately 24° SSW over a length of about 15 km, consistent with seismic reflection data propagating downward from the mainshock hypocenter, confirming the ongoing seismotectonic activity of this sector of the Apennines that is still propagating towards the foreland, approximately in a piggy-back thrust sequence. The area of greatest impact is along the coastal stretch between Fano and Ancona, where a maximum intensity of 5 European Macroseismic Scale (EMS-98) [25] has been estimated. Within this zone, very slight sporadic damages have been observed. Particularly, the district outskirts of Ancona have reported damages to recent reinforced concrete buildings, likely due to local amplification effects. The estimated maximum intensity reached 5 EMS-98 in some locations; however, macroseismic effects rapidly decreased to 4 EMS-98 inland, at a short distance from the coast [26].

This study aims to provide a comprehensive view of the effects observed on various geological and atmospheric layers within the framework of lithosphere–atmosphere–ionosphere coupling (LAIC) models, during the preparatory phase of the seismic event recorded offshore the coast of Marche on 9 November 2022. Previous studies related to this seismic event [24,27–31] have predominantly focused on its tectonic and seismological aspects, while we adopted the LAIC approach as a novelty for this EQ. Additionally, the characteristics of this seismic event, such as the magnitude slightly below 6 and its occurrence at sea, piqued our interest in investigating whether this approach could yield fruitful results. In particular, the presence of the sea conductive layer could limit the occurrence of one kind of LAIC instead of another.

Figure 1 describes the parameters studied across different geological layers from bottom to top, looking for anomalous perturbations potentially associated with the preparation phase of the $M_L = 5.7$ EQ. Specifically, seismicity acceleration preceding the mainshock is investigated in the lithosphere using the revised accelerated moment release (R-AMR) method. Moving upwards, the atmosphere is studied through parameters such as Skin Temperature (SKT) and Outgoing Longwave Radiation (OLR). Lastly, ionospheric analysis includes data from the Rome AIS-INGV ionosonde, the European Space Agency (ESA) Swarm satellite mission, and the National Oceanic and Atmospheric Administration (NOAA) satellites.

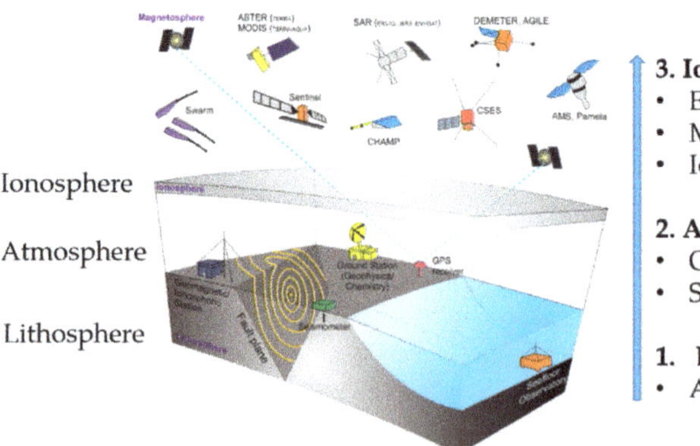


**Figure 1.** Summary map of the study conducted: the different layers analyzed are observed, from bottom to top. For each layer, the types of parameters considered are indicated.

In the next section, the data used will be introduced, followed by a presentation of the applied methods alongside the main results. Finally, the work will conclude with a discussion and a conclusion.

Given the proximity of Fano town to the epicenter, i.e., approximately 29 km, this event will be simply referred to as the "Fano EQ".

## 2. Seismotectonic Settings

The Adriatic Sea appears as a result of a good variety of structural and stratigraphic processes (Figure 2) guided by fault-related anticlines formed in the Plio–Miocene connected to the main Apennine thrust chain, deeper carbonate structures developed in the south, and a very shallow structure in the Late Pliocene to Quaternary in the central area [32]. During the Mesozoic, this area was affected by an extensional tectonic phase in the Middle Liassic and a compressional paleoinversion in the Lowermost Cretaceous [33,34]. The development of the Alps and the Apennines started from the Middle Eocene onwards in the African continental margin [35–39]. Subsequently, a flexure of the lithosphere belonging to the Adria margin concerned the most internal areas and migrated eastward through time, forming foredeep basins oriented sub-parallel to the belts. The Adriatic domain corresponds to the youngest part of the belt, strictly connected to the evolution of the Apennine fold and thrust belt and to the interaction with the Dinarides, sub-parallel orogenic belts with opposing vergences [32]. In detail, this thrust front is buried beneath Early Pliocene–Quaternary synorogenic deposits.

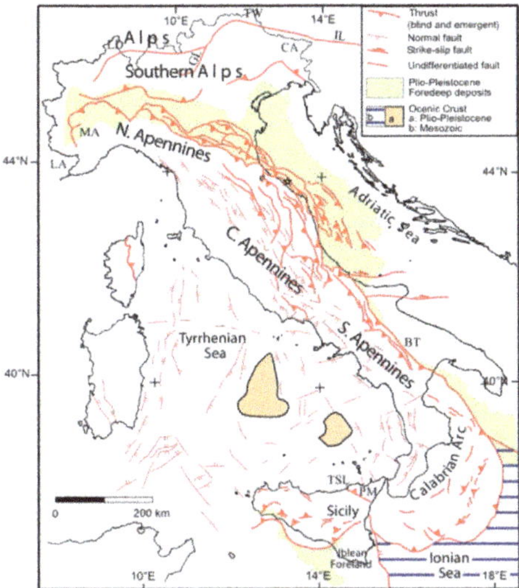


**Figure 2.** Simplified structural map of Italy, with the epicenter of the 9 November 2022 Fano EQ indicated by a small yellow star (modified from [40]).

The entire seismic sequence of November 2022 unfolds along the outermost structure of the Apennine orogeny, characterized by a series of NW–SE trending, NE verging folds forming the easternmost edge of the Apennine thrust front [31,41–43]. Over the Tertiary–Quaternary period, this front has gradually migrated towards the east-northeast (e.g., [44]). Geological evidence suggests the ongoing growth of these folds, indicating the continued activity of blind thrust fronts (e.g., [41]).

The presence of historical and instrumental earthquakes with Mw $\geq$ 5.5, as shown in Figure 3, suggests that thrust faults are also seismogenic [30].

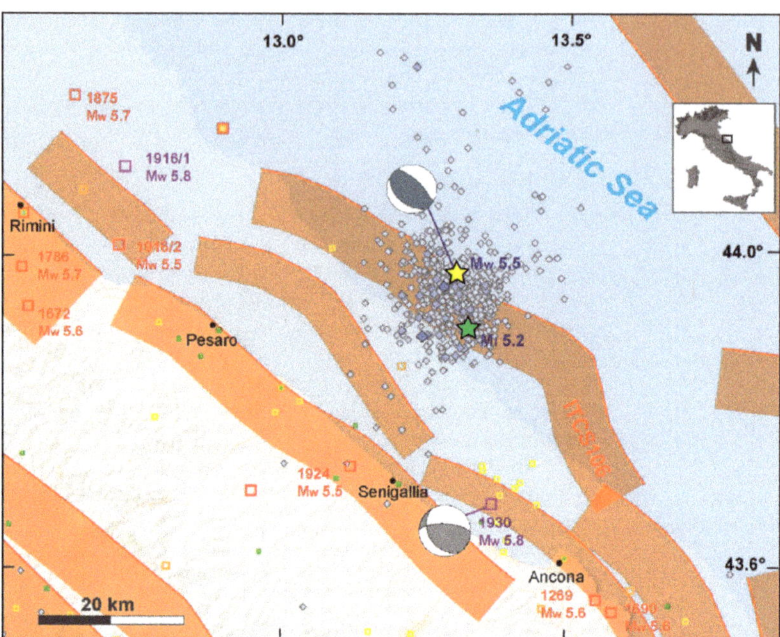



**Figure 3.** Seismotectonic framework of the coastal area of the Marche region. The light blue squares represent the seismic sequence from 9 November 2022 to 14 February 2023; the first event is marked with a yellow star and the second event is shown with a green star. Historical and instrumental earthquakes from CPTI15 [45] are indicated with colored squares, with earthquakes of Mw ≥ 5.5 highlighted in red. The surface projections of seismogenic zones are depicted with orange ribbons [41]. The focal mechanisms of the 9 November 2022 earthquake and the event of 30 October 1930, represented by the grey and white balls, come from TDMT (Time Domain Moment Tensor) and Vannoli et al. [46], respectively (modified from [30]).

The events recorded on 9 November 2022 represent a manifestation of the ongoing contraction between the Apennine chain, moving towards the northeast, and the Balkan area, which is experiencing a similar but opposite (southwestward) movement. The two hypocenters belong to the blind thrust fault system running parallel to the Marche Coast. This compressive front is located approximately 25–35 km offshore, with a length of about 70 km, and it is referred to as ITCS10 in the Database of Individual Seismogenic Sources (DISS) [41]. This thrust system has therefore been identified as responsible for this seismic sequence (see Figure 3).

## 3. Data and Methods

To study the effects of LAIC, several datasets are required since each geolayer being investigated demands specific data from various sources. As the analysis is conducted separately in each layer, there are different resolutions both in time and in space, which will be described below.

### 3.1. Lithospheric Data

From a lithospheric point of view, a seismological analysis was carried out to characterize the seismicity of the area affected by the $M_L = 5.7$ EQ under inspection, paying particular attention to the area mainly affected by the imminent seismic event. The seismic data provided by the Italian INGV Catalog [47] were used by selecting a circular area with a radius of 150 km around the epicenter and imposing a depth limit of 100 km, covering the period from 1 January 2012 to 8 November 2022 (Figure 4). The temporal and spatial resolutions of the used catalog in terms of earthquake detection (with associated information) are of the order of seconds (or even fractions of a second) and a few km, respectively.

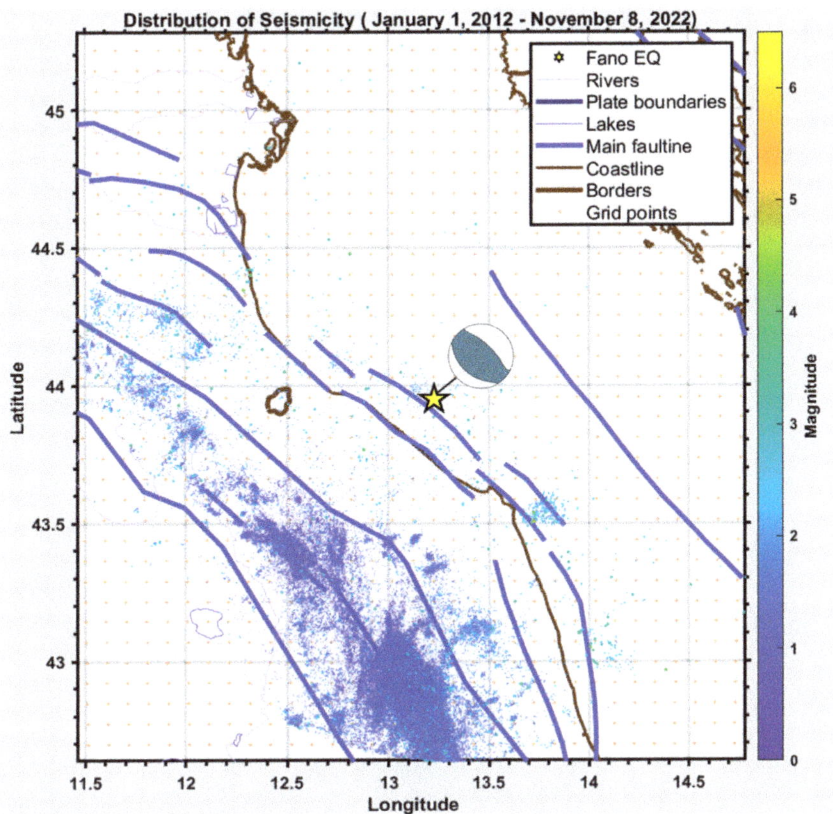



**Figure 4.** Spatial distribution of the 174.723 events extracted from the INGV Catalog during the period 2012–2022 within a circular radius of 150 km from the epicenter of the main EQ, highlighted by the yellow star. The grey and white sphere represents the focal mechanism of the earthquake on 9 November 2022. The chosen radius includes the Central Italy sequence (2016), identifiable by the cluster of events to the south near the edge of the area.

To evaluate R-AMR [48,49], the magnitude completeness (Mc) of the catalog, which represents the minimum value of magnitude for detection, was estimated using the method of maximum curvature [50]. The estimation of Mc was performed as a function of time by sliding time windows, each containing 150 EQs and stepping by five events. The uneven distribution of events within the selected circular area, due to the lack of seismometers on the Adriatic seabed, was taken into consideration. However, the presence of a seismic station located in Banja Luka (Bosnia and Herzegovina), named BLY, allowed for precise event localization. Based on these considerations, an Mc = 2.2 was chosen, resulting in a catalog of 9.099 events. Figure 5 depicts the events (blue and red points) that contributed to the acceleration identified by the R-AMR analysis.

Several studies (e.g., [51,52]) suggested that before significant EQs, there is an accelerated seismic activity under specific conditions. This phenomenon, explained by the Critical Point Theory, likens the main EQ to a phase transition occurring at a "time-to-failure", $t_f$.

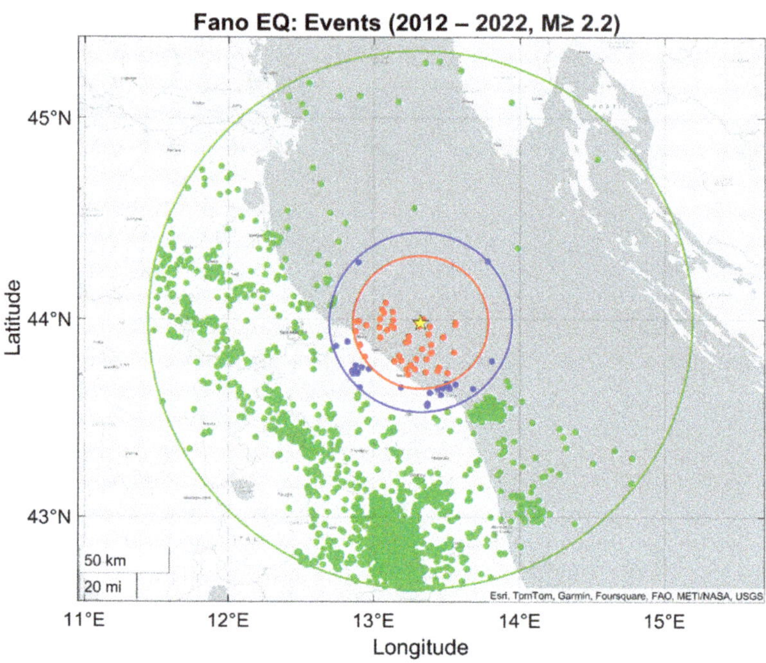


**Figure 5.** Spatial distribution of ten years of seismicity around the mainshock, in a radius of 150 km from the epicenter (largest green circle). Blue and red dots (confined within blue and red circles, respectively) represent the events contributing to the acceleration found by the R-AMR analysis. The red dots are the events closer to the seismogenic fault.

The seismicity preceding the mainshock, often hidden in catalogs, can be revealed through methods like Accelerated Moment Release (AMR), particularly its revised version, known as R-AMR [48,49]. The R-AMR algorithm was applied to EQ data in the Marche region before the Fano EQ. Acceleration in seismicity is measured by examining the accumulation of seismic Benioff strain $s_i = \sqrt{E_i}$, where each event releases strain proportional to the square root of its energy $E_i = 10^{(1.5M_i+4.8)}$ J. The cumulative strain, $s(t) = \sum s_i$, is known as Cumulative Benioff Strain. The regional increase in the cumulative Benioff deformation before a large shock is expressed by a power-law time-to-failure functional relation:

$$s(t) = A + B\left(t_f - t\right)^m \tag{1}$$

where $t_f$ is the time-to-failure (i.e., the occurrence of the mainshock) and $m$ is an inverse measure of how quickly the acceleration grows around $t_f$. To evaluate the quality of seismic acceleration compared to a linear trend representing the background seismicity, Bowman et al. (1998) [51] introduced the C factor, which is the ratio of the sum of the squares of the residuals of the fit of $s = s(t)$ and the same quantity of a linear fit; a C value less than 1 indicates acceleration, with lower values indicating more prominent acceleration. De Santis et al. [48] improved the technique by focusing on strain deposited on the mainshock fault by surrounding seismicity, corrected by a damping function with distance. Cianchini et al. [49] further enhanced the algorithm based on 14 case studies, confirming its ability to reveal hidden acceleration in seismic sequences and provide estimates of $t_f$ and expected magnitude based on parameters A and B of the functional relation. In this analysis, the algorithm is automatic: considering the $s = s(t)$ time series backward and excluding the mainshock, it detects the time when the acceleration starts and the minimum (no attenuation) and maximum circles (with some attenuation). Therefore, for the characterization of seismicity, the R-AMR algorithm was applied to the seismic catalog without taking into account any potential heterogeneity.

### 3.2. Atmospheric Data

For the atmospheric analyses, data from the ECMWF (European Center for Medium-range Weather Forecasts) ERA5 climatological reanalysis dataset were utilized. This dataset provides comprehensive reanalysis from 1940 up to 5 days prior to the current date, assimilating a wide range of observations in the upper atmosphere and near-surface regions. In our study, however, we focused on parameters dating back to 1980 [53]. This dataset is known for its consistent coverage in both space and time and is minimally affected by observational conditions such as cloud cover in satellite observations. Nighttime values were specifically considered due to their reduced susceptibility to local meteorological changes. Specifically, we analyzed the parameters of SKT and OLR, which are typically reported to be influenced by impending earthquakes. Several studies [54–56] have demonstrated how these parameters are directly affected by the "thermodynamic channel" in LAIC models. The ECMWF time series for each atmospheric parameter were collected and pre-processed to apply the Climatological Analysis for Seismic Precursor Identification (CAPRI) algorithm [54,55]. This algorithm compares daily time series of the current year with a historical dataset spanning forty-two years (1980–2021), within a temporal window preceding the seismic event, in our case of 90 days. An anomaly is identified if the observed value persistently exceeds the mean of the historical series by two standard deviations. However, for the current study, anomalies were defined as values exceeding 1.5 standard deviations, considering that the earthquake magnitude was below 6. Additionally, the geographic area investigated was determined based on the circular earthquake preparation region (or Dobrovolsky area) centered on the epicenter [57], resulting in the selection of a geographical area of 2° latitude and 2° longitude.

### 3.3. Ionospheric Data

To investigate the ionosphere for potential disturbances associated with the Fano EQ, data from ionosonde and satellite sources were analyzed. From the ground, ionosonde measurements can detect the critical frequency of the F2 layer (foF2), the height (hmF2) of the main electron density peak, and also the information of the sporadic E (Es) layer (such as its height, h'Es; its critical frequency, foEs; and the blanketing frequency, fbEs), which can represent the variations of the corresponding layers. Low Earth Orbit (LEO) satellites can detect the in situ plasma parameters and electromagnetic field. The ionospheric station of Rome (43.98° N; 13.32° E) is located 251.44 km from the epicenter, so within the EQ preparation zone according to the formula by Dobrovolsky et al. [57]. Hourly data manually scaled by an experienced operator from the ionograms recorded with the Advanced Ionospheric Sounder (AIS-INGV) [58] were used in this study [59]. Ionospheric anomalies are defined by significant deviations in the parameters h'Es, fbEs, and foF2 compared to a specified background level determined by 27-day hourly running medians centered on the observation day. Specifically, these deviations are calculated according to the following expressions and must satisfy the following criteria, provided they occur within a few hours under geomagnetically quiet conditions specified by daily geomagnetic index values Ap < 9 nT:

$$\Delta h'Es = \left( h'Es - \left( h'Es \right)_{med} \right) \geq 10 \text{ km} \tag{2}$$

$$\delta f_b Es = \left( f_b Es - \left( f_b Es \right)_{med} \right) / \left( f_b Es \right)_{med} \geq 0.2 \tag{3}$$

$$\delta f_o F2 = \left( f_o F2 - \left( f_o F2 \right)_{med} \right) / \left( f_o F2 \right)_{med} \geq 0.1 \tag{4}$$

where $\Delta$ indicates absolute deviations and $\delta$ indicates relative deviations.

The satellite data, on the other hand, are collected by the ESA Swarm three-satellite constellation (Alpha, Bravo, and Charlie). Alpha and Charlie orbit side by side at an altitude of around 460 km, while Bravo orbits at 510 km. Satellite magnetic data have been analyzed using the MASS (MAgnetic Swarm anomaly detection by Spline analysis) methodology for the magnetic data method (see, e.g., [60]). This technique is used to detect electromagnetic anomalies from Swarm magnetic field data (Level 1B, low resolution of 1 Hz) from the analysis of the three components of the geomagnetic field (X, Y, and Z) and intensity (F) for every track of each satellite (Swarm A, Swarm B, and Swarm C) recorded over the EQ preparation area [57]. The analysis consists of the determination of the first differences of the time series (dX/dt, dY/dt, dZ/dt, and dF/dt) and the removal of the long trend using a cubic spline. Moreover, it is important to evaluate the quality of the data using the quality flags provided by ESA, considering only magnetic quiet times ($|Dst| \leq 20$ nT and ap $\leq 10$ nT) and excluding polar regions ($\pm50°$ geomagnetic latitudes) because they are very disturbed. After these steps, the root mean square (rms) of sliding windows of $7°$ shifting every $1.4°$ was compared with the whole root mean square of the track (RMS). When rms is greater than kt times the value of RMS, the corresponding window is classified as anomalous. For satellite magnetic data, kt is established as 2.5 [60]. This analysis covered a time period ranging from 90 days before the earthquake occurrence to 10 days after and was confined in an area comparable with the EQ preparation region [57].

Since the 1980s, NOAA satellites have continuously circled Earth's orbit, employing a shared instrument for detecting charged particles since 1998. This instrument, the Medium-Energy Proton and Electron Detector (MEPED), is integrated into NOAA satellites and features eight solid-state detectors meticulously crafted to gauge proton and electron counting rates (CRs) within the 30 keV–200 MeV range with a sampling of 2s [61]. These measurements provide invaluable insights into various phenomena, including radiation belt populations, energetic solar proton events, and the low-energy segment of the galactic cosmic-ray population. Specifically tailored for electron detection, the two telescopes within MEPED delineate three energy bands ranging from 30 keV to 2.5 MeV. The first telescope, angled at $9°$ towards the local zenith, captures one perspective, while the second telescope, positioned orthogonally at $90°$ along the satellite's motion, provides a complementary view. Each compact solid-state detector boasts nominal geometric acceptances of 0.1 cm$^2$sr and opening angle apertures of $\pm15°$.

Statistical correlations between strong EQs and electron bursts (EBs) detected by NOAA were mainly observed considering seismic events around the equator in both the West and East Pacific [62,63]. It was in agreement with the bouncing points of the inner Van Allen Belt (VAB), points where electrons go down close to the earth's surface, which are also around the equator and cross the ring of fire twice. However, the inner VAB does not generally overhang latitudes above $35°$, while mid-latitude EQs are located below the slot region between the internal and external VABs. Seismic events occurring at mid latitudes have therefore apparently minor probability to interact with trapped particles. The representation of a possible interaction scenario between mid-latitude seismic events and trapped electrons is depicted in Figure 6, where the possible area to observe EBs connected to the Fano EQ is shown in pink color. These areas have a longitude range of a few tenths of degrees around those of the seismic event. In fact, the longitude of the seismic event is close to the South Atlantic Anomaly (SAA) border longitude, and the electrons' mirror points that are far from the SAA longitude will hardly descend to the satellite altitude. Thus, the probability of bouncing electrons to cross the satellite decreases getting away from the SAA. For what concerns the lower latitude EBs, electrons are thought to be coming (direction indicated with pink arrow) from the inner VAB (in yellow), with the interaction (in blue) running along the magnetic field lines. Instead, for what concerns mid–high latitude EBs, electrons are thought to be escaping the trapped conditions (direction indicated with another pink arrow) from the external VAB (still in yellow), with the interaction (still in blue) connecting the two phenomena along the minimum path. The geomagnetic lines (in green) of the internal VAB can cross the lithosphere up to $30°$–$35°$ in latitude, so being unlikely to be affected by tectonic activity, whereas the radial propagation of LAIC (in blue) should be able to intercept with greater probability the external belts.

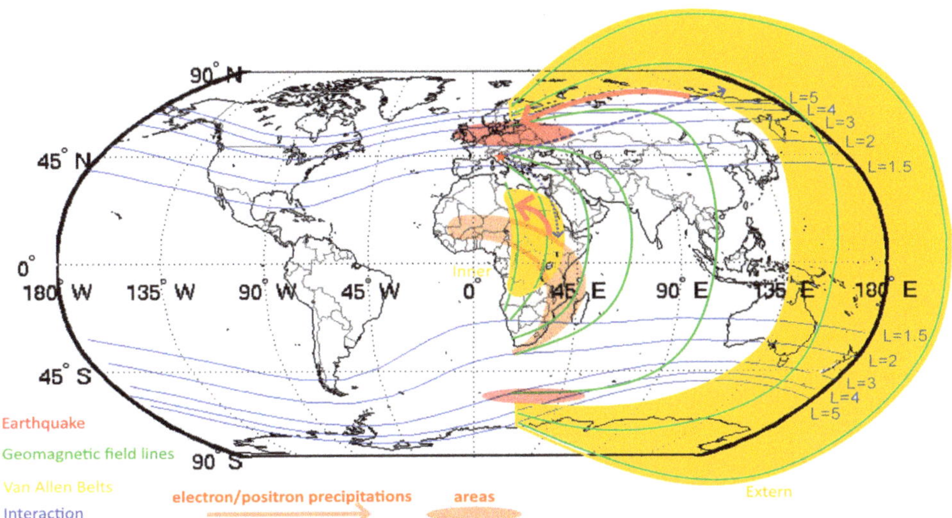


**Figure 6.** Scenario of hypothesized pre-EQ coupling processes between the lithosphere of Central Italy and areas where possible EBs could be detected by LEO satellites.

Given the pivotal role of geomagnetic activity in perturbing the ionosphere, instances where the ap and Dst indices exceeded some predefined thresholds were excluded from the ionospheric data analysis. We also verified that no X-class flares occurred during the detected ionospheric anomalies.

## 4. Results

### 4.1. Lithospheric Data Analysis

The acceleration of seismicity prior to the mainshock was analyzed by applying the R-AMR method [48,49] to the INGV Catalog, in order to highlight a diverging power-law function over time for the cumulative value of Benioff strain. The focal search area was centered on the epicenter of the event, whose responsible fault system has a length of approximately 70 km [41]. Figure 7 shows the result of the R-AMR analysis, excluding the mainshock and its aftershocks. An evident acceleration is observed, characterized by the C value indicating the onset of "critical" behavior relative to the background, which is 0.598; however, the estimated critical time $t_f$ is 200 days after the mainshock. This accelerated behavior was observed in the fault area within a radius ranging from 0 to 50 km from the epicenter. Additionally, the application of this method provides two expected magnitudes, M(A) = 4.8 and M(B) = 4.7, which, although underestimating the real mainshock magnitude, predict an impending EQ that significantly exceeds the background seismicity. The result of this automatic analysis allows us to place the initial progression of seismic acceleration in the lithosphere around September 2019, i.e., 1155 days before the mainshock, resuming in August 2021, i.e., 455 days before the event. These accelerations are visible from the change in slope in the cumulative curve. Around these dates, it is possible to attempt to identify the establishment of different energy transmission channels towards the ionosphere, as hypothesized in LAIC models.

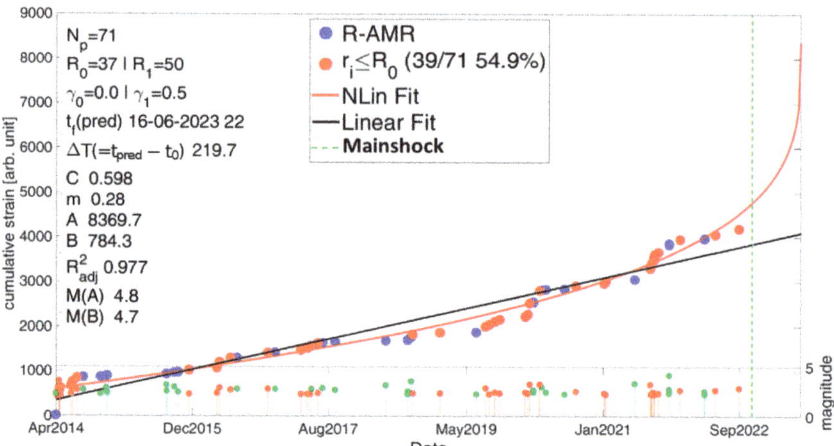



**Figure 7.** Outcome of the R-AMR algorithm applied to the extracted seismic dataset. The red points represent EQs that are closer to the fault (within 37 km) than those represented by the blue points. At the bottom of the main figure, the magnitudes of the involved events are represented: red is used for EQs falling within 37 km from the fault and green those outside that limit.

## 4.2. Atmospheric Data Analysis

SKT and OLR were analyzed to conduct the climatological study. We examined 90 days of ECMWF data preceding the Fano EQ, comparing it with a historical time series spanning the previous 42 years, i.e., from 1980 to 2021. The analysis revealed two highly anomalous days for each atmospheric parameter where the 2022 time series reached the limit of the two standard deviation bands of the historical series. The first anomalous day for SKT occurred on 18 August, i.e., 82 days before the EQ, while the second occurred on 15 September, i.e., 54 days before the EQ (as shown in Figure 8). Furthermore, Figure 9 shows the spatial distribution of these anomalous values, confirming their proximity to the epicenter, especially for the main anomaly. Please note that SKT is defined only on land.

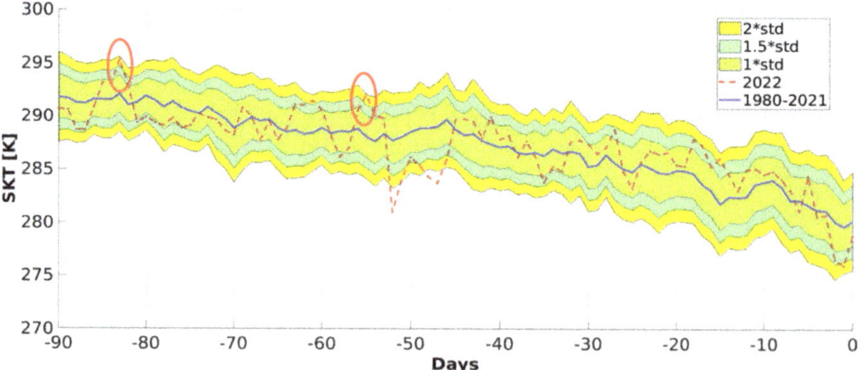



**Figure 8.** Analysis of the SKT parameter for the Fano EQ with comparison between the 2022 time series (dashed red line) and the historical time series (1980–2021, blue line). Evidenced by red circles there are two quite anomalous values near the second standard deviations from the mean: the first one refers to 18 August, and the second one to 15 September.

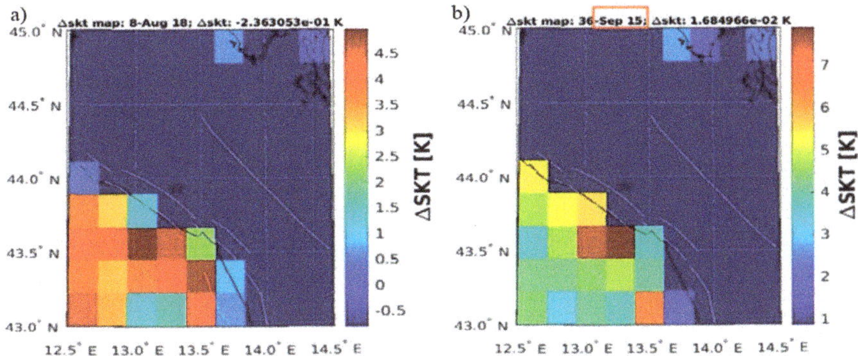

**Figure 9.** Maps of the SKT anomalous days in terms of difference with respect to the historical mean: (**a**) 18 August; (**b**) 15 September. The EQ epicenter is indicated by the central star. SKT is defined only on land.

Regarding the OLR parameter, Figure 10 shows two anomalies exceeding the historical mean by two standard deviations on 5 and 12 September 2022, i.e., respectively, 65 and 58 days before the Fano EQ. Interestingly, both anomalies occurred after the first SKT anomaly but before the second one. Figure 11 depicts the spatial distribution of these OLR anomalies, with the first map (**a**) revealing an extended structure located to the north of the EQ epicenter.

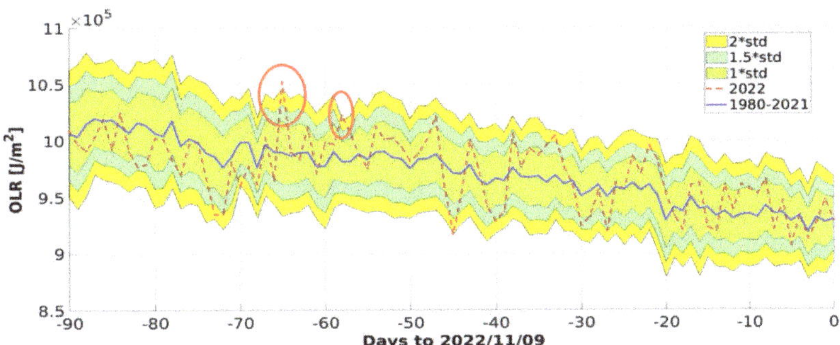

**Figure 10.** Analysis of the OLR for the Fano EQ with the identification of two anomalous days that exceed the historical average calculated from 1980 to 2021 by two standard deviations.

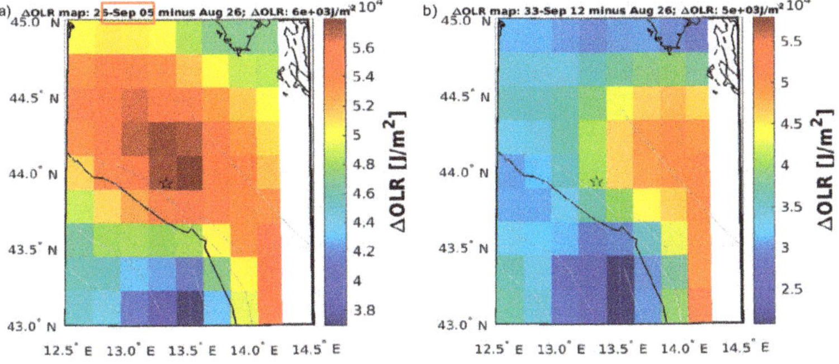

**Figure 11.** Maps of OLR anomalous days maps in terms of difference with respect to the historical mean: (**a**) 5 September; (**b**) 12 September. The epicenter is indicated by the central star.

*4.3. Ionospheric Data Analysis*

4.3.1. Ionosonde Data Analysis

The application of the ionosonde multiparametric approach [64,65], that takes into account the variations of three ionosonde characteristics, h'Es, fbEs, and foF2, manually scaled from hourly ionograms of the AIS-INGV ionosonde of Rome [66], revealed a single anomaly occurring 9 days prior to the Fano EQ (i.e., on 31 October) at 06:00 UT (Table 1; Figure 12). Since the anomaly occurred on a day with the geomagnetic index Ap = 11 nT, the criterion was not strictly satisfied; however, we preferred to take this anomaly into consideration, given the moderate magnitude of the Fano EQ.

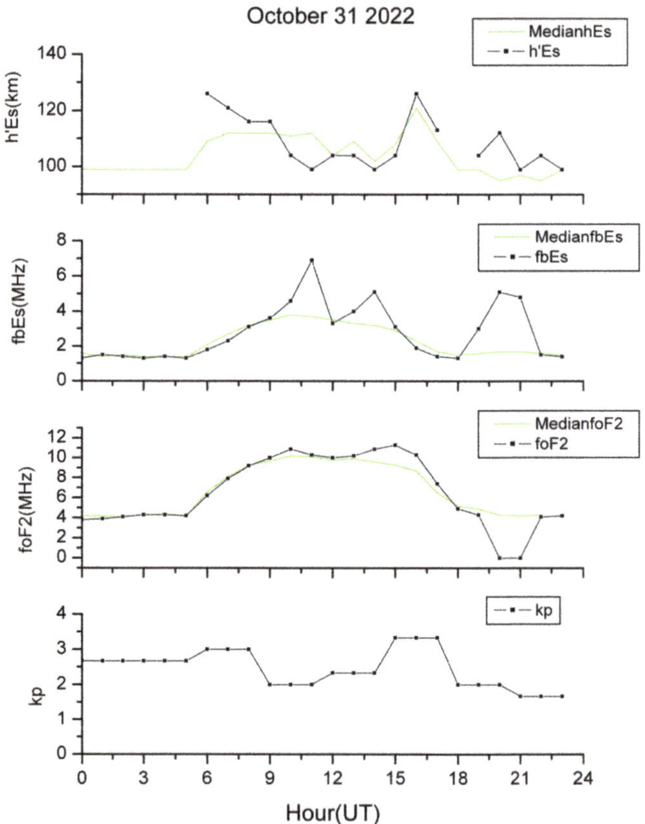


**Figure 12.** The anomaly observed 9 days before the 9 November 2022 Fano EQ using Δh'Es, δfbEs, and δfoF2 variations, along with 3 h Kp index values given as a reference of geomagnetic activity.

**Table 1.** Anomaly detected at the ionospheric station of Rome from ionosonde measurements and possibly related to the Fano EQ.

| Date | Hour(UT) | Δh'Es | δfbEs | δfoF2 | ΔT(Days) | Ap Index | AE Index |
|---|---|---|---|---|---|---|---|
| 31 October 2022 | 06–07 | 06–07 | 0.22 | 0.21 | 9 | =11 nT | <300 nT |

As shown in Figure 13, it is worth noting that the anomaly is consistent with the relationship between ΔT·R and M previously found by the analysis of the most powerful Central Italian EQs since 1984 [66], with ΔT representing the anticipation time (in days), R the distance (in km) between the epicenter and the ionosonde, and M the EQ magnitude.

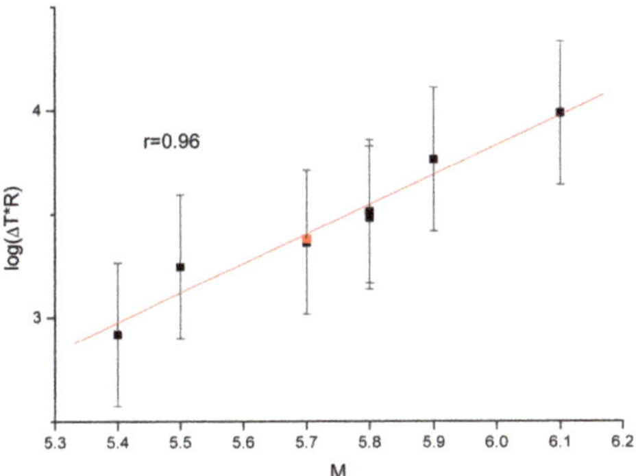



**Figure 13.** Ionosonde anomaly for the 9 November 2022 M5.7 Fano EQ (red square), compared to the relationship between ΔT·R and M previously found by the analysis of the most powerful Central Italian EQs since 1984 (red line and black squares).

### 4.3.2. Swarm Satellite Data Analysis

The analysis of satellite magnetic field data from the Swarm constellation has reported two electromagnetic anomalies, which could be correlated with the Fano EQ. Specifically, applying the MASS algorithm to the Swarm A satellite, considering the 90 days before the EQ and 10 days after, an anomaly was detected 4 days after the EQ (Figure 14a), potentially associated with the aftershock period. Another anomaly was also detected 75 days before the EQ, very close to the edge of the analyzed track (Figure 14b).

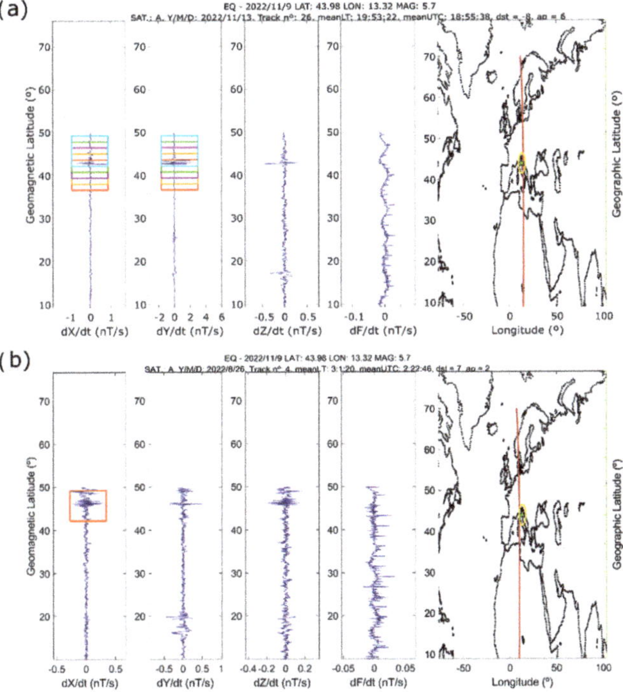



**Figure 14.** Anomalies found 4 days after (**a**) and 75 days before the Fano EQ (**b**) by means of an automatic search for magnetic anomalies 90 days before and 10 days after the EQ; MASS algorithm (kt = 2.5) applied to Swarm A satellite. The anomalies are evidenced by coloured rectangles. The vertical red line on the geographical map represents the satellite track.

### 4.3.3. NOAA Satellite Data Analysis: Electron Burst Data Analysis

NOAA electron fluxes were analyzed two days before the EQ and on the event day. Looking at the geomagnetic indices, the geomagnetic activity of these days was quite calm. Then, some candidate EBs were observed over the expected regions. For example, a flux of about 600 electrons cm$^{-2}$ s$^{-1}$ str$^{-1}$ was observed in the NOAA-15 track on November 8 at a latitude of 50–55° around 20:40 UT, as shown by a red circle in Figure 15. Slight perturbations of around 400 electrons cm$^{-2}$ s$^{-1}$ str$^{-1}$ were also observed at the previous eastward satellite trajectory, all occurring at around 19:00 UT. Thus, these two perturbations anticipated the Central Italy EQ by 9.5 and 11.2 h, respectively.

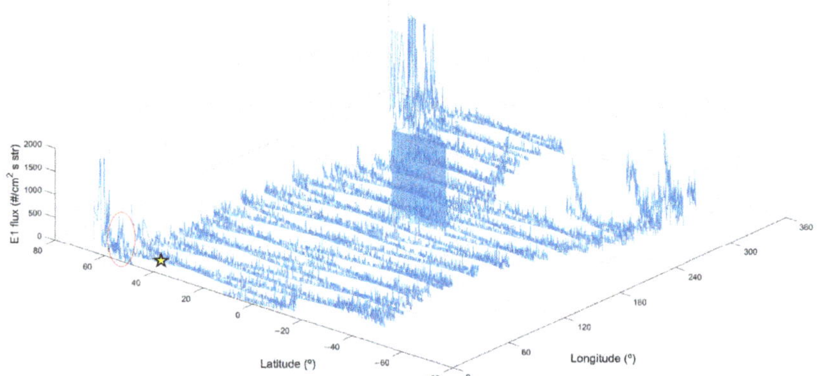

**Figure 15.** Three-dimensional representation of the NOAA-15 semi-orbits on 8 November 2022; EB evidenced by a red circle while the star identified the EQ epicenter.

## 5. Discussion

In this paper, precursor anomalies possibly associated with the 2022 $M_L$ = 5.7 Fano EQ (Marche, Italy) were studied using a multiparametric and multilayer approach, including seismic, atmospheric, and ionospheric parameters. The purpose of this multidisciplinary approach is to gather various contributions and connect them to identify the best LAIC model.

Although the resolutions of data are different, we are confident that, when we assimilate them at daily intervals (see Table 2), we can reconstruct a reliable cumulative number of anomalies. The tracking of the cumulative number of anomalies in chronological order reveals a distinctive behavior, as illustrated in Figure 16.

**Table 2.** List of anomalies detected for the case study of the Fano EQ. From right to left, it shows the analyzed parameter, the day when the anomaly was identified, and the days of occurrence relative to the mainshock. All anomalies appear from bottom (lithosphere) to top (atmosphere and ionosphere). There is only one exception (indicated in bold): a satellite anomaly appears among the atmospheric anomalies.

| Parameter | Date | Days to Mainshock |
|---|---|---|
| R-AMR | 11 September 2019 | −1155 |
| R-AMR | 11 August 2021 | −455 |
| SKT | 18 August 2022 | −82 |
| **Swarm A mag. field** | **26 August 2022** | **−75** |
| OLR | 05 September 2022 | −65 |
| OLR | 12 September 2022 | −58 |
| SKT | 15 September 2022 | −54 |
| Ionosonde | 31 October 2022 | −9 |
| Electron bursts | 08 November 2022 | −1 |
| Swarm A mag. field | 13 November 2022 | +4 |

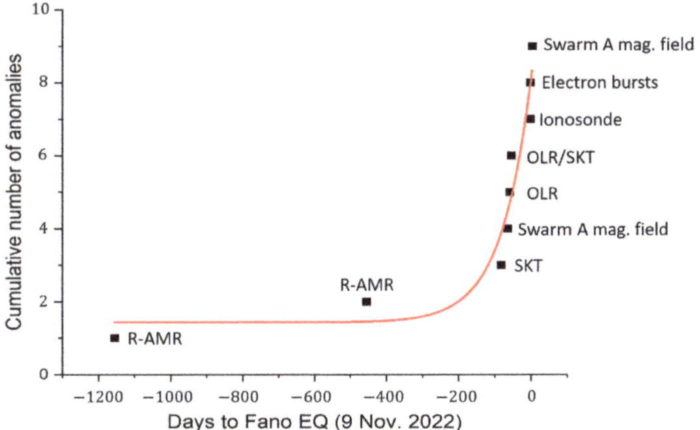


**Figure 16.** In a comprehensive approach of the anomalies, the cumulative number of anomalies for Fano EQ is shown here. It is possible to notice that the anomalies appear in time mostly from below (seismic data in the lithosphere) to above (atmosphere and ionosphere). The red curve is an exponential fit of the data.

An exponential fit (indicated by the red curve) accurately represents the overall acceleration of anomalies. This collective progression of anomalies from different geolayers cannot simply be attributed to chance and is probably associated with the preparation phase of the Fano EQ. Furthermore, as highlighted in Table 2, most anomalies appear chronologically from the lithosphere to the atmosphere and ionosphere. This pattern suggests these can be defined as "thermodynamic anomalies", related to a diffusive–delayed coupling model likely driven by thermodynamic processes. Notably, there is an ionospheric satellite anomaly (indicated in bold in Table 2) amidst the atmospheric anomalies, which could be due to direct electromagnetic coupling between the lithosphere and the ionosphere.

Based on the multiparametric and multilayer analysis, a long-term precursor in the lithosphere was identified. For example, the R-AMR value showed its anomalous acceleration starting about three years before the EQ. From the analysis of atmospheric and ionospheric parameters, anomalies were detected starting 82 days before the Fano EQ. Specifically, the energetic particle signal from NOAA showed an anomaly one day before the seismic event. The Swarm satellites detected an anomaly 75 days before the event and another 4 days after, likely associated with aftershocks. An anomaly was recorded in the ionosonde 9 days before the EQ, and atmospheric anomalies were mainly detected at −82 days and −54 days. The number of anomalies in the atmosphere and ionosphere for this EQ is comparable. This similarity suggests a LAIC behavior of the "thermodynamic" or "diffusive–delayed" coupling, with progression from the lithosphere through the atmosphere to the ionosphere. This progression might correspond to the chemical channel or the acoustic gravity channel, as described in the Hayakawa model [1].

## 6. Conclusions

In conclusion, this study presents results from the retrospective analysis of the major earthquake that occurred in Italy in November 2022, applying a multiparametric and multilayer approach. This approach involved analyzing data from different geophysical layers (lithosphere, atmosphere, and ionosphere) engaged in the coupling process during the earthquake preparation phase. In the ca. 1200 days preceding the Fano EQ, anomalies appeared primarily from the lowest level (seismic data in the lithosphere) to the higher levels (atmosphere and ionosphere), following an overall acceleration pattern. This confirms that the observed anomalies, which originated during the EQ preparation phase and progressed thermodynamically from the lithosphere to the atmosphere and ionosphere, seem to be consistent with the delayed coupling model. However, the presence of a satellite anomaly between other atmospheric anomalies also seems to confirm the possibility of another direct coupling. Therefore, the overall results would confirm a two-way LAIC model.

**Author Contributions:** Conceptualization, M.O. and A.D.S.; Data curation, M.O., A.D.S., M.D.C., L.P., S.A.C., G.C., A.P., S.D., M.C., C.F., D.S. and M.S.; Formal analysis, M.O., A.D.S., M.D.C., L.P., S.A.C., G.C., A.P., S.D., M.C., C.F., D.S. and M.S.; Funding acquisition, A.D.S. and L.P.; Investigation, M.O., A.D.S., M.D.C., L.P., S.A.C., G.C., A.P., S.D., M.C., C.F., A.N., D.S. and M.S.; Methodology, M.O., A.D.S., M.D.C., L.P., S.A.C., G.C., A.P., S.D., M.C., C.F., A.N., D.S. and M.S.; Project administration, A.D.S. and L.P.; Resources, M.O., A.D.S., M.D.C., L.P., S.A.C., G.C., A.P., S.D., M.C., C.F., A.N., D.S. and M.S.; Software, L.P., S.A.C., G.C., A.P. and C.F.; Supervision, M.O., A.D.S., M.D.C. and L.P.; Validation, M.O., A.D.S., M.D.C., L.P., S.A.C., G.C., A.P., S.D., M.C., C.F. and D.S.; Visualization, M.O., A.D.S., M.D.C., L.P., S.A.C., G.C., A.P., S.D., M.C., C.F., A.N., D.S. and M.S.; Writing—original draft, M.O., A.D.S., M.D.C., L.P., S.A.C., G.C., A.P., S.D., C.F. and D.S.; Writing—review and editing, M.O., A.D.S., M.D.C., L.P., S.A.C., G.C., A.P., S.D., C.F., A.N., D.S. and M.S. All authors have read and agreed to the published version of the manuscript.

**Funding:** The authors acknowledge funding provided by the Ministry of University and Research (MUR) for the Project Unitary of "Pianeta Dinamico" ("Working Earth"), INGV for the Project "Further", and the Italian Space Agency (ASI) for the Project "Limadou Scienza +".

**Data Availability Statement:** Publicly available datasets were analyzed in this study. The seismic data provided by the Italian INGV Catalog are available at https://terremoti.ingv.it/en/. The atmospheric data were furnished by the European Centre for Medium-range Weather Forecasts (ECMWF) available at https://www.ecmwf.int/. The analysis of the ionosonde data was possible thanks to the ionosonde dataset of the AIS-INGV in Rome, available at the following link: http://eswua.ingv.it/. The Swarm satellite data are available in the ESA Swarm FTP and HTTP Server https://swarm-diss.eo.esa.int/. Finally, we retrieved the electron flux data from the National Oceanic and Atmospheric Administration, available at https://www.ngdc.noaa.gov/stp/satellite/poes/dataaccess.html. All indicated websites have been accessed on 15 January 2024.

**Acknowledgments:** We would like to thank INGV, ECMWF, ESA, and NOAA for the free availability of their data.

**Conflicts of Interest:** The authors declare no conflicts of interest.

# References

1. Hayakawa, M. Electromagnetic phenomena associated with earthquakes: A frontier in terrestrial electromagnetic noise environment. *Recent Res. Dev. Geophys.* **2004**, *6*, 81–112.
2. Freund, F. Pre-earthquake signals: Underlying physical processes. *J. Asian Earth Sci.* **2011**, *41*, 383–400. [CrossRef]
3. Dahlgren, R.P.; Johnston, M.J.S.; Vanderbilt, V.C.; Nakaba, R.N. Comparison of the Stress-Stimulated Current of Dry and Fluid-Saturated Gabbro Samples. *Bull. Seismol. Soc. Am.* **2014**, *104*, 2662–2672. [CrossRef]
4. Freund, F. Charge generation and propagation in igneous rock. *J. Geodyn.* **2002**, *33*, 543–570. [CrossRef]
5. Freund, F. Electric currents streaming out of stressed igneous rocks: A step towards understanding pre-earthquake low frequency EM emissions. *Phys. Chem. Earth* **2006**, *31*, 389–396. [CrossRef]
6. Freund, F. Pre-earthquake signals, Part I: Deviatoric stresses turn rocks into a source of electric current. *Nat. Hazards Earth Syst. Sci.* **2007**, *7*, 535–541. [CrossRef]
7. Scoville, J.; Heraud, J.; Freund, F. Pre-earthquake magnetic pulses. *Nat. Hazards Earth Syst. Sci.* **2015**, *15*, 1873–1880. [CrossRef]
8. Pulinets, S.; Ouzounov, D. Lithosphere–Atmosphere–Ionosphere Coupling (LAIC) model—An unified concept for earthquake precursors validation. *J. Southeast Asian Earth Sci.* **2011**, *41*, 371–382. [CrossRef]
9. Surkov, V.V.; Pilipenko, V.A.; Silina, A.S. Can Radioactive Emanations in a Seismically Active Region Affect Atmospheric Electricity and the Ionosphere? *Izv. Phys. Solid Earth* **2022**, *58*, 297–305. [CrossRef]
10. Schekotov, A.; Hayakawa, M.; Potirakis, S.M. Does air ionization by radon cause low-frequency atmospheric electromagnetic earthquake precursors? *Nat. Hazards* **2021**, *106*, 701–714. [CrossRef]
11. Hayakawa, M.; Molchanov, O.A.; NASDA/UEC Team. Summary report of NASDA's earthquake remote sensing frontier project. *Phys. Chem. Earth* **2004**, *29*, 617–625. [CrossRef]
12. Molchanov, O.A.; Hayakawa, M. *Seismo Electromagnetics and Related Phenomena: History and Latest Results*; Terra Sceintific Publishing: Tokyo, Japan, 2008; Volume 189.

13. Korepanov, V.; Hayakawa, M.; Yampolski, Y.; Lizunov, G. AGW as a seismo-ionospheric coupling responsible agent. *Phys. Chem. Earth* **2009**, *34*, 485–495. [CrossRef]
14. Akhoondzadeh, M. Investigation of the LAIC mechanism of the Haiti earthquake (August 14, 2021) using CSES-01 satellite observations and other earthquake precursors. *Adv. Space Res.* **2024**, *73*, 672–684. [CrossRef]
15. Sasmal, S.; Chowdhury, S.; Kundu, S.; Politis, D.Z.; Potirakis, S.M.; Balasis, G.; Hayakawa, M.; Chakrabarti, S.K. Pre-seismic irregularities during the 2020 Samos (Greece) earthquake (M = 6.9) as investigated from multi-parameter approach by ground and space-based techniques. *Atmosphere* **2021**, *12*, 1059. [CrossRef]
16. Nayak, K.; Romero-Andrade, R.; Sharma, G.; Cabanillas Zavala, J.L.; Làopez Urias, C.; Trejo Soto, M.E.; Aggarwal, S.P. A combined approach using b-value and ionospheric GPS-TEC for large earthquake precursor detection: A case study for the Colima earthquake of 7.7 Mw, Mexico. *Acta Geod. Geophys.* **2023**, *58*, 515–538. [CrossRef]
17. Nayak, K.; Làopez-Uràias, C.; Romero-Andrade, R.; Sharma, G.; Guzmàan-Acevedo, G.M.; Trejo-Soto, M.E. Ionospheric Total Electron Content (TEC) anomalies as earthquake precursors: Unveiling the geophysical connection leading to the 2023 Moroccan 6.8 Mw earthquake. *Geoscience* **2023**, *13*, 319. [CrossRef]
18. Shah, M.; Calabria Aiber, A.; Arslan Tariq, M.; Ahmed, J.; Ahmed, A. Possible ionosphere and atmosphere precursory analysis related to Mw > 6.0 earthquakes in Japan. *Remote Sens. Environ.* **2020**, *239*, 111620. [CrossRef]
19. Sharma, G.; Nayak, K.; Romero-Andrade, R.; Mohammed Aslam, M.A.; Sarma, K.K.; Aggarwal, S.P. Low ionosphere density above the earthquake epicenter region of Mw 7.2, El Mayor-Cucapah earthquake evident from dense CORS data. *J. Indian Soc. Remote Sens.* **2024**, *52*, 543–555. [CrossRef]
20. Tachema, A. Identifying pre-seismic ionospheric disturbances using space geodesy: A case study of the 2011 Lorca earthquake (Mw 5.1), Spain. *Earth Sci. Inform.* **2024**, *17*, 2055–2071. [CrossRef]
21. De Santis, A.; Cianchini, G.; Marchetti, D.; Piscini, A.; Sabbagh, D.; Perrone, L.; Campuzano, S.A.; Inan, S. A Multiparametric Approach to Study the Preparation Phase of the 2019 M7.1 Ridgecrest (California, United States) Earthquake. *Front. Earth Sci.* **2020**, *8*, 40398. [CrossRef]
22. Wang, Y.; Ma, W.; Zhao, B.; Yue, C.; Zhu, P.; Yu, C.; Yao, L. Responses to the preparation of the 2021 M7.4 Madoi earthquake in the Lithosphere-Atmosphere-Ionosphere System. *Atmosphere* **2023**, *14*, 1315. [CrossRef]
23. Akhoondzadeh, M.; Marchetti, D. Study of the preparation phase of Turkey's powerful earthquake (6 February 2023) by a geophysical multi-parametric fuzzy interference system. *Remote Sens.* **2023**, *15*, 2224. [CrossRef]
24. Pezzo, G.; Billi, A.; Carminati, E.; Conti, A.; De Gori, P.; Devoti, R.; Lucente, F.P.; Palano, M.; Petracchini, L.; Serpelloni, E.; et al. Seismic source identification of the 9 November 2022 Mw 5.5 offshore Adriatic sea (Italy) earthquake from GNSS data and aftershock relocation. *Sci. Rep.* **2023**, *13*, 11474. [CrossRef]
25. Grunthal, G. *European Macroseismic Scale 1998 (EMS-98). Cahiers du Centre Europeen de Ge ´odynamique et de Seismologie 15*; Centre Européen de Geodynamique et de Seismologie: Luxembourg, 1998; 99p.
26. Tertulliani, A.; Antonucci, A.; Berardi, M.; Borghi, A.; Brunelli, G.; Caracciolo, C.H.; Castellano, C.; D'Amico, V.; Del Mese, S.; Ercolani, E.; et al. (Gruppo Operativo Quest). Rapporto Macrosismico sul Terremoto del 9/11/2022 Della Costa Marchigiana [Data set]. Istituto Nazionale di Geofisica e Vulcanologia (INGV). 2022. Available online: https://quest.ingv.it/rilievi-macrosismici (accessed on 16 May 2024).
27. Maesano, F.E.; Buttinelli, M.; Maffucci, R.; Toscani, G.; Basili, R.; Bonini, L.; Burrato, P.; Fedorik, J.; Fracassi, U.; Panara, Y.; et al. Buried alive: Imaging the 9 November 2022, Mw 5.5 earthquake source on the offshore Adriatic blind thrust front of the Northern Apennines (Italy). *Geophys. Res. Lett.* **2023**, *50*, e2022GL102299. [CrossRef]
28. Dezi, F.; Merli, A. Chiaradonna A. Soil liquefaction potential of the central Adriatic Coast after the 9th of November 2022 earthquake (Italy). *Jpn. Geotech. Soc. Spec. Publ.* **2024**, *10*, 597–602. [CrossRef]
29. Montone, P.; Mariucci, M.T. Deep well new data in the area of the 2022 Mw 5.5 earthquake, Adriatic Sea, Italy: In situ stress state and P-velocities. *Front. Earth Sci.* **2023**, *11*, 1164929. [CrossRef]
30. Console, R.; Vannoli, P.; Carluccio, R. The 2022 Seismic Sequence in the Northern Adriatic Sea and Its Long-Term Simulation. *Appl. Sci.* **2023**, *13*, 3746. [CrossRef]
31. Vannoli, P.; Basili, R.; Valensise, G. New geomorphic evidence for anticlinal growth driven by blind-thrust faulting along the northern Marche coastal belt (central Italy). *J. Seismol.* **2004**, *8*, 297–312. [CrossRef]
32. Casero, P.; Bigi, S. Structural setting of the Adriatic basin and the main related petroleum exploration plays. *Mar. Pet. Geol.* **2013**, *42*, 135–147. [CrossRef]
33. Ziegler, P.A. Compressional Irma plate deformation in the Alpine foreland. *Tectonophysics* **1987**, *137*, 420.
34. Ziegler, P.A.; Cloetingh, S.; van Wees, J.D. Dynamics of intra-plate compressional deformation: The Alpine foreland and other examples. *Tectonophysics* **1995**, *252*, 7–59. [CrossRef]
35. Doglioni, C. A proposal for the kinematic modelling of w-dipping subductions—Possible applications to the Tyrrhenian-Apennines system. *Terra Nova* **1991**, *3*, 423–434. [CrossRef]
36. Bertotti, G.; Picotti, V.; Chilovi, C.; Fantoni, R.; Merlini, S.; Mosconi, A. Neogene to quaternary sedimentary basins in the South Adriatic (Central Mediterranean): Foredeeps and lithospheric buckling. *Tectonics* **2001**, *20*, 771–787. [CrossRef]
37. Faccenna, C.; Jolivet, L.; Piromallo, C.; Morelli, A. Subduction and the depth of convection in the Mediterranean mantle. *J. Geophys. Res.* **2003**, *108*, 2099. [CrossRef]
38. Doglioni, C.; Carminati, E.; Cuffaro, M. Simple kinematics of subduction zones. *Int. Geol. Rev.* **2006**, *48*, 479–493. [CrossRef]

39. Patacca, E.; Scandone, P.; Di Luzio, E.; Cavinato, G.P.; Parotto, M. Structural architecture of the central Apennines. Interpretation of the CROP 11 seismic profile from the Adriatic coast to the orographic divide. *Tectonics* **2008**, *27*, TC3006. [CrossRef]
40. *Consiglio Nazionale delle Ricerche, Structural Model of Italy and Gravity Map*; Quaderni de "La Ricerca Scientifica"; Progetto Finalizzato Geodinamica: Roma, Italy, 1992; Volume 3.
41. DISS_Working_Group. *Database of Individual Seismogenic Sources (DISS), Version 3.3.0: A Compilation of Potential Sources for Earthquakes Larger Than M 5.5 in Italy and Surrounding Areas*; Istituto Nazionale di Geofisica e Vulcanologia: Roma, Italy, 2021.
42. Fantoni, R.; Franciosi, R. Tectono-sedimentary setting of the Po Plain and Adriatic foreland. *Rend. Fis. Acc. Lincei* **2010**, *21*, 197–209. [CrossRef]
43. Kastelic, V.; Vannoli, P.; Burrato, P.; Fracassi, U.; Tiberti, M.M.; Valensise, G. Seismogenic sources in the Adriatic Domain. *Mar. Pet. Geol.* **2013**, *42*, 191–213. [CrossRef]
44. Elter, P.; Giglia, G.; Tongiorgi, M.; Trevisan, L. Tensional and Compressional Areas in the Recent (Tortonian to Present) Evolution of the Northern Apennines. *Boll. Geofis. Teor. Appl.* **1975**, *17*, 3–18.
45. Rovida, A.N.; Locati, M.; Camassi, R.D.; Lolli, B.; Gasperini, P.; Antonucci, A. *Catalogo Parametrico dei Terremoti Italiani CPTI15; Versione 4.0*; Istituto Nazionale di Geofisica e Vulcanologia (INGV): Roma, Italy, 2022.
46. Vannoli, P.; Vannucci, G.; Bernardi, F.; Palombo, B.; Ferrari, G. The source of the 30 October 1930 M w 5.8 Senigallia (Central Italy) earthquake: A convergent solution from instrumental, macroseismic, and geological data. *Bull. Seismol. Soc. Am.* **2015**, *105*, 1548–1561. [CrossRef]
47. Istituto Nazionale di Geofisica e Vulcanologia (INGV). Earthquake List with Real Time Updates. Available online: https://terremoti.ingv.it/en/ (accessed on 15 January 2024).
48. De Santis, A.; Cianchini, G.; Di Giovambattista, R. Accelerating moment release revisited: Examples of application to Italian seismic sequences. *Tectonophysics* **2015**, *639*, 82–98. [CrossRef]
49. Cianchini, G.; De Santis, A.; Di Giovambattista, R.; Abbattista, C.; Amoruso, L.; Campuzano, S.A.; Carbone, M.; Cesaroni, C.; De Santis, A.; Marchetti, D.; et al. Revised Accelerated Moment Release Under Test: Fourteen Worldwide Real Case Studies in 2014–2018 and Simulations. *Pure Appl. Geophys.* **2020**, *177*, 4057–4087. [CrossRef]
50. Wiemer, S.; Wyss, M. Minimum magnitude of complete reporting in earthquake catalogs: Examples from alaska, the western United States, and Japan. *Bull. Seismol. Soc. Am.* **2000**, *90*, 859–869. [CrossRef]
51. Bowman, D.D.; Ouillon, G.; Sammis, C.G.; Sornette, A.; Sornette, D. An observational test of the critical earthquake concept. *J. Geophys. Res.* **1998**, *103*, 24359–24372. [CrossRef]
52. Jaume´, S.; Sykes, L.R. Evolving towards a critical point: A review of accelerating seismic moment/energy release prior to large and great earthquakes. *Pure Appl. Geophys.* **1999**, *155*, 279–305. [CrossRef]
53. Hersbach, H.; Bell, B.; Berrisford, P.; Biavati, G.; Horányi, A.; Muñoz Sabater, J.; Nicolas, J.; Peubey, C.; Radu, R.; Rozum, I.; et al. *ERA5 Hourly Data on Single Levels from 1940 to Present*; Copernicus Climate Change Service (C3S) Climate Data Store (CDS): Reading, UK, 2018. [CrossRef]
54. Piscini, A.; De Santis, A.; Marchetti, D.; Cianchini, G. A Multi-parametric Climatological Approach to Study the 2016 Amatrice–Norcia (Central Italy) Earthquake Preparatory Phase. *Pure Appl. Geophys.* **2017**, *174*, 3673–3688. [CrossRef]
55. Piscini, A.; Marchetti, D.; De Santis, A. Multi-Parametric Climatological Analysis Associated with Global Significant Volcanic Eruptions During 2002–2017. *Pure Appl. Geophys.* **2019**, *176*, 3629–3647. [CrossRef]
56. Ouzounov, D.; Liu, D.; Chunli, K.; Cervone, G.; Kafatos, M.; Taylor, P. Outgoing long wave radiation variability 1086 from IR satellite data prior to major earthquakes. *Tectonophysics*, **2007**; *431*, 211–220. [CrossRef]
57. Dobrovolsky, I.R.; Zubkov, S.I.; Myachkin, V.I. Estimation of the size of earthquake preparation zones. *Pure Appl. Geophys.* **1979**, *117*, 1025–1044. [CrossRef]
58. Zuccheretti, E.; Tutone, G.; Sciacca, U.; Bianchi, C.; Arokiasamy, B.J. The new AIS-INGV digital ionosonde. *Ann. Geophys.* **2003**, *46*, 647–659. [CrossRef]
59. Upper Atmosphere Physics and Radiopropagation Working Group; Cossari, A.; Fontana, G.; Marcocci, C.; Pau, S.; Pezzopane, M.; Pica, E.; Zuccheretti, E. *Electronic Space Weather Upper Atmosphere Database (eSWua)—HF Validated Data (1.0)*; Istituto Nazionale di Geofisica e Vulcanologia (INGV): Roma, Italy, 2020.
60. De Santis, A.; Marchetti, D.; Spogli, L.; Cianchini, G.; Pavón-Carrasco, F.J.; De Franceschi, G.; Di Giovambattista, R.; Perrone, L.; Qamili, E.; Cesaroni, C.; et al. Magnetic field and electron density data analysis from Swarm satellites searching for ionospheric effects by great earthquakes: 12 case studies from 2014 to 2016. *Atmosphere* **2019**, *10*, 371. [CrossRef]
61. Evans, D.S.; Greer, M.S. *Polar Orbiting Environmental Satellite Space Environment Monitor—2: Instrument Descriptions and Archive Data Documentation, NOAA Technical Memorandum January*; Version 1.4; NOAA National Geophysicial Data Centre: Boulder, CO, USA, 2004; p. 155. Available online: https://ngdc.noaa.gov/stp/satellite/poes/docs/SEM2v1.4b.pdf (accessed on 16 May 2024).
62. Fidani, C. West Pacific Earthquake Forecasting Using NOAA Electron Bursts With Independent LShells and Ground-Based Magnetic Correlations. *Front. Earth Sci.* **2021**, *9*, 673105. [CrossRef]
63. Fidani, C. The Conditional Probability of Correlating East Pacific Earthquakes with NOAA Electron Bursts. *Appl. Sci.* **2022**, *12*, 10528. [CrossRef]
64. Perrone, L.; Korsunova, L.P.; Mikhailov, A.V. Ionospheric precursors for crustal earthquakes in Italy. *Ann. Geophys.* **2010**, *28*, 941–950. [CrossRef]

65. Perrone, L.; De Santis, A.; Abbattista, C.; Alfonsi, L.; Amoruso, L.; Carbone, M.; Cesaroni, C.; Cianchini, G.; De Franceschi, G.; De Santis, A.; et al. Ionospheric Anomalies Detected by Ionosonde and Possibly Related to Crustal Earthquakes in Greece. *Ann. Geophys.* **2018**, *36*, 361–371. [CrossRef]
66. Ippolito, A.; Perrone, L.; De Santis, A.; Sabbagh, D. Ionosonde Data Analysis in Relation to the 2016 Central Italian Earthquakes. *Geosciences* **2020**, *10*, 354. [CrossRef]

*geoscasciences*

MDPI

*Article*

# On the Impact of Geospace Weather on the Occurrence of M7.8/M7.5 Earthquakes on 6 February 2023 (Turkey), Possibly Associated with the Geomagnetic Storm of 7 November 2022

**Dimitar Ouzounov [1],* and Galina Khachikyan [2]**

[1]  Institute for Earth, Computing, Human and Observing, Chapman University, Orange, CA 92866, USA
[2]  National Scientific Center for Seismological Observations and Research, Almaty 050060, Kazakhstan; galina.khachikyan@seismology.kz
*   Correspondence: ouzounov@chapman.edu

**Abstract:** A joint analysis of solar wind, geomagnetic field, and earthquake catalog data showed that before the catastrophic M = 7.8 and M = 7.5 Kahramanmaras earthquake sequence on 6 February 2023, a closed strong magnetic storm occurred on 7 November 2022, SYM/H = −117 nT. The storm started at 08:04 UT. At this time, the high-latitudinal part of Turkey's longitudinal region of future epicenters was located under the polar cusp, where the solar wind plasma would directly access the Earth's environment. The time delay between storm onset and earthquake occurrence was ~91 days. We analyzed all seven strong (M7+) earthquakes from 1967 to 2020 to verify the initial findings. A similar pattern has been revealed for all events. The time delay between magnetic storm onset and earthquake occurrence varies from days to months. To continue these investigations, a retrospective analysis of seismic and other geophysical parameters just after preceded geomagnetic storms in the epicenter areas is desirable.

**Keywords:** solar wind; geomagnetic storm onset; polar cusp; geospace weather; magnetic local time; earthquake

**Citation:** Ouzounov, D.; Khachikyan, G. On the Impact of Geospace Weather on the Occurrence of M7.8/M7.5 Earthquakes on 6 February 2023 (Turkey), Possibly Associated with the Geomagnetic Storm of 7 November 2022. *Geosciences* **2024**, *14*, 159. https://doi.org/10.3390/geosciences14060159

Academic Editors: Jesus Martinez-Frias and Masashi Hayakawa

Received: 30 April 2024
Revised: 28 May 2024
Accepted: 4 June 2024
Published: 7 June 2024



## 1. Introduction

It has been found in some papers, for example [1–5] and references therein, that earthquake occurrence may be preceded by a geomagnetic storm, which is one of Earth's most striking manifestations of solar wind activity. A lag time between a magnetic storm onset and an earthquake occurrence varies; it could be about 2–6 days [1], 12–14 days [2], 26–27 days [3], several months [4], and for very large earthquakes (M7.5+), it may reach up to multiple years [5]. Nevertheless, the idea of a relationship between earthquake occurrence and a magnetic storm has been considered controversial up to now. In this direction, the most famous and cited work is [6], where at a high statistical level, a hypothesis was analyzed to see if the solarterrestrial interaction, as measured by sunspots, solar wind velocity, and geomagnetic activity, might play a role in triggering earthquakes. The authors of [6] counted the number of earthquakes occurring globally with magnitudes exceeding chosen thresholds in calendar years, months, and days. Then, they ordered these counts by the corresponding rank of the solarterrestrial variables' annual, monthly, and daily averages. A statistical significance of the difference between the earthquake number distributions below and above the median of the solar–terrestrial averages was estimated by $\chi^2$ and Student's $t$-tests. They found no consistent and statistically significant distributional differences. When they introduced time lags (±5 days) between the solar–terrestrial variables and the number of earthquakes, again, no statistically significant distributional difference was found. Since they could not reject the null hypothesis of no solar–terrestrial triggering of earthquakes, they concluded that there was insignificant solar–terrestrial triggering of earthquakes. However, the results of [6] can be interpreted so that the Earth's crust does not respond to solar events immediately or over the next five days. In [7], a similar result was obtained for a limited territory (Anatolian Peninsula). This work investigated the ratio of earthquakes that occurred during geomagnetically disturbed days (Dst $\leq -30$ nT) to those that occurred on geomagnetically quiet days using data on 122,838 events with magnitudes from 3.0 to 7.9 that occurred from 1965 to 2005 in the Anatolian Peninsula. It was concluded: "As a result of all these data, a hypothesis cannot be put forward which suggests that geomagnetic storms trigger earthquakes in the Anatolian peninsula". In the same work [7], the author concluded, "However, these results should not hinder the conduct of further research. A global study on this subject can potentially provide new approaches". Indeed, a new, to some extent, approach (discussed below) gives us a hint that the solar–terrestrial interaction has an impact on earthquake occurrence, but its nature may not be a trigger. The basis for this approach is our previous results [4], which showed that sometimes earthquakes can look like targeted earthquakes. Namely, they may occur near the footprint in the Earth's crust of those geomagnetic lines, which were populated by high-energy electrons pouring out from the outer radiation belt downwards at the time of geomagnetic storms, in other words, when conductivity in surrounding media is increased. Increasing conductivity in the ionosphere, for example, is a key parameter in the mathematical model [8], which considers the hypothesis of the electromagnetic generation of earthquakes due to the influence of solar flare energy on the ionosphere–atmosphere–lithosphere system and the intensification of telluric currents in the lithosphere, including around tectonic faults. The essence of the model [8] is that the absorption of solar flare radiation in the ionosphere creates additional ionization in it, which is accompanied by the appearance of an additional electric current and an additional electric field, which will ultimately lead to an increase in the telluric current in the Earth's crust. It has been well known for many decades that at the dayside of the high latitudes, there is a funnel-shaped area (polar cusp) where the solar wind plasma has direct access to the atmosphere. As summarized in [9], the polar cusp is a vital connection point for the solar wind–magnetosphere–ionosphere interaction, where the plasma density irregularities have a wide range of spatial scales. The reconnection of the solar wind magnetic field with the geomagnetic field at the dayside magnetopause impacts the polar cusp through flux transfer events that enhance ionospheric flow, input to the appearance of the field-aligned currents, and auroral particle precipitation. Often, polar-orbiting spacecraft observe Alfven waves with scale sizes perpendicular to the geomagnetic field of the order of an electron skin depth [10]. In the cusp, the density of the neutral atmosphere is always increased, on average, by one and a half times, relative to the density in neighboring areas [11,12]. From a magnetic field point of view, the polar cusp is a funnel-shaped region where the high-latitude dayside (compressed) and nightside (elongated) magnetic field lines converge toward the geomagnetic poles [13]. Cusps have small sizes and small ionospheric footprints; nevertheless, they are essential in transferring solar wind energy, momentum, and plasma to the atmosphere. The investigation of the penetration through the polar cusps of the shocked solar wind is one of the primary science objectives of the Cluster mission, which is composed of four identical spacecraft flying around the Earth [14]. The Cluster data revealed [15] that most of the time, the cusp is located at magnetic latitudes ~75–80° and in a longitudinal region of 10–14 h magnetic ocal time (MLT). However, sometimes its longitudinal extension may be more comprehensive, depending, for example, on the length of the reconnection X-line at the magnetopause [16]. So, data from the Polar satellite demonstrate [17] that the polar cusp may be between 8 and 16 h MLT. Considering the above, it can be assumed that strong earthquakes can occur "targeted" in the longitudinal region, in which its high-latitude area was located under the polar cusp at the time of the arrival of the shocked solar wind (at the time of geomagnetic storm onset). Below, we test this assumption.

## 2. Materials and Methods

This study investigates earthquakes from the United States Geological Survey (USGS) global seismological catalog (https://earthquake.usgs.gov/earthquakes/search/ accessed on 20 February 2023), which presents the moment magnitude (Mw, hereafter referred to as M for simplicity). The data on the solar wind parameters are taken from the OMNI database (http://cdaweb.gsfc.nasa.gov, accessed on 20 February 2023), which was obtained from current and past space missions and projects. From the World Data Center for Geomagnetism, Kyoto (https://wdc.kugi.kyoto-u.ac.jp/ accessed on 20 February 2023), the onset and intensity of the geomagnetic storms were revealed using the 1 h Dst (Disturbance Storm Time Index) before 1981 and the 1 min "SYM-H" index after 1981, since the latter is absent before 1981. The "SYM-H" index, measured at the Earth's surface, is, in fact, the high-resolution Dst index [18], allowing one to determine the onset and intensity of a magnetic storm more precisely. According to [19], depending on the Dst value, geomagnetic storms are classified into weak (Dst from −30 to −50 nT), moderate (Dst from −50 to −100 nT), strong (Dst from −100 to −200 nT), powerful (Dst from −200 to −350 nT), and extreme (Dst below −350 nT). Also, we used data on the storm sudden commencement—SSC, which were obtained by the Observatorio del Ebro, Roquetes, Spain, from the web page (ftp://ftp.ngdc.noaa.gov/STP/SOLARDATA/suddencommencements/storm2.SSC, accessed on 20 February 2023). We suggest that the occurrence of a strong earthquake may follow a particular geomagnetic storm in only a specific longitudinal region (depending on the time of geomagnetic storm onset). At this step, the method of investigation was manual (visual). Since we know the date of the earthquake and the coordinates of its epicenter, we visually determine those preceded by geomagnetic storms, during the beginning of which (in time of SSC) the longitude of the earthquake epicenter was located under the polar cusp; that is, the magnetic local time at the point of the future epicenter was within 8–16 h. The MLT is often used to describe the position in near-Earth space because the longitude, which rotates with the Earth, is not a helpful parameter for this (https://ecss.nl/item/?glossary_id=1619, accessed on 20 February 2023). The MLT has a value of 0 (midnight) in the anti-sunward direction, noon in the sunward direction, and 6 (dawn) and 18 (dusk) perpendicular to the sunward/anti-sunward line. This system is essential for understanding phenomena in geospace, such as the aurora, plasma motion, ionospheric currents, and associated magnetic field disturbances, which are highly organized by Earth's main magnetic field. The MLT values were estimated using the online program (https://omniweb.gsfc.nasa.gov/vitmo/cgm.html, accessed on 20 February 2023). In Table 1, the data on the nine analyzed earthquakes are presented, and in Figure 1, the location of their epicenters is shown.

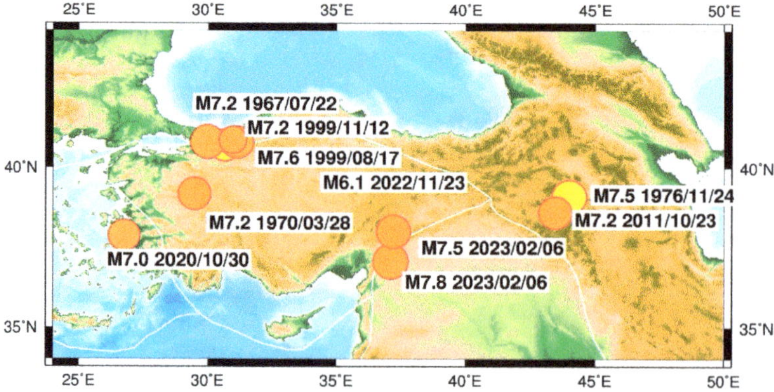


**Figure 1.** A map of strong earthquakes M ≥ 7.0 that occurred inside and around the Anatolian Plate in 1967–2023. The orange indicates a depth of <30 km, and the yellow indicates earthquakes with a depth of >30 km. An added epicenter of M = 6.1 occurred on 23 November 2022 and is discussed in the text.

**Table 1.** Data on strong M ≥ 7.0 earthquakes inside and around the Anatolian Plate, preceding geomagnetic storms, magnetic local time at the area of an epicenter at the time of geomagnetic storm onset, and lag time between storm onset and earthquake occurrence. Shades of gray indicate two Prime events. in this manuscript. Everything starts with these two events.

| | Earthquake Catalog (USGS) | | | | Geomagnetic Storm (Date, Intensity, Time of Storm Onset, and Positive SYM/H Index (nT) | | | | Magnetic Local Time in the Epicenter at Time of Storm Onset (Hour) | Time Lag between Storm Onset and EQ (Days) |
| | Date Time (UTC) | Lat./Long. | H (km) | M | Date | Intensity (nT) | Class | Onset (UTC) | | |
|---|---|---|---|---|---|---|---|---|---|---|
| 1 | 6 February 2023, 01:17:34 | 37.226° N, 37.014° E | 10 | 7.8 | 7 November 2022 | −117 | Strong | +10 nT at 08:04 | 10.96 | 91 |
| 2 | 6 February 2023, 10:24:48 | 38.011° N, 37.196° E | 7.4 | 7.5 | 7 November 2022 | −117 | Strong | +10 nT at 08:04 | 10.97 | 91 |
| 3 | 30 October 2020, 11:51:27 | 37.897° N 26.784° E | 21 | 7 | 2 August 2020 | −39 | Small | +27 nT at 09:23 | 11.66 | 89 |
| | | | | | 5 October 2020 | −37 | Small | +1 nT at 08.12 | 10.48 | 25 |
| 4 | 23 October 2011, 10:41:23 | 38.721° N, 43.508° E | 18 | 7.1 | 9 September 2011 | −77 | Moderate | +74 nT at 13:16 | 16.63 | 44 |
| | | | | | 17 September 2011 | −43 | Small | +61 nT at 8:10 | 11.53 | 36 |
| | | | | | 26 September 2011 | −111 | Strong | +62 nT at12:38 | 15.99 | 27 |
| 5 | 12 November 1999, 16:57:19 | 40.758° N, 31.161° E | 10 | 7.2 | 22 September 1999 | −166 | Strong | +33 nT at 12:34 | 15.27 | 51 |
| | | | | | 21 October 1999 | −211 | Powerful | +42 nT at 7:06 * | 9.8 | 21 |
| 6 | 17 August 1999, 00:01:39 | 40.748° N, 29.864° E | 17 | 7.6 | 16 April 1999 | −123 | Strong | +10 nT at 11:25 | 14.04 | 123 |
| | | | | | 15 August 1999 | −44 | Small | +36 nT at 11:52 | 14.49 | 1.5 |
| 7 | 24 November 1976, 12:22:18 | 39.121° N, 44.029° E | 36 | 7.3 | 30 October 1976 | −57 | Moderate | +18 nT at 10:30 | 14.03 | 25 |
| 8 | 28 March 1970, 21:02:26 | 39.098° N 29.570° E | 25 | 7.2 | 15 January 1970 | −51 | Small | +20 nT at 9:30 | 12.2 | 72 |
| | | | | | 27 March 1970 | −52 | Small | +44 nT at 8:30 | 11.2 | 1.5 |
| 9 | 22 July 1967, 16:57:00 | 40.751° N, 30.8° E | 30 | 7.3 | 25 May 1967 | −387 | Extreme | +55 nT at 12:30 | 15.29 | 58 |

* marks the time of a dense solar wind flux arrival that resulted in a sharp increase in the solar wind dynamic pressure at the magnetopause.

### 2.1. Case Study for M = 7.8 and M = 7.5 of 6 February 2023

Two catastrophic earthquakes in Turkey on 6 February 2023, M = 7.8 at 01:17:34 UTC with epicenter 37.226° N, 37.014° E, 10.0 km depth, and M = 7.5 at 10:24:48 UTC with epicenter 38.011° N, 37.196° E, 7.4 km depth, were preceded by a strong geomagnetic storm on 7 November 2022, with the most significant negative "SYM/H" = −117 nT (Figure 2).

This storm started at 08:04 UT with positive peak "SYM/H" = +10 nT. At this time, in the area of the future M = 7.8 earthquake epicenter (37.226° N, 37.014° E), the magnetic local time was equal to MLT = ~10.96 h, and in the area of the future M = 7.5 earthquake epicenter (38.011° N, 37.196° E), it was equal to MLT = ~10.97 h. Thus, the storm on 7 November 2022, met a given criterion: at the time of its onset, the high-latitude part of the longitudinal region where, in the future, the Kahramanmaras earthquake sequence occurred was located under the polar cusp.

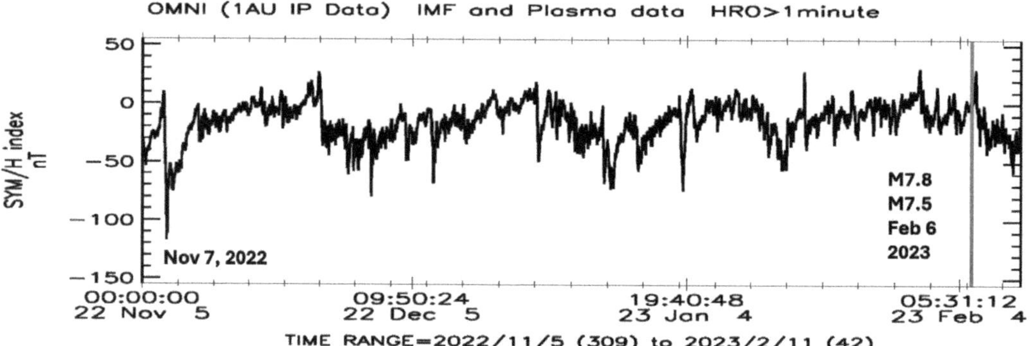


**Figure 2.** The 1 min data on the geomagnetic "SYM/H" index from 1 November 2022 to 10 February 2023; on the right, a Red line marks the date of the M = 7.8 and M = 7.5 Kahramanmaras earthquake sequence on 6 February 2023; on the left, a preceding geomagnetic storm on 7 November 2022, is indicated.

To consider in more detail the solar wind (space weather) parameters that provoked this geomagnetic storm and then earthquakes (as we suppose), let us consider Figure 3, which presents not only the "SYM/H" index but also the solar wind proton density, solar wind dynamic pressure at the magnetopause, and the vertical component of the solar wind magnetic field (interplanetary magnetic field) in the GSM coordinate system (Bz_GSM), which is a critical parameter for development of a magnetic storm. In this plot, we show data from the OMNI database for only a short time interval (6–9 November 2022) for better visualization. It can be seen from Figure 3 that on 7 November 2022, at 06:25 UT, the "SYM/H" index changed its value from negative to positive and reached its positive peak at 08:04 UT. The solar wind flux density (n) and, accordingly, the dynamic pressure of the solar wind on the dayside magnetopause (P) also started to increase and reached their peaks on 7 November 2022, at 10:23 UT (n = 27.88 cm$^{-3}$, P = 8.81 nPa). A change in the orientation of the vertical component of the interplanetary magnetic field Bz_GSM from positive to negative (at which an effective reconnection of the solar wind magnetic lines with geomagnetic lines occurs) began at 09:31 UT. At 10:37 UT, the "SYM/H" index changed its value from positive to negative (the main phase of the geomagnetic storm started). Thus, the initial phase of the magnetic storm on 7 November 2022 lasted from ~08:04 UT to ~10:37 UT. At 10:37 UT, the magnetic local time in the areas of the future Kahramanmaras earthquake sequence was equal to 13.51 h and 13.52 h, respectively. This means that during the initial storm phase, the high-latitudinal area of the Kahramanmaras longitudinal region was located under the polar cusp. The time delay between geomagnetic storm onset and earthquake occurrence equals ~91 days (Table 1).

To check and confirm that the result obtained for the Kahramanmaras earthquake sequence is not random, we carried out a similar analysis for the other seven strong M ≥ 7.0 earthquakes inside and around the Anatolian Plate (Table 1, Figure 1).

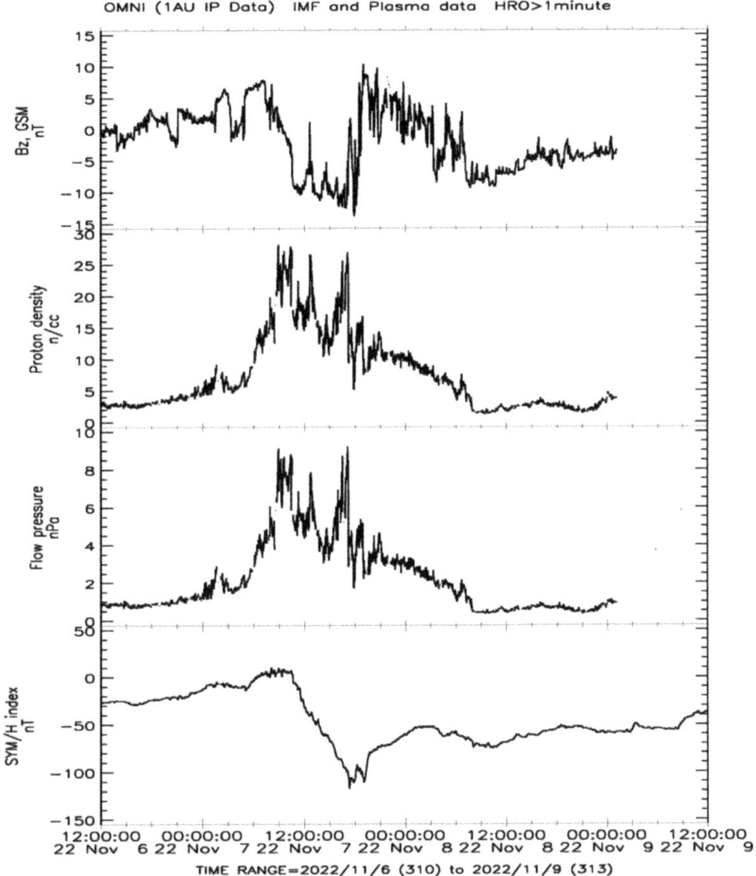



**Figure 3.** rom bottom to top: The 1 min data on the geomagnetic "SYM/H" index, the pressure of the solar wind at the magnetopause, solar wind protons density, and the vertical component of the interplanetary magnetic field in the GSM coordinate system (Bz_GSM) in the Earth's orbit for 6–9 November 2022, from the OMNI database.

## 2.2. Case Study of M = 7.0 of 30 October 2020

On 30 October 2020, a strong M = 7.0 earthquake occurred in the Aegean Sea at 11:51:27 UTC with coordinates of the epicenter 37.897° N, 26.784° E, at a depth of 21.0 km. This earthquake was preceded by two sequential small magnetic storms (Figure 4), which satisfied the chosen criteria: at the time of geomagnetic storm onset, the high-latitudinal area of the longitudinal region of the earthquake epicenter was located under the polar cusp.

The first such minor geomagnetic storm started on 2 August 2020, at 09:23 UT with a positive "SYM/H" = +27 nT and reached its most significant negative "SYM/H" = −39 nT on 3 August at 03:31 UT (Figure 5). In the initial phases of this storm, the magnetic local time around the future epicenter (37.897° N, 26.784° E) was equal to MLT = ~11.66 h. Thus, at the time of this storm's onset, the high-latitude part of the longitudinal region, where the M = 7.0 earthquake later occurred, was located under the polar cusp. The delay between this magnetic storm onset and earthquake occurrence equals ~89 days (Table 1).

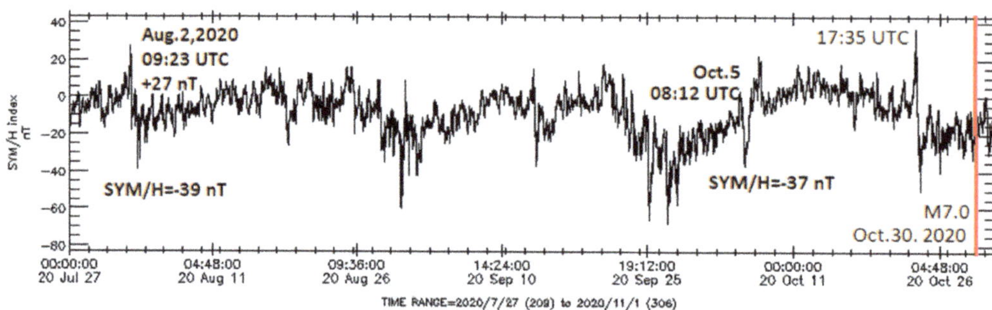



**Figure 4.** The 1 min data on the geomagnetic "SYM/H" index for 27 July–30 October 2020; on the right, a red line marks the date of the M = 7.0 earthquake on 30 October 2020 (37.897° N, 26.784° E); on the left, parameters of two small geomagnetic storms, which satisfied the given criteria, are indicated.

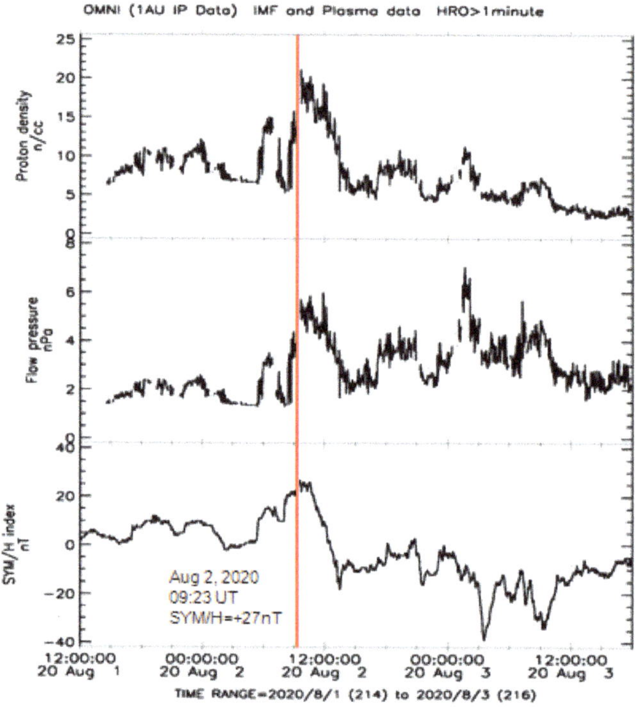



**Figure 5.** From bottom to top: The 1 min data on the geomagnetic "SYM/H" index, the dynamic pressure of the solar wind at the magnetopause, and the solar wind proton density for 1–3 August 2020; a red line marks the positive peaks in the solar wind and geomagnetic field parameters on 2 August 2020, at 09:23 UT.

The second small geomagnetic storm "SYM/H" = −37 nT on 5 October 2020, at 22:38 UT, started on 5 October at 08:12 UT with a sudden jump in the "SYM/H" index from −9 nT to +1 nT (Figure 6). In the initial phases of this storm (08:12 UT), the magnetic local time in the area of the future epicenter (37.897° N, 26.784° E) was equal to MLT = 10.48 h. Thus, this storm met a given criterion: at the time of its onset, the high-latitude part of the longitudinal region, where in the future the M = 7.0 earthquake occurred, was located under the polar cusp. The delay between this magnetic storm onset and earthquake occurrence equals ~25 days (Table 1).

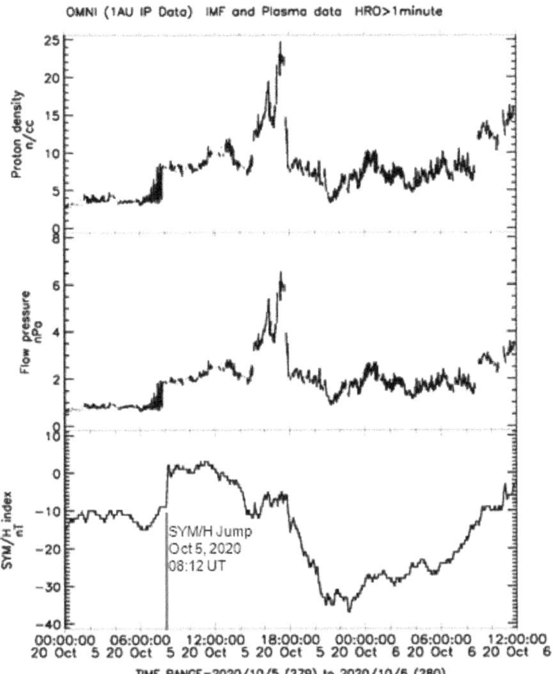



**Figure 6.** rom bottom to top: The 1 min data on the geomagnetic SYM/H index, the dynamic pressure of the solar wind at the magnetopause, and the solar wind proton density for 5–6 October 2020; a red line marks a sharp positive jump in the solar wind and geomagnetic field parameters on 5 October 2020, at 08:12 UT.

### 2.3. Case Study of M = 7.1 of 23 October 2011

On 23 October 2011, a strong M = 7.1 earthquake occurred at 10:41:23 UTC with coordinates of the epicenter 38.721° N, 43.508° E, at a depth of 18.0 km. This event was preceded by three geomagnetic storms in September 2011 with clear, sudden onsets (Figure 7).

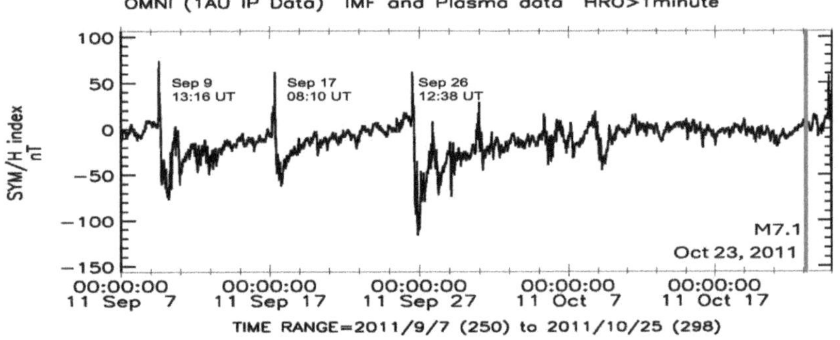



**Figure 7.** The 1 min data on the geomagnetic "SYM/H" index from 7 September 2011 to 24 October 2011; on the right, a red line marks the date of the M7.1 earthquake on 23 October 2011 (38.721° N, 43.508° E); on the left, sudden onsets of the three preceding geomagnetic storms are indicated.

The first moderate storm, "SYM/H" = −77 nT, started on 9 September at 13:16 UT with positive "SYM/H" = +74 nT; the second weak storm, "SYM/H"= −43 nT, started on 17 September at 08:10 UT with positive "SYM/H" = +61 nT; and the third strong storm,

"SYM/H" = −111 nT, started on 26 September at 12:38 UT with positive "SYM/H" = +62 nT. In the initial phases of these three geomagnetic storms, the MLT values in the area of the future epicenter (38.721° N, 43.508° E) were equal to ~16.63 h, 11.53 h, and 15.99 h, respectively. Thus, in the initial phase of these geomagnetic storms, the high-latitude area of the longitudinal region where the M = 7.1 earthquake occurred was located near the cusp at the time of the first and the third magnetic storms, while being strictly under the cusp at the time of the second storm. The lag times between the magnetic storm onsets and earthquake occurrence are equal to ~44, ~36, and ~27 days, respectively (Table 1).

### 2.4. Case Study for M = 7.2 of 12 November 1999

On 12 November 1999, a strong M = 7.2 earthquake occurred at 16:57:19 UTC with coordinates of the epicenter 40.758° N, 31.161° E, at a depth of 10.0 km. It was preceded by a strong geomagnetic storm on 22 September 1999, "SYM/H" = −166 nT, and a powerful one on 21 October 1999, "SYM/H" = −211 nT (Figure 8).

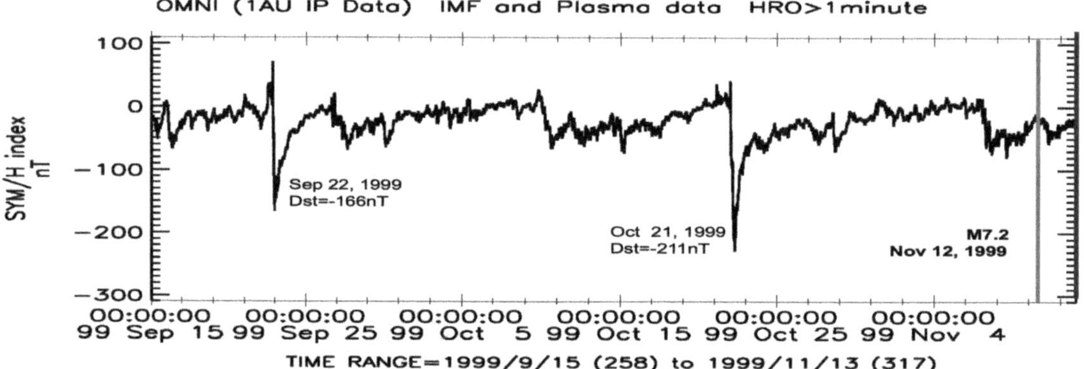


**Figure 8.** The 1 min data on the geomagnetic "SYM/H" index from 15 September 1999 to November 15, 1999; on the right, a red line marks the date of the M = 7.2 earthquake on 12 November 1999 (40.758° N, 31.161° E); on the left, two preceding geomagnetic storms are indicated.

Below, we analyze these two storms in more detail. Figure 9 shows that the first storm started on 22 September 1999, at 12:34 UT, due to the arrival of a dense solar wind proton flux (upper panel), which increased dynamic pressure at the magnetopause (middle panel). At the time of storm onset, a magnetic local time around the future epicenter (40.758° N, 31.161° E) was equal to MLT = 15.27 h; that is, the high-latitude part of the M = 7.2 epicenter longitudinal region (31.161° E) was under the cusp. The delay time between the geomagnetic storm onset and earthquake occurrence equals ~51 days (Table 1).

Besides this, Figure 9 shows that during this storm, the second arrival of a dense solar wind flux occurred, which resulted in increasing positive "SYM/H" up to +71 nT at 20:12 UT. It is not difficult to estimate that this time, the high-latitude part of an American longitudinal region was located under the polar cusp; one could expect a strong earthquake in this region. Indeed, according to the USGS seismic catalog, three strong earthquakes occurred here, namely, M = 7.5 in Mexico (16.059° N, 96.931° W) on 30 September 1999, with a time lag of ~8 days; M = 7.1 in California (34.603° N, 116.265° W) on 16 October 1999 with a time lag of ~24 days, and M = 7.0 at Alaska (57.342° N, 154.347° W) on 6 December 1999 with a time lag of ~75 days.

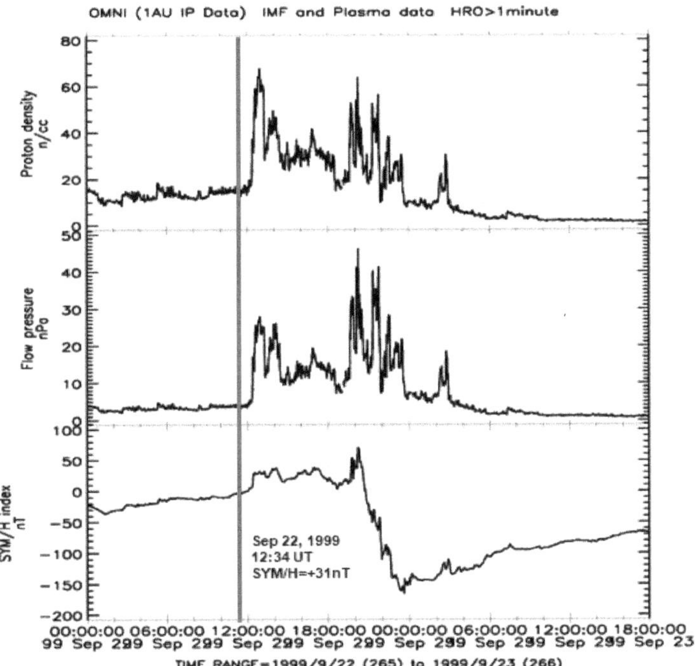



**Figure 9.** From bottom to top: The 1 min data on the geomagnetic "SYM/H" index, the dynamic pressure of the solar wind at the magnetopause, and the solar wind proton density for 22–23 September of 1999; a red line marks a sharp increase in these parameters, which started on 22 September 1999, at 12:34 UT.

In Figure 10, we show the solar wind parameters for the powerful geomagnetic storm with "SYM/H" = −219 nT, which started on 21 October 1999, at 23:41 UT, with positive "SYM/H" = +42 nT. It is not difficult to calculate that at this time (23:41 UT), the high-latitude part of the longitudinal region ~125–245 E could be located under the polar cusp if it is between 8 and 16 h. Again, one could expect strong earthquakes in this region, which is indeed what happened. According to the USGS seismological catalog, on 19 November 1999, an M = 7.0 event occurred in Papua New Guinea (6.351° S, 148.763° E), a time lag of ~29 days; on 26 November 1999, an M = 7.5 event occurred in Vanuatu (16.423° S, 168.214° E), a time lag of ~36 days; and on 6 December 1999, an M = 7.0 event occurred in Alaska (57.342° N, 154.347° W = 205.653 E), a time lag of ~46 days.

Besides this, Figure 10 shows that at the negative peak of the main phase of this powerful magnetic storm, the arrival of a dense solar wind flux occurred, which resulted in a sharp increase in the solar wind dynamic pressure at the magnetopause up to 35 nPa on 22 October 1999, at 07:06 UT. At this time, a magnetic local time in the area of the future epicenter (40.758° N, 31.161° E) was equal to MLT = 9.8 h; that is, the high-latitude part of the M = 7.2 epicenter longitudinal region (31.161° E) was under the polar cusp. The arrival of this dense solar wind flux could add some energy to the region of earthquake preparation. The delay between the arrival of a dense solar wind flux on 22 October 1999 at 07:06 UT and the M = 7.2 earthquake occurrence on 12 November 1999 was equal to ~21 days (Table 1).

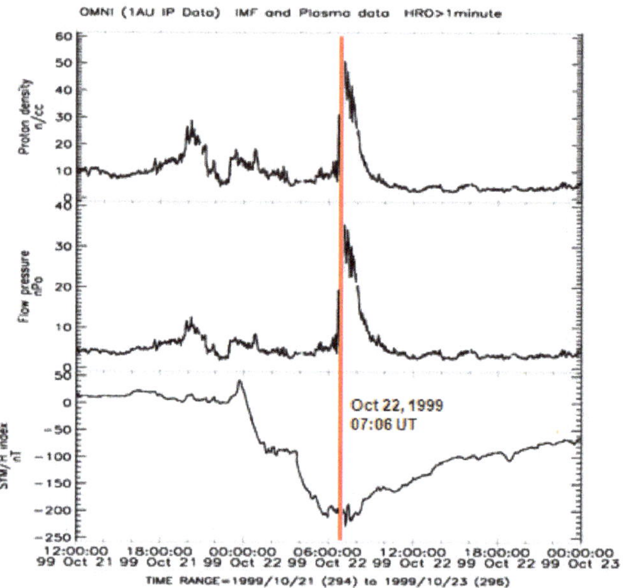



**Figure 10.** From bottom to top: The 1 min data on the geomagnetic "SYM/H" index, the dynamic pressure of the solar wind at the magnetopause, and the solar wind proton density in the Earth's orbit for 21–22 October 1999. The red line marks the arrival of a dense solar wind flux.

### 2.5. Case Study for M = 7.6 of 17 August 1999

On 17 August 1999, a catastrophic M = 7.6 earthquake occurred at 00:01:39 UT with epicenter coordinates 40.748° N, 29.864° E, at a depth of 17 km. This earthquake was preceded by a strong magnetic storm "SYM/H" = −123 nT on 16 April 1999 (Figure 11). According to the Observatorio del Ebro, Roquetes, Spain, the sudden onset of this storm (SSC) occurred at ~11:25 UT on 16 April 1999, with a positive "SYM/H" = +10 nT, which then increased up to "SYM/H" = +63 nT at ~14:49 UT. During geomagnetic storm onset (SSC at ~11:25 UT), the magnetic local time at the territory of the future epicenter (40.748° N, 29.864° E) was equal to MLT = ~14.04 h. The high-latitude zone of the longitudinal region in which the M = 7.6 occurred was under the polar cusp.

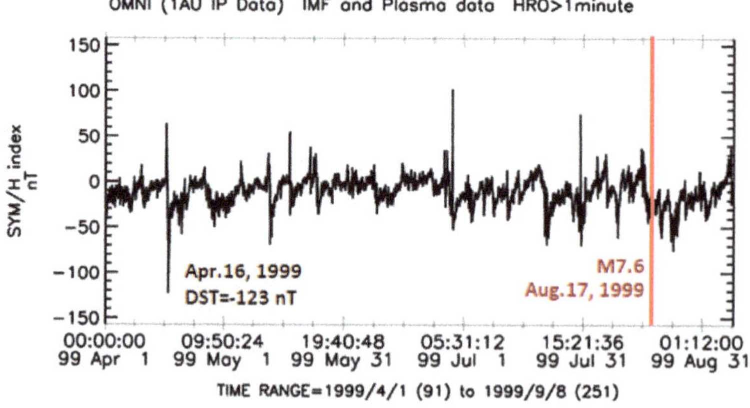



**Figure 11.** The 1 min data on the geomagnetic "SYM/H" index from 1 April 1999 to 31 August 1999; on the right, a red line marks the date of the M = 7.6 earthquake on 17 August 1999 (40.748° N, 29.864° E); on the left, a preceding geomagnetic storm on 16 April 1999 is indicated.

Besides this, an analysis of the solar wind and geomagnetic field behavior before the earthquake on 17 August 1999 has revealed (Figure 12) that on the eve of the M = 7.6 event, the positive value of the "SYM/H" index sharply increased up to 36 nT on 15 August 1999, at 11:52 UT (almost the same time as for SSC on 16 April 1999). A small geomagnetic storm ("SYM/H" = −44 nT) started. At 11:52 UT, the magnetic local time at the territory of the future epicenter (40.748° N, 29.864° E) was equal to MLT = ~14.49 h; that is, the high-latitude area of the epicenter longitudinal region was under the polar cusp. Considering two geomagnetic storms that preceded the M = 7.6 earthquake, one may conclude that the delay times between storm onsets and earthquake occurrence were equal to ~123 and ~1.5 days, respectively (Table 1). Again, it seems that the arrival of the dense solar wind flux on 15 August 1999, at 11:52 UT, could add some energy to the region of earthquake preparation, which could start after a magnetic storm onset on 16 April 1999, at ~11:25 UT.

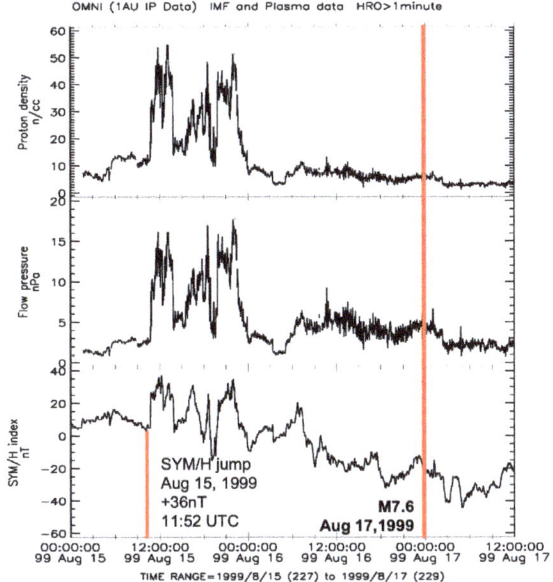


**Figure 12.** From bottom to top: The 1 min data on the "SYM/H" index, the dynamic pressure of the solar wind at the magnetopause, and the solar wind proton density in the Earth's orbit for 15–17 August 1999; on the right, a red line marks the date of the M = 7.6 earthquake on 17 August 1999 (40.748° N, 29.864° E); on the left, the date of a sharp jump in the solar wind and geomagnetic field parameters on 15 August at 11:52 UTC is indicated.

### 2.6. Case Study for M = 7.3 of 24 November 1976

On 24 November 1976, a strong M = 7.3 earthquake occurred at 12:22:18 UT with coordinates of the epicenter 39.121° N, 44.029° E, at a depth of 36.0 km. This event was preceded by a moderate geomagnetic storm (Dst= −57 nT) that started on 30 October 1976, at 10:30 UTC, with positive Dst = +18 nT (Figure 13). At this time, the magnetic local time in the area of the future epicenter (39.121° N, 44.029° E) was equal to MLT = 14.03 h. The high-latitude zone of the longitudinal region where the M = 7.3 earthquake occurred was located under the polar cusp. The delay between the magnetic storm onset and earthquake occurrence equals ~25 days (Table 1).

**Figure 13.** The 1 h data on the Dst index in 1976 from 28 October to 29 November; on the right, a black line marks the date of the M = 7.3 earthquake on 24 November 1976 (39.121° N, 44.029° E); on the left, the start of a preceding geomagnetic storm on 30 October 1976 at 10:30 UTC is indicated.

### 2.7. Case Study for M = 7.2 28 March 1970

On 28 March 1970, a strong M = 7.2 earthquake occurred at 21:02:26 UT with coordinates of the epicenter 39.098° N, 29.570° E, at a depth of 25.0 km. This event was preceded by three geomagnetic storms in January–March 1970 (Figure 14).

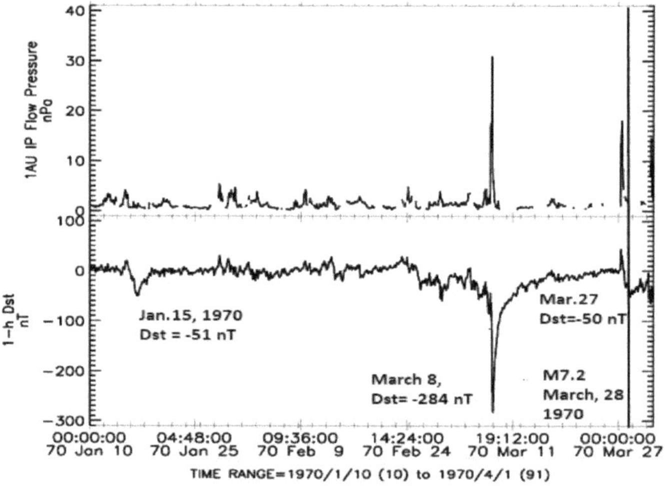



**Figure 14.** From bottom to top: The 1-h data on the geomagnetic Dst index, the dynamic pressure of the solar wind at the magnetopause, and the solar wind proton density from 10 January to 1 April 1970; on the right, a black line marks the date of the M = 7.2 earthquake on 28 March 1970 (39.098° N, 29.570° E); on the left, preceding geomagnetic storms are indicated.

The first weak geomagnetic storm (Dst = −51 nT) started on 15 January at about 09:30 UT (positive Dst = +20 nT). At this time, the magnetic local time in the area of the future epicenter (39.098° N 29.570° E) was equal to MLT = ~12.2 h. Thus, this storm met a given criterion: at the time of its onset, the high-latitude part of the longitudinal region where the M = 7.2 earthquake later occurred was located under the polar cusp. The delay between this magnetic storm onset and earthquake occurrence equals ~72 days.

The second powerful storm (Dst = −284 nT) had no clear initial phase. However, at the negative peak of the main phase of this powerful magnetic storm, the arrival of a dense solar wind flux occurred, which resulted in a sharp increase in the solar wind dynamic pressure at the magnetopause up to 32 nPa on 8 March at 19:30 UT. It is not difficult to understand that at this time, the high-latitude area of the American longitudinal region was under the polar cusp. Three strong earthquakes occurred here with a time lag of 52, 84, and 145 days. The first M = 7.3 occurred in Mexico on 29 April 1970, and the other two occurred in Peru: M = 7.9 on 31 May and M = 8.0 on 31 July.

The third weak magnetic storm (Dst= −50 nT) started on 27 March at 08:30 UT (positive Dst = +44 nT). At this time, the magnetic local time in the area of a considered epicenter (39.098° N, 29.570° E) was equal to MLT = ~11.2 h. The high-latitude area of the longitudinal region in which the M = 7.2 earthquake occurred was located under the polar cusp. The delay times between two small magnetic storms' onset (15 January and 27 March 1970) and earthquake occurrence on 28 March equal ~72 and ~1.5 days, respectively (Table 1). Again, it seems that as a small magnetic storm started on 27 March 1970, at 08:30 UT, it could add some energy to an area of M = 7.2 earthquake preparation, which could start after a small magnetic storm on 15 January 1970, at ~09:30 UT.

### 2.8. Case Study for M = 7.3 of 22 July 1967

On 22 July 1967, a strong M = 7.3 earthquake occurred at 16:57 UT with coordinates of the epicenter 40.751° N, 30.8° E, at a depth of 30.0 km. This seismic event was preceded by an extreme geomagnetic storm (Dst = −387 nT) starting on 25 May 1967, at 12:30 UT, from a sudden positive increase in the Dst index to +55 nT. Figure 15 presents the 1 h Dst data for 10 May–25 July 1967. At the time of the magnetic storm onset (12:30 UT), the magnetic local time in the area of the future epicenter (40.751° N, 30.8° E) was equal to MLT = ~15.29 h. Thus, this storm met a given criterion: at the time of its onset, the high-latitude part of the longitudinal region where the M = 7.3 earthquake later occurred was located under the polar cusp. The time delay between an extreme magnetic storm on 25 May 1967 and the M = 7.3 earthquake on 22 July 1967 was equal to ~58 days (Table 1).

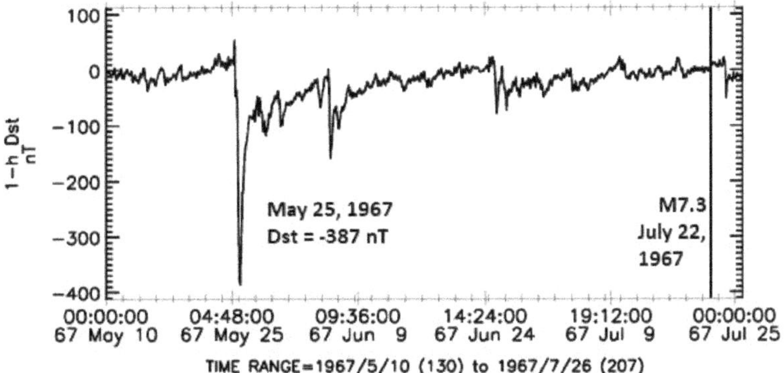


**Figure 15.** The 1 h data on the geomagnetic Dst index for 10 May–25 July 1967; on the right, a black line marks the date of the M = 7.3 earthquake on 22 July 1967 (40.751° N, 30.8° E); on the left, a preceding geomagnetic storm on 25 May 1967 is indicated.

## 3. Discussion and Conclusions

One of the still-open questions to the Earth and space community is how the energy of the geospace environment impacts lithospheric processes. Some papers show that solar flare X-ray radiation, coronal mass ejections, and geomagnetic storms may precede the occurrence of earthquakes ([1–8,20–23] and references therein). Many years of statistical searching in this direction have led to a mathematical model [8] that considers a hypothesis of electromagnetic earthquakes being triggered by a sharp rise of telluric currents in the lithosphere, including crust faults, due to the interaction of solar flare X-ray radiation with the ionosphere–atmosphere–lithosphere system. Recently, the authors of [24] have investigated the level of the geomagnetically induced current index (GIC) in the Mediterranean region during the strongest magnetic storms of solar cycle 24 (2008–2019). The GIC index is a proxy of the geoelectric field, calculated entirely from geomagnetic field variations. Their results showed that the GIC index increased during the magnetic storm events in the 24th solar cycle, and its increase appears simultaneously with the SSC occurrence, which agrees with other GIC studies for low and middle latitudes, e.g., [25]. In the mathematical model [8], a critical point is the increase in the ionosphere's radiation and conductivity. In the approach to explain an increase in global seismic activity in the solar minimums, again, a critical point is an increase in radiation and conductivity of the upper troposphere and lower stratosphere, produced by the galactic cosmic rays [26,27], whose intensity increases in solar minimums. It has been found recently [4] that strong earthquakes may appear addressed (targeted) because they occur near the footprints of certain geomagnetic lines belonging to a newly created radiation belt into the lower magnetosphere when the high-energy electrons in the outer radiation belt spill down due to a geomagnetic storm. Again, the critical point for this effect may be an increase in radiation and conductivity in the mesosphere and upper stratosphere due to the precipitation of energetic electrons from the radiation belt up to the stratopause, as shown in [28]. The above suggests that active space weather can provoke strong earthquakes in those longitudinal regions above which a near-space environment, including geomagnetic lines, can be sufficiently populated with charged particles (be conductive). On Earth, there are two places where geomagnetic lines may constantly be filling with charged particles. These are the polar cusps where the solar wind plasma would directly access the atmosphere [9–17]. The magnetic local time determines the length of the polar cusps in longitude, and its largest extension is expected to be between 8 and 16 h MLT [17]. Due to the Earth's rotation, different longitudinal regions are located under the polar cusps during the arrival of the shocked solar wind flows. Thus, an idea was formed to check whether, after the magnetic storm, strong earthquakes will be more likely to occur in that longitudinal region, where its high-latitudinal area was located under the cusp when a shocked solar wind arrived (in time of SSC). Our analyses of nine strong ($M \geq 7.0$) earthquakes inside and around the Anatolian Plate, including the $M = 7.8$ and $M = 7.5$ Kahramanmaras earthquake sequence on 6 February 2023, proved this suggestion. Moreover, for each of the other seven earthquakes, which occurred here from 1967, we could also identify preceding geomagnetic storms that met a given criterion: the magnetic local time (MLT) at an area of a future epicenter was between 8 and 16 h at the time of geomagnetic storm onset. The results showed (Table 1) that before four earthquakes ($M = 7.8$ and $M = 7.5$ on 6 February 2023, $M = 7.3$ on 24 November 1976, and $M = 7.3$ on 22 July 1967), there was only one geomagnetic storm with a given criterion (strong, moderate, and extreme, respectively). Before four earthquakes ($M = 7.0$ on 30 October 2020; $M = 7.2$ on 12 November 1999; $M = 7.6$ on 17 August 1999; and $M = 7.2$ on 28 March 1970), there were two magnetic storms with a given criterion (small + small, strong + powerful, strong + small, and small + small, respectively). Before one seismic event ($M = 7.1$ on 23 October 2011), there were three consecutive magnetic storms with a given criterion (moderate, small, and strong). The lag time between a magnetic storm onset and earthquake occurrence varied from ~1.5 days (two cases: $M = 7.6$ on 17 August 1999, and 28 March 1970, when two magnetic storms preceded earthquakes) to 123 days (one case for the $M = 7.6$ on 17 August 1999, preceded by two magnetic storms). On average, the delay is equal to ~50 days.

Observed long delays between geomagnetic storm onset and earthquake occurrence may tell us that space weather phenomena do not trigger earthquakes immediately. Instead, the solar wind stimulates the lithosphere processes that we see, which then, with a time delay, result in earthquakes. These processes may be related to geoelectricity, for example. So, the authors of [29] investigated seismicity and geoelectric potential changes, possibly associated with the seismic swarm activity in the Izu Island region, Japan, which took place in June–September of 2000. They showed that a swarm activity was preceded by pronounced electrical activity with innumerable signals that started two months before the swarm onset. The authors [30] also revealed a time delay for the ULF geomagnetic anomaly associated with the 2000 Izu Islands earthquake swarm. Namely, about three months before the beginning of swarm activity, the local signal level around the magnetometer associated with crustal activity (such as the vibration of the ground) was slightly enhanced. A similar result was obtained in [31] for the M = 7.8 earthquake at the Anatolian Fault on 6 February 2023. The authors of [31] performed a natural time analysis (NTA) of the seismicity preceding the Kahramanmaras earthquake doublet. They revealed a minimum fluctuation of the order parameter of seismicity that ended on 18 October 2022, pointing to the initiation of seismic electrical activity. This occurred almost three and a half months before the M = 7.8 earthquake. Table 1 shows that this earthquake occurred three months (91 days) after the start of the geomagnetic storm on 7 November 2022. There may be a connection between the initiation of seismic and electrical activity detected in [31] and the geomagnetic storm that occurred. If such a connection could be strictly established, it could provide direct evidence of the contribution of space weather to the earthquake preparation process.

Also, a time delay could appear if the shocked solar wind influences the upward lifting of fluids, which are active participants in tectonic earthquakes [32]. We cannot know what happens to fluids at a depth of the earthquake source, but we can expect that sometimes (in exceptional situations), the effects of fluids can manifest themselves at the Earth's surface. It is possible that one of these exceptions was the sharp emanation of radon before the M = 6.9 earthquake in Kobe, Japan, on 16 January 1995 [33].

If the idea with fluids is true, then, depending on the fragmentation of the lithosphere in different places, the time duration of fluid rise could vary. It is possible that such a situation occurred after the magnetic storm on 7 November 2022. It is shown above that the Kahramanmaras earthquake doublet occurred 91 days after the magnetic storm. However, 16 days after the storm, on 23 November 2022, an earthquake M = 6.1 occurred in the north of the Anatolian Fault, with epicenter coordinates 40.836° N, 30.983° E (Figure 1). Here, in the recent past, three strong events occurred, namely, M = 7.6 on 17 August 1999, M = 7.2 on 12 November 1999, and M = 7.2 on 22 July 1967. It is possible that previous strong earthquakes caused the conditions created in this fault area, and the rise of fluids was facilitated here.

If this proves to be true, in the frame of our investigation, the next question could be: "How is it possible that the fluid uplifting happens very efficiently only in the time sector of a cusp and is not efficient in all other longitudinal sectors?" To our mind, the processes into the cusp could give us a small hint. So, the CHAMP and DMSP satellites discovered that the density of the neutral atmosphere in the cusp funnel is always increased, on average, by one and a half times relative to the density in neighboring areas [11,12]. A leading candidate driver mechanism for explaining the density anomaly in the cusp involves soft auroral precipitation driving neutral upwelling. The authors of [12] say: "Our preferred explanation is that dayside reconnection fuels Joule heating of the thermosphere causing air upwelling and at the same time heating of the electron gas that pulls up ions along affected flux tubes". Thus, in the outer geospheres (magnetosphere, ionosphere), the rise of thermospheric gases occurs most effectively in the polar cusp funnel. In the middle geosphere (troposphere), the upward rise of air masses occurs, as we know, most effectively in the funnel of an atmospheric tornado. Suppose the processes in different geospheres are electrically interconnected. In that case, it is possible that the solar wind energy received by a beam of geomagnetic lines in the polar cusp will be delivered to the inner geospheres precisely along these geomagnetic lines (via Alfven waves, for example). Then, the upward rise of intra-terrestrial gases (fluids) can also occur along the bundle of these specific geomagnetic lines, and earthquakes will occur "targeted" at the footprint of these geomagnetic lines. The journey of intra-terrestrial fluids from the deep Earth is a fascinating process. As they move upward, they create fractures, mainly in the upward direction, as the rock overburden finishes and becomes less resistant to gas-pressure fracturing. These fractures then serve as conduits through the solid lithosphere [34]. By holding the faces apart, gas inflow into fault lines significantly reduces internal friction, thereby facilitating earthquakes. When these fluids reach near-surface pressure, they transform into invisible gases such as methane, carbon dioxide, hydrogen sulfide, hydrogen, nitrogen, helium, and various trace gases like radon [35,36].

Electrical coupling between an external and internal geosphere is expected; it is suggested in [37] that the process of earthquake preparation and realization could be related to the functioning of the global electric circuit (GEC). One of the main characteristics of the GEC is a unitary variation—a dependence of the fair-weather electric field on universal time, called the Carnegie curve, which demonstrates a steady increase of the electric field in fair-weather regions at ~19:00 UT. It was revealed in [38,39] that the global seismic activity also shows a unitary variation, which correlates relatively strongly with the Carnegie curve. It is shown in [40] that statistically significant anomalous changes appear simultaneously before major earthquakes in independent datasets of different geophysical observables. Therefore, the authors of [37,39] noted that electromagnetic monitoring of earthquake-prone areas is needed to identify and verify these precursory changes in future major earthquakes. Besides this, a retrospective analysis of the solid-earth parameters in the epicenter areas after identifying preceding geomagnetic storms is desirable.

**Author Contributions:** D.O. and G.K. provided the concepts for the manuscript. G.K. organized and wrote the manuscript. All authors provided critical feedback and helped shape the research, analysis, and manuscript. All authors have read and agreed to the published version of the manuscript.

**Funding:** The research of G.K. is partly funded by the Science Committee of the Ministry of Science and Higher Education of the Republic of Kazakhstan (Grant No. AP19677977).

**Data Availability Statement:** The original contributions presented in the study are included in the article: further inquiries can be directed to the corresponding author.

**Acknowledgments:** We thank the US Geological Survey and European–Mediterranean Seismological Centre for providing earthquake information services and data. We acknowledge the use of the NASA/GSFC's Space Physics Data Facility's CDAWeb service, OMNIdata. We are also very grateful to all anonymous reviewers for their valuable comments, which helped improve the work.

**Conflicts of Interest:** The authors declare that the research was conducted without any commercial or financial relationships that could be construed as a potential conflict of interest.

## References

1. Sobolev, G.A.; Zakrzhevskaya, N.A.; Kharin, E.P. On the relation between seismicity and magnetic storms. *Phys. Solid Earth* **2001**, *37*, 917–927.
2. Urata, N.; Duma, G.; Freund, F. Geomagnetic Kp Index and Earthquakes. *Open J. Earthq. Res.* **2018**, *7*, 39–52. [CrossRef]
3. Chen, H.; Wang, R.; Miao, M.; Liu, X.; Ma, Y.; Hattori, K.; Han, P. A Statistical Study of the Correlation between Geomagnetic Storms and M $\geq$ 7.0 Global Earthquakes during 1957–2020. *Entropy* **2020**, *22*, 1270. [CrossRef]
4. Ouzounov, D.; Khachikyan, G. Studying the Impact of the Geospace Environment on Solar Lithosphere Coupling and Earthquake Activity. *Remote Sens.* **2024**, *16*, 24. [CrossRef]
5. Marchetti, D.; DeSantis, A.; Campuzano, S.A.; Zhu, K.; Soldani, M.; D'Arcangelo, S.; Orlando, M.; Wang, T.; Cianchini, G.; Di Mauro, D.; et al. Worldwide Statistical Correlation of Eight Years of Swarm Satellite Data with M5.5+ Earthquakes: New Hints about the Preseismic Phenomena from Space. *Remote Sens.* **2022**, *14*, 2649. [CrossRef]
6. Love, J.J.; Thomas, J.N. Insignificant solar-terrestrial triggering of earthquakes. *Geophys. Res. Lett.* **2013**, *40*, 1165–1170. [CrossRef]
7. Yesugey, S.C. Comparative evaluation of the influencing effects of geomagnetic solar storms on earthquakes in Anatolian peninsula. *Earth Sci. Res. J.* **2009**, *13*, 82–89.
8. Sorokin, V.; Yaschenko, A.; Mushkarev, G.; Novikov, V. Telluric Currents Generated by Solar Flare Radiation: Physical Model and Numerical Estimations. *Atmosphere* **2023**, *14*, 458. [CrossRef]
9. Ivarsen, M.F.; Jin, Y.; Spicher, A.; St-Maurice, J.-P.; Park, J.; Billett, D. GNSS scintillations in the cusp, and the role of precipitating particle energy fluxes. *J. Geophys. Res. Space Phys.* **2023**, *128*, e2023JA031849. [CrossRef]
10. Goertz, C.K. Kinetic Alfven waves on auroral field lines. *Planet. Space Sci.* **1984**, *32*, 1387–1392. [CrossRef]
11. Lühr, H.; Rother, M.; Köhler, W.; Ritter, P.; Grunwaldt, L. Thermospheric up-welling in the cusp region: Evidence from CHAMP observations. *Geophys. Res. Lett.* **2004**, *31*, L06805. [CrossRef]
12. Kervalishvili, G.N.; Lühr, H. The relationship of thermospheric density anomaly with electron temperature, small-scale FAC, and ion up-flow in the cusp region, as observed by CHAMP and DMSP satellites. *Ann. Geophys.* **2013**, *31*, 541–554. [CrossRef]
13. Pitout, F.; Bogdanova, Y.V. The polar cusp seen by Cluster. *J. Geophys. Res. Space Phys.* **2021**, *126*, e2021JA029582. [CrossRef]
14. Escoubet, C.P.; Fehringer, M.; Goldstein, M. The Cluster mission. *Ann. Geophys.* **2001**, *19*, 1197–1200. [CrossRef]

15. Pitout, F.; Escoubet, C.P.; Klecker, B.; Rème, H. Cluster survey of the middle altitude cusp: 1. size, location, and dynamics. *Ann. Geophys.* **2006**, *24*, 3011–3026. [CrossRef]
16. Crooker, N.U. Dayside merging and cusp geometry. *J. Geophys. Res.* **1979**, *84*, 951–959. [CrossRef]
17. Russell, C.T. Polar Eyes the Cusp Cluster-II Workshop: Multiscale/Multipoint Plasma Measurements. In Proceedings of the Workshop Held at Imperial College, London, UK, 22–24 September 1999; European Space Agency (ESA), ESA-SP: Paris, Italy, 2000; p. 47, ISBN 9290927968. Available online: https://articles.adsabs.harvard.edu//full/2000ESASP.449...47R/0000050.000.html (accessed on 20 February 2023).
18. Wanliss, J.A.; Showalter, K.M. High-resolution global storm index: Dst versus SYM-H. *J. Geophys. Res.* **2006**, *111*, A02202. [CrossRef]
19. Loewe, C.A.; Prolss, G.W. Classification and Mean Behavior of Magnetic Storms. *J. Geophys. Res.* **1997**, *102*, 14209. [CrossRef]
20. Novikov, V.; Ruzhin, Y.; Sorokin, V.; Yaschenko, A. Space weather and earthquakes: Possible triggering of seismic activity by strong solar flares. *Ann. Geophys.* **2020**, *63*, PA554. [CrossRef]
21. Shestopalov, I.P.; Kharin, E.P. Secular variations of solar activity and seismicity of the Earth. *Geophys. J.* **2006**, *28*, 59–70.
22. Zhang, G.Q. Relationship between global seismicity and solar activities. *Acta Seismol. Sin.* **1998**, *11*, 495–500. [CrossRef]
23. Huzaimy, J.M.; Yumoto, K. Possible correlation between solar activity and global seismicity. In Proceeding of the 2011 IEEE International Conference on Space Science and Communication (IconSpace), Penang, Malaysia, 12–13 July 2011; pp. 138–141.
24. Boutsi, A.Z.; Balasis, G.; Dimitrakoudis, S.; Daglis, I.A.; Tsinganos, K.; Papadimitriou, C.; Giannakis, O. Investigation of the geomagnetically induced current index levels in the Mediterranean region during the strongest magnetic storms of solar cycle 24. *Space Weather* **2023**, *21*, e2022SW003122. [CrossRef]
25. Zhang, J.J.; Wang, C.; Sun, T.R.; Liu, C.M.; Wang, K.R. GIC due to storm sudden commencement in low-latitude high-voltage power network in China: Observation and simulation. *Space Weather* **2015**, *13*, 643–655. [CrossRef]
26. Bazilevskaya, G.A.; Usoskin, I.G.; Flückiger, E.O.; Harrison, R.G.; Desorgher, L.; Bütikofer, R.; Krainev, M.B.; Makhmutov, V.S.; Stozhkov, Y.I.; Svirzhevskaya, A.K.; et al. Cosmic ray induced ion production in the atmosphere. *Space Sci. Rev.* **2008**, *137*, 149–173. [CrossRef]
27. Phillips, T.; Johnson, S.; Koske-Phillips, A.; White, M.; Yarborough, A.; Lamb, A.; Schultz, J. Space weather ballooning. *Space Weather* **2016**, *14*, 697–703. [CrossRef]
28. Kavanagh, A.J.; Cobbett, N.; Kirsch, P. Radiation Belt Slot Region Filling Events: Sustained Energetic Precipitation Into the Mesosphere. *J. Geophys. Res. Space Phys.* **2018**, *123*, 7999–8020. [CrossRef]
29. Uyeda, S.; Kamogawa, M.; Tanaka, H. Analysis of electrical activity and seismicity in the natural time domain for the volcanic-seismic swarm activity in 2000 in the Izu Island region, Japan. *J. Geophys. Res.* **2009**, *114*, B02310. [CrossRef]
30. Hattori, K.; Serita, A.; Gotoh, K.; Yoshino, C.; Harada, M.; Isezaki, N.; Hayakawa, M. ULF geomagnetic anomaly associated with 2000 Izu Islands earthquake swarm, Japan. *Phys. Chem. Earth Parts A/B/C* **2004**, *29*, 425–435. [CrossRef]
31. Nicholas, V.S.; Efthimios, S.S.; Stavros-Richard, G.C.; Panayiotis, K.V. Identifying the Occurrence Time of the Destructive Kahramanmaraş-Gazientep Earthquake of Magnitude M7.8 in Turkey on 6 February 2023. *Appl. Sci.* **2024**, *14*, 1215. [CrossRef]
32. Miller, S.A. The Role of Fluids in Tectonic and Earthquake Processes. *Adv. Geophys.* **2013**, *54*, 1–38. [CrossRef]
33. Yasuoka, Y.; Igarashi, G.; Ishikawa, T.; Tokonami, S.; Shinogi, M. Evidence of precursor phenomena in the Kobe earthquake obtained from atmospheric radon concentration. *Appl. Geochem.* **2006**, *21*, 1064–1072. [CrossRef]
34. Pulinets, S.A. Physical mechanism of the vertical electric field generation over active tectonic faults. *Adv. Space Res.* **2009**, *44*, 767–773. [CrossRef]
35. Gold, T. The Deep Hot Biosphere. In *The Myth of Fossil Fuels*; Springer: Heidelberg, Germany, 1998; p. 243.
36. Soter, S.; Gold, T. The Deep-Earth-Gas Hypothesis. *Sci. Am. Mag.* **1980**, *242*, 154. [CrossRef]
37. Ouzounov, D.; Pulinets, S.; Liu, J.-Y.; Hattori, K.; Han, P. Multiparameter Assessment of Pre-Earthquake Atmospheric Signals. In *Pre-Earthquake Processes*; Ouzounov, D., Pulinets, S., Hattori, K., Taylor, P., Eds.; American Geophysical Union; John Wiley & Sons: Hoboken, NJ, USA, 2018; 385p. [CrossRef]
38. Pulinets, S.A.; Khachikyan, G.Y. Unitary Variation in the Seismic Regime of the Earth: Carnegie-Curve Matching. *Geomagn. Aeron* **2020**, *60*, 787–792. [CrossRef]
39. Pulinets, S.; Khachikyan, G. The Global Electric Circuit and Global Seismicity. *Geosciences* **2021**, *11*, 491. [CrossRef]
40. Varotsos, P.A.; Sarlis, N.V.; Skordas, E.S. Direct interconnection of seismicity with variations of the Earth's electric and magnetic field before major earthquakes. *Europhys. Lett.* **2024**, *146*, 22001. [CrossRef]

*Article*

# Possible Interrelations of Space Weather and Seismic Activity: An Implication for Earthquake Forecast

Valery Sorokin [1] and Victor Novikov [2,3,*]

[1] Pushkov Institute of Terrestrial Magnetism, Ionosphere, and Radio Wave Propagation, Russian Academy of Sciences, Moscow 108840, Russia; sova@izmiran.ru
[2] Joint Institute for High Temperatures, Russian Academy of Sciences, Moscow 125412, Russia
[3] Sadovsky Institute of Geospheres Dynamics, Russian Academy of Sciences, Moscow 119334, Russia
* Correspondence: novikov@ihed.ras.ru

**Abstract:** The statistical analysis of the impact of the top 50 X-class solar flares (1997–2024) on global seismic activity as well as on the earthquake preparation zones located in the illuminated part of the globe and in an area of 5000 km around the subsolar point was carried out. It is shown by a method of epoch superposition that for all cases, an increase in seismicity is observed, especially in the region around the subsolar point (up to 33%) during the 10 days after the solar flare in comparison with the preceding 10 days. The case study of the aftershock sequence of a strong $M_w = 9.1$ earthquake (Sumatra–Andaman Islands, 26 December 2004) after the solar flare of X10.16 class (20 January 2005) demonstrated that the number of aftershocks with a magnitude of $M_w \geq 2.5$ increases more than 17 times after the solar flare with a delay of 7–8 days. For the case of the Darfield earthquake ($M_w = 7.1$, 3 September 2010, New Zealand), it was shown that X-class solar flares and M probably triggered two strong aftershocks ($M_w = 6.1$ and $M_w = 5.9$) with the same delay of 6 days on the Port Hills fault, which is the most sensitive to external electromagnetic impact from the point of view of the fault electrical conductivity and orientation. Based on the obtained results, the possible application of natural electromagnetic triggering of earthquakes is discussed for the earthquake forecast using confidently recorded strong external electromagnetic triggering impacts on the specific earthquake preparation zones, as well as ionospheric perturbations due to aerosol emission from the earthquake sources recorded by satellites.

**Keywords:** solar flare; geomagnetic field variations; geomagnetically induced currents; electromagnetic earthquake triggering; aerosol emission; short-term earthquake forecast

**Citation:** Sorokin, V.; Novikov, V. Possible Interrelations of Space Weather and Seismic Activity: An Implication for Earthquake Forecast. *Geosciences* **2024**, *14*, 116. https://doi.org/10.3390/geosciences14050116

Academic Editors: Dimitrios Nikolopoulos and Jesus Martinez-Frias

Received: 23 February 2024
Revised: 21 April 2024
Accepted: 24 April 2024
Published: 25 April 2024

## 1. Introduction

The problem of a possible relation between solar activity and the Earth's seismicity has been discussed over 170 years [1–9], with references therein. Despite a fairly large number of publications devoted to research on the possible influence of the Sun on seismic processes, a final conclusion about the possibility of earthquakes (EQs) being triggered by solar flares (SFs) or geomagnetic storms has not yet been reached. The results obtained to date are fuzzy and contradictive, and some authors deny the real existence of interrelationships between the processes in the Sun and in the lithosphere that resulted in the occurrence of EQs [10,11].

It should be noted that all the studies mentioned above employed only a statistical approach to the analysis of geophysical and seismological data when the hypothesis of the presence or absence of a possible correlation (positive or negative) between solar activity and Earth's seismicity was tested. The physical mechanisms of solar–terrestrial relations that resulted in the possible triggering of EQs were not considered in detail, and their possible existence was only indicated phenomenologically when a statistically significant relationship has been found between solar activity and the response of the Earth's seismicity.

Such a simplified approach to the study of the solar–terrestrial relationships may provide false results and incorrect conclusions, and no practical recommendations may be proposed for seismic risk mitigation.

In contrast to a pure statistical approach, the study of Sorokin et al. [12] considers a possible physical mechanism of earthquake (EQ) triggering by electromagnetic (EM) impacts on the area of EQ preparation due to X-ray radiation from SFs. This idea has been proposed in Sorokin et al. [5] and Novikov et al. [7], when it was numerically demonstrated that due to the interaction of SF X-ray radiation with the ionosphere–atmosphere–lithosphere system, strong geomagnetic field pulsations occur, resulting in the sharp rise of geomagnetically induced currents (GICs) in the conductive crust faults. It is known that EQs can be triggered by strong variations in both natural and artificial electric currents in the Earth's crust as a result of the interaction of EM and electric fields with rocks and faults under subcritical stress–strain state [13].

The results of numerical studies obtained in Sorokin et al. [12] using the developed physical model and computer code indicate that after an X-class SF (with peak radiation flux $\geq 10^{-4}$ W/m$^2$), geomagnetic field pulsations up to 100 nT can occur, and the density of GICs in the conductive layer of the lithosphere can rise to $10^{-8}$–$10^{-6}$ A/m$^2$. In this case, the current pulse duration is about 100 s and the duration of the current rise front is ~10 s. These values are 2–3 orders of magnitude higher than the average density of telluric currents in the lithosphere [14] and they are comparable with the parameters of electric current pulses generated in the lithosphere ($10^{-7}$–$10^{-8}$ A/m$^2$) by artificially pulsed sources of electrical energy [13]. It should be noted that the injections of electrical impulses into the Earth's crust in seismically active regions result in the EM triggering of weak EQs and the regional spatiotemporal redistribution of seismicity in the Pamirs and Northern Tien Shan. This means that strong SFs providing an energy flux density above 0.005 J/m$^2$ are also capable of triggering EQs in seismically hazardous regions, as was assumed in [7,12,15]. This conclusion is confirmed by cases of observation of magnetic pulses before an EQ [16,17] similarly to the obtained numerical estimates of magnetic pulses generated by X-rays of SF provided telluric current pulses in the conductive layer of the lithosphere, as well as the case of observation of a sharp increase in global and regional seismicity (Greece) after the SF of X13.37 class that occurred on 6 September 2017 [7].

For additional verification of numerical results obtained with the application of the physical model of the Sun–Earth interaction [12], we carried out the statistical analysis of the impact of the top 50 X-class SFs on global seismic activity, the EQs located on the illuminated part of the globe and the EQs located in an area of 5000 km around the subsolar point (SSP). We demonstrated that in all cases, an increase in seismicity is observed, especially in the region around the SSP (up to 33%) during the 6–8 days after the SF. Moreover, we found that the maximum seismic sensitivity to the SF impact is observed in the aftershock area of the strong EQ. The case study of the aftershock sequence behavior of strong $M_w = 9.1$ EQ (Sumatra–Andaman Islands, 26 December 2004) after the X10.16-class SF (20 January 2005, peak radiation flux is $10.16 \times 10^{-4}$ W/m$^2$) demonstrated that starting from the 7th day after the SF, the number of aftershocks with a magnitude of $M_w \geq 2.5$ increased more than 20 times by the 8th day and returned to the background level within the following two days. In addition, we consider the case of the Darfield EQ (4 September 2010, $M_w = 7.1$, New Zealand) with two strong aftershocks ($M_w \sim 6$) occurring in the Port Hills fault, which were most sensitive to external EM impact from the point of view of the fault electrical conductivity and orientation, with a delay of 6 days after strong X-class SFs and M.

Finally, based on the obtained results, we discuss the possibility of applying natural EM triggering of EQs for the EQ forecast, using confidently recorded strong external EM triggering impacts on the specific electromagnetically sensitive EQ preparation zones.

## 2. Methods of Verification of Hypothesis of Electromagnetic Earthquake Triggering by Strong X-Class SFs

### *2.1. Testable Hypothesis of Earthquake Triggering by Strong SFs*

In Sorokin et al. [5], Novikov et al. [7], and Sorokin et al. [12], a possible mechanism of EQ triggering by the ionizing radiation of SFs has been proposed. The theoretical model of Sorokin et al. [12] considered a disturbance of electric field, electric current, and heat release in the lithosphere associated with a variation in ionosphere conductivity caused by the absorption of ionizing radiation from SFs. The model predicted the generation of geomagnetic field disturbances in a range of seconds to tens of seconds as a result of a large-scale perturbation of the conductivity of the bottom part of the ionosphere in a horizontal direction in the presence of an external electric field. Amplitude-time characteristics of the geomagnetic disturbance depend upon a perturbation of the integral conductivity of the ionosphere [12]. Numerical calculations demonstrated that, depending on the relationship between the integral Hall and Pedersen conductivities of the disturbed ionosphere, oscillating and aperiodic modes of magnetic disturbances may be observed. For strong perturbations of ionosphere conductivities, the amplitude of pulsations may be ~$10^2$ nT. In this case, the amplitude of the horizontal component of the electric field on the Earth's surface will be 0.01 mV/m, and the electric current density in the lithosphere may reach $10^{-6}$ A/m$^2$ [12]. Thus, it was shown that the absorption of ionizing radiation from SFs can result in variations of a density of telluric currents in seismogenic faults comparable with a current density of $10^{-8}$–$10^{-7}$ A/m$^2$ generated in the Earth's crust by artificial pulsed power systems (geophysical MHD generator "Pamir-2" and electric pulsed facility "ERGU-600"), which provided regional EQ triggering and spatiotemporal variation of seismic activity [13]. Therefore, we can expect that the triggering of seismic events is possible not only due to the EM impact of the artificial pulsed power sources on the lithosphere but also due to SFs. Based on the theoretical study above mentioned and the numerical results obtained with the employment of the theoretical model [12], the hypothesis was put forward about the triggering of EQs by strong SFs under certain favorable conditions (the electrical conductivity of the fault in the Earth's crust, its orientation relative to the direction of the GIC density vector, and the level of stress–strain state of the fault, e.g., its maturity for the dynamic rupture). This study is directed at the verification of the hypothesis by the analysis of variations in the geomagnetic field recorded at INTERMAGNET observatories during strong X-class SFs, as well as by analysis of seismicity behavior after the flares.

### *2.2. Analysis of Geomagnetic Field Variations and Seismic Activity during Strong X-Class SFs*

According to the theoretical model [12], SF X-ray radiation will be absorbed in the ionosphere, resulting in a short-term increase in its conductivity (Figure 1, reproduced from [12]) and disturbances of the geomagnetic field in various ranges of periods in the presence of an external electric field. It was assumed that the maximal currents in the lithosphere are induced by short-period oscillations of the geomagnetic field. The "earth-ionosphere" resonator generates the geomagnetic field oscillations with periods of 1 to 100 s in the process of ionosphere ionization by SF radiation with a short-term increase in its amplitude.

Considering that the increase in ionosphere conductivity occurs on the illuminated part of the globe, we analyzed the records of the INTERMAGNET observatories [18] that were located there during SFs from the catalog of the 50 strongest X-class SFs with a peak X-ray intensity of above $10^{-4}$ W/m$^2$ [18]. The coordinates of the SSP and the area of the illuminated part of the globe were determined by a solar calculator [19].

For the verification of the proposed model and the obtained numerical results [12] on the possible triggering of EQs by SFs, an analysis of the Earth's seismicity before and after the strongest X-class SFs was carried out. A representative part of the US Geological Survey (USGS) EQ catalogue ($M_w \geq 4.5$) [20] and a catalogue of the 50 strongest X-class SFs [21] for the period 1997–2024 were used.

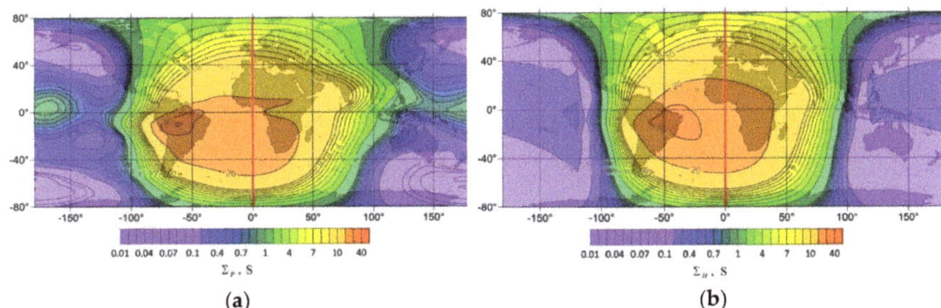

(a)                                                                                              (b)

**Figure 1.** An example of the spatial distribution of the integrated Pedersen (**a**) and Hall (**b**) conductivities in the −80–80° latitude range for universal time 12:00 UT [12]. The red vertical line shows the position of the subsolar meridian.

The analysis of the possible correlation between SFs and EQs employed the epoch superposition method, when for the time windows of 10 days before and after the arrival of X-rays from the SF to the Earth (~8 min), all EQs that occurred in the selected region of the Earth's crust were summed up for each day. In accordance with the physical model [12], according to which the maximum burst of telluric currents in the Earth's crust should occur on the illuminated part of the globe, the seismicity of two regions with a center in the SSP and radii of 5000 km and 10,000 km was analyzed. The SSP coordinates were determined by the date and time of the SF occurrence.

## 3. Results of Verification of Hypothesis of Earthquake Triggering by Strong Solar Flares

*3.1. Response of Geomagnetic Field to Strong Solar Flare: The Case Study of Solar Flare X5.01 of 31 December 2023*

We considered the behavior of the geomagnetic field ($Bx$) during strong SF X5.01 on 31 December 2023. The SF occurred at 21:36 UTC, the maximal X-ray flux was reached at 21:55 UTC, and the end of the SF was at 22:08 UTC (Figure 2).

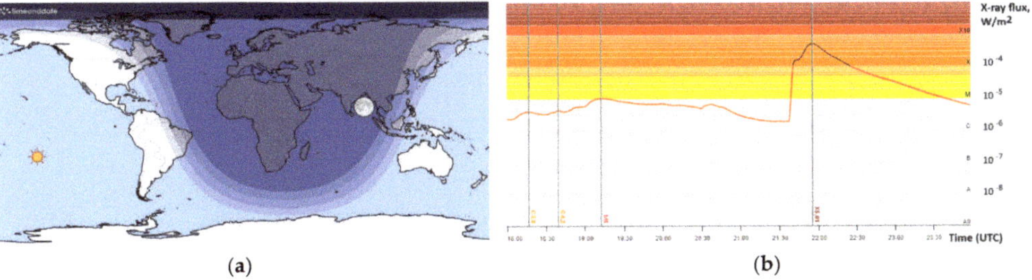



(a)                                                                                              (b)

**Figure 2.** (**a**) Location of the SSP (the Sun sign) for the X5.01 SF of 31 December 2023 with the coordinates 23.067° S, 147.733° W during peak radiation on 21:55 UTC; (**b**) 1 min solar X-ray average in the 1–8 Angstrom passband (red line) recorded by the GOES X-ray satellite [21].

According to the model [12], we analyzed the recordings of geomagnetic observatories located just near the SSP (23.067° S, 147.733° W) at a distance of 640.96 km and at distances of 4287.45 to 10,003.97 km (on the illuminated part of the globe at the moment of the SF) and on the non-illuminated part of the globe at a distance of 15,780.24 km from the SSP (Table 1). We considered records of the horizontal component $Bx$ of the geomagnetic field and its derivative $dBx/dt$, keeping in mind their maximal contribution to GICs in the lithosphere [22]. The magnetograms for different geomagnetic observatories at different distances R from the SSP $Bx$ (a), (c), I, (g), (i), (k), (m), and $dBx/dt$ (b), (d), (f), (h), (j), (l), (n) for the time around the SF occurrence downloaded from the INTERMAGNET site [18] for 31 December 2023 are shown in Figure 3. The SF duration is depicted by a shadowed rectangle.

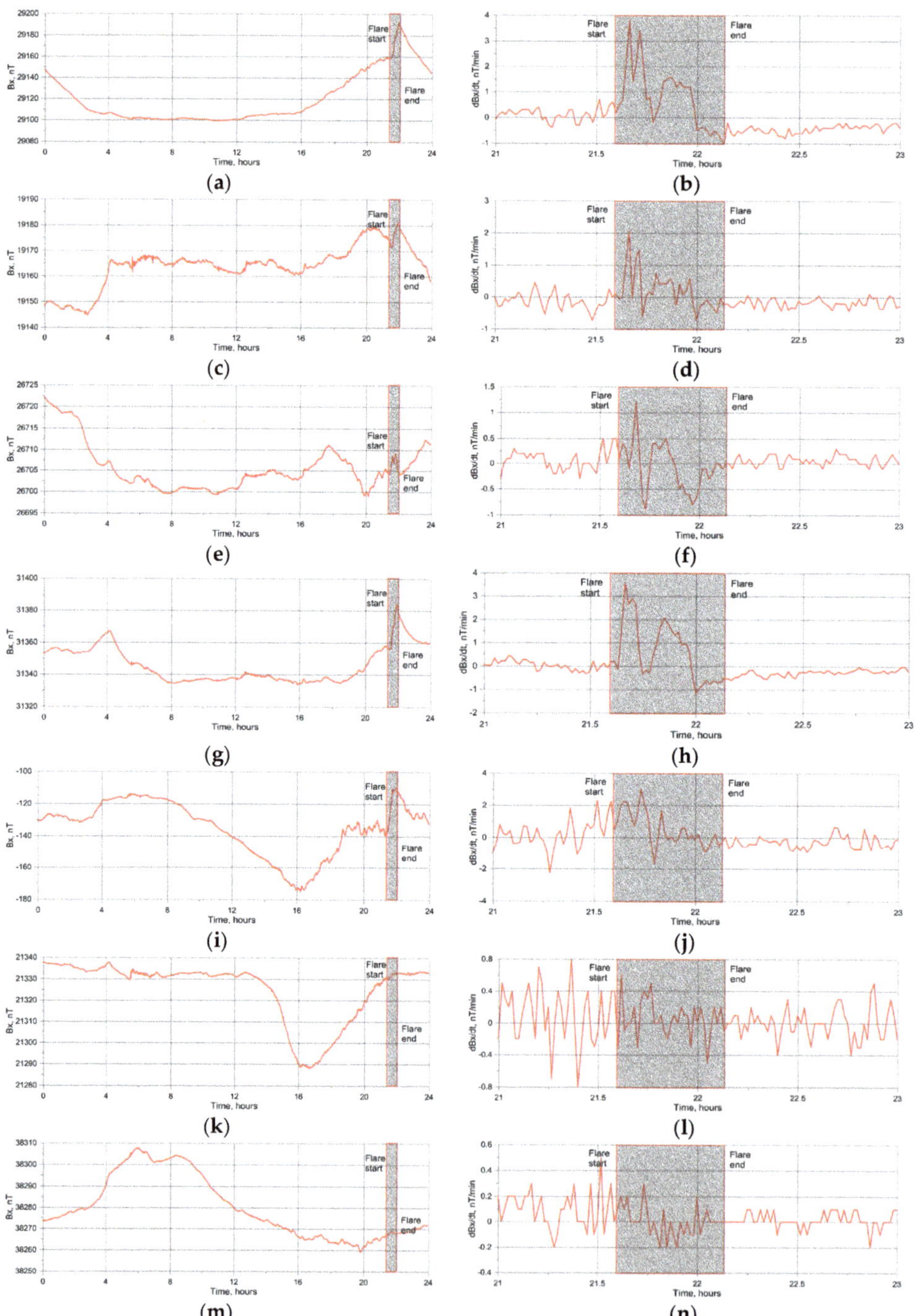

**Figure 3.** Variations of geomagnetic field $Bx$ (left panels) and $dBx/dt$ (right panels) recorded at various geomagnetic observatories (Table 1) during the X5.01 class SF of 31 December 2023: (**a**,**b**)—PPT, R = 640.92 km; (**c**,**d**)—EYR, R = 4287.45 km; (**e**,**f**)—HON, R = 5059.38 km; (**g**,**h**)—CTA, R = 6774.29 km; (**i**,**j**)—AIA, R = 7388.67 km; (**k**,**l**)—FRD, R = 10,003.97 km; (**m**,**n**)—ABG, R = 15,780.24 km. The SF duration is depicted by a shadowed rectangle.

**Table 1.** Location of INTERMAGNET observatories with IAGA codes used for analysis of *Bx* and *dBx/dt* variations during the X5.01 SF on 31 December 2023.

| IAGA Code | Latitude | Longitude | Distance to Subsolar Point R, km |
|---|---|---|---|
| PPT | −17.567 | 210.426 | 640.92 |
| EYR | −43.474 | 172.393 | 4287.45 |
| HON | 21.320 | 202.000 | 5059.38 |
| CTA | −20.090 | 146.264 | 6774.29 |
| AIA | −65.245 | 295.742 | 7388.67 |
| FRD | 38.210 | 282.633 | 10,003.97 |
| ABG | 18.638 | 72.872 | 15,780.24 |

The analysis of the recorded variations in the geomagnetic field in different parts of the globe (illuminated and non-illuminated) at various distances from the SSP demonstrated that the *Bx* and *dBx/dt* pulses predicted by the model [12] during the SF were observed on the illuminated part of the globe.

At the same time, there were no geomagnetic field pulses during the X-class SF on the border of the illuminated part (FRD observatory) and on the non-illuminated part. A sharp increase in *Bx* during the SF totaled 20–25 nT, and *dBx/dt* pulsations totaled 1–4 nT/min. These observations confirmed the numerical results obtained with the employment of the model [12], that the pulses of the geomagnetic field were generated by the X-ray radiation of the SF and resulted in GIC occurrence in the lithosphere were not anticipated in the non-illuminated part of the globe. Thus, the analysis of the response of seismic activity to SFs should consider only the illuminated part of the globe. The next question arises: "Are the observed pulsed variations in the geomagnetic field capable of provoking EQs according to the hypothesis of the EM triggering of EQs by strong SFs?"

*3.2. Seismic Activity before and after Strong X-Class SFs*

For the analysis of the possible response of seismic activity to the impact of strong SFs, we used the representative part ($M_w \geq 4.5$) of the USGS earthquake catalog [21] and a catalogue of the 50 strongest X-class SFs [19]. The analysis of the possible correlation between SFs and EQs employed the epoch superposition method, when for the time windows of 10 days before and after the arrival of X-rays from the SF to the Earth, all EQs occurring in the selected region of the Earth's crust were summed up for each day. In accordance with the physical model [12], when the maximum burst of telluric currents in the Earth's crust was anticipated in the illuminated part of the globe, we considered seismic activity in two circular regions with a center in the SSP and radii of 5000 km and 10,000 km (the illuminated part of the globe) and compared the results with the Earth's overall seismicity. The detailed list of specific SFs and the analyzed possible seismic responses is given in Appendix A (see Table A1).

The summary results based on the detailed Table A1 are shown in Table 2, where $\Sigma_{R=5000}$ is the sum of the EQs in an area with a radius of 5000 km around the SSP, $\Sigma_{R=10,000}$ is the sum of the EQs in an area with a radius of 10,000 km around the SSP, $\Sigma_{global}$ is the sum of EQs for the whole globe, *a* is the cumulative number of EQs occurring within 10 days after the SF, *b* is the cumulative number of EQs occuring within 10 days before the SF, and $\Delta$EQ is the increase in the number of EQs after the SF (%).

**Table 2.** The number of EQs ($M_w \geq 4.5$) after (*a*) and before (*b*) the SF at a distance of 5000 km ($\Sigma_{R=5000}$) and 10,000 km ($\Sigma_{R=10,000}$) from the SSP, as well for the whole globe ($\Sigma_{global}$).

| | $\Sigma_{R=5000}$ | | $\Sigma_{R=10,000}$ | | $\Sigma_{global}$ | |
|---|---|---|---|---|---|---|
| | *a* | *b* | *a* | *b* | *a* | *b* |
| | 1696 | 1276 | 4565 | 4113 | 8629 | 7987 |
| $\Delta$EQ, % | 32.92 | | 10.99 | | 8.04 | |

The analysis of the values in Table 2 demonstrated that, according to the physical model and the results of the numerical studies obtained with its application [12], there was a significant seismic response in the illuminated part of the Earth. The maximum increase in the number of EQs (32.92%) was observed in the region with a radius of 5000 km around the SSP. As the radius of the region increased to 10,000 km, which is equivalent to the entire illuminated area of the globe, seismic growth decreased to 10.99%, similar to a decrease in geomagnetic field variations (see Section 3.1). In comparison with the global seismicity, the $\Delta$EQ for the illuminated part was one and a half times higher. The histogram of the distribution of the daily number of EQs before and after the SF is shown in Figure 4. It should be noted that the increase in seismicity for the illuminated area of the globe was observed with a delay of 7–8 days after SF occurrence (Figure 4a,b) similar to a few-day delay in the response of seismic activity to the impact of artificial EM on the Earth's crust [13].

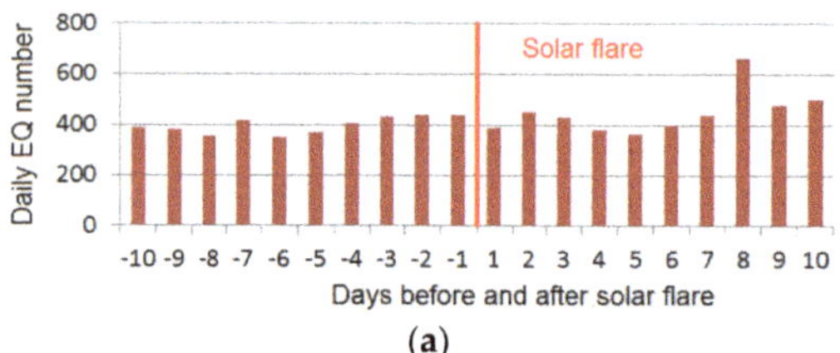


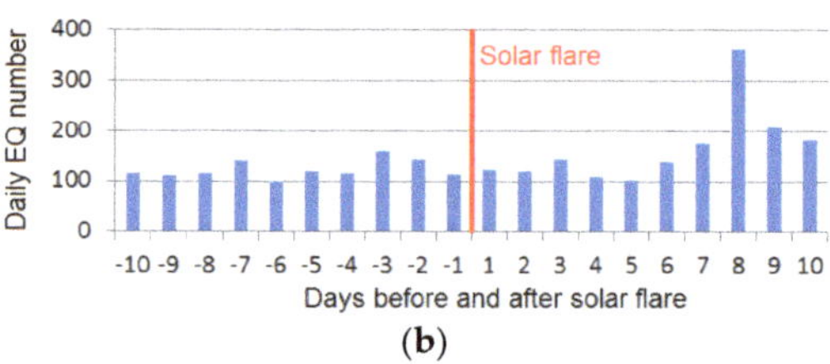


**Figure 4.** Seismic activity before and after SF for (**a**) the illuminated part of the globe and (**b**) the area with a radius of 5000 km around the SSP. The red vertical line denotes a moment of SF.

One of the general conditions for the triggering effect of EM on the earthquake preparation zone is the level of stress–strain state of rocks and the crust fault, which, according to the results of laboratory modeling [13], should be 0.98–0.99 of the critical stresses when the dynamic rupture of the fault occurs. The current level of stresses in a particular region of the globe can be roughly estimated by indirect signs, e.g., by the current seismic activity that is used by methods of mid-term earthquake forecasting [23].

However, there are situations when it is possible, with a high degree of probability, to determine areas where subcritical stresses arise regularly in the Earth's crust. Such areas are the aftershock zones of strong EQs. Aftershocks are a sequence of seismic events that occur after a larger main earthquake around the fault zone of the main shock. Aftershocks are a consequence of the stress field redistribution in the Earth's crust after the main displacement along the fault as the result of the main shock. In the aftershock zone, areas occur constantly where the stresses in the Earth's crust will be close to critical values when the rock rupture (aftershock) occurs. Aftershocks become less frequent over time, although they may continue for days, weeks, months, or even years [24]. Thus, due to the indicated stress redistribution in the aftershock zone, areas with a subcritical stress–strain state always appear, which are most sensitive to triggering impacts. In this case, when such an area is located near the SSP during the occurrence of a strong SF, it is possible to anticipate the EM triggering of aftershocks. In this regard, to verify the hypothesis of the EM triggering of earthquakes by SFs, it is quite reasonable to consider the seismic activity in the aftershock zones located near the SSP at the time of the SF occurrence.

As a case study, we considered the impact of the X10.16 SF of 20 January 2005 on the aftershock zone of the Sumatra–Andaman $M_w = 9.1$ EQ that occurred on 26 December 2004. The aftershock zone [25] is covered by the area of the 5000 km radius around the SSP of the X10.16-class SF (20.083° S, 77.767° E) (Figure 5). The distance from the SSP to the epicenter of the $M_w = 9.1$ EQ is 3306.36 km.

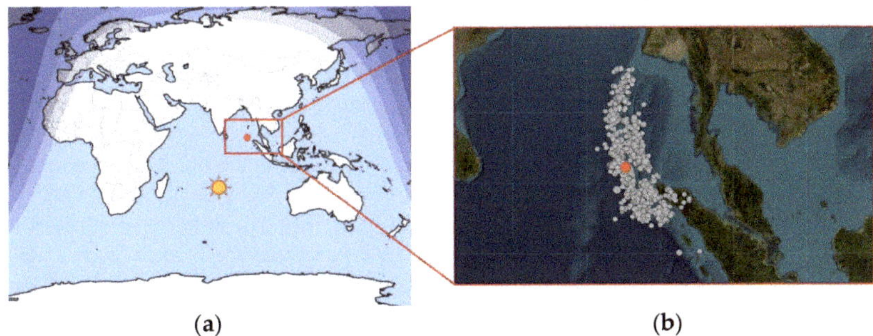

(a)              (b)

**Figure 5.** (**a**) SSP location (20.083° S, 77.767° E) for the X10.16 SF of 20 January 2005; the red circle is the Sumatra–Andaman $M_w$ = 9.1 EQ epicenter location (3.295° N, 95.982° E) [21]; (**b**) enlarged aftershock zone of the Sumatra–Andaman $M_w$ = 9.1 EQ of 26 December 2004; the red circle is the EQ epicenter located at a distance of 2717.6 km from the SSP; the grey circles are the epicenters of the aftershocks ($M_w \geq$ 2.5) [21].

The histogram of the daily distribution of the aftershocks after the main shock of 26 December 2004 is shown in Figure 6. The redistribution of stresses in the crust just after the $M_w$ = 9.1 EQ triggered many aftershocks during the first two days (269 and 202), then the daily aftershock number was reduced up to an average number of about 14, and after the X10.16 SF, with a delay of 6 days, we observed a sharp increase in seismic activity that lasted four days, with two peaks of 244 and 240 aftershocks during 27–28 January 2005. Thus, for favorable conditions from the point of view of the generation of geomagnetic field pulsations (for this case recorded at the PHU observatory located at a distance of 2933.4 km from the SSP and 2248.2 km from the $M_w$ = 9.1 EQ epicenter, Figure 7), as well as for the zone with subcritical stress values close to the fault rupture (aftershock zone), we observed a clear EQ triggering effect of the X10.16-class SF.

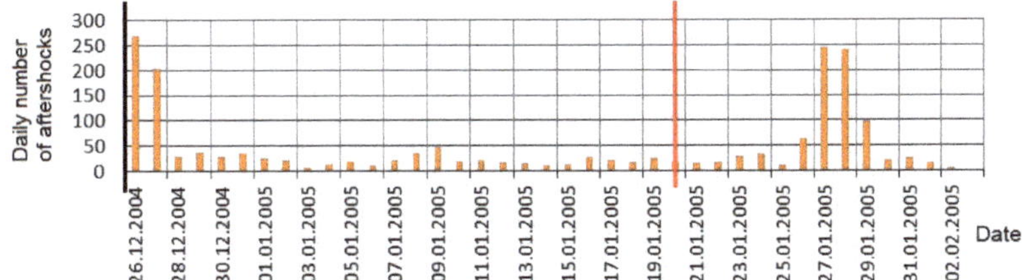

**Figure 6.** Daily distribution of the aftershocks of the Sumatra–Andaman $M_w$ = 9.1 EQ (depicted on the left by a thick black vertical line). The date of the X10.16 SF (20 January 2005) is depicted by a thick red vertical line.

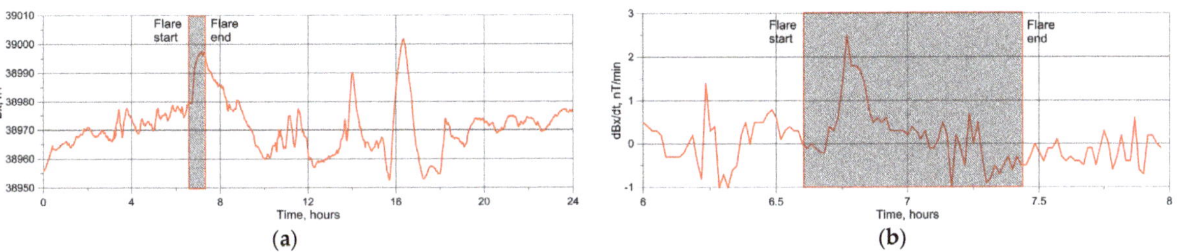

(a)              (b)

**Figure 7.** variations in (**a**) geomagnetic field $Bx$ and (**b**) $dBx/dt$ recorded at the PHU observatory located at a distance of 2933.4 km from the SSP and 2248.2 km from the $M_w$ = 9.1 EQ epicenter during the X10.16-class SF of 20 January 2005. The SF duration is depicted by the shadowed rectangle.

It should be noted that during the month before the sharp intensification of the aftershock sequence in the period of 27–29 January 2005, there were no strong EQs in the immediate vicinity of the region under study (and throughout the entire globe) with a magnitude of $M_w \geq 7$, which could provide a distant dynamic triggering effect in the considered aftershock area.

There were a few space weather disturbances due to seven SFs that occurred after the Sumatra–Andaman $M_w = 9.1$ EQ (see Table 3). Nevertheless, the EM impact on the aftershock zone of the strong EQ can only be attributed to the X10.16 SF of 20 January 2005 (Figure 7), due to its X-ray radiation that exceeded the other SFs 5 to 10 times and the distance from its SSP to the epicenter of the main shock that was two times less than the similar distance of the strong X5.51 SF of 17 January 2005. Nevertheless, for this case, the combined action of both SFs (X10.16 and X5.51) may be considered, which can explain the beginning of the increase in aftershocks on 26 January 2005 (see Figure 6).

**Table 3.** Solar activity (X-class SFs) before a splash of the aftershock sequence of the Sumatra–Andaman EQ $M_w = 9.1$ (20 January 2005), possibly triggered by a strong X10.16-class SF.

| SF Date | SF Class | Time of Max X-ray Flux (UT) | SSP Latitude | SSP Longitude | Distance to Sumatra–Andaman EQ Epicenter, km |
|---|---|---|---|---|---|
| 1 January 2005 | 2.49 | 0:31 | 23°01′ S | 172°52′ E | 8815.447 |
| 15 January 2005 | 1.79 | 0:43 | 21°08′ S | 171°20′ E | 8628.460 |
| 15 January 2005 | 1.21 | 4:26 | 21°07′ S | 115°51′ E | 3471.743 |
| 15 January 2005 | 1.24 | 5:54 | 21°06′ S | 93°52′ E | 2722.371 |
| 15 January 2005 | 3.79 | 23:02 | 20°58′ S | 163°05′ W | 7782.587 |
| 17 January 2005 | 5.51 | 9:52 | 20°41′ S | 34°33′ E | 7201.479 |
| 19 January 2005 | 2.00 | 8:22 | 20°17′ S | 57°12′ E | 4977.074 |
| 20 January 2005 | 10.16 | 7:02 | 20°05′ S | 77°16′ E | 3306.360 |

Thus, this case study of aftershock sequence variation due to the impact of a strong SF indicates that, if the aftershock zone is located near the SSP during the occurrence of a strong SF (Figure 5), we can expect the EM triggering of aftershocks, which can have a magnitude comparable to that of the main shock, and they are also dangerous, especially during rescue operations after a catastrophic EQ. For the Sumatra–Andaman $M_w = 9.1$ EQ, the maximal magnitude of the aftershocks was $M_w = 6.3$.

The obtained results are supported by the occurrence of strong aftershocks after X- and M-class SFs in the New Zealand region in 2011. A strong $M_w = 7.1$ EQ occurred on 3 September 2010 near Darfield on the South Island of New Zealand [26], which injured about 100 people. After this EQ, an aftershock sequence began, which included a strong $M_w = 6.1$ aftershock that occurred on 21 February 2011 and killed 185 people. It should be noted that 6 days before this strong aftershock, an X2.3-class SF occurred (15 February 2011, 01:44–02:06 UTC) when New Zealand was in the central zone of the illuminated part of the globe at a distance of 3853.7 km from the SSP to the $M_w = 6.1$ aftershock epicenter. The EYR observatory (New Zealand) recorded geomagnetic field pulsations during the SF of about 20–25 nT.

According to the results of calculations using the model [12], the GIC density vector in this region had a southeastern direction and was at a level of $10^{-7}$ A/m$^2$, which was comparable to the GIC density created in the Earth's crust by an MHD generator and resulted in the triggering of EQs in the northern Tien Shan [13]. In the case of New Zealand, the GIC density vector coincided with the direction of the strike of the Port Hills fault [27], where a strong aftershock occurred. Thus, for an EQ to be triggered by a SF, the presence of all three conditions of the EM triggering effect was ensured: a subcritical stress–strain state fault (EQ aftershock zone), the required level of GICs ($10^{-7}$ A/m$^2$), and the optimal direction of the GIC density vector (parallel to the direction of the Port Hill fault). It should be noted that this $M_w = 6.1$ aftershock occurred with a delay of 6 days after the SF, which is similar to the seismic response to artificial EM impacts [13], as well as aftershock activity for the Sumatra–Andaman $M_w = 9.1$ EQ (the delay of the aftershock response to the SF was 6–8 days). The relationship between these two events (SF and aftershock) seems not accidental, since on the same fault there was a repeated strong $M_w = 5.9$ aftershock on 13 June 2011 after an M3.64 SF (7 June 2011, peak radiation flux was $3.64 \times 10^{-5}$ W/m$^2$) with the same delay of 6 days.

## 4. Discussion

During the verification of the physical model [12] by the comparison of numerical results on the possible triggering of EQs by strong SFs with field observations of variations in the geomagnetic field and seismic activity during and after the top 50 strong X-class SFs, we obtained the following results:

(1)  Pulsations in the geomagnetic field predicted by the model [12] due to the interaction of X-ray radiation from SFs with the ionosphere were observed during the SF on the illuminated part of the globe. The maximal $Bx$ and $dBx/dt$ pulsations were observed in an area of 5000 km around the SSP at the time of SF occurrence. With an increasing area radius, $Bx$ and $dBx/dt$ pulsations decreased and practically disappeared at the border of the illuminated part. Such pulsations were not observed on the non-illuminated part of the globe. These results are consistent with those obtained earlier and presented in a review by Curto [28] and a study by Grodji et al. [29].

(2)  The observed sharp variations in the geomagnetic field were capable of generating GICs in the conductive elements of the lithosphere, including seismogenic faults. According to the model [12], these GICs are comparable to a splash of telluric currents generated by artificial pulsed power systems, which resulted in the EQ triggering and spatiotemporal redistribution of seismicity in the northern Tien Shan and Pamir [13]. Our analysis of seismicity after strong SFs supported the hypothesis of Sorokin et al. [12] of the EM triggering of EQs by SFs (Table 2). For the illuminated part, within 10 days after an X-class SF, the seismicity increased in comparison with 10 days before the SF by ~11 to ~33%, depending on the distance from the SSP. It significantly exceeded the Earth's overall seismic response. This result positively estimates the hypothesis proposed in Sorokin et al. [12] on EQ triggering by the X-ray radiation of SFs and indicates the incorrectness of a purely statistical approach to the study of the interrelationship of solar and seismic activities without any physical model, explaining a possible relationship between the processes on the Sun and within the Earth. For further study, it is reasonable to consider the solar–terrestrial relations based on the physical model [12], or any models considering another physical mechanism of these relations, provided that a refined approach is used to select the data for statistical analysis. In other words, "Physics should be ahead of Statistics".

(3)  The next finding of the presented analysis was the response of the aftershock area of a strong EQ to the impact of a SF, where areas with a subcritical stress–strain state appeared constantly due to the redistribution of the stresses in the crust after the main shock. Based on two case studies of the aftershock zones of a strong magnitude $M_w = 7.1$ EQ in New Zealand and a strong $M_w = 9.1$ EQ in Indonesia, a clear response of the aftershock sequences to X-class SFs was discovered (Figure 6). The general feature of this response is a delay of 6 to 8 days, which may indicate a multi-stage physical mechanism triggering processes in the crust fault, including fluid migration under EM impact, that require some time for fluid diffusion into the fault, reducing its frictional properties and strength.

In our opinion, the presented results of the analysis of field observations not only indicate the possibility of EQs being triggered by strong SFs, but also, taking into account the delay of several days in the seismic response to the EM impact, point to the possibility of the application of natural EM triggering effects as additional prognostic information for EQ forecasting methods along with other known precursors of strong EQs.

In this case, the concept of EQ predictability based on triggering phenomena, which was formulated by Sobolev [30], may be used. Based on observations of seismic behavior before strong EQs as well as laboratory studies of the response of acoustic emissions (crack formation) from rock samples in a subcritical stress–strain state under external triggering impacts, the following algorithm for short-term EQ forecasting based on triggering phenomena was proposed [30]: (a) determination of the volume of the unstable zone (a system of unstable zones of various scales); (b) monitoring of the triggering effects and assessing their impact on the unstable areas; and (c) assessment of the probability of the location, time, and magnitude of the impending EQ.

The first step (a) of this concept can be performed based on various methods for the selection of regions with an impending EQ (e.g., [31]). For the case of an EM impact on the EQ preparation zone in these regions, it is necessary to additionally select faults in the Earth's crust, taking into account their orientation and electrical conductivity, where the generation of maximum GICs can be expected. It is obvious that the maximum GICs in the fault will be generated when the GIC density vector is parallel to the fault direction, which will contribute to the GIC concentration in the fault and increase the efficiency of its impact on rocks.

Numerical results [12] demonstrated that the maximum GIC density values should be observed in the southern hemisphere when the SSP is located in the northern hemisphere. Thus, the response of seismic activity to a strong SF can be expected with a higher probability in the southern hemisphere. The GIC density vectors in the northern hemisphere at low and middle latitudes are oriented mainly in the latitudinal direction, and at high latitudes—in the meridional direction. In the southern hemisphere, they are usually oriented in the meridional direction. When choosing regions and faults, where the response of seismic activity to SFs will be statistically analyzed is important. In this case, the selection of EQs from seismic catalogs in step (b) should be made only for faults where the expected triggering effect from the EM impact will be maximal. To increase the reliability of the statistical analysis, taking into account the numerical results obtained by Sorokin et al. [12], only those regions should be selected where the orientations of the faults in the Earth's crust approximately coincide with the direction of the GIC density vector. Otherwise, the density of the GICs generated in the fault may be insufficient for the EM triggering of an EQ, resulting in false statistical results and conclusions about solar–terrestrial relationships when seismic activity is analyzed for an entire region with faults of different orientations and electrical conductivities.

Another important aspect when selecting the crustal faults that are sensitive to strong variations in space weather is their electrical resistivity, which is usually determined by magnetotelluric (MT) sounding [32]. The results of field studies showed that the San Andreas fault (California, USA) [33] and other major faults, such as the Alpine fault in New Zealand [34] and the Fraser River fault in British Columbia, Canada [35], have conducting zones with a resistivity of 0.8 to 50 Ohm·m. At the same time, some large faults demonstrate the presence of both conductive and resistive zones. For example, the MT sounding of the Tintina fault in the northern Cordillera [32] showed that the fault is associated with a 20 km wide resistive zone (>400 Ohm·m) at depths of more than 5 km. The Denali fault in Alaska also has a relatively resistive structure in the upper layers of the Earth's crust [36], and the San Andreas fault in the Carrizo Plain region has a resistive zone in the mid-deep region [37]. The resistivity of these faults varies in the range of ~250 to 10,000 Ohm·m, and, therefore, the generation of a GIC pulse resulting from strong variations in space weather parameters that is sufficient to trigger an EQ is unlikely or impossible.

Keeping in mind the obtained numerical estimates [12] and their field verification during this study, further detailed and correct statistical analysis of solar and seismic activities should first be based on a physical model of the mechanism of solar–terrestrial relationships, and second, the specific model of the EM triggering of EQs should be prepared out as follows:

(a) determination of an unstable area (a fault section in the Earth's crust), where strong EQs are expected based on existing mid-term methods for selecting seismic-prone regions [31];

(b) selection of crustal faults in the regions identified in step (a) that are the most sensitive to EM impact in terms of their orientation and close to the direction of the GIC density vector, as well as based on their electrical conductivity;

(c) selection of EQs that occurred in the faults identified in step from regional seismic catalogs (b);

(d) correlation analysis of EQ occurrence times and variations in space weather parameters to determine the delay time of EQ triggering and the threshold values of space weather parameters that resulted in the triggering effect in the EQ preparation zone.

As mentioned above, in our opinion, the results obtained, both using the model [12] and field observations employed in this study, can be applied to EQ forecasting methods. The algorithm for such a forecast can be as follows: After a strong X-class SF, the areas of the possible triggering of EQs should be determined, where the impact of the SF can be the greatest in terms of the medium-term forecast of seismic hazards, the electrical conductivity of faults where EQs are expected, and their orientation. In these areas, immediately after the SF, the monitoring of other known EQ precursors should also be carried out. Based on a multi-parameter approach for the analysis of possible precursors, a seismic alarm can be issued, taking into account the delay of several days (6–8 days, according to the analysis results obtained above) in the response of the EQ preparation zone to the EM triggering impact. Obviously, this algorithm is not universal and can only be used in seismically hazardous areas that are sensitive to EM impacts.

As one of the possible precursors related to space weather that can be considered after a SF in the areas of possible EQ triggering, anomalous variations in ionospheric parameters caused by the emission of aerosols from the EQ preparation zone may be considered physically validated in Sorokin et al. [38].

## 5. Conclusions

The statistical analysis of the impact of the top 50 X-class SFs (1997–2024) on the seismic activity carried out for the verification of the physical model of the EM triggering impact of strong SFs on EQ preparation zones indicated a possibility of EQ triggering on the illuminated part of the globe. It was shown that an increase in seismicity was observed, especially in the region around the SSP with a radius of 5000 km (up to 33%) during the 10 days after the SF. The case study of the aftershock sequence behavior of a strong $M_w = 9.1$ EQ (Sumatra–Andaman Islands, 26 December 2004) after an X10.16-class SF (20 January 2005) demonstrated that the number of aftershocks with a magnitude of $M_w \geq 4.5$ increased more than 17 times with a delay of 7–8 days. For the case of the Darfield EQ ($M_w = 7.1$, New Zealand, 3 September 2010), it was shown that strong X-class (peak radiation flux is $\geq 10^{-4}$ W/m$^2$) and M-class (peak radiation flux is $\geq 10^{-5}$ W/m$^2$ and $< 10^{-4}$ W/m$^2$) SFs probably triggered two strong aftershocks ($M_w = 6.1$, $M_w = 5.9$) with an observed delay of 6 days on the Port Hills fault, which is the most sensitive to an external EM impact from the point of view of the fault's electrical conductivity and orientation. Based on the obtained results, we concluded that data on the possible natural EM triggering of EQs may be applied as supporting information in addition to known EQ precursors for EQ forecasting using confidently recorded strong external EM triggering impacts on the specific EQ preparation zones as well as well-known EQ precursors, including satellite-recorded ionospheric perturbations due to aerosol emissions from the EQ preparation zone. It is obvious that such an approach, based on EM-triggering phenomena, has limited use and may be applied only to earthquake-prone regions with conductive crustal faults, the orientation of which roughly coincides with the direction of the GIC density vector excited by the SF. In this case, the expected EQ should be located in the illuminated part of the globe as close as possible to the SSP, where maximum pulsations in the geomagnetic field were observed.

**Author Contributions:** Conceptualization, V.S. and V.N.; methodology, V.N.; software, V.S.; formal analysis, V.N.; writing—original draft preparation, V.N.; writing—review and editing, V.S. All authors have read and agreed to the published version of the manuscript.

**Funding:** The study was supported by the Ministry of Science and Higher Education of the Russian Federation (State Assignments of IZMIRAN No. 1021060808637-6-1.3.8, JIHT RAS No. 075-00270-24-00, and IDG RAS No. 122032900178-7).

**Data Availability Statement:** All data used in the study are available online (see refs. [20,21]).

**Acknowledgments:** The authors would like to thank the anonymous reviewers and the academic editor for their valuable comments and suggestions, which undoubtedly improved the manuscript.

**Conflicts of Interest:** The authors declare no conflicts of interest.

## Appendix A

**Table A1.** Parameters of SFs [21], the number of EQs (M $\geq$ 4.5) within 10 days after (*a*) and before (*b*) an SF at a distance of 5000 km ($\Sigma_{R=5000}$) and 10,000 km ($\Sigma_{R=10,000}$) from the SSP, as well for the whole globe ($\Sigma_{global}$) [20].

| No. | SF Date | SF Class | Time of Max X-ray Flux (UT) | SSP LAT | SSP LONG | $\Sigma_{R=5000}$ | | $\Sigma_{R=10,000}$ | | $\Sigma_{global}$ | |
|---|---|---|---|---|---|---|---|---|---|---|---|
| | | | | | | *a* | *b* | *a* | *b* | *a* | *b* |
| 1 | 6 November 1997 | X12.97 | 11:55 | 16°04′ S | 2°35′ W | 7 | 0 | 35 | 50 | 151 | 136 |
| 2 | 6 May 1998 | X3.81 | 8:09 | 16°31′ N | 57°10′ E | 13 | 13 | 66 | 61 | 121 | 96 |
| 3 | 18 August 1998 | X7.03 | 22:19 | 12°56′ N | 153°33′ W | 15 | 14 | 85 | 71 | 114 | 106 |
| 4 | 18 August 1998 | X4.03 | 8:24 | 13°07′ N | 54°59′ E | 11 | 18 | 52 | 57 | 112 | 106 |
| 5 | 19 August 1998 | X5.57 | 21:45 | 12°37′ N | 145°06′ W | 7 | 11 | 88 | 67 | 116 | 100 |
| 6 | 22 November 1998 | X5.37 | 6:42 | 20°06′ S | 76°01′ E | 8 | 7 | 72 | 68 | 121 | 100 |
| 7 | 28 November 1998 | X4.77 | 5:52 | 21°16′ S | 88°58′ E | 36 | 17 | 94 | 84 | 129 | 108 |
| 8 | 14 July 2000 | X8.21 | 10:24 | 21°36′ N | 25°28′ E | 9 | 6 | 44 | 52 | 183 | 202 |
| 9 | 26 November 2000 | X5.83 | 16:48 | 21°05′ S | 75°07′ W | 14 | 12 | 26 | 19 | 133 | 376 |
| 10 | 2 April 2001 | X28.57+ | 21:51 | 5°13′ N | 146°38′ W | 13 | 9 | 82 | 82 | 117 | 115 |
| 11 | 6 April 2001 | X8.08 | 19:21 | 6°42′ N | 109°25′ W | 12 | 13 | 46 | 61 | 109 | 132 |
| 12 | 15 April 2001 | X20.67+ | 13:50 | 9°55′ N | 27°30′ W | 1 | 8 | 27 | 40 | 134 | 115 |
| 13 | 25 August 2001 | X7.7 | 16:45 | 10°34′ N | 70°30′ W | 9 | 18 | 19 | 24 | 148 | 173 |
| 14 | 11 December 2001 | X4.02 | 8:08 | 23°01′ S | 56°19′ E | 8 | 3 | 59 | 61 | 113 | 109 |
| 15 | 13 December 2001 | X8.9 | 14:30 | 23°11′ S | 38°55′ W | 14 | 17 | 31 | 29 | 106 | 119 |
| 16 | 28 December 2001 | X4.99 | 20:45 | 23°15′ S | 130°33′ W | 12 | 10 | 85 | 54 | 143 | 108 |
| 17 | 15 July 2002 | X4.39 | 20:08 | 21°27′ N | 120°30′ W | 1 | 3 | 46 | 45 | 108 | 116 |
| 18 | 20 July 2002 | X4.74 | 21:30 | 20°34′ N | 140°54′ W | 1 | 1 | 58 | 75 | 97 | 122 |
| 19 | 23 July 2002 | X6.98 | 0:35 | 20°09′ N | 173°07′ E | 37 | 43 | 74 | 85 | 102 | 110 |
| 20 | 24 August 2002 | X4.54 | 1:12 | 11°13′ N | 162°38′ E | 82 | 91 | 105 | 118 | 141 | 135 |
| 21 | 28 May 2003 | X5.17 | 0:27 | 21°22′ N | 172°47′ E | 56 | 45 | 124 | 120 | 154 | 160 |
| 22 | 23 October 2003 | X7.77 | 8:35 | 11°18′ S | 47°36′ E | 2 | 5 | 45 | 47 | 132 | 130 |
| 23 | 28 October 2003 | X24.57+ | 11:10 | 13°04′ S | 8°28′ E | 1 | 1 | 24 | 24 | 152 | 129 |
| 24 | 29 October 2003 | X14.36 | 20:49 | 13°32′ S | 136°04′ W | 25 | 15 | 77 | 78 | 163 | 121 |
| 25 | 2 November 2003 | X11.96 | 17:25 | 14°47′ S | 30°08′ E | 4 | 1 | 35 | 30 | 160 | 134 |
| 26 | 3 November 2003 | X5.61 | 9:55 | 14°60′ S | 27°24′ E | 4 | 1 | 41 | 28 | 164 | 135 |
| 27 | 3 November 2003 | X3.88 | 1:30 | 14°53′ S | 153°24′ E | 82 | 66 | 125 | 109 | 162 | 133 |
| 28 | 4 November 2003 | X40+ | 19:53 | 15°26′ S | 122°06′ W | 6 | 5 | 64 | 62 | 174 | 139 |
| 29 | 16 July 2004 | X5.24 | 13:55 | 21°15′ N | 26°58′ W | 5 | 4 | 37 | 27 | 117 | 120 |
| 30 | 17 January 2005 | X5.52 | 9:52 | 20°41′ S | 34°33′ E | 2 | 6 | 239 | 164 | 376 | 265 |
| 31 | 20 January 2005 | X10.16 | 7:01 | 20°05′ S | 77°46′ E | 524 | 127 | 621 | 223 | 679 | 275 |
| 32 | 7 September 2005 | X24.42+ | 17:40 | 5°50′ N | 85°32′ W | 14 | 19 | 31 | 37 | 164 | 184 |
| 33 | 8 September 2005 | X7.77 | 21:06 | 5°24′ N | 137°07′ W | 2 | 11 | 97 | 102 | 158 | 192 |
| 34 | 9 September 2005 | X8.87 | 20:04 | 5°02′ N | 121°42′ W | 11 | 13 | 77 | 79 | 147 | 204 |
| 35 | 9 September 2005 | X5.17 | 9:59 | 5°12′ N | 29°50′ E | 5 | 6 | 40 | 72 | 144 | 207 |
| 36 | 5 December 2006 | X12.95 | 10:35 | 22°23′ S | 19°08′ E | 3 | 3 | 25 | 35 | 152 | 210 |
| 37 | 6 December 2006 | X9.4 | 18:47 | 22°32′ S | 103°43′ W | 13 | 22 | 40 | 73 | 149 | 208 |
| 38 | 13 December 2006 | X4.88 | 2:40 | 23°08′ S | 138°29′ E | 61 | 66 | 123 | 131 | 150 | 157 |

**Table A1.** *Cont.*

| No. | SF Date | SF Class | Time of Max X-ray Flux (UT) | SSP LAT | SSP LONG | $\Sigma_{R=5000}$ | | $\Sigma_{R=10,000}$ | | $\Sigma_{global}$ | |
|---|---|---|---|---|---|---|---|---|---|---|---|
| | | | | | | *a* | *b* | *a* | *b* | *a* | *b* |
| 39 | 9 August 2011 | X9.96 | 8:05 | 15°55′ N | 60°24′ E | 15 | 20 | 101 | 109 | 176 | 191 |
| 40 | 7 March 2012 | X7.79 | 0:24 | 5°12′ S | 176°46′ E | 70 | 87 | 187 | 217 | 249 | 265 |
| 41 | 13 May 2013 | X4.11 | 16:05 | 18°32′ N | 61°55′ W | 18 | 16 | 39 | 36 | 263 | 163 |
| 42 | 14 May 2013 | X4.64 | 1:11 | 18°38′ N | 161°35′ E | 172 | 77 | 211 | 101 | 266 | 162 |
| 43 | 5 November 2013 | X4.93 | 22:12 | 15°56′ S | 157°05′ W | 51 | 52 | 147 | 151 | 196 | 199 |
| 44 | 25 February 2014 | X7.13 | 0:49 | 9°11′ S | 171°17′ E | 49 | 60 | 101 | 110 | 174 | 159 |
| 45 | 24 October 2014 | X4.58 | 21:41 | 11°58′ S | 148°57′ W | 34 | 38 | 120 | 157 | 200 | 256 |
| 46 | 5 May 2015 | X3.93 | 22:11 | 16°22′ N | 153°20′ W | 19 | 11 | 222 | 139 | 285 | 202 |
| 47 | 6 September 2017 | X13.37 | 12:02 | 6°15′ N | 0°55′ W | 4 | 9 | 43 | 59 | 231 | 170 |
| 48 | 10 September 2017 | X11.88 | 16:06 | 4°41′ N | 62°17′ W | 68 | 104 | 100 | 140 | 206 | 229 |
| 49 | 31 December 2023 | X5.01 | 21:55 | 23°04′ S | 147°44′ W | 31 | 49 | 122 | 148 | 216 | 204 |
| 50 | 22 February 2024 | X6.3 | 22:34 | 10°07′ S | 155°08′ W | 35 | 23 | 113 | 102 | 172 | 173 |
| | | | Total EQs after and before 50 SFs | | | 1696 | 1276 | 4565 | 4113 | 8629 | 7987 |
| | | | Total difference in EQ amount after and before 50 SFs, $\Delta EQ = \sum a - \sum b$ | | | 420 | | 452 | | 642 | |
| | | | Total variation in EQs amount after 50 SFs, $\Delta EQ$, % | | | 32.92 | | 10.99 | | 8.04 | |

# References


1. Sobolev, G.A. The effect of strong magnetic storms on the occurrence of large earthquakes. *Izv. Phys. Solid Earth* **2021**, *57*, 20–36. [CrossRef]
2. Rabeh, T.; Miranda, M.; Hvozdara, M. Strong earthquakes associated with high amplitude daily geomagnetic variations. *Nat. Hazards* **2010**, *53*, 561–574. [CrossRef]
3. Tavares, M.; Azevedo, A. Influence of solar cycles on earthquakes. *Nat. Sci.* **2011**, *3*, 436–443. [CrossRef]
4. Urata, N.; Duma, G.; Freund, F. Geomagnetic Kp Index and Earthquakes. *Open J. Earthq. Res.* **2018**, *7*, 39–52. [CrossRef]
5. Sorokin, V.M.; Yashchenko, A.K.; Novikov, V.A. A possible mechanism of stimulation of seismic activity by ionizing radiation of solar flares. *Earthq. Sci.* **2019**, *32*, 26–34. [CrossRef]
6. Marchitelli, V.; Harabaglia, P.; Troise, C.; De Natale, G. On the Correlation between Solar Activity and Large Earthquakes Worldwide. *Sci. Rep.* **2020**, *10*, 11495. [CrossRef] [PubMed]
7. Novikov, V.; Ruzhin, Y.; Sorokin, V.; Yaschenko, A. Space weather and earthquakes: Possible triggering of seismic activity by strong solar flares. *Ann. Geophys.* **2020**, *63*, PA554. [CrossRef]
8. Tarasov, N.T. Effect of Solar Activity on Electromagnetic Fields and Seismicity of the Earth. *IOP Conf. Ser. Earth Environ. Sci.* **2021**, *929*, 012019. [CrossRef]
9. Anagnostopoulos, G.; Spyroglou, I.; Rigas, A.; Preka-Papadema, P.; Mavromichalaki, H.; Kiosses, I. The sun as a significant agent provoking earthquakes. *Eur. Phys. J. Spec. Top.* **2021**, *230*, 287–333. [CrossRef]
10. Love, J.J.; Thomas, J.N. Insignificant solar-terrestrial triggering of earthquakes. *Geophys. Res. Lett.* **2013**, *40*, 1165–1170. [CrossRef]
11. Akhoondzadeh, M.; De Santis, A. Is the Apparent Correlation between Solar-Geomagnetic Activity and Occurrence of Powerful Earthquakes a Casual Artifact? *Atmosphere* **2022**, *13*, 1131. [CrossRef]
12. Sorokin, V.; Yaschenko, A.; Mushkarev, G.; Novikov, V. Telluric Currents Generated by Solar Flare Radiation: Physical Model and Numerical Estimations. *Atmosphere* **2023**, *14*, 458. [CrossRef]
13. Zeigarnik, V.A.; Bogomolov, L.M.; Novikov, V.A. Electromagnetic Earthquake Triggering: Field Observations, Laboratory Experiments, and Physical Mechanisms—A Review. *Izv. Phys. Solid Earth* **2022**, *58*, 30–58. [CrossRef]
14. Helman, D.S. Earth electricity: A review of mechanisms which cause telluric currents in the lithosphere. *Ann. Geophys.* **2014**, *56*, G0564. [CrossRef]
15. Han, Y.; Guo, Z.; Wu, J.; Ma, L. Possible triggering of solar activity to big earthquakes ($Ms \geq 8$) in faults with near west-east strike in China. *Sci. China Ser. G Phys. Mech. Astron.* **2004**, *47*, 173–181. [CrossRef]
16. Scoville, J.; Heraud, J.; Freund, F. Pre-earthquake magnetic pulses. *Nat. Hazards Earth Syst. Sci.* **2015**, *15*, 1873–1880. [CrossRef]
17. Guglielmi, A.V.; Zotov, O.D. Magnetic perturbations before the strong earthquakes. *Izv. Phys. Solid Earth* **2012**, *48*, 171–173. [CrossRef]
18. INTERMAGNET Data Viewer. Available online: https://imag-data.bgs.ac.uk/GIN_V1/GINForms2 (accessed on 10 April 2024).
19. Day and Night World Map. Available online: https://www.timeanddate.com/worldclock/sunearth.html (accessed on 10 April 2024).
20. Search Earthquake Catalog. Available online: https://earthquake.usgs.gov/earthquakes/search/ (accessed on 10 April 2024).
21. Real-Time Data and Plots Auroral Activity. Available online: https://www.spaceweatherlive.com/en/solar-activity/top-50-solar-flares.html (accessed on 1 February 2024).

22. Thomson, A.W.P.; McKay, A.J.; Viljanen, A. A Review of Progress in Modelling of Induced Geoelectric and Geomagnetic Fields with Special Regard to Induced Currents. *Acta Geophys.* **2009**, *57*, 209–219. [CrossRef]
23. Zavialov, A.D.; Morozov, A.N.; Aleshin, I.M.; Ivanov, S.D.; Kholodkov, K.I.; Pavlenko, V.A. Medium-term Earthquake Forecast Method Map of Expected Earthquakes: Results and Prospects. *Izv. Atmos. Ocean. Phys.* **2022**, *58*, 908–924. [CrossRef]
24. Sedghizadeh, M.; Shcherbakov, R. The Analysis of the Aftershock Sequences of the Recent Mainshocks in Alaska. *Appl. Sci.* **2022**, *12*, 1809. [CrossRef]
25. Araki, E.; Shinohara, M.; Obana, K.; Yamada, T.; Kaneda, Y.; Kanazawa, T.; Suyehiro, K. Aftershock distribution of the 26 December 2004 Sumatra-Andaman earthquake from ocean bottom seismographic observation. *Earth Planets Space* **2006**, *58*, 113–119. [CrossRef]
26. Potter, S.H.; Becker, J.S.; Johnston, D.M.; Rossiter, K.P. An overview of the impacts of the 2010–2011 Canterbury earthquakes. *Int. J. Disaster Risk Reduct.* **2015**, *14*, 6–14. [CrossRef]
27. New Zealand Active Faults Database. Available online: https://data.gns.cri.nz/af/ (accessed on 10 April 2024).
28. Curto, J.J. Geomagnetic solar flare effects: A review. *J. Space Weather. Space Clim.* **2020**, *10*, 27. [CrossRef]
29. Grodji, O.D.F.; Doumbia, V.; Amaechi, P.O.; Amory-Mazaudier, C.; N'guessan, K.; Diaby, K.A.A.; Zie, T.; Boka, K. A Study of Solar Flare Effects on the Geomagnetic Field Components during Solar Cycles 23 and 24. *Atmosphere* **2022**, *13*, 69. [CrossRef]
30. Sobolev, G.A. Seismicity dynamics and earthquake predictability. *Nat. Hazards Earth Syst. Sci.* **2011**, *11*, 445–458. [CrossRef]
31. Dzeboev, B.A.; Gvishiani, A.D.; Agayan, S.M.; Belov, I.O.; Karapetyan, J.K.; Dzeranov, B.V.; Barykina, Y.V. System-Analytical Method of Earthquake-Prone Areas Recognition. *Appl. Sci.* **2021**, *11*, 7972. [CrossRef]
32. Ledo, J.; Jones, A.G.; Ferguson, I.J. Electromagnetic images of a strike-slip fault: The Tintina fault-Northern Canadian. *Geophys. Res. Lett.* **2002**, *29*, 1225. [CrossRef]
33. Unsworth, M.J.; Malin, P.E.; Egbert, G.D.; Booker, J.R. Internal Structure of the San Andreas Fault Zone at Parkfield, California. *Geology* **1997**, *25*, 359–362.
34. Ingham, M.; Brown, C. A magnetotelluric study of the Alpine Fault, New Zealand. *Geophys. J. Int.* **1998**, *2*, 542–552. [CrossRef]
35. Jones, A.G.; Kurtz, R.D.; Boerner, D.E.; Craven, J.A.; McNeice, G.W.; Gough, D.I.; DeLaurier, J.M.; Ellis, R.G. Electromagnetic constraints on strike-slip fault geometry—The Fraser River fault system. *Geology* **1992**, *20*, 561–564.
36. Stanley, W.D.; Labson, V.F.; Nokleberg, W.J.; Csejtey, B.; Fisher, M.A. The Denali fault system and Alaska Range of Alaska: Evidence for underplated Mesozoic flysch from magnetotelluric surveys. *Geol. Soc. Am. Bull.* **1990**, *102*, 160–173.
37. Mackie, R.L.; Livelybrooks, D.W.; Madden, T.R.; Larsen, J.C. A magnetotelluric investigation of the San Andreas Fault at Carrizo Plain, California. *Geoph. Res. Lett.* **1997**, *24*, 1847–1850. [CrossRef]
38. Sorokin, V.M.; Chmyrev, V.M.; Hayakawa, M. A Review on Electrodynamic Influence of Atmospheric Processes to the Ionosphere. *Open J. Earthq. Res.* **2020**, *9*, 113–141. [CrossRef]

*Article*

# Observation of the Preparation Phase Associated with Mw = 7.2 Haiti Earthquake on 14 August 2021 from a Geophysical Data Point of View

**Dedalo Marchetti**

Independent Researcher, 00145 Rome, Italy; dedalo.marchetti@ingv.it or dedalo.marchetti.work@gmail.com;
Tel.: +86-18936866083

**Abstract:** On 14 August 2021, an earthquake of moment magnitude Mw = 7.2 hit Haiti Island. Unfortunately, it caused several victims and economic damage to the island. While predicting earthquakes is still challenging and has not yet been achieved, studying the preparation phase of such catastrophic events may improve our knowledge and pose the basis for future predictions of earthquakes. In this paper, the six months that preceded the Haiti earthquake are analysed, investigating the lithosphere (by seismic catalogue), atmosphere (by climatological archive) and ionosphere by China Seismo-Electromagnetic Satellite (CSES-01) and Swarm satellites, as well as Total Electron Content (TEC) data. Several anomalies have been extracted from the analysed parameters using different techniques. A comparison, especially between the different layers, could increase or decrease the probability that a specific group of anomalies may be (or not) related to the preparation phase of the Haiti 2021 earthquake. In particular, two possible coupling processes have been revealed as part of the earthquake preparation phase. The first one was only between the lithosphere and the atmosphere about 130 days before the mainshock. The second one was about two months before the seismic event. It is exciting to underline that all the geo-layers show anomalies at that time: seismic accumulation of stress showed an increase of its slope, several atmospheric quantities underline abnormal atmospheric conditions, and CSES-01 Ne depicted two consecutive days of ionospheric electron density. This suggested a possible coupling of lithosphere–atmosphere and ionosphere as a sign of the increased stress, i.e., the impending earthquake.

**Keywords:** Haiti; earthquake; LAIC; CSES; swarm; atmosphere; ionosphere

**Citation:** Marchetti, D. Observation of the Preparation Phase Associated with Mw = 7.2 Haiti Earthquake on 14 August 2021 from a Geophysical Data Point of View. *Geosciences* **2024**, *14*, 96. https://doi.org/10.3390/geosciences14040096

Academic Editors: Masashi Hayakawa and Jesus Martinez-Frias

Received: 10 February 2024
Revised: 17 March 2024
Accepted: 26 March 2024
Published: 30 March 2024

## 1. Introduction

This paper applies a multilayer geophysical investigation to the recent earthquake that occurred on 14 August 2021 in Haiti Island. Earthquakes are potentially one of our planet's most significant natural phenomena, releasing huge amounts of energy in a few seconds. While most earthquakes are due to the known tectonic movements of plates on the Earth, their time, location and magnitude of occurrence are uncertain [1]. In fact, predicting an earthquake means providing, in advance, space–time coordinates and the magnitude of an incoming event [2]. It is still debated whether the prediction of an earthquake would ever be possible, but that is not the topic of this paper. However, it is pretty unrealistic that such a large natural event could suddenly occur without influencing the geo-system. In fact, for several decades, many reports have claimed pre-earthquake phenomena called precursors [3,4]. In some cases, abnormal animal behaviour has also been reported, for example, before the Mw = 6.2 L'Aquila (Italy) 2009 devasting seismic event [5]. The abnormal behaviour of the animals could be due to the perception of some substances or properties (such as electrical or magnetic fields) that humans cannot depict but the instrumentation could monitor. Consequently, efforts have been made to study the properties of the lithosphere, atmosphere and ionosphere before the earthquakes.

In addition, on 2 February 2018, China launched a satellite in partnership with Italy, entirely dedicated to investigating possible effects in the ionosphere induced by the earthquake: China Seismo-Electromagnetic Satellite (CSES-01) and planning the launch of the second satellite in the current year [6]. The satellite provides excellent results in different geophysical topics, including the study of geomagnetic field, geomagnetic storms and ionospheric pulsations as Pi2 [7–12].

Research on the study of the preparation of earthquakes has been dedicated, for example, to the investigation of seismic acceleration before the earthquake in the approach known as accelerated moment release [13–15]. The atmospheric parameters have been explored using multiple techniques. For example, the Robust Satellite Technique (RST) showed the appearance of thermal infrared anomalies with a statistically significant number of earthquakes [16–19] or other parameters, such as ozone [20]. Tronin [21] provided a review of different parameters retrieved from remote sensing data. Electromagnetic precursors have been searched historically from ground geomagnetic observatories in different areas of the world [22–27] and, more recently, from space satellites [28–32]. Comprehensive recent reviews of seismo-electromagnetic phenomena have been provided by Chen et al. [33] and Hayakawa et al. [34,35]. A comparison between surface and land temperature values and Swarm magnetic anomalies has been supplied by Ghamry et al. [36]. Several studies proved that Swarm and CSES-01 electromagnetic anomalies statistically preceded several shallow M5.5+ earthquakes [37–40]. A recent study by Chen et al. [41] investigated the relationship between space–time distance and magnitude of the incoming earthquake for the first four years of CSES-01 Ne satellite data. Alterations of the ionosphere before the earthquakes are also commonly investigated using Total Electron Content (TEC), which can be retrieved from different instruments, generally the Ground Global Navigational Satellite System (GNSS) detectors [42–45]. Other ways of investigating ionosphere include the study of ionosonde and Very Low-Frequency (VLF) transmitter–receiver alterations for changing the altitude of vertical reflection of the ionosphere's electromagnetic signals [46–51].

The $Mw = 7.2$ Haiti earthquake occurred on 14 August 2021 at 12:29:08 Universal Time (UT). Its hypocentre was localised at a depth of about 10 km, and the inverse solution of its moment is compatible with a strike–slip focal mechanism with a small thrust component. The approximate fault plane interested in the slip was more than 70 km long along the strike and more than 40 km deep along the fault dip. The maximum slip overpassed 2.5 m at a depth of about 10 km and for a length of about 15 km. Unfortunately, there were at least 2,248 victims, 12,763 injured and 329 missing persons and huge economic losses as reported by USGS (https://earthquake.usgs.gov/earthquakes/eventpage/us6000f65h, last access 9 February 2024).

The location of the $Mw = 7.2$ Haiti 2021 earthquake, together with tectonic settings, is shown in the map in Figure 1. The represented active fault system of the Caribbean and Central American area has been retrieved in the dataset [52] developed by Styron et al. [53]. Some geological information reported in the map has been acquired from [54–56]. In particular, the place of the earthquake is well located in the strike-slip Enriquillo–Plantain Garden (EPGFZ) fault, which crossed Jamaica Island, the offshore section and the Southern side of Haiti/Hispaniola Island.

A previous work on this earthquake was conducted by Akhoondzadeh [57], which analysed 75 days before up to 15 days after the $Mw = 7.2$ Haiti 2021 earthquake using CSES-01, Swarm Alpha Bravo and Charlie satellites and atmospheric data from the Giovanni web-portal. He identified an increase in anomalies in the last 20 days before the earthquake, similar to the result of the recent catastrophic earthquake in Turkey [58]. Another work by Khan et al. [59] used a machine learning technique to analyse satellite atmospheric data 30 days before and up to 15 days after the mainshock. They identified an anomaly increase ten days before the mainshock. Chen et al. [60] analysed GNSS TEC data from 17 days before up to 7 days after the mainshock, claiming possible seismo-induced anomalies 13 days before the earthquake. However, those studies analysed a limited time before the earthquake, which is selected equal to 6 months in the present paper. A possible

LAIC related to the Haiti 2021 earthquake was investigated by D'Angelo et al. [61], but the investigation was focused on the co-seismic effect detecting Acoustic Gravity Waves produced by the shaking induced by the seismic event. This phenomenon has been theorised and detected in several cases [62–64] and even identified before several earthquakes [65–67] with a source mechanism based on a thermal gradient as proposed by Hayakawa [68].

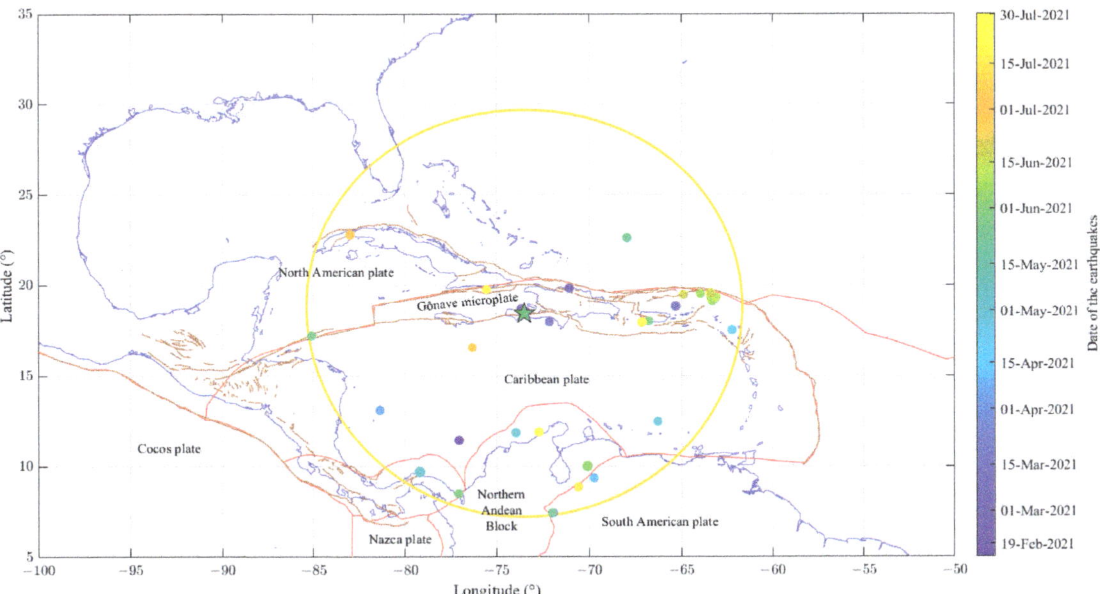


**Figure 1.** Geographical and tectonic context of the Mw = 7.2 Haiti 14 August 2021 earthquake. The epicentre is shown as a green star. The yellow circle depicts Dobrovoslky's area. Blue lines represent the coasts, brown lines represent the active faults (shown only for the Central American and Caribbean regions) and red lines represent the plate borders. Earthquakes of magnitude equal to or greater than 4.2 with a maximum depth of 50 km that occurred in the six months before the Haiti in the Dobrovolsky have been visualised as a filled dot with size proportional to their magnitude and colour to their origin time.

## 2. Materials and Methods

In this paper, we applied several methods to investigate the lithosphere, atmosphere and ionosphere, following the approach we had successfully used before several earthquakes and some volcano eruptions. These cases include among all, the Mw = 6.7 Lushan (China) 2013 [69], Mw = 7.8 Ecuador 2016 [70], Mw = 6.0 and Mw = 6.5 Amatrice-Norcia 2016 [71], Mw = 7.5 Indonesia 2018 [72], Mw = 7.2 Kermadec Islands (New Zealand) 2019 [73] and Mw = 7.8 and Mw 7.5 Turkey 2023 [58] earthquakes. In all of these mentioned cases, a possible lithosphere–atmosphere and ionosphere coupling (LAIC) has been identified in the form of a chain of anomalies and, in some cases (as Lushan 2013), even more possible couplings with different physical mechanisms have been proposed by Zhang et al. [69]. A sort of confutation analysis showed a lack of lithosphere–atmosphere and ionosphere coupling before a small earthquake, such as the ML = 3.3 Guidonia (Rome, Italy) 2023 event [74]. Still, a seismic acceleration before such a small earthquake has been clearly detected, supporting the concept that LAIC requires a higher magnitude. Still, seismological analyses may be effective even in lower magnitude events depending on the characteristics of the specific fault.

A period of six months before the Mw = 7.2 Haiti 2021 earthquake was selected. The time period to search for possible precursors is likely to increase with earthquake magnitude, as proposed by Rikitake [75,76] and Scholz et al. [77]. In addition, recent work on satellite data confirmed such a relationship initially developed for ground observations [37,38]. All of these relationships affirm that the logarithm of the anticipation time in days of the precursor is directly proportional to the magnitude of the incoming earthquake:

$$\log_{10}(\Delta T) = a - b \cdot M$$

where "*a*" and "*b*" are two coefficients that, according to Rikitake [75,76], are different for each precursor. However, he tried to classify them into macro-categories with common anticipation time but great dispersion of the coefficient. For example, for an earthquake of magnitude 7.2 (like Haiti) and geophysical precursors of the quasi-first kind, the anticipation would be about 72 days. If all the disciplines are involved, the same calculus can bring it to about 1400 days. A part of this, there is another important consideration, i.e., the recharge on the fault is a very long process (see Appendix C, where the time between two events of magnitude greater or equal to 7.2 has been estimated as two decades and this is much less than recharge time of a single fault). In addition, the preparation phase of the earthquake is considered a process that involves different stages according to the "Dilatancy Model" of Scholz [77,78]. Considering this, the investigated time before the earthquake affects the explored stages. From a practical point of view, another seismicity in the area must be checked, excluding other large (or even larger) seismic events and the eventual time to recover from seismic sequence. In fact, for the case study of this earthquake, a large earthquake occurred on 28 January 2020, not so far (closer to Jamaica), with magnitude 7.7. Finally, a time of six months for this case study is a balance of these considerations, and it is expected to cover medium and short-term precursors [79]. The following briefly illustrates the datasets used and their processing methods.

### 2.1. Data and Methods for Lithosphere

The lithosphere has been investigated using the USGS earthquake catalogue. All the events localised six months before the Haiti earthquake and inside Dobrovolsky's area (see yellow circle in Figure 1) have been retrieved. The Dobrovolsky's area defined by Dobrovolsky's radius ($R[km] = 10^{0.43 \times Mw}$) [80] is an estimation of the possible area of preparation for the earthquake under study. It scales exponentially with the moment magnitude (Mw) of the same event. Only earthquakes with a magnitude greater or equal to the completeness magnitude have been selected. The completeness magnitude has been estimated with the ZMap software, version 7.1 [81] with the "best fit" option, and more details are provided in Appendix C. In addition, only events with a depth equal to or less than 50 km have been selected, and their epicentres are shown in Figure 1 as coloured filled dots. The colour indicates the occurrence time according to the colour bar. The cumulated Benioff stress $S(t)$ has been estimated using the following equation:

$$S(t) = \sum_i \sqrt{10^{(1.5 \cdot M_i)}} \tag{1}$$

where "$M_i$" is the magnitude of the $i$-th event that occurred before time "$t$".

### 2.2. Data and Methods for Atmosphere

The atmosphere has been investigated using the climatological archive MERRA-2 provided by NASA with data from 1980 to the present [82]. The dataset is based on real observation from the ground and remote sensing (satellites, airborne, aviation, etc.), constituting the input of a chemical–physical model of the atmosphere with a space resolution of 0.625° longitude, 0.5° latitude and 1 h of time resolution. The analysis has been conducted with the "MErra-2 ANalysis to search Seismic precursors" (MEANS) algorithm developed for the first time to study volcano eruptions and subsequently applied to several earthquakes [69,71,72,83] and, recently, La Palma 2021 volcano eruption [84]. Some improvements to the method have been applied over the years. The selected version was previously used to analyse the Lushan (China) 2013 earthquake [69]. In particular, global warming is removed from all the parameters as a linear fit shown in Figure 2. It is interesting to note that SO$_2$ showed a high peak in 1991 due to Pinatubo's great volcano eruption (Volcano Explosive Index = 6).

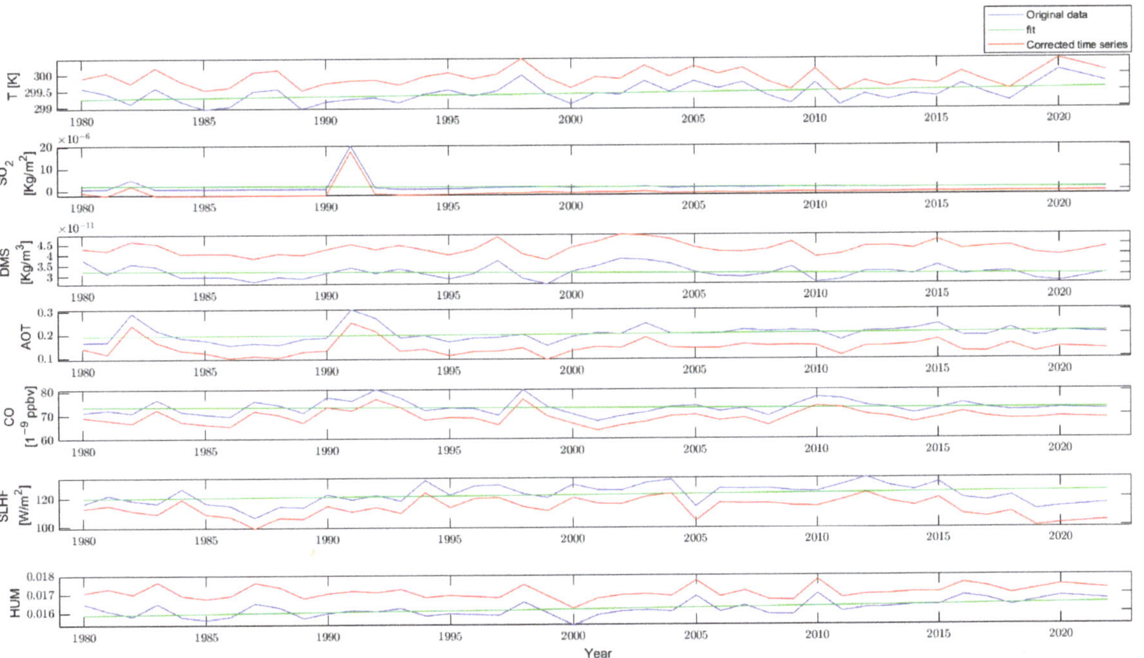

**Figure 2.** Yearly average of the selected atmospheric parameters in the investigated six-month period (blue line). The linear fit is shown as a green line and the residual after removing the fit as a red line.

For each day, the data closer to the local midnight were selected, and the spatial average in the investigated areas was computed. The historical value was characterised as a function of mean and standard deviations of the same day for other years, excluding the year of the earthquake occurrence (i.e., as an example, the typical value of 15 February, the first day of the time series is the mean of 15 February 1980, 15 February 1981, . . . 15 February 2019, 15 February 2020, 15 February 2022). The atmospheric analyses were performed in two distinct areas: a square box centred on the epicentre and 3° side and a rectangular area of 2.2° longitude and 1.6° latitude drawn around the seismic displacement. Finally, the year of the earthquake (in this case, 2021) was compared with the typical values and the one outside the two standard deviations were classified as anomalous.

Regarding methane, data was directly acquired by the Atmospheric INfrared Sounder (AIRS) dataset of the AQUA satellite. For the first time in this approach, the version 7.0 dataset has been explored instead of version 6.0. For this reason, Appendix A shows the analysis of the previous version, 6.0, for comparison. In particular, the whole time series of the available values at a maximum distance of 3.0° from the epicentre from 2002 up to 2023 has been shown in Figure 3a. A linear fit and an "exponential–linear" one were calculated and overplotted on the time series. The data were normalised, removing the exponential–linear fit but maintaining the same absolute value. The histogram of the distribution of the values before and after the detrending is plotted in Figure 3b. It is possible to note that the dispersion of the values was reduced after detrending them as expected. The multi-year trend is likely due to global warming produced by increased methane concentration in the atmosphere.

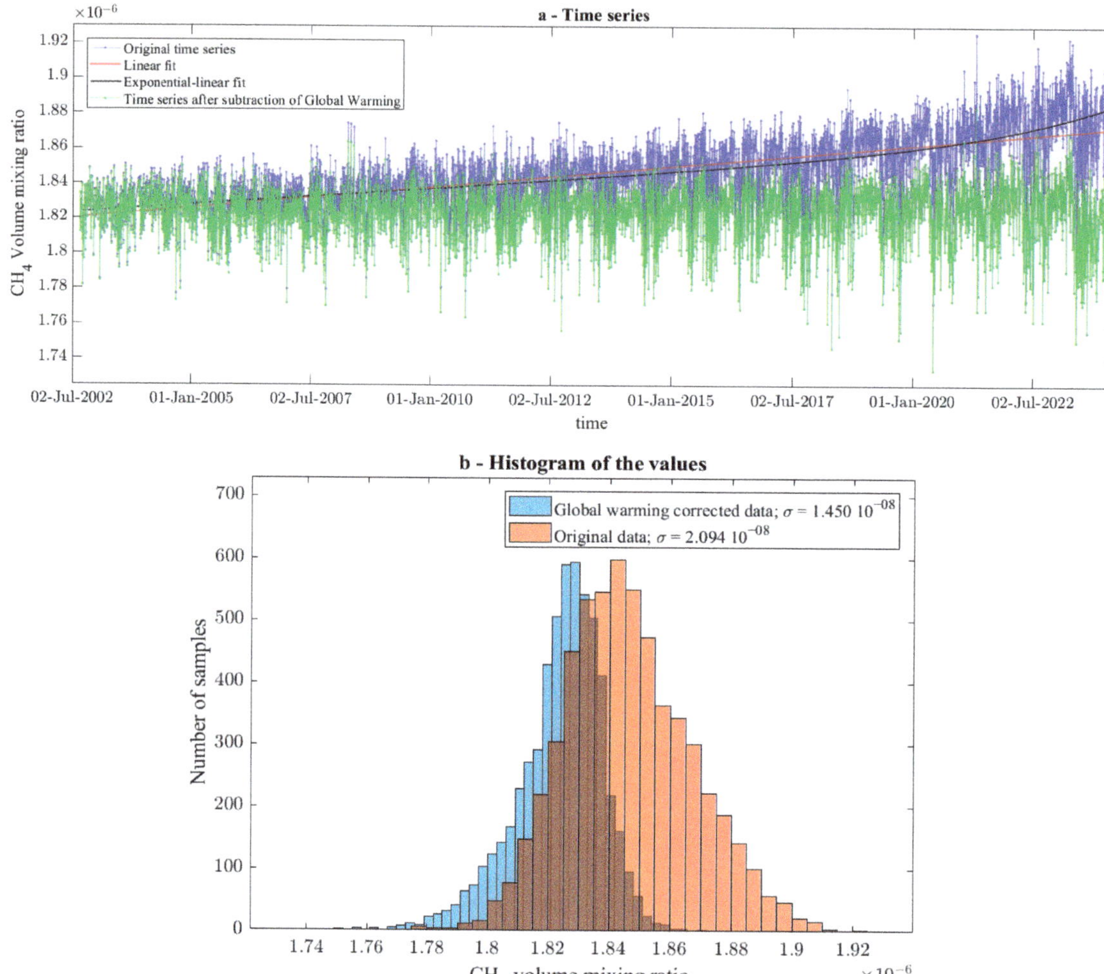

**Figure 3.** Time series (**a**) and histogram (**b**) of the CH4 values estimated by the instrument AIRS onboard the AQUA satellite provided in version 7.0.

In atmospheric analyses, years after the earthquake (2022 and 2023 for methane) are used to improve the background definition. In fact, the scope of this work is still not to predict the earthquake but to understand as much as possible what happened before the event.

### 2.3. Data and Methods for the Ionosphere

### 2.3.1. CSES-01, Ne

Several methodologies have been applied to the data of CSES-01 and Swarm satellites to investigate the ionosphere. In particular, the data of electron density (Ne) of CSES-01 recorded in the Dobrovolsky's area in the six months preceding the Mw = 7.2 Haiti 2021 earthquake have been collected, and the daily mean values have been estimated separately for night-time (2 A.M.) and daytime (2 P.M.). The time series was detrended by seasonal variation by a polynomial fit. The values outside the median plus or minus 1.5 times the interquartile ranges were classified as anomalous. The technique has already been applied to Mw = 7.6 Papua New Guinea 2019 [85].

A second investigation compares the identical orbit that CSES-01 flies every five days. Figure 4 shows the two selected orbits for day and night times as the closest to the epicentre of the Mw = 7.2 Haiti 2021 earthquake. The method, already applied to La Palma 2021 volcano eruption [84], compares the complete latitudinal profiles of Ne of CSES-01 acquired in geomagnetic quiet conditions ($|\text{Dst}| \leq 20$ nT and ap $\leq 10$ nT). It then characterises the distribution of the Ne samples in windows of 1 degree of latitude in terms of interquartile range and outliers, i.e., the values that are 1.5 times the interquartile range out of the upper or lower percentiles (25% and 75%), i.e., the whiskers. The outlier values for each data distribution are marked with a red cross in the graph. If the phenomena occurred at the latitude of the incoming earthquake, it is further discussed as possibly caused by the seismic event.

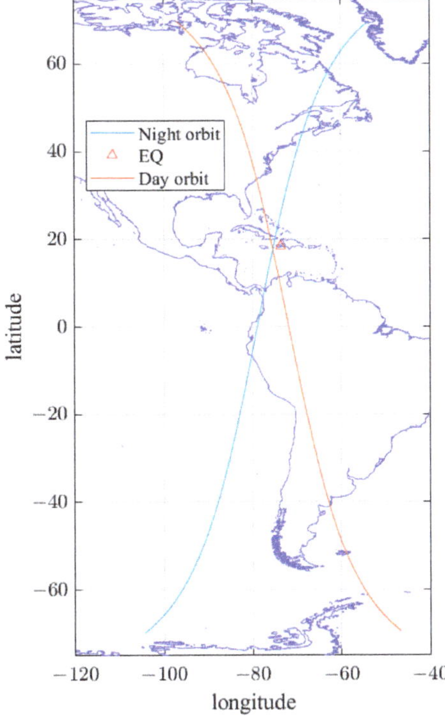


**Figure 4.** Selected orbits of CSES-01 as the closest to the Mw = 7.2 Haiti 2021 epicentre.

### 2.3.2. Swarm, Magnetic Field

The three Swarm satellites (Alpha Bravo and Charlie) have been analysed by the MASS (MAgnetic Swarm anomaly detection by Spline analysis) algorithm, widely utilised to investigate single earthquakes [28,70,72,74,86], as well as large statistical studies [37,38]. The MASS algorithm first removed the main field by a numerical derivative of the signal and subtracting a cubic spline. For the residuals, it compared the root mean square (rms) inside a moving window of 3° latitude within the Root Mean Square (RMS) of the whole track between −50° and +50° geomagnetic latitude. If the rms > kt × RMS, the window is defined as anomalous. This approach extracts anomalous signals, but they are not necessarily induced by an earthquake. Previous statistical works, in fact, demonstrate that the earthquake likely induces a significant number of these anomalies, but they are a few per cent of the whole ensemble of extracted anomalies [37,38]. So, to further discriminate possible pre-earthquake signals, the approach proposed for the Ridgecrest 2019 [53] earthquake has been followed here. It is supposed to compare the research area with a comparison one at the same magnetic latitude and a similar ratio between land and sea. In this case, a comparison area at longitude 15.69° W has been selected. Only the anomalies that exclusively appear in the earthquake area are further discussed as possibly related to the incoming seismic event.

### 2.3.3. TEC

The Global Ionospheric Maps (GIMs) of Total Electron Content (TEC) have been investigated using a method previously applied to volcano eruption and earthquake occurrence [69,84]. GIM-TEC has a space resolution of 5° longitude and 2.5° latitude and a time resolution of 2 h. The method selects the four nearest neighbour values for the previous and following UT available hours. Firstly, a bidimensional cubic interpolation of the value above the epicentre is calculated, and then a linear interpolation at the selected local time (LT in this paper is set as 2 A.M.) is performed for each day in the six months before the mainshock. For each year, the yearly median TEC is calculated and linearly fitted versus the mean Sunspot number. The median is generally preferred over the mean in ionospheric studies with TEC and ionosonde to avoid the effect of the outliers and intrinsic strong variability [87,88]. The normalised TEC is calculated by dividing the time series of each year by the value obtained by the previous fit. After this passage, the background, i.e., the typical TEC value for the specific day, is calculated as the median, the median and interquartile range of the yearly values and thresholds to extract anomalies defined as the median $\pm$ 1.5 times interquartile range. The value of the year of the earthquake (2021, excluded from the background) is graphically superposed to check if any of the values are outside the thresholds, i.e., anomalous. A second threshold (represented as green horizontal lines) was calculated as the median and interquartile range of the 2021 normalised TEC values. The values that are outside such a threshold are anomalous for 2021. Still, if inside the previous background, they can be considered inside the historical variation of TEC at the epicentre location, so they are not strictly anomalous.

## 3. Results

### 3.1. Lithosphere

The analysis of the earthquake catalogue is presented in Figure 5. A combined fit of a sigmoidal curve with a linear growth has been performed (green line).

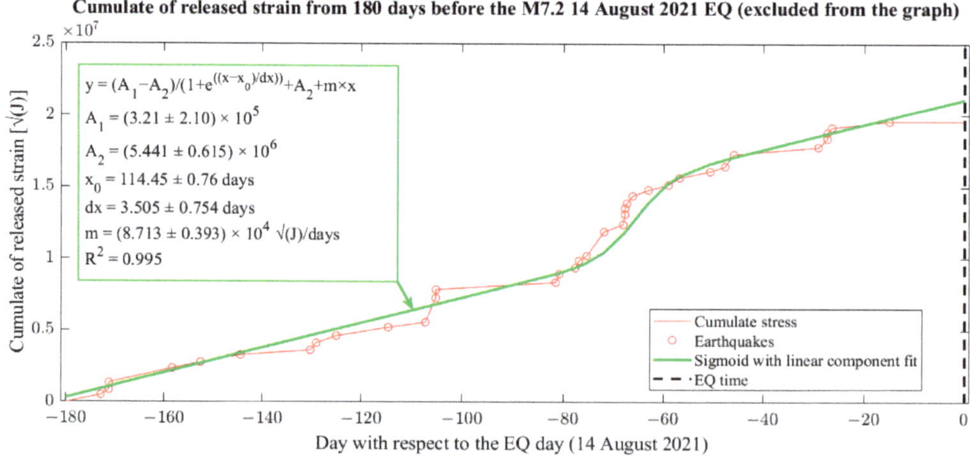


**Figure 5.** Cumulative seismic stress calculated with earthquakes in Dobrovolsky's area and with a magnitude equal to or greater than completeness magnitude 4.3. The fit is shown as a green line, and its coefficients are presented in the box inside the figure (days in the box are counted from 0 to 180 days, in particular, $x_0 = -65.03$ in days with respect to the earthquake day).

The investigation of the cumulate of released strain underlined a trend variation 66 days before the mainshock ($x_0$ parameter of sigmoidal fit). This means that from a tectonic point of view, Dobrovolsky's area experienced higher seismicity about two months before the Haiti 2021 earthquake. In addition, the $E_S$ parameter defined by Hattori [23,89] has been estimated and reported in Figure 6. The cited papers considered the region as active when $E_S \geq 10^8$. This occurred twice: 171 and 130 days before the Mw = 7.2 Haiti 2021 earthquake.

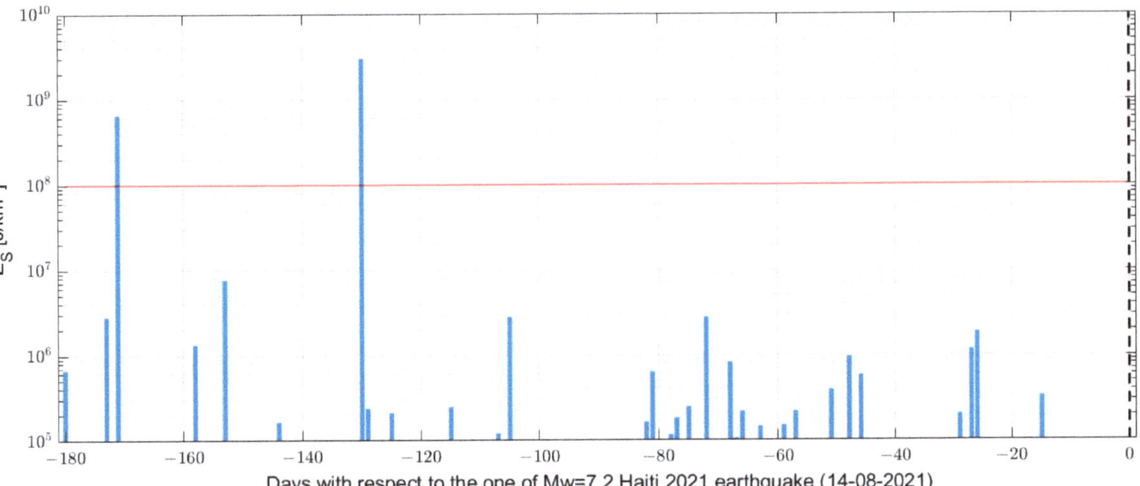


**Figure 6.** Daily seismicity analysis made by the daily $E_S$ parameter (shown as blue bars). According to the previous literature, the red line represents the threshold to consider active the region on a specific day. The dashed black vertical line indicates the day of Mw = 7.2 Haiti 2021 earthquake.

### 3.2. Atmosphere

Various atmospheric physical and chemical parameters extracted from MERRA2 were analysed six months before the Mw = 7.2 Haiti 2021 earthquake. The time series of Aerosol (as Aerosol Optical Thickness-AOT), the surface concentration of Carbon Monoxide (CO), the surface mass concentration of Dimethyl Sulphide and the surface mass concentration of Sulphur Dioxide ($SO_2$) have been plotted in Figures 7–11. Each parameter is analysed in a 3° side square area centred on the epicentre (above) and a rectangular box of 2.2° longitude and 1.6° latitude drawn around seismic displacement (below). Using the area that included the seismic displacement information generally provides more reliable pre-earthquake anomalies. In fact, this analysis aims to discriminate among all atmospheric anomalies, which ones have more chances of being induced by seismic activity. Consequently, the anomalies confirmed by this investigation are marked with red circles, while the others with orange circles are excluded for further consideration in the discussion section.

Specific humidity, surface air temperature and surface latent heat flux have also been investigated, and their time series are presented in Supplementary Materials. For Specific humidity, according to the LAIC model of Pulinets and Ouzounov [90], only a reduction (decrease) of humidity has been considered. In fact, the hydration of aerosol particles in the atmosphere is considered the cause of the drop in humidity, as further explained in the recent book by the same authors and colleagues [91].

In addition to the previous atmospheric parameters from the MERRA-2 climatological archive, the methane measured by AIRS has been investigated and shown in Figure 11.

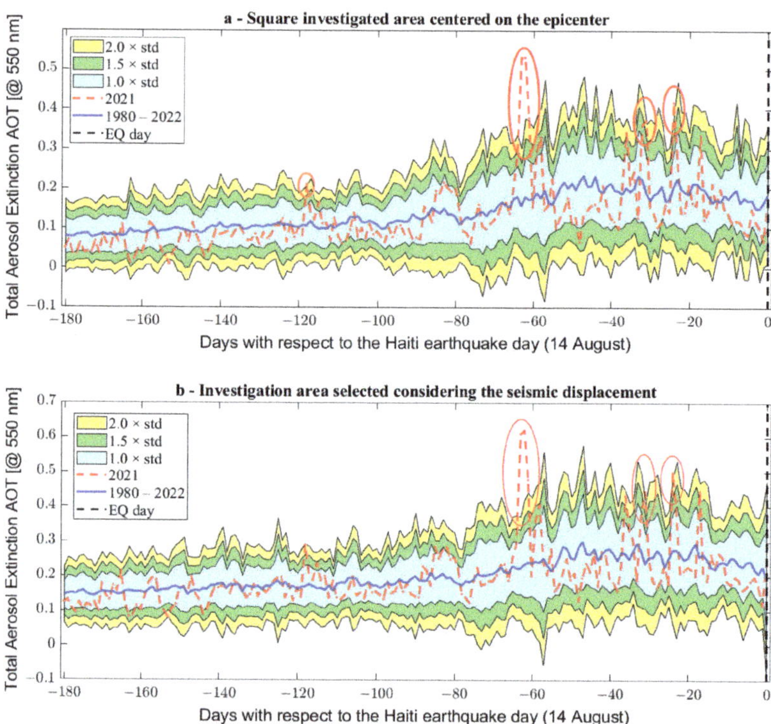

**Figure 7.** Aerosol investigation in the six months before the Haiti 2021 earthquake (vertical dashed black line) using (**a**) a symmetric square area centred on the epicentre or (**b**) a rectangular area around the seismic displacement. The years 1980 (only **a**), 1991, 2015 and 2020 have been excluded for outlier values. Red circles underline anomalies in the lower panel and the one in the upper panel, confirmed by investigations in both areas, while orange circles underline anomalies in the upper panel only.

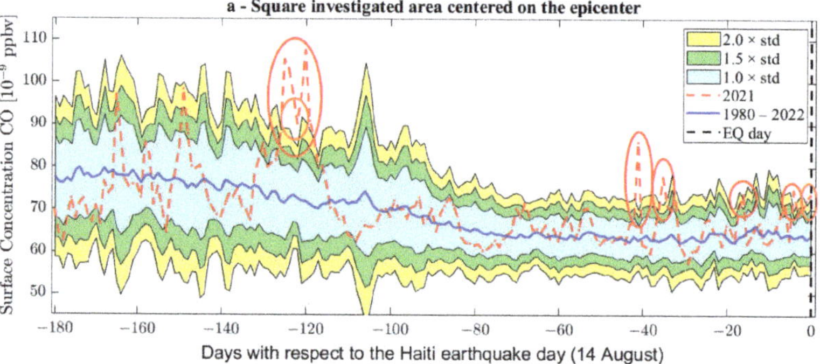

**Figure 8.** *Cont.*

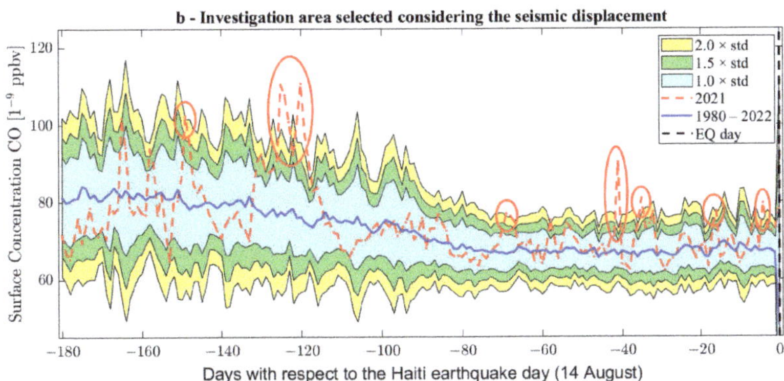

**Figure 8.** Carbon monoxide (CO) investigation in the six months before the Haiti 2021 earthquake (vertical dashed black line) using (**a**) a symmetric square area centred on the epicentre or (**b**) a rectangular area around the seismic displacement. The years 1981, 1982, 1983, 1989, 1992 (only **b**) and 1998 have been excluded for outlier values. Red circles underline anomalies in the lower panel and the one in the upper panel, confirmed by investigations in both areas, while orange circles underline anomalies in the upper panel only.

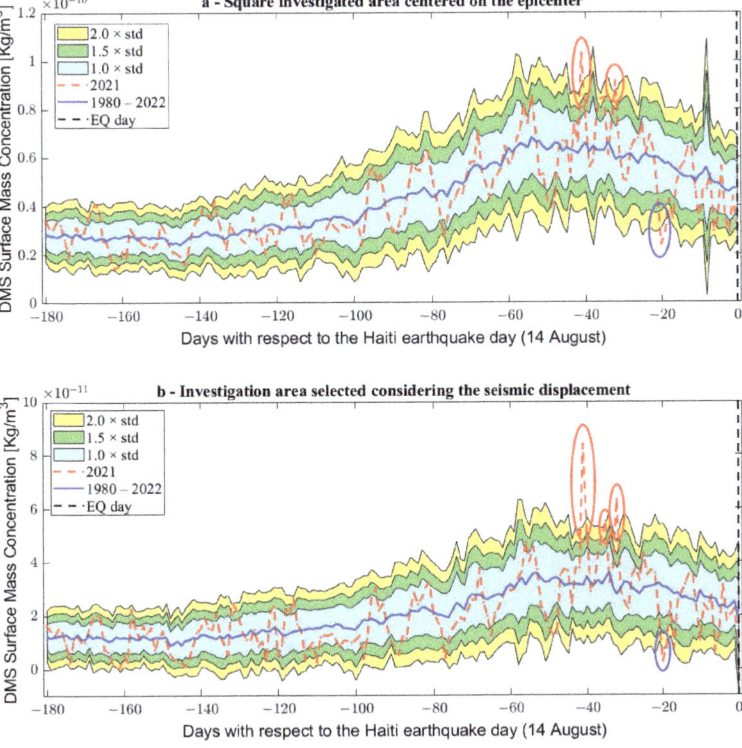

**Figure 9.** Dimethyl sulphide (DMS) investigation in the six months before the Haiti 2021 earthquake (vertical dashed black line) using (**a**) a symmetric square area centred on the epicentre or (**b**) a rectangular area around the seismic displacement. The years 1980 and 2005 have been excluded for outlier values. Red circles underline positive anomalies, while blue circle underlines a negative anomaly.

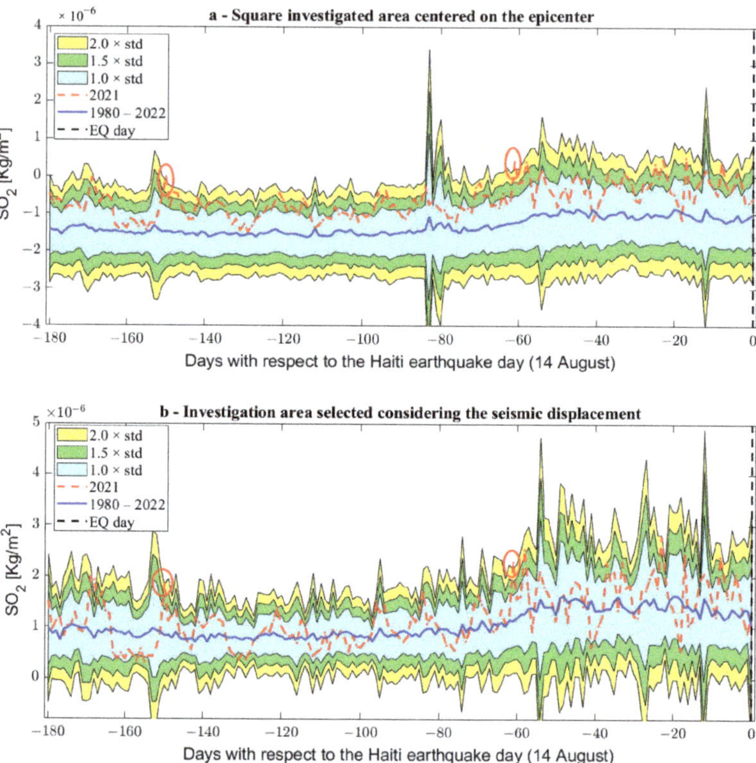

**Figure 10.** Sulphur dioxide (SO$_2$) investigation in the six months before the Haiti 2021 earthquake (vertical dashed black line) using (**a**) a symmetric square area centred on the epicentre or (**b**) a rectangular area around the seismic displacement. The years 1982, 1991, 1992, 2003 (only **a**) and 2006 (only **b**) have been excluded for outlier values. Red circles underline anomalies.

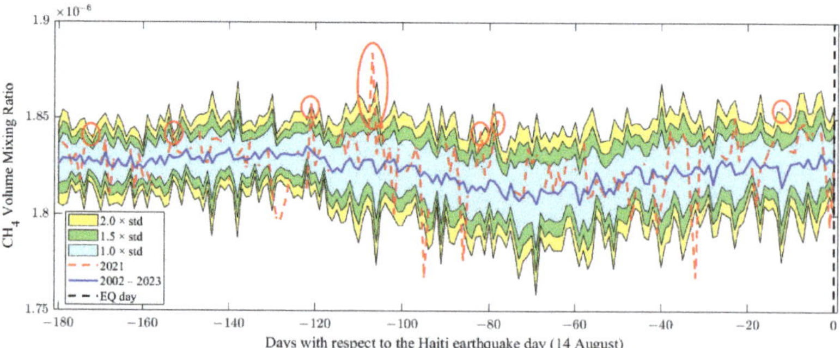

**Figure 11.** Methane (CH$_4$) investigation in the six months before the Haiti 2021 earthquake (vertical dashed black line). The data source is the AISR instrument, so the background has been calculated over a shorter period (from 2002 to 2023, excluding the earthquake year 2021). Red circles underline anomalies.

Figure 12 shows a couple of maps of two anomalous days for aerosol content: 13 June and 14 July 2021. In the first map, the aerosol concentration is quite close and unique to the epicentre; overall, it is aligned with the main plate boundary. This supports a possible release of some substances, such as radon, from the active fault due to increased stress from the impending earthquake. On the second map on 14 July 2021, aerosol concentration is not unique to the epicentre, so some doubt exists about a possible link with seismic activity. It cannot be excluded that more phenomena coexist at the same time. For example, the large concentration in Colombia/Venezuela was surely not due to the preparation phase of the Haiti earthquake. However, it seems clear that South American aerosol emissions are a separate phenomenon from the concentration in the Caribbean Sea. Consequently, considering the close-to-epicentre aerosol concentration as a possible pre-earthquake phenomenon cannot be wholly excluded or confirmed.

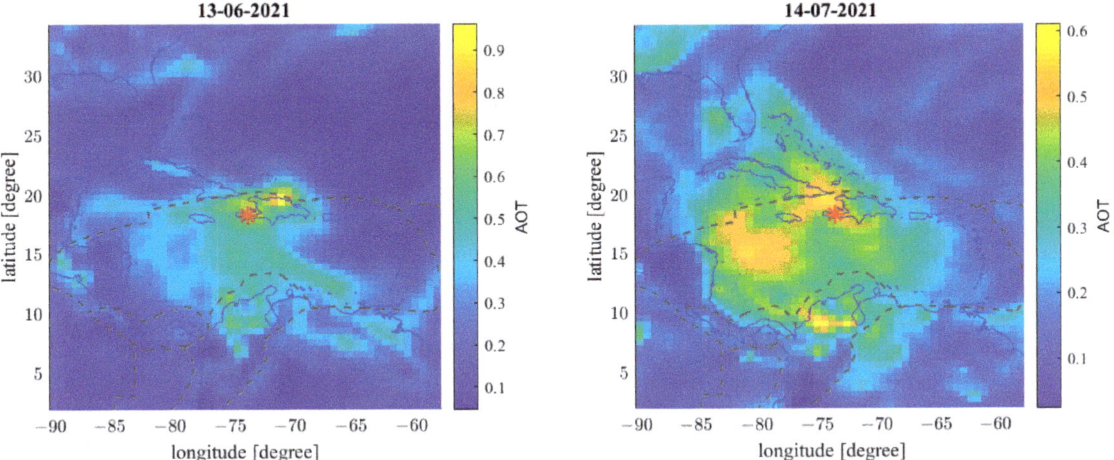


**Figure 12.** Maps of Aerosol (AOT) on 13 June and on 14 July 2021. A red asterisk marks the epicentre. Dashed grey lines represent the main plate boundaries. The blue lines represent coastlines.

*3.3. Ionosphere*

The ionosphere has been investigated by two satellite missions: CSES and Swarm. Figure 13 presents the time series of CSES-01 Ne values acquired in Dobrovoslky's area six months before the Mw = 7.2 Haiti 2021 earthquake during geomagnetic quiet conditions.

It is possible to note that there are more CSES Ne anomalies as the earthquake approaches. In addition, two consecutive anomalous days with a similar amount of high Ne appeared about two months (64 days) before the earthquake. In some of the previous research on pre-earthquake phenomena, the persistence or duration of the anomaly is considered (and statically proved) to be more effective [92,93].

Another analysis with CSES-01 Ne nighttime data is presented in Figure 14. In this case, only one orbit is selected, and its values are characterised in geomagnetic quiet conditions, and eventual outliers (marked as red crosses) are extracted. To select the most probable anomalies that may be related to the preparation for the earthquake, I focused on the latitude of the earthquake. The other part of the profile can be used as a comparison to see if the phenomena are local or affect the entire orbit, making it less likely to be associated with seismic activity. The most anomalous increase in electron density occurred on 29 July 2021, 16 days before the mainshock. The peak is slightly North of the incoming epicentre, well inside the Dobrovoslky's area. Another important increase in electron density occurred on 4 June 2021, 71 days before the Mw = 7.2 Haiti 2021 earthquake. It is worth noting that almost at the same time, the seismic trend accelerates, supporting a possible coupling between the lithosphere and the ionosphere. The corresponding analysis with CSES-01 Ne daytime data is reported in Appendix B.

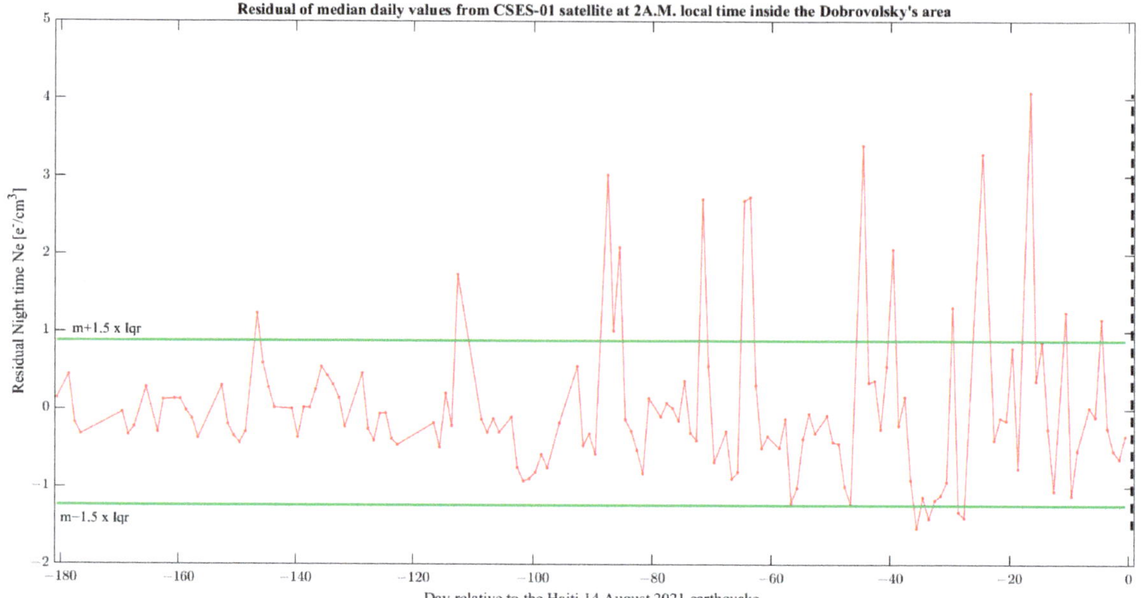

**Figure 13.** Time series of CSES-01 Ne night-time values (red dot/line) acquired in the Dobrovolsky's area in the six months before the Mw = 7.2 Haiti 2021 earthquake (vertical dashed black line). Green lines indicate thresholds to define anomalies based on median (m) and interquartile range (Iqr).

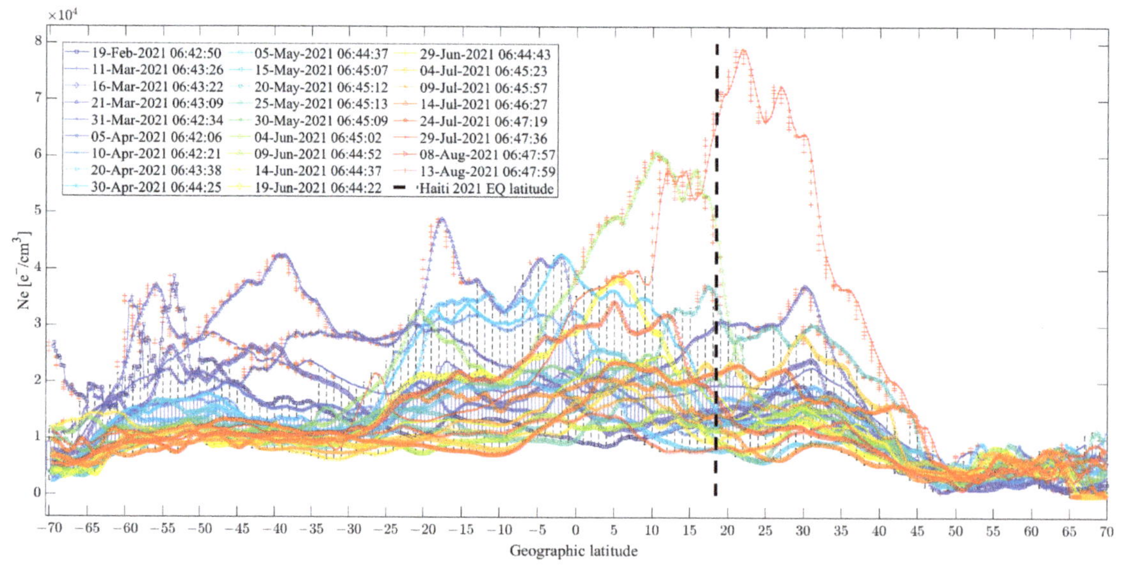

**Figure 14.** CSES-01 Ne night-time latitudinal profiles acquired in geomagnetic quiet time. The blue boxes and dashed black lines represent the standard ranges of the values for each degree of latitude. The red crosses indicate the outlier values of Ne. The latitude of the Haiti earthquake (EQ) is represented as a vertical black line.

The Swarm magnetic Y-East component has been objectively analysed in the area of a 3.32° radius around the epicentre, according to a previous study [86]. All the tracks that have at least one 3° latitude window with a root mean square (rms) greater or equal to 3.5 times the rms of the whole track (between −50° and +50° geomagnetic latitude) are counted as anomalous, and their cumulative trend is shown in Figure 15a. The same approach was applied to a comparison area shown in Figure 15d centred at a longitude of 22.5° W and the same magnetic latitude of the epicentre, obtaining the cumulative trend plotted in Figure 15b. Finally, the difference between the two trends has been estimated in Figure 15e. Whenever such a trend is positively increasing, it means that in the epicentral area, more anomalies are recorded than in the comparison area. Especially 80 days before the Mw = 7.2 Haiti 2021 earthquake, a higher rate of anomalous tracks were recorded in the epicentral area. However, at this time, the trend also increased in the comparison area. A combination of a global geomagnetic phenomenon that interacted with an altered electromagnetic field at the epicentral area cannot be excluded, for example, due to the release of positive charges (p-holes) for the increase of the seismic stress as proposed by Freund [94,95].

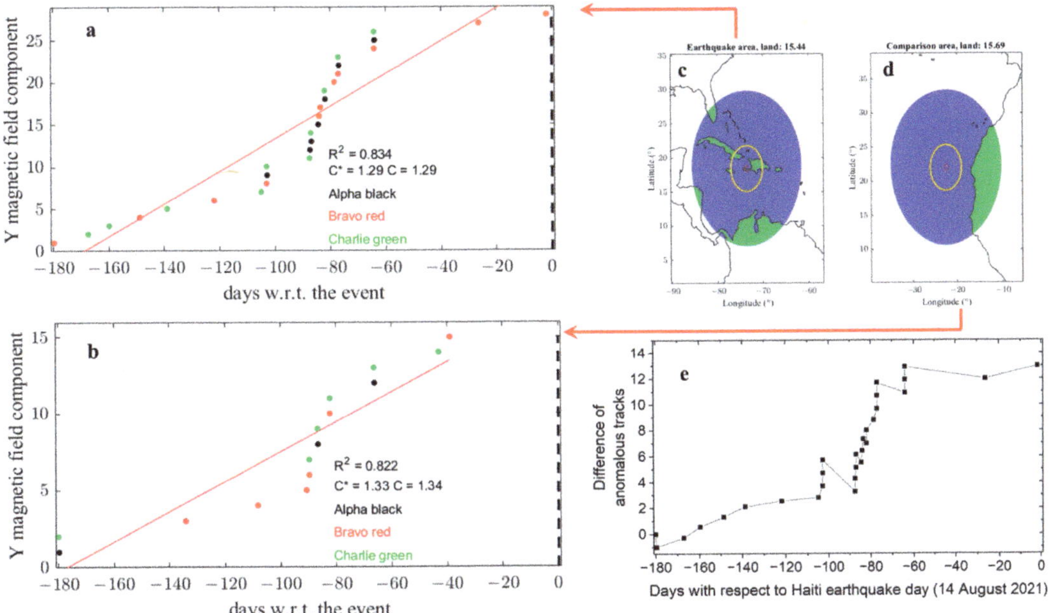



**Figure 15.** Cumulate of the Swarm Y-East component of magnetic field anomalies in the six months before Mw = 7.2 Haiti 2021 earthquake in (**a**) epicentral area shown in (**c**); (**b**) in a comparison area shown in (**d**). A linear fit of the cumulative anomaly trend has been performed and shown as red line in (**a**,**b**). The maps in (**c**,**d**) of the Dobrovolsky's and comparison areas underline the sea part with blue patches and land by green patches. The epicentre and centre of comparison area is represented by a red and the investigation areas with yellow circles. (**e**) Differences in the trends to extract the anomalies are more likely related to the preparation phase of the earthquake.

In addition to the previous investigation, the TEC at 2 a.m. local time has been estimated, interpolating spatially and temporarily the values reported by Global Ionospheric Maps of Total Electron Content (GIM-TEC). The chosen local time is the one of night-time passages of the CSES-01 satellite to allow a better comparison. The same technique was already applied to La Palma 2021 volcano eruption [84] and the Lushan 2013 earthquake [69] with a slightly different approach to take into account the cyclic variation of TEC due to solar activity. The first work used a sinusoidal fit of TEC over 20 years, while the second one performed an exponential fit with the Sun Spot Number (SSN). The advantage of using SSN is to consider that each cycle of solar activity has a specific intensity both in minimum and maximum value, and SSN is one of the indices that can measure this feature that directly influences the ionisation of Earth's environment, so TEC. This work uses the SSN again, and a linear fit between TEC and SSN is performed. Then, the TEC values for each year are normalised versus the mean SSN value in the analysed six months. Despite the implemented normalisation, it is necessary to note that 2021 is a year of particularly low solar activity, which may affect the extraction of anomalies. In fact, as you can see from Figure 16, there are only two anomalous days, which are 1 and 2 July 2021, i.e., about 44 and 43 days before the mainshock. At that time, the geomagnetic conditions were quiet, as underlined by the following indexes: Dst = −8 nT, ap = 5 nT and AE = ~107 nT.

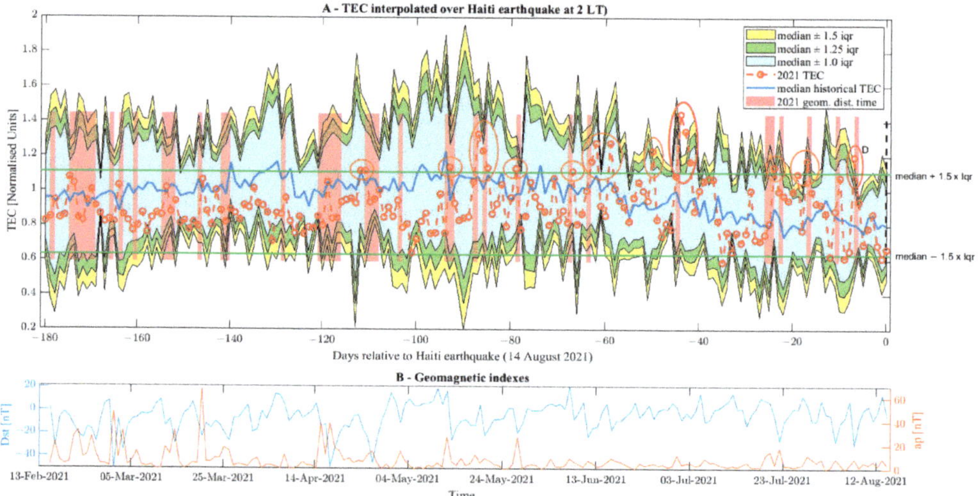

**Figure 16.** (**A**)Time series of the Total Electron Content (TEC) residuals estimated from the Global Ionospheric Maps of Total Electron Content (GIM-TEC) maps and interpolated above the earthquake epicentre and at 2 LT. The green horizontal lines indicate thresholds for anomalous values in the year of the earthquake (2021). (**B**) Geomagnetic activity represented by Dst (blue line) and ap (orange line) indexes.

## 4. Discussion

Atmospheric and ionospheric analyses depicted several anomalies. Regarding the investigation of the atmospheric parameter, if the broken segment by the incoming earthquake is considered, there is an improvement in the results. However, this point is more important for larger-magnitude events, as the approximation of the source as a point is more accurate for smaller-magnitude events.

An external and occasional hazard that may affect the results, especially atmospheric analyses, is the eventual presence of extreme weather events, such as storms or hurricanes. In the literature, the identified pre-earthquake anomalies have been compared with hurricanes to exclude the ones probably not induced by earthquakes [96]. In the investigated time, NOAA reported several storms and hurricanes in the Atlantic Ocean sector (https://www.nhc.noaa.gov/data/tcr/index.php?season=2021&basin=atl, last access 10 March 2024) as reported in the following list.

1. Tropical Storm ANA from 22 to 23 May 2021. This storm affected the North-Center Atlantic Ocean, which is very far from the investigated area.
2. Tropical Storm BILL from 14 to 15 June 2021. This storm was generated in the North Atlantic Ocean and arrived close to the Northern coast of Florida, which is still far from the investigated area, affecting the results of this work.
3. Tropical Storm CLAUDETTE from 19 to 22 June 2021. This storm was generated in the North Atlantic Ocean, close to the Nova Scotia (Canada) coast, travelled inside the

USA and ended in the Gulf of Mexico. Despite passing around the investigated area, it was completely outside the Dobrovoslkys's area.

4. Tropical Storm DANNY from 27 to 29 June 2021. This storm was generated in the Central Atlantic Ocean and arrived at the coast border between Florida and Georgia States. It was far from the investigated area.
5. Hurricane ELSA from 30 June to 9 July 2021. This hurricane passed in the investigated area, which is further discussed below.
6. Tropical Storm FRED from 11 to 17 August 2021. This storm passed inside the investigated area, which is further discussed below.
7. Hurricane GRACE from 13 to 21 August 2021. This hurricane crossed the investigated area, but mainly after the Haiti 2021 earthquake.

Considering the above list, three events are selected as potentially affecting the analyses of this paper: Hurricane Elsa [97], Tropical Storm Fred [98] and Hurricane Grace [99]. The positions of the centre of Elsa and Grace hurricanes and Fred storm are reported in Figure 17, along with the date reported by the National Hurricane Center [97–99].

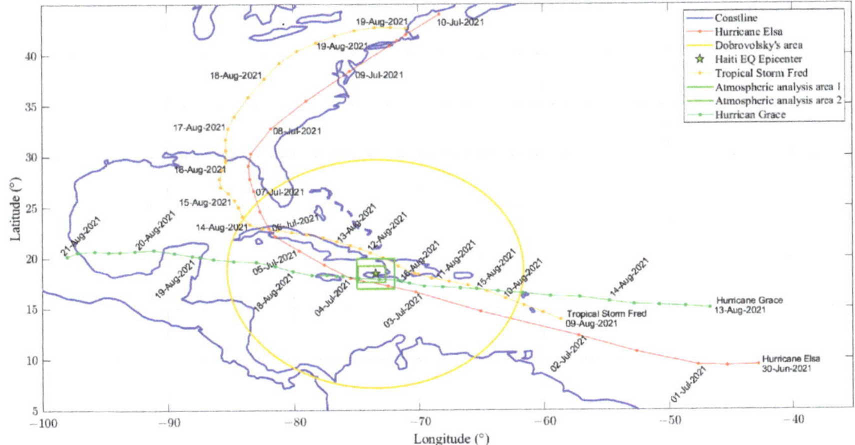


**Figure 17.** Map of positions of Elsa and Grace hurricanes and Tropical Storm Fred that crossed the investigated area during the six months before the Haiti 2021 earthquake occurrence. The yellow circle represents the Dobrovolsky's area and the green boxes shows the atmospheric research areas implemented in MEANS algorithm for this earthquake.

Some atmospheric anomalies in carbon monoxide in the days immediately following the extreme weather events may be related to the increase of using power generators as emergency recovery solutions, as reported in previous cases dealing with a higher risk of CO poisoning for this reason [100]. Sulphur dioxide increase due to the same reason was reported in the aftermath of the previous Hurricane Maria, which hit Puerto Rico in 2017 [101].

Figure 18 provides a summary of all the identified anomalies in the atmosphere and ionosphere compared with the seismicity inside Dobrovolsky's area. Several possible interactions could be inferred by analysing the different trends in geo-layers, which are underlined by red ellipses and dashed red arrows. The four numbers in grey circles indicate important moments that these analyses can depict.

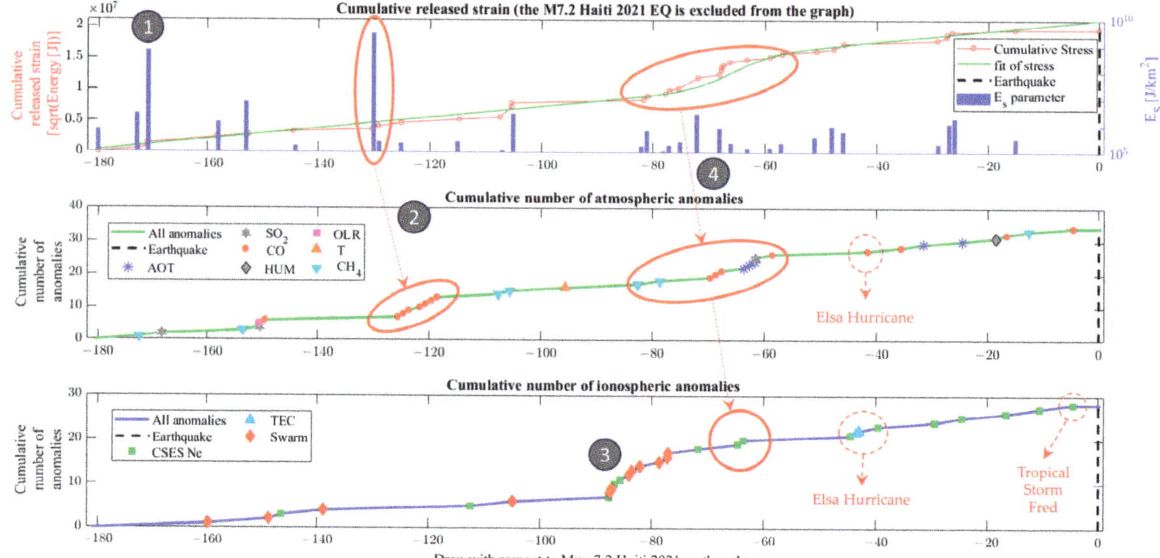



**Figure 18.** Cumulative trends in the lithosphere (earthquakes), atmosphere and ionosphere (anomalies). The number inside grey circles indicates the most important stages depicted by this summary graph: (1) particular high seismicity, (2) high seismicity immediately followed by aerosol release, (3) increase of ionospheric anomalies (especially Swarm magnetic field) and (4) variation of seismic trend synchronous to increase of atmospheric anomalies followed by two consecutive days of CSES-01 electron density anomalies. These important group of anomalies possibly related to the incoming earthquake has been underlined by bold red circles. Anomalies possibly related to weather hurricanes and storms have been underlined by dashed red circles.

1.  High seismicity in the Dobrovolsky area 171 days before the mainshock.
2.  Possible coupling between lithosphere and atmosphere occurred 130~120 days before the mainshock. The increase in seismicity could lead to the release of CO and carbon dioxide (or monoxide), which are proven to be linked to seismic activity [102,103].
3.  Increased ionospheric anomalies from 88 to 77 days before the mainshock. This group of anomalies is detected mainly by the Swarm magnetic field, but CSES-01 Ne also depicted some anomalies. It is the most abundant increase in ionospheric anomalies before the Mw = 7.2 Haiti 2021 earthquake, but it is not preceded by significant anomalies in the lithosphere or ionosphere, so its link with the incoming earthquake is uncertain.
4.  Increased seismic rate from about 80 to 60 days before the mainshock, synchronous with anomalies in the atmosphere and two consecutive days of CSES-01 Ne anomalies. This is a possible complete coupling between the lithosphere–atmosphere and ionosphere. It is still possible to describe it with the Pulinets and Ouzounov LAIC model, as in addition to carbon monoxide release, there are also aerosol anomalies and higher concentrations of surface SO$_2$, which may also be released due to the stress increase. The first atmospheric anomalous parameter in this time range (−80 days) was methane, which could have acted as a gas carrier to bring radioactive substances into the atmosphere, inducing ionisation, as proposed by Etiope and Martinelli [104]. This may lead to an increase of Ne recorded by CSES-01, as simulated by Kuo [105]. This is also somehow compatible with the LAIC theory of Pulinets and Ouzonouv [90], despite the lack of evidence (such as missing OLR, humidity drop, etc.) to confirm this explanation. This may be due to particular local conditions or observation limitations, especially for ground instrumentation.

A specific list of anomalies with the values of the investigated parameters is provided in Table 1. The value is the absolute one and not the exceedance of the threshold.

**Table 1.** List of lithosphere, atmosphere and ionosphere anomalies in the six months before Mw = 7.2 Haiti 2021 earthquake occurrence.

| Parameter | Anomalous Day [2021] | Anticipation Time [Day] | Value |
|---|---|---|---|
| | **Lithosphere** | | |
| $E_S$ | 24 February | −171 | $6.5 \times 10^8$ J/km$^2$ |
| | 6 April | −130 | $3.0 \times 10^9$ J/km$^2$ |
| | **Atmosphere** | | |
| Aerosol | 12 June | −63 | 0.61 |
| | 13 June | −62 | 0.62 |
| | 14 June | −61 | 0.48 |
| | 14 July | −31 | 0.44 |
| | 21 July | −24 | 0.50 |
| SO$_2$ | 27 February | −168 | $1.93 \times 10^{-6}$ kg/m$^2$ |
| | 17 March | −150 | $1.86 \times 10^{-6}$ kg/m$^2$ |
| | 14 June | −61 | $2.29 \times 10^{-6}$ kg/m$^2$ |
| CO | 18 March | −149 | $102 \times 10^{-9}$ ppbv |
| | From 11 to 18 April (excluding 14 April) | From −125 to −118 (excluded −122) | $97 \div 111 \times 10^{-9}$ ppbv |
| | From 6 to 8 June | From −69 to −67 | $75 \div 77 \times 10^{-9}$ ppbv |
| | 17 June | −58 | $75 \times 10^{-9}$ ppbv |
| | 4 July [1] | −41 | $90 \times 10^{-9}$ ppbv |
| | 10 July | −35 | $81 \times 10^{-9}$ ppbv |
| | 29 July | −16 | $78 \times 10^{-9}$ ppbv |
| | 10 August | −4 | $80 \times 10^{-9}$ ppbv |
| Humidity | 27 July | −18 | 0.0164 |
| OLR | 17 March | −150 | 187 W/m$^2$ |
| Temperature | 11 May | −95 | 301 K |
| CH$_4$ | 24 February | −171 | 2.24 $\sigma$ |
| | 15 April | −153 | 2.40 $\sigma$ |
| | 17 April | −107 | 3.76 $\sigma$ |
| | 1 May | −105 | 2.29 $\sigma$ |
| | 14 May | −82 | 2.18 $\sigma$ |
| | 28 May | −78 | 2.71 $\sigma$ |
| | 7 July | −12 | 2.11 $\sigma$ |
| | **Ionosphere** | | |
| CSES-01 Ne | 20 March | −147 | $1.232 \times 10^4$ e$^-$/cm$^3$ |
| | 23 April | −113 | $1.737 \times 10^4$ e$^-$/cm$^3$ |
| | 18–20 May | From −88 to −86 | $(1.006 \div 3.019) \times 10^4$ e$^-$/cm$^3$ |
| | 3 June | −72 | $2.704 \times 10^4$ e$^-$/cm$^3$ |
| | 10–11 June | −65 | $2.7 \times 10^4$ e$^-$/cm$^3$ |
| | 30 June | −45 | $3.398 \times 10^4$ e$^-$/cm$^3$ |
| | 5 July | −40 | $2.065 \times 10^4$ e$^-$/cm$^3$ |
| | 15 July | −30 | $1.309 \times 10^4$ e$^-$/cm$^3$ |
| | 20 July | −25 | $3.288 \times 10^4$ e$^-$/cm$^3$ |
| | 28 July | −17 | $4.076 \times 10^4$ e$^-$/cm$^3$ |
| | 3 August | −11 | $1.246 \times 10^4$ e$^-$/cm$^3$ |
| | 9 August | −5 | $1.155 \times 10^4$ e$^-$/cm$^3$ |
| TEC | 1 July | −44 | 1.44 TECU [2] |
| | 2 July | −43 | 1.35 TECU [2] |

[1] On this day, hurricane Elsa passed inside the atmospheric research area, so this anomaly is likely due to weather perturbation. [2] The value is in normalised TECU units.

Comparing the results of this work with previous ones [57,59,60], it is possible to confirm the presence of sparse anomalies in the last weeks and days before the earthquake, but it is possible to depict more interesting periods with higher anticipation time not investigated in previous work. This is not surprising, considering that a single fault's seismic recharge time could be hundreds or thousands of years [106].

Comparing the hurricanes and storms with the list and location of the anomalies, it is possible to note that the CO anomaly recorded on 4 July 2021 is likely caused by the Elsa hurricane that crossed the atmospheric investigation area exactly on the same day. The same hurricane is likely the source of the two TEC anomalies on 1st and 2nd July 2021 and the CSES-01 Ne anomaly recorded on 5 July 2021 inside the Dobrovolsky area. The anomaly of CO recorded on 10 August 2021 could be due to tropical storm Fred, but on the same day, the storm was still far from the area of investigation for atmospheric anomalies. On the other hand, the CSES-01 Ne anomalies recorded on 9 August 2021 (the last one before the earthquake) could be due to the starting of the perturbation of the tropical storm Fred on the same day. The presence of anomalies likely associated with the Elsa Hurricane may still be a sign of possible geophysical coupling induced by the strong weather, especially the ionospheric perturbation in Ne and its vertically integrated quantity TEC identified from CSES-01 and Global Navigation Satellite System (GNSS) detectors. LAIC induced by other natural hazards has also been proposed, but it is beyond the scope of this paper [90,107–109].

In Supplementary Materials, a comparison of electron density measured by Swarm satellites and CSES-01 is provided during two nights with anomalous features close to the epicentre. Figure S4 represents Ne on 5 July 2021 and Figure S5 on August 2021, i.e., 40 and 8 days before the Mw = 7.2 Haiti 2021 earthquake, respectively. Despite the fact that an increase in electron density seems present inside Dobrovoslky's area, the tracks acquired at the same local time eastward and outside the earthquake preparation area also show high electron density values. In the electron density profile of 6 August 2021, a double peak above and symmetric to the epicentre is present, as visible in Figure S5, which is the ochre colour track. This track is close to the TEC anomaly extracted by Chen et al. [60] (1 August 2021). A similar result was presented in the analysis of Mw = 7.2 Kermadec Island (New Zealand) 2019 earthquake [73] in Figure 11 of the cited paper, with the same problem as the present work, i.e., the Eastward track outside Dobrovolsky's area also shows the high value of Ne. The comparison is interesting as the magnitude is the same for both earthquakes, but the tectonic context and region differ. Unfortunately, due to the unclearness of the results, it is not easy to draw any conclusions.

Previous works identified an important feature: the spatial organisation of the anomalies before the earthquake [69,110], identifying a pattern that started far from the epicentre and gradually approached the incoming earthquake. However, in this work, only some considerations about the seismic patterns of Figure 1 can be discussed, as the anomalies in the atmosphere and ionosphere have been extracted in areas closer to the epicentre than the entire Dobrovolsky area. An important difference between the cited works and the present one is that they analysed continental earthquakes (Mw6.7 Lushan, China, 2013 and Mw = 7.8 Nepal 2015) while Mw = 7.2 Haiti 2021 epicentre is surrounded by the Caribbean Sea. The sea and the great variability of the area could make it more difficult to extract pre-earthquake anomalies far from the epicentre (still inside Dobrovolsky's area) due to contamination of other phenomena. However, regarding the seismicity, it is possible to note that the first earthquakes six months before Mw = 7.2 Haiti 2021 occurred close to the future mainshock. Then, several seismicities were recorded far in the three months before the earthquake gradually approached the incoming earthquake, leaving a gap in the middle at the mainshock future location. This is a well-known phenomenon that, from a seismological point of view, can be explained as the seismogenic fault being locked before greater earthquakes [111]. Some simulations of this phenomenon, also known as seismic gap theory, have been performed in the literature [112].

## 5. Conclusions

In this paper, the lithosphere (earthquake catalogue), atmosphere (climatological dataset MERRA2 and AIRS methane data) and ionosphere (CSES and Swarm missions + TEC) geo-layers have been investigated in the six months before the Mw = 7.2 Haiti 2021 earthquake. Geomagnetic disturbed time has been skipped to avoid confusing external source ionospheric anomalies with a possible seismo-induced phenomenon. In addition, extreme weather events have been scrutinised, and some anomalies have been identified as possibly caused by such events. Several anomalies have been extracted by applying specific methods for the particular parameter, and finally, they have been discussed together. Here, an approach of separating the geo-layer is preferred to the one that combines all the anomalies in one (used, for example, for Ridgecrest 2019, New Zealand 2019 and a previous one about Haiti 2021 [57,73,83]) because, in this way, possible transfer of anomalies from bottom layers to the upper ones can be inferred. In particular, a coupling between the lithosphere and atmosphere may occur 130 days before the mainshock, and another one between the lithosphere, atmosphere and ionosphere may occur 80 to 60 days before the mainshock. Both require some days of propagation from one geo-layer to the next and are compatible with the LAIC model of Pulinets and Ouzonov [90]. These results cannot be used directly to make predictions or directly implement in a prediction system. Still, they provide crucial knowledge that is necessary but insufficient to make a scientifically reliable prediction of an earthquake. However, this further confirms that this method of studying multiple layers and parameters permits the depiction of alteration of the geosystem possibly associated with the preparation phase of the earthquake and supports possible LAIC and specific models. Further studies are necessary, especially to discuss why the patterns of anomalies are often different case by case and also to statistically assess the validity of an identified pattern on a large number of medium/large earthquakes. The final stage of constructing a prediction system may be building a platform based on the accumulated knowledge to be tested in a forward way and from an independent organisation.

**Supplementary Materials:** The following supporting information can be downloaded at: https://www.mdpi.com/article/10.3390/geosciences14040096/s1, Figure S1: Atmospheric analysis of humidity; Figure S2: Atmospheric analysis of temperature; Figure S3: Atmospheric analysis of OLR; Figure S4: Swarm Bravo and CSES-01 on the night of 5 July 2021; Figure S5: Swarm Alpha and CSES-01 on the night of 6 August 2021.

**Funding:** This research received no external funding.

**Data Availability Statement:** The earthquake catalogue has been retrieved from the USGS free archive (https://earthquake.usgs.gov/earthquakes/search/ last access 26 January 2024). MERRA-2 data can be downloaded from https://disc.gsfc.nasa.gov/datasets?project=MERRA-2 (last accessed 23 July 2022) with Earth Observation NASA free credential. Swarm data are freely available via ftp and http at swarm-diss.eo.esa.int server (last accessed 27 July 2022). China Seismo Electromagnetic Satellite data are freely available at https://www.leos.ac.cn/#/home (last access on 8 March 2024, a warning about "not secure" website could appear in certain browsers) upon registration and approval.

**Acknowledgments:** The author acknowledges ISSI/ISSI-BJ for supporting the International Team 553 "CSES and Swarm Investigation of the Generation Mechanisms of Low Latitude Pi2 Waves" led by Essam Ghamry and Zeren Zhima, and International Team 23-583 (57) "Investigation of the Lithosphere Atmosphere Ionosphere Coupling (LAIC) Mechanism before the Natural Hazards" led by Dedalo Marchetti and Essam Ghamry. A special acknowledgement to Guido Ventura, Essam Ghamry, Xuhui Shen, Rui Yan, Zeren Zhima, Alessandro Piscini, Loredana Perrone, Saioa Arquero Campuzano, Francisco Javier Pavón-Carrasco, Maurizio Soldani, Angelo De Santis, Yiqun Zhang, Ting Wang, Wenqi Chen, Donghua Zhang and Hanshuo Zhang, for re-use of some codes and discussions.

**Conflicts of Interest:** The author declares no conflicts of interest. The funders had no role in the design of the study; in the collection, analyses or interpretation of data; in the writing of the manuscript; or in the decision to publish the results.

## Appendix A. Analysis of Methane Data Version 6 of AIRS

Here, the results are presented using the previous methane data version, 6.0, from AIRS. In addition, the original data before and after detrending are shown in Figure A1 as a time series (a) and histogram (b). It is possible to note, comparing the same figure in the manuscript (see Figure 3), that the time series of the new version 7 is a bit changed with a more accentuated global warming trend. The CH4 anomalies using this version are shown in Figure A2. This trend and the extracted anomalies are compatible with the ones obtained with the new version, even though some minor differences exist. As the new version is provided as a revised and improved data calibration from the team mission, it is also considered more reliable for investigating the study of possible methane emissions before earthquake occurrence.

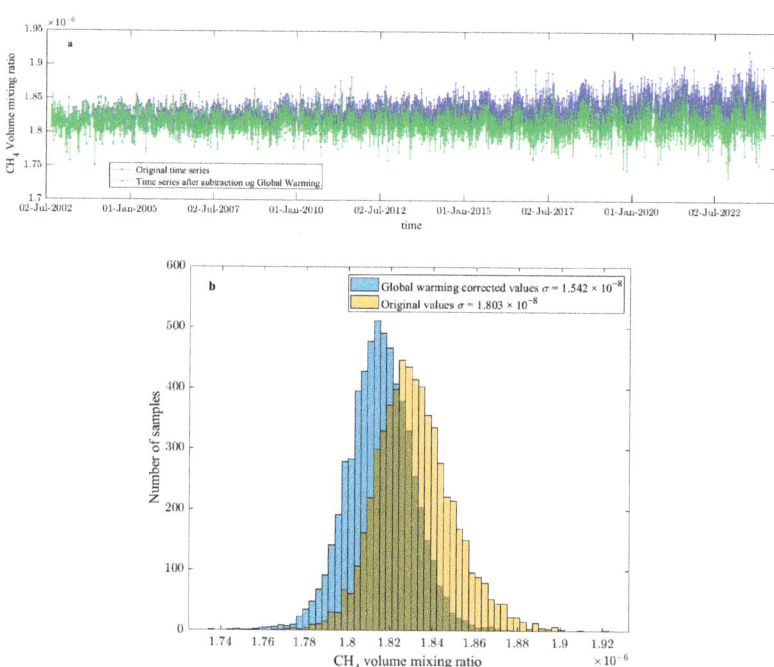


**Figure A1.** Methane (CH$_4$) version 6 data in the form of time series (**a**) and histogram (**b**) before and after the global warming detrending.

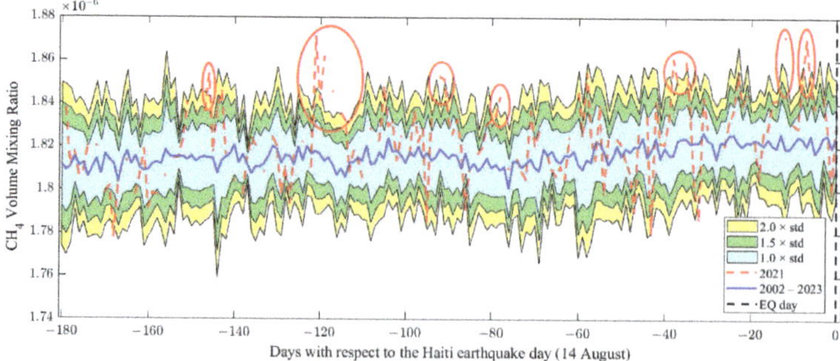


**Figure A2.** Methane (CH$_4$) version 6 investigation in the six months before the Haiti 2021 earthquake. The data source is AISR instrument, and so the background has been calculated on a shorter time period (from 2002 to 2023, excluding the year of the earthquake–2021). Red circles underline anomalies.

### Appendix B. CSES-01 Ne Daytime Analysis

In this appendix, the analysis of CSES-01 Ne daytime data is presented. In particular, all of the same orbits (see Figure 4) acquired during geomagnetic quiet conditions are superposed and visualised in Figure A3. The general shape is the typical one for the altitude of the satellite (~510 km) with only one Equatorial Ionospheric Anomaly (EIA) peak. Some outliers are depicted as such peaks that are significantly higher for some specific orbits. There is also an orbit with lower Ne values detected as an outlier (2 February 2021). At the latitude of the Haiti 2021 earthquake, no anomalous values were extracted. Despite some theories proposing the variations of EIA as one of the pre-earthquake phenomena [113], in this case, it looks more like ionospheric intrinsic variability, especially due to the very long anticipation time. However, it is quite interesting to note that on 19 March 2021, double EIA peaks were formed, with maximum values on 24 March 2021. These anomalous tracks with double EIA peaks preceded from 18 to 8 days, the highest seismic day in the area (excluding the Haiti event) according to the $E_S$ parameter ($E_S = 3.0 \times 10^9$ J/km$^2$ on 6 April 2021).

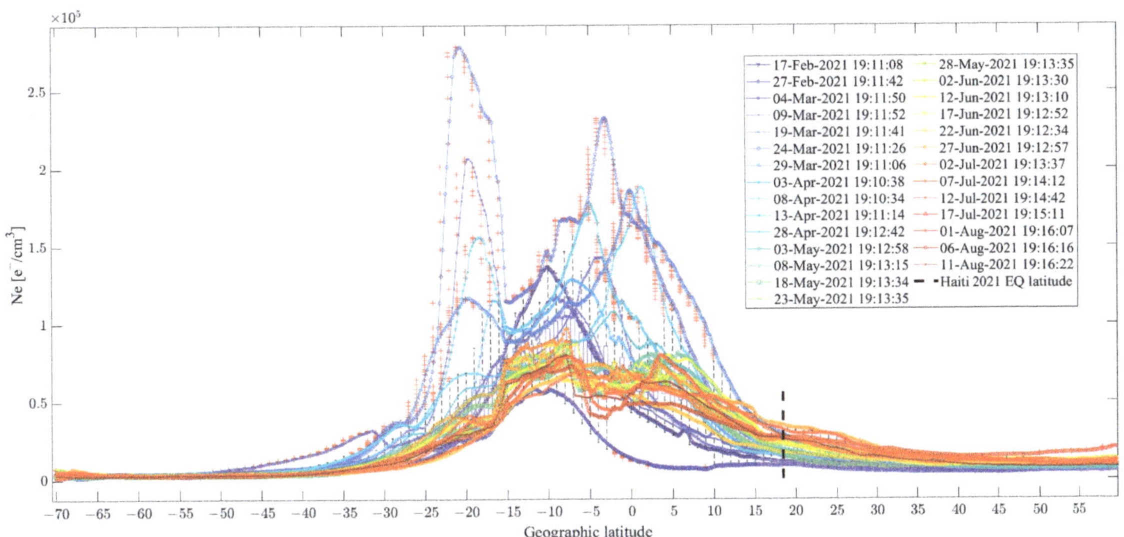


**Figure A3.** CSES-01 Ne daytime latitudinal profiles acquired in geomagnetic quiet time. The blue boxes and dashed black lines represent the standard ranges of the values for each degree of latitude. The red crosses indicate the outlier values of Ne. The latitude of the Haiti earthquake (EQ) is represented as a vertical black line.

### Appendix C. Magnitude of Completeness of Earthquake Catalogue

In this appendix, the magnitude of completeness of the USGS earthquake catalogue in the investigated area and time is checked. It was noted that the catalogue seems to have a geographical area (Dominican Republic) with higher completeness, probably due to a dense network of seismic stations. So, the completeness was separately checked into two distinct areas represented in Figure A4.

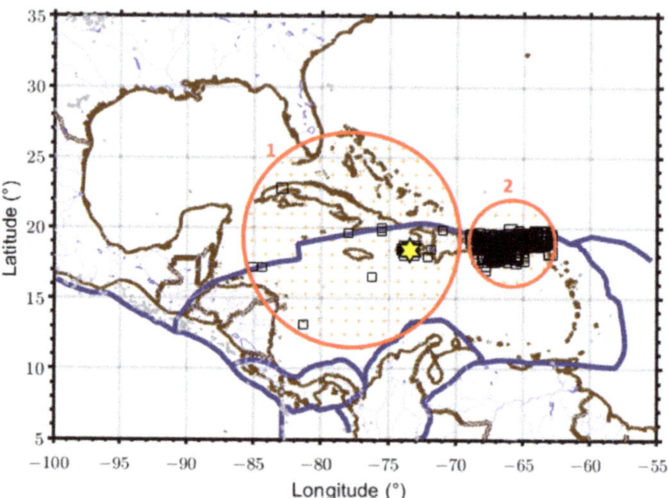

**Figure A4.** Map of the two selected areas (red circles) marked with numbers "1" and "2". The earthquakes are represented as black boxes and the Mw = 7.2 Haiti 2021 earthquake epicentre as yellow star. Main plate boundaries are depicted with bold dark blue lines.

The Gutenberg–Richter (G-R) [114] distribution into the two selected areas is shown in Figure A5 "a" and "b". The graph shows the number of events as a function of earthquake magnitude. Both the binned value (grey box) and cumulative values of the number of earthquakes with magnitude equal to or greater than the corresponding value on the abscissa are shown by white boxes. It is possible to confirm that in Area 2, the completeness magnitude Mc is better and equal to 2.6. However, in Area 1, the Mc = 4.2. Consequently, to be sure not to lose events in the whole of Dobrovolsky's area, the most conservative value of Mc = 4.2 was selected. Using this value and selecting the two years of the earthquake catalogue before the Mw = 7.2 Haiti 2021 earthquake, the G-R distribution is calculated again and shown in Figure A5c. Using such a final solution, it is possible to calculate how many events of magnitude equal or greater than 7.2 occurred in this area by simply using the a and b coefficient of best fit:

$$N(M \geq 7.2) = 10^{6.484 - 1.07 \times 7.2} = 0.060 \text{ events/year}$$

The inverse of the above number represents the mean interseismic time $\tau$ of occurrence of events equal to or greater than magnitude 7.2 in the area:

$$\tau = \frac{1}{N(M \geq 7.2)} = 16.6 \text{ years}$$

It is possible to say that statistically, the time between an earthquake of at least magnitude 7.2 in this area is of the order of two decades. This time must not be taken as rigorous, considering that the empirical distribution of interseismic times is generally considered to follow a Poissonian statistic, even though some debate about this assumption exists [115–117].

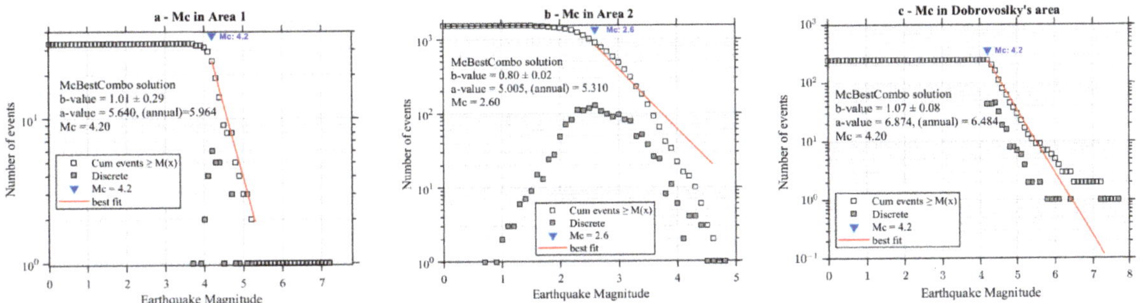

**Figure A5.** Gutenberg–Richter distributions in (**a**) selected Area 1 of Figure A4; (**b**) selected Area 2 and (**c**) the Dobrovolsky's area. The last case considers two years of data before the Haiti earthquake.

# References


1. Ze, Z.; Guojie, M.; Xiaoning, S.; Jicang, W.; Xiaojing, L.J. Global Crustal Movement and Tectonic Plate Boundary Deformation Constrained by the ITRF2008. *Geod. Geodyn.* **2012**, *3*, 40–45. [CrossRef]
2. Geller, R.J. Earthquake Prediction: A Critical Review. *Geophys. J. Int.* **1997**, *131*, 425–450. [CrossRef]
3. Conti, L.; Picozza, P.; Sotgiu, A. A Critical Review of Ground Based Observations of Earthquake Precursors. *Front. Earth Sci.* **2021**, *9*, 676766. [CrossRef]
4. Picozza, P.; Conti, L.; Sotgiu, A. Looking for Earthquake Precursors From Space: A Critical Review. *Front. Earth Sci.* **2021**, *9*, 676775. [CrossRef]
5. Fidani, C.; Freund, F.; Grant, R. Cows Come Down from the Mountains before the (Mw = 6.1) Earthquake Colfiorito in September 1997; A Single Case Study. *Animals* **2014**, *4*, 292–312. [CrossRef] [PubMed]
6. Shen, X.; Zong, Q.-G.; Zhang, X. Introduction to Special Section on the China Seismo-Electromagnetic Satellite and Initial Results. *Earth Planet. Phys.* **2018**, *2*, 439–443. [CrossRef]
7. Ghamry, E.; Marchetti, D.; Yoshikawa, A.; Uozumi, T.; De Santis, A.; Perrone, L.; Shen, X.; Fathy, A. The First Pi2 Pulsation Observed by China Seismo-Electromagnetic Satellite. *Remote Sens.* **2020**, *12*, 2300. [CrossRef]
8. Yang, Y.; Hulot, G.; Vigneron, P.; Shen, X.; Zhima, Z.; Zhou, B.; Magnes, W.; Olsen, N.; Tøffner-Clausen, L.; Huang, J.; et al. The CSES Global Geomagnetic Field Model (CGGM): An IGRF-Type Global Geomagnetic Field Model Based on Data from the China Seismo-Electromagnetic Satellite. *Earth Planets Space* **2021**, *73*, 45. [CrossRef]
9. Yang, Y.-Y.; Zhima, Z.-R.; Shen, X.-H.; Chu, W.; Huang, J.-P.; Wang, Q.; Yan, R.; Xu, S.; Lu, H.-X.; Liu, D.-P. The First Intense Geomagnetic Storm Event Recorded by the China Seismo-Electromagnetic Satellite. *Space Weather* **2020**, *18*, e2019SW002243. [CrossRef]
10. Yan, R.; Zhima, Z.; Xiong, C.; Shen, X.; Huang, J.; Guan, Y.; Zhu, X.; Liu, C. Comparison of Electron Density and Temperature From the CSES Satellite With Other Space-Borne and Ground-Based Observations. *J. Geophys. Res. Space Phys.* **2020**, *125*, e2019JA027747. [CrossRef]
11. Zhou, B.; Cheng, B.; Gou, X.; Li, L.; Zhang, Y.; Wang, J.; Magnes, W.; Lammegger, R.; Pollinger, A.; Ellmeier, M.; et al. First In-Orbit Results of the Vector Magnetic Field Measurement of the High Precision Magnetometer Onboard the China Seismo-Electromagnetic Satellite. *Earth Planets Space* **2019**, *71*, 119. [CrossRef]
12. Gou, X.; Li, L.; Zhou, B.; Zhang, Y.; Xie, L.; Cheng, B.; Feng, Y.; Wang, J.; Miao, Y.; Zhima, Z.; et al. Electrostatic Ion Cyclotron Waves Observed by CSES in the Equatorial Plasma Bubble. *Geophys. Res. Lett.* **2023**, *50*, e2022GL101791. [CrossRef]
13. Mignan, A. The Stress Accumulation Model: Accelerating Moment Release and Seismic Hazard. In *Advances in Geophysics*; Elsevier: Amsterdam, The Netherlands, 2008; Volume 49, pp. 67–201. ISBN 978-0-12-374231-5.
14. De Santis, A.; Cianchini, G.; Di Giovambattista, R. Accelerating Moment Release Revisited: Examples of Application to Italian Seismic Sequences. *Tectonophysics* **2015**, *639*, 82–98. [CrossRef]
15. Kato, A.; Fukuda, J.; Kumazawa, T.; Nakagawa, S. Accelerated Nucleation of the 2014 Iquique, Chile Mw 8.2 Earthquake. *Sci. Rep.* **2016**, *6*, 24792. [CrossRef] [PubMed]
16. Filizzola, C.; Corrado, A.; Genzano, N.; Lisi, M.; Pergola, N.; Colonna, R.; Tramutoli, V. RST Analysis of Anomalous TIR Sequences in Relation with Earthquakes Occurred in Turkey in the Period 2004–2015. *Remote Sens.* **2022**, *14*, 381. [CrossRef]
17. Tramutoli, V.; Cuomo, V.; Filizzola, C.; Pergola, N.; Pietrapertosa, C. Assessing the Potential of Thermal Infrared Satellite Surveys for Monitoring Seismically Active Areas: The Case of Kocaeli (İzmit) Earthquake, August 17, 1999. *Remote Sens. Environ.* **2005**, *96*, 409–426. [CrossRef]
18. Tramutoli, V.; Bello, G.D.; Pergola, N.; Piscitelli, S. Robust Satellite Techniques for Remote Sensing of Seismically Active Areas. *Ann. Geophys.* **2001**, *44*, 295–312. [CrossRef]
19. Genzano, N.; Filizzola, C.; Hattori, K.; Pergola, N.; Tramutoli, V. Statistical Correlation Analysis Between Thermal Infrared Anomalies Observed From MTSATs and Large Earthquakes Occurred in Japan (2005–2015). *J. Geophys. Res. Solid Earth* **2021**, *126*, e2020JB020108. [CrossRef]

20. Jing, F.; Singh, R.P. Changes in Tropospheric Ozone Associated With Strong Earthquakes and Possible Mechanism. *IEEE J. Sel. Top. Appl. Earth Obs. Remote Sens.* **2021**, *14*, 5300–5310. [CrossRef]
21. Tronin, A.A. Remote Sensing and Earthquakes: A Review. *Phys. Chem. Earth Parts A/B/C* **2006**, *31*, 138–142. [CrossRef]
22. Molchanov, O.A.; Hayakawa, M.; Rafalsky, V.A. Penetration Characteristics of Electromagnetic Emissions from an Underground Seismic Source into the Atmosphere, Ionosphere, and Magnetosphere. *J. Geophys. Res.* **1995**, *100*, 1691. [CrossRef]
23. Han, P.; Hattori, K.; Hirokawa, M.; Zhuang, J.; Chen, C.-H.; Febriani, F.; Yamaguchi, H.; Yoshino, C.; Liu, J.-Y.; Yoshida, S. Statistical Analysis of ULF Seismomagnetic Phenomena at Kakioka, Japan, during 2001–2010: Ulf Seismo-Magnetic Phenomena at Kakioka. *J. Geophys. Res. Space Phys.* **2014**, *119*, 4998–5011. [CrossRef]
24. Cianchini, G.; De Santis, A.; Barraclough, D.R.; Wu, L.X.; Qin, K. Magnetic Transfer Function Entropy and the 2009 Mw = 6.3 L'Aquila Earthquake (Central Italy). *Nonlin. Process. Geophys.* **2012**, *19*, 401–409. [CrossRef]
25. Fidani, C.; Orsini, M.; Iezzi, G.; Vicentini, N.; Stoppa, F. Electric and Magnetic Recordings by Chieti CIEN Station During the Intense 2016–2017 Seismic Swarms in Central Italy. *Front. Earth Sci.* **2020**, *8*, 536332. [CrossRef]
26. Fraser-Smith, A.C.; Bernardi, A.; McGill, P.R.; Ladd, M.E.; Helliwell, R.A.; Villard, O.G. Low-Frequency Magnetic Field Measurements near the Epicenter of the $M_s$ 7.1 Loma Prieta Earthquake. *Geophys. Res. Lett.* **1990**, *17*, 1465–1468. [CrossRef]
27. Uyeda, S.; Hayakawa, M.; Nagao, T.; Molchanov, O.; Hattori, K.; Orihara, Y.; Gotoh, K.; Akinaga, Y.; Tanaka, H. Electric and Magnetic Phenomena Observed before the Volcano-Seismic Activity in 2000 in the Izu Island Region, Japan. *Proc. Natl. Acad. Sci. USA* **2002**, *99*, 7352–7355. [CrossRef] [PubMed]
28. De Santis, A.; Balasis, G.; Pavón-Carrasco, F.J.; Cianchini, G.; Mandea, M. Potential Earthquake Precursory Pattern from Space: The 2015 Nepal Event as Seen by Magnetic Swarm Satellites. *Earth Planet. Sci. Lett.* **2017**, *461*, 119–126. [CrossRef]
29. Christodoulou, V.; Bi, Y.; Wilkie, G. A Tool for Swarm Satellite Data Analysis and Anomaly Detection. *PLoS ONE* **2019**, *14*, e0212098. [CrossRef] [PubMed]
30. Xie, T.; Chen, B.; Wu, L.; Dai, W.; Kuang, C.; Miao, Z. Detecting Seismo-Ionospheric Anomalies Possibly Associated With the 2019 Ridgecrest (California) Earthquakes by GNSS, CSES, and Swarm Observations. *JGR Space Phys.* **2021**, *126*, e2020JA028761. [CrossRef]
31. Athanasiou, M.A.; Anagnostopoulos, G.C.; Iliopoulos, A.C.; Pavlos, G.P.; David, C.N. Enhanced ULF Radiation Observed by DEMETER Two Months around the Strong 2010 Haiti Earthquake. *Nat. Hazards Earth Syst. Sci.* **2011**, *11*, 1091–1098. [CrossRef]
32. Ouyang, X.Y.; Parrot, M.; Bortnik, J. ULF Wave Activity Observed in the Nighttime Ionosphere Above and Some Hours Before Strong Earthquakes. *J. Geophys. Res. Space Phys.* **2020**, *125*, e2020JA028396. [CrossRef]
33. Chen, H.; Han, P.; Hattori, K. Recent Advances and Challenges in the Seismo-Electromagnetic Study: A Brief Review. *Remote Sens.* **2022**, *14*, 5893. [CrossRef]
34. Hayakawa, M. Seismo Electromagnetics and Earthquake Prediction: History and New Directions. *IJEAR* **2019**, *6*, 1–23. [CrossRef]
35. Hayakawa, M.; Schekotov, A.; Izutsu, J.; Nickolaenko, A.P.; Hobara, Y. Seismogenic ULF/ELF Wave Phenomena: Recent Advances and Future Perspectives. *Open J. Earthq. Res.* **2023**, *12*, 45–113. [CrossRef]
36. Ghamry, E.; Mohamed, E.K.; Abdalzaher, M.S.; Elweekeil, M.; Marchetti, D.; De Santis, A.; Hegy, M.; Yoshikawa, A.; Fathy, A. Integrating Pre-Earthquake Signatures From Different Precursor Tools. *IEEE Access* **2021**, *9*, 33268–33283. [CrossRef]
37. De Santis, A.; Marchetti, D.; Pavón-Carrasco, F.J.; Cianchini, G.; Perrone, L.; Abbattista, C.; Alfonsi, L.; Amoruso, L.; Campuzano, S.A.; Carbone, M.; et al. Precursory Worldwide Signatures of Earthquake Occurrences on Swarm Satellite Data. *Sci. Rep.* **2019**, *9*, 20287. [CrossRef]
38. Marchetti, D.; De Santis, A.; Campuzano, S.A.; Zhu, K.; Soldani, M.; D'Arcangelo, S.; Orlando, M.; Wang, T.; Cianchini, G.; Di Mauro, D.; et al. Worldwide Statistical Correlation of Eight Years of Swarm Satellite Data with M5.5+ Earthquakes: New Hints about the Preseismic Phenomena from Space. *Remote Sens.* **2022**, *14*, 2649. [CrossRef]
39. He, Y.; Zhao, X.; Yang, D.; Wu, Y.; Li, Q. A Study to Investigate the Relationship between Ionospheric Disturbance and Seismic Activity Based on Swarm Satellite Data. *Phys. Earth Planet. Inter.* **2022**, *323*, 106826. [CrossRef]
40. Zhima, Z.; Yan, R.; Lin, J.; Wang, Q.; Yang, Y.; Lv, F.; Huang, J.; Cui, J.; Liu, Q.; Zhao, S.; et al. The Possible Seismo-Ionospheric Perturbations Recorded by the China-Seismo-Electromagnetic Satellite. *Remote Sens.* **2022**, *14*, 905. [CrossRef]
41. Chen, W.; Marchetti, D.; Zhu, K.; Sabbagh, D.; Yan, R.; Zhima, Z.; Shen, X.; Cheng, Y.; Fan, M.; Wang, S.; et al. CSES-01 Electron Density Background Characterisation and Preliminary Investigation of Possible Ne Increase before Global Seismicity. *Atmosphere* **2023**, *14*, 1527. [CrossRef]
42. Perevalova, N.P.; Sankov, V.A.; Astafyeva, E.I.; Zhupityaeva, A.S. Threshold Magnitude for Ionospheric TEC Response to Earthquakes. *J. Atmos. Sol. Terr. Phys.* **2014**, *108*, 77–90. [CrossRef]
43. Cicerone, R.D.; Ebel, J.E.; Britton, J. A Systematic Compilation of Earthquake Precursors. *Tectonophysics* **2009**, *476*, 371–396. [CrossRef]
44. Lin, J.-W.; Chiou, J.-S. Detecting Total Electron Content Precursors Before Earthquakes by Examining Total Electron Content Images Based on Butterworth Filter in Convolutional Neural Networks. *IEEE Access* **2020**, *8*, 110478–110494. [CrossRef]
45. Liu, J.; Wang, W.; Zhang, X.; Wang, Z.; Zhou, C. Ionospheric Total Electron Content Anomaly Possibly Associated with the April 4, 2010 Mw7.2 Baja California Earthquake. *Adv. Space Res.* **2022**, *69*, 2126–2141. [CrossRef]
46. Perrone, L.; Korsunova, L.P.; Mikhailov, A.V. Ionospheric Precursors for Crustal Earthquakes in Italy. *Ann. Geophys.* **2010**, *28*, 941–950. [CrossRef]

47. Sabbagh, D.; Orlando, M.; Perrone, L.; Cianchini, G.; De Santis, A.; Piscini, A. Analysis of the Ionospheric Perturbations Prior to the 2009 L'Aquila and 2002 Molise Earthquakes from Ground- and Space-Based Observations. *URSI Radio Sci. Lett.* **2021**, *3*, 1–4. [CrossRef]
48. Korsunova, L.P.; Khegai, V.V. Medium-Term Ionospheric Precursors to Strong Earthquakes. *Int. J. Geomagn. Aeron.* **2006**, *6*, GI3005. [CrossRef]
49. Kumar, A.; Kumar, S.; Hayakawa, M.; Menk, F. Subionospheric VLF Perturbations Observed at Low Latitude Associated with Earthquake from Indonesia Region. *J. Atmos. Sol. Terr. Phys.* **2013**, *102*, 71–80. [CrossRef]
50. Hayakawa, M.; Kasahara, Y.; Nakamura, T.; Muto, F.; Horie, T.; Maekawa, S.; Hobara, Y.; Rozhnoi, A.A.; Solovieva, M.; Molchanov, O.A. A Statistical Study on the Correlation between Lower Ionospheric Perturbations as Seen by Subionospheric VLF/LF Propagation and Earthquakes: Seismo-ionospheric perturbations. *J. Geophys. Res.* **2010**, *115*, A09305. [CrossRef]
51. Politis, D.Z.; Potirakis, S.M.; Contoyiannis, Y.F.; Biswas, S.; Sasmal, S.; Hayakawa, M. Statistical and Criticality Analysis of the Lower Ionosphere Prior to the 30 October 2020 Samos (Greece) Earthquake (M6.9), Based on VLF Electromagnetic Propagation Data as Recorded by a New VLF/LF Receiver Installed in Athens (Greece). *Entropy* **2021**, *23*, 676. [CrossRef]
52. Styron, R.; García-Pelaez, J.; Pagani, M. GEM Central America and Caribbean Active Faults Database. 2018. Available online: https://github.com/GEMScienceTools/central_am_carib_faults (accessed on 9 February 2024).
53. Styron, R.; García-Pelaez, J.; Pagani, M. CCAF-DB: The Caribbean and Central American Active Fault Database. *Nat. Hazards Earth Syst. Sci.* **2020**, *20*, 831–857. [CrossRef]
54. Burbach, G.V.; Frohlich, C.; Pennington, W.D.; Matumoto, T. Seismicity and Tectonics of the Subducted Cocos Plate. *J. Geophys. Res. Solid Earth* **1984**, *89*, 7719–7735. [CrossRef]
55. Cordani, U.; Ramos, V.; Fraga, L.; Cegarra, M.; Delgado, I.; de Souza, K.G.; Gomes, F.E.; Schobbenhaus, C. *Tectonic Map of South America at 1:5,900,000 Scale*, 600th ed.; Commission for the Geological Map of the World (CGMW): Paris, France, 2019.
56. Cediel, F.; Shaw, R.P.; Cáceres, C. Tectonic Assembly of the Northern Andean Block. In *The Circum-Gulf of Mexico and the Caribbean: Hydrocarbon Habitats, Basin Formation and Plate Tectonics*; Bartolini, C., Buffler, R.T., Blickwede, J.F., Eds.; American Association of Petroleum Geologists: Tulsa, OK, USA, 2003; Volume 79, ISBN 978-1-62981-054-6.
57. Akhoondzadeh, M. Investigation of the LAIC Mechanism of the Haiti Earthquake (14 August 2021) Using CSES-01 Satellite Observations and Other Earthquake Precursors. *Adv. Space Res.* **2024**, *73*, 672–684. [CrossRef]
58. Akhoondzadeh, M.; Marchetti, D. Study of the Preparation Phase of Turkey's Powerful Earthquake (6 February 2023) by a Geophysical Multi-Parametric Fuzzy Inference System. *Remote Sens.* **2023**, *15*, 2224. [CrossRef]
59. Khan, M.M.; Ghaffar, B.; Shahzad, R.; Khan, M.R.; Shah, M.; Amin, A.H.; Eldin, S.M.; Naqvi, N.A.; Ali, R. Atmospheric Anomalies Associated with the 2021 Mw 7.2 Haiti Earthquake Using Machine Learning from Multiple Satellites. *Sustainability* **2022**, *14*, 14782. [CrossRef]
60. Chen, D.; Meng, D.; Wang, F.; Gou, Y. A Study of Ionospheric Anomaly Detection before the August 14, 2021 Mw7.2 Earthquake in Haiti Based on Sliding Interquartile Range Method. *Acta Geodaetica et Geophysica* **2023**, *58*, 539–551. [CrossRef]
61. D'Angelo, G.; Piersanti, M.; Battiston, R.; Bertello, I.; Carbone, V.; Cicone, A.; Diego, P.; Papini, E.; Parmentier, A.; Picozza, P.; et al. Haiti Earthquake (Mw 7.2): Magnetospheric–Ionospheric–Lithospheric Coupling during and after the Main Shock on 14 August 2021. *Remote Sens.* **2022**, *14*, 5340. [CrossRef]
62. Rapoport, Y.G.; Gotynyan, O.E.; Ivchenko, V.M.; Kozak, L.V.; Parrot, M. Effect of Acoustic-Gravity Wave of the Lithospheric Origin on the Ionospheric F Region before Earthquakes. *Phys. Chem. Earth Parts A/B/C* **2004**, *29*, 607–616. [CrossRef]
63. Godin, O.A. Finite-Amplitude Acoustic-Gravity Waves: Exact Solutions. *J. Fluid Mech.* **2015**, *767*, 52–64. [CrossRef]
64. Yeh, K.C.; Liu, C.H. Acoustic-Gravity Waves in the Upper Atmosphere. *Rev. Geophys.* **1974**, *12*, 193. [CrossRef]
65. Yang, S.-S.; Hayakawa, M. Gravity Wave Activity in the Stratosphere before the 2011 Tohoku Earthquake as the Mechanism of Lithosphere-Atmosphere-Ionosphere Coupling. *Entropy* **2020**, *22*, 110. [CrossRef] [PubMed]
66. Yang, S.; Asano, T.; Hayakawa, M. Abnormal Gravity Wave Activity in the Stratosphere Prior to the 2016 Kumamoto Earthquakes. *JGR Space Phys.* **2019**, *124*, 1410–1425. [CrossRef]
67. Kundu, S.; Chowdhury, S.; Ghosh, S.; Sasmal, S.; Politis, D.Z.; Potirakis, S.M.; Yang, S.-S.; Chakrabarti, S.K.; Hayakawa, M. Seismogenic Anomalies in Atmospheric Gravity Waves as Observed from SABER/TIMED Satellite during Large Earthquakes. *J. Sens.* **2022**, *2022*, 1–23. [CrossRef]
68. Hayakawa, M.; Kasahara, Y.; Nakamura, T.; Hobara, Y.; Rozhnoi, A.; Solovieva, M.; Molchanov, O.; Korepanov, V. Atmospheric Gravity Waves as a Possible Candidate for Seismo-Ionospheric Perturbations. *J. Atmos. Electr.* **2011**, *31*, 129–140. [CrossRef]
69. Zhang, Y.; Wang, T.; Chen, W.; Zhu, K.; Marchetti, D.; Cheng, Y.; Fan, M.; Wang, S.; Wen, J.; Zhang, D.; et al. Are There One or More Geophysical Coupling Mechanisms before Earthquakes? The Case Study of Lushan (China) 2013. *Remote Sens.* **2023**, *15*, 1521. [CrossRef]
70. Akhoondzadeh, M.; De Santis, A.; Marchetti, D.; Piscini, A.; Cianchini, G. Multi Precursors Analysis Associated with the Powerful Ecuador (MW = 7.8) Earthquake of 16 April 2016 Using Swarm Satellites Data in Conjunction with Other Multi-Platform Satellite and Ground Data. *Adv. Space Res.* **2018**, *61*, 248–263. [CrossRef]
71. Marchetti, D.; De Santis, A.; D'Arcangelo, S.; Poggio, F.; Piscini, A.; Campuzano, S.A.; De Carvalho, W.V.J.O. Pre-Earthquake Chain Processes Detected from Ground to Satellite Altitude in Preparation of the 2016–2017 Seismic Sequence in Central Italy. *Remote Sens. Environ.* **2019**, *229*, 93–99. [CrossRef]

72. Marchetti, D.; De Santis, A.; Shen, X.; Campuzano, S.A.; Perrone, L.; Piscini, A.; Di Giovambattista, R.; Jin, S.; Ippolito, A.; Cianchini, G.; et al. Possible Lithosphere-Atmosphere-Ionosphere Coupling Effects Prior to the 2018 Mw = 7.5 Indonesia Earthquake from Seismic, Atmospheric and Ionospheric Data. *J. Asian Earth Sci.* **2020**, *188*, 104097. [CrossRef]

73. De Santis, A.; Perrone, L.; Calcara, M.; Campuzano, S.A.; Cianchini, G.; D'Arcangelo, S.; Di Mauro, D.; Marchetti, D.; Nardi, A.; Orlando, M.; et al. A Comprehensive Multiparametric and Multilayer Approach to Study the Preparation Phase of Large Earthquakes from Ground to Space: The Case Study of the June 15 2019, M7.2 Kermadec Islands (New Zealand) Earthquake. *Remote Sens. Environ.* **2022**, *283*, 113325. [CrossRef]

74. Marchetti, D.; Zhu, K.; Marchetti, L.; Zhang, Y.; Chen, W.; Cheng, Y.; Fan, M.; Wang, S.; Wang, T.; Wen, J.; et al. Quick Report on the ML = 3.3 on 1 January 2023 Guidonia (Rome, Italy) Earthquake: Evidence of a Seismic Acceleration. *Remote Sens.* **2023**, *15*, 942. [CrossRef]

75. Rikitake, T. Classification of Earthquake Precursors. *Tectonophysics* **1979**, *54*, 293–309. [CrossRef]

76. Rikitake, T. Earthquake Precursors in Japan: Precursor Time and Detectability. *Tectonophysics* **1987**, *136*, 265–282. [CrossRef]

77. Scholz, C.H.; Sykes, L.R.; Aggarwal, Y.P. Earthquake Prediction: A Physical Basis. *Science* **1973**, *181*, 803–810. [CrossRef] [PubMed]

78. Scholz, C.H. Earthquakes and Friction Laws. *Nature* **1998**, *391*, 37–42. [CrossRef]

79. Console, R.; Yamaoka, K.; Zhuang, J. Implementation of Short- and Medium-Term Earthquake Forecasts. *Int. J. Geophys.* **2012**, *2012*, 1–2. [CrossRef]

80. Dobrovolsky, I.P.; Zubkov, S.I.; Miachkin, V.I. Estimation of the Size of Earthquake Preparation Zones. *Pageoph* **1979**, *117*, 1025–1044. [CrossRef]

81. Wiemer, S. A Software Package to Analyze Seismicity: ZMAP. *Seismol. Res. Lett.* **2001**, *72*, 373–382. [CrossRef]

82. Gelaro, R.; McCarty, W.; Suárez, M.J.; Todling, R.; Molod, A.; Takacs, L.; Randles, C.A.; Darmenov, A.; Bosilovich, M.G.; Reichle, R.; et al. The Modern-Era Retrospective Analysis for Research and Applications, Version 2 (MERRA-2). *J. Climate* **2017**, *30*, 5419–5454. [CrossRef] [PubMed]

83. De Santis, A.; Cianchini, G.; Marchetti, D.; Piscini, A.; Sabbagh, D.; Perrone, L.; Campuzano, S.A.; Inan, S. A Multiparametric Approach to Study the Preparation Phase of the 2019 M7.1 Ridgecrest (California, United States) Earthquake. *Front. Earth Sci.* **2020**, *8*, 540398. [CrossRef]

84. Marchetti, D.; Zhu, K.; Zhang, H.; Zhima, Z.; Yan, R.; Shen, X.; Chen, W.; Cheng, Y.; He, X.; Wang, T.; et al. Clues of Lithosphere, Atmosphere and Ionosphere Variations Possibly Related to the Preparation of La Palma 19 September 2021 Volcano Eruption. *Remote Sens.* **2022**, *14*, 5001. [CrossRef]

85. Akhoondzadeh, M.; De Santis, A.; Marchetti, D.; Shen, X. Swarm-TEC Satellite Measurements as a Potential Earthquake Precursor Together With Other Swarm and CSES Data: The Case of Mw7.6 2019 Papua New Guinea Seismic Event. *Front. Earth Sci.* **2022**, *10*, 820189. [CrossRef]

86. Marchetti, D.; De Santis, A.; Campuzano, S.A.; Soldani, M.; Piscini, A.; Sabbagh, D.; Cianchini, G.; Perrone, L.; Orlando, M. Swarm Satellite Magnetic Field Data Analysis Prior to 2019 Mw = 7.1 Ridgecrest (California, USA) Earthquake. *Geosciences* **2020**, *10*, 502. [CrossRef]

87. Jakowski, N.; Béniguel, Y.; De Franceschi, G.; Pajares, M.H.; Jacobsen, K.S.; Stanislawska, I.; Tomasik, L.; Warnant, R.; Wautelet, G. Monitoring, Tracking and Forecasting Ionospheric Perturbations Using GNSS Techniques. *J. Space Weather Space Clim.* **2012**, *2*, A22. [CrossRef]

88. Badeke, R.; Borries, C.; Hoque, M.M.; Minkwitz, D. Empirical Forecast of Quiet Time Ionospheric Total Electron Content Maps over Europe. *Adv. Space Res.* **2018**, *61*, 2881–2890. [CrossRef]

89. Hattori, K.; Serita, A.; Gotoh, K.; Yoshino, C.; Harada, M.; Isezaki, N.; Hayakawa, M. ULF Geomagnetic Anomaly Associated with 2000 Izu Islands Earthquake Swarm, Japan. *Phys. Chem. Earth Parts A/B/C* **2004**, *29*, 425–435. [CrossRef]

90. Pulinets, S.; Ouzounov, D. Lithosphere–Atmosphere–Ionosphere Coupling (LAIC) Model–An Unified Concept for Earthquake Precursors Validation. *J. Asian Earth Sci.* **2011**, *41*, 371–382. [CrossRef]

91. Pulinets, S.; Ouzounov, D.; Karelin, A.; Boyarchuk, K. Near-Ground Processes as a Result of Air Ionization. In *Earthquake Precursors in the Atmosphere and Ionosphere: New Concepts*; Pulinets, S., Ouzounov, D., Karelin, A., Boyarchuk, K., Eds.; Springer: Dordrecht, The Netherlands, 2022; pp. 1–60. ISBN 978-94-024-2172-9.

92. Piscini, A.; De Santis, A.; Marchetti, D.; Cianchini, G. A Multi-Parametric Climatological Approach to Study the 2016 Amatrice–Norcia (Central Italy) Earthquake Preparatory Phase. *Pure Appl. Geophys.* **2017**, *174*, 3673–3688. [CrossRef]

93. Liu, S.; Cui, Y.; Wei, L.; Liu, W.; Ji, M. Pre-Earthquake MBT Anomalies in the Central and Eastern Qinghai-Tibet Plateau and Their Association to Earthquakes. *Remote Sens. Environ.* **2023**, *298*, 113815. [CrossRef]

94. Freund, F. Pre-Earthquake Signals: Underlying Physical Processes. *J. Asian Earth Sci.* **2011**, *41*, 383–400. [CrossRef]

95. Freund, F.; Ouillon, G.; Scoville, J.; Sornette, D. Earthquake Precursors in the Light of Peroxy Defects Theory: Critical Review of Systematic Observations. *Eur. Phys. J. Spec. Top.* **2021**, *230*, 7–46. [CrossRef]

96. Politis, D.Z.; Potirakis, S.M.; Kundu, S.; Chowdhury, S.; Sasmal, S.; Hayakawa, M. Critical Dynamics in Stratospheric Potential Energy Variations Prior to Significant (M > 6.7) Earthquakes. *Symmetry* **2022**, *14*, 1939. [CrossRef]

97. John, P.C.; Sandy, D.; Robbie, B. *Hurricane Elsa*; National Hurricane Center Tropical Cyclone Report; National Hurricane Center: Miami, FA, USA, 2022.

98. Robbie, B. *Tropical Storm Fred*; National Hurricane Center Tropical Cyclone Report; National Hurricane Center: Miami, FA, USA, 2021.

99. Brad, J.R.; Amanda, R.; Robbie, B. *Hurricane Grace*; National Hurricane Center Tropical Cyclone Report; National Hurricane Center: Miami, FA, USA, 2022.
100. Alfaro, S.; Sen-Crowe, B.; Autrey, C.; Elkbuli, A. Trends in Carbon Monoxide Poisoning Deaths in High Frequency Hurricane States from 2014–19: The Need for Prevention Intervention Strategies. *J. Public Health* **2023**, *45*, e250–e259. [CrossRef] [PubMed]
101. Subramanian, R.; Ellis, A.; Torres-Delgado, E.; Tanzer, R.; Malings, C.; Rivera, F.; Morales, M.; Baumgardner, D.; Presto, A.; Mayol-Bracero, O.L. Air Quality in Puerto Rico in the Aftermath of Hurricane Maria: A Case Study on the Use of Lower Cost Air Quality Monitors. *ACS Earth Space Chem.* **2018**, *2*, 1179–1186. [CrossRef]
102. Chiodini, G.; Cardellini, C.; Amato, A.; Boschi, E.; Caliro, S.; Frondini, F.; Ventura, G. Carbon Dioxide Earth Degassing and Seismogenesis in Central and Southern Italy: Carbon Dioxide Earth Degassing and Seismogenesis. *Geophys. Res. Lett.* **2004**, *31*, L07615. [CrossRef]
103. Chiodini, G.; Cardellini, C.; Di Luccio, F.; Selva, J.; Frondini, F.; Caliro, S.; Rosiello, A.; Beddini, G.; Ventura, G. Correlation between Tectonic $CO_2$ Earth Degassing and Seismicity Is Revealed by a 10-Year Record in the Apennines, Italy. *Sci. Adv.* **2020**, *6*, eabc2938. [CrossRef] [PubMed]
104. Etiope, G.; Martinelli, G. Migration of Carrier and Trace Gases in the Geosphere: An Overview. *Phys. Earth Planet. Inter.* **2002**, *129*, 185–204. [CrossRef]
105. Kuo, C.L.; Lee, L.C.; Huba, J.D. An Improved Coupling Model for the Lithosphere-Atmosphere-Ionosphere System. *J. Geophys. Res. Space Phys.* **2014**, *119*, 3189–3205. [CrossRef]
106. Mirzaee, A.; Estekanchi, H.E.; Vafai, A. Improved Methodology for Endurance Time Analysis: From Time to Seismic Hazard Return Period. *Sci. Iran.* **2012**, *19*, 1180–1187. [CrossRef]
107. Pulinets, S.; Krankowski, A.; Hernandez-Pajares, M.; Marra, S.; Cherniak, I.; Zakharenkova, I.; Rothkaehl, H.; Kotulak, K.; Davidenko, D.; Blaszkiewicz, L.; et al. Ionosphere Sounding for Pre-Seismic Anomalies Identification (INSPIRE): Results of the Project and Perspectives for the Short-Term Earthquake Forecast. *Front. Earth Sci.* **2021**, *9*, 610193. [CrossRef]
108. Pulinets, S.A.; Ouzounov, D. Multi-Instrument Observations and Validation of Laic. American Geophysical Union, Fall Meeting 2014, abstract id NH21C-03. 2014. Available online: https://www.researchgate.net/publication/283515340_Multi-Instrument_Observations_and_Validation_of_Laic (accessed on 29 March 2024).
109. Tramutoli, V.; Marchese, F.; Falconieri, A.; Filizzola, C.; Genzano, N.; Hattori, K.; Lisi, M.; Liu, J.-Y.; Ouzounov, D.; Parrot, M.; et al. Tropospheric and Ionospheric Anomalies Induced by Volcanic and Saharan Dust Events as Part of Geosphere Interaction Phenomena. *Geosciences* **2019**, *9*, 177. [CrossRef]
110. Wu, L.; Qi, Y.; Mao, W.; Lu, J.; Ding, Y.; Peng, B.; Xie, B. Scrutinizing and Rooting the Multiple Anomalies of Nepal Earthquake Sequence in 2015 with the Deviation–Time–Space Criterion and Homologous Lithosphere–Coversphere–Atmosphere–Ionosphere Coupling Physics. *Nat. Hazards Earth Syst. Sci.* **2023**, *23*, 231–249. [CrossRef]
111. Ben-Zion, Y.; Zaliapin, I. Localization and Coalescence of Seismicity before Large Earthquakes. *Geophys. J. Int.* **2020**, *223*, 561–583. [CrossRef]
112. Console, R.; Carluccio, R.; Papadimitriou, E.; Karakostas, V. Synthetic Earthquake Catalogs Simulating Seismic Activity in the Corinth Gulf, Greece, Fault System: Corinth Earthquakes Simulations. *J. Geophys. Res. Solid Earth* **2015**, *120*, 326–343. [CrossRef]
113. Parrot, M.; Li, M. Demeter Results Related to Seismic Activity. *URSI Radio Sci. Bull.* **2015**, *2015*, 18–25. [CrossRef]
114. Gutenberg, B. *Seismicity of the Earth and Associated Phenomena*; Read Books Ltd.: Redditch, UK, 2013.
115. Gardner, J.K.; Knopoff, L. Is the Sequence of Earthquakes in Southern California, with Aftershocks Removed, Poissonian? *Bull. Seismol. Soc. Am.* **1974**, *64*, 1363–1367. [CrossRef]
116. Peresan, A.; Gorshkov, A.; Soloviev, A.; Panza, G.F. The Contribution of Pattern Recognition of Seismic and Morphostructural Data to Seismic Hazard Assessment. *Bull. Geophys. Oceanogr.* **2015**, *56*, 295–328. [CrossRef]
117. Panza, G.F.; Bela, J. NDSHA: A New Paradigm for Reliable Seismic Hazard Assessment. *Eng. Geol.* **2020**, *275*, 105403. [CrossRef]

*geosciences*

*Article*

# The Lithosphere-Atmosphere-Ionosphere Coupling of Multiple Geophysical Parameters Approximately 3 Hours Prior to the 2022 M6.8 Luding Earthquake

Chieh-Hung Chen [1,*], Shengjia Zhang [2], Zhiqiang Mao [1], Yang-Yi Sun [1], Jing Liu [3], Tao Chen [4], Xuemin Zhang [3], Aisa Yisimayili [5], Haiyin Qing [6], Tianya Luo [7], Yongxin Gao [8] and Fei Wang [9]

[1] School of Geophysics and Geomatics, China University of Geosciences, Wuhan 430074, China; jaycobmao@cug.edu.cn (Z.M.); sunyy@cug.edu.cn (Y.-Y.S.)
[2] College of Computer Science and Cyber Security, Chengdu University of Technology, Chengdu 610059, China; sjzhang@cdut.edu.cn
[3] Institute of Earthquake Forecasting, China Earthquake Administration, Beijing 100036, China; liujing@ief.ac.cn (J.L.); xm@ief.ac.cn (X.Z.)
[4] National Space Science Center, Chinese Academy of Sciences, Beijing 100049, China; tchen@nssc.ac.cn
[5] Earthquake Agency of Xinjiang Uygur Autonomous Region, Urumqi 830011, China; aisa@cug.edu.cn
[6] School of Electronic Information and Artificial Intelligence, Leshan Normal University, Leshan 614000, China; qinghaiyin123@whu.edu.cn
[7] Guangxi Transportation Science and Technology Group Co., Ltd., Nanning 530007, China; luotianya@cug.edu.cn
[8] Institute of Applied Mechanics, School of Civil Engineering, Hefei University of Technology, Hefei 230009, China; gaoyx@hfut.edu.cn
[9] School of Environment and Civil Engineering, Chengdu University of Technology, Chengdu 610059, China; wangfei19@cdut.edu.cn
* Correspondence: zjq02010@cug.edu.cn or nononochchen@gmail.com

**Citation:** Chen, C.-H.; Zhang, S.; Mao, Z.; Sun, Y.-Y.; Liu, J.; Chen, T.; Zhang, X.; Yisimayili, A.; Qing, H.; Luo, T.; et al. The Lithosphere-Atmosphere-Ionosphere Coupling of Multiple Geophysical Parameters Approximately 3 Hours Prior to the 2022 M6.8 Luding Earthquake. *Geosciences* **2023**, *13*, 356. https://doi.org/10.3390/geosciences13120356

Academic Editors: Masashi Hayakawa and Jesus Martinez-Frias

Received: 6 October 2023
Revised: 18 November 2023
Accepted: 20 November 2023
Published: 21 November 2023

**Abstract:** Investigating various geophysical parameters from the Earth's crust to the upper atmosphere is considered a promising approach for predicting earthquakes. Scientists have observed that changes in these parameters can occur days to months before earthquakes. Understanding and studying the impending abnormal phenomena that precede earthquakes is both urgent and challenging. On 5 September 2022, a magnitude 6.8 earthquake occurred in Sichuan, China, at 4:52:18 (Universal Time). The earthquake happened approximately 175 km away from an instrumental array established in 2021 for monitoring vibrations and perturbations in the lithosphere, atmosphere, and ionosphere (MVP-LAI). This array consisted of over 15 instruments that regularly monitor changes in various geophysical parameters from the subsurface up to an altitude of approximately 350 km in the ionosphere. Its purpose was to gain insights into the mechanisms behind the coupling of these different geospheres during natural hazards. The seven geophysical parameters from the MVP-LAI system simultaneously exhibited abnormal behaviors approximately 3 h before the Luding earthquake. These parameters include ground tilts, air pressure, radon concentration, atmospheric vertical electric field, geomagnetic field, wind field, and total electron content. The abnormal increase in radon concentration suggests that the chemical channel could be a promising mechanism for the coupling of geospheres. Furthermore, air pressure, the geomagnetic field, and total electron content exhibited abnormal characteristics with similar frequencies. Horizontal wind experienced temporary cessation or weakening, while vertical wind displayed frequent reversals. These anomalies may be attributed to atmospheric resonance before the earthquake. The results demonstrate that the coupling of geospheres, as indicated by the anomalous phenomena preceding an earthquake, could be influenced by multiple potential mechanisms. The multiple anomalies observed in this study provided approximately 3 h of warning for people to prepare for the seismic event and mitigate hazards.

**Keywords:** lithosphere–atmosphere–ionosphere coupling; pre-earthquake anomaly phenomena; MVL-LAI system; Luding earthquake

## 1. Introduction

Investigations into multiple anomalous phenomena in the lithosphere–atmosphere–ionosphere (LAI) coupling before major earthquakes are considered an important aspect of predicting earthquake occurrences [1–6]. While earthquake prediction remains a significant challenge, integrating various geophysical parameters can help us understand the evolution of pre-earthquake anomalous phenomena and uncover the causal mechanisms involved [7]. Typically, the evolution of pre-earthquake anomalous phenomena starts with crustal deformation and seismicity, and culminates in electromagnetic anomalies near the Earth's surface [7] and total electron contents (TECs) in the ionosphere [8–10]. Here, we take the 1999 M7.6 Chi-Chi earthquake and the 2011 M9.0 Tohoku-Oki earthquake as examples to demonstrate the evaluation of pre-earthquake anomalies. Anomalous surface deformation, which triggers unusual changes in groundwater levels [11], can be observed, starting approximately one year before the Chi-Chi earthquake [12]. Groundwater levels in the footwall of the Chelungpu fault began to change accordingly and remained at a low stage for around three months (from five months to two months before the Chi-Chi earthquake). Two months before the Chi-Chi earthquake, groundwater levels significantly increased due to noticeable compression stress loading in the Earth's crust [12]. Meanwhile, the stress loading in the crust near the epicenter caused variations in the magnetic field's daily pattern [13], as demonstrated by raw data showing several strong magnetic disturbances [14]. Additionally, a significant decrease in total electron contents (TECs) was observed over a wide area covering the epicenter approximately two to five days before the earthquake [15]. Similarly, researchers such as Chen et al. [16], Han et al. [17], Orihara et al. [18], Varotsos et al. [19], and Sarlis et al. [20] observed abnormal surface movements, changes in groundwater levels, diurnal variations in the geomagnetic field, and unusual seismic activity and slow slips approximately two months before the Tohoku-Oki earthquake, respectively. Abnormal crustal deformation was noted by Chen et al. [16] about 1–1.5 months before the earthquake, along with abnormal changes in the foreshock sequence, slow slip events, and ground motion reported by Kato et al. [21], Ito et al. [22], and Hattori et al. [23], respectively. Moreover, seismo-TEC anomalies were observed four days before the Tohoku-Oki earthquake [24]. Heki [25] found TEC enhancements approximately 40 min before earthquake occurrences. These anomalies in seismo-crustal deformation, groundwater levels, and seismic activity often act as early indicators of major earthquakes. The TEC anomalies typically emerge 2–5 days before the earthquake, surpassing these other anomalies. On the other hand, numerous previous studies have revealed statistical anomalies in the ionosphere within approximately 2 weeks and a few days before the occurrence of earthquakes by utilizing the nighttime fluctuation method and the terminator time method [26–30]. However, it is important to note that most pre-earthquake anomalies span from days [8,9] to months [31–33]. Detecting impending anomalies associated with earthquakes remains a significant challenge that requires immediate attention.

In 2021, an instrumental array was established in Leshan, Sichuan, China (103.91° E, 29.60° N) to address one of the main challenges in detecting impending anomalies related to earthquakes [34]. The instrumental array consists of more than 15 distinct instruments that monitor over 15 different geophysical parameters, ranging from the subsurface to the ionosphere (for more details, refer to Chen et al. [34]). These instruments are primarily installed within a 20 m × 20 m area to monitor vibrations and perturbations in the lithosphere, atmosphere, and ionosphere (MVP-LAI), with a predominant focus on vertical propagation. It is important to note that geostationary satellites operated by the BeiDou Navigation Satellite System serve the MVP-LAI system for continuously monitoring changes in TECs at fixed ionospheric pierce points (IPPs) over specific locations on the Earth's surface. To monitor TEC changes across the MVP-LAI system, ground-based GNSS (Global Navigation Satellite System) receivers were installed at two substations: YADU (103.4° E, 31.9° N) and CAXI (105.8° E, 31.8° N), located approximately 200 km north of the MVP-LAI system. The IPPs for the YADU and CAXI substations, corresponding to the BeiDou geostationary satellites (G3 and G2), are positioned directly over the MVP-LAI system at an altitude of approximately 350 km (Figure 1). Thus, changes in the ionosphere over the MVP-LAI system are monitored by calculating TEC values from electromagnetic signals transmitted by the associated BeiDou geostationary satellites [35,36] received by ground-based GNSS stations at YADU and CAXI. The MVP-LAI system detected resident signals prior to the Maduo and Yangbi earthquakes [37] and observed interactions between the lithosphere, atmosphere, and ionosphere associated with the Lamb waves triggered by the Tonga volcano eruption [38]. Furthermore, Chen et al. [39] identified and proposed four distinct coupling and/or interactions between the LAI triggered by the Lamb waves associated with the Tonga volcano eruption. Although the MVP-LAI system is insufficient for monitoring changes in Very Low-Frequency (VLF) data [40], the observational results related to earthquakes and the Tonga volcano eruption have demonstrated the capability of the MVP-LAI system to monitor phenomena during natural hazards.

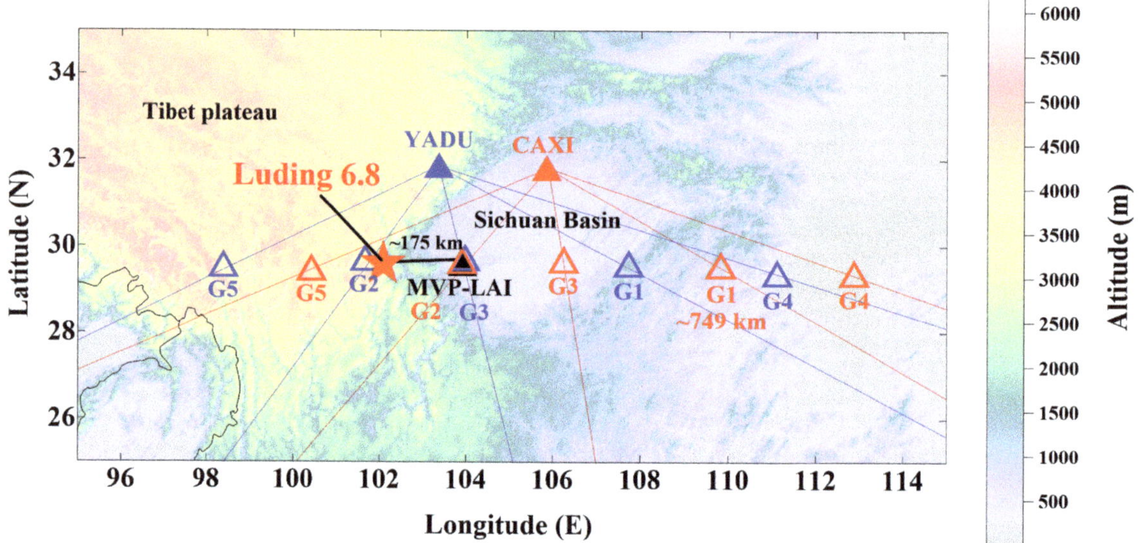


**Figure 1.** The locations of various observational points related to the 2022 M6.8 Luding earthquake. The red star represents the epicenter of the Luding earthquake. The black triangle indicates the instrumental array for monitoring vibrations and perturbations in the lithosphere, atmosphere, and ionosphere (MVP-LAI). The blue and red solid triangles denote the YADU and CAXI ground-based GNSS (Global Navigation Satellite System) stations, respectively. The blue and red open triangles show the ionospheric pierce points (IPPs) corresponding to the BeiDou geostationary (G1–G5) satellites. The approximate distances from the MVP-LAI system and the IPPs on the CAXI-G1 route to the epicenter of the Luding earthquake are 175 km and 749 km, respectively.

On 5 September 2022, at 4:52:18 Universal Time (UT), an earthquake, 6.8 in magnitude, struck near Luding County in Sichuan, China, at the coordinates 102.08° E, 29.59° N [41,42]. The earthquake reached a ground intensity level of nine, resulting in the loss of 97 lives, with 20 people reported as missing. In a study conducted by Liu et al. [43], various geophysical parameters in the ionosphere, infrared radiation, atmospheric electrostatic field, and hot spring ions in the seismogenic region were examined. Pre-earthquake anomalies were observed in the lithosphere–atmosphere–ionosphere coupling approximately 10 days before the Luding earthquake. However, the specific impending anomalies prior to the Luding earthquake remain unclear. The estimated radius of the earthquake preparation zone, calculated using a formula proposed in Dobrovolsky et al. [44], is approximately 840 km. The epicenter of the earthquake is approximately 175 km away from the MVP-

LAI system. This suggests that the MVP-LAI system has a high potential for detecting multiple anomalous phenomena before an earthquake occurs. As part of this study, data on ground tilts, air pressure, radon concentration, atmospheric vertical electric field near the Earth's surface, wind field up to an altitude of approximately 5000 m, magnetic field corresponding to the ionospheric current at around 100 km altitude, and TECs from the YADU and CAXI substations at approximately 350 km altitude were collected from 4 to 5 September 2022. Since there was no precipitation during this period, data from the rain gauge in the MVP-LAI system were not included.

## 2. Materials and Methods

Daily variations in data from the MVP-LAI system have been routinely shown at http://geostation.top (accessed on 4 October 2023; please also visit the website for references). Data from 4 September to 5 September 2022 were retrieved to illustrate various factors, including the atmospheric vertical electric field near the Earth's surface (Figure 2a,b), radon concentration (Figure 2c,d), air pressure (Figure 2e,f), and ground tilts (Figure 2g–j). Here, we examine this diverse dataset, spanning from the lithosphere to the ionosphere, in order to investigate potential abnormal phenomena preceding the Luding earthquake. It is important to note that the tilts embedded in the magnetometers in the MVP-LAI system were used as seismic data in this study due to the seismometers breaking down during the Luding earthquake. Significant ground tilts were observed at 4:52:48 UT in the tilt data shown in Figure 2g–j. These vibrations were triggered by the seismic waves of the Luding earthquake and appeared approximately 30 s later. The velocity of the ground vibration propagation was approximately ~5.8 (=175/30) km/s, corresponding to the seismic primary waves. Around three hours before the Luding earthquake, there was a temporal change in the east-downward tilts, with persistent east-downward declinations shifting to east-upward declinations (Figure 2i,j). Roughly one hour after the earthquake, the east-upward declination reverted back to east-downward declinations. Similar changes were observed in the north-downward tilts, where the steady north-downward declination temporarily shifted to north-upward declination from three hours to two hours before the earthquake (Figure 2g,h).

Barometers monitor changes in air pressure near the Earth's surface with a sampling interval of 2 s. Figure 2e illustrates air pressure changes with semi-diurnal variations. Additionally, Figure 2f shows a small enhancement coinciding with the semi-diurnal variations around 1:56 UT, with an amplitude of approximately 0.3 hPa. The emanometer records variations in air radon with a sampling interval of 10 min. The radon concentration (Figure 2c) remained low, between 0 Bq/m$^3$ and 30 Bq/m$^3$, but notably increased to around 50 Bq/m$^3$ near 1:50 UT (Figure 2d). Vertical changes in the atmospheric electric field near the Earth's surface are monitored using the atmospheric electric meter with a sampling interval of 1 s. Significant drops in the atmospheric electric field occurred around 3:00 UT on 4 September 2022, and 1:30 UT on 5 September 2022 (Figure 2a). No abnormal phenomenon was observed in the other geophysical parameters around 3:00 UT on 4 September 2022. However, a drop in the atmospheric electric field, corresponding to the anomalous changes in the other geophysical parameters around 1:50 UT, can be seen (Figure 2b). In summary, abnormal changes were detected in multiple geophysical parameters near the Earth's surface in the MVP-LAI system at approximately 1:50 UT, approximately 3 h before the Luding earthquake.

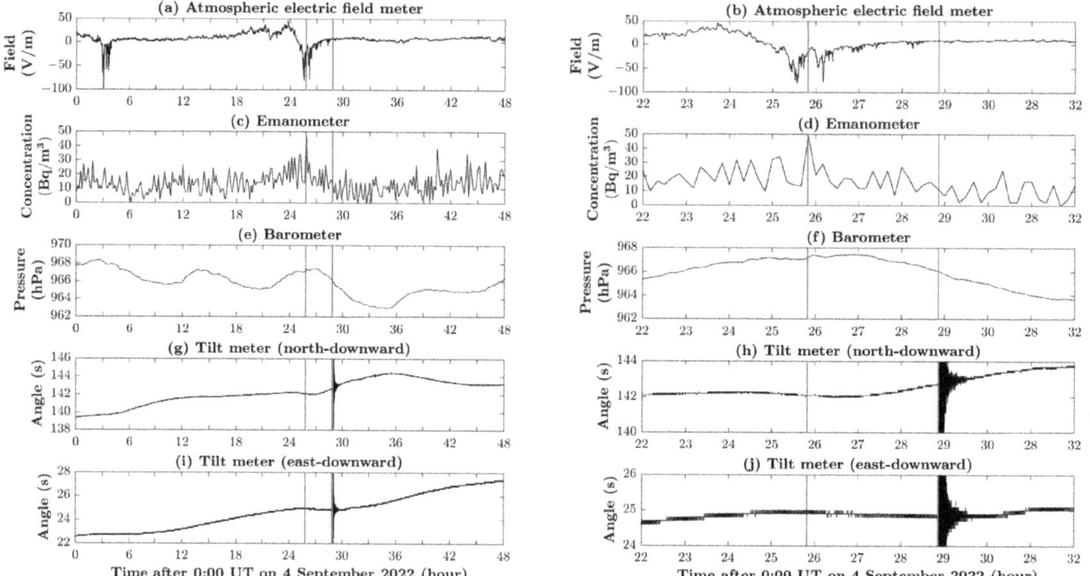

**Figure 2.** The changes in five geophysical parameters from the MVP-LAI system between the 4 and 5 September 2022. The variations in the atmosphere's electric field are depicted in (**a**,**b**), while the changes in radon concentration are shown in (**c**,**d**). The alterations in air pressure are displayed in (**e**,**f**), and the ground tilts for the north downward are illustrated in (**g**,**h**). Furthermore, the ground tilts for the east downward can be seen in (**i**,**j**). The changes in geophysical parameters over a time span from the 4 to the 5 September 2022 are shown in (**a**,**c**,**e**,**g**,**i**). Additionally, the changes in geophysical parameters over a shorter time span from 22:00 UT on the 4 September 2022 to 8:00 UT on the 5 September 2022 are shown in (**b**,**d**,**f**,**h**,**j**). The red vertical lines indicate the occurrence of the Luding earthquake, while the blue lines highlight the abnormal changes in geophysical parameters occurring three hours prior to the earthquake.

Figure 3 illustrates changes in the wind field, geomagnetic field, and TECs during the Luding earthquake. The radar wind profile in the MVP-LAI system monitors the three-dimensional wind field from the Earth's surface up to an altitude of approximately 5000 m, with a 2 min sampling interval. Between approximately 12:00 UT and 23:00 UT on 4 September 2022, an intense horizontal wind with a speed of around 15 m/s was observed at altitudes ranging from approximately 2000 m to 4000 m (Figure 3c). After approximately 23:00 UT on 4 September 2022, and until around 13:00 UT on 5 September 2022, this intense horizontal wind either ceased or weakened. However, after 13:00 UT on 5 September 2022, the horizontal wind reversed from weakening to intensifying, with a speed of approximately 10 m/s at altitudes ranging from approximately 3000 m to 4000 m. Simultaneously, the vertical wind frequently changed direction, exhibiting intense upward wind speeds exceeding approximately 3 m/s when the intense horizontal wind either ceased or weakened (Figure 3d). These intense upward winds coincided with abnormal tilt changes, indicating that the wind field was affected by abnormal ground vibrations over a wide area. A magnetometer monitored changes in the geomagnetic field at the north–south, east–west, and vertical components with a 0.1 s sampling interval. Changes in the north–south and east–west components correspond to ionospheric currents at an altitude of approximately 100 km above the Earth's surface [45]. Variations in the vertical geomagnetic field are attributed to underground currents [46,47]. No significant abnormal phenomena were detected in the north–south and vertical geomagnetic field. In contrast, periodic magnetic disturbances with an amplitude of approximately 10 nT were observed from around 0:00 UT to 3:30 UT on 5 September 2022 (Figure 3b). TEC variations over the MVP-LAI system were obtained from ground-based GNSS receivers using electromagnetic signals transmitted from the BeiDou geostationary G3 satellites. To remove long-term background variations (e.g., diurnal variation) in TECs, a moving average computed with a 1 h window was subtracted, resulting in residual TEC (dTEC) values. The dTEC variations with an amplitude of approximately 1.5 TECU (TEC unit = $10^{16}$ el/m$^2$) were observed around the same time as abnormal ground vibrations (Figure 2g–j), perturbations in air pressure (Figure 2e,f), increases in radon concentration (Figure 2c,d), decreases in atmospheric electric field (Figure 2a,b), mitigation of horizontal wind (Figure 3c), enhancements of upward wind (Figure 3d), and periodic magnetic perturbations (Figure 3b). This suggests that dTECs in the ionosphere may have been influenced by abnormal ground vibrations approximately 3 h before the Luding earthquake.

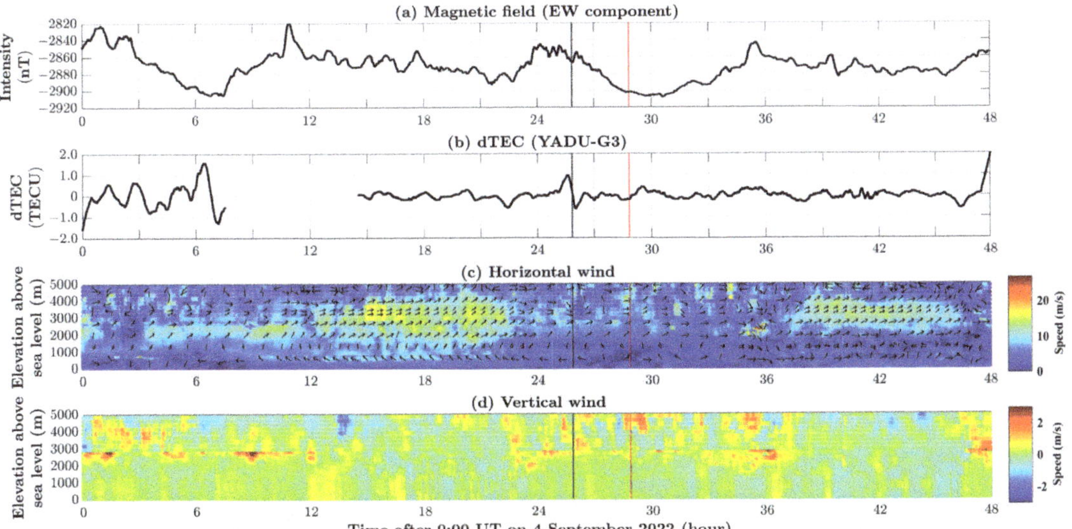

**Figure 3.** The changes in the magnetic field at the east–west component, residual total electron content (dTEC) at the ionospheric pierce point (IPP) for the YADU-G3 route, and wind field from the MVP-LAI system between the 4 and 5 September 2022. The variations in the magnetic field at the east–west component are presented in (**a**). The residual TEC (dTEC) at the IPP for the YADU-G3 route is displayed in (**b**). The horizontal and vertical winds from the surface up to an altitude of 5000 m can be seen in (**c,d**), respectively. In (**c**), the arrows indicate the azimuths of the wind. The red vertical lines signify the occurrence of the Luding earthquake, while the black lines highlight the abnormal changes in geophysical parameters occurring three hours before the earthquake.

## 3. Discussion

Figures 2 and 3 depict abnormal changes in multiple geophysical parameters at approximately 1:50 UT, which occurred about 3 h before the Luding earthquake. In addition, we gathered TEC data from the YADU and CAXI stations to investigate the spatial distribution of TEC abnormal changes for further clarifying the potential source locations. These data were obtained through the reception of electromagnetic signals emitted by the BeiDou geostationary satellites (G1, G2, G3, and G5) using the standard method [37,48]. The assumed altitude of the IPPs for the collected TEC data is 350 km above the Earth's surface. Figure 1 shows the approximate latitude alignment of the IPPs at 29.6° N, which is due to the geostationary satellites being stationary and positioned around the equator in space. However, it is worth noting that a series of discontinuities was observed in the electromagnetic signals emitted by the BeiDou G4 satellite. As a result, the TEC data from the BeiDou G4 satellite were not considered in this study.

The dTEC values shown in Figure 4 were computed using TEC data collected from the YADU and CAXI stations, employing the same method of removing a 1 h moving average. Generally, dTEC values from both stations exhibited synchronized changes. This synchronization can be attributed to the fact that the IPPs associated with the TEC measurements were located within a radius of approximately 850 km. Notably, abnormal changes occurring 3 h prior to the Luding earthquake were consistently observed throughout the entire dTEC dataset, except for the dTEC derived from signals emitted by the G1 satellite at the CAXI station. The delayed abnormal changes in dTEC for the CAXI-G1 route compared to others suggest that the sources of TEC perturbations may exist on the west side of this route. Additionally, the IPP for the CAXI-G1 route was approximately 749 km away from the epicenter, a distance that roughly aligns with the earthquake preparation zone proposed in Dobrovolsky et al. [44].

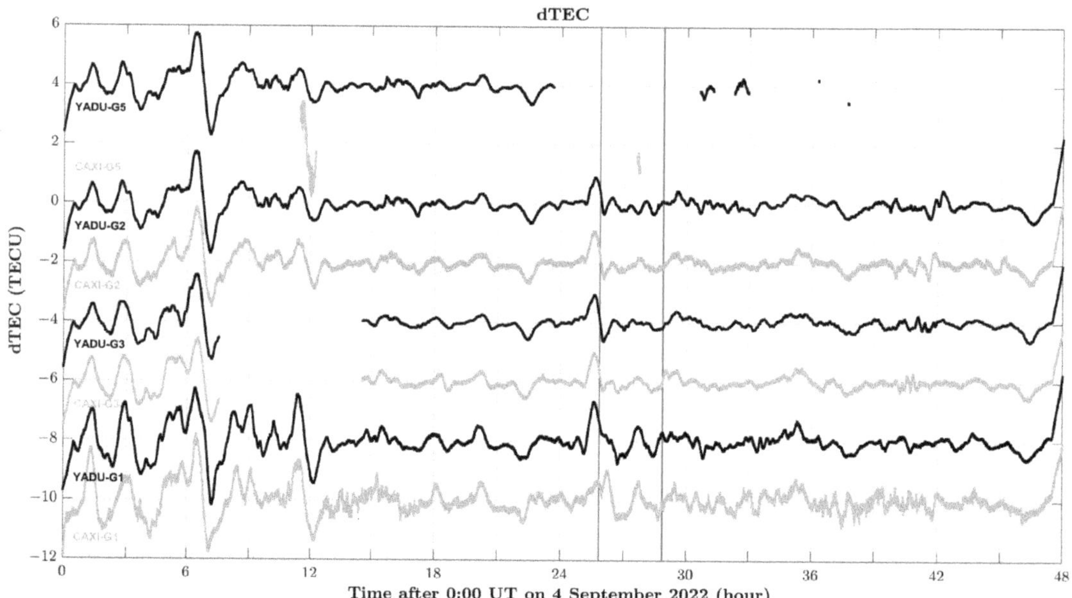


**Figure 4.** The changes in residual total electron content (dTEC) at the ionospheric pierce points (IPPs) for ground-based GNSS receivers at the YADU and CAXI substations corresponding to the BeiDou geostationary satellites (i.e., G1, G2, G3, and G) between the 4 and 5 September 2022. The residual TEC (dTEC) at the IPPs, as depicted in Figure 1, for ground-based GNSS receivers at the YADU and CAXI substations corresponding to the BeiDou geostationary satellites, are represented by the black and gray curves. The routes (station code–satellite code) are denoted below the curves with the same color. The red vertical line indicates the occurrence of the Luding earthquake, while the blue line highlights the abnormal changes in other geophysical parameters obtained three hours before the earthquake.

The second question pertains to the relatively small abnormal signals observed in the geomagnetic field and air pressure, which may not, by themselves, be convincing. To further analyze these signals, we employed the Morlet wavelet transform [49] to convert the magnetic and air pressure data into the frequency domain shown in Figure 5. We examined the magnetic field, specifically the north–south component (Figure 5a) and the vertical component (Figure 5c), no significant enhancements were detected around 3 h before the Luding earthquake that numerous anomalies were observed. Conversely, in the east–west component, enhancements with a period of 14–32 min were noted from 6 h to 3.5 h before the earthquake (Figure 5b). These enhancements temporarily subsided and then reversed around 3 h before the earthquake. The period of the secondary enhancements was similar to the first but slightly longer (16–32 min). In the air pressure data, Figure 5d displays a clear enhancement with a period of approximately 16–40 min around 3 h before the earthquake. The presence of these enhancements in both air pressure and the east–west component of the geomagnetic field, occurring at the same time and with similar periods, suggests that they may have a common underlying physical mechanism. These abnormal changes in air pressure and the geomagnetic field at the east-west component could provide valuable clues for identifying other abnormal parameters.

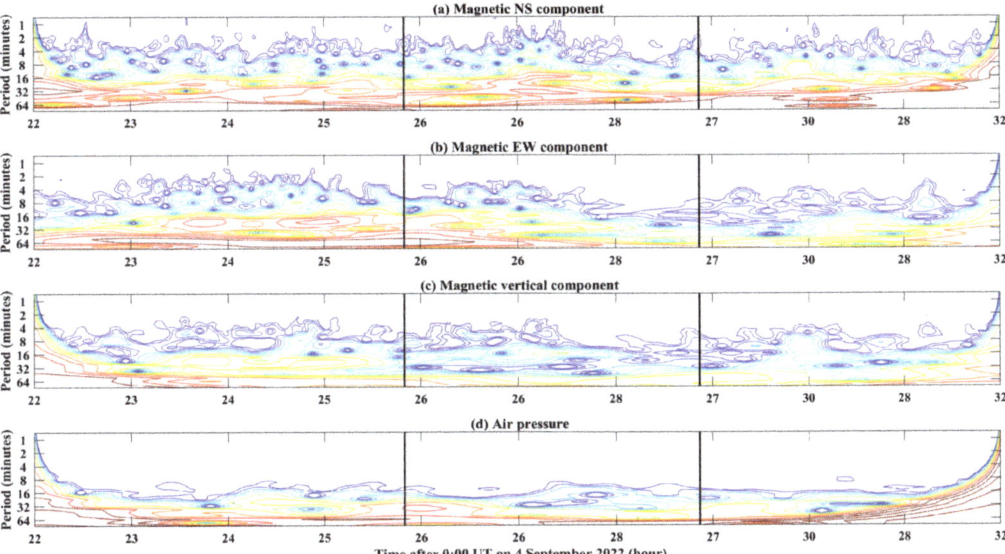

**Figure 5.** The time-period–energy distribution of the geomagnetic field and the air pressure from the MVP-LAI system between the 4 and 5 September 2022. The geomagnetic and air pressure data, obtained from the MVP-LAI system, are transformed into the frequency domain using the Morlet wavelet transform [44]. The color changing from blue to red suggests that the energy goes from weak to strong. Panels (**a–c**) display the time-period-energy distribution of the geomagnetic field for the north-south, east-west, and vertical components, respectively. Panel (**d**) presents the time-period-energy distribution of the air pressure data. The black vertical lines indicate occurrences three hours before the Luding earthquake, which were determined based on abnormal changes in other geophysical parameters and the earthquake occurrence time.

Four potential channels have been proposed to explain the coupling between the lithosphere, atmosphere, and ionosphere for multiple geophysical parameters before earthquakes in Hayakawa [50,51]. These channels include the chemical channel, conductivity channel, acoustic-gravity channel, and electromagnetic channel. In the chemical channel, the concentration of radon plays a key role in triggering anomalies in various geophysical parameters [52–55]. The decay of radon radiation leads to electromagnetic anomalies. The conductivity channel is associated with an increase in conductivity, which can result in more upward lightning strikes, causing heating in the ionosphere [56–60]. The acoustic-gravity channel involves thermal anomalies and pre-earthquake ground vibrations that generate upward acoustic-gravity waves, leading to changes in TECs in the ionosphere [61–69]. In the electromagnetic channel, electromagnetic waves emitted from the lithosphere can modify the ionosphere [70–73].

Figure 2c,d show an abnormal increase in radon concentration approximately 3 h before the Luding earthquake. This suggests that the chemical channel dominates the seismo-anomalies in multiple geophysical parameters. Additionally, the decay of radon radiation leads to a decrease in the atmospheric electric field, as demonstrated in Figure 2a,b. This observation coincides with the statistical results in Chen et al. [74], which indicate that negative abnormal signals in the atmospheric electric field usually appear 2–48 h before major earthquakes. However, the periodic changes in atmospheric pressure, electromagnetic field, and TEC anomalies cannot be fully explained by the chemical channel alone.

Chen et al. [39] observed that Lamb waves changed air pressure by approximately 1 hPa, causing variations of about 10 nT in the geomagnetic field and approximately 2 TECU in the TEC due to the upward propagation of the acoustic waves during the Tonga volcano eruption. The perturbations of air pressure, magnetic field, and TEC share a frequency at around 0.002 Hz. Meanwhile, once the coupling is dominated by the acoustic-gravity waves, abnormal changes in the lithosphere can lead to corresponding changes in the ionosphere within a few minutes to hours [75].

Abnormal changes in the amplitudes of air pressure, magnetic field, and TEC are approximately 0.3 hPa, 10 nT, and 1.5 TECU in this study, respectively. These abnormal changes are comparable to those observed during the Tonga volcano eruption. Abnormal changes in the geomagnetic field components and air pressure exhibit frequency characteristics of 0.0005–0.001 Hz, approximately corresponding to periods of 15–30 min in Figure 5b,d. The TEC variations, with a clear amplitude of about 1.5 TECU (TEC unit) approximately 3 h before the earthquake, have a period of approximately 30 min. This suggests that the coupling between air pressure near the Earth's surface, the magnetic field at the east–west component influenced by ionospheric currents at around 100 km altitude, and TEC at approximately 350 km altitude may be dominated by the acoustic-gravity channel due to the observed abnormal amplitudes and frequency characteristics.

Nevertheless, once the coupling is dominated by the acoustic-gravity waves, it contradicts the observations in this study that abnormal changes in the lithosphere, atmosphere, and ionosphere exhibit no significant time difference. Although the frequencies were slightly lower than the atmospheric resonance frequency, which is around 0.005 Hz from the surface to an altitude of 500 km [76], the lack of significant time discrepancy between these anomalies suggests a promising mechanism of double resonance [77]. This mechanism implies that a resonance in the ground [32] triggers a resonance in the atmosphere before major earthquakes. Additionally, Figure 3c,d show that the horizontal wind temporarily decreases, and the vertical wind frequently reverses during the Luding earthquake. The frequent reversal of the vertical wind replaces the horizontal winds, providing further evidence of atmospheric resonance. Therefore, it can be concluded that at least two mechanisms dominate the coupling between the lithosphere, atmosphere, and ionosphere before the Luding earthquake, including the chemical channel and double resonance, which simultaneously affect various geophysical parameters.

## 4. Conclusions

Multiple geophysical parameters exhibit anomalous phenomena in a sequence spanning from months to days before major earthquakes. This has been widely reported in numerous previous studies. This is the first instance where seven distinct geophysical parameters distributed in the lithosphere, atmosphere, and ionosphere all show abnormal phenomena occurring nearly simultaneously, approximately 3 h before the major earthquake. The increase in radon concentration supports the idea that the seismo-LAI coupling is primarily influenced by the chemical channel. However, the chemical channel alone cannot explain the shared frequency characteristics observed among air pressure, the geomagnetic field, and TEC data. A novel abnormal phenomenon involving vertical winds frequently reversing direction, corresponding to the cessation or weakening of horizontal winds, was detected by the radar wind profile in the MVP-LAI system. The characteristics of these reverse wind directions and shared frequencies suggest the presence of promising atmospheric resonance in the LAI coupling. It appears that multiple channels and/or mechanisms play a dominant role in the LAI coupling of various parameters before the Luding earthquake. Additionally, our study's results confirm the effectiveness of the MVP-LAI system in distinguishing these distinct LAI coupling mechanisms once again.

**Author Contributions:** Conceptualization, C.-H.C. and Z.M.; methodology, C.-H.C., Z.M. and Y.-Y.S.; software, S.Z.; validation, J.L., Z.M. and Y.-Y.S.; formal analysis, C.-H.C. and Z.M.; investigation, A.Y., T.C., H.Q., T.L. and F.W.; writing—original draft preparation, C.-H.C. and Z.M.; writing—review and editing, Z.M., Y.-Y.S., J.L., T.C., X.Z., T.L., Y.G. and F.W.; visualization, C.-H.C. and Z.M.; funding acquisition, C.-H.C., X.Z., A.Y. and Y.G. All authors have read and agreed to the published version of the manuscript.

**Funding:** This research was funded by the Special Fund of China Seismic Experimental Site, grant number CEAIEF20230401, National Natural Science Foundation of China (NNSFC), grant number 42230207, the Special Training Project of National Science Foundation of Xinjiang Uygur Autonomous Region, grant number 2022D03031, and the State Key Laboratory of Geodesy and Earth's Dynamics, Innovation Academy for Precision Measurement Science and Technology, grant number SKLGED2023-3-4.

**Data Availability Statement:** Daily data have been shown on the website: http://geostation.top (assessed on 20 November 2023). Data are available by directly contacting the first author, Chieh-Hung Chen, through the E-mail: nononochchen@gmail.com.

**Acknowledgments:** The authors thank the people who established and maintain the MVP-LAI system and the substations providing numerous geophysical data.

**Conflicts of Interest:** Author Tianya Luo was employed by Guangxi Transportation Science and Technology Group Co., Ltd. The remaining authors declare that the research was conducted in the absence of any commercial or financial relationships that could be construed as a potential conflict of interest.

## References

1. Hayakawa, M.; Molchanov, O.A. *Seismo Electromagnetics, Lithospheric-Atmospheric-Ionospheric Coupling*; Terra Science Publication Co.: Tokyo, Japan, 2002.
2. Pulinets, S.; Ouzounov, D. Lithosphere–Atmosphere–Ionosphere Coupling (LAIC) model–An unified concept for earthquake precursors validation. *J. Asian Earth Sci.* **2011**, *41*, 371–382. [CrossRef]
3. Mehdi, S.; Shah, M.; Naqvi, N.A. Lithosphere atmosphere ionosphere coupling associated with the 2019 Mw 7.1 California earthquake using GNSS and multiple satellites. *Environ. Monit. Assess.* **2021**, *193*, 501.
4. Adil, M.A.; Şentürk, E.; Pulinets, S.A.; Amory-Mazaudier, C. A lithosphere–atmosphere–ionosphere coupling phenomenon observed before M 7.7 Jamaica Earthquake. *Pure Appl. Geophys.* **2021**, *178*, 3869–3886. [CrossRef]
5. Parrot, M.; Tramutoli, V.; Liu, T.J.Y.; Pulinets, S.; Ouzounov, D.; Genzano, N.; Lisi, M.; Hattori, K.; Namgaladze, A. Atmospheric and ionospheric coupling phenomena associated with large earthquakes. *Eur. Phys. J. Spec. Top.* **2021**, *230*, 197–225. [CrossRef]
6. Shen, X.; Zhang, X.; Wang, L.; Chen, H.; Wu, Y.; Yuan, S.; Shen, J.; Zhao, S.; Qian, J.; Ding, J. The earthquake-related disturbances in ionosphere and project of the first China seismo-electromagnetic satellite. *Earthq. Sci.* **2011**, *24*, 639–650. [CrossRef]
7. Reid, H.F. The Mechanics of the Earthquake. In *The California Earthquake of April 18, 1906: Report of the State Earthquake Investigation Commission*; Carnegie Institution of Washington Publication: Washington, DC, USA, 1910.
8. Liu, J.Y.; Chen, Y.I.; Pulinets, S.A.; Tsai, Y.B.; Chuo, Y.J. Seismo-ionospheric signatures prior to M ≥ 6.0 Taiwan earthquakes. *Geophys. Res. Lett.* **2000**, *27*, 3113–3116. [CrossRef]
9. Liu, J.Y.; Chen, Y.I.; Jhuang, H.K.; Lin, Y.S. Ionospheric foF2 and TEC anomalous days associated with M ≥ 5.0 earthquake in Taiwan during 1997–1999. *Terr. Atmo. Oceanic Sci.* **2004**, *15*, 371–383. [CrossRef]
10. Oyama, K.-I.; Kakinami, Y.; Liu, J.-Y.; Kamogawa, M.; Kodama, T. Reduction of electron temperature in low-latitude ionosphere at 600 km before and after large earthquakes. *J. Geophys. Res.* **2008**, *113*, A11317. [CrossRef]
11. Chen, C.H.; Wang, C.H.; Wen, S.; Yeh, T.K.; Lin, C.H.; Liu, J.Y.; Yen, H.Y.; Lin, C.; Rau, R.J.; Lin, T.W. Anomalous frequency characteristics of groundwater level before major earthquake in Taiwan. *Hydrol. Earth Syst. Sci.* **2013**, *17*, 1693–1703. [CrossRef]
12. Chen, C.H.; Tang, C.C.; Cheng, K.C.; Wang, C.H.; Wen, S.; Lin, C.H.; Wen, Y.Y.; Meng, G.; Yeh, T.K.; Jan, J.C.; et al. Groundwater-strain coupling before the 1999 Mw 7.6 Taiwan Chi-Chi earthquake. *J. Hydrol.* **2015**, *524*, 378–384. [CrossRef]
13. Liu, J.Y.; Chen, C.H.; Chen, Y.I.; Yen, H.Y.; Hattori, K.; Yumoto, K. Seismo-geomagnetic anomalies and M ≥ 5.0 earthquakes observed in Taiwan during 1988–2001. *Phy. Chem. Earth.* **2006**, *30*, 215–222. [CrossRef]
14. Yen, H.Y.; Chen, C.H.; Yeh, Y.H.; Liu, J.Y.; Lin, C.R.; Tsai, Y.B. Geomagnetic fluctuations during the 1999 Chi-Chi earthquake in Taiwan. *Earth Planets Space* **2004**, *56*, 39–45. [CrossRef]

15. Liu, J.Y.; Chen, Y.I.; Chuo, Y.J.; Tsai, H.F. Variations of ionospheric total electron content during the Chi-Chi earthquake. *Geophys. Res. Lett.* **2001**, *28*, 1383–1386. [CrossRef]
16. Chen, C.H.; Wen, S.; Liu, J.Y.; Hattori, K.; Han, P.; Hobara, Y.; Wang, C.H.; Yeh, T.K.; Yen, H.Y. Surface displacements in Japan before the 11 March 2011 M9.0 Tohoku-Oki earthquake. *J. Asian Earth Sci.* **2014**, *80*, 165–171. [CrossRef]
17. Han, P.; Hattori, K.; Huang, Q.; Hirooka, S.; Yoshino, C. Spatiotemporal characteristics of the geomagnetic diurnal variation anomalies prior to the 2011 Tohoku earthquake (Mw 9.0) and the possible coupling of multiple pre-earthquake phenomena. *J. Asian Earth Sci.* **2016**, *129*, 13–21. [CrossRef]
18. Orihara, Y.; Kamogawa, M.; Nagao, T. Preseismic Changes of the level and temperature of confined groundwater related to the 2011 Tohoku earthquake. *Sci. Rep.* **2014**, *4*, 6907. [CrossRef]
19. Varotsos, P.A.; Sarlis, N.V.; Skordas, E.S.; Lazaridou, M.S. Seismic electric signals: An additional fact showing their physical interconnection with seismicity. *Tectonophysics* **2013**, *589*, 116–125. [CrossRef]
20. Sarlis, N.V.; Skordas, E.S.; Varotsos, P.A.; Nagao, T.; Kamogawa, M.; Tanaka, H.; Uyeda, S. Minimum of the order parameter fluctuations of seismicity before major earthquakes in Japan. *Proc. Natl. Acad. Sci. USA* **2013**, *110*, 13734–13738. [CrossRef]
21. Kato, A.; Obara, K.; Igarashi, T.; Tsuruoka, H.; Nakagawa, S.; Hirata, N. Propagation of slow slip leading up to the 2011 Mw 9.0 Tohoku-Oki earthquake. *Science* **2012**, *335*, 705–708. [CrossRef]
22. Ito, Y.; Hino, R.; Kido, M.; Fujimoto, H.; Osada, Y.; Inazu, D.; Ohta, Y.; Iinuma, T.; Ohzono, M.; Miura, S.; et al. Episodic slow slip events in the Japan subduction zone before the 2011 Tohoku-Oki earthquake. *Tectonophysics* **2013**, *600*, 14–26. [CrossRef]
23. Hattori, K.; Han, P. Investigation on preparation process of the 2011 off the Pacific Coast of Tohoku Earthquake (Mw 9.0) by GPS data. In Proceedings of the 2014 American Geophysics Union Fall Meeting, San Francisco, CA, USA, 15–19 December 2014.
24. Le, H.; Liu, L.; Liu, J.-Y.; Zhao, B.; Chen, Y.; Wan, W. The ionospheric anomalies prior to the M9.0 Tohoku-Oki earthquake. *J. Asian Earth Sci.* **2013**, *62*, 476–484. [CrossRef]
25. Heki, K. Ionospheric electron enhancement preceding the 2011 Tohoku-Oki earthquake. *Geophys. Res. Lett.* **2011**, *38*, 312. [CrossRef]
26. Hayakawa, M. Probing the Lower Ionospheric Perturbations Associated with Earthquakes by Means Of Sub-Ionospheric VLF/LF Propagation. *Earthq. Sci.* **2011**, *24*, 609–637. [CrossRef]
27. Hayakawa, M.; Molchanov, O.A.; Ondoh, T.; Kawai, E. The Precursory Signature Effect of the Kobe Earthquake on VLF Sub-Ionospheric Signals. *J. Comm. Res. Lab.* **1996**, *43*, 169–180.
28. Hayakawa, M. VLF/LF Radio Sounding of Ionospheric Perturbations Associated with Earthquakes. *Sensors* **2007**, *7*, 1141–1158. [CrossRef]
29. Molchanov, O.; Hayakawa, M.; Oudoh, T.; Kawai, E. Precursory Effects in the Sub-Ionospheric VLF Signals for the Kobe Earthquake. *Phys. Earth Planet. Inter.* **1998**, *105*, 239–248. [CrossRef]
30. Politis, D.Z.; Potirakis, S.M.; Contoyiannis, Y.F.; Biswas, S.; Sasmal, S.; Hayakawa, M. Statistical and Criticality Analysis of the Lower Ionosphere Prior to the 30 October 2020 Samos (Greece) Earthquake (M6.9), Based on VLF Electromagnetic Propagation Data as Recorded by a New VLF/LF Receiver Installed in Athens (Greece). *Entropy* **2021**, *23*, 676. [CrossRef]
31. Chen, C.H.; Yeh, T.K.; Liu, J.Y.; Wang, C.H.; Wen, S.; Yen, H.Y.; Chang, S.H. Surface Deformation and Seismic Rebound: Implications and applications. *Surv. Geophys.* **2011**, *32*, 291–313. [CrossRef]
32. Chen, C.H.; Lin, L.-C.; Yeh, T.-K.; Wen, S.; Yu, H.; Yu, C.; Gao, Y.; Han, P.; Sun, Y.-Y.; Liu, J.-Y.; et al. Determination of epicenters before earthquakes utilizing far seismic and GNSS data, Insights from ground vibrations. *Remote Sens.* **2020**, *12*, 3252. [CrossRef]
33. Bedford, J.R.; Moreno, M.; Deng, Z.; Oncken, O.; Schurr, B.; John, T.; Báez, J.C.; Bevis, M. Months-long thousand-kilometre-scale wobbling before great subduction earthquakes. *Nature* **2020**, *580*, 628–635. [CrossRef]
34. Chen, C.H.; Sun, Y.Y.; Lin, K.; Zhou, C.; Xu, R.; Qing, H.Y.; Gao, Y.; Chen, T.; Wang, F.; Yu, H.; et al. A new instrumental array in Sichuan, China, to monitor vibrations and perturbations of the lithosphere, atmosphere and ionosphere. *Surv. Geophys.* **2021**, *42*, 1425–1442. [CrossRef]
35. Su, X.; Meng, G.; Sun, H.; Wu, W. Positioning performance of BDS observation of the crustal movement observation network of China and its potential application on crustal deformation. *Sensors* **2018**, *18*, 3353. [CrossRef]
36. Wang, F.; Zhang, X.; Dong, L.; Liu, J.; Mao, Z.; Lin, K.; Chen, C.-H. Monitoring Seismo-TEC perturbations utilizing the Beidou geostationary satellites. *Remote Sens.* **2023**, *15*, 2608. [CrossRef]
37. Chen, C.H.; Sun, Y.-Y.; Xu, R.; Lin, K.; Wang, F.; Zhang, D.; Zhou, Y.; Gao, Y.; Zhang, X.; Yu, H.; et al. Resident waves in the ionosphere before the M6.1 Dali and M7.3 Qinghai earthquakes of 21–22 May 2021. *Earth Space Sci.* **2022**, *9*, e2021EA002159. [CrossRef]
38. Chen, C.-H.; Zhang, X.; Sun, Y.-Y.; Wang, F.; Liu, T.-C.; Lin, C.-Y.; Gao, Y.; Lyu, J.; Jin, X.; Zhao, X. Individual Wave Propagations in Ionosphere and Troposphere Triggered by the Hunga Tonga-Hunga Ha'apai Underwater Volcano Eruption on 15 January 2022. *Remote Sens.* **2022**, *14*, 2179. [CrossRef]
39. Chen, C.-H.; Sun, Y.-Y.; Zhang, X.; Wang, F.; Lin, K.; Gao, Y.; Tang, C.-C.; Lyu, J.; Huang, R.; Huang, Q. Far-field coupling and interactions in multiple geospheres after the Tonga volcano eruptions. *Surv. Geophys.* **2023**, *44*, 587–601. [CrossRef]
40. Hayakawa, M.; Schekotov, A.; Izutsu, J.; Nickolaenko, A.P.; Hobara, Y. Seismogenic ULF/ELF Wave Phenomena: Recent Advances and Future Perspectives. *Open J. Earthq. Res.* **2023**, *12*, 45–133. [CrossRef]
41. Fan, X.; Wang, X.; Dai, L.; Fang, C.; Deng, Y.; Zou, C.; Tang, M.; Wei, Z.; Dou, X.; Zhang, J.; et al. Characteristics and spatial distribution pattern of Ms 6.8 Luding earthquake occurred on September 5, 2022. *J. Eng. Geol.* **2022**, *30*, 1504–1516. [CrossRef]

42. An, Y.; Wang, D.; Ma, Q.; Xu, Y.; Li, Y.; Zhang, Y.; Liu, Z.; Huang, C.; Su, J.; Li, J.; et al. Preliminary report of the September 5, 2022 Ms 6.8 Luding earthquake, Sichuan, China. *Earthq. Res. Adv.* **2023**, *3*, 100184. [CrossRef]
43. Liu, J.; Zhang, X.; Yang, X.; Yang, M.; Zhang, T.; Bao, Z.; Wu, W.; Qiu, G.; Yang, X.; Lu, Q. The Analysis of Lithosphere–Atmosphere–Ionosphere Coupling Associated with the 2022 Luding Ms6.8 Earthquake. *Remote Sens.* **2023**, *15*, 4042. [CrossRef]
44. Dobrovolsky, I.P.; Zubkov, S.I.; Miachkin, V.I. Estimation of the size of earthquake preparation zones. *Pure Appl. Geophys.* **1979**, *117*, 1025–1044. [CrossRef]
45. Yamazaki, Y.; Maute, A. Sq and EEJ—A review on the daily variation of the geomagnetic field caused by ionospheric dynamo currents. *Space Sci. Rev.* **2017**, *206*, 299–405. [CrossRef]
46. Campbell, W.H. *Introduction to Geomagnetic Fields*; Cambridge University Press: Cambridge, UK, 2003.
47. Rasson, J.L.; Lévêque, J.J. *The Earth's Magnetic Field: Its History, Origin and Planetary Perspective*; Springer: Berlin/Heidelberg, Germany, 2004.
48. Liu, J.Y.; Tsai, H.F.; Jung, T.K. Total electron content obtained by using the global positioning system. *Terr. Atmos. Ocean Sci.* **1996**, *7*, 107–117. [CrossRef]
49. Torrence, C.; Compo, G.P. A practical guide to wavelet analysis. *Bull Am. Meteor Soc.* **1998**, *79*, 61–78. [CrossRef]
50. Hayakawa, M. *Earthquake Prediction with Radio Techniques*; John Wiley & Sons: Hoboken, NJ, USA; Singapore Pte Ltd.: Singapore, 2015; p. 294.
51. Hayakawa, M. Earthquake Prediction with Electromagnetic Phenomena. In *AIP Conference Proceedings*; AIP Publishing: Melville, NY, USA, 2016; Volume 1705, p. 20002.
52. Pulinets, S.A.; Boyarchuk, K. *Ionospheric Precursors of Earthquakes*; Springer: Berlin/Heidelberg, Germany, 2004.
53. Sorokin, V.; Chemyrev, V.; Hayakawa, M. *Electrodynamic Coupling of Lithosphere-Atmosphere-Ionosphere of the Earth*; Nova Science Publishing: Hauppauge, NY, USA, 2015.
54. Harrison, R.G.; Aplin, K.L.; Rycroft, M.J. Atmospheric electricity coupling between earthquake regions and the ionosphere. *J. Atmos. Sol. Terr. Phys.* **2010**, *72*, 376–381. [CrossRef]
55. Harrison, R.; Aplin, K.; Rycroft, M. Brief Communication: Earthquake-cloud coupling through the global atmospheric electric circuit. *Nat. Hazards Earth Syst. Sci.* **2014**, *14*, 773–777. [CrossRef]
56. Inan, U.S.; Bell, T.F.; Rodriguez, J.V. Heating and ionization of the lower ionosphere by lightning. *Geophys. Res. Lett.* **1991**, *18*, 705–708. [CrossRef]
57. Pasko, V.P.; Inan, U.S.; Bell, T.F.; Taranenko, Y.N. Sprites produced by quasi-electrostatic heating and ionization in the lower ionosphere. *J. Geophys. Res. Space Phys.* **1997**, *102*, 4529–4561. [CrossRef]
58. Cho, M.; Rycroft, M.J. Computer simulation of the electric field structure and optical emission from cloud-top to the ionospheres. *J. Atmos. Solar.-Terr. Phys.* **1998**, *60*, 871–888.
59. Rodger, C.J. Red sprites, upward lighting and VLF perturbations. *Rev. Geophys.* **1999**, *37*, 317–336. [CrossRef]
60. Takahashi, Y.; Miyasato, R.; Adachi, T.; Adachi, K.; Sera, M.; Uchida, A.; Fukunishi, H. Activities of sprites and elves in the winter season, Japan. *J. Atmos. Solar.-Terr. Phys.* **2003**, *65*, 551–560. [CrossRef]
61. Molchanov, O.A.; Hayakawa, M.; Miyaki, K. VLF/LF sounding of the lower ionosphere to study the role of atmospheric oscillations in the lithosphere-ionosphere coupling. *Adv. Polar Upper Atmos. Res. Tokyo* **2001**, *15*, 146–158.
62. Sun, Y.-Y.; Oyama, K.I.; Liu, J.Y.; Jhuang, H.K.; Cheng, C.Z. The neutral temperature in the ionospheric dynamo region and the ionospheric F region density during Wenchuan and Pingtung Doublet earthquakes. *Nat. Hazards Earth Syst. Sci.* **2011**, *11*, 1759–1768. [CrossRef]
63. Oyama, K.I.; Devi, M.; Ryu, K.; Chen, C.H.; Liu, J.Y.; Liu, H.; Bankov, L.; Kodama, T. Modifications of the ionosphere prior to large earthquakes: Report from the ionospheric precursor study group. *Geosci. Lett.* **2016**, *3*, 6. [CrossRef]
64. Yang, S.-S.; Asano, T.; Hayakawa, M. Abnormal gravity wave activity in the stratosphere prior to the 2016 Kumamoto earthquakes. *J. Geophys. Res. Space Phys.* **2019**, *124*, 1410–1425. [CrossRef]
65. Liu, J.Y.; Chen, C.H.; Sun, Y.Y.; Chen, C.H.; Tsai, H.F.; Yen, H.Y.; Chum, J.; Lastovicka, J.; Yang, Q.S.; Chen, W.S.; et al. The vertical propagation of disturbances triggered by seismic waves of the 11 March 2011 M9.0 Tohoku earthquake over Taiwan. *Geophys. Res. Lett.* **2016**, *43*, 1759–1765. [CrossRef]
66. Chou, M.Y.; Cherniak, I.; Lin, C.C.H.; Pedatella, N.M. The persistent ionospheric responses over Japan after the impact of the 2011 Tohoku earthquake. *Space Weather* **2020**, *18*, e2019SW002302. [CrossRef]
67. Yang, S.S.; Hayakawa, M. Gravity wave activity in the stratosphere before the 2011 Tohoku earthquake as the mechanism of lithosphere-atmosphere-ionosphere coupling. *Entropy* **2020**, *22*, 110. [CrossRef]
68. Yang, S.S.; Potirakis, S.M.; Sasmal, S.; Hayakawa, M. Natural time analysis of global navigation satellite system surface deformation: The case of the 2016 Kumamoto earthquakes. *Entropy* **2020**, *22*, 674. [CrossRef]
69. Chen, C.-H.; Oyama, K.; Jhuang, H.-K.; Das, U. Driving Source of Change for Ionosphere before Large Earthquake -Vertical Ground Motion. *Remote Sens.* **2023**, *15*, 4556. [CrossRef]
70. Fraser-Smith, A.C.; Bernardi, A.; McGill, P.R.; Ladd, M.E.; Helliwell, R.A.; Villard, O.G., Jr. Low-frequency magnetic field measurements near the epicenter of the Ms 7.1 Loma Prieta Earthquake. *Geophys. Res. Lett.* **1990**, *17*, 1465–1468. [CrossRef]
71. Molchanov, O.A.; Mazhaeva, O.A.; Goliavin, A.N.; Hayakawa, M. Observations by the intercosmos-24 satellite of ELF-VLF electromagnetic emissions associated with earthquakes. *Ann. Geophys.* **1993**, *11*, 431–440.

72. Molchanov, O.A.; Hayakawa, M.; Rafalsky, V.A. Penetration characteristics of electromagnetic emissions from an under-ground seismic source into the atmosphere, ionosphere and magnetosphere. *J. Geophys. Res.* **1995**, *100*, 1691–1712. [CrossRef]
73. Molchanov, O.A.; Hayakawa, M. Generation of ULF electromagnetic emissions by microfracturing. *Geophys. Res. Lett.* **1995**, *22*, 3091–3094. [CrossRef]
74. Chen, T.; Li, L.; Zhang, X.; Wang, C.; Jin, X.; Wu, H.; Ti, S.; Wang, S.; Song, J.; Li, W.; et al. Possible Locking Shock Time in 2–48 Hours. *Appl. Sci.* **2023**, *13*, 813. [CrossRef]
75. Sun, Y.Y.; Chen, C.H.; Zhang, P.; Li, S.; Xu, H.R.; Yu, T.; Lin, K.; Mao, Z.; Zhang, D.; Lin, C.Y.; et al. Explosive eruption of the Tonga underwater volcano modulates the ionospheric E-region current on 15 January 2022. *Geophys. Res. Lett.* **2022**, *49*, e2022GL099621. [CrossRef]
76. Dautermann, T.; Calais, E.; Lognonné, P.; Mattioli, G.S. Lithosphere—Amosphere—Ionosphere coupling after the 2003 explosive eruption of the Soufriere Hills Volcano, Montserrat. *Geophys. J. Int.* **2009**, *179*, 1537–1546. [CrossRef]
77. Chen, C.-H.; Sun, Y.-Y.; Zhang, X.; Gao, Y.; Yisimayili, A.; Qing, H.; Yeh, T.-K.; Lin, K.; Wang, F.; Yen, H.-Y.; et al. Double resonance in seismo-lithosphere-atmosphere-ionosphere coupling. *Ann. Geophys.* **2023**, *accepted*.

 *geosciences*

*Article*

# Topside Ionospheric Structures Determined via Automatically Detected DEMETER Ion Perturbations during a Geomagnetically Quiet Period

**Mei Li *, Hongzhu Yan and Yongxian Zhang**

Institute of Earthquake Forecasting, China Earthquake Administration, Beijing 100036, China; m13453460853@163.com (H.Y.); yxzhang@ief.ac.cn (Y.Z.)
* Correspondence: mei_seis@163.com

**Abstract:** In this study, 117,718 ionospheric perturbations, with a space size ($t$) of 20–300 s but no amplitude ($A$) limit, were automatically globally searched via software utilizing ion density data measured by the DEMETER satellite for over 6 years. The influence of geomagnetic storms on the ionosphere was first examined. The results demonstrated that storms can globally enhance positive ionospheric irregularities but rarely induce plasma variations of more than 100%. The probability of PERs with a space size falling in 200–300 s (1400–2100 km if a satellite velocity of 7 km/s is considered) occurring in a geomagnetically perturbed period shows more significance than that in a quiet period. Second, statistical work was performed on ion PERs to check their dependence on local time, and it was shown that 24.8% of the perturbations appeared during the daytime (10:30 LT) and 75.2% appeared during the nighttime (22:30 LT). Ionospheric fluctuations with an absolute amplitude of $A < 10\%$ tend to be background variations, and the percentages of positive perturbations with a small $A < 20\%$ occur at an amount of 64% during the daytime and 26.8% during the nighttime, but this number is reversed for mid–large-amplitude PERs. Large positive PERs with $A > 100\%$ mostly occurred at night and negative ones with $A < -100\%$ occurred entirely at night. There was a demarcation point in the space size of $t = 120$ s, and the occurrence probabilities of day PERs were always higher than that of nighttime ones before this point, while this trend was contrary after this point. Finally, distributions of PERs according to different ranges of amplitude and space scale were characterized by typical seasonal variations either in the daytime or nighttime. EIA only exists in the dayside equinox and winter, occupying two low-latitude crests with a lower Np in both hemispheres. Large WSAs appear within all periods, except for dayside summer, and are full of PERs with an enhanced amplitude, especially on winter nights. The WN-like structure is obvious during all seasons, showing large-scale space. On the other hand, several magnetically anomalous zones of planetary-scale non-dipole fields, such as the SAMA, Northern Africa anomaly, and so on, were also successfully detected by extreme negative ion perturbations during this time.

**Keywords:** ionospheric structures; ion perturbations; automatic detection method

**Citation:** Li, M.; Yan, H.; Zhang, Y. Topside Ionospheric Structures Determined via Automatically Detected DEMETER Ion Perturbations during a Geomagnetically Quiet Period. *Geosciences* **2024**, *14*, 33. https://doi.org/10.3390/geosciences14020033

Academic Editors: Dimitrios Nikolopoulos and Jesus Martinez-Frias

Received: 7 December 2023
Revised: 6 January 2024
Accepted: 19 January 2024
Published: 28 January 2024



## 1. Introduction

As a conductive component of layers of the atmosphere, investigations of various properties of the ionosphere have always been a controversial issue due to different sources from above and below that contribute to ionospheric variations. Solar activity mainly includes sunspots, solar winds, solar flares, etc. These activities can release a large number of energy particles, causing a direct or indirect impact on the formation and characteristics of the Earth's ionosphere, while during geomagnetic storms, vast amounts of energy and momentum in the form of increased particle precipitation and Joule heating from solar wind and the magnetosphere have been deposited into the Earth's upper atmosphere and ionosphere, causing global disturbances of the ionosphere. On the other hand, the ionosphere is subject to tremendous responses to natural and artificial events, such as earthquakes, volcanoes, tsunamis, communication engineering, etc. Ionospheric irregularities caused by inhomogeneous ionization density in the topside ionosphere have been described as early as the 1930s [1–3]. Following several decades of research, ionospheric structures, and even their inner various features, have gradually gained more and more clear configurations from ground-based remote ionospheric responding and satellite in situ instruments [4–7]. Some, such as the equatorial ionization anomaly (EIA) and mid-latitude ionospheric trough (MIT), have been well reconstructed by different parameters, although parts of the potential mechanisms of these various ionospheric irregularities are yet to be fully understood [8–12].

Plasma density is the primary parameter that is commonly utilized to reconstruct the ionosphere configuration and test the reliability of the data measured from different instruments. Electron density (Ne) is a key parameter to characterize the status of ionospheric plasma, and $O^+$ is the main component among the ions, although this depends on certain factors such as local time and altitude [7,13–15]. Electron density measured employing in situ DEMETER (Detection of Electro-Magnetic Emissions Transmitted from Earthquake Regions) and Swarm satellites or from FORMOSAT-3/COSMIC radio occultation measurements were employed to investigate the properties of ionospheric irregularities, with variations such as EIA, MIT, middle latitudinal band structures, Wedell Sea Anomaly, etc. [11,16–21]. Other parameters such as $O^+$, $H^+$, and $He^+$ densities and total electron densities are also used to research the ionospheric characteristics of background, seasonal, and day-to-day variations, as well as the large-scale depletion of oxygen ions [22–24].

However, it must be mentioned that variations in ionospheric parameters are not only due, in large part, to sources from above, such as solar and geomagnetic activity variations, but also natural hazards and meteorological sources from below [25]. As the Earth's observation from space has developed, seismo-ionospheric influences have been highlighted as a possible candidate for earthquake forecasting and potential seismogenic electromagnetic energy transmitted along the lithosphere, atmosphere, ionosphere, and even magnetosphere [26–31]. As a weak factor of strong ionospheric background variations, weak information potentially associated with seismic activities is always submerged in other enhanced irregularities. Aside from a case study in a relatively small region and within a specified short time, statistical investigations on seismo-ionospheric influence have always been a way to distinguish anomalous features of earthquake precursors [32–41]. With an alternative statistical method, Parrot [31] correlated DEMETER ion perturbations (PERs), automatically searched via software, with strong earthquake events occurring during the DEMETER period, and the results have shown that the number and intensity of the ionospheric PERs are a little larger prior to earthquakes than prior to random events. A similar method has also been utilized by Li et al. [42,43] to correlate ionospheric PERs measured by the DEMETER satellite with strong seismic activities occurring within a corresponding period. They found that the obtained results are inadequate, because not all ionospheric PERs are caused by EQs, and the number of false alarms is large, even with the detection range reaching up to 1500 km. Furthermore, Li et al. [15] conveyed that CSES (China Seismo-Electromagnetic Satellite) ion PERs, with a space size of 200–300 s, are located collectively in the equatorial area, with no specified correlation to the main seismic zones of the world. Contrastingly, Li et al. [44] found that there are different properties for ion PERs and electron ones obtained from a CSES satellite for more than three years and ionospheric PERs with large amplitudes and space sizes tend to collocate with large-scale ionospheric structures such as EIA and MIT.

Therefore, all evidence indicates that it will be of great significance to explore ionospheric PERs caused by various interferences. A comprehensive investigation of the properties of ionospheric PERs with different amplitudes and space sizes can help us fully understand their producing mechanisms and correctly distinguish earthquake precursors. Hence, in this paper, DEMETER data and the data processing method are introduced in Section 2. In Section 3, different seasonal and local time characteristics of ion PERs are comparatively exhibited. Finally, the discussion and conclusions are presented in Sections 4 and 5, respectively.

## 2. Data and Data Processing Method

### 2.1. Dataset

DEMETER was launched in June 2004, measuring electromagnetic waves and plasma parameters around the globe, except in the auroral zones [45]. It is a low-orbit satellite with an altitude of 710 km, decreased to 660 km in December 2005. The orbit of DEMETER is nearly sun-synchronous, and the upgoing and downgoing half-orbits correspond to nighttime (22:30 LT) and daytime (10:30 LT), respectively. Its payload IAP (plasma analyzer instrument) outputs the plasma density of ion density with data resolutions of 4 s in the survey mode and 2 s in the burst mode for all data. This satellite's science mission stopped measuring at the end of December 2010. More details can be found in the study by Parrot et al. [45].

This investigation is based on the ion ($O^+$) density (Ni) data from IAP onboard the DEMETER in situ measurements of over 6 years (from November 2004 to August 2010). All data were transmitted into the resolution of 4 s for ion density issued by the original recordings. At the same time, the SAVGOL method was employed to smooth the data, eliminating pulse-like peaks before searching for PERs. The SAVGOL function returns the coefficients of the Savitzky–Golay smoothing filter [46].

### 2.2. Automatic Search for Ionospheric PERs

The plasma data processing method employed here is similar to the one used by Li et al. [45]. Ionospheric PERs in the ion dataset were automatically searched by software (code) globally rather than around the main seismic zones, as in our previous work [42,43]. Here, two key parameters are specified: the PER spatial scale ($t$, in situ measurement time) was kept within the 20–300 s range (140–2100 km if a satellite velocity of 7 km/s is considered), and there were still no limits for the value of various amplitudes ($A$) in the PER database. The minimum size was also set to 20 s in order to avoid spurious impulsions caused by one or two points [42]. Each ionospheric disturbance is described by several parameters, such as peak appearing time, orbit number, location (latitude and longitude), amplitude, spatial scale, etc. Eight three-hourly averaged Kp index values each day were available from Li and Parrot [42], and the Kp value was also examined when each ionospheric PER detected occurred.

Figure 1a represents the upgoing orbit of 12545 (black line) measured by DEMETER on 8 November 2006. The $O^+$ density recorded along this orbit during nighttime is denoted by a blue line (data-ion) in Figure 1b, and its smoothed data using the SAVGOL function is lined in red (Smo-ion).

Two ionospheric perturbations were automatically detected within this Smo-ion line by the software, with a defined space size of 20–300 s and without amplitude limits. The corresponding parameters of these two PERs are presented in Table 1.

**Table 1.** Parameters of two PERs detected by the software along the DEMETER orbit of 12545 at nighttime.

|  | **PER1** | **PER2** |
|---|---|---|
| Date (y m d) | 8 November 2006 | 8 November 2006 |
| Time (h m s ms) | 12 41 3 797 | 12 46 0 527 |
| Orbit | 12545_1 | 12545_1 |
| Latitude (°) | −28.5122 | −10.5681 |
| Longitude (°) | 148.210 | 144.081 |
| BkgdIon (cm$^{-3}$) | 4649.35 | 8378.52 |
| Amplitude (cm$^{-3}$) | 6738.61 | 14,763.6 |
| Trend | Increase | Increase |
| Percent (%) | 44.9 | 76.2 |
| Time_width (m s ms) | 1 37 433 | 4 56 113 |
| Extension (km) | 669 | 2036 |

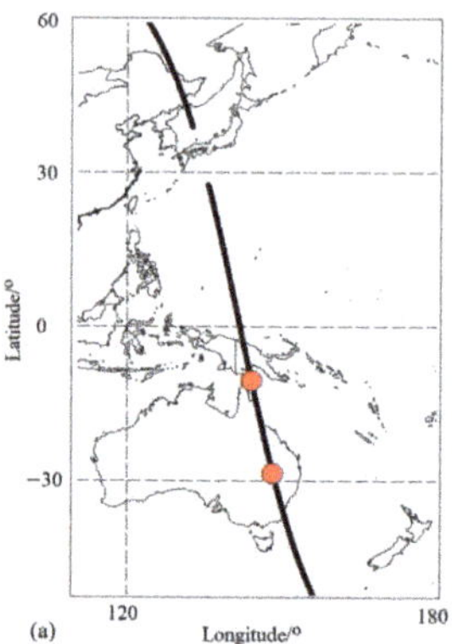
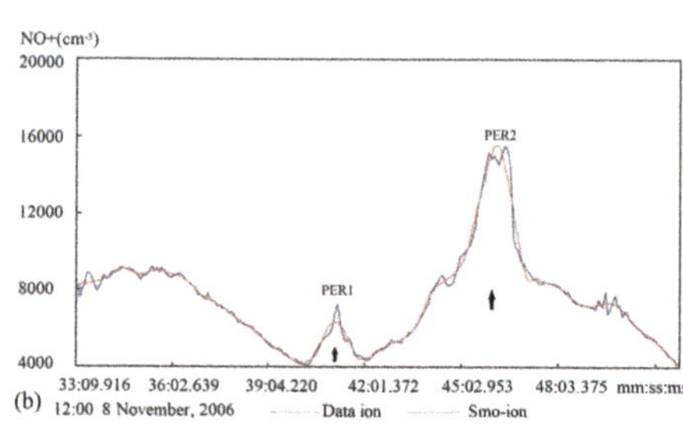



**Figure 1.** (**a**) Location of the flight path (upgoing orbit 12545) conducted by DEMETER on 8 November 2006. Two red dots represent the peak values of (**b**) two ionospheric PERs (indicated by two black arrows) detected by the software (the code) in smoothed ion data of the DEMETER orbit 12545 at night.

In total, 117,718 ion PERs were attained from the ion density dataset measured by the DEMETER satellite during this considered period.

## 3. Properties of Ionospheric PERs

### 3.1. Effect of a Geomagnetic Storm on the Ionosphere

In this study, the Kp index was first used to examine the geomagnetic effect on plasma densities. Figure 2 shows the distributions of all 117,718 ion PERs with an increasing value of corresponding Kp indexes as one PER occurs. In Figure 2, it is clear that the distributions of plasma variations in light of the Kp index present no special areas where PERs with large Kp values appear collectively, although solar activity tends to have a slightly heavier effect on equatorial and high-latitude ionospheric areas. The response of the ionosphere to solar and geomagnetic activity variations depends on the season, latitude, and storm time occurrence [15,32,44,47,48].

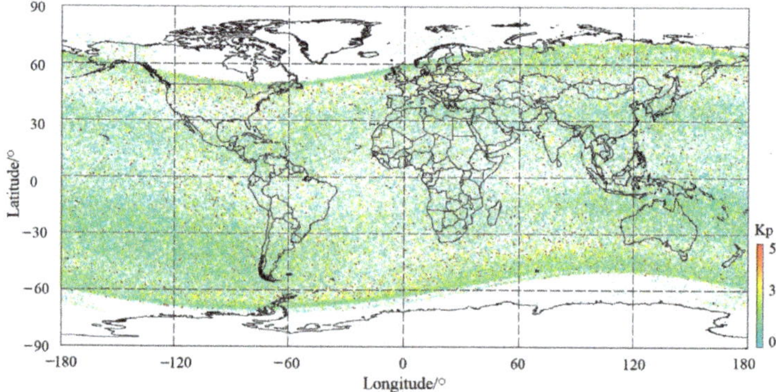



**Figure 2.** Distribution of 117,718 ion PERs with respect to different ranges of the Kp index during the DEMETER satellite period considered in this paper.

Usually, Kp ≥ 3 implies that there is an effect on space weather from geomagnetic storms. To further examine the exact influence of the geomagnetic storm on the amplitude and space size of ionospheric PERs and attain obviously comparable results, PERs with Kp > 4 within the perturbed period and Kp ≤ 2 within the quiet period were selected from all 117,718 ion PERs to form two new groups of PERs: 87,057 with Kp ≤ 2 and 4263 with Kp > 4. Second, for each group of PERs, the number for certain PERs (*n*, the same in the following parts) and their occurrence probability ($p$, the same in the following parts) were calculated as a function of different ranges of amplitude ($A$): $\geq$100, 90–100, 80–90, 70–80, 60–70, 50–60, 40–50, 30–40, 20–30, 10–0, 0––100, and <−100; and space size ($t$): 20–40, 40–60, 60–80, 80–100, 100–120, 120–140, 140–160, 160–180, 180–200, and 200–300. The results are listed in Tables 2 and 3 as the number of each sub-group of PERs and its corresponding percentage.

**Table 2.** Number and its corresponding percentage of sub-group PERs with different ranges of amplitude ($A$) for two-group ion PERs (Kp > 4 and Kp $\leq$ 2).

| | Kp > 4 | | Kp $\leq$ 2 | |
| --- | --- | --- | --- | --- |
| $A$/% | $n$ | $p$ | $n$ | $p$ |
| $\geq$100 | 510 | 12.0 | 12,014 | 13.8 |
| 90–100 | 100 | 2.3 | 1543 | 1.8 |
| 80–90 | 112 | 2.6 | 1872 | 2.2 |
| 70–80 | 131 | 3.1 | 2292 | 2.6 |
| 60–70 | 151 | 3.5 | 2915 | 3.3 |
| 50–60 | 211 | 4.9 | 3623 | 4.2 |
| 40–50 | 281 | 6.6 | 4850 | 5.6 |
| 30–40 | 388 | 9.1 | 6890 | 7.9 |
| 20–30 | 479 | 11.2 | 9379 | 10.8 |
| 10–20 | 684 | 16.0 | 13,881 | 15.9 |
| 0–10 | 776 | 18.3 | 17,529 | 20.1 |
| −100–0 | 436 | 10.3 | 10,111 | 11.6 |
| <−100 | 4 | 0.1 | 158 | 0.2 |

**Table 3.** Number and its corresponding percentage of sub-group PERs with different ranges of space size ($t$) for two-group ion PERs (Kp > 4 and Kp $\leq$ 2).

| | Kp > 4 | | Kp $\leq$ 2 | |
| --- | --- | --- | --- | --- |
| $t$/s | $n$ | $p$ | $n$ | $p$ |
| 20–40 | 773 | 18.1 | 16,045 | 18.4 |
| 40–60 | 469 | 11.0 | 9121 | 10.5 |
| 60–80 | 646 | 15.2 | 13,553 | 15.6 |
| 80–100 | 536 | 12.6 | 12,594 | 14.5 |
| 100–120 | 482 | 11.3 | 10,200 | 11.7 |
| 120–140 | 287 | 6.7 | 6098 | 7.0 |
| 140–160 | 263 | 6.2 | 5439 | 6.2 |
| 160–180 | 164 | 3.8 | 3255 | 3.7 |
| 180–200 | 158 | 3.7 | 2994 | 3.5 |
| 200–300 | 485 | 11.4 | 7758 | 8.9 |

For an easy comparison, data from Tables 2 and 3 are represented as polylines in Figure 3.

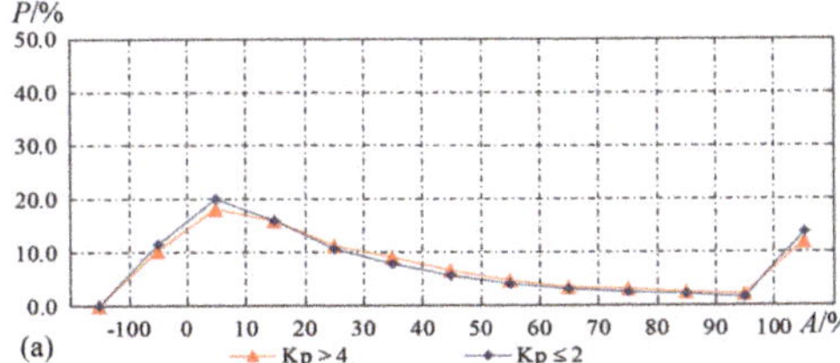


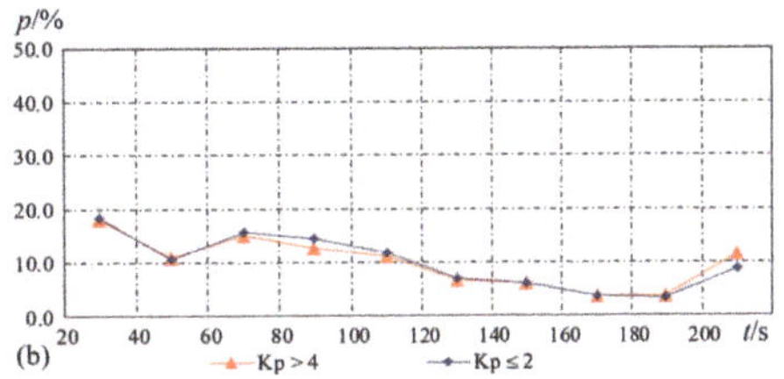


**Figure 3.** Polyline of percentages for different groups of perturbations: PERs for Kp > 4 in red, ones for Kp $\leq$ 2 in blue line, as a function of (**a**) an amplitude ($A$) and (**b**) a space size ($t$).

Table 2 and Figure 3a show the percentage values corresponding to different scales of amplitude during a geomagnetically perturbed period ($Kp > 4$) and a quiet period ($Kp \leq 2$); for positive PERs, the occurrence probabilities for plasma variations with a magnitude covering the range $A = 0$–10% is 18.3% and 20.1%, for $10\% < A < 100\%$ it is 59.3% and 54.3%, and for $A > 100\%$ is 12% and 13.8%, respectively. From this point, it is easy to infer that geomagnetic storms can accelerate ionospheric variations but rarely induce plasma irregularities of more than 100%. On the other hand, it is possible, but unlikely, for magnetic storms to give rise to negative ionospheric variations even beyond 100% (see Table 2 and Figure 3a), which is coincident with the results attained by Prölss [49], who presented that the increase in geomagnetic activity gave rise to negative ionospheric variations at the nightside during the summer season. Xiong et al. [50] also reported that positive and negative ionospheric responses were observed during the recovery phase of a geomagnetic storm on the dayside.

In Table 3 and Figure 3b, there is no obvious discrepancy between the space sizes of PERs appearing during the disturbed period and quiet time when $t < 200$ s. Contrastingly, there is a relatively obvious gap between the percentages during $Kp > 4$ and $Kp \leq 2$ when $t = 200$–300 s, which indicates that geomagnetic storms tend to induce larger ionospheric disturbances. Alternatively, the space size defined here does not cover the range of geomagnetically perturbed ionospheric variations well. Large-scale ionospheric density enhancements are frequently observed during geomagnetic storms [50].

### 3.2. Local Time Discrepancy of Ionospheric PERs

The DEMETER measurement heavily depends on two local times: 22:30 LT for nighttime and 10:30 LT for the morning. For all 117,718 ion PERs, 29,226 occurred during daytime and 88,492 for nighttime, accounting for 24.8% and 75.2%, respectively. To check the dependence of the occurrence of plasma PERs on local time, PERs were separated into two groups according to their occurrence time of daytime and nighttime. Then, for each group of PERs (dayside or nightside), the percentage for different ranges of amplitude, $A$, and space size, $t$, were calculated and are presented in Tables 4 and 5.

**Table 4.** Number and its corresponding percentage of sub-group PERs with different ranges of amplitude ($A$) for two-group ion PERs (daytime and nighttime).

| $A$/% | Daytime | | Nighttime | |
|---|---|---|---|---|
| | $n$ | $p$ | $n$ | $p$ |
| $\geq 100$ | 249 | 0.9 | 15,650 | 17.7 |
| 90–100 | 73 | 0.2 | 2054 | 2.4 |
| 80–90 | 116 | 0.4 | 2486 | 2.8 |
| 70–80 | 148 | 0.6 | 3011 | 3.4 |
| 60–70 | 332 | 1.1 | 3716 | 4.2 |
| 50–60 | 451 | 1.5 | 4562 | 5.2 |
| 40–50 | 844 | 2.9 | 5868 | 6.6 |
| 30–40 | 1699 | 5.8 | 7718 | 8.7 |
| 20–30 | 3104 | 10.6 | 9651 | 10.9 |
| 10–20 | 6712 | 23.0 | 12,179 | 13.8 |
| 0–10 | 11,983 | 41.0 | 11,541 | 13.0 |
| $-100$–0 | 3515 | 12.0 | 9860 | 11.1 |
| $< -100$ | 0 | 0.0 | 196 | 0.2 |

Data from Tables 4 and 5 are also presented as polylines in Figure 4 for a better comparison.

From Table 4 and Figure 4a, the percentages for all amplitude segments varied widely, and this property was more obvious during the daytime. The significant feature is that plasma PERs with small amplitudes <20% made up a prominent proportion of 64% on the dayside. Contrastingly, this parameter only accounts for 26.8% on the nightside. The number of each group of PERs with an amplitude between 0 and $-10\%$ was also checked: 3406 and 5603 for day and night ion PERs, accounting for 11.7% and 6.3% of all 29,226 day ion PERs and 88,492 night ones, respectively. That means that most ionospheric variations with a positive and negative magnitude of less than 10% tend to be background irregularities. This conclusion seems more correct during nighttime than daytime, when sunlight can speed the ionization of plasma, giving rise to more positive ionospheric irregularities. The occurrence probabilities of mid–large-amplitude perturbations, for instance, >20%, account for 24% and 61.9% for daytime and nighttime, respectively. Comparatively, the ionosphere can enhance amplitudes to more than 100% easily during nighttime but rarely decrease to 100%. From the space size, the probability for various sections keeps a relative balance, but there is a demarcation point, $t = 120$ s. The occurrence probabilities of day PERs were always higher than that of nighttime before this point, while this result was reversed after this point (See Table 5 and Figure 4b). A primary conclusion was almost attained on the basis of these statistical results: relatively, the ionosphere varies more frequently and more violently during nighttime but with a relatively small space size.

**Table 5.** Number and its corresponding percentage of sub-group PERs with different ranges of space size ($t$) for two-group ion PERs (daytime and nighttime).

| $t$/s | Daytime | | Nighttime | |
|---|---|---|---|---|
| | $n$ | $p$ | $n$ | $p$ |
| 20–40 | 5521 | 18.9 | 15,731 | 17.8 |
| 40–60 | 2580 | 8.8 | 9714 | 11.0 |
| 60–80 | 3599 | 12.3 | 14,535 | 16.4 |
| 80–100 | 3342 | 11.4 | 13,478 | 15.2 |
| 100–120 | 3127 | 10.7 | 10,501 | 11.9 |
| 120–140 | 2150 | 7.4 | 6130 | 6.9 |
| 140–160 | 2058 | 7.0 | 5302 | 6.0 |
| 160–180 | 1419 | 4.9 | 3130 | 3.5 |
| 180–200 | 1373 | 4.7 | 2830 | 3.2 |
| 200–300 | 4057 | 13.9 | 7141 | 8.1 |

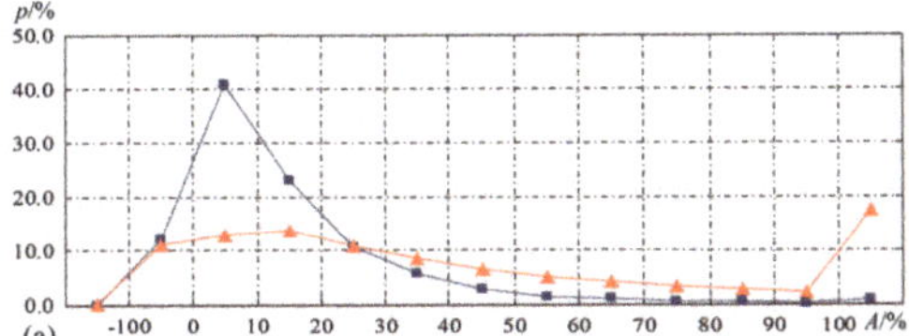


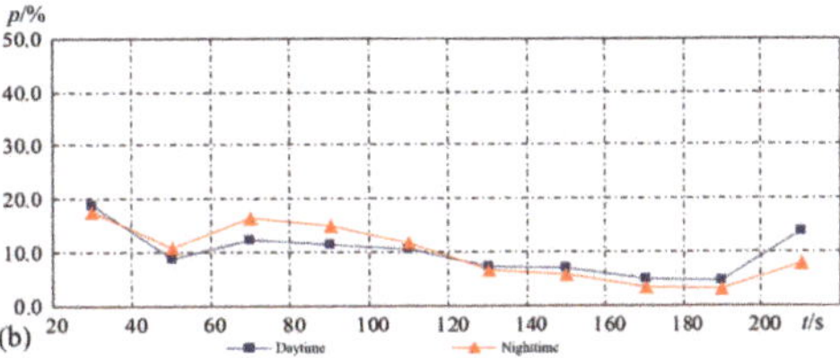


**Figure 4.** Polyline of percentages for different groups of perturbations: PERs during nighttime in red and daytime in blue as a function of (**a**) amplitude range ($A$) and (**b**) space size ($t$).

### 3.3. Seasonal Variation in Ionospheric PERs

The ionosphere is also characterized by local time [19,23], as well as seasonal variations [23,51]. According to the Lloyd criteria [52], the months of winter for the Northern Hemisphere (summer for the Southern Hemisphere) include November, December, January, and February; equinox covers March, April, September, and October; and summer (winter for the Southern Hemisphere) contains the months of May, June, July, and August.

On the basis of dividing ion PERs into different groups according to their occurrence at local time (dayside and nightside), plasma PERs issued by ion density in this section were further divided into different groups by seasons. However, before that, we eliminated PERs with Kp $\geq$ 3 (18,169 PERs) in order to eliminate the global influence from solar activities and only keep PERs with Kp < 3 (99,549 PERs) to a statistical in this part. For all 99,549 PERs with Kp < 3, 25,780 PERs occurred on the dayside and 73,769 on the nightside. Furthermore, each group of these PERs was separated into three sub-groups according to different seasons: 3862 for summer, 6680 for equinox, and 15,238 for winter on the dayside; and 22,852, 22,389, and 28,528 on the nightside. Their distributions corresponding to various amplitudes, *A*, as well as space size, *t*, are exhibited in Figures 5 and 6, respectively. In Figure 5, the left panels represent the distributions of ion PERs appearing on the dayside in summer, equinox, and winter from the top to the bottom, and the right panels correspond to ones on the nightside. Figure 6 shows the same arrangement as Figure 5.

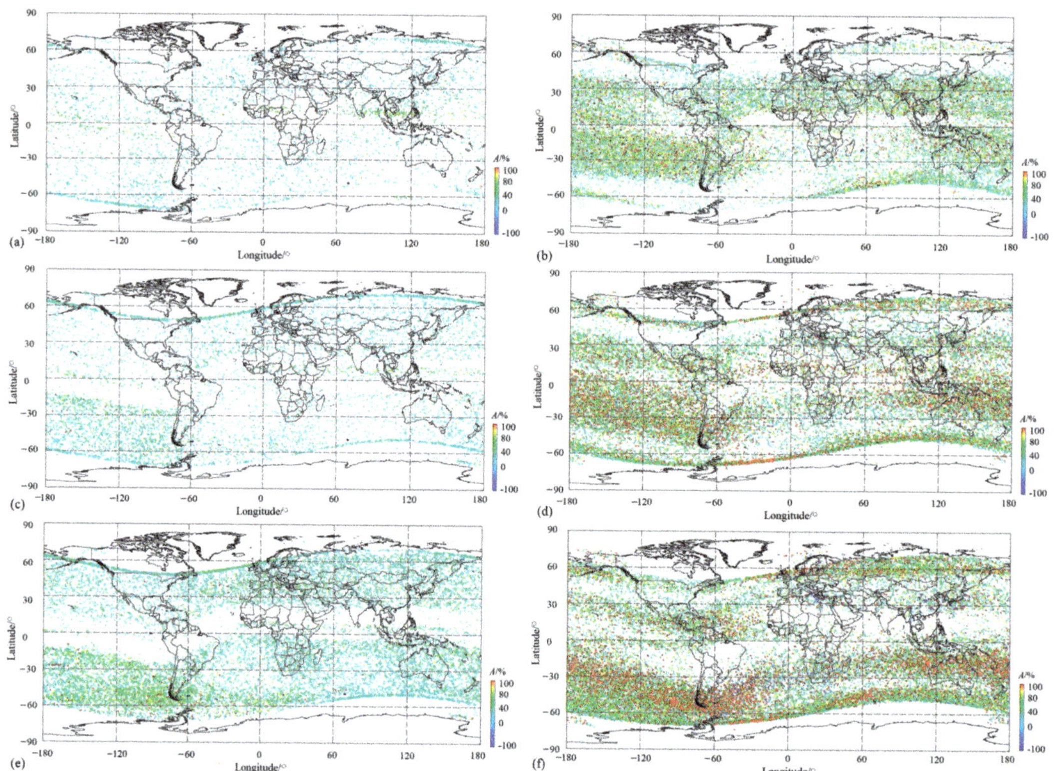

**Figure 5.** Distribution of various amplitude ion PERs with respect to different seasons and local times. (**a**) Daytime in summer; (**b**) nighttime in summer; (**c**) daytime in equinox; (**d**) nighttime in equinox; (**e**) daytime in winter; and (**f**) nighttime in winter.

The equatorial ionization anomaly (EIA) is one of the ionospheric phenomena during daytime occurring in a low-latitude F-region and characterized by an electron density trough above the magnetic equator and double crests of enhanced plasma density at approximately 15° north and south of the magnetic equator [4,53]. The longitudinal arranged wavenumber-4 (WN4) in the summer and autumn and wavenumber-3 (WN3) in the winter of plasma density also developed in the morning within the equator [54]. On the dayside, the EIA structure was exhibited well in equinox and winter, as shown in Figure 5c,e, but not in the summer, as shown in Figure 5a. This structure displays a typical feature of low-plasma PER density (Np) on both sides of the magnetic equator at a low latitude and a sudden enhancement of Np at about 15° on both sides of magnetic latitudes. At the same time, this daytime anomaly is also clearly presented with the simultaneous enhancement both in Np and space size, even beyond 200 s (Figure 6c,e). Except for this, the WN4 structure arranged longitudinally was outlined from the distributions of O+ PERs with a relatively obvious Np during equinox daytime either according to the amplitude (Figure 5c) or strengthened large space size (Figure 6c), and in winter, this structure gives way to the WN3 (See Figures 5e and 6e). Additionally, a WN-like structure along the east longitude ~60° E–120° E on the magnetic equator was discovered on the dayside in summer, with an obviously enhanced amplitude (Figure 5a) and space size (Figure 6a), which seems the most outstanding phenomenon occurring on the dayside in summer. On the other hand, on the dayside in winter, the symmetric structure of both sides of the North and South Hemispheres is also significant for the distributions of O$^+$ PERs as a function of amplitude, as well as space size (see Figures 5e and 6e).

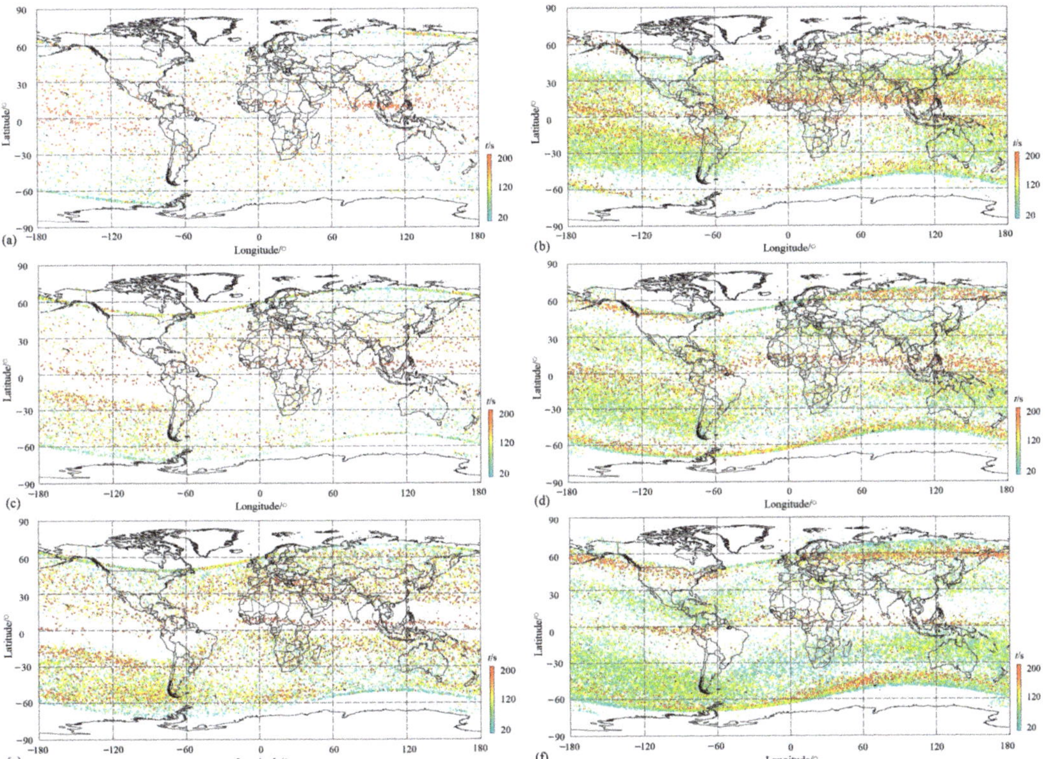

**Figure 6.** Distribution of various-space size ion PERs with respect to different seasons and local times. (**a**) Daytime in summer; (**b**) nighttime in summer; (**c**) daytime in equinox; (**d**) nighttime in equinox; (**e**) daytime in winter; and (**f**) nighttime in winter.

On the topside of the ionosphere, there is a large Wedell Sea Anomaly (WSA) zone (30° W–180° W and 30° S–75° S) [16], a summer ionospheric anomaly, which is characterized by a greater nighttime ionospheric density than that in daytime in the region near the Weddell Sea (20° W–150° W and 40° S–70° S) [55]. However, during this time, this WSA structure appeared at all times, except daytime in summer (left panels in Figures 5 and 6). On the dayside amplitude shown in Figure 5, the WSA was found both in the equinox and winter, occupying a zone (60° W–180° W and 15° S–60° S) with an enhanced Np, as well as moderate magnitude (Figure 5c,e) and large space size (Figures 5e and 6c), but this was not the case for daytime in summer (Figures 5a and 6a). Comparatively, the WSA shows its pattern more clearly during nighttime than daytime, covering a zone (30° W–180° W, 100° E–180° E, and 20° S–50° S) (See Figures 5 and 6 comparing right panels with the left ones). Contrastingly, during the daytime, with the exception of a high Np in the WSA area during nighttime, PERs with large amplitudes, for example, $A > 100\%$, displayed their significance (as can be seen in the right panels in Figure 5), especially in winter during the night (Figure 5f), but not so much ones with a large space size, for example, $t > 200$ s (as seen from the right panels in Figure 6).

The mid-latitude ionospheric trough during nighttime was characterized by a lower Np than around the area centered at 50° latitude in both hemispheres from the right panels in Figures 5 and 6. A thin WN-like structure constructed by lager-space size PERs clearly runs longitudinally above the magnetic equator on the nightside in summer (Figure 6b) and in equinox (Figure 6b). However, this pattern is slightly confusing in winter (Figure 6f) regarding space size, and the amplitude completely disappears in winter (Figure 5f), possibly due to the winter oxygen ion ($O^+$)-depleted (WOD) region at a latitude about 20°–60° at different longitudes during nighttime [23].

The nighttime South Atlantic Magnetic Anomaly (SAMA) developed mainly in equinox, especially in winter, with a high Np but without an outstanding amplitude (Figure 5d,f) or space size (Figure 6d,f) due to its negative varying properties [16,20,23,56]. At the same time, we found the phenomenon of large negative-amplitude PERs collecting locally during the night equinox and winter seasons (Figure 5d,f). For a desirable statement, we distributed all negative PERs as a function of magnitude in the map in Figure 7. In Figure 7, it can be seen that PERs with a magnitude of less than −100% gather mainly in Northern Africa, Southeast China to the Japanese Ocean, and the South Atlantic Magnetic Anomaly area, as well as a few anomalies such as the Eurasian continent and Australia. Among these, the negative ionospheric anomaly seems to be stronger in Northern Africa and Eastern Asia than in other areas.

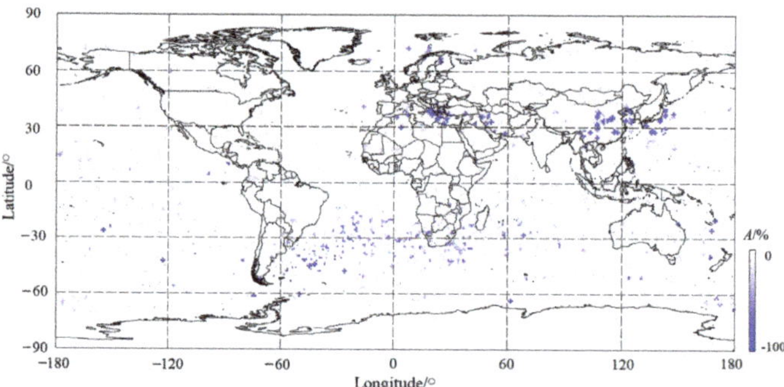

**Figure 7.** Distribution of night PERs with respect to various negative-amplitude ranges.

We examined all night PERs and found that there are 196 with an amplitude less than −100% in total, and they occurred mostly in equinox and winter, with a few occurring in summer. The occurrence probability in winter was 82.7%. The key point is that these PERs are of similar space size at ~100 s, 700 km, if a velocity of the DEMETER satellite of 7 km/s is considered, possibly illustrating the outcome of the satellite flying over the same region at different times. Xu et al. [57] reported that there are mainly five planetary-scale geomagnetic anomalies worldwide: Australia, Africa, the Southern Atlantic Ocean, the Eurasian continent, and Northern America. However, in these areas, a strong negative ionospheric anomaly was not detected in Northern America but in Eastern Asia, including Southeast China and Japan during this time.

## 4. Discussion

Ionospheric irregularity has always been a main topic of investigations of ionospheric dynamics systems, and various data from ground-based sensors and in situ measurement onboard satellites have been utilized to establish ionospheric models and main large-scale ionospheric structures [2–7,9–13,16–24], as well as their inner structure and property [58]. Unlike previous works using continuous data, during this time, automatically searched ion PERs were investigated for properties such as varied amplitude, space scale, location, and occurrence time. During this period, the influence of geomagnetic storms on the ionospheric ion density was first examined by analyzing two-group PERs occurring at a perturbed period when Kp > 4 and quiet time Kp $\leq$ 2, respectively. The results indicate that strong storms can enhance overall ionospheric irregularities more in amplitudes and less in space sizes. The impact of geomagnetic storms tends to be global, although this effect in equatorial and subauroral regions exhibits its significance more than in other areas. Geomagnetic storms generally give rise to global disturbances in the ionosphere, but ionospheric irregularities perform quite differently from one to another, and a strong spatial and local time dependency from either model or observational results mainly due to complex coupling mechanisms of the Earth's magnetosphere–ionosphere–thermosphere system [48,50]. In a geomagnetic storm, the solar wind and magnetosphere output a large amount of energy and momentum into the Earth's upper layers of atmosphere and ionosphere via enhanced particle precipitation and Joule heating. The enhanced electric fields at high latitudes can penetrate the equatorial region almost instantaneously, causing equatorial ionospheric disturbances [59,60]. Accompanying this process, the expansion of the neutral atmosphere via enhanced Joule heating at the auroral region can further drive traveling atmospheric/ionospheric disturbances [61,62]. Therefore, the influence of geomagnetic storms on ionospheric perturbance on the basis of this automatic detection method needs further research.

The dependence of PERs on local time has also been checked, and the appearance probability was 24.8% during daytime and 75.2% during nighttime, respectively. The statistical results, with respect to various amplitudes and space scales of PERs occurring at different local times, also revealed that the ratio of PERs with small amplitudes ($A < 20\%$) accounts for 64% on the dayside and 26.8% on the nightside, respectively. On the other hand, large-amplitude PERs ($A > 100\%$) occurred entirely at night. Nevertheless, the condition associated with space sizes has presented a contrary conclusion: PERs with space size $t > 120$ s occurred more frequently on the dayside than the nightside. This conclusion was also drawn from the left panels shown in Figure 6 of the dayside: PERs with a large space size appear collectively in the EIA area, which is the typical ionospheric feature during the daytime. Daytime EIA is driven by the equatorial plasma fountain effect. In the equatorial region, magnetic field lines are primarily horizontal, pointing northward, and the daytime eastward electric field drives the plasma upward via $E \times B$ drift. Under the combined force of gravity and pressure gradient, the up-lifted plasma diffuses poleward and sediments downward along the geomagnetic field lines into both hemispheres, forming two density crests alongside the magnetic equator [63,64]. The interhemispheric asymmetry EIA phenomena were constructed via daytime equinox and winter $O^+$ PERs, having two clear crests with a higher density of large-space size PERs but a lower density between them. Moreover, the WN running along the magnetic equator longitude was also established during all periods considered in this timeframe. Therefore, the equatorial plasma fountain effect can simultaneously lead to large-scale ionospheric fluctuations.

The Wedell Sea Anomaly (WSA) displays its clear boundary within the area of 60° W–180° W longitude and 15° S–60° S latitude on the dayside in equinox and winter seasons from distributions of PERs, both on amplitude and on space size. Comparatively, this phenomenon occurs during nighttime and all seasons and is characterized typically by a high density of PERs with large positive amplitudes, especially in winter during nighttime (see the right panels in Figure 5), but without obviously increased space sizes (see Figure 6). Another feature of this structure is an expanded occupation area, covering a longitude of almost 30°–180° in the West Hemisphere, 100°–180° in the East Hemisphere, and a latitude of 15° S–60° S during the night in the winter. Chen et al. [65] have presented that the WSA can extend from South America and Antarctica to the Central Pacific. The major physical mechanisms of its formation involve equatorward neutral wind, an electric field, photoionization, and downward diffusion from the plasmasphere [55,65]. In this research, an enhanced PER density also appeared at a middle latitude of 20° N–40° N and along the entire longitude in the Northern Hemisphere in summer (See Figures 5a and 6a). Horvath and Lovell [56] reported a WSA-like feature with an electron density enhancement occurring near Northeast Asia in the Northern Hemisphere. During a mid-latitude night, a plasma density enhancement exists in both hemispheres [66,67], which is known as the Mid-Latitude Summer Nighttime Anomaly (MSNA) [55]. On the other hand, the mid-latitude ionospheric trough (MIT) can also be understood based on the night seasons shown in Figures 5 and 6; this phenomenon presents a narrow latitudinal extension of several degrees ($\pm 50°$–$55°$) with a lower Np than the surroundings in both hemispheres. Its formation mechanism is the plasma "stagnation" mechanism with the interaction of high-latitude plasma convection and mid-latitude corotation flow, as well as other forces, such as a subauroral ion drift, subauroral electron temperature enhancement, and frictional heating [68–70].

Ion PERs with large negative amplitudes, for example, $A < -100\%$, were successfully detected predominantly in winter during the night in most mentioned magnetic anomalies of the world, like the Africa anomaly and SAMA (Figure 7). Xu and Bai [57] concluded that these planetary-scale non-dipole fields could heavily control the secular variation in geomagnetic fields, and the combined effect from the Africa anomaly and SAMA have tremendously modified the shape and position of the magnetic equator. Unexpectedly, we also found a large negative ion density anomaly from Southeast China to the Japanese Ocean (Figure 7), but its formation mechanism remains unknown.

## 5. Conclusions

In this research, ion density measured by the DEMETER satellite for nearly 6 years was first collected. Then, software was utilized to automatically search ion perturbations globally, and 117,718 ion PERs were attained in total. All PERs were distributed on the map with various Kp indexes to examine the effects of solar activity. The effect of the geomagnetic storm was exhibited globally, although there were regions surrounding the equator of $0°$ and mid-latitude of $50°$ in both hemispheres where this effect showed more prominently. The occurrence probabilities of PERs appearing during the disturbed period (Kp > 4) and quiet period (Kp $\leq$ 2) were checked as functions of various amplitudes and space sizes, and the results present that geomagnetic storms can completely enhance ionospheric-positive variations but rarely beyond 100% and can sometimes also induce negative ones. On the other hand, the statistical results show no clear discrepancy between the space sizes of two-group PERs occurring during the disturbed time and quiet time, although the geomagnetic storm tends to induce ionospheric irregularities with relatively larger space sizes to some degree.

Statistical work was also performed on ion PERs occurring at different local times on the dayside (10:30 LT) and nightside (22:30 LT) for the DEMETER satellite. It was testified that ionospheric variations depend heavily on the local time: 24.8% during the day and 75.2% during the night, respectively. The statistical results of the PERs occurring during daytime or nighttime according to different scales of amplitude and space size indicated that I ionospheric fluctuations with an absolute $A < 10\%$ tend to be background variations; II PERs on the dayside with a small positive amplitude (< 20%) show a significance of 64%, while this number is only 26.8% for the nightside, but there is a contrary conclusion for ones with a mid–large-amplitude of $A > 20\%$; III large positive PERs ($A > 100\%$) predominantly occurred on the nightside but rarely the dayside, and large negative ones ($A < -100\%$) only occurred on the nightside; IV there is a critical point for space size $t = 120$ s and occurrence probabilities of day PERs are always higher than that of night ones before this point, while this result is reverse after this point, which indicates that, comparatively, the ionosphere varies more frequently and more violently during the nighttime, causing relatively small-scale perturbations.

The distributions of seasonal PERs on the dayside and at nightside have displayed that there are more complex regional collections during the nighttime. These zonal collections generally show different aspects of main ionospheric structures during various seasons and local times. The EIA only exists on the dayside in equinox and winter, occupying two low-latitude crests with a lower Np in both hemispheres. The huge WSA appears during all periods except for on the dayside in summer, and is full of PERs with an enhanced amplitude, especially in winter during night. The WN-like structure can be clearly found in all seasons, showing absolutely large-scale spaces. Furthermore, several magnetically anomalous zones of planetary-scale non-dipole fields were also successfully detected by extremely negative ion PERs in this study. Therefore, the inner properties and formation physical mechanisms of these phenomena determined via automatically searched ion PERs will be the main focus of future investigations.

**Author Contributions:** Conceptualization, writing—original draft preparation, review, and editing, M.L.; validation and visualization, H.Y.; investigation and funding acquisition, Y.Z. All authors have read and agreed to the published version of the manuscript.

**Funding:** This work was supported by the Special Expenses for Basic Scientific Research under grant no. CEAIEF2022030206 and the National Natural Science Foundation of China (NSFC) under grant no. 41774084.

**Data Availability Statement:** Ion density data utilized here are available by directly contacting the first author, Mei Li, via the email mei_seis@163.com.

**Acknowledgments:** This work was supported by the Centre National d'Etudes Spatiales. It is based on observations with the plasma analyzer IAP embarked on DEMETER.

**Conflicts of Interest:** The authors declare that the research was conducted in the absence of any commercial or financial relationships that could be construed as potential conflicts of interest.

# References

1. Eckersley, T.L. Studies in radio transmission. *Inst. Electr. Eng.-Proc. Wirel. Sect. Inst.* **1932**, *71*, 405–459. [CrossRef]
2. Eckersley, T.L. Irregular ionic clouds in the E layer of the ionosphere. *Nature* **1937**, *140*, 846–847. [CrossRef]
3. Booker, H.G.; Wells, H.W. Scattering of radio waves by the F-region of the ionosphere. *J. Geophys. Res.* **1938**, *43*, 249. [CrossRef]
4. Appleton, E.V. Two anomalies in the ionosphere. *Nature* **1946**, *157*, 691. [CrossRef]
5. Aarons, J. Global morphology of ionospheric scintillations. *Proc. IEEE* **1982**, *70*, 360–378. [CrossRef]
6. Basu, S. VHF ionospheric Scintillationsat L = 2.8 and formation of stable auroral red arcs by magnetospheric heat conduction. *J. Geophys. Res.* **1974**, *79*, 3160. [CrossRef]
7. Hargreaves, J.K. Principles of the ionosphere at middle and low latitude. In *The Solar-Terrestrial Environment. An Introduction to Geospace-the Science of the Terrestrial Upper Atmosphere, Ionosphere, and Magnetosphere*; Cambridge University Press: Cambridge, UK; New York, NY, USA, 1992; pp. 208–210. [CrossRef]
8. Houminer, Z.; Aarons, J. Production and dynamics of high-latitude irregularities during magnetic storms. *J. Geophys. Res.* **1981**, *86*, 9939–9944. [CrossRef]
9. Krankowski, A.; Shagimuratov, I.I.; Ephishov, I.I.; Krypiak-Gregorczyk, A.; Yakimova, G. The occurrence of the mid-latitude ionospheric trough in GPS-TEC measurements. *Adv. Space Res.* **2009**, *43*, 1721–1731. [CrossRef]
10. Xiong, C.; Stolle, C.; Lühr, H.; Park, J.; Fejer, B.G.; Kervalishvili, G.N. Scale analysis of the equatorial plasma irregularities derived from Swarm constellation. *Earth Planets Space* **2016**, *68*, 1–12. [CrossRef]
11. Matyjaslak, B.; Przepiorka, D.; Rothkaehl, H. Seasonal Variations of Mid-Latitude Ionospheric Trough Structure Observed with DEMETER and COSMIC. *Acta Geophys.* **2016**, *64*, 2734–2747. [CrossRef]
12. Xiong, C.; Park, J.; Lühr, H.; Stolle, C.; Ma, S. Comparing plasma bubble occurrence rates at CHAMP and GRACE altitudes during high and low solar activity. *Ann. Geophys.* **2010**, *28*, 1647–1658. [CrossRef]
13. Lomidze, L.; Knudsen, D.J.; Burchill, J.; Kouznetsov, A.; Buchert, S.C. Calibration and validation of Swarm plasma densities and electron temperatures using ground-based radars and satellite radio occultation measurements. *Radio Sci.* **2018**, *53*, 15–36. [CrossRef]
14. Schunk, R.W.; Nagy, A.F. Introduction. In *Ionospheres: Physics, Plasma Physics, and Chemistry*; Cambridge University Press: New York, NY, USA, 2009; pp. 1–10. [CrossRef]
15. Li, M.; Shen, X.; Parrot, M.; Zhang, X.; Zhang, Y.; Yu, C.; Yan, R.; Liu, D.; Lu, H.; Guo, F.; et al. Primary joint statistical seismic influence on ionospheric parameters recorded by the CSES and DEMETER satellites. *J. Geophys. Res. Space Phys.* **2020**, *125*, e2020JA028116. [CrossRef]
16. Li, L.; Yang, J.; Cao, J.; Lu, L.; Wu, Y.; Yang, D. Statistical backgrounds of topside ionospheric electron density and temperature and their variations during geomagnetic activity. *Chin. J. Geophys.* **2011**, *54*, 2437–2444. [CrossRef]
17. Zhong, J.; Lei, J.; Yue, X.; Luan, X.; Dou, X. Middle-latitudinal band structure observed in the nighttime ionosphere. *J. Geophys. Res. Space Phys.* **2019**, *124*, e2018JA026059. [CrossRef]
18. Ren, Z.; Wan, W.; Liu, L.; Zhao, B.; Wei, Y.; Yue, X.; Heelis, R. Longitudinal variations of electron temperature and total ion density in the sunset equatorial topside ionosphere. *Geophys. Res. Lett.* **2008**, *35*. [CrossRef]

19. Li, Q.; Hao, Y.; Zhang, D.; Xiao, Z. Nighttime enhancements in the midlatitude ionosphere and their relation to the plasmasphere. *J. Geophys. Res. Space Phys.* **2018**, *123*, 7686–7696. [CrossRef]
20. Li, L.; Cao, J.; Yang, J.; Berthelier, J.; Lebreton, J.-P. Semiannual and solar activity variations of daytime plasma observed by demeter in the ionosphere-plasmasphere transition region. *J. Geophys. Res. Space Phys.* **2016**, *120*, e2015JA021102. [CrossRef]
21. Zhang, Y.; Paxton, L.; Kil, H. Nightside midlatitude ionospheric arcs: TIMED/GUVI observations. *J. Geophys. Res. Space Phys.* **2013**, *118*, 3584–3591. [CrossRef]
22. Shen, X.; Zhang, X. The spatial distribution of hydrogen ions at topside ionosphere in local daytime. *Terr. Atmos. Ocean. Sci.* **2017**, *28*, 1009–1017. [CrossRef]
23. Li, L.; Zhou, S.; Cao, J.; Yang, J.; Berthelier, J. Large-scale depletion of nighttime oxygen ions at the low and middle latitudes in the winter hemisphere. *J. Geophys. Res. Space Phys.* **2022**, *127*, e2022JA030688. [CrossRef]
24. Kuai, J.; Li, Q.; Zhong, J.; Zhou, X.; Liu, L.; Yoshikawa, A.; Hu, L.; Xie, H.; Huang, C.; Yu, X.; et al. The ionosphere at middle and low latitudes under geomagnetic quiet time of December 2019. *J. Geophys. Res. Space Phys.* **2021**, *126*, e2020JA028964. [CrossRef]
25. Parrot, M. Statistical analysis of the ion density measured by the satellite DEMETER in relation with the seismic activity. *Earthq. Sci.* **2011**, *24*, 513–521. [CrossRef]
26. Stangl, G.; Boudjada, M.Y. Investigation of TEC and VLF space measurements associated to L'Aquila (Italy) earthquakes. *Nat. Hazards Earth Syst. Sci.* **2011**, *11*, 1019–1024. [CrossRef]
27. Li, M.; Lu, J.; Zhang, X.; Shen, X. Indications of Ground-based Electromagnetic Observations to A Possible Lithosphere-Atmosphere-Ionosphere Electromagnetic Coupling before the 12 May 2008 Wenchuan $M_S$ 8.0 Earthquake. *Atmosphere* **2019**, *10*, 355. [CrossRef]
28. Hayakawa, M. VLF/LF Radio Sounding of Ionospheric Perturbations Associated with Earthquakes. *Sensors* **2007**, *7*, 1141–1158. [CrossRef]
29. Pulinets, S.A.; Boyarchuk, K.A.; Hegai, V.V.; Kim, V.P.; Lomonosov, A.M. Quasielectrostatic model of atmosphere-thermosphere-ionosphere coupling. *Adv. Space Res.* **2000**, *26*, 1209–1218. [CrossRef]
30. Hayakawa, M.; Molchanov, O.A. (Eds.) *Seismo-Electromagnetics: Lithosphere-Atmosphere-Ionosphere Coupling*; Terrapub: Tokyo, Japan, 2002.
31. Parrot, M.; Li, M. Statistical analysis of the ionospheric density recorded by the DEMETER satellite during seismic activity. In *Pre-Earthquake Processes: A Multi-Disciplinary Approach to Earthquake Prediction Studies*; Ouzounov, D., Pulinets, S., Hattori, K., Taylor, P., Eds.; Geophysical Monograph Series 234; AGU and John Wiley & Sons Inc.: Hoboken, NJ, USA, 2018; pp. 319–328. [CrossRef]
32. Pulinets, S.A.; Legen, A.D.; Gaivoronskaya, T.V.; Depuev, V.K. Main phenomenological features of ionospheric precursors of strong earthquakes. *J. Atmos. Sol.-Terr. Phys.* **2003**, *65*, 1337–1347. [CrossRef]
33. Dobrovolsky, I.R.; Zubkov, S.I.; Myachkin, V.I. Estimation of the size of earthquake preparation zones. *Pure Appl. Geophys.* **1979**, *117*, 1025–1044. [CrossRef]
34. Chen, Y.I.; Chuo, Y.J.; Liu, J.Y.; Pulinets, S.A. Statistical study of ionospheric precursors of strong Earthquakes at Taiwan area. In Proceedings of the XXVIth General Assembly of The International Union of Radio Science, Toronto, ON, Canada, 13–21 August 1999.
35. Píša, D.; Němec, F.; Parrot, M.; Santolík, O. Attenuation of electromagnetic waves at the frequency ~1.7 kHz in the upper ionosphere observed by the DEMETER satellite in the vicinity of earthquakes. *Ann. Geophys.* **2012**, *55*, 157–163. [CrossRef]
36. Píša, D.; Němec, F.; Santolík, O.; Parrot, M.; Rycroft, M. Additional attenuation of natural VLF electromagnetic waves observed by the DEMETER spacecraft resulting from preseismic activity. *J. Geophys. Res. Space Phys.* **2013**, *118*, 5286–5295. [CrossRef]
37. Němec, F.; Santolík, O.; Parrot, M.; Berthelier, J.J. Spacecraft observations of electromagnetic perturbations connected with Seismic Activity. *Geophys. Res. Lett.* **2008**, *35*, L05109. [CrossRef]
38. Němec, F.; Santolík, O.; Parrot, M. Decrease of intensity of ELF/VLF Waves observed in the upper ionosphere close to earthquakes: A statistical study. *J. Geophys. Res. Space Phys.* **2009**, *114*, A04303. [CrossRef]
39. Liu, J.; Qiao, X.; Zhang, X.; Wang, Z.; Zhou, C.; Zhang, Y. Using a spatial analysis method to study the Seismo-Ionospheric Disturbances of Electron Density Observed by China Seismo-Electromagnetic Satellite. *Front. Earth Sci.* **2022**, *10*, 811658. [CrossRef]
40. Korsunova, L.P.; Khegai, V.V. Analysis of seismo-ionospheric disturbances at the chain of Japanese stations for vertical sounding of the ionosphere. *Geomagn. Aeron.* **2008**, *48*, 392–399. [CrossRef]
41. Ippolito, A.; Perrone, L.; Santis, A.D.; Sabbagh, D. Ionosonde data analysis in relation with the 2016 central italian earthquakes. *Geosciences* **2020**, *10*, 354. [CrossRef]
42. Li, M.; Parrot, M. "Real time analysis" of the ion density measured by the satellite DEMETER in relation with the seismic activity. *Nat. Hazards Earth Syst. Sci.* **2012**, *12*, 2957–2963. [CrossRef]
43. Li, M.; Parrot, M. Statistical analysis of an ionospheric parameter as a base for earthquake prediction. *J. Geophys. Res. Space Phys.* **2013**, *118*, 3731–3739. [CrossRef]
44. Li, M.; Jiang, X.; Li, J.; Zhang, Y.; Shen, X. Temporal-spatial characteristics of seismo-ionospheric influence observed by the CSES satellite. *Adv. Space Res.* **2024**, *73*, 607–623. [CrossRef]
45. Parrot, M.; Berthelier, J.J.; Lebreton, J.P.; Sauvaud, J.A.; Santolík, O.; Blecki, J. Examples of unusual ionospheric observations made by the DEMETER satellite over seismic regions. *Phys. Chem. Earth* **2006**, *31*, 486–495. [CrossRef]

46. Savitzky, A.; Golay, M.J.E. Smoothing and differentiation of data by simplified least squares procedures. *Anal. Chem.* **1964**, *36*, 1627–1639. [CrossRef]
47. Gou, X.; Li, L.; Zhang, Y.; Zhou, B.; Feng, Y.; Cheng, B.; Raita, T.; Liu, J.; Zhima, Z.; Shen, X. Ionospheric Pc1 waves during a storm recovery phase observed by the China Seismo-Electromagnetic Satellite. *Ann. Geophys.* **2020**, *38*, 775–787. [CrossRef]
48. Wan, X.; Zhong, J.; Xiong, C.; Wang, H.; Liu, Y.; Li, Q.; Kuai, J.; Weng, L.; Cui, J. Persistent occurrence of strip-like plasma density bulges at conjugate lower-mid latitudes during the September 8–9, 2017 geomagnetic storm. *J. Geophys. Res. Space Phys.* **2021**, *126*, e2020JA029020. [CrossRef]
49. Prölss, G.W. *Ionospheric F-Region Storms*; CRC Press: Boca Raton, FL, USA, 2017.
50. Xiong, C.; Lühr, H.; Yamazaki, Y. An opposite response of the low-latitude ionosphere at Asian and American sectors during storm recovery phases: Drivers from below or above. *J. Geophys. Res. Space Phys.* **2019**, *124*, 6266–6280. [CrossRef]
51. Yan, R.; Zhima, Z.; Xiong, C.; Shen, X.; Huang, J.; Guan, Y.; Zhu, X.; Liu, C. Comparison of electron density and temperature from the CSES satellite with other space-borne and ground-based observations. *J. Geophys. Res. Space Phys.* **2020**, *125*, e2019JA027747. [CrossRef]
52. Lloyd, H. On earth-currents, and their connexion with the diurnal changes of the horizontal magnetic needle. *Trans. R. Ir. Acad.* **1861**, *24*, 115–141.
53. Liang, P.H. F2 ionization and geomagnetic latitudes. *Nature* **1947**, *160*, 642–643. [CrossRef]
54. He, M.; Liu, L.; Wan, W.; Lei, J.; Zhao, B. Longitudinal modulation of the O/N2 column density retrieved from TIMED/GUVI measurement. *Geophys. Res. Lett.* **2010**, *37*, L20108. [CrossRef]
55. Zakharenkova, I.; Cherniak, I.; Shagimuratov, I. Observations of the Weddell Sea Anomaly in the ground-based and space-borne TEC measurements. *J. Atmos. Sol.-Terr. Phys.* **2017**, *161*, 105–117. [CrossRef]
56. Horvath, I.; Lovell, B.C. Investigating the relationships among the South Atlantic magnetic Anomaly, southern nighttime midlatitude trough, and nighttime Weddell Sea Anomaly during southern summer. *J. Geophys. Res.* **2009**, *114*, A02306. [CrossRef]
57. Xu, W.; Bai, C. Role of the African magnetic anomaly in controlling the magnetic configuration and its secular variation. *Chin. J. Geophys* **2009**, *52*, 1985–1992. [CrossRef]
58. Liu, Y.; Xiong, C.; Wan, X.; Lai, Y.; Wang, Y.; Yu, X.; Ou, M. Instability mechanisms for the F-region plasma irregularities inside the midlatitude ionospheric trough: Swarm observations. *Space Weather* **2021**, *19*, e2021SW002785. [CrossRef]
59. Kikuchi, T.; Luhr, H.; Kitamura, T.; Saka, O.; Schlegel, K. Direct penetration of the polar electric field to the equator during a DP 2 event as detected by the auroral and equatorial magnetometer chains and the EISCAT radar. *J. Geophys. Res.* **1996**, *101*, 17161–17173. [CrossRef]
60. Nishida, A. Coherence of geomagnetic DP 2 fluctuations with interplanetary magnetic variations. *J. Geophys. Res.* **1968**, *73*, 5549–5559. [CrossRef]
61. Hocke, K.; Schlegel, K. A review of atmospheric gravity waves and traveling ionospheric disturbances: 1982–1995. *Ann. De Geophys.* **1996**, *14*, 917–940. [CrossRef]
62. Richmond, A.D.; Matsushita, S. Thermospheric response to a magnetic substorm. *J. Geophys. Res.* **1975**, *80*, 2839–2850. [CrossRef]
63. Duncan, R.A. The equatorial F-region of the ionosphere. *J. Atmos. Terr. Phys.* **1960**, *18*, 89–100. [CrossRef]
64. Xiong, C.; Rang, X.Y.; Huang, Y.Y.; Jiang, G.Y.; Hu, K.; Luo, W.H. Latitudinal four-peak structure of the nighttime F region ionosphere: Possible contribution of the neutral wind. *Rev. Geophys. Planet. Phys.* **2024**, *55*, 94–108. [CrossRef]
65. Chen, C.H.; Huba, J.D.; Saito, A.; Lin, C.H.; Liu, J.Y. Theoretical study of the ionospheric Weddell sea anomaly using SAMI2. *J. Geophys. Res.* **2011**, *116*, A04305. [CrossRef]
66. Luan, X.; Wang, W.; Burns, A.; Solomon, S.C.; Lei, J. Midlatitude nighttime enhancement in F region electron density from global COSMIC measurements under solar minimum winter condition. *J. Geophys. Res. Sp. Phys.* **2008**, *113*, 1–13. [CrossRef]
67. Lin, C.H.; Liu, C.H.; Liu, J.Y.; Chen, C.H.; Burns, A.G.; Wang, W. Midlatitude summer nighttime anomaly of the ionospheric electron density observed by FORMOSAT-3/COSMIC. *J. Geophys. Res.* **2010**, *115*, A03308. [CrossRef]
68. Kelley, M.C. *The Earth' Ionosphere, Plasma Physics and Electrodynamics*; Academic Press: Cambridge, MA, USA, 1989.
69. Anderson, P.C.; Heelis, R.A.; Hanson, W.B. The ionospheric signatures of rapid subauroral ion drifts. *J. Geophys. Res.* **1991**, *96*, 5785–5792. [CrossRef]
70. Ishida, T.; Ogawa, Y.; Kadokura, A.; Hiraki, Y.; Haggstrom, I. Seasonal variation and solar activity dependence of the quiet-time ionospheric trough. *J. Geophys. Res. Sp. Phys.* **2014**, *119*, 6774–6783. [CrossRef]

Article

# A PLL-Based Doppler Method Using an SDR-Receiver for Investigation of Seismogenic and Man-Made Disturbances in the Ionosphere

Nazyf Salikhov [1], Alexander Shepetov [1,2,*], Galina Pak [1], Vladimir Saveliev [1,3], Serik Nurakynov [1], Vladimir Ryabov [2] and Valery Zhukov [2]

[1] Institute of Ionosphere, Almaty 050020, Kazakhstan; snurakynov@ionos.kz (S.N.)
[2] P. N. Lebedev Physical Institute of the Russian Academy of Sciences, Moscow 119991, Russia
[3] V. G. Fesenkov Astrophysical Institute, Almaty 050020, Kazakhstan
[*] Correspondence: ashep@tien-shan.org

**Abstract:** The article describes in detail the equipment and method for measuring the Doppler frequency shift (DFS) on an inclined radio path, based on the principle of the phase-locked loop using an SDR receiver for the investigation of seismogenic and man-made disturbances in the ionosphere. During the two M7.8 earthquakes in Nepal (25 April 2015) and Turkey (6 February 2023), a Doppler ionosonde detected co-seismic and pre-seismic effects in the ionosphere, the appearances of which are connected with the various propagation mechanisms of seismogenic disturbance from the lithosphere up to the ionosphere. One day before the earthquake in Nepal and 90 min prior to the main shock, an increase in the intensity of Doppler bursts was detected, which reflected the disturbance of the ionosphere. A channel of geophysical interaction in the system of lithosphere–atmosphere–ionosphere coupling was traced based on the comprehensive monitoring of the DFS of the ionospheric signal, as well as of the flux of gamma rays in subsoil layers of rocks and in the ground-level atmosphere. The concept of lithosphere–atmosphere–ionosphere coupling, where the key role is assigned to ionization of the atmospheric boundary layer, was confirmed by a retrospective analysis of the DFS records of an ionospheric signal made during underground nuclear explosions at the Semipalatinsk test site. A simple formula for reconstructing the velocity profile of the acoustic pulse from a Dopplerogram was obtained, which depends on only two parameters, one of which is the dimension of length and the other the dimension of time. The reconstructed profiles of the acoustic pulses from the two underground nuclear explosions, which reached the height of the reflection point of the sounding radio wave, are presented.

**Keywords:** Doppler frequency shift; phase-locked loop; PLL; SDR receiver; M7.8 earthquakes; acoustic wave; lithosphere-atmosphere-ionosphere coupling; seismogenic effects; underground nuclear explosion

**Citation:** Salikhov, N.; Shepetov, A.; Pak, G.; Saveliev, V.; Nurakynov, S.; Ryabov, V.; Zhukov, V. A PLL-Based Doppler Method Using an SDR-Receiver for Investigation of Seismogenic and Man-Made Disturbances in the Ionosphere. *Geosciences* **2024**, *14*, 192. https://doi.org/10.3390/geosciences14070192

Academic Editors: Jesus Martinez-Frias and Masashi Hayakawa

Received: 5 June 2024
Revised: 26 June 2024
Accepted: 27 June 2024
Published: 16 July 2024

## 1. Introduction

One of the generally accepted methods for studying the connection between ionospheric disturbances and helio–geophysical processes is the radiophysical method of Doppler sounding of the ionosphere in the shortwave frequency range. The essence of the method is that when a radio wave propagates through the ionosphere, its frequency changes due to the time variability of electron concentration, and the Doppler effect arises. The higher the variation rate of the electron concentration, the larger the observed value of Doppler frequency. An important stage in the development of Doppler observations is associated with the use of publicly available cesium or rubidium frequency standards, which makes it possible to determine the phase and frequency of oscillations with great accuracy. Doppler frequency measurements are taken both with vertical and inclined sounding of the ionosphere. At present, both methods are widely used in studies of the response of the ionosphere to various disturbances.

Doppler frequency measurements on an inclined radio path were carried out for the first time by Essen in 1935, and then they were actively developed in the 1960s [1,2]. Of fundamental importance and impact for further research were the data of Davis and Baker [3] on the possible connection between the disturbances in the atmosphere and ionosphere around the time of the Alaskan earthquake on 28 March 1964. Over the next half-century and beyond, significant progress was made in the study of the state of the ionosphere disturbed by various sources of natural and artificial origin, which is summarized in a number of reviews [4–12]. A significant part of the works is devoted to the application of Doppler methods for studying the ionosphere under various helio–geophysical and man-made disturbances [13–18].

However, the first studies revealed a number of limitations for the use of the Doppler sounding technique of the ionosphere and showed that, due to the presence of ionospheric heterogeneities, reflected signals arrive at a radio-receiving device in different ways. As a result, a multipath of radio signals is observed, and the problem of interference distortion arises. Measuring Doppler frequencies in multipath conditions is a most difficult issue. In many Doppler installations, spectrum analyzers are required to select individual beams, although using spectrum analyzers, in turn, leads to limitations in the accuracy of measurements of both time and frequency. As is known, the accuracy of spectral analysis depends on the duration of continuous measurement. This is why, for example, to ensure the measurement accuracy of the Doppler frequency shift of 0.01 Hz, a time of about 100 s or more is required. Then, by constructing a time plot of Doppler frequency variations, a detailed description of the disturbances with periods of less than 100 s will be impossible in many cases.

Since the 1970s, the Doppler method of ionospheric sounding has been actively applied for the research of the ionosphere and radio wave propagation in the Ionosphere Sector of the Academy of Sciences of the Kazakh SSR and then at the Institute of Ionosphere of the Republic of Kazakhstan. At first, only vertical sounding was used, and then sounding on an inclined radio path. However, though the accuracy of Doppler frequency measurement with a single-path signal was very high, in the case of the multipath propagation of radio waves, certain difficulties arose in determining the Doppler frequencies and interpreting research results. It was found that the percentage of single-path communication sessions was small, usually 7–9% of their total number, and varied for different radio paths, from 7% to 30% on average [19]. Therefore, the following tasks were assigned by the employers of the Institute of Ionosphere: to ensure the possibility of the continuous registration of the Doppler frequency shift of ionospheric signals, to solve the problem of interference distortions caused by multipath signals, and to increase the accuracy of the measurements.

Further development of the Doppler method at the Institute of Ionosphere was associated with the application of the phase-locked loop (PLL) system. This allowed the registration of the Doppler frequency of a larger-amplitude beam when there is the multipath propagation of radio waves. With this method, the responses of the ionosphere to industrial explosions and underground nuclear explosions have been studied, and the ionospheric effects have been recorded over time intervals from a few tens of seconds to several hours [20–24]. Using the Doppler method with application of PLL has significantly expanded the capabilities of recording ionospheric responses to various natural sources of disturbances, including earthquakes [25,26], volcanic eruptions [27], X-ray and ultraviolet solar flares [28], and the launches of carrier rockets from the cosmodromes "Baikonur" and "Vostochny" [29].

This article describes the method and equipment for measuring the Doppler frequency shift at an inclined sounding of the ionosphere, which is based on the PLL operation principle using a Software-Defined Radio (SDR) receiver. To process the data received and identify the signatures of disturbances, a special program was developed that uses digital filters, correlation, and spectral methods. Then, some examples of the application of the hardware–software complex of Doppler measurements for studying seismogenic disturbances in the ionosphere are considered.

## 2. The Hardware-Software Complex for Doppler Sounding of the Ionosphere

For the registration of the ionospheric response to disturbances propagating from the lithosphere to the height of the ionosphere, we use the method of Doppler sounding on an inclined radio path with continuous carrier in conjunction with a PLL system and an SDR receiver. The sounding radio wave is reflected from the ionosphere and received by the Doppler radio receiver on the ground with some frequency shift. The receiving Doppler equipment is located in the two measuring points: Institute of Ionosphere (43.17594° N, 76.95342° E) and Radiopolygon Orbita(43.05831° N, 76.97361° E). The measurements of Doppler frequency are carried out on a 13.2 km long low-inclined radio path divided by a mountain ridge, which reduces the passage of the direct radio wave from the transmitter to the Doppler receiver. A special Doppler radio transmitter is used, synchronized by the rubidium frequency standard FE 5650A from "MORION", INC. (St. Petersburg, Russia).

When measuring the Doppler frequencies on a longer inclined radio path with the length in the range of (200–3000) km, the transmitters of broadcasting stations are employed, as selected from the BBC Short-Wave catalog [30] for different times of day and seasons. The radio paths used in Doppler measurements are listed in Table 1, and the diagram of their relative layout is shown in Figure 1.

**Table 1.** The list of radio transmitters whose signals were used for the Doppler sounding of the ionosphere.

| Transmitter Point | Geographic Coordinates | Length of the Radio Path, km | Frequency, kHz | Power of the Transmitter, kW |
|---|---|---|---|---|
| Kuwait | 29.51306° N, 47.67306° E | 3010 | 5860 | 250 |
| Urumqi (China) | 44.15944° N, 86.89917° E | 808 | 5960/9560 | 100 |
| Dushanbe-orzu (Tajikistan) | 37.53778° N, 68.79389° E | 908 | 7245 | 100 |
| Beijing (China) | 39.74750° N, 116.81361° E | 3326 | 7210/7215 | 500 |
| Kashi-Saibagh (China) | 39.36444° N, 75.71611° E | 423 | 7205 | 100 |
| Kujang (DPRK) | 40.07833° N, 126.10861° E | 4056 | 7570 | 200 |
| Institute of Ionosphere, Almaty (Kazakhstan) | 43.17594° N, 76.95342° E | 13.2 | 2963/5121 | 0.3 |

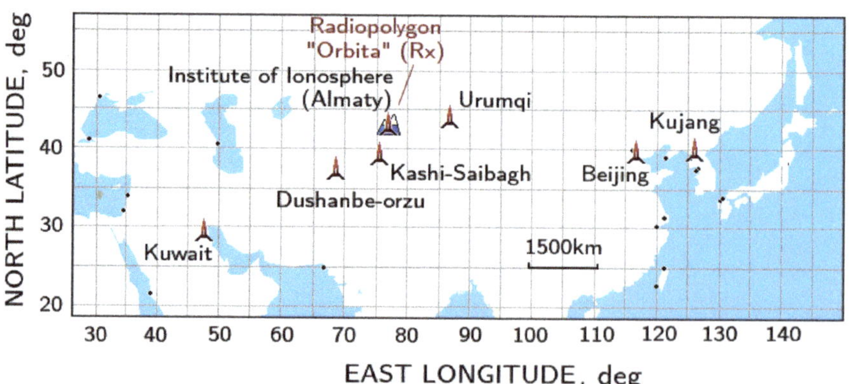


**Figure 1.** The disposition scheme of radio transmitters whose signals are used for the Doppler sounding of the ionosphere, relative to the radio receivers (RX) in the points with Doppler installations at the Institute of Ionosphere and Radiopolygon Orbita.

For the measurement of Doppler frequencies in expeditionary mobile conditions, a receiving Doppler installation was developed on the basis of SDR technology. The hardware-software complex provides a round-the-clock monitoring of the shift of Doppler frequency with remote control and access to the data over the Internet.

*2.1. The Doppler Receiver*

The operation of the receiving Doppler installation utilizes the principle of the PLL based synchronous-phase detector. The main advantages of the application of a PLL are the high sensitivity to small disturbances, the high time resolution, and the ability to operate in multipath conditions which result from the spatiotemporal heterogeneity of the ionosphere and specific propagation features of shortwave radio waves [21].

Previously, in the Doppler installation, we used a radio receiver of the R-399 type, built in accordance with superheterodyne circuitry with a synthesizer-type local oscillator, where a rubidium frequency standard was used as a reference oscillator. At present, with development of the Software-Defined Radio technology [31], a possibility appeared to include an SDR receiver into our Doppler installation instead of the R-399 radio receiver.

The operation principle of SDR is based on the real-time digitization of received radio signal and its further processing in digital form, such that the main treatment of the signal is performed by a computer. A broadband, full-featured 14-bit SDR receiver of RSPdx type [32] covers the radio frequency spectrum from 1 kHz to 2 GHz, providing the spectrum width of up to 10 MHz. The RSPdx receiver operates together with the SDRuno's own SDRplay software application designed for the Windows operating system. The important condition for the use of the RSPdx receiver in the Doppler installation is the ability to connect to it an external highly stable rubidium frequency standard.

The functional diagram of the receiving part of the Doppler installation based on the SDR receiver is shown in Figure 2. The signal from a highly stable ($\Delta f / f \simeq 10^{-9}$) radio transmitter, after reflection from the ionosphere, arrives via an antenna to the input of the SDR receiver of RSPdx type, which is connected to a computer through the USB port. The SDRuno program, designed for processing the signal from the RSPdx receiver, is set to the CW telegraph signals receiving mode in the band of 150 Hz. As a result, at the output of the computer sound card the signal appears with the frequency of about 1 kHz, which contains the information on the Doppler frequency of the ionospheric signal.

Next, a PLL is organized for selection of the Doppler frequency shift of the ionospheric signal. For this purpose, the 1 kHz signal is transferred by a mixer to a frequency of 215 kHz and then, after filtering with a narrow-band electromechanical filter ( *Filter 300 Hz*), it is fed to one of the inputs of a phase detector. The signal of a voltage-controlled oscillator (VCO) comes to the other input of the phase detector. The output signal of the phase detector is connected to a low-pass filter of proportional-integrating type (*Filter #1*), which determines the parameters of the PLL. This signal acts on the VCO and adjusts it to the frequency of the received signal, such that the change in the voltage at the output of the phase detector is proportional to the Doppler shift. After the passage of a DC amplifier and another low-pass filter (*Filter #2*), the voltage is digitized by an analogue-to-digital converter (ADC) of E-154 type (L-card, Moscow, Russia), and the results are kept as a file in the computer memory.

The selection principle of the Doppler frequency of a larger-amplitude beam in the case of mutipath signal is described below in Section 2.2.

For remote control, the computer of the Doppler installation is connected to the Internet via a Wi-Fi modem. The on-board time of the computer is synchronized over the network from an NTP server of an atomic frequency and time standard. A special computer program was designed to automate the measurement process, to organize the collection of ADC data, and to visualize the results of the measurement.

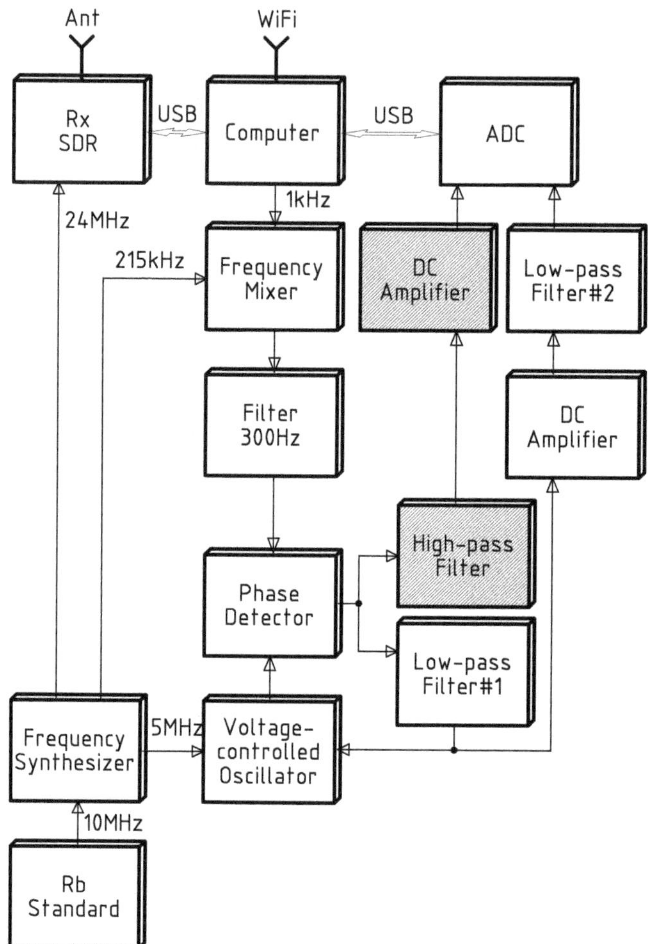



**Figure 2.** Functional scheme of the receiving part of the hardware-software complex for measuring the Doppler frequency shift of the ionospheric radio signal using the SDR receiver. See the text for explanations.

The non-linearity of the PLL based conversion of Doppler frequency into voltage was verified by a reference oscillator with hopped adjustment of the output frequency. A highly stable signal was tuned in steps of 0.5 Hz within a frequency range of 5 Hz and fed to the input of the RSPdx receiver. Next, a graph was drawn of the dependence of frequency change on the number of tunings $N$. As a result, the real transformation characteristic was obtained, denoted as $R$ in Figure 3. The analytic linear approximation of this characteristic looks as $f(N) = 1.012 \cdot N - 3.396$.

As it follows from Figure 3, some non-linearity was observed by comparison of the ideally linear characteristic $I$ with the real conversion characteristic $R$. Within the PLL hold band, $f_{hold} = 15$ Hz, the non-linearity of the real characteristic was less than 1%, and had the value of 0.46%. This is quite sufficient for the high quality measurement of the Doppler frequency shift of the radio signal reflected from the ionosphere in the short-wave frequency range.

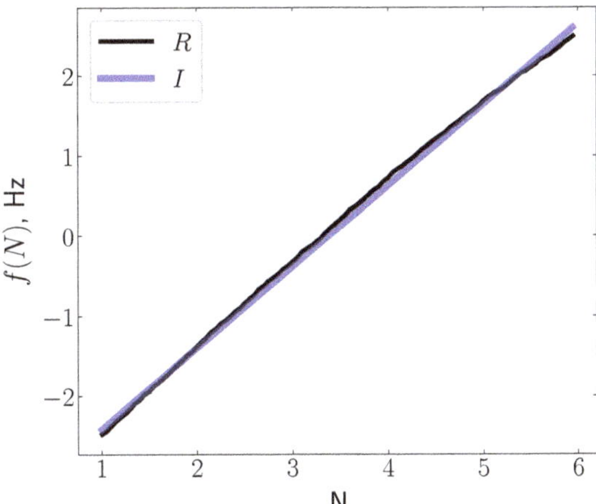

**Figure 3.** Comparison of the real characteristic ($R$) of the PLL conversion of the Doppler equipment with the ideal linear characteristic ($I$). The units along the horizontal axis are the number $N$ of the succeeding frequency tunings (see text). The plot is taken from the publication Salikhov and Somsikov, 2014 [33].

### 2.2. Measuring Doppler Frequencies in Multipath Conditions

As noted above, with single-beam receiving, the accuracy of the Doppler frequency measurement is very high, and is determined only by the frequency instability of reference generator and by the characteristics of ADC equipment. In contrast, by receiving a multipath signal, the interference beats occur between the incoming beams. In dependence on the ratio between the amplitudes of interfering signals, the instantaneous values of frequency spikes can reach tens of Hz, significantly exceeding the Doppler frequencies.

With multipath propagation of radio waves, it is often the case that one of the beams has a higher amplitude than the others. By tracking with a PLL based synchronous-phase detector the behavior of a larger-amplitude beam and averaging its instantaneous values with a low-pass filter, one can quite accurately measure the Doppler frequency of the larger-amplitude beam even in multipath signal.

The measurement accuracy of the Doppler shift of ionospheric signal was determined in the measurements at a special test facility which imitated the conditions of multipath receiving. For the check of the Doppler installation, it was sufficient to use a pair of highly stable oscillators. As shown in Figure 4, two signals from the generators #1 and #2 come, through attenuators, to a frequency mixer, and the interference signal from the mixer output is fed to the input of the PLL-based Doppler equipment. Then, the signal at the output of the PLL is digitized by an ADC and recorded in the computer memory as a text file. The resulting signal of the beating between two frequencies can be simulated with the help of this test facility.

The signal at the output of the mixer is illustrated by the vector diagram shown in Figure 5, which represents most clearly the physical essence of the phenomenon. As known, on a vector diagram any number of the beams received can be reduced to the two vectors with time-varying amplitude and frequency [19]. With account to this fact, it is sufficient to consider in detail the arrival of only two beams at the input of a radio receiver device. In the simplest case we have the beating of two sinusoids, $A1$ and $A2$, and their resulting oscillation can be determined using the vector diagram of Figure 5. The diagram is built for the case when the circular frequency $\omega_2$ is higher then $\omega_1$, and the amplitude of the beam $A1$ significantly exceeds that of $A2$, $K = A_2/A_1 \ll 1$.

Let's suppose that the vector $A1$ revolves with the frequency $\omega_1$ together with the drawing plane, such that it remains immovable on the diagram, while the vector $A2$ rotates with the velocity equal to the circular frequency of the beating, $\omega_d = \omega_2 - \omega_1$. Then the resulting vector $A3$, which represents the geometric sum $A1 + A2$, turns out to be modulated in its phase (frequency) and amplitude with the frequency of the beating $\omega_d$. The frequency of the vector $A3$ oscillates around the frequency of the larger-amplitude beam ($A1$) at all values $K < 1$. Consequently, the smaller $K$, the less the instantaneous value of the resulting frequency differs from the frequency of the beam with the larger amplitude.

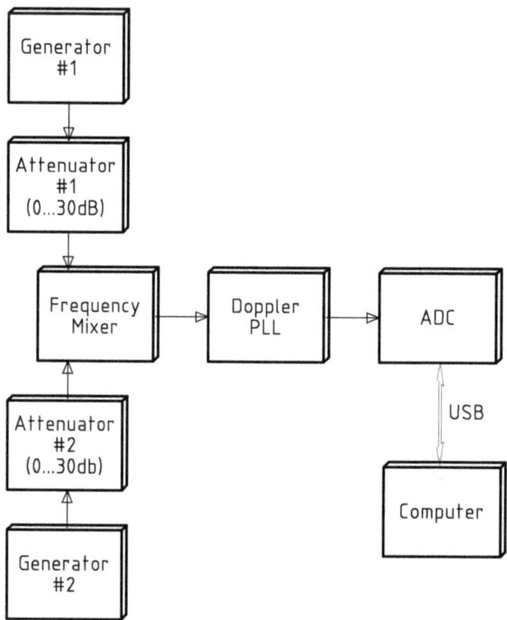


**Figure 4.** Schematic of the testing facility to check the operation of the hardware-software complex of Doppler measurements in the imitated conditions of multipath receive.

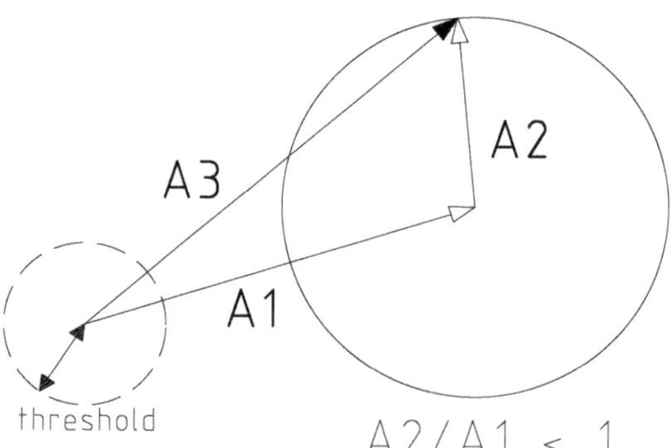


**Figure 5.** The vector diagram of the resulting oscillation in a sum of two sinusoidal signals. Designations: $A1$—the vector of the larger-amplitude beam, $A2$—the vector of the smaller beam, $A3$—the resulting vector, which equals to the geometric sum of the vectors $A1$ and $A2$, modulated in phase (frequency) and amplitude by the frequency of the beating.

In Figure 6 an experimental record is shown of the beating of two signals with different amplitude ratios, $K = 0.2$, $K = 0.5$, and $K = 0.85$, as registered by the Doppler installation at the output of the frequency mixer (see Figure 4). It is seen that in the case of $K = 0.85$ (when the beam $A1$ is only 15% larger than $A2$), the resulting frequency deviates from the frequency of the larger-amplitude beam and reaches $-4\,\text{Hz}$. With the amplitude ratios of $K = 0.5$ and $K = 0.2$, the deviation of the resulting frequency is significantly lower.

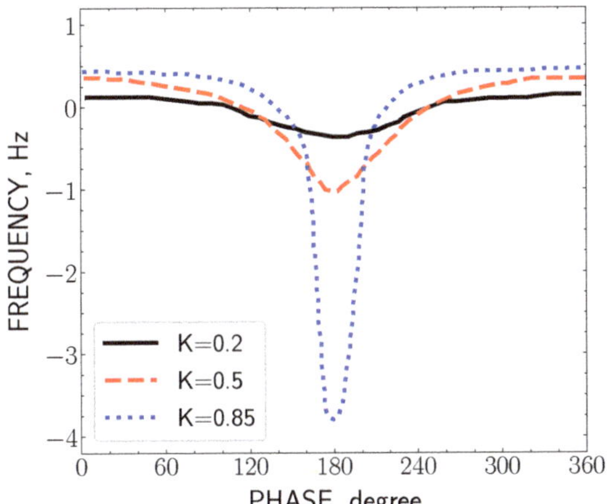


**Figure 6.** The change in the resulting frequency, per a cycle of the smaller-amplitude vector *A2*, as measured with the different ratios *K* between the amplitudes of the two sinusoids *A1* and *A2*.

According to the mathematical analysis of beats [34], in an ideal case the areas under the positive and negative parts of the curves in Figure 6 are equal and opposite in sign. When integrating the oscillations of the difference frequency with a low-pass filter over a single period, a straight line will be obtained coinciding on the graph with the X axis. If the frequency of the beam with larger amplitude varies, the graph obtained after application of integrating filter to the oscillations of difference frequency will repeat the frequency variation of the larger amplitude beam. In the Doppler installation, the averaging of the difference frequency signal proceeds due to the integrating filter (*Low-pass Filter #2* in Figure 2). By proper choice of the time constant of this filter, it is possible to adapt the installation to different conditions of multipath receive.

By the conversion of signals in the PLL system, some errors may appear caused both by the operation of electronic devices and the conditions of radio wave propagation. To reveal such errors, the special test measurements were made using the testing facility from Figure 4 with two highly stable generators. One generator was always tuned to the frequency of 5 MHz, the frequency of the other was changed between the limits of 5 MHz + 0.5 Hz and 5 MHz + 7 Hz, which corresponds to the change in the difference frequency $F_d$ from 0.5 Hz to 7 Hz. The parameters of the PLL of the Doppler installation, both the double width of the hold band $2 \cdot \Delta f_{hold} = 3.2\,\text{Hz}$ and the time constant of the low-pass filter $2 \cdot \tau = 20\,\text{s}$, remained the same throughout entire measuring time.

In Figure 7 a family of curves is shown which demonstrate the dependence of the measurement error of difference frequency for the various ratios *K* between the amplitudes of the summed signals: $K = 0.85$, $K = 0.5$, and $K = 0.2$. As it follows from this plot, the measurement error of Doppler frequency increases with the rise of the difference frequency $F_d$. For any value of $K < 0.85$ and $F_d < 8\,\text{Hz}$, the error always remains below 0.04 Hz. With decreasing $K$, the value of errors diminishes also, and at $K = 0.2$ the error does not exceed 0.002 Hz. Also, a characteristic segment is seen in the graph of Figure 7, where all curves run parallel to the abscissa axis, and their bend begins at a value of $F_d \simeq 2\,\text{Hz}$.

Therefore, with optimal choice of the PLL double hold band $2 \cdot \Delta f_{hold}$, the accuracy of the Doppler shift measurement with multipath signal can be better than 0.01 Hz, which is 1.5–2 orders of magnitude below the background variations level of the Doppler frequency shift of ionospheric signal.

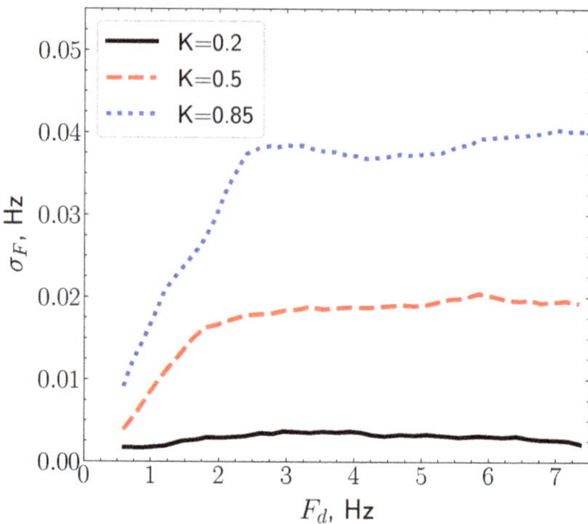



**Figure 7.** The measurement error of the difference frequency $F_d$ determined for the various amplitude ratios $K$.

In Figure 8 an example is shown of the Doppler frequency shift registration by receiving of a real two-beam ionospheric signal on the inclined radio path "Urumqi—Institute of Ionosphere" at a frequency of 9560 kHz. It can be seen there that by selection of the interference signal with the low-pass filter, the Doppler frequency of the larger-amplitude beam was clearly distinguished.

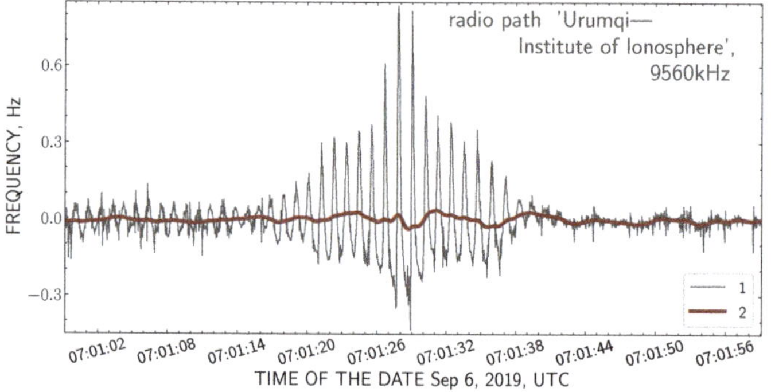



**Figure 8.** Receiving a two-beam ionospheric signal by the hardware-software complex of Doppler measurements in realistic conditions. *1*—the interfering signal at the output of PLL, *2*—the variation of the Doppler shift of the larger amplitude signal selected by the low-pass filter (*Filter #2 in Figure 2*).

### 2.3. Registration of Short-Term Ionospheric Bursts in the Records of Doppler Frequency

In the monograph [35] Ya. L. Alpert mentioned the existence of unusual transient ionization flashes which appear and shortly vanish in small regions of the ionosphere. When analyzing the records of the Doppler ionosonde, our attention was drawn by the presence of short-term Doppler bursts with the amplitudes essentially above the background. For further analysis, the detection of such isolated bursts was selected into a special channel of the Doppler installation. A sample of the data recorded in this channel is shown in Figure 9. It was found that the intensity of such bursts was increasing, sometimes considerably, under the influence on the ionosphere of various disturbing factors as, e.g., the sunrise and sunset periods and solar flares. This observation gave the ground to suppose that the intensity of the bursts of Doppler frequency should reflect the state of disturbance of the ionosphere. In this regard, an idea has arisen to measure the intensity of the ionization flashes during Doppler measurements, and the principle of the counting of short-term Doppler bursts was realized at the modification of the PLL based Doppler equipment.

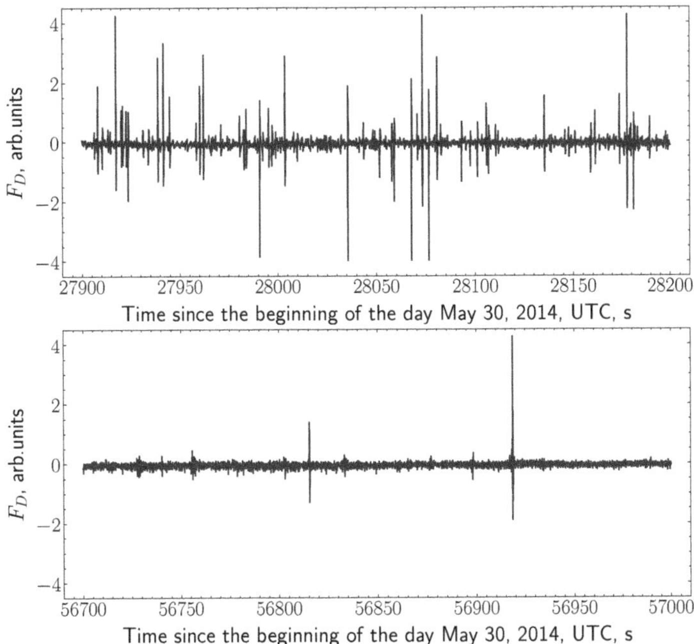

**Figure 9.** Two samples of the short-term Doppler bursts reflecting the ionization flashes in the ionosphere. The vertical axis is expressed in the units of Doppler frequency $F_D$.

For identification of the short-term Doppler bursts, additional units were incorporated into the functional scheme of the hardware-software complex of Doppler measurements: a high-pass frequency filter and a DC amplifier with the regulated gain. In Figure 2 above these two blocks are indicated by hatching. The signal from the output of the phase detector comes to the filter, where its high-frequency component is selected. Then, after amplification, this signal is digitized with a periodicity of 25 Hz by the second ADC channel, and the results are kept in the computer memory. The calculation and graphical visualization of the variations in the intensity of short-term Doppler bursts is made by a specially developed program.

## 3. Response of the Ionosphere in the Doppler Frequency Shift to Disturbances of Seismogenic Origin

Each strong earthquake presents unique opportunity to study the response of the ionosphere both prior and during the earthquake, and thereby better understand the lithosphere-atmosphere-ionosphere coupling and identify the anomalies which could be used for earthquakes prediction. With the development of the Doppler method of radio sounding of the ionosphere, significant progress was made in the investigation of the ability of acoustic waves to transmit energy from the ground to the uppermost layers of the atmosphere. It was shown that the Doppler method is highly effective for detecting the impact of acoustic waves on the ionosphere even at weak earthquakes and at the earthquakes with the large distance to the epicenter [17,36].

*3.1. The Ionospheric Response in the Doppler Frequency Shift to the M7.8 Earthquake in Nepal on 25 April 2015*

The M7.8 25 April 2015 earthquake in Nepal occurred at 06:11:23 UTC. The epicenter was located at 28.231° N, 84.731° E, and the depth of its focus was of 8.2 km [37]. The disposition of the hardware-software complex of Doppler measurements relative to the epicenter of the earthquake is presented in the diagram of Figure 10. In these measurements, we used the transmitter of a broadcast radio station in Urumqi, which operated at a frequency of 5960 kHz. The radio receiver of the Doppler installation is placed in the Radiopolygon Orbitapoint, at a distance of 805 km from the transmitter. The projection of the radio wave reflection point on the ground (sub-ionospheric point) was located at a distance of 1732 km from the epicenter of the earthquake. The Dobrovolsky radius ($R_D = 10^{0.43M}$ km), which determines the size of the zone of deformation processes in the period of earthquake preparation [38], equals to 2259 km for a M7.8 earthquake. Therefore, the monitoring results of Doppler frequency obtained in the said sub-ionospheric point can be used for an objective estimation of the state of the ionosphere at this earthquake.

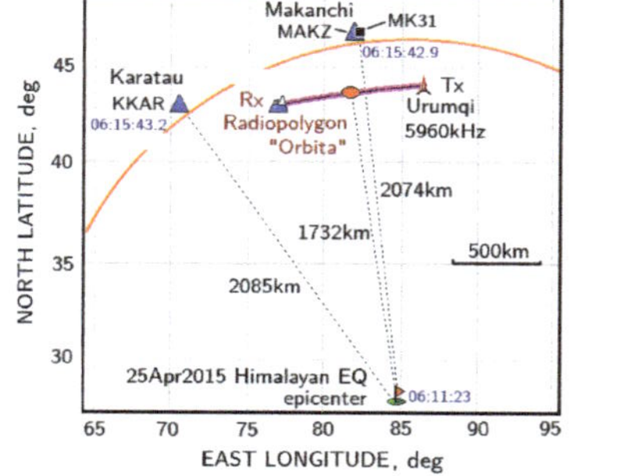


**Figure 10.** The disposition scheme of the radio path of Doppler measurements and of the seismic network stations of the National Nuclear Center in Makanchi (MAKZ, MK31) and Karatau (KKAR), in relation to the epicenter of the 25 April 2015 M7.8 earthquake. *Tx*—the radio transmitter (44.15944° N, 86.89917° E), *Rx*—the receiver (43.05831° N, 76.97361° E). The red oval indicates the projection of the reflection point of the sounding radio wave (sub-ionospheric point).

An important advantage of Doppler method is its high sensitivity and ability to detect the disturbances in the ionosphere even at considerable distance from an earthquake. Thus, as shown in Figure 11, the seismogenic disturbance from the Nepal earthquake was distinctly observed in the Doppler frequency shift at a distance of 1732 km from the epicenter. The disturbance, with an amplitude above 2 Hz, was detected at 06:28:35 UTC, approximately 17 min (1032 s) after the main shock, and lasted about 91 s.

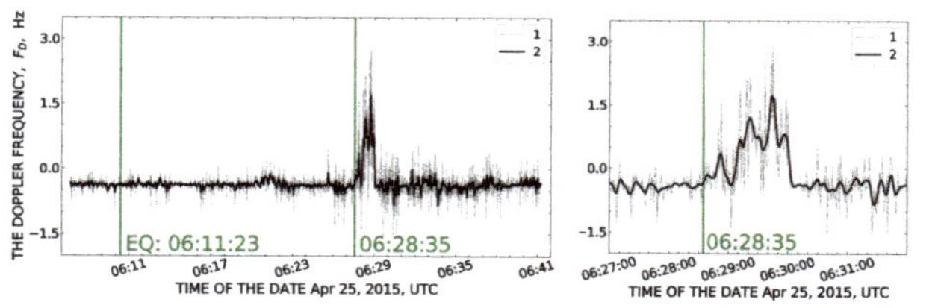


**Figure 11.** The response of the ionosphere to the M7.8 earthquake in Nepal on 25 April 2015 registered by the Doppler ionosonde on the "Urumqi—Radiopolygon Orbita" radio path. *1*—the original measurement data; *2*—same data after application of the 10 points running average filter.

### 3.1.1. Determination of the Arrival Time of the Rayleigh Wave to the Sub-Ionospheric Point and the Reflection Height of the Sounding Radio Wave

The acoustic waves generated by the propagation of surface Rayleigh waves cause disturbances in the ionosphere and can be detected by a Doppler ionosonde. The time of response appearing in the Doppler shift record is the sum of the propagation time of the seismic wave to the sub-ionospheric point, plus the time necessary for an acoustic wave to spread from the ground up to the reflection point of the sounding radio wave in the ionosphere.

In the case of the 25 April 2015 Nepal earthquake, no seismic station was situated near the sub-ionospheric point, where the arrival time of the Rayleigh seismic wave could be directly determined. For this purpose, the information was used gained at the closest seismic stations of the National Nuclear Center of the Republic of Kazakhstan: Karatau (KKAR), Makanchi (MAKZ, MK31) and KNDC (Almaty). The data of these stations are accessible at the site [39].

In the seismograms of the vertical Z-component recorded on 25 April 2015 and shown Figure 12, it is clearly seen the successive passage of the Rayleigh wave preceding the appearance of the disturbance in the Doppler shift. It is observed also a certain similarity between the shape of the Doppler shift record and the waveform of the Z-component of seismic wave at the station MAKZ, which was the closest to the sub-ionospheric point.

The information necessary for the calculation of the speed and arrival time of the Rayleigh wave to the sub-ionospheric point is listed in Table 2. It is seen, that the seismic stations were located at an approximately same distance from the epicenter of the earthquake. The arrival time of the seismic surface wave to the sub-ionospheric point, 630.5 s, was deduced as an arithmetic mean of the indications of nearest seismic stations.

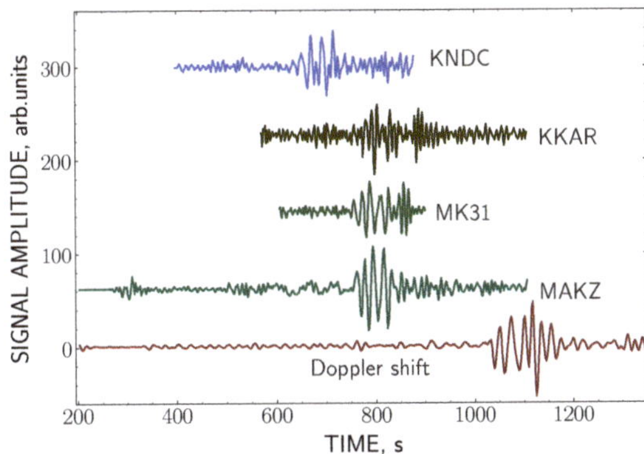


**Figure 12.** The fragments of the Doppler shift record and of the seismograms of Z-component written on 25 April 2015 at different distances from the epicenter, at the stations in Makanchi (MAKZ, MK31), Karatau (KKAR), and Almaty (KNDC). The scale of the horizontal axis is expressed in seconds passed since the moment of the earthquake.

Now, knowing the appearance time of the disturbance in the ionosphere, 1032 s, and the arrival moment of the Rayleigh wave to the sub-ionospheric point, 630.5 s, the time required for an acoustic wave to propagate from the earth's surface to the reflection point of the sounding radio wave in the ionosphere can be calculated as $1032 - 630.5 = 401.5$ s.

To determine the propagation trajectory and the reflection height of radio waves, we used the profile of the electron concentration calculated on the basis of the IRI2016 model [40] for the middle point of the radio path, $43.57\,°$ N, $81.75\,°$ E. The calculation was made directly by the program resided at the model web site [40]. The parameters necessary for the calculation: the index of solar activity $F10.7 = 126$ and the index of magnetic activity $Ap = 3$ were taken for the date of interest, April 25th, from the sites, correspondingly, [41] and [42].

**Table 2.** Parameters of the seismological stations used for calculation of the arrival time of the Rayleigh wave to the sub-ionospheric point.

| Point | Geographic Coordinates | Epicenter Distance, km | Arrival Time of the Rayleigh Wave, s | The Rayleigh Wave Propagation Speed, km/s |
|---|---|---|---|---|
| KNDC (Almaty) | 43.21710° N, 76.96710° E | 1796 | 642 | 2.797 |
| MAKZ (Makanchi) | 46.80800° N, 81.97700° E | 2069 | 752 | 2.742 |
| MK31 (Makanchi) | 46.79370° N, 82.29040° E | 2066 | 750 | 2.754 |
| KKAR (Karatau) | 43.10340° N, 70.51150° E | 2087 | 774 | 2.706 |
| sub-ionospheric point | 43.67529° N, 81.72963° E | 1732 | 630.5 [1] | 2.747 [1] |

[1] calculated value.

Then, using the calculated profile of the electron concentration, the path of radio wave propagation was defined by a special program, which took into account the geomagnetic field in accordance with the IGRF13 model for the ordinary component. The result of the calculation is shown in Figure 13, where it is especially indicated the reflection height of the sounding radio wave at 126.0 km.

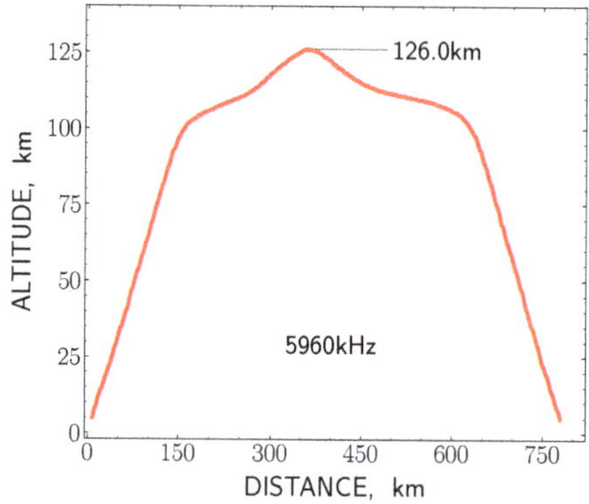


**Figure 13.** The propagation trajectory of the sounding wave of Doppler measurements on the radio path "Urumqi—Radiopolygon Orbita" in the time of the M7.8 Nepal earthquake.

### 3.1.2. The Propagation Time of Infrasonic Waves to the Reflection Point of the Sounding Radio Wave

In this section, to determine the height in the ionosphere, which an acoustic wave propagating vertically upward from the earth's surface reaches in 401.5 s, we use the calculation of an altitude profile of the sound speed with account to the viscosity and thermal conductivity of the atmosphere. The calculation method, based on the international atmosphere model NRLMSISE-00, is described in [43,44]. The calculation was made for the coordinates of the sub-ionospheric point in the earthquake time, and for an accepted index of solar activity $F10.7 = 126$. The result is shown in the left frame of Figure 14 as a plot of the altitude profile of the sound speed.

Taking into account the profile of the sound speed, the dependence of the impact time of acoustic wave at the various heights in the ionosphere was calculated, which is presented in the right frame of Figure 14. According to this plot, for the time of 401.5 s the acoustic wave reached a height of 124.9 km, which may be compared with the previous estimation made by the calculation of the trajectory of radio wave propagation—126.0 km. The smallness of the difference between both estimates obtained by the two independent methods, $126.0 - 124.9 = 1.1$ km, indicates a rather good agreement between the experimental and calculated data.

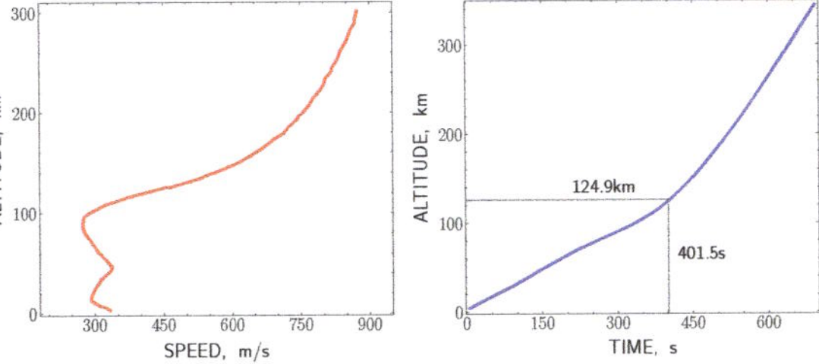


**Figure 14.** The estimated altitude profile of the sound speed (**left**), and the calculated arrival moment of an acoustic wave at the different heights in the ionosphere (**right**) in the time of the M7.8 Nepal earthquake.

Since geomagnetic storms and substorms induce strong disturbing effect on the ionosphere, taking into account the geomagnetic situation is important for the analysis of seismogenic anomalies in the ionosphere. Note, that during the period from 18 April to 30 April 2015, the geomagnetic situation remained calm, with $Kp$ index below 3. Only on April 21, it was observed for several hours a magnetosphere disturbance with $Kp = 4$ [45]. The $Dst$-index value during the specified period did not exceed the negative value of $-39$ nTl, and equaled to $-11$ nTl on April 25 [42].

Similar co-seismic response of the ionosphere was detected during the catastrophic earthquake in Turkey on 6 February 2023, at a distance of 1591 km from the earthquake epicenter. Monitoring of the Doppler frequency shift was carried out at a frequency of 5860 kHz on the inclined two-hop radio path "Kuwait—Institute of Ionosphere (Almaty)". The length of the radio path equals to 3010 km, and the distance between the sub-ionospheric point and the earthquake epicenter was 1591 km [26]. As shown in Figure 15, approximately 17 min (998 s) after the main shock, a disturbance was registered in the Doppler shift record, which lasted about 160 s. The double amplitude of the disturbance was more than 2 Hz, which significantly exceeded the background variation of Doppler frequency. The geomagnetic situation during the period from 21 January to 7 February 2023 was calm, such that the Dst-index did not exceed the negative values of $-27$ nTl, $Kp \leqslant 3$ [42]. Just as in the case with M7.8 earthquake in Nepal, the effect observed in the Doppler frequency was a result of the penetration into the ionosphere of the acoustic waves generated by the propagating Rayleigh wave.

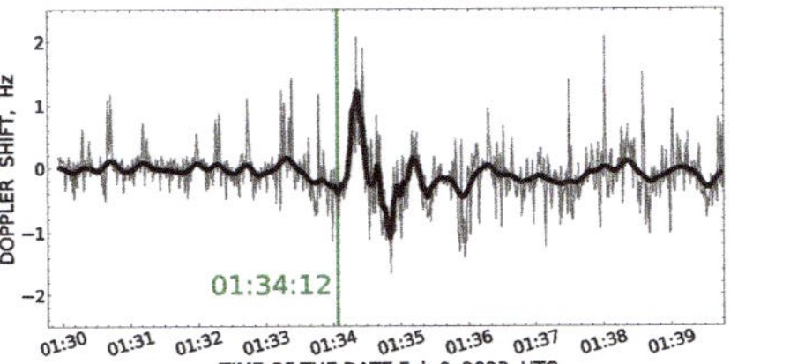


**Figure 15.** The response in the Doppler frequency shift of ionospheric signal to the *M7.8* earthquake in Turkey on 6 February 2023, as detected at the radio path "Kuwait—Institute of Ionosphere (Almaty)". Thin curve—original measurements, bold curve—same data smoothed by a 10 points running average filter. The vertical line indicates the beginning of the ionospheric disturbance in the reflection point of radio waves at 01:34:12 UTC. The plot is taken from the publication Salikhov et al., 2023 [26].

## 4. The Doppler Observations of Pre-Seismic Disturbances in the Ionosphere before the M7.8 Earthquakes

Taking into account catastrophic consequences of large earthquakes, one of the most pressing issues of modern geophysics is the detection of ionospheric anomalies which could be precursors of seismic events. In recent decades, for investigation of ionospheric anomalies before earthquakes most studies used the data of the Global Navigation Satellite System (GNSS). Thus, intensive studies of seismogenic disturbances in the ionosphere were carried out using the data from the electromagnetic satellites DEMETER and CSES [45,46]. In refs. [47–49], brief reviews are given of seismogenic phenomena interpreted as possible precursors of earthquakes, which were recorded in the ionosphere by the ground-based and satellite measurement methods. Up to the present, we have not met any examples of identifying pre-seismic anomalies with application of the Doppler sounding method of the ionosphere.

*Short-Term Doppler Bursts before and during the M7.8 Earthquake in Nepal on 25 April 2015*

The registration method of short-term Doppler bursts as indicators of ionization flashes was used to study the response of the ionosphere to the earthquake in Nepal and to the preparation process of this catastrophic event.

In Figure 16 the day-to-day records are presented of the variations in the amount of the Doppler bursts. It is seen that the intensity of the bursts significantly increases in the periods of sunrise and sunset, i.e., respectively, around the left and right margin of each plot. During the daytime, the intensity of Doppler bursts remains, as a rule, minimal, provided that the geomagnetic environment is calm and there are no other factors disturbing the ionosphere. In Figure 16 this is clearly seen in the plots built for the days of 25 April 2014 and 15 April 2015.

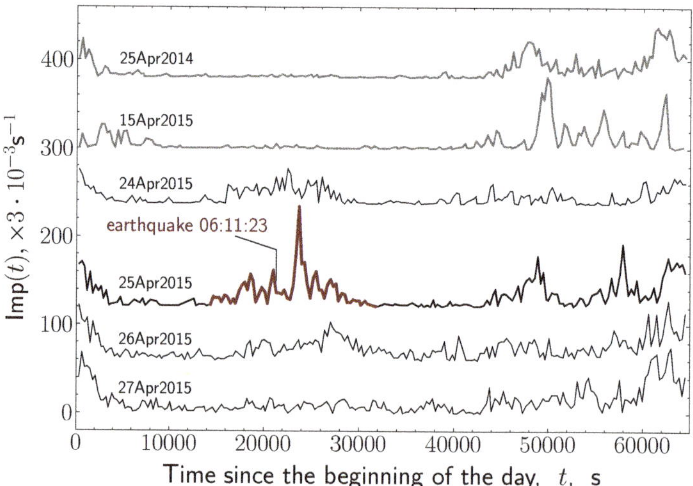


**Figure 16.** Intensity of the short-term Doppler bursts detected by the modified method of Doppler measurements before and in the time of the M7.8 Nepal earthquake. The scale of the abscissa axis is expressed in thousands of the seconds passed since the beginning of the day in local time; each distribution covers the time interval of ~18 h. Along the ordinate axis, the counting rate of short-term Doppler bursts Imp($t$) is recorded; the curves are displaced in vertical direction for convenience of comparison.

As noted above, the geomagnetic situation in the period from 18 April to 30 April 2015 was calm. The earthquake occurred at noon, at 12:11:26 of the local time (06:11:26 UTC), which made it possible to register the seismogenic effects in the ionosphere. A noticeable increase in the intensity of Doppler bursts was detected on 24 April, the day before the earthquake, and on 25 April, approximately 90 min before the main shock. The maximum intensity was reached 40 min after the earthquake. Further on, from day to day, the disturbances with decreasing trend were observed at the approximately same time on 26 and 27 April, which is apparently associated with aftershock activity after the earthquake. Thus, the registration of short-term Doppler bursts made it possible to detect the seismogenic disturbances in the ionosphere not only during the earthquake itself, but also in the period of its preparation.

The precursor effects in the ionosphere were also identified at the 6 February 2023 M7.8 earthquake in Turkey. Using the PLL based hardware-software complex of Doppler measurements, the Doppler frequency shift of ionospheric signal was monitored on an inclined, 3010 km long, two-hop radio path "Kuwait— Institute of Ionosphere (Almaty)". In the variation record of Doppler signal, an expressed increase of the Doppler frequency shift was found, which started approximately 8 days before the main shock, while the maximum values of Doppler frequency were reached at the day of the earthquake [26]. According to the calculated Dobrovolsky radius, the sub-ionospheric point at the first hop of the radio path used was within the limits of the action zone of deformation processes in the lithosphere. This permits to associate the appearance of anomalous effects in the ionosphere with the processes which took place in the lithosphere region of the earthquake preparation.

## 5. The Lithosphere-Atmosphere-Ionosphere Coupling

*5.1. Lithosphere-Atmosphere-Ionosphere Coupling on Example of the M4.2 Earthquake on 30 December 2017*

Experimental data on the conjugacy and transmission sequence of seismogenic disturbances in geophysical fields from the lithosphere to the ionospheric heights are of undeniable importance for the study of anomalous phenomena arising at the final stage of earthquake preparation. An important assumption by development of the methods of seismological forecast is the expectation that the preparation process of a strong earthquake should impact, at least, the local geophysical fields. This section presents the examples of ionospheric signatures which have appeared simultaneously with the disturbances in the flux of gamma-rays measured both in the subsoil layers of rock and in the ground-level atmosphere 7 days before a M4.2 earthquake on 30 December 2017.

The measurements of the Doppler frequency shift of ionospheric signal were carried out on a low-inclined radio path between the Institute of Ionosphere (43.17594° N 76.95342° E) and Radiopolygon Orbita(43.05831° N 76.97361° E). The flux of gamma rays, as an exhalation marker of radon and gamma-radioactive products of its decay, was measured at a depth of 40 m inside a borehole, where a gamma detector on the basis of a $\varnothing$40 mm NaI(Tl) crystal was installed. Another gamma detector, with a NaI(Tl) crystal of 150 mm in diameter, was placed close to the borehole on the surface of the ground. Both detectors were operating in the continuous monitoring mode. It should be stressed, that the measuring equipment was situated at a distance of only 5.3 km from the epicenter of the earthquake.

On 23 December, seven days before the earthquake, an anomalous intensity outburst of the flux of low-energy, (50–200) keV, gamma rays was registered in the borehole. Presumably, the cause of this effect was a change in the stress-strain state of the rocks before the earthquake, resulting in increased release of radon and its decay products. At the same time, an acoustic microphone with a sensitivity of 25 mV/Pa, also installed in the borehole [50], registered two episodes of the increasing geoacoustic emission, and the thermosensors placed there detected the rising trend of local temperature [25]. As seen in Figure 17, an enhanced level of gamma ray intensity was kept for 3 days, and then during the whole subsequent period until the earthquake.

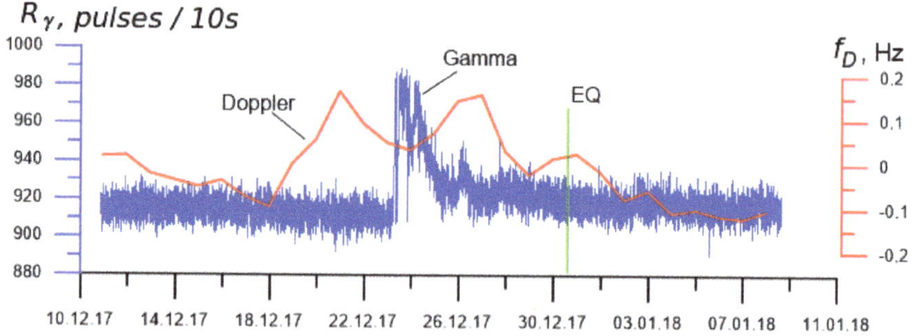



**Figure 17.** The outburst of the flux of (50–200) keV gamma rays detected 40 m underground in the borehole prior to the 30 December 2017 M4.2 earthquake, together with the simultaneous negative anomaly in the Doppler frequency shift of ionospheric signal. The moment of the earthquake (EQ) is indicated by a vertical line. The counting rate of gamma rays, $R_\gamma$, is expressed in the units of the amount of pulses obtained from the gamma detector in 10 s. The data on the Doppler frequency $f_D$, in Hz, are presented with daily averaging.

In the ground-level atmosphere, an increase of the gamma radiation flux was also observed by the on-ground gamma detector, as illustrated by Figure 18. In this plot, the similar parts in the two records of radiation intensity both above and under the surface of the ground are marked by vertical lines. It is seen that in the ground-level atmosphere the maximum of the radiation flux was reached approximately 5 h later than inside the borehole, which presumably may be stipulated by the sequence of the moving of radioactive substances from the depth of the earth's crust up to the surface.

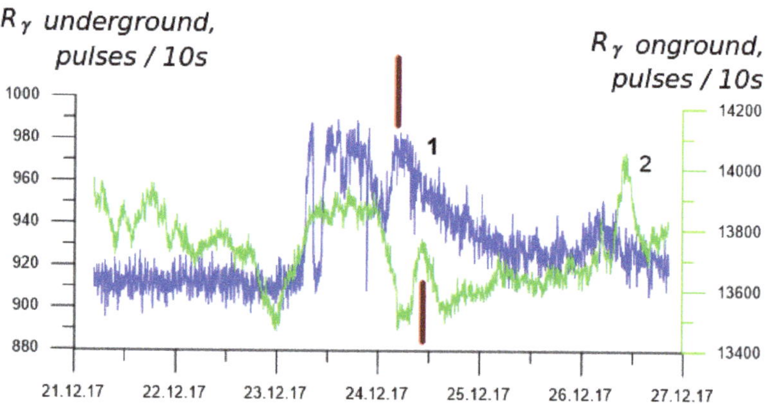



**Figure 18.** Comparison of the gamma ray flux variations measured during the period of 22–27 December 2017 at a depth of 40 m in the borehole (*1*), and at the surface of the ground (*2*). The counting rates of gamma radiation $R_\gamma$ are expressed as the amount of detector pulses per one second. Two vertical lines mark the mutually corresponding bursts of gamma ray intensity, which were successively appearing, at first in the borehole and then in the ground-level atmosphere.

An advantage of the gamma ray registration 40 m deep in the borehole is the temperature stability, as well as the absence of any influence of atmospheric precipitation to the radiation background. This is why the disturbance in the gamma radiation flux has revealed itself more distinctly in the borehole, so that not only the burst itself was well registered there, but also the fine structure of the response to the earthquake preparation process.

Simultaneously with the appearance of anomalous effects in the borehole, we registered the disturbances in the ionosphere. In Figure 17 it is seen that 7 days before the earthquake a decrease of Doppler frequency occurred, which happened in the time of a sharp increase in the gamma ray intensity. In the day of the earthquake, 30 December 2017, it was also observed small growing in the Doppler frequency shift. It should be noted, that the geomagnetic conditions remained quite at that time, and any geomagnetic storms were absent during the period since 18 until 31 December, which is an important condition for revealing ionospheric response to earthquake preparation.

Thus, our experimental data have shown, that the ionization process of the ground-level atmosphere which occurred during the short-term period of earthquake preparation is an important link in realization of the lithosphere-atmosphere-ionosphere coupling and in appearance of the precursor disturbance in the ionosphere.

### 5.2. Lithosphere-Atmosphere-Ionosphere Coupling on Example of Underground Nuclear Explosion

In the works of Pulinets et al. [51,52], convincing examples are presented of disturbances appearing in the ionosphere at the growth of the ionization in the boundary layer of the atmosphere, including the cases of underground nuclear explosion. According to the authors, during the underground nuclear explosion in North Korea on 12 February 2013, an earthquake with a magnitude of M4.9 was registered. One hour after the explosion, when a small quantity of radioactive components seeped out onto the surface of the Earth, the GPS satellites registered a negative anomaly in the ionosphere. This observation confirms the concept of the authors, that the burst of ionization may be accompanied by a local increase of atmospheric conductivity, and by a change of electron concentration in the ionosphere above the region with intensified conductivity.

Similar examples of disturbance appearance in the ionosphere at the release of radioactive components of underground nuclear explosion on daytime surface we have observed in 1980th during the tests at the Semipalatinsk test site. To detect the response of the ionosphere, it was used the PLL-based complex of Doppler measurements, and its ability to measure the Doppler frequency shift of a larger amplitude beam was realized [21,53]. In Figure 19 an original record of the Doppler frequency shift is presented, which was obtained at the last underground nuclear explosion on 19 October 1989. The explosion, with a power of 75 kt of TNT equivalent, was executed at 09:50:00 UTC in the borehole #1365 at the Balapan area of the Semipalatinsk test site. The radio-sounding of the ionosphere was proceeding on the radio-path "Kurchatov (50.715° N, 78.621° E)—Sarzhal (49.6° N, 78.683° E)" at a frequency of 7727 kHz.

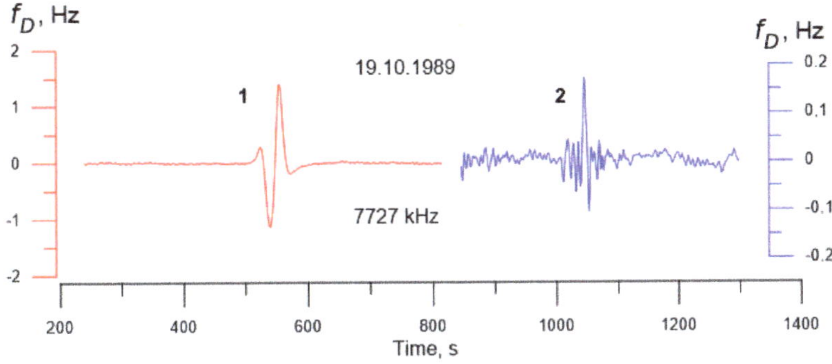


**Figure 19.** Two responses in the Doppler frequency shift $f_D$ of ionospheric signal detected at the 75 kt underground nuclear explosion on 19 October 1989 at the Semipalatinsk test site. *1*—510 s after the explosion, *2*—1005 s after the explosion. The scale of the horizontal axis is expressed in the seconds passed since the moment of the explosion.

It is seen in the plots, that 510 s (8.5 min) after the explosion a disturbance appears in the record of the Doppler frequency shift of ionospheric signal. This effect is caused by the penetration into the ionosphere of the acoustic waves, the source of which is the movement of the Earth's surface in the spallation zone (Figure 19, left). Then, 1005 s (16.75 min) after the explosion, another frequency burst is observed, which has an essentially smaller amplitude than that caused by the acoustic impact. From the comparison of experimental records of the Doppler frequency shift it follows, that not only the appearance time, but also the shape and amplitude of the secondary effect differ from the effect connected with the impact of the acoustic waves from the explosion on the ionosphere. As a rule, in our experiments the registration of the Doppler frequency shift proceeded no more than (30–40) min; nevertheless, this occurred sufficient for the detection of the secondary disturbance in the ionosphere. In average, the secondary effects in the records of the Doppler frequency shift of ionospheric signal were registered (15–18) min since explosion.

We noted, that the appearance time of ionospheric anomalies after explosion was comparable with the time of radioactivity enhancement in the atmosphere as revealed by direct measurements of the radioactive background in the atmosphere from helicopters and planes [54,55]. Thus, on 19 October 1989, in the borehole #1365 an exit of the products of nuclear explosion onto the ground surface was observed, with a rise of radioactive background up to (15–40) R/h in the epicenter zone [56]. According to statistics, explosions of incomplete camouflage accompanied by the small exhalation of radioactive gases composed a 45% share of the whole number of tests at the Semipalatinsk site [55], and 67% of the tests at the Novaya Zemlya site [54]. The practice of the tests at the Arctic nuclear polygon shows, that by all camouflage nuclear explosions an escape of radioactive products of underground explosion onto the surface occurred, as a result of filtration through the zones of melting, crushing, micro- and macro-fissuring. If the release happened less than one minute after the explosion, the isotopes Sr-89, Sr-90, Ba-140, Cs-137 were detected in the atmosphere. If the time delay between the explosion and the gas exit was above one, but below 33 min, the isotopes Sr-89 and Cs-137 were detected, and after 33 min only Cs-137 remained.

As a result of the seismic impact of the energy of nuclear explosion on rock massif, large masses of natural Rn, Th and their daughter products are also thrown out into the atmosphere [54]. The ingress of radionuclides in the atmosphere leads to ionization of its ground-level layer, which may be one of the links of disturbance transmission in the lithosphere-atmosphere-ionosphere system.

According to modern concepts, the key factor of anomalies generation in the ionosphere is the change in the conductivity of the atmospheric boundary layer and the complex of physical and chemical processes proceeding under the influence of ionization [52,57]. Then, the appearance of anomalous effects in the ionosphere both after an underground nuclear explosion and in the short-term period of earthquake preparation can be explained on the basis of a general mechanism consisting of the processes which take place at the ionization stage of the ground-level atmosphere. In the case of an underground nuclear explosion, the exit of radioactive substances occurs after explosion, while by earthquakes the ionization of the near-earth atmosphere, as an exhalation result of radon, thoron, and their daughter products, is detected before the main shock.

Experimental records of the response of the ionosphere to underground nuclear explosions were a basis for an estimation of the velocity profile of an acoustic pulse by the Doppler frequency shift of the sounding radio wave.

## 6. A Simple Formula for Estimating the Profile of an Acoustic Pulse Velocity by the Doppler Shift of the Frequency of Sounding Radio Wave

In this section, a possibility is considered to restore the profile of the acoustic pulses which have reached the reflection point of radio waves in the ionosphere from the two underground nuclear explosions. In principle, such formula can be only derived under exclusive conditions of a powerful acoustic impact on the ionosphere, and on the basis of reliable data of Doppler sounding.

It should be noted, that proceeding of any direct acoustic observations at the height of upper atmosphere (ionosphere) seems impossible. This is why we turned to an analysis of the historical records of the Doppler frequency shift obtained by the Doppler method using PLL, the effectiveness of which was demonstrated during underground nuclear explosions at the Semipalatinsk test site. Measuring the Doppler frequency shift by the short-wave radio sounding of the ionosphere is a widely used instrument for studying the disturbances in the thermosphere of both natural and artificial origin. Usually, the Doppler frequency shift is calculated in the approximation of geometric optics, and its value is determined by the rate with which the electron concentration changes in time along the trajectory of radio beam [58].

Here, a simple formula is obtained which allows to acquire a good estimate for the relationship between the velocity field of acoustic disturbances and the measured Doppler frequency. The results obtained may be a reference point in the development of numerical models of the propagation of an acoustic pulse from an underground nuclear explosion into the upper layers of atmosphere.

Under disturbances that do not affect the ionization balance, the change in electron concentration over time is entirely determined by the velocity field of the charged component. The velocity of the charged component can be related to the velocity of the neutral gas and, as a result, the Doppler frequency shift can be represented as a functional of the velocity field of the neutral component of the ionospheric plasma. This representation is convenient, since calculating the Doppler shift no longer requires preliminary labor-intensive reconstruction of the electron density profile from the field of disturbance velocities.

For the acoustic disturbances, the characteristic spatial size of which exceeds the size of the region of significant interaction of the radio wave with the ionospheric plasma, it is possible to analytically invert the expression for the Doppler frequency shift, and to obtain a fairly simple formula for the velocity of the neutral gas, suitable for practical processing of Dopplergrams.

In the isotropic case, the phase of the radio wave at the receiving point relative to the transmission point is given by integrating the refractive index $n = \sqrt{1 - N/N_w}$ along the radio beam trajectory

$$\phi = -\frac{\omega}{c} \int_L n \, dl, \tag{1}$$

where $\omega$ is the circular frequency of the sounding radio wave; $c$ is the speed of light. By virtue of Fermat's principle:

$$\delta \int_L n \, dl = \int_{L+\delta L} n \, dl - \int_L n \, dl = 0, \tag{2}$$

the perturbation of the radio wave phase corresponding to the perturbation of the refractive index can be expressed, accurate to linear terms, by an integral along an unperturbed trajectory

$$\phi + \delta\phi = -\frac{\omega}{c} \int_{L+\delta L} (n + \delta n) \, dl = -\frac{\omega}{c} \int_L (n + \delta n) \, dl. \tag{3}$$

This leads to the determination of the Doppler frequency shift [58]:

$$\omega_D = \frac{\partial}{\partial t}\phi = -\frac{\omega}{c} \int_{L+\delta L} \frac{\partial}{\partial t}\delta n \, dl = -\frac{\omega}{c} \int_{L+\delta L} \frac{\partial}{\partial N}\frac{\partial}{\partial t}\delta N \, dl, \tag{4}$$

where $\delta N$ is the disturbance of the electron density; $dl$ is an element of the length along the trajectory. Using the linearized continuity equation

$$\frac{\partial}{\partial t}\delta N + \nabla \cdot N v = 0, \tag{5}$$

the rate of change in the electron concentration can be expressed in terms of the velocity field of the charged component of the plasma. As a result, we get:

$$\omega_D = \frac{\omega}{c} \int_{L+\delta L} \frac{\partial n}{\partial N}(\boldsymbol{v} \cdot \nabla N + N \nabla \cdot \boldsymbol{v})\, dl = \frac{\omega}{c} \int_{L+\delta L} \nabla n \cdot (\boldsymbol{v} + H_e \nabla \cdot \boldsymbol{v})\, dl, \tag{6}$$

where $H_e = \nabla \ln N / (\nabla \ln N)^2$ is the height vector of the homogeneous ionosphere. The equations for the change along the trajectory of the wave vector of a radio wave $\boldsymbol{k}, k = n\frac{\omega}{c}$,

$$\frac{\partial}{\partial l}\boldsymbol{k} = \frac{\omega}{c}\nabla n; \quad \frac{\partial}{\partial l}\boldsymbol{r} = \frac{\boldsymbol{k}}{k}, \tag{7}$$

allow us to express the Doppler frequency shift (4) through the parameters of the unperturbed trajectory

$$\omega_D = \int_L \boldsymbol{u} \cdot d\boldsymbol{k}(l), \quad \boldsymbol{u}(\boldsymbol{r},t) = \boldsymbol{v} + H_e \nabla \cdot \boldsymbol{v}. \tag{8}$$

Under the magnetization conditions of plasma ions, the charged component is entrained by the neutral component only along the geomagnetic field

$$\boldsymbol{v} = \frac{\boldsymbol{B}}{B^2}\boldsymbol{B} \cdot \boldsymbol{v}_n, \tag{9}$$

where $\boldsymbol{v}_n(\boldsymbol{r},t)$ is the perturbation of the velocity field of the neutral component, $\boldsymbol{B}$ is the induction of the magnetic field. Formulas (8) and (9) make it possible to efficiently calculate the magnitude of the Doppler frequency shift in the numerical simulation of the unperturbed beam trajectory $\boldsymbol{r}(l)$, $\boldsymbol{k}(l)$, which occurs due to the motion of the neutral component of the ionospheric plasma. A noticeable change in the wave vector of the radio wave, and consequently the contribution to the integral (8), occurs only in the regions of significant interaction of the radio wave with the ionospheric plasma. Usually, it is sufficient to consider a small neighborhood of the reflection point as such an area.

If the characteristic spatial scale of the acoustic disturbance exceeds the size of the region of significant interaction of the radio wave with the ionospheric plasma, the integral (8) can be approximated by the first terms of its asymptotic expansion at $H_e \to 0$. To do this, following the usual method of obtaining asymptotic expansions of this kind of integrals, we decompose the integral expression into the Taylor series at the reflection point $z_0$, and the integration region is limited to the section of the radio beam trajectory in the interval of altitudes from $z^* = z_0 - H_e Z_0$ to $z_0$. In this region, the change in the wave vector of the sounding radio wave constitutes a $\sim 0.8$ share of its complete change.

$$\omega_D(t) = \int_L u_z\, dk_z(z) \approx \oint_{z^*} \left[ u_z(\boldsymbol{r}_0) + (\boldsymbol{r} - \boldsymbol{r}_0)\frac{\partial}{\partial \boldsymbol{r}_0}u_z \right] dk_z(z) =$$

$$\Delta k_z \left[ u_z(\boldsymbol{r}_0) + \langle z - z_0 \rangle \cdot \frac{\partial}{\partial z_0}u_z \right], \tag{10}$$

where

$$z^* = z_0 - H_e, \quad k_z(z) = \frac{\omega}{c}\cos\theta_{kz}\sqrt{1 - \frac{N(z)}{N(z_0)}}; \tag{11}$$

$$\Delta k_z = -2k_z(z^*) = -2\frac{\omega}{c}\cos\theta_{kz}\sqrt{1 - \frac{N(z)}{N(z_0)}} \approx -2\frac{\omega}{c}\cos\theta_{kz} \times 0.795; \tag{12}$$

$$\langle z - z_0 \rangle = \frac{1}{k_z(z^*)}\int_{z^*}^{z_0}(z - z_0)\, dk_z(z) = -\int_{z^*}^{z_0}\left(1 - \sqrt{1 - \frac{N(z_0) - N(z)}{N(z_0) - N(z^*)}}\right) dz \approx -0.271 H_e, \tag{13}$$

where $z_0$ is the altitude of the reflection point; $\theta_{kz}$ is the initial value of the angle between the wave vector and the vertical; $x_+(z)$, $x_-(z)$ are the values of x-coordinates on the ascending and descending branches of the trajectory. The numerical values of the quantities are obtained by approximating the electron concentration profile in the vicinity of the reflection point by the exponent $N(z)/N(z_0) = \exp[(z-z_0)/H_e]$. For formula (10) to be applicable, it is necessary that the vertical linearity of the velocity profile was not less than $h \sim 0.271 H_e$, and the horizontal dimension was not less than $\langle |x - x_0| \rangle \approx \tan\theta_{kz} H_e$.

Let us consider the problem of the Doppler frequency shift caused by the disturbance from an acoustic wave. We approximate the velocity field of the neutral component $v_n(\mathbf{r}, t)$ in the vicinity of the reflection point of sounding radio wave by a longitudinal pulse propagating with the local sound velocity $c_s$ and with exponential change of the amplitude with altitude. The vector $\mathbf{H}_a$ is directed vertically upwards.

$$v_n = \frac{c_s}{c_s} v_n = c_s \exp\left(\frac{\mathbf{H}_a \cdot \mathbf{r}}{H_a^2}\right) w(\tau), \quad \tau = t - \frac{c_s \cdot \mathbf{r}}{c_s^2}. \tag{14}$$

By differentiating (14), we obtain the evolutionary equation for each of the components of $v_n$:

$$\left(\frac{c_s}{c_s^2}\frac{\partial}{\partial t} + \frac{\partial}{\partial r} - \frac{\mathbf{H}_a}{H_a^2}\right) \circ v_n = 0. \tag{15}$$

Let us express, with the help of (15), the spatial derivatives of the velocity field in terms of temporal ones, for example:

$$\frac{\partial}{\partial r} \cdot v_n = \left(\frac{\cos\theta_{kz}}{H_a} - \frac{1}{c_s}\frac{\partial}{\partial t}\right) v_n;$$

$$\frac{\partial}{\partial z} v_n = \left(-\frac{\cos\theta_{cz}}{c}\frac{\partial}{\partial t} - \frac{1}{H_a}\right) v_n. \tag{16}$$

By substituting (16) into (10) and neglecting the second derivative of time, we get the final formula:

$$\omega_D(t) = \frac{1}{L_0}\left(T\frac{\partial}{\partial t} - 1\right) v_a(r_a, t), \tag{17}$$

where

$$T = \frac{\cos\theta_{Bc}}{\cos\theta_{Bz}}\frac{H_e}{c_s}\left(1 + \frac{H_e}{H_a}\right)^{-1} - \cos\theta_{cz}\frac{h}{c_s}\left(1 - \frac{h}{H_a}\right)^{-1}; \tag{18}$$

$$L_0^{-1} = \Delta k_z \cos\theta_{Bz} \cos\theta_{Bc}\left(1 - \frac{h}{H_a}\right)\left(1 + \frac{H_e}{H_a}\right); \tag{19}$$

$$h = -\langle z - z_0 \rangle \approx 0.271 H_e; \quad H_a = 2H. \tag{20}$$

$\theta_{Bc}, \theta_{B_z}, \theta_{cz}$ are the angles between the magnetic field and the direction of acoustic pulse propagation, the magnetic field and the vertical, the direction of acoustic pulse propagation and the vertical; $H$ is the height of the homogeneous atmosphere at the reflection point; $c_s$ is the speed of sound at the reflection point. When $H_e \to 0$, formula (17) turns into the formula for the Doppler frequency shift for a radio wave reflected from a moving mirror, which is often used for estimates (though not always justified).

Formula (17) gives a solution to the problem of determining the Doppler frequency shift from the time profile of the velocity of wave disturbance in the neutral gas at the reflection point of radio wave. In addition, it is not difficult to reverse the formula by solving differential equation, and thereby to solve the inverse problem of determining the velocity profile from the measured Doppler frequency shift. The treatment will be unambiguous if the physically necessary conditions of boundedness are imposed on the velocity disturbance,

$$|v_a(r_a, t)| < \infty; \, -\infty < t < \infty, \tag{21}$$

since Equation (17) for $v_n(\mathbf{r}_0, t)$ at $\omega_D(t) = 0$ has no non-zero bounded solutions. From (17) and (21), taking into account the relation $1 - T\frac{\partial}{\partial t} = -\exp\left(\frac{t}{T}\right)T\frac{\partial}{\partial t}\exp\left(-\frac{t}{T}\right)$, the desired inversion formula follows (it is assumed that $T > 0$):

$$v_n(\mathbf{r}_0, t) = -L_0 \int_0^\infty \frac{dt'}{T} \exp\left(-\frac{t'}{\tau}\right)\omega_D(t + t'). \tag{22}$$

Let us dwell in more detail on the case when the Doppler frequency shift occurs only starting from the zero moment of time:

$$\omega_D(t) = 0 \text{ when } t < 0. \tag{23}$$

In this case, as follows from (22),

$$v_n(\mathbf{r}_0, t) = v_n(\mathbf{r}_0, 0)\exp\left(\frac{t}{T}\right) \text{ when } t \leqslant 0. \tag{24}$$

From (24) we conclude that if the acoustic pulse is non-zero in the interval $t \leqslant 0$, then it must have a leading edge of a special shape ($v_n(\mathbf{r}_0, t) \sim \exp\left(\frac{t}{T}\right)$, $t \leqslant 0$). In a real experiment, such situation can only arise if additional measures were taken. We do not consider this option, so we believe that $v_n(\mathbf{r}_0, 0) = 0$. Thus, Dopplerograms that originate at $t = 0$ must satisfy the condition

$$\int_0^\infty \frac{t'}{T}\exp\left(-\frac{t'}{\tau}\right)\omega_D(t') = 0 \tag{25}$$

Equation (25) can be used to adjust the parameter $T$ if the available ionospheric data is insufficiently reliable.

Our Equation (22) for reconstructing the velocity profile of an acoustic pulse depends on only two parameters: $L_0$, which has the dimension of length, and $T$, which has the dimension of time. If we set the parameter $T$ equal to zero, we arrive at the case of the reflection of radio wave from a moving mirror, which is often used to estimate the velocity profile of disturbance. It is important to note, that in this case, the restoring error of the velocity profile can distort the result several times and distort the shape of the profile.

Figures 20 and 21 show the records of the Doppler frequency shift and reconstructed velocity profiles in the reflection point of sounding radio wave for the two acoustic pulses which were propagating from the underground nuclear explosions carried out at the Semipalatinsk nuclear test site on 17 December 1988 and 19 October 1989.

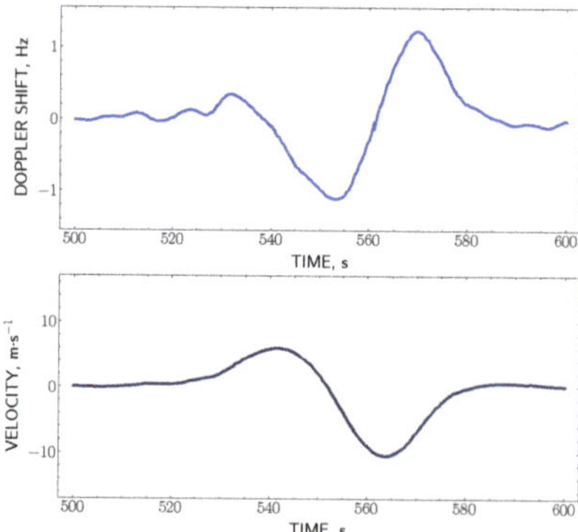

**Figure 20.** The record of the Doppler shift measured at the underground nuclear explosion on 17 December 1988, and the restored velocity profile $v_n(\mathbf{r}_0, t)$. The frequency of the sounding radio wave $\omega = 7.7$ MHz, the altitude of the reflection point $z = 237$ km, $L_0 = 225$ m, $T = 21.3$ s.

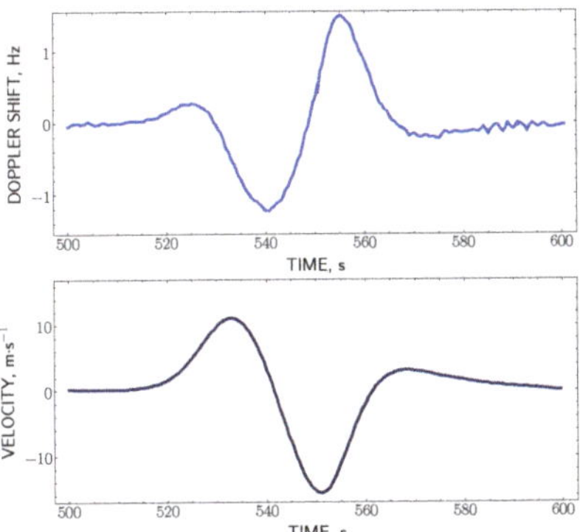

**Figure 21.** The record of the Doppler shift measured at the underground nuclear explosion on 19 October 1988, and the restored velocity profile $v_n(r_0, t)$. The frequency of the sounding radio wave $\omega = 7.7$ MHz, the altitude of the reflection point $z = 225$ km, $L_0 = 232$ m, $T = 20.0$ s.

The obtained results can serve as a reference point in development of the numerical propagation models of the acoustic pulses which arise in the upper layers of the atmosphere from underground nuclear explosions.

## 7. Conclusions

The article describes in detail the method and equipment for measuring the Doppler frequency shift by inclined sounding of the ionosphere on the basis of PLL operation principle. The task was to expand the capabilities of the Doppler method for registration of the ionospheric response to seismogenic and anthropogenic impacts, and to detect the appearance of pre-seismic signatures in the ionosphere prior to large earthquakes. As a result, it became possible to register continuously, with good accuracy and reliability, both the Doppler frequency shift and the short-term Doppler bursts which reflect disturbances in the ionosphere.

1. Application of an SDR receiver using the digital technology of Software-Defined Radio in the Doppler installation ensured the high characteristics of the radio receiving tract.
2. The use of the PLL system permitted to carry out the continuous measurement of Doppler frequency and to measure the Doppler frequency shift of larger-amplitude beam under multipath conditions; this is an evident advantage of the considered method.
3. With an optimal choice of the PLL hold band, it was achieved an accuracy of $\leqslant 0.01$ Hz in the measurement of Doppler frequency shift, which is 1.5–2 orders of magnitude below the background variations of Doppler frequency in the $F$-region of the ionosphere.
4. A modification of the Doppler installation was made for the registration of short-term ionization bursts in the ionosphere. For this purpose, two additional blocks were included into the functional scheme of the Doppler installation: the high-pass filter and DC amplifier, a separate channel of signal digitization was organized, and a special program was developed for operation of the data obtained.

The radiophysical method of Doppler sounding of the ionosphere using PLL was applied for investigation of seismogenic and artificial disturbances in the ionosphere. During the two M7.8 earthquakes, in Nepal (25 April 2015) and Turkey (6 February 2023), the pre- and co-seismic effects in the ionosphere were revealed, the appearance of which relates to different propagation mechanisms of seismogenic disturbance from the lithosphere up to height of the ionosphere.

1. The co-seismic effects, which arose as a result of the penetration into the ionosphere of the acoustic waves caused by propagation of Rayleigh wave, were detected by the Doppler ionosonde at a distance of 1732 km from the epicenter of Nepal earthquake, and at 1591 km from the epicenter of the earthquake in Turkey.

2. Pre-seismic effects, as noticeable increase in the intensity of Doppler bursts reflecting the disturbance state of the ionosphere, were registered one day before the earthquake in Nepal, as well as 90 min prior to the main shock. The intensity of Doppler bursts had a constant rising trend, and its maximum was achieved 40 min after the earthquake.
3. A pre-seismic effect in the ionosphere, as noticeable increase of the Doppler frequency shift, was detected in the records of Doppler frequency 8 and 3 days before the earthquake in Turkey, and the maximum value of Doppler frequency was achieved on the day of the earthquake.
4. A channel of geophysical interaction in the system of lithosphere-atmosphere-ionosphere coupling was traced, when 7 days before a M4.2 earthquake the disturbances in the ionosphere were detected simultaneously with an intensity increase of the flux of gamma-rays both in the borehole, under the surface of the ground, and in the ground-level atmosphere, which resulted in the ionization of the ground-level atmosphere.
5. The concept of lithosphere-atmosphere-ionosphere coupling, where the key role is assigned to the ionization of the atmospheric boundary layer, found confirmation in an retrospective analysis of the records of Doppler frequency shift of ionospheric signal made during the underground nuclear explosions at the Semipalatinsk test site in the late 1980s. It was established, that after nuclear explosion the Doppler ionosonde registered first the distinct penetration signature of an acoustic wave into the ionosphere, then a disturbance in the ionosphere coinciding with the rise of radioactivity in the atmospheric boundary layer.
6. A simple formula for reconstructing the velocity profile of an acoustic pulse from Dopplerogram was obtained, which depends on only two parameters, one of which has the dimension of length, and the other the dimension of time. The article presents the reconstructed profiles of the acoustic pulses from the two underground nuclear explosions, on 17 December 1988 and 19 October 1989, which have reached the reflection points of sounding radio wave in the ionosphere.

**Author Contributions:** Conceptualization, N.S.; methodology, N.S. and V.S.; software, N.S. and A.S.; formal analysis, N.S., V.S. and G.P.; data curation, N.S. and A.S.; writing—original draft preparation, N.S., G.P. and V.S.; writing—review and editing, N.S., V.S., G.P., S.N. and A.S.; project administration, N.S., S.N., V.R. and V.Z.; funding acquisition, N.S. and S.N. All authors have read and agreed to the published version of the manuscript.

**Funding:** 1. This research was funded by the Science Committee of the Ministry of Science and Higher Education of the Republic of Kazakhstan, grant number AP19678127 "Development of the methods for geophysical fields monitoring in the lithosphere-atmosphere-ionosphere-magnetosphere system in a seismically hazardous region of Northern Tien Shan". 2. This research was funded by the Science Committee of the Ministry of Science and Higher Education of the Republic of Kazakhstan, grant number BR20280979 "Comprehensive study of the impact of the solar origin sources on the near-Earth space state".

**Data Availability Statement:** The data presented in this study are available on request from the corresponding author.

**Conflicts of Interest:** The authors declare no conflicts of interest.

## References


1. Watts, J.M.; Davies, K. Rapid frequency analysis of fading radio signals. *J. Geophys. Res.* **1960**, *65*, 2295–2301. [CrossRef]
2. Davies, K.; Watts, J.M.; Zacharisen, D.H. A study of F2-layer effects as observed with a Doppler technique. *J. Geophys. Res.* **1962**, *65*, 601–609. [CrossRef]
3. Davies, K.; Baker, D. Ionospheric effects observed around the time of the Alaskan earthquake of March 28, 1964. *J. Geophys. Res.* **1965**, *70*, 2251–2253. [CrossRef]
4. Afraimovich, E.L. *Interference Methods of Radio Sounding of the Ionosphere*; Nauka: Moscow, Russia, 1982; p. 198. (In Russian)
5. Blanc, E. Observations in the upper atmosphere of infrasonic waves from natural or artificial sources: A summary. *Ann. Geophys.* **1985**, *3*, 673–688.
6. Mikhailov, A.V. The geomagnetic control concept of the F2-layer parameter long-term trends. *Phys. Chem. Earth* **2002**, *27*, 595–606. [CrossRef]

7. Buonsanto, M.J. Ionospheric storms—A Review. *Space Sci. Rev.* **1999**, *88*, 563–601. [CrossRef]
8. Laštovička, J. Lower ionosphere response to external forcing: A brief review. *Adv. Space Res.* **2009**, *43*, 1–14. [CrossRef]
9. Danilov, A.D. Ionospheric F-region response to geomagnetic disturbances (Review). *Adv. Space Res.* **2013**, *52*, 343–366. [CrossRef]
10. Astafyeva, E. Ionospheric detection of natural hazards. *Rev. Geophys.* **2019**, *57*, 1265–1288. [CrossRef]
11. Danilov, A.D.; Konstantinova, A.V. Long-term variations of the parameters of the middle and upper atmosphere and ionosphere (review). *Geom. Aeron.* **2020**, *60*, 397–420. [CrossRef]
12. Laštovička, J. Long-term changes in ionospheric climate in terms of f0F2. *Atmosphere* **2022**, *13*, 110. [CrossRef]
13. Pokhotelov, O.A.; Parrot, M.; Fedorov, E.N.; Pilipenko, V.A.; Surkov, V.V.; Gladychev, V.A. Response of the ionosphere to natural and man-made acoustic sources. *Ann. Geophys.* **1995**, *13*, 1197–1210. [CrossRef]
14. Artru, J.; Farges, T.; Lognonne, P. Acoustic waves generated from seismic surface waves: Propagation properties determined from Doppler sounding observations and normal-mode modelling. *J. Int.* **2004**, *158*, 1067–1077. [CrossRef]
15. Krasnov, V.M.; Drobzheva, Y.V.; Laštovička, J. Recent advances and difficulties of infrasonic wave investigation in the ionosphere. *Surv. Geophys.* **2006**, *27*, 169–209. [CrossRef]
16. Petrova, I.R.; Bochkarev, V.V.; Teplov, V.Y.; Sherstyukov, O.N. Response of the ionosphere to natural and man-made acoustic sources. *Adv. Space Res.* **2007**, *40*, 825–834. [CrossRef]
17. Chum, J.; Hruška, F.; Zedník, J.; Laštovička, J. Ionospheric disturbances (infrasound waves) over the Czech Republic exited by the 2011 Tohoku earthquake. *J. Geophys. Res.* **2012**, *117*, A08319. [CrossRef]
18. Laštovička, J.; Chum, J. A review of results of the international ionospheric Doppler sounder network. *Adv. Space Res.* **2017**, *60*, 1629–1643. [CrossRef]
19. Krasnov, V.M. Method of selecting one-beam communication sessions. *Radio Eng. Electron.* **1976**, *XXI*, 600–602. (In Russian)
20. Drobzhev, V.I.; Krasnov, V.M.; Salikhov, N.M. Temporal variations of ionospheric waves in the D- and F-regions. *J. Atmos. Terr. Phys.* **1979**, *41*, 1011–1013. [CrossRef]
21. Salikhov, N.M. *Response of the Ionosphere to Acoustic Sources of Natural and Artificial Origin. Abstract Diss. Candidate of Physics and Mathematics Sciences*; Tomsk State University: Tomsk, Russia, 1985; 17p. (In Russian)
22. Alperovich, L.S.; Afraimovich, E.L.; Vugmeister, V.O.; Gokhberg, I.B.; Drobzhev, V.I.; Erushchenkov, A.I.; Ivanov, E.A.; Kalikhman, A.D.; Kudryavtsev, V.P.; Kulichkov, S.N.; et al. Acoustic wave explosion. *Phys. Solid Earth* **1985**, *1*, 32–42. (In Russian)
23. Alebastrov, V.A.; Bezruchenko, L.I.; Belenky, M.I.; Borisov, B.B.; Drobzhev, V.I.; Kaliev, M.Z.; Kiselev, V.F.; Krasnov, V.M.; Liadze, Z.L.; Litvinov, Y.G.; et al. Ionospheric response of disturbances initiated by an industrial explosion. In *Ionospheric Researches*; Nauka: Moscow, Russia, 1986; pp. 61–68. (In Russian)
24. Drobzhev, V.I.; Zheleznyakov, E.V.; Idrisov, I.K.; Kaliev, M.Z.; Kazakov, V.V.; Krasnov, V.M.; Pelenitsyn, G.M.; Savel'ev, V.L.; Salikhov, N.M.; Shingarkin, A.D. Ionospheric effects of the acoustic wave above the epicenter of an industrial explosion. *Radiophys. Quant. Electr.* **1987**, *30*, 1047–1051. [CrossRef]
25. Salikhov, N.; Shepetov, A.; Pak, G.; Nurakynov, S.; Ryabov, V.; Saduyev, N.; Sadykov, T.; Zhantayev, Z.; Zhukov, V. Monitoring of gamma radiation prior to earthquakes in a study of lithosphere-atmosphere-ionosphere coupling in Northern Tien Shan. *Atmosphere* **2022**, *13*, 1667. [CrossRef]
26. Salikhov, N.; Shepetov, A.; Pak, G.; Nurakynov, S.; Kaldybayev, A.; Ryabov, V.; Zhukov, V. Investigation of the pre- and co-seismic ionospheric effects from the 6 February 2023 M7.8 Turkey earthquake by a Doppler ionosonde. *Atmosphere* **2023**, *14*, 1483. [CrossRef]
27. Salikhov, N.; Shepetov, A.; Pak, G.; Saveliev, V.; Nurakynov, S.; Ryabov, V.; Zhukov, V. Disturbances of Doppler frequency shift of ionospheric signal and of telluric current caused by atmospheric waves from explosive eruption of Hunga Tonga volcano on January 15, 2022. *Atmosphere* **2023**, *14*, 245. [CrossRef]
28. Salikhov, N.M.; Pak, G.D. Ionospheric effects of solar flares and earthquake according to Doppler frequency shift on an inclined radio path. *Bull. Nat. Acad. Sci. Kazakhstan Repub.* **2020**, *331*, 108–115. [CrossRef]
29. Salikhov, N.M.; Pak, G.D.; Kryakunova, O.N.; Milyutin, V.I.; Mayevskaya, V.I.; Nikolayevskiy, N.F.; Tsepakina, I.L. Effects of launch vehicles from "Baikonur" and "Vostochny" spaceports on the surface atmosphere and ionosphere and its ecological significance. *The Bulletin of the National Nuclear Center of the Republic of Kazakhstan* **2016**, *2*, 135–145. ISSN 1729-7516. (In Russian)
30. BBC Frequencies and Sites. Available online: https://www.short-wave.info (accessed on 1 March 2024).
31. SDR-Radio.com. Available online: https://www.sdr-radio.com (accessed on 1 March 2024).
32. SDRplay.com. Available online: https://www.sdrplay.com/ (accessed on 1 March 2024).
33. Salikhov, N.M.; Somsikov, V.M. The program- and hardware complex for registration of the Doppler frequency shift of ionosphere radio-signal over earthquake epicenters. *Bull. Nat. Acad. Sci. Kazakhstan Repub.* **2014**, *296*, 115–121. (In Russian)
34. Popov, A.N. *Mathematical Analysis of Beats*; Gosenergoizdat: Moscow, Russia, 1956. (In Russian)
35. Alpert, Y.L. *Propagation of Electromagnetic Waves and the Ionosphere*, 2nd ed.; Nauka: Moscow, Russia, 1972; p. 564. (In Russian)
36. Chum, J.; Liu, J.Y.; Laštovička, J.; Fišer, J.; Mošna, Z.; Baše, J.; Sun, Y.Y. Ionospheric signatures of the April 25, 2015 Nepal earthquake and the relative role of compression and advection for Doppler sounding of infrasound in the ionosphere. *Earth Planets Space* **2016**, *68*, 24. [CrossRef]
37. Earthquake Hazards Program. Available online: https://earthquake.usgs.gov/earthquakes (accessed on 1 April 2024).
38. Dobrovolsky, I.P.; Zubkov, S.I.; Miachkin, V.I. Estimation of the size of earthquake preparation zones. *Pure Appl. Geophys.* **1979**, *117*, 1025–1044. [CrossRef]

39. Kazakhstan National Data Center. Available online: https://www.kndc.kz (accessed on 1 April 2024).
40. Community Coordinated Modeling Center. Available online: https://ccmc.gsfc.nasa.gov (accessed on 1 April 2024).
41. Laboratory of X-ray Astronomy of the Sun. Available online: https://tesis.xras.ru/en/ (accessed on 1 April 2024).
42. World Data Center for Geomagnetism, Kyoto. Available online: https://wdc.kugi.kyoto-u.ac.jp (accessed on 1 April 2024).
43. Krasnov, V.M.; Drobzheva, Y.V. Nonlinear acoustics in the in homogeneous atmosphere within the limits of analytical solutions. In *The Printery KROM*; KROM: St. Petersburg, Russia, 2018; p. 172.
44. Krasnov, V.M.; Kuleshov, Y.V. Variation of infrasonic signal spectrum during wave propagation from Earth's surface to ionospheric altitudes. *Acoust. Phys.* **2014**, *60*, 19–28. [CrossRef]
45. Liu, J.; Zhang, X.; Wu, W.; Chen, C.; Wang, M.; Yang, M.; Guo, Y.; Wang, J. The seismo-ionospheric disturbances before the 9 June 2022 Maerkang $M_s$6.0 earthquake swarm. *Atmosphere* **2022**, *13*, 1745. [CrossRef]
46. Zhang, X.; Chen, C.H. Lithosphere–atmosphere–ionosphere coupling processes for pre-, co-, and post-earthquakes. *Atmosphere* **2023**, *14*, 4. [CrossRef]
47. Oyama, K.I.; Devi, M.; Ryu, K.; Chen, C.H.; Liu, J.Y.; Liu, H.; Bankov, L.; Kodama, T. Modifications of the ionosphere prior to large earthquakes: Report from the Ionosphere Precursor Study Group. *Geosci. Lett.* **2016**, *3*, 6. [CrossRef]
48. Conti, L.; Picozza, P.; Sotgiu, A. A critical review of ground based observations of earthquake precursors. *Front. Earth Sci.* **2021**, *9*, 676766. [CrossRef]
49. Picozza, P.; Conti, L.; Sotgiu, A. Looking for earthquake precursors from space: A critical review. *Front. Earth Sci.* **2021**, *9*, 676775. [CrossRef]
50. Mukashev, K.M.; Sadykov, T.K.; Ryabov, V.A.; Shepetov, A.L.; Khachikyan, G.Y.; Salikhov, N.M.; Muradov, A.D.; Novolodskaya, O.A.; Zhukov, V.V.; Argynova, A.K. Investigation of acoustic signals correlated with the flow of cosmic ray muons in connection with seismic activity of Northern Tien Shan. *Acta Geophys.* **2019**, *67*, 1241–1251. [CrossRef]
51. Pulinets, S.; Davidenko, D. Ionospheric precursors of earthquakes and Global Electric Circuit. *Adv. Space Res.* **2014**, *53*, 709–723. [CrossRef]
52. Pulinets, S.A.; Ouzounov, D.P.; Karelin, A.V.; Davidenko, D.V. Physical bases of the generation of short-term earthquake precursors: A complex model of ionization-induced geophysical processes in the lithosphere-atmosphere-ionosphere-magnetosphere system. *Geomagn. Aeron.* **2015**, *55*, 521–538. [CrossRef]
53. Krasnov, V.M. Remote monitoring of nuclear explosions during ratio sounding of ionosphere over explosion site: Report. In Proceedings of the 16th National Radio Science Conference, NRSC'99, Cairo, Egypt, 30 May–1 June 1999; pp. INV2/1–INV2/7. [CrossRef]
54. Mikhailov, V.N. (Ed.) The radioecological situation in the areas of underground nuclear tests. In *Nuclear Tests in the Arctic*; Institute for Strategic Stability (Rosatom): Moscow, Russia, 2004; Volume 2, Chapter 2.5.
55. Logachev, V.A. (Ed.) *Semipalatinsk Test Site: Ensuring the General and Radiation Safety of Nuclear Tests*; IGEM RAS: Moscow, Russia, 1997; p. 347. (In Russian)
56. Smagulov, S.G. *Signs of Fate. Memoirs of a Nuclear Tester*; RFNC-VNIIEF: Sarov, Russia, 2012; p. 212. (In Russian)
57. Pulinets, S.A. Physical bases of the short-term forecast of earthquakes. *Astron. Astroph. Trans.* **2024**, *34*, 65–84. [CrossRef]
58. Ginzburg, V.L. *The Propagation of Electromagnetic Waves in Plasmas*; Pergamon Press: Oxford, UK, 1964.

*Article*

# Thermal Anomalies Observed during the Crete Earthquake on 27 September 2021


Soujan Ghosh [1,*], Sudipta Sasmal [2], Sovan K. Maity [2,3], Stelios M. Potirakis [4,5,6] and Masashi Hayakawa [7,8]

[1] National Atmospheric Research Laboratory, Department of Space, Government of India, Gadanki 517112, India

[2] Institute of Astronomy Space and Earth Science, Kolkata 700054, India; meet2ss25@gmail.com (S.S.)

[3] Uttar Amtala Gita Rani Vidyabhawan (H.S.), Contai 721427, India

[4] Department of Electrical and Electronics Engineering, University of West Attica, Ancient Olive Grove Campus, Egaleo, GR-12244 Athens, Greece; spoti@uniwa.gr

[5] Institute for Astronomy, Astrophysics, Space Applications and Remote Sensing, National Observatory of Athens, Metaxa and Vasileos Pavlou, Penteli, GR-15236 Athens, Greece

[6] Department of Electrical Engineering, Computer Engineering and Informatics, School of Engineering, Frederick University, CY-1036 Nicosia, Cyprus

[7] Hayakawa Institute of Seismo-Electromagnetics Co., Ltd. (Hi-SEM), UEC (University of Electro-Communications), Alliance Center 521, 1-1-1 Kojima-cho, Chofu, Tokyo 182-0026, Japan; hayakawa@hi-seismo-em.jp

[8] Advanced Wireless and Communications Research Center (AWCC), UEC, 1-5-1 Chofugaoka, Chofu, Tokyo 182-8585, Japan

* Correspondence: soujanghosh89@gmail.com



**Abstract:** This study examines the response of the thermal channel within the Lithosphere–Atmosphere–Ionosphere Coupling (LAIC) mechanism during the notable earthquake in Crete, Greece, on 27 September 2021. We analyze spatio-temporal profiles of Surface Latent Heat Flux (SLHF), Outgoing Longwave Radiation (OLR), and Atmospheric Chemical Potential (ACP) using reanalysis data from the National Oceanic and Atmospheric Administration (NOAA) satellite. Anomalies in these parameters are computed by removing the background profile for a non-seismic condition. Our findings reveal a substantial anomalous increase in these parameters near the earthquake's epicenter 3 to 7 days before the main shock. The implications of these observations contribute to a deeper understanding of the LAIC mechanism's thermal channel in seismic events.



**Keywords:** lithospheric–atmospheric–ionospheric coupling; thermal anomaly; surface latent heat flux; outgoing longwave radiation; atmospheric chemical potential



**Citation:** Ghosh, S.; Sasmal, S.; Maity, S.K.; Potirakis, S.M.; Hayakawa, M. Thermal Anomalies Observed during the Crete Earthquake on 27 September 2021. *Geosciences* **2024**, *14*, 73. https://doi.org/10.3390/geosciences14030073

Academic Editors: Jesus Martinez-Frias and Rosa Nappi

Received: 6 November 2023
Revised: 19 February 2024
Accepted: 4 March 2024
Published: 9 March 2024

## 1. Introduction

Various complicated and nonlinear processes govern the interaction of the lithosphere, atmosphere, and ionosphere. Studying various channels, like chemical, thermal, acoustics, and electromagnetic channels, it was found that precursory earthquake activities begin around a month before the actual catastrophe within the area known as the "preparation zone". These channels cover a wide range of parameters that can be treated as potential tools to understand the seismogenic effects from beneath the earth's surface to the magnetospheric heights [1]. These pre-earthquake activities involve complex, heterogeneous, multidimensional, and multi-parametric physical and chemical processes [1–5]. So, the precursory framework of seismic hazards includes a variety of chemical, thermal, acoustic, mechanical, and electromagnetic events, each with its own set of characteristics. Based on these characteristics and their interconnecting properties, a unified theory called Lithospheric–Atmospheric–Ionospheric Coupling (LAIC) has been proposed to explain the coupling that occurs across different layers of the atmosphere and ionosphere during seismic events through a variety of channels, e.g., [1,6,7]. This idea has been addressed and validated in light of the significant variabilities in several atmospheric and ionospheric parameters. This mechanism establishes the inter-relationships between various geochemical and geophysical processes related to earthquake occurrences. Radon emanation [8] is the most commonly accepted possible primary cause of air ionizations in the near-surface area, where the early geochemical and physical processes begin. Ion clusters are created when the ions generated by this mechanism interact with water molecules. These nanometer-sized cluster ions are then released into the atmosphere, changing the natural profile of air conductivity and, as a result, the ionospheric potential also changes [7,9]. These ongoing changes are therefore reflected in the ionosphere as seismogenic ionospheric disturbances, as shown by [10–21]. In the lower atmosphere, the condensation of water molecules owing to a change in the relative humidity of air eventually leads to latent heat release. The latent heat release raises in the lower atmosphere are the total energy budget, which initiates the emission of the Outgoing Longwave Radiation flux (W/m$^2$) as measured on the topside of the atmosphere during and prior to seismic events. These thermal anomalies are observed from both ground-based and satellite observations [14,22,23].

It is well established that during the final stage of the earthquake preparation process, the system tends to move to a self-organizing state from a chaotic state to reach the critical state where one can expect the maximum instability in the strain energy profile that is about to go through an avalanche, and after that it passes the system to another quality [24]. Using nonlinear thermodynamics and synergetics, these processes can be described [25]. Some integral parameters with threshold values mainly describe the system. When the parameter crosses the threshold value, the system approaches the critical state. This threshold is the final stage of the earthquake preparation and rupture processes in the seismogenic process. Furthermore, the integral parameter describes several simultaneous processes that lead the system to the critical state. The chemical potential of water vapor at a high ionization level is one of the probable candidates to be the integral parameter during the seismogenic process, as its values are derived from the anomalies in both the atmospheric temperature and humidity.

Pulinets et al. [26] presented that during the phase transition, the latent heat required by the water molecule is equivalent to its chemical potential. In multicomponent media with external forcing, the relative humidity can be represented as

$$H(t) = \frac{exp(-U(t)/kT)}{exp(-U_0/kT)} = exp\left(\frac{U_0 - U(t)}{kT}\right) = exp\left(-\frac{0.032\Delta U cos^2 t}{(kT)^2}\right), \quad (1)$$

where $U(t) = U_0 + \Delta U cos^2 t$, and $\Delta U$ is the volume averaged correction of chemical potential, arising from external forcing of the environment. The square of the cosine of time represents the diurnal variation in the intensity of the solar radiation, and $U_0$ represents the boiling temperature. $\Delta U$ is a complex parameter reflecting the formation of cluster ions. A higher value of $\Delta U$ indicates that the energy of the water molecule binding with ions is also higher, and the cluster ions are more stable. These stable cluster ions have a much longer lifetime and grow larger due to the attachment of water molecules.

Gorny et al. [27] first observed prominent signatures for thermal anomalies in the Central Asian region. They analyzed a satellite image of the earth's surface within the spectral range of 10.5–11.3 *mcm* over some linear structures of the Middle Asian seismically active region (Kopet Dagh, Talasso-Ferghana, and other faults) and observed positive anomalies of IR radiation along the point of intersection of the Talasso-Ferghana and Tamdy-Tokrauss faults. After that, several researchers started to study thermal anomalies associated with large earthquakes using various parameters in China, Japan, India, Iran, and Algeria [28–33]. Pulinets et al. [34] presented that Atmospheric Chemical Potential (ACP) anomalies are generated in the last stage of the seismic cycle. The temporal variation in the ACP is very similar to the radon activity in the preparation zone. The magnitude of the ACP drops close to the moment of the main shock. This magnitude of the ACP variation is weaker over the ocean. It was also found that there are areas where ACP anomalies are very weak or negative variation is observed. Pulinets et al. [35] presented the importance of ACP anomalies before seismogenic activities and other atmospheric phenomena.

In our study, we chose the Crete earthquake of 2021 to investigate the possible pre- and co-seismic thermal anomalies from the satellite data and present them with the conventional spatio-temporal variation in the anomalies by removing the background variation in the non-seismic condition. Our main aim is to find the anomaly variation of various parameters related to thermal channels before earthquakes to establish the hypothesis of a possible coupling mechanism during pre-seismic conditions. This work emphasizes one of the most important channels of the LAIC mechanism. The earthquake under study was chosen in such a way that evidence of such seismogenic impression is already established by using other parameters of other channels of LAIC [36]. This will also enable us to study and compare a multi-parametric approach of LAIC mechanisms [37–41]. The plan of this paper is as follows: In Section 2, we present our observations and the methods of extracting anomalies; in Section 3, we present the results and discuss our results; and finally, in Section 4, we make concluding remarks.

## 2. Data and Methodologies

In our study, we chose the Crete earthquake of 2021 to investigate the possible pre- and co-seismic irregularities of thermal parameters. On 27 September 2021, at 06:17 UT, an earthquake of $M_w = 6$ magnitude struck the island of Crete, Greece, at a depth of 6 km. In Figure 1, we present the epicenter of the earthquake, marked with a tiny red circle; the earthquake preparation zone, marked with a black circle; and the earthquake fault line, marked with a red curve. We present the details of the earthquake in Table 1.

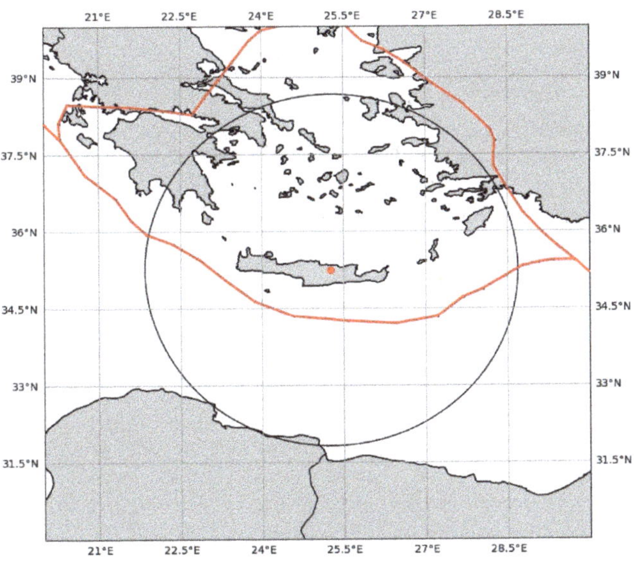

**Figure 1.** The location of the earthquake epicenter (red circle), the earthquake preparation zone (black circle), and the local earthquake fault lines for the Crete, Greece, earthquake.

**Table 1.** Earthquake details.

| Earthquake | Location of Epicenter | Magnitude ($M_w$) | Depth (km) | Date and Time (UT) | Radius of Preparation Zone (km) |
|---|---|---|---|---|---|
| Crete Earthquake | 35.244° N 25.27° E | 6 | 6 | 27 September 2021 06:17:21 | 380.189 |

### 2.1. Surface Latent Heat Flux Anomaly

To identify thermal anomalies related to earthquakes, we first removed the diurnal, seasonal variation, and other meteorological variations. We computed the background data using the seismically quiet period for the same grid area. To compute the background data, we used the following equation:

$$G_{bac}(x,y,t) = \frac{1}{N} \sum_{i=1}^{n} G_i(x,y,t). \tag{2}$$

Here, $G_i(x,y,t)$ is the background SLHF as a function of latitude $(x)$, longitude $(y)$, and time $(t)$ and $G_bac$ is the background flux. For the studied case, the value of $N$ is 3. We computed the background data for the Crete earthquake using the years 2015, 2017, and 2019, which were seismically quiet. We removed the background variation from the seismically active time period to compute the anomaly as

$$Anomaly(x,y,t) = [G(x,y,t) - G_{bac}(x,y,t)], \tag{3}$$

where $G(x,y,t)$ is the data obtained for the seismically active time period.

### 2.2. Outgoing Longwave Radiation Data Analysis

Several researchers have used techniques to analyze Outgoing Longwave Radiation (OLR) data; one of the most significant ones is the Eddy Field Calculation Mean method. This method detects the presence of any singularities in OLR data between adjacent points within the epicenter region [14,42–44]. The Eddy Field Calculation Mean is defined as the "total sum of the difference value" of the "measured value" of OLR between adjacent points [45]. This method compares the parameters for one particular point over the grid data with their nearest adjacent grid locations (latitudinal and longitudinal directions). The sum of the difference values of all measured values gives the parameter value for that particular point. Finally, an interpolated values gives the spatio-temporal profile of the said parameter. This method of computation can be expressed as follows:

$$S_d^*(x_{i,j}, y_{i,j}) = 4S(x_{i,j}, y_{i,j}) - [S(x_{i-1,j}, y_{i,j}) + S(x_{i,j}, y_{i,j-1}) + S(x_{i+1,j}, y_{i,j}) + S(x_{i,j}, y_{i,j+1})], \tag{4}$$

where $S_d^*(x_{i,j}, y_{i,j})$ is the daily Eddy field, $S(x_{i,j}, y_{i,j})$ is the daily mean, and $x$ and $y$ are the latitude and longitude, respectively. $i$ and $j$ are the integers representing the number of grids.

### 2.3. Atmospheric Chemical Potential Analysis

To compute the ACP, we followed the method suggested by Boyarchuk et al. [46]. The authors expressed $\Delta U$ with the air temperature at the earth's surface and relative humidity. $\Delta U$ is expressed as

$$\Delta U = 5.8 \times 10^{-10}(20T_g + 5463)^2 ln(100/H), \tag{5}$$

where $\Delta U$ is in eV.

In this study, we used data obtained from the National Oceanic and Atmospheric Administration (NOAA) reanalysis dataset. NOAA reanalysis data were taken from https://www.esrl.noaa.gov (accessed on 11 March 2023). These datasets data formats and classes were previously discussed in several publications, e.g., [22]. For the Crete earthquake, we used the latitude range 31° N to 39° N and the longitude range of 20° E to 30° E for the spatio-temporal variation. We computed the anomalies in all these parameters and overlaid the world map over the results to recognize the spatio-temporal variation regarding the epicenter of the earthquake. This showed a systematic variation over both space and time of the analyzed parameters to find out their temporal evolution and the most affected area that shows the maximum perturbation around the EQ epicenter.

## 3. Results

### 3.1. Surface Latent Heat Flux Observational Results

In Figures 2 and 3, we present the spatio-temporal variation in Surface Latent Heat Flux (SLHF) anomalies in $W/m^2$. This spatio-temporal variation gives direct evidence of the increase in the SLHF. We present the SLHF anomalies from 13 September 2021 to 12 October 2021, from longitudinal and latitudinal spans from 20° E to 30° E and 31° N to 39° N, respectively. The epicenter is indicated with a red dot and marked with the letter "C". Figures 2 and 3 show that the sudden intensification of the SLHF anomaly was observed near the epicenter on 23 September 2021. The next day, the maximum intensification of the SLHF anomaly was observed over the epicenter. The anomaly also propagated towards the southeast direction on that particular day. From 25 September, no such anomalous behavior of the SLHF was observed near the epicenter. From 30 September to 6 October, a slight increase in the SLHF anomaly is observed within the preparation zone of the earthquake. These increments in the SLHF are due to aftershocks followed by the mainshock and also due to the combined effect of the earthquake of 12 October, which occurred very close to the epicenter of the first earthquake.

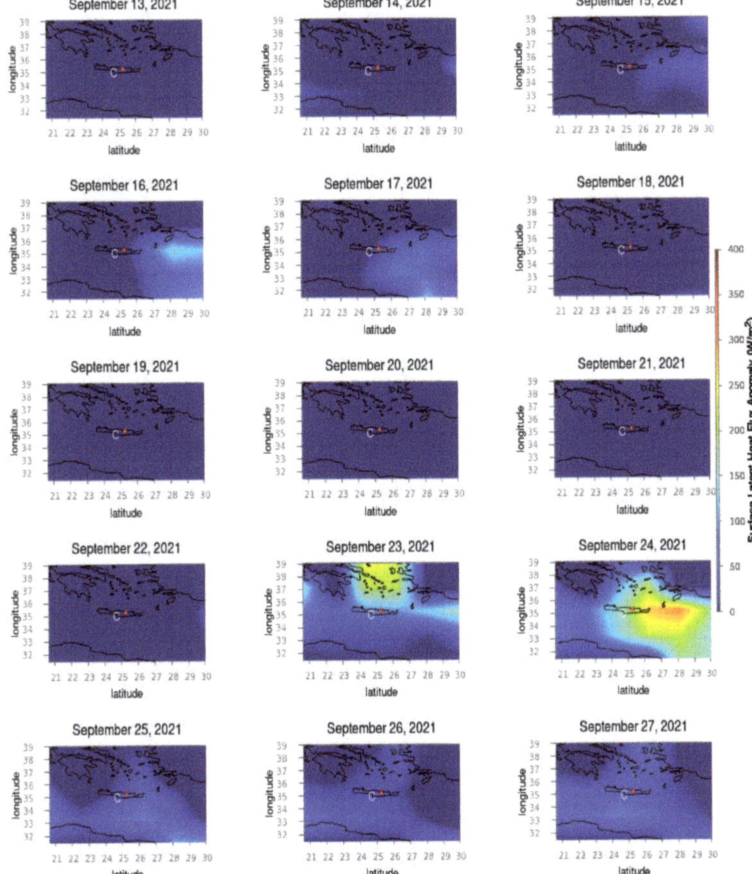

**Figure 2.** Variation in the SLHF anomaly from 13 to 27 September 2021 for the Crete earthquake. Along the X and Y axes, we present the geographic longitude and latitude, respectively. The black outlines are the country border, and the red dot marks the epicenter of the Crete earthquake, also denoted by the letter "C".

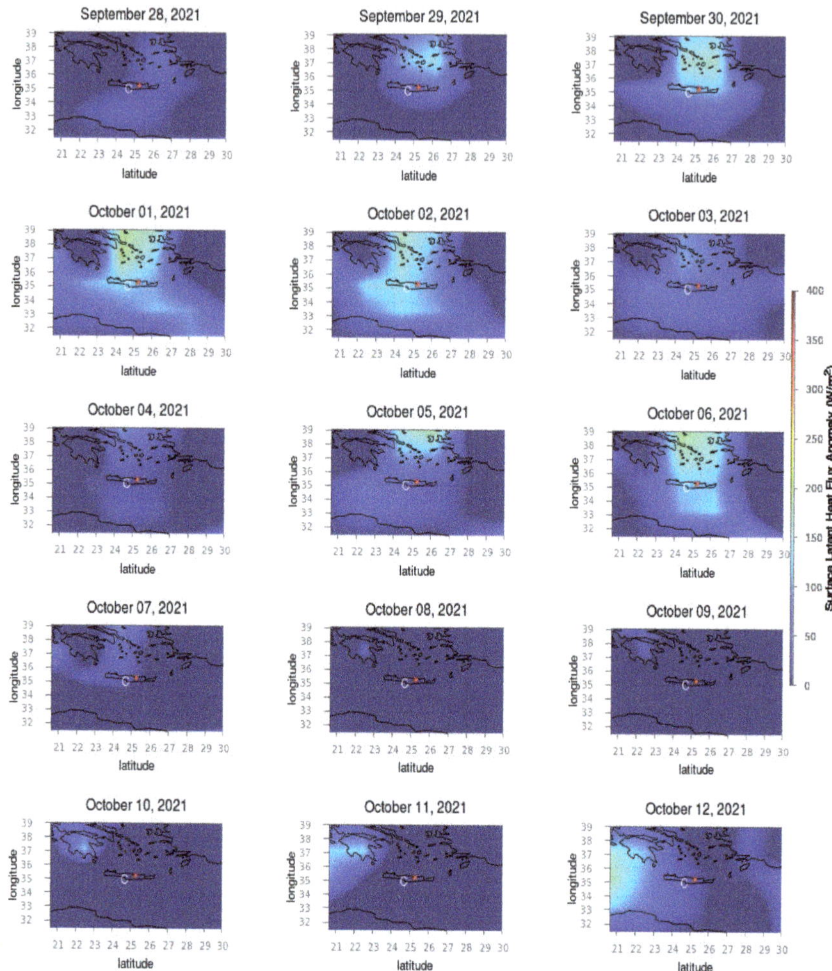

**Figure 3.** SLHF anomaly variation from 28 September to 12 October 2021. The figure format follows the format of Figure 2. The black outlines are the country border, and the red dot marks the epicenter of the Crete earthquake, also denoted by the letter "C".

### 3.2. OLR Results

Figures 4 and 5 represent the daily OLR variation around the earthquake epicenter. Figure 4 shows the OLR variation from 13 to 27 September 2021, whereas in Figure 4, we present the same from 28 September to 12 October 2021. From Figure 4, it is clearly found that the OLR variation was low from 13 to 17 September. On 18 September, the OLR increased near the northern part of the earthquake preparation zone. The maximum intensification was observed on 20 September. The OLR variation decreased in the next few days and again started increasing on 23 September near the earthquake's epicenter. From 24 September, it started decreasing and completely vanished on 25 September. In Figure 5, we see that the OLR again increased during the first week of October due to aftershocks and the combined effect of another mainshock that occurred on 12 October, south-southeast of Crete island [12 October 2021, 09:24:04.8 UT, (34.894° N, 26.472° N), $M_w = 6.4$, depth = 10 km], near the epicenter of the earthquake under study. It is also found that on the day of the second earthquake, the maximum intensification of OLR variation was observed within the region close to the epicenter.

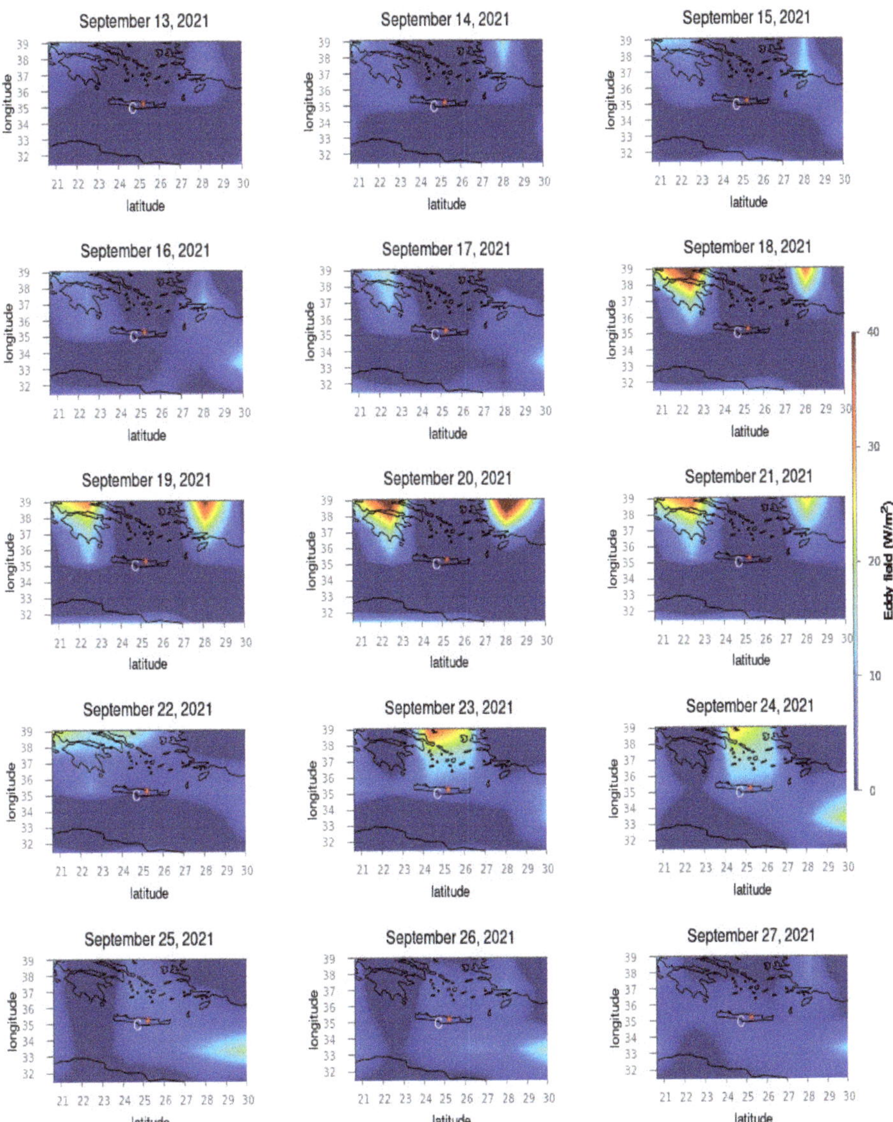

**Figure 4.** Eddy Field OLR variations around the Crete earthquake epicenter during 13–27 September 2021 with a spatial span of latitudes 30° N to 39° N and longitudes 20° E to 30° E. The red dot and the letter "C" indicate the epicenter of the Crete earthquake. The country boundaries are indicated by black lines. The color bar represents the intensity of the mean Eddy Field in W/m$^2$.

### 3.3. ACP Variation

In Figures 6 and 7, we present the spatio-temporal variation in the ACP within the earthquake preparation zone. The ACP value started showing an anomalous increase from 16 September, and the maximum ACP value near the epicenter of the studied earthquake was observed on 21 September. Again, on 26 September, the ACP value increased rapidly near the epicenter. From Figure 7, it was found that during October, the ACP value remained normal, and no rapid increase was observed near the epicenter.

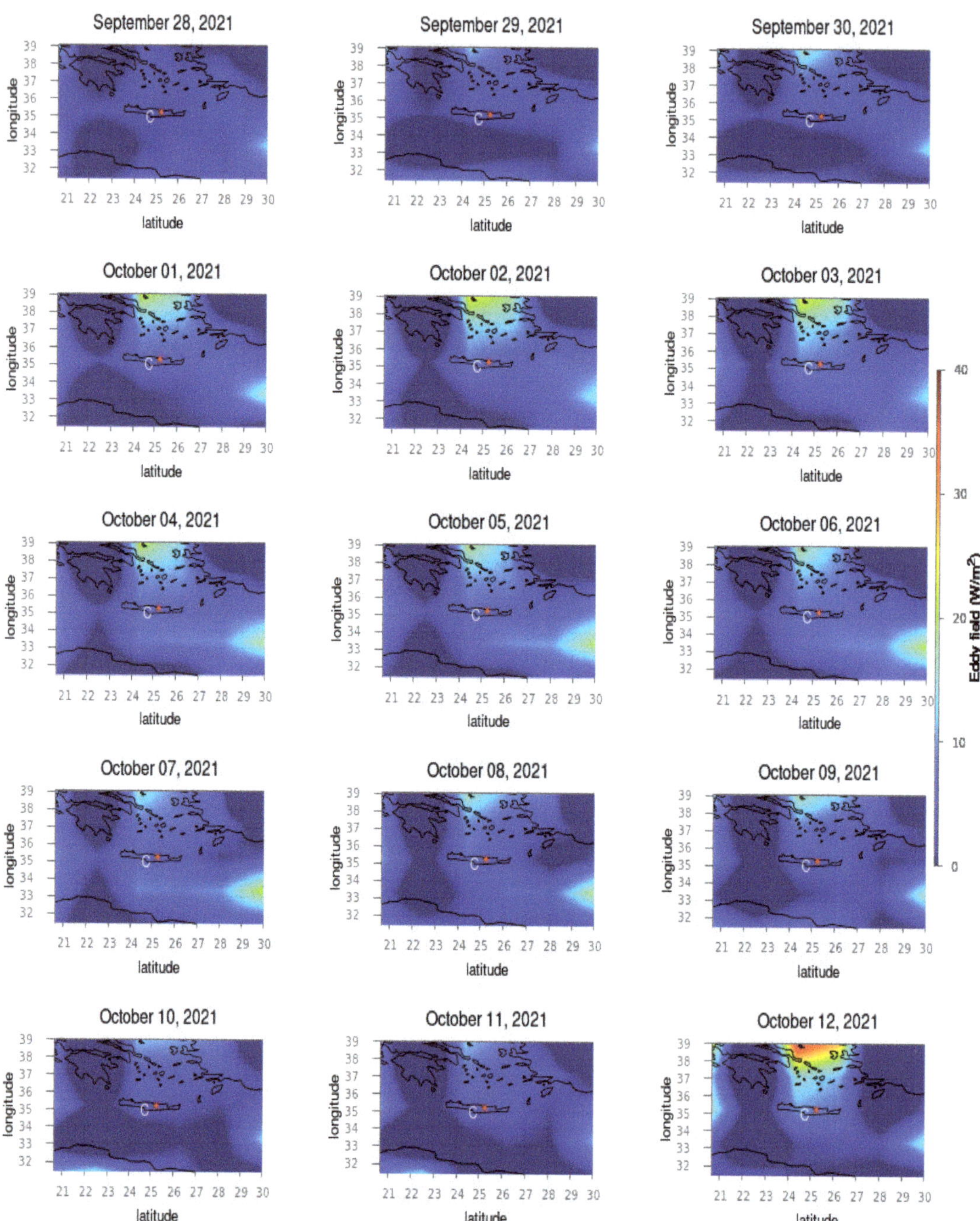

**Figure 5.** Same as Figure 4 for 28 September to 12 October 2021. The black outlines are the country border, and the red dot marks the epicenter of the Crete earthquake, also denoted by the letter "C".

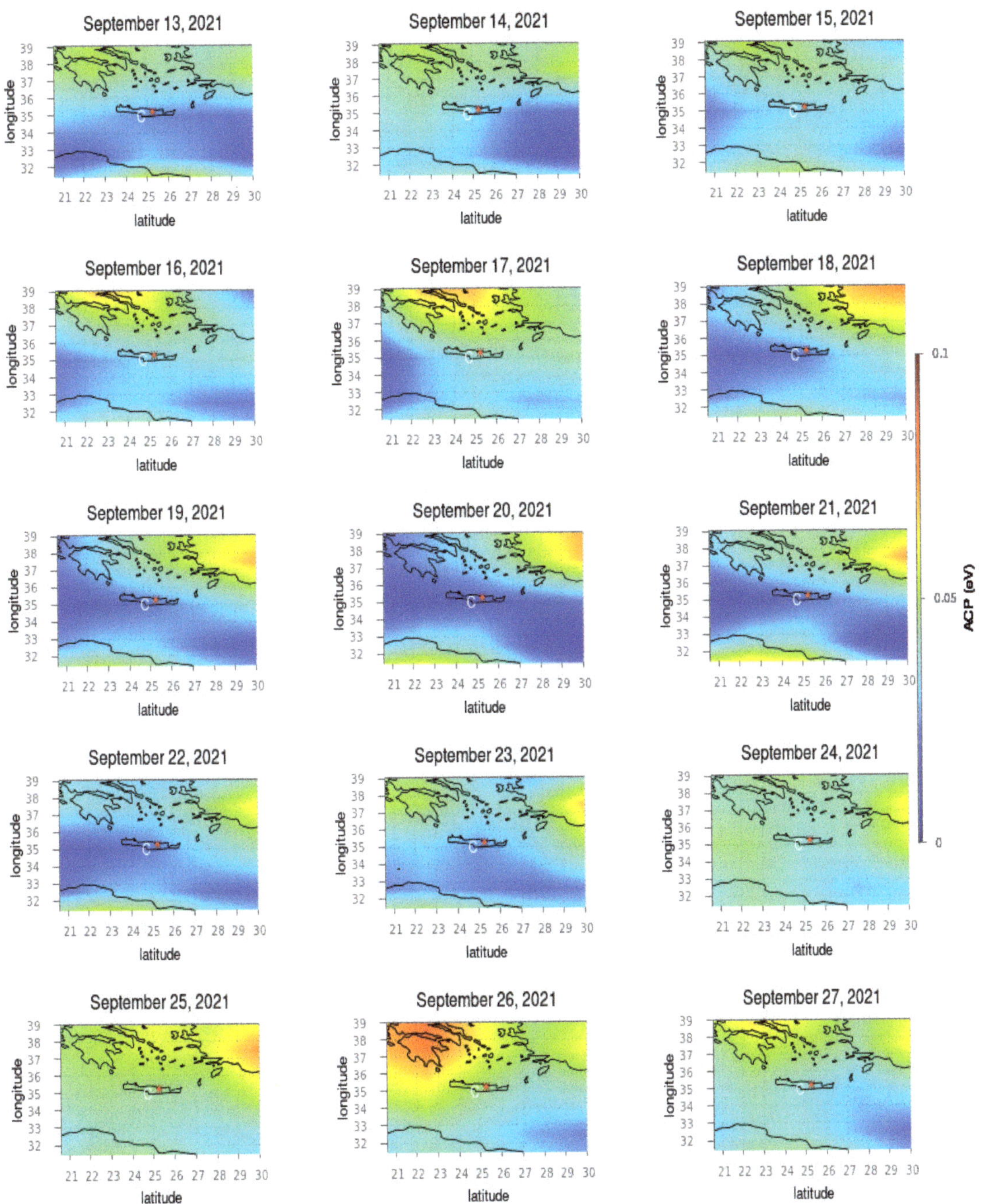

**Figure 6.** ACP distribution around the epicenter of the Crete earthquake from 28 September to 12 October 2021. The black outlines are the country border, and the red dot marks the epicenter of the Crete earthquake, also denoted by the letter "C".

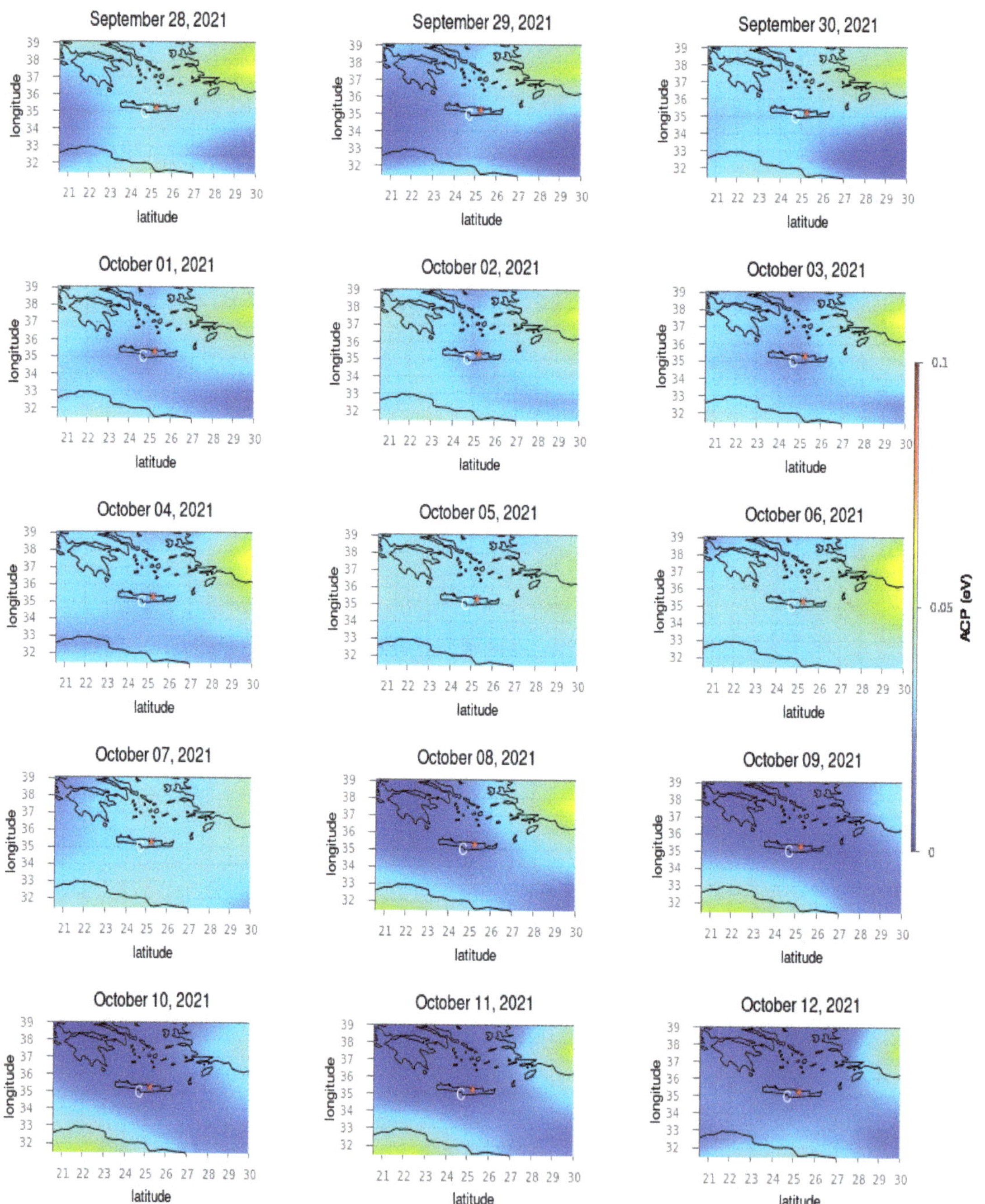

**Figure 7.** Same as Figure 6 for 28 September to 12 October 2021. The black outlines are the country border, and the red dot marks the epicenter of the Crete earthquake, also denoted by the letter "C".

## 4. Discussion

This work uses space-based observation to present the lower atmospheric thermal anomalies associated with the $M_w = 6$ Crete earthquake that took place on 27 September 2021. We investigated the anomalies in the SHLF, OLR, and ACP by using the NOAA reanalysis data. The spatio-temporal profiles of these parameters were studied around the earthquake occurrence day after the removal of background profiles by choosing a non-seismic period of observation.

(i)   It is evident from Figure 2 that the intensification in the SLHF is observed on 23 and 24 September 2021 over the epicenter region. The intensified SLHF shows a longitudinal spread over the earthquake epicenter and migrates towards the southeast direction. After the earthquake, comparatively less intensification is observed from 29 September to 6 October 2021 (Figure 2). For all these days, the anomalies are observed mostly in the north–south direction over the epicenter. This can be attributed to the combined effects of a series of aftershocks of the mainshock of 27 September and a second mainshock that took place on 12 October 2021, south-southeast of Crete island, the epicenter of which was in close vicinity of the first mainshock.

(ii)  In contrast to the SLHF, the intensification in the OLR Eddy Field was observed from 18 to 21 September 2021, in two different patches in the northeast and northwest directions of the epicenter, which lie within the earthquake preparation zone (Figure 4). On 23 and 24 September, it became a single intensification over the epicenter, similar to the SLHF variation. After the Crete earthquake, a similar post-earthquake OLR Eddy field enhancement was observed with much less intensity (Figure 5). On 12 September 2021, the Eddy Field again increased in the northern direction of the epicenter, possibly due to the second mainshock mentioned above.

(iii) The ACP variation shows an anomalous increase from 16 September, and the maximum enhancement took place on 21 September 2021 (Figure 6). During this period, the intensification of the ACP is found to be a bit away from the epicenter of the 2021 Crete earthquake. On 26 September, a secondary enhancement was observed near the earthquake epicenter. For the ACP, no such post-earthquake enhancement is observed, in contrast to the SLHF and OLR.

The analysis of very-low-frequency (VLF) sub-ionospheric propagation data received at multiple receivers during the 2021 Crete earthquake shows a moderate shift in electron density variation during the sunrise and sunset terminator times observed on 24, 25, and 27 September 2021 for the ISR-UWA propagation path [36]. Significant changes were not observed during both terminator times for the ISR-GER propagation path. However, for the 12 October earthquake, a significant change in the electron density profile was observed. Furthermore, some intermediate change in the electron density profile was observed on 11 October [36]. In the present study, we also found thermal anomalies during the same time period, which indicates that during the pre-seismic process, the lithospheric perturbations percolate to the troposphere and lower ionosphere through various channels according to the LAIC mechanism. The air ionization creates cluster hydration that releases latent heat and increases the air temperature. According to the LAIC hypothesis, this may create an ion uplift by the formation of an EQ cloud, and that reduces the air conductivity. Thus, the ground–ionosphere potential differences are modulated. This changes the horizontal electric field in the ionosphere, leading to ion drifts, and therefore, the electron and ion concentration may become perturbed over the EQ preparation zone. So, a lithospheric phenomenon can be coupled with ionospheric irregularities, as the LAIC theory prescribes. Our study also found that ACP variation is much weaker for the 2021 Crete earthquake than for the other studied case. It was previously reported that near the oceanic earthquake, the ACP values were lower than those of the land earthquakes. We observed a similar variation from the studied case.

Our manuscript shows that the OLR and SHLF give comparatively clearer indications than the ACP. It needs to be noted the ACP is not a fundamental parameter but is derived from two important parameters, viz., air temperature and relative humidity. It is commonly found that during or prior to seismic activities, the air temperature shows an increment, and the relative humidity experiences a decrease in the profiles. Equation (5) in the manuscript shows that the overall ACP values generally show an increase in nature due to the increase and decrease in air temperature and relative humidity, respectively. As OLR and SLHF are not derived parameters, the variabilities in these parameters are different in comparison to the ACP. Furthermore, for a consistent change in the ACP, one can expect equal proportionate changes in air temperature and relative humidity that can satisfy the equation, which is possibly not always true. Thus, the intensification of ACP patches may have more dynamic characteristics in comparison to the other two parameters in this manuscript.

It is very important to accept that based on the research on the LAIC mechanism, a variety of parameters are involved in each channel of LAIC (chemical, thermal, acoustic, and electromagnetic). It needs to be remembered that the pre-seismic processes, as mentioned in LAIC coupling, are highly nonlinear, anisotropic, and multi-parametric in nature. Refs. [38,41] gave a detailed description of this anisotropy and dependency of multi-parameters. The works on surface deformation characteristics [38,47] before strong earthquakes indicate vital information about the frictional force that leads to generating enhanced thermal profiles around an earthquake's epicenter that sometimes follow the earthquake fault lines. Furthermore, as the friction-generated heat energy is not very regular during the earthquake preparation process [48,49], the increased energy budget due to friction may not be synchronized with the heat energy originated due to geochemical mechanisms. Therefore, there is always a possibility of breaking this synchronization, and it is difficult to identify any particular thermal anomaly that will temporally dominate every time. This is the same for the other parameters in other channels in LAIC. For example, for the well-known Nepal earthquake in April and May 2015, refs. [14,22] reported the OLR and SLHF separately. It is found that the temporal variation does not show proper synchronization. The OLR intensification shows a maximum anomaly 4 to 5 days before the earthquake, whereas the SLHF shows the maximum anomaly 4 to 5 days before the earthquake. Most interestingly, the same parameter (OLR/SLHF) shows different time frames to show the maximum anomaly before the earthquake. Therefore, this is not a linear problem where one can expect similar outcomes every time. The synchronization of all the parameters in individual channels may not happen every time. The anisotropy in the various parameters may bar achieving uniform directionality. For example, even though the spatio-temporal profile of OLR intensification followed the direction of the Himalayan fault line (east to west), a significant amount of the OLR energy budget also went to the north-to-south direction where there was no such fault line [14]. In the LAIC mechanism, the generation of any anomaly is equally important as the migration of such anomalies for the source location due to the presence of other atmosphere dynamics. Therefore, there is quite a possibility that one cannot expect the most intense anomaly over the epicenter every time. In our case, we experienced similar features for OLR anomalies.

## 5. Conclusions

During the pre-seismic process, the immediate after-effects are observed over the surface and lower tropospheric regions. In this region, the observed effects are mostly in the form of thermal anomalies, which makes it an important channel in the LAIC mechanism. In this study, we used space-based observations of thermal excitation during pre-seismic processes. The Outgoing Longwave Radiation from the fault lines changes the normal temperature trends and relative humidity near the epicenter, which eventually contributes to the Surface Latent Heat Flux anomaly. So, all of these parameters are considered significant parameters in the LAIC mechanism. In this study, we tried to investigate these parameters for the same earthquake and find the contrasting behavior of the parameters due to the geographic location and the parameter with the most potential for studying the precursory effects. We used parameters like the SLHF anomaly, which is originated from the coagulation of the water molecules to ions, whereas the OLR anomaly is directly emerging thermal waves in the infrared range. On the other hand, the ACP depicts the measure of these variations and the extent of the process criticality during pre-seismic conditions. We found that all these parameters exhibit significant increases 3 to 7 days prior to the mainshock of the earthquake near the epicenter. To date, we have been able to show the changes related to pre-seismic processes. So many significant parameters are associated with the LAIC mechanism's thermal channel and play various roles in seismogenic conditions. We do not have a clear idea of how this complex interplay is taking place and which are the most significant parameters in this interplay. We must study all possible parameters for different geographic and climatic conditions to understand the entire process. Multi-parametric and multidisciplinary approaches will help us understand the LAIC process, starting from the lower atmosphere to the upper ionosphere. A further approach to coordinate the observed anomaly with the ionospheric and magnetospheric perturbations using a multidimensional approach will help us to understand the entire LAIC process. These new approaches to solving the unanswered question of the LAIC mechanism will be studied in the future and will be published elsewhere.

**Author Contributions:** Conceptualization, S.G., S.S., S.M.P. and M.H.; methodology, S.G. and S.S.; software, S.G., S.S. and S.K.M.; validation, S.G., S.S. and S.M.P.; formal analysis, S.G.; investigation, S.G., S.S. and S.M.P.; resources, S.G. and S.S.; data curation, S.G.; writing—original draft preparation, S.G.; writing—review and editing, S.S., S.M.P. and M.H.; visualization, S.S. and S.M.P.; supervision, S.S., S.M.P. and M.H.; project administration, S.S. All authors have read and agreed to the published version of the manuscript.

**Funding:** The authors have not received any external funding for this research.

**Data Availability Statement:** The data are freely available in the NOAA Reanalysis repository https://psl.noaa.gov/data/gridded/data.ncep.reanalysis.html (accessed on 26 April 2023).

**Acknowledgments:** The authors acknowledge the NOAA for providing satellite data.

**Conflicts of Interest:** Author Masashi Hayakawa was employed by the company Hayakawa Institute of Seismo-Electromagnetics Co., Ltd. The remaining authors declare that the research was conducted in the absence of any commercial or financial relationships that could be construed as a potential conflict of interest.

# References

1. Pulinets, S.; Boyarchuk, K. *Ionospheric Precursors of Earthquakes*; Springer: Berlin/Heidelberg, Germany; New York, NY, USA, 2004; ISBN 3-540-20839-9. [CrossRef]
2. Molchanov, O.A.; Hayakawa, M. *Seismo Electromagnetics and Related Phenomena: History and Latest Results*; TERRAP: Tokyo, Japan, 2008; 189p.
3. Ouzounov, D.; Pulinets, S.; Hattori, K.; Taylor, P. (Eds.) *Pre-Earthquake Processes: A Multidisciplinary Approach to Earthquake Prediction Studies*; AGU Geophysical Monograph; Wiley: Hoboken, NJ, USA, 2018; Volume 234, 365p.
4. Nayak, K.; López-Urías, C.; Romero-Andrade, R.; Sharma, G.; Guzmán-Acevedo, G.M.; Trejo-Soto, M.E. Ionospheric Total Electron Content (TEC) Anomalies as Earthquake Precursors: Unveiling the Geophysical Connection Leading to the 2023 Moroccan 6.8 Mw Earthquake. *Geosciences* **2023**, *13*, 319. [CrossRef]
5. Sharma, G.; Saikia, P.; Walia, D.; Banerjee, P.; Raju, P.L.N. TEC anomalies assessment for earthquakes precursors in North-Eastern India and adjoining region using GPS data acquired during 2012–2018. *Quat. Int.* **2021**, *575*, 120–129. [CrossRef]
6. Molchanov, O.A. Lithosphere-Atmosphere-Ionosphere Coupling due to Seismicity. *Electromagn. Phenom. Assoc. Earthquakes* **2009**, 255–279.
7. Pulinets, S.; Ouzonov, D. Lithosphere-Atmosphere-Ionosphere Coupling (LAIC) model—An unified concept for earthquake precursors validation. *J. Asian Earth Sci.* **2011**, *41*, 371–382. [CrossRef]
8. Omori, Y.; Yasuoka, Y.; Nagahama, H.; Kawada, Y.; Ishikawa, T.; Tokonami, S.; Shinogi, M. Anomalous radon emanation linked to preseismic electromagnetic phenomena. *Nat. Hazards Earth Syst. Sci.* **2007**, *7*, 629–635. [CrossRef]
9. Klimenko, M.V.; Klimenko, V.V.; Zakharenkova, I.E.; Pulinets, S.A.; Zhao, B.; Tsidilina, M.N. Formation mechanism of great positive TEC disturbances prior to Wenchuan earthquake on May 12, 2008. *Adv. Space Res.* **2011**, *48*, 488–499. [CrossRef]
10. Asano, T.; Hayakawa, M. On the tempo-spatial evolution of the lower ionospheric perturbation for the 2016 Kumamoto earthquakes from comparisons of VLF propagation data observed at multiple stations with wave-hop theoretical computations. *Open J. Earthq. Res. (OJER)* **2018**, *7*, 161–185. [CrossRef]

11. Biswas, S.; Kundu, S.; Chowdhury, S.; Ghosh, S.; Yang, S.; Hayakawa, M.; Chakraborty, S.; Chakrabarti, S.; Sasmal, S. Contaminated effect of geomagnetic storm on pre-seismic atmospheric and ionospheric anomalies during imphal earthquake. *Open J. Earthq. Res. (OJER)* **2020**, *9*, 383–402. [CrossRef]

12. Chakrabarti, S.K.; Sasmal, S.; Chakrabarti, S. Ionospheric anomaly due to seismic activities-II: Possible evidence from D-layer preparation and disappearance times. *Nat. Hazards Earth Syst. Sci.* **2010**, *10*, 1751–1757. [CrossRef]

13. Chakraborty, S.; Sasmal, S.; Basak, T.; Ghosh, S.; Palit, S.; Chakrabarti, S.K.; Ray, S. Numerical modeling of possible lower ionospheric anomalies associated with Nepal earthquake in May, 2015. *Adv. Space Res.* **2017**, *60*, 1787–1796. [CrossRef]

14. Chakraborty, S.; Sasmal, S.; Chakrabarti, S.K.; Bhattacharya, A. Observational signatures of unusual outgoing longwave radiation (OLR) and atmospheric gravity waves (AGW) as precursory effects of May 2015 Nepal earthquakes. *J. Geodyn.* **2018**, *113*, 43–51. [CrossRef]

15. Ghosh, S.; Sasmal, S.; Midya, S.; Chakrabarti, S. Unusual Change in Critical Frequency of F2 Layer during and Prior to Earthquakes. *J. Open Earthq. Res. (OJER)* **2017**, *6*, 191–203. [CrossRef]

16. Ghosh, S.; Chakraborty, S.; Sasmal, S.; Basak, T.; Chakrabarti, S.K.; Samanta, A. Comparative study of the possible lower ionospheric anomalies in Very Low Frequency (VLF) signal during Honshu, 2011 and Nepal, 2015 earthquakes. *Geomat. Nat. Hazards Risk* **2019**, *10*, 1596–1612. [CrossRef]

17. Hayakawa, M.; Molchanov, O.A.; Ondoh, T.; Kawai, E. Anomalies in the sub-ionospheric VLF signals for the 1995 Hyogo-ken Nanbu earthquake. *J. Phys. Earth* **1996**, *44*, 413–418. [CrossRef]

18. Hayakawa, M.; Molchanov, O.A.; Ondoh, T.; Kawai, E. The precursory signature effect of the Kobe earthquake on VLF subionospheric signals. *J. Comm. Res. Lab. Tokyo* **1996**, *43*, 169–180.

19. Hayakawa, M.; Potirakis, S.M.; Saito, Y. Possible relation of air ion density anomalies with earthquakes and the associated precursory ionospheric perturbations: An analysis in terms of criticality. *Int. J. Electron. Appl. Res.* **2018**, *5*, 56–75. [CrossRef]

20. Sasmal, S.; Chakrabarti, S.K. Ionosperic anomaly due to seismic activities—Part 1: Calibration of the VLF signal of VTX 18.2 kHz station from Kolkata and deviation during seismic events. *Nat. Hazards Earth Syst. Sci.* **2009**, *9*, 1403–1408. [CrossRef]

21. Sasmal, S.; Chakrabarti, S.K.; Ray, S. Unusual behavior of Very Low Frequency signal during the earthquake at Honshu/Japan on 11 March, 2011. *Indian J. Phys.* **2014**, *88*, 1013–1019. [CrossRef]

22. Ghosh, S.; Chowdhury, S.; Kundu, S.; Sasmal, S.; Politis, D.Z.; Potirakis, S.M.; Hayakawa, M.; Chakraborty, S.; Chakrabarti, S.K. Unusual Surface Latent Heat Flux Variations and Their Critical Dynamics Revealed before Strong Earthquakes. *Entropy* **2022**, *24*, 23. [CrossRef]

23. Ghosh, S.; Sasmal, S.; Naja, M.; Potirakis, S.; Hayakawa, M. Study of aerosol anomaly associated with large earthquakes (M > 6). *Adv. Space Res.* **2022**, *71*, 129–143. [CrossRef]

24. Pulinets, S.; Ouzounov, D.; Karelin, A.; Davidenko, D. Physical Bases of the Generation of Short Term Earthquake Precursors: A Complex Model of Ionization Induced Geophysical Processes in the Lithosphere–Atmosphere–Ionosphere–Magnetosphere System. *Geomagn. Aeron.* **2015**, *55*, 521–538. [CrossRef]

25. Kondepudi, D.; Prigogine, I. *Modern Thermodynamics: From Heat Engines to Dissipative Structures*; Wiley: Chichester, UK, 1998.

26. Pulinets, S.A.; Ouzounov, D.; Karelin, A.V.; Boyarchuk, K.A.; Pokhmelnykh, L.A. The physical nature of the thermal anomalies observed before strong earthquakes. *Phys. Chem. Earth* **2006**, *31*, 143–153. [CrossRef]

27. Gorny, V.I.; Salman, A.G.; Tronin, A.A.; Shilin, B.V. The Earth's outgoing IR radiation as an indicator of seismic activity. *Proc. Acad. Sci. USSR* **1988**, *301*, 67–69.

28. Ouzounov, D.; Freund, F. Mid-infrared emission prior to strong earthquakes analyzed by remote sensing data. *Adv. Space Res.* **2004**, *33*, 268–273. [CrossRef]

29. Qiang, Z.J.; Xu, X.D.; Dian, C.G. Abnormal infrared thermal satellite-forewarning of earthquakes. *Chin. Sci. Bull.* **1990**, *35*, 1324–1327.

30. Saraf, A.K.; Choudhury, S. Satellite detects surface thermal anomalies associated with the Algerian earthquakes of May 2003. *Int. J. Remote Sens.* **2004**, *26*, 2705–2713. [CrossRef]

31. Tronin, A.A. Satellite thermal survey—A new tool for the study of seismo active regions. *Int. J. Remote Sens.* **1996**, *41*, 1439–1455. [CrossRef]

32. Tronin, A.A.; Hayakawa, M.; Molchanov, O.A. Thermal IR satellite data application for earthquake research in Japan and China. *J. Geodyn.* **2002**, *33*, 519–534. [CrossRef]

33. Xu, X.; Qiang, Z.; Dian, C. Thermal infrared image of Meteosat and prediction of impending earthquake–conclusion of studying thermal infrared image before Lanhang-Genma earthquake occurring. *Remote Sens. Environ.* **1991**, *6*, 261–266.

34. Pulinets, S.; Ouzounov, D.; Karelin, A.; Davidenko, D. Lithosphere-Atmosphere-Ionosphere-Magnetosphere Coupling—A Concept for Pre-Earthquake Signals Generation. In *Pre-Earthquake Processes: A Multidisciplinary Approach to Earthquake Prediction Studies*; Ouzounov, D., Pulinets, S., Hattori, K., Taylor, P., Eds.; AGU/Wiley: Washington, DC, USA, 2018; pp. 77–98.

35. Pulinets, S.; Budnikov, P. Atmosphere Critical Processes Sensing with ACP. *Atmosphere* **2022**, *13*, 1920. [CrossRef]

36. Politis, D.Z.; Potirakis, S.M.; Sasmal, S.; Malkotsis, F.; Dimakos, D.; Hayakawa, M. Possible Pre-Seismic Indications Prior to Strong Earthquakes That Occurred in Southeastern Mediterranean as Observed Simultaneously by Three VLF/LF Stations Installed in Athens(Greece). *Atmosphere* **2023**, *14*, 673. [CrossRef]

37. Biswas, S.; Chowdhury, S.; Sasmal, S.; Politis, D.Z.; Potirakis, S.M.; Hayakawa, M. Numerical modelling of sub-ionospheric Very Low Frequency radio signal anomalies during the Samos (Greece) earthquake (M = 6.9) on October 30, 2020. *Adv. Space Res.* **2022**, *70*, 1453–1471. [CrossRef]

38. Chowdhury, S.; Kundu, S.; Ghosh, S.; Hayakawa, M.; Schekotov, A.; Potirakis, S.M.; Chakrabarti, S.K.; Sasmal, S. Direct and indirect evidence of pre-seismic electromagnetic emissions associated with two large earthquakes in Japan. *Nat. Hazards* **2022**, *112*, 2403–2432. [CrossRef]

39. Hayakawa, M.; Schekotov, A.; Izutsu, J.; Yang, S.-S.; Solovieva, M.; Hobara, Y. Multi-Parameter Observations of Seismogenic Phenomena Related to the Tokyo Earthquake (M = 5.9) on 7 October 2021. *Geosciences* **2022**, *12*, 265. [CrossRef]

40. Kundu, S.; Chowdhury, S.; Ghosh, S.; Sasmal, S.; Politis, D.Z.; Potirakis, S.M.; Yang, S.S.; Chakrabarti, S.K.; Hayakawa, M. Seismogenic Anomalies in Atmospheric Gravity Waves as Observed from SABER/TIMED Satellite during Large Earthquakes. *J. Sens.* **2022**, *2022*, 3201104. [CrossRef]

41. Sasmal, S.; Chowdhury, S.; Kundu, S.; Politis, D.Z.; Potirakis, S.M.; Balasis, G.; Hayakawa, M.; Chakrabarti, S.K. Pre-Seismic Irregularities during the 2020 Samos (Greece) Earthquake (M = 6.9) as Investigated from Multi-Parameter Approach by Ground and Space-Based Techniques. *Atmosphere* **2021**, *12*, 1059. [CrossRef]

42. Kang, C.; Liu, D. The applicability of satellite remote sensing in monitoring earthquake. *Sci. Surv. Mapp.* **2001**, *26*, 46–48.

43. Liu, D.; Peng, K.; Liu, W.; Li, L.; Hou, J. Thermal omens before earthquake. *Acta Seismol. Sin.* **1999**, *12*, 710–715. [CrossRef]

44. Liu, D.; Anomalies analyses on satellite remote sensing OLR before Jiji earthquake of Taiwan Province. *Geo-Inf. Sci.* **2000**, *2*, 33–36.

45. Xiong, P.; Shen, X.H.; Bi, Y.X.; Kang, C.L.; Chen, L.Z.; Jing, F.; Chen, Y. Study of outgoing longwave radiation anomalies associated with Haiti earthquake. *Nat. Hazards Earth Syst. Sci.* **2010**, *10*, 2169–2178. [CrossRef]

46. Boyarchuk, K.A.; Karelin, A.V.; Shirokov, R.V. *Bazovaya Model' Kinetiki Ionizirovannoi Atmosfery (The Reference Model of Ionized Atmospheric Kinetics)*; VNIIEM: Moscow, Russia, 2006.

47. Yang, S.S.; Potirakis, S.M.; Sasmal, S.; Hayakawa, M. Natural Time Analysis of Global Navigation Satellite System Surface Deformation: The Case of the 2016 Kumamoto Earthquakes. *Entropy* **2020**, *22*, 674. [CrossRef]

48. Aubry, J.; Passelègue, F.X.; Deldicque, D.; Girault, F.; Marty, S.; Lahfid, A.; Bhat, H.S.; Escartin, J.; Schubnel, A. Frictional Heating Processes and Energy BudgetDuring Laboratory Earthquakes. *Geophys. Res. Lett.* **2018**, *12*, 274.

49. Coffey, J.L.; Savage, H.M.; Polissar, P.J. Estimates of earthquake temperature rise, frictional energy, and implications to earthquake energy budgets. *Seismica* **2023**, *2*. [CrossRef]

*geosciences*

*Article*

# Feasibility of Principal Component Analysis for Multi-Class Earthquake Prediction Machine Learning Model Utilizing Geomagnetic Field Data

**Kasyful Qaedi [1], Mardina Abdullah [1,2,*], Khairul Adib Yusof [1,3,*] and Masashi Hayakawa [4,5]**

[1]  Space Science Center, Institute of Climate Change, Universiti Kebangsaan Malaysia (UKM), Bangi 43600, Malaysia; qaedi96@gmail.com
[2]  Department of Electrical, Electronic and Systems Engineering, Faculty of Engineering and Built Environment, Universiti Kebangsaan Malaysia (UKM), Bangi 43600, Malaysia
[3]  Department of Physics, Faculty of Science, Universiti Putra Malaysia (UPM), Seri Kembangan 43400, Malaysia
[4]  Hayakawa Institute of Seismo Electromagnetics Co., Ltd. (Hi-SEM), UEC Alliance Center, 1-1-1 Kojimacho, Chofu 182-0026, Japan; hayakawa@hi-seismo-em.jp
[5]  Advanced & Wireless Communications Research Center (AWCC), The University of Electro-Communications, 1-5-1 Chofugaoka, Chofu 182-8585, Japan
*   Correspondence: mardina@ukm.edu.my (M.A.); adib.yusof@upm.edu.my (K.A.Y.)

**Abstract:** Geomagnetic field data have been found to contain earthquake (EQ) precursory signals; however, analyzing this high-resolution, imbalanced data presents challenges when implementing machine learning (ML). This study explored feasibility of principal component analyses (PCA) for reducing the dimensionality of global geomagnetic field data to improve the accuracy of EQ predictive models. Multi-class ML models capable of predicting EQ intensity in terms of the Mercalli Intensity Scale were developed. Ensemble and Support Vector Machine (SVM) models, known for their robustness and capabilities in handling complex relationships, were trained, while a Synthetic Minority Oversampling Technique (SMOTE) was employed to address the imbalanced EQ data. Both models were trained on PCA-extracted features from the balanced dataset, resulting in reasonable model performance. The ensemble model outperformed the SVM model in various aspects, including accuracy (77.50% vs. 75.88%), specificity (96.79% vs. 96.55%), F1-score (77.05% vs. 76.16%), and Matthew Correlation Coefficient (73.88% vs. 73.11%). These findings suggest the potential of a PCA-based ML model for more reliable EQ prediction.

**Keywords:** principal component analysis (PCA); ensemble; machine learning (ML); earthquake (EQ) prediction

**Citation:** Qaedi, K.; Abdullah, M.; Yusof, K.A.; Hayakawa, M. Feasibility of Principal Component Analysis for Multi-Class Earthquake Prediction Machine Learning Model Utilizing Geomagnetic Field Data. *Geosciences* **2024**, *14*, 121. https://doi.org/10.3390/geosciences14050121

Academic Editors: Dimitrios Nikolopoulos and Jesus Martinez-Frias

Received: 25 March 2024
Revised: 24 April 2024
Accepted: 26 April 2024
Published: 29 April 2024

## 1. Introduction

The non-linear, chaotic, scale-invariant phenomena of earthquakes (EQs) have led some researchers to conclude that predicting EQs in the conventional sense is inherently impossible due to complex interactions involving plate tectonics, fault mechanics, and material properties within the Earth's crust [1]. EQ precursor studies have shown that many short-term precursors are non-seismic, with the ionosphere, atmosphere, and lithosphere being perturbed prior to an EQ [2]. Various methods can be employed for EQ prediction, including the study of precursor phenomena such as fluctuations in electric and magnetic fields [3], variations in the total electron content of the ionosphere [4], observations of animal behavior [5], and the use of multiple remote sensing data sources such as electron and ion density data [6,7]. Hattori et al. [8] and Ouyang et al. [9], in their studies, observed distinctive perturbations in the spectral density ratio between the horizontal and vertical components of Ultra-Low-Frequency (ULF) geomagnetic field measurements. ULF magnetic data can provide useful EQ precursory information with optimal prediction performance depending on the distance and event size [10].

The dynamic nature of seismic events poses a challenge for traditional prediction methods based on historical and empirical observations. These methods often struggle to account for the complex factors that trigger EQs, leading to limitations in accuracy and reliability. However, machine learning (ML) algorithms like the Support Vector Machine (SVM), decision trees, and ensemble have demonstrated promising results in EQ forecasting [11–14]. ML classifiers have also shown potential for making accurate EQ magnitude predictions, which could significantly improve seismic risk assessment and preparedness efforts [15]. EQ prediction using geomagnetic data faces a significant challenge in large classification datasets. As data dimensions multiply, issues such as overfitting lead to increased computational costs and decreased model stability, which become major concerns. The dynamic nature and large coverage of the global geomagnetic field, including both spatial and temporal variations, translates into a large number of variables within the dataset [16]. Chen et al. [17] emphasize the need for effective dimensionality reduction techniques to alleviate these challenges, recommending methods like principal component analysis (PCA) or feature selection strategies to extract relevant information while handling large dimensionality.

PCA is a widely utilized technique for dimensionality reduction in high-dimensional datasets, including the electromagnetic or geomagnetic data that are consecutively applied in EQ predictions [18,19]. Hattori et al. [20] demonstrated the effectiveness of PCA in extracting the ULF signals associated with potential EQ precursors. Their study showcased PCA's ability to unravel the essential patterns within geomagnetic data, particularly those linked to ULF phenomena indicative of EQs. Ensemble methods like bagging and boosting have emerged as powerful tools for EQ prediction [21]. A study by Mukherjee et al. [22] demonstrated that ensemble models not only capture complex spatiotemporal patterns in seismic data but also exhibit a superior generalization performance compared to individual models. The ensemble approach leverages diverse learning strategies and mitigates the risk of overfitting, providing a robust framework for addressing the inherent uncertainties and dynamic nature of seismic processes.

This study applies a PCA with ensemble and SVM models to enhance EQ prediction using geomagnetic data categorized by the Mercalli Intensity Scale. Utilizing global geomagnetic data spanning from 1970 to 2021, sourced from SuperMAG (Laurel, MA, USA), alongside EQ records from the USGS and focusing on events with magnitudes M5.0 and above, this approach emphasizes dimensionality reduction via PCA to manage complex datasets for ML. Model efficacy is evaluated through accuracy, precision, recall, F1-score, and the Matthew Correlation Coefficient (MCC). By identifying key data components that correlate with seismic activity, the integration of a PCA with the ensemble and SVM algorithms aims to advance seismic risk mitigation by improving EQ prediction studies.

## 2. Data and Methods

This study utilized low-frequency 1 min global geomagnetic field data sourced from the SuperMAG database [23], combined with EQ data from the USGS [24], covering the period from 1970 to 2021. The dataset was filtered to include only EQs with a magnitude equal to or exceeding M5.0 and hypocentral locations situated within a radius of 200 km from their corresponding geomagnetic observatories, as can be observed in Figure 1 [25]. The study focused on earthquakes occurring within a seven-day window prior to significant seismic events, coinciding with the availability of station data [26]. The length of the observation period was chosen to maximize the number of constructed datasets as well as to balance between model optimization and computational cost. A total of 7525 EQs that met the criteria were selected. To refine the analysis, the Ap index was applied, using values below 27 to eliminate and exclude periods of geomagnetic quiescence, which represent geomagnetically quiet conditions. This ensures that the analysis is focused on more dynamic conditions [27]. Additionally, a Dst index cutoff of −30, which is commonly used to filter out instances of severe magnetic field disturbances, was applied [28]. The EQ magnitude scale was categorized according to the Mercalli

Intensity Scale to allow for a more refined multi-class model, encompassing distinct seismic intensities ranging from Non (non-seismic days) to VI (M5.0 to M5.5), VII (M5.5 and M6.0), VIII (M6.0 to M6.5), IX (M6.5 to M7.0), X (M7.0 to M7.5), XI (M7.5 and M8.0), and XII (>M8.0). The scale, which is based on observed effects and damage, offers a complementary perspective that can potentially help mitigate these limitations, therefore providing a more comprehensive picture for prediction purposes. The Mercalli Intensity Scale allows for a more refined categorical classification (in this case, 8 classes) compared to the Richter Scale, which uses a more generalized single-integer scale and could potentially increase computational costs [29].

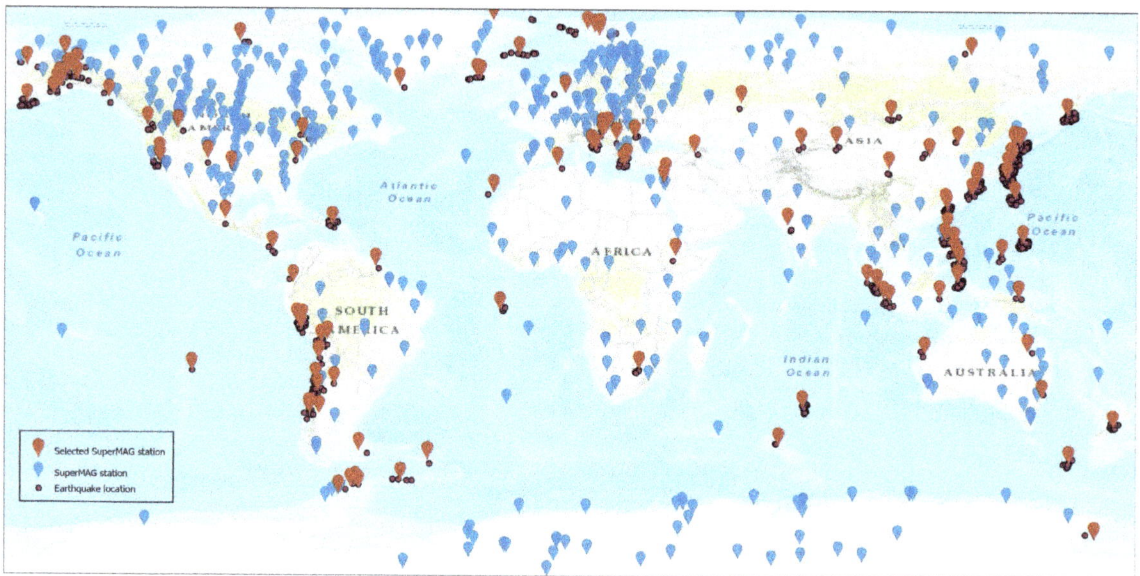


**Figure 1.** SuperMAG geomagnetic observatory locations around the world (blue pins) and filtered geomagnetic observatory locations based on selected EQ events (red pins).

The resolution of SuperMAG data (1 min sampling period) into 7-day windows resulted in a complex dataset, even with only three features (X, Y, Z). These features exhibited intricate relationships and variations over time, crucial for understanding EQ precursors. Applying a PCA addressed the complexity of the task by extracting the most informative temporal patterns and reducing dimensionality, while preserving the key interactions among the features. This approach facilitated the simplification of the data for analysis, allowing the extraction of the most pertinent information for the EQ prediction models. These components revealed key insights, including projected data points that represent observations in the reduced space, the variance explained by each component, and the contributions of features as indicated by the coefficient. While the coefficient provided interpretability, the projected data points served as the primary input for subsequent ML models. By choosing a cumulative explained variance threshold that captured 87% of the data's variance (number of components retained) based on a combination of grid and random search, the approach ensured that most of the relevant information was preserved while maintaining model flexibility. PCA proved to be a valuable tool in navigating the challenges of high-dimensional data, facilitating further analysis and model development.

To address the class imbalance caused by low-magnitude EQs, Synthetic Minority Oversampling Technique (SMOTE) was employed [30] to potentially improve EQ prediction accuracy. Bao et al. [31] successfully addressed the data imbalance issue in their EQ prediction model by employing SMOTE. This technique augmented the minority class within the dataset, enabling the model to learn their characteristics more effectively. This improvement did not compromise the model's sensitivity to smaller EQs, for example, class VII to IX, ensuring their proper identification and prediction. By oversampling the minority, SMOTE created a more balanced dataset, allowing the model to learn equally from both positive and negative examples. This reduced bias towards the majority class. A new synthetic instance, $x_{new}$, can be generated using the following formula:

$$x_{new} = x_i + \lambda \times \left( x_j - x_i \right)$$

where $x_i$ represents a minority class instance and $x_j$ represents its randomly selected neighbor, while $0 \leq \lambda \leq 1$ controls the proportion of synthetic samples created.

Leveraging the dimensionality reduction achieved through PCA and the balanced dataset obtained via oversampling, a 10-fold k-fold cross-validation was implemented. This approach iteratively trained and tested the models on various data subsets, providing a more reliable estimate of their generalizability compared to a simple train–test split and mitigating potential biases specific to individual data distributions. Subsequently, two models—an SVM and an ensemble model—were developed on the full dataset. Each model underwent hyperparameter tuning through a grid search method, optimizing their key settings to maximize their predictive power. The details of this hyperparameter selection process are further discussed in Section 3.2. This comprehensive approach ensured the models were not only accurate on the specific training data but also generalizable to unseen examples, providing a more reliable estimate of their generalizability.

Model evaluation was conducted using the following multi-class classification metrics:

$$Accuracy = \frac{TP + TN}{\text{Total samples}}$$

$$Sensitivity \ (Recall) = \frac{TP}{TP + FN}$$

$$Specificity = \frac{TN}{TN + FP}$$

$$Precision = \frac{TP}{TP + FP}$$

$$F = 2\frac{(\text{Precision} \times \text{Recall})}{(\text{Precision} + \text{Recall})}$$

$$MCC = \frac{(TP \times TN - FP \times FN)}{\sqrt[2]{(TP + FP) \times (TP + FN) \times (FP + TN) \times (TN + FN)}}$$

where TP = True Positive, TN = True Negative, FP = False Positive, and FN = False Negative.

Given the multi-class nature of the EQ prediction model, the evaluation employed metrics that provided a comprehensive understanding of its performance across all EQ intensity levels. Metrics like precision, recall, and F1-score were utilized to assess the model's ability to correctly identify different EQ intensities, balancing the trade-off between true positives and false positives/negatives. Additionally, the MCC offered a balanced perspective on overall model performance by considering all true and false classifications. The detailed workflow is shown in Figure 2.

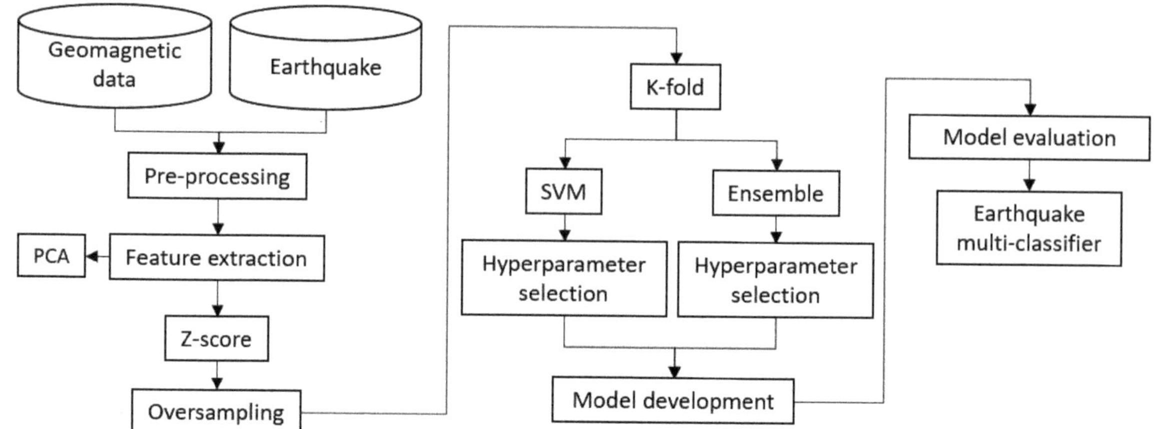

**Figure 2.** Illustration of the detailed workflow of a PCA-based approach for feature extraction and dimensionality reduction, leading to the construction of a multi-class model for EQ prediction.

## 3. Results and Discussion

### 3.1. PCA Scores for Model Development

In the PCA results, each principal component was plotted with all three of its original geomagnetic components. This approach was adopted to visually ascertain the relationships and correlations of each PCA result with the geomagnetic components to determine which components are most suitable for feature extraction. The position of each point on the first principal component (PC1) in Figure 3a indicates its similarity to its geomagnetic X component. Negative values aligned strongly with PC1, with a minimum of $-2017.8$ nT, and positive values also showed strong alignment, reaching a maximum of 1334.9 nT. The spread of points around zero values, highlighted by the yellow dashed box, reflects the correlation between PC1 and the X component. A tight cluster, as shown in the red dashed box, suggests a linear relationship, while a wider spread indicates a weaker or non-linear connection. The interpretation of PC1 relied on its correlation with other variables. In this case, its strong correlation with the X component signified northward variations in the Earth's magnetic field. The statistical values presented in Table 1 justified the resemblance between PC1 and the X component, indicating a minimal trade-off between the X component and PC1 when compared to the Y and Z components. The PC1 had a variance of 598.34 nT, slightly lower than the X component, with its variance of 624.26 nT. Similarly, the standard deviation for PC1 was 24.46 nT, closely matching that of the X component, which was 24.98 nT.

**Table 1.** Statistical value of geomagnetic components (X, Y, and Z) and principal components (PC1, PC2, and PC3).

| | X (nT) | Y (nT) | Z (nT) | PC1 (nT) | PC2 (nT) | PC3 (nT) |
|---|---|---|---|---|---|---|
| Mean | 8.24 | $-0.35$ | 1.46 | $7.33 \times 10^{-8}$ | $-5.80 \times 10^{-9}$ | $4.14 \times 10^{-8}$ |
| Median | $-4.75$ | $-0.10$ | 1.21 | 1.13 | $-0.09$ | 0.02 |
| Variance | 624.26 | 132.58 | 169.87 | 598.34 | 138.28 | 52.12 |
| Standard deviation | 24.98 | 11.51 | 13.03 | 24.46 | 11.75 | 7.21 |
| Range | 3343.75 | 3275.55 | 2711.71 | 3352.75 | 2429.60 | 1670.76 |
| Min | $-1924.85$ | $-1809.16$ | $-1561.00$ | $-2017.81$ | $-1660.95$ | $-926.62$ |
| Max | 1418.90 | 1466.38 | 1150.70 | 1334.93 | 768.65 | 744.13 |

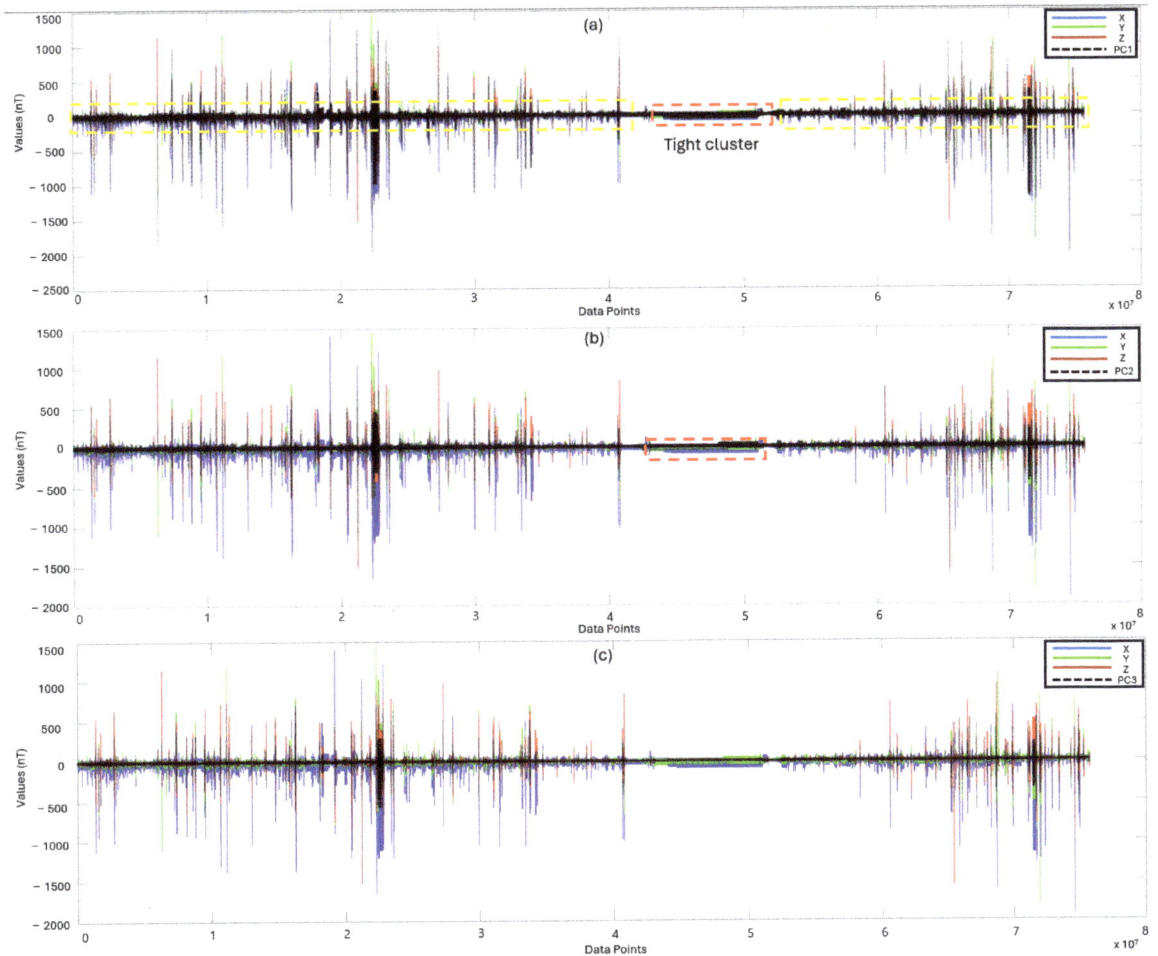



**Figure 3.** Comparative plots of principal component and geomagnetic field components (X (blue), Y (green), Z (red)) over data points. The y axis represents geomagnetic field values, and the x axis enumerates data points. Subfigures: (**a**) PC1 against X, Y, and Z; (**b**) PC2 against X, Y, and Z; (**c**) PC3 against X, Y, and Z, with the PCs depicted by black lines. The unit values are nanoTesla (nT).

Similarly, points on the second principal component (PC2) axis in Figure 3b illustrate their alignment within the Y axis. PC2 had a broader spread compared to PC1, capturing a wider range of geomagnetic variability. The clustering of points slightly above zero for PC2 indicated a correlation with the Y component. The statistical values for PC2 revealed a similarity with the Y component. Specifically, PC2 exhibited a similar variance of 138.28 nT compared to the variance of the Y component, which was 132.58 nT. Furthermore, the standard deviation of PC2 was 11.75 nT, compared to the standard deviation of the Y component, which was 11.51 nT.

The third principal component (PC3), as shown in Figure 3c, had a spread comparable to PC2, suggesting that it captured a similar level of variability. However, its correlations with geomagnetic components were even weaker than for PC2, indicating that PC3 most likely captured subtle or complex variations influenced by multiple factors or smaller-scale fluctuations. The statistical results showed no correlation with any component.

Understanding PC3 might require additional context such as location, time, or specific geomagnetic events. Therefore, PC3 was not included in the model training.

### 3.2. Hyperparameter Tuning and Algorithm Selection

This study evaluated Random Undersampling Boosting (RUSBoost), AdaBoostM2, bagging, and SVM algorithms for multi-class EQ prediction. Despite being compared to a baseline model, boosting methods like RUSBoost and AdaBoostM2 demonstrated poor predictive accuracy. In contrast, bagging achieved a good performance across all EQ classes. This finding underscores the importance of careful algorithm selection for multi-class problems, as distinct methodologies exhibit varying sensitivities to class imbalance and data complexity.

The optimization of SVM hyperparameters in Table 2 shows that the Gaussian kernel function was selected for its effectiveness in handling non-linear data relationships. The box constraint, set at 50, and the kernel scale, chosen as 0.5, were pivotal in balancing the trade-off between model complexity and overfitting, ensuring its robust predictive capability. The Nu parameter, fixed at 0.01, regulated the model's margin of error in classification, fine-tuning its sensitivity to seismic activity indicators. Subsequent hyperparameter tuning further optimized the bagging model, as shown in Table 2. Two hundred base learners were identified as offering a balance between model complexity and computational efficiency. A split size of 13,000 facilitated effective data partitioning, enhancing the model's ability to capture underlying patterns. Additionally, a minimum leaf size of 0.01 prevented overfitting while maintaining optimized model performance. The predictor selection strategy focusing on curvature had a minimal impact on performance.

**Table 2.** Optimized hyperparameter selection for the SVM and ensemble models.

| SVM Hyperparameter | SVM | Ensemble Hyperparameter | Ensemble |
|---|---|---|---|
| Kernel function | Gaussian | Method | bagging |
| Box constraint | 50 | Num of learners | 200 |
| Kernel | 0.5 | Split size | 13,000 |
| Nu | 0.01 | Leaf size | 0.01 |
| | | Predictor selection | curvature |

The accuracy of the models in Table 3, which represents the overall correctness of the predictions, showed that the ensemble model's algorithm outperformed the SVM with 77.50% accuracy compared to 75.88%. Sensitivity, which measures the ability to correctly identify positive instances, also favored the ensemble model at 77.50%, surpassing the SVM's performance of 75.88%. Both models exhibited high specificity, with the SVM at 96.55% and the ensemble model at 96.79%, indicating that both models correctly identified negative cases and rarely predicted an EQ when none actually occurred. High specificity might indicate inherent biases in the models due to their architecture, the potential over-sampling of negative data instances, and the imbalanced nature of the EQ data itself, as negative cases greatly outnumbered positive classes. Precision, which reflects the accuracy of positive predictions, was slightly higher for the SVM, at 77.56%, compared to the ensemble model, at 76.69%. However, the F1-score, which considers both precision and sensitivity, favored the ensemble model at 77.06% against the SVM at 76.16%. The MCC values for both models were almost identical at 73.88% for the ensemble and 73.11% for the SVM, suggesting a balanced performance in capturing true and false positives and negatives. Overall, the ensemble model demonstrated superior predictive capabilities for EQ prediction in this multi-class model, showcasing its effectiveness across multiple performance metrics.

**Table 3.** Performance measurements demonstrate that the ensemble model outperforms the SVM model.

| Model | SVM | Ensemble |
|---|---|---|
| Accuracy | 75.88% | 77.50% |
| Sensitivity | 75.88% | 77.50% |
| Specificity | 96.55% | 96.79% |
| Precision | 77.56% | 76.69% |
| F1-score | 76.16% | 77.05% |
| MCC | 73.11% | 73.88% |

### 3.3. Handling Imbalanced Data Using SMOTE

The implementation of SMOTE successfully mitigated the imbalance challenge by oversampling the underrepresented high-magnitude events. SMOTE's effectiveness is reflected in the showcased model's performance, as shown in the confusion matrix presented in Figure 4. The model achieved high precision and recall values for low-magnitude EQs, indicating its accurate identification of both positive and negative cases. Furthermore, for high-magnitude EQs exceeding scale VII, the model demonstrated near-perfect accuracy. By oversampling the scarce high-magnitude data, the model received more training examples to learn patterns specific to these critical events. However, it is important to acknowledge the potential limitations of SMOTE. While oversampling increases the representation of the minority class, it is crucial to ensure the introduced synthetic data points maintain proximity to their original distribution. Otherwise, overfitting or biased predictions could occur. In this case, the quality of the synthetic data generated was carefully monitored and its impact on model performance was evaluated through cross-validation techniques. Despite oversampling, the overall EQ data might still be limited, particularly for rare events like class XI and XII EQs. This limitation could restrict the generalizability of the study's findings and potentially lead to the models' overfitting to the specific dataset used. While the employed models offered good overall performance, their "black-box" nature presents another challenge. The lack of interpretability makes it difficult to fully understand their decision-making process, potentially hindering the evaluation of their prediction validity and identification of potential biases or inaccuracies.

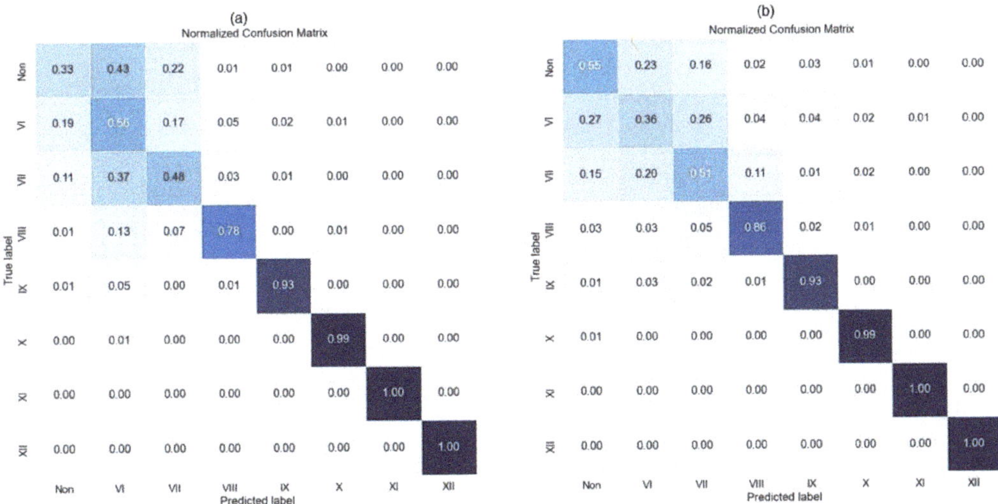

**Figure 4.** SVM (**a**) and ensemble (**b**) tend to produce confusion matrices that are susceptible to non- and low-magnitude EQs.

### 3.4. Ensemble Model Performance Based on PCA

This study explored EQ prediction using various ML models and addressed challenges like imbalanced data through oversampling. Both the ensemble and SVM models benefited from using a reduced feature set derived from PCA. This mitigates the risk of overfitting on the limited EQ data, especially for rare events like "XII" EQs, where overfitting can lead to unreliable predictions. By focusing on the most significant features extracted through PCA, both models can generalize better and potentially improve their performance on unseen data. This improvement can be attributed to two key factors. First, the improved separability of EQ classes: reduced dimensionality helps emphasize the essential features that distinguish different EQ categories, leading to more accurate classifications. Second, enhanced computational efficiency: working with fewer features reduces training time and complexity, which is particularly beneficial for complex models like SVMs. The ensemble model's advantage lies in its inherent diversity. Combining multiple decision tree models captures different perspectives on the data, which is particularly valuable in complex, non-linear domains like EQ prediction, where SVMs, with their single hyperplane approach, might struggle. This aligns with previous findings by Cui et al. [32], where stacking ensembles outperformed individual models, including SVMs, in EQ magnitude prediction. Furthermore, ensembles exhibit greater resilience to data imbalances compared to individual models like SVMs. This advantage stems from their ability to collectively learn from scarce data points across multiple models, potentially addressing the imbalanced classes suggested by the oversampling used in these models.

## 4. Conclusions

As a conclusion, this study investigated the feasibility of ML models for EQ prediction based on the Mercalli Intensity Scale, while simultaneously addressing the challenge of imbalanced data. PCA proved valuable in reducing the dimensionality of geomagnetic data and as feature extraction, potentially mitigating overfitting and improving model performance. Among the evaluated models, the ensemble approach achieved the highest performance across multiple metrics (accuracy: 77.50%, sensitivity: 77.50, precision: 76.69%, F1-score: 77.05%, and MCC: 73.88%). This suggests a significant potential for accurate EQ prediction, reflecting the method's effectiveness despite the fundamental challenges of this field. These results suggest the promising potential for integrating such techniques into existing earthquake monitoring systems to enhance their prediction capabilities and disaster risk reduction. Overall, this study has demonstrated the feasibility of utilizing ML techniques for EQ prediction based on the Mercalli Intensity Scale. This study is part of the ongoing challenges we face in understanding earthquakes, and in specifically aiming to minimize false alarms. Further research exploring new dimensionality reduction methods and interpretable models could pave the way for even more accurate and reliable predictions, ultimately contributing to enhanced EQ preparedness and risk mitigation.

**Author Contributions:** K.Q. and K.A.Y., methodology; K.Q. and K.A.Y., formal analysis; K.Q. and K.A.Y., resources; K.Q. and K.A.Y., software; K.Q., writing—original draft; M.A., K.A.Y. and M.H. supervision; M.A., K.A.Y. and M.H., writing—review and editing. All authors have read and agreed to the published version of the manuscript.

**Funding:** This research was funded by the Malaysia Ministry of Higher Education (MOHE); the support of this work is under the FRGS/1/2020/TK0/UKM/01/1.

**Data Availability Statement:** Restrictions apply to the availability of these data. Data were obtained from SuperMAG database and USGS, accessible at https://supermag.jhuapl.edu/ and www.earthquake.usgs.gov respectively.

**Acknowledgments:** This study is supported by the Fundamental Research Grant Scheme, Ministry of Higher Education, Malaysia (FRGS/1/2020/TK0/UKM/01/1) and the research was made possible through the generous contribution of geomagnetic field data by SuperMAG and its collaborators. Sincere gratitude is also extended to the United States Geological Survey (USGS) for providing critical earthquake data. We utilized AI tools for English language corrections (Quillbot and Grammarly) and code debugging (ChatGPT). No text entirely generated by AI is included in the manuscript; all content, analyses, and methodologies are original and were authored by us.

**Conflicts of Interest:** Masashi Hayakawa was employed by the company Hayakawa Institute of Seismo Electromagnetics Co., Ltd. The remaining authors declare that the research was conducted in the absence of any commercial or financial relationships that could be construed as a potential conflict of interest. The funding sponsors had no role in the design of the study; in the collection, analyses, or interpretation of data; in the writing of the manuscript, and in the decision to publish the results.

# References


1. Wang, Z. Predicting or Forecasting Earthquakes and the Resulting Ground-Motion Hazards: A Dilemma for Earth Scientists. *Seismol. Res. Lett.* **2015**, *86*, 1–5. [CrossRef]
2. Ghamry, E.; Mohamed, E.K.; Abdalzaher, M.S.; Elwekeil, M.; Marchetti, D.; de Santis, A.; Hegy, M.; Yoshikawa, A.; Fathy, A. Integrating Pre-Earthquake Signatures from Different Precursor Tools. *IEEE Access* **2021**, *9*, 33268–33283. [CrossRef]
3. Han, R.; Cai, M.; Chen, T.; Yang, T.; Xu, L.; Xia, Q.; Jia, X.; Han, J. Preliminary Study on the Generating Mechanism of the Atmospheric Vertical Electric Field before Earthquakes. *Appl. Sci.* **2022**, *12*, 6896. [CrossRef]
4. Yue, Y.; Koivula, H.; Bilker-Koivula, M.; Chen, Y.; Chen, F.; Chen, G. TEC Anomalies Detection for Qinghai and Yunnan Earthquakes on 21 May 2021. *Remote Sens.* **2022**, *14*, 4152. [CrossRef]
5. Zöller, G.; Hainzl, S.; Tilmann, F.; Woith, H.; Dahm, T. Comment on "Potential short-term earthquake forecasting by farm animal monitoring" by Wikelski, Mueller, Scocco, Catorci, Desinov, Belyaev, Keim, Pohlmeier, Fechteler, and Mai. *Ethology* **2021**, *127*, 302–306. [CrossRef]
6. Moro, M.; Saroli, M.; Stramondo, S.; Bignami, C.; Albano, M.; Falcucci, E.; Gori, S.; Doglioni, C.; Polcari, M.; Tallini, M.; et al. New insights into earthquake precursors from InSAR. *Sci. Rep.* **2017**, *7*, 12035. [CrossRef] [PubMed]
7. Asaly, S.; Gottlieb, L.-A.; Inbar, N.; Reuveni, Y. Using Support Vector Machine (SVM) with GPS Ionospheric TEC Estimations to Potentially Predict Earthquake Events. *Remote Sens.* **2022**, *14*, 2822. [CrossRef]
8. Hattori, K.; Han, P. Statistical Analysis and Assessment of Ultralow Frequency Magnetic Signals in Japan As Potential Earthquake Precursors: 13. In *Pre-Earthquake Processes*; American Geophysical Union (AGU): Washington, DC, USA, 2018; pp. 229–240. ISBN 9781119156949.
9. Ouyang, X.-Y.; Parrot, M.; Bortnik, J. ULF Wave Activity Observed in the Nighttime Ionosphere above and Some Hours before Strong Earthquakes. *J. Geophys. Res. Space Phys.* **2020**, *125*, e2020JA028396. [CrossRef]
10. Han, P.; Zhuang, J.; Hattori, K.; Chen, C.-H.; Febriani, F.; Chen, H.; Yoshino, C.; Yoshida, S. Assessing the Potential Earthquake Precursory Information in ULF Magnetic Data Recorded in Kanto, Japan during 2000–2010: Distance and Magnitude Dependences. *Entropy* **2020**, *22*, 859. [CrossRef]
11. Asim, K.M.; Martínez-Álvarez, F.; Basit, A.; Iqbal, T. Earthquake magnitude prediction in Hindukush region using machine learning techniques. *Nat. Hazards* **2017**, *85*, 471–486. [CrossRef]
12. Asim, K.M.; Idris, A.; Iqbal, T.; Martínez-Álvarez, F. Earthquake prediction model using support vector regressor and hybrid neural networks. *PLoS ONE* **2018**, *13*, e0199004. [CrossRef] [PubMed]
13. Chang, X.; Zou, B.; Guo, J.; Zhu, G.; Li, W.; Li, W. One sliding PCA method to detect ionospheric anomalies before strong Earthquakes: Cases study of Qinghai, Honshu, Hotan and Nepal earthquakes. *Adv. Space Res.* **2017**, *59*, 2058–2070. [CrossRef]
14. Gitis, V.G.; Derendyaev, A.B. Machine Learning Methods for Seismic Hazards Forecast. *Geosciences* **2019**, *9*, 308. [CrossRef]
15. Debnath, P.; Chittora, P.; Chakrabarti, T.; Chakrabarti, P.; Leonowicz, Z.; Jasinski, M.; Gono, R.; Jasińska, E. Analysis of Earthquake Forecasting in India Using Supervised Machine Learning Classifiers. *Sustainability* **2021**, *13*, 971. [CrossRef]
16. Arafa-Hamed, T.; Khalil, A.; Nawawi, M.; Marzouk, H.; Arifin, M. Geomagnetic Phenomena Observed by a Temporal Station at Ulu-Slim, Malaysia during The Storm of March 27, 2017. *Sains Malays.* **2019**, *48*, 2427–2435. [CrossRef]
17. Chen, B.H. Minimum standards for evaluating machine-learned models of high-dimensional data. *Front. Aging* **2022**, *3*, 901841. [CrossRef] [PubMed]
18. Liu, Y.; Yong, S.; He, C.; Wang, X.; Bao, Z.; Xie, J.; Zhang, X. An Earthquake Forecast Model Based on Multi-Station PCA Algorithm. *Appl. Sci.* **2022**, *12*, 3311. [CrossRef]
19. Li, J.; Li, Q.; Yang, D.; Wang, X.; Hong, D.; He, K. Principal Component Analysis of Geomagnetic Data for the Panzhihua Earthquake (Ms 6.1) in August 2008. *Data Sci. J.* **2011**, *10*, IAGA130–IAGA138. [CrossRef]
20. Hattori, K.; Serita, A.; Gotoh, K.; Yoshino, C.; Harada, M.; Isezaki, N.; Hayakawa, M. ULF geomagnetic anomaly associated with 2000 Izu Islands earthquake swarm, Japan. *Phys. Chem. Earth Parts A/B/C* **2004**, *29*, 425–435. [CrossRef]
21. Fernández-Gómez, M.; Asencio-Cortés, G.; Troncoso, A.; Martínez-Álvarez, F. Large Earthquake Magnitude Prediction in Chile with Imbalanced Classifiers and Ensemble Learning. *Appl. Sci.* **2017**, *7*, 625. [CrossRef]
22. Mukherjee, S.; Gupta, P.; Sagar, P.; Varshney, N.; Chhetri, M. A Novel Ensemble Earthquake Prediction Method (EEPM) by Combining Parameters and Precursors. *J. Sens.* **2022**, 5321530. [CrossRef]
23. SuperMAG Database. Available online: https://supermag.jhuapl.edu/ (accessed on 9 October 2023).
24. United States Geological Survey (USGS) Database. Available online: www.earthquake.usgs.gov (accessed on 9 October 2023).

25.  Yusof, K.A.; Abdullah, M.; Hamid, N.S.A.; Ahadi, S.; Ghamry, E. Statistical Global Investigation of Pre-Earthquake Anomalous Geomagnetic Diurnal Variation Using Superposed Epoch Analysis. *IEEE Trans. Geosci. Remote Sens.* **2022**, *60*, 1–13. [CrossRef]
26.  Yusof, K.A.; Mashohor, S.; Abdullah, M.; Rahman, M.A.A.; Hamid, N.S.A.; Qaedi, K.; Matori, K.A.; Hayakawa, M. Earthquake Prediction Model Based on Geomagnetic Field Data Using Automated Machine Learning. *IEEE Geosci. Remote Sens. Lett.* **2024**, *21*, 1–5. [CrossRef]
27.  Ismail, N.H.; Ahmad, N.; Mohamed, N.A.; Tahar, M.R. Analysis of Geomagnetic $A_p$ Index on Worldwide Earthquake Occurrence using the Principal Component Analysis and Hierarchical Cluster Analysis. *Sains Malays.* **2021**, *50*, 1157–1164. [CrossRef]
28.  Xu, G.; Han, P.; Huang, Q.; Hattori, K.; Febriani, F.; Yamaguchi, H. Anomalous behaviors of geomagnetic diurnal variations prior to the 2011 off the Pacific coast of Tohoku earthquake (Mw9.0). *J. Asian Earth Sci.* **2013**, *77*, 59–65. [CrossRef]
29.  Alvarez, D.A.; Hurtado, J.E.; Bedoya-Ruíz, D.A. Prediction of modified Mercalli intensity from PGA, PGV, moment magnitude, and epicentral distance using several nonlinear statistical algorithms. *J. Seismol.* **2012**, *16*, 489–511. [CrossRef]
30.  Elreedy, D.; Atiya, A.F. A Comprehensive Analysis of Synthetic Minority Oversampling Technique (SMOTE) for handling class imbalance. *Inf. Sci.* **2019**, *505*, 32–64. [CrossRef]
31.  Bao, Z.; Zhao, J.; Huang, P.; Yong, S.; Wang, X. A Deep Learning-Based Electromagnetic Signal for Earthquake Magnitude Prediction. *Sensors* **2021**, *21*, 4434. [CrossRef]
32.  Cui, S.; Yin, Y.; Wang, D.; Li, Z.; Wang, Y. A stacking-based ensemble learning method for earthquake casualty prediction. *Appl. Soft Comput.* **2021**, *101*, 107038. [CrossRef]

*Article*

# Global Rayleigh Wave Attenuation and Group Velocity from International Seismological Centre Data

**Thomas Martin Hearn**

Physics Department, New Mexico State University, Las Cruces, NM 88003, USA; thearn@nmsu.edu

**Abstract:** This paper presents a study of global Rayleigh wave attenuation and group velocity at a period of around 20 s using data from the International Seismological Centre (ISC) bulletin. Rayleigh waves at this period are sensitive to the crustal structure beneath continents and the uppermost mantle beneath oceans. Tomographic imaging reveals strong continental-ocean contrasts due to this. Oceanic group velocities are high but vary with seafloor depth, while oceanic attenuation shows mid-ocean ridges. Subduction zone regions display high attenuation but little velocity reduction, indicating scattering attenuation. Low attenuation regions are associated with the Earth's major cratonic regions, but there are no associated velocity changes. This implies that intrinsic attenuation is low and scattering dominates. Cratonic crustal scatterers have been annealed. A new surface wave magnitude scale is constructed that is valid from near-source to near-antipode distances.

**Keywords:** Rayleigh waves; attenuation tomography; velocity tomography

**Citation:** Hearn, T.M. Global Rayleigh Wave Attenuation and Group Velocity from International Seismological Centre Data. *Geosciences* **2024**, *14*, 50. https://doi.org/10.3390/geosciences14020050

Academic Editors: Masashi Hayakawa and Jesus Martinez-Frias

Received: 18 December 2023
Revised: 2 February 2024
Accepted: 8 February 2024
Published: 10 February 2024



## 1. Introduction

Global Rayleigh wave attenuation and velocity play a crucial role in understanding the behavior of seismic waves propagating through the Earth's interior. By studying the attenuation and velocity of these waves, we can gain insights into the composition, structure, and temperature of the crust and mantle. These properties play a significant role in understanding the dynamic processes occurring within the Earth, such as mantle convection, plate tectonics, and the formation of geological features. The amplitudes and velocity of Rayleigh waves are also crucial in assessing seismic hazards since understanding and accurately estimating the locations and magnitudes of seismic events is essential for characterizing global seismicity and improving our ability to predict earthquake-related ground motion. Furthermore, researching the worldwide attenuation and velocity of Rayleigh waves is crucial for seismic monitoring. Amplitudes play a key role in discerning the Mb/Ms ratio for discrimination and assessing yield in seismic events.

Most previous global tomographic studies of global Rayleigh waves used waveform and ambient noise data for periods of 40 s and above to image phase velocity and attenuation [1–7]. At these long periods, the mantle is imaged and shows both higher attenuation and lower velocity beneath the oceans. Midocean ridges and subduction zones show high attenuation, while continental cratons have low attenuation. At periods nearer to 20 s, midocean ridges, subduction zones, and cratons are still visible [2,8–15].

In this study, the International Seismological Centre (ISC) bulletin surface wave amplitude and group travel time data [16] collected at periods of around 20 s were used to image both global attenuation and group velocity using standard tomographic methods. For the continents, the wave velocities and attenuation are sensitive to crustal attenuation and velocity structure, but for the oceans, the sensitivity is within the top of the mantle. Ocean ridges, subduction zones, and cratons are imaged. Furthermore, the simultaneous use of both velocity and attenuation tomography allows intrinsic attenuation to be distinguished from scattering attenuation.

## 2. Methods

The basic tomographic method here uses great circle raypaths along with both event and station terms [17]. Geodesic rays on the ellipsoid were traced using the Vincenty [18] algorithm. For velocity tomography, the data is first inverted for the average slowness using the following equation:

$$t_i = a + \Delta_i \left( \frac{1}{\overline{v}} \right)$$ (1)

where $t_i$ is the travel time for event $i$ and $\Delta_i$ is the distance. The equation is solved for the intercept, $a$, and the average slowness, $1/\overline{v}$, using standard least squares.

The residuals are then inverted using the following equation:

$$t_{ij} = a_i + b_j + \sum_k \Delta_{ijk} \delta \left( \frac{1}{v_k} \right)$$ (2)

where $t_{ij}$ is the residual travel time from station $i$ to event $j$ and $\Delta_{ijk}$ is the distance the ray travels in cell $k$. The unknowns are the station delays, $a_i$, the event delays $b_j$, and $\delta \left( \frac{1}{v_k} \right)$ is the slowness perturbation in cell $k$. The sum is over all cells crossed by the raypath, and in this study, five-degree cells are used. These equations are inverted on a latitude-longitude grid using great-circle raypaths. On a latitude-longitude grid, the cell sizes differ, especially near the poles. This is compensated for by using the following damped least squares solution:

$$m = \left( A^T A + D^T W^T W D \right)^{-1} A^T d$$ (3)

where the model vector, $m$, is composed of the station delays, event delays, and slownesses, and the data vector, $d$, is composed of the travel times. Smoothing is performed with a Laplacian operator, $D$, and $W$ is a diagonal weighting matrix with elements proportional to the area of the cell sizes. This equalizes the weighting per unit area. Without this, the solution will be overdamped at the poles. The LSQR conjugate-gradient algorithm was used to invert the data [19,20].

Tomography using the amplitudes proceeds much the same way as with velocity tomography. The operative attenuation equation is [21] as follows:

$$A = \frac{1}{(\sin \Delta)^{\frac{1}{2}} \Delta^k} 10^a 10^{Ms} e^{\frac{-\pi \Delta}{vTQ}}$$ (4)

where $A$ is the amplitude, $\Delta$ is the distance in degrees, $v$ is the velocity, $T$ is the period, and $Q$ is the attenuation quality factor. The sine term is the geometrical spreading factor on a sphere. The factor $\Delta^k$ represents the temporal spreading due to dispersion. According to Ewing et al. [21], $k$ should be one-half for Rayleigh waves and one-third for Airy waves.

First, the log-amplitude data are fit to an attenuation function using the following equation:

$$\log A_i - Ms_i + 0.5\log \sin \Delta_i + k\log \Delta_i = a - \frac{\log e \pi \Delta_i}{vT_i} \frac{1}{Q}$$ (5)

Here, $A_i$ is the amplitude of arrival $i$, and $Ms_i$ is its event magnitude. The unknowns are the intercept, $a$, and the average attenuation, $1/Q$. The log of the amplitude is corrected for event size by subtracting the magnitude, and then a spherical geometric spreading correction is added.

The residuals for this equation are then used for the tomographic equation

$$\log A_{ij} = a_i + b_j - \sum_k \frac{\ln 10 \pi \Delta_{ijk}}{vT_i} \delta \left( \frac{1}{Q_k} \right)$$ (6)

This is applied to the five-degree grid of cells, and the sum is over a great circle raypath. Here $A_{ij}$ is the residual reduced amplitude between event $i$ and station $j$, and $\Delta_{ijk}$ is the distance the ray travels in each cell. The unknowns are the station and event gains, $a_i$ and $b_j$, and the attenuation in each cell $k$, $\delta\left(\frac{1}{Q_k}\right)$. Because the station and event terms trade off, the average of the station terms is set to zero. The least-squared equations are then again solved using the LRSQ algorithm.

The formula used for converting the attenuation perturbation, $\Delta\left(\frac{1}{Q}\right)$ to $\Delta Q$ is given by solving $\frac{1}{Q} = \frac{1}{Q_0} + \Delta\frac{1}{Q}$ for $\Delta Q = Q - Q_0$:

$$\Delta Q = \frac{-Q_0^2 \Delta\left(\frac{1}{Q}\right)}{1 + Q_0 \Delta\left(\frac{1}{Q}\right)} \tag{7}$$

where $Q_0$ is the Q value from the initial line fit to the data. Traditional checkerboard tests are performed to evaluate resolution and variance. The checkerboards consist of alternating squares of low and high. Rays are traced through this, an appropriate amount of noise is added, and the results are inverted in the same manner as the data. An approximate resolution width can be found by finding the smallest square size that can be reasonably imaged.

One source of error is the potential for focusing. Unfortunately, phase velocity is not available for these data, and thus focusing cannot be estimated using the methods of Dahlen and Tromp [22] and Dalton and Ekstrom [5]. This could create noise in the image. Another source of error is using great circle raypaths. As shall be seen, there are significant variations in velocity that cause refraction, especially along the coastlines. However, Dalton and Ekstrom [5] argue that the refraction effect is negligible at global scales. Bao et al. [23] compared different methods of accounting for focusing. They found that methods other than great-circle paths did improve the image, but the major features were visible without. Chen et al. [24] examined deviations from great circle raypaths and found errors in the phase velocity tomography of 1.5% at 30 s, with the effect increasing for shorter periods. This is a relatively small source of error. A final source of error includes not correcting for focal mechanisms, but with a minimum of 10 arrivals per event and station, this should average out. The above error sources are not directly represented in the checkerboard tests except as being included as random errors.

## 3. Results

Data is from the ISC bulletin [16] (International Seismological Centre, 2023) up until February 2020 for events of Ms > 4 and above and depths less than 50 km. There are several different designations for surface wave amplitude measurements in the bulletin for different stations at different times. This study used the measurements labeled LR, MLR, AMS, and IAMs_20. Only vertical components were used.

The time-distance curve for the travel time data is in Figure 1. Unlike most travel time data, the data fans out with distance. This is because of the dichotomy between the low continental velocities and the high oceanic velocities. In addition, there are mispicked Love wave phases forming a line at 4.0 km/s apparent velocity. These were cut out of the data set for both velocity and attenuation tomography. Travel time data were restricted to distances of 2 to 160 degrees based on Figure 1. Ten arrivals at each station and event were required, with a maximum residual of 1000 s. A total of 1,044,918 arrivals were used. The initial fit gave an average group velocity of 3.2 km/s but had a very large standard deviation of 65 s. The final inversion used cell sizes of five degrees, and the results are in Figure 2. There are small artifacts at the poles because the raypath tracing step size becomes comparable to the width of the cell; nevertheless, structure in the Arctic and Antarctica is still resolvable. Checkerboard test results show that 10-degree squares can be imaged (Figure S1).

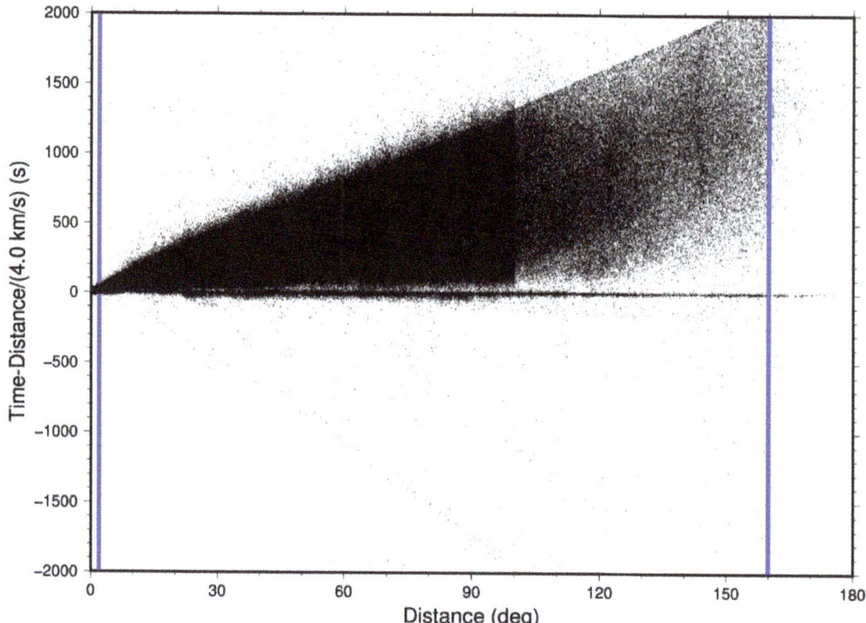

**Figure 1.** Travel time data from ISC data. The apparent cutoff at 99 degrees corresponds to the far body-wave cutoff distance at the shadow zone. Many stations do not pick arrivals past this distance. The cutoff at 160 degrees corresponds to the far cutoff from the surface-wave magnitude definition. Only a few stations pick arrivals past this distance. Blue lines indicate the range of data used in the inversion (2 to 160 degrees). The Love wave arrivals at 4 km/s were cut out of the data.

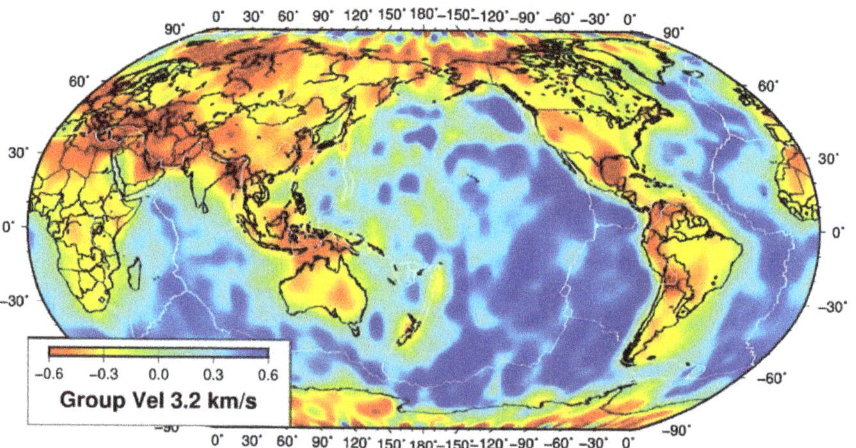

**Figure 2.** Results of velocity tomography.

A normalized amplitude-distance curve after the $\frac{1}{2}\log(2\pi R sin\Delta)$ geometrical spreading correction is shown in Figure 3. There is linearity in the data with no indication of the $1/(\sqrt{\Delta})$ curvature predicted by dispersion. Furthermore, adding the dispersion term gave average Q values that seemed too high. As a result, k was set to zero, and it is concluded that dispersion does not play a role in determining amplitude within the instrument bandwidth. Hearn et al. (2008) [25] made a similar conclusion for six-second period surface waves in China; however, this contrasts with Rezapour and Pearce (1998) [26], who did find a $1/\sqrt[3]{\Delta}$ dependence in early ISC data. Amplitude data were restricted to 2 to 99 degrees, the body-wave cutoff, based on Figure 3. Only events with reported surface wave magnitudes were used. Amplitude data had a minimum of 10 arrivals at each station and event and a maximum residual of two magnitude units. The modern definition of Ms amplitude uses periods between 18 and 22 s, so data was restricted to this range. However, it was found that using the measured period in the inversion made no improvement in the fit, so 20 s was used in the inversion. The initial fit for the attenuation tomography gave an average 1/Q value of 1/275 with an rms of 0.23 magnitude units for 844,079 arrivals. The intercept was 0.75 magnitude units.

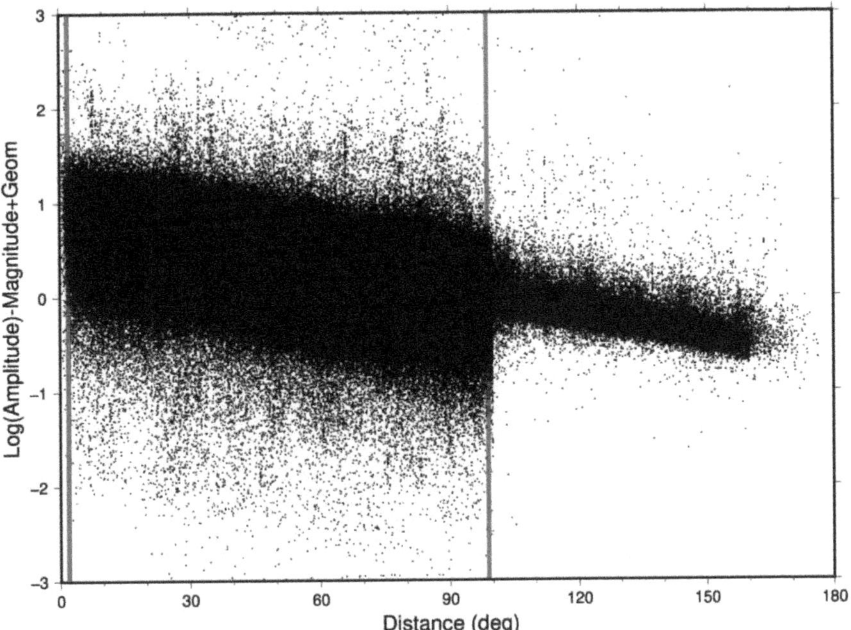

**Figure 3.** Amplitude data from ISC data after correction for magnitude and geometric spreading. Note the apparent linearity. The discontinuity in the data at 100 degrees corresponds to the far Mb-magnitude cutoff distance. Blue lines indicate the range of data used (2 to 99 degrees).

From these results, a physics-based magnitude relation can be defined based on the following vertical amplitudes:

$$Msp = \log A + 0.5\log(2\pi R \sin\Delta) + 0.0086\Delta - 0.75; 18s \le T \le 22s; \; 2° \le \Delta \le 178°; h < 40 \,\text{km} \tag{8}$$

where R is the Earth's radius of 6371 km. The factor 0.0086 is equal to $\frac{\log(e)\pi}{vTQ}$ times 111.19 km/deg. Its residual plot is shown in Figure 4a. As can be seen, the formula is good for the nearly entire measured distance range. One-degree averages were made, and the first two bins had averages of over 0.1 magnitude units, so the formula was truncated at two degrees to avoid source and antipodal effects. The period is not required in this formula. The current effective ISC Ms magnitude formula is (isc.ac.uk/standards/magnitudes) as follows for vertical components:

$$Ms = \log \frac{A}{T} + 1.66\log\Delta + 0.3 + 0.074; 18s \le T \le 22s; \; 20° \le \Delta \le 160°. \tag{9}$$

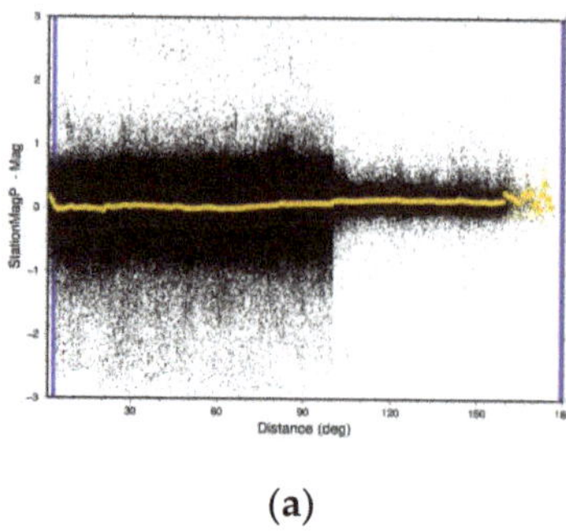
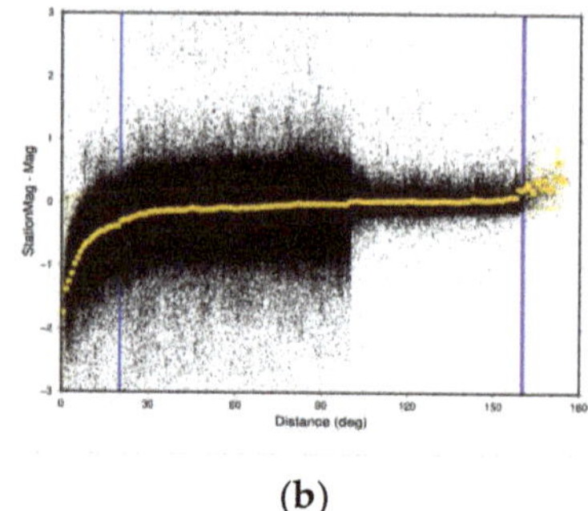

(a)          (b)

**Figure 4.** (**a**) Magnitude residuals with distance for the physics-based magnitude scale Msp. Yellow dots represent 1-degree averages with standard deviations. (**b**) Magnitude residuals using the effective ISC Ms formula; note that these are only valid in the prescribed 20 to 160 degree range. Blue lines indicate the range of data for the magnitude scale.

The last term corrects for a slight measured bias of the average station ISC Ms magnitude residuals in the 20 to 60 degree range. This probably occurs because ISC is not using the exact same set of data to compute magnitude. Residuals for the Ms formula are plotted in Figure 4b. The standard Ms expression has a slight bias within its range, and it is clearly invalid for distances of less than 20 degrees. The two expressions yield the same value at 56 degrees for a 20 s period. The physics-based magnitude scale is calibrated to the current scale but with an extended range (Figure S2).

Attenuation tomography results are shown in Figure 5. Contrasts between the continents and oceans, subduction zones, and cratons can be observed. The checkerboard tests show resolution is at best 15 degrees wide (Figure S3). There are some places where focusing could occur; in particular, there are some patchy low attenuation zones in the Pacific that seem odd, and Russia seems patchy too; however, the overall results clearly correspond to major tectonic features.

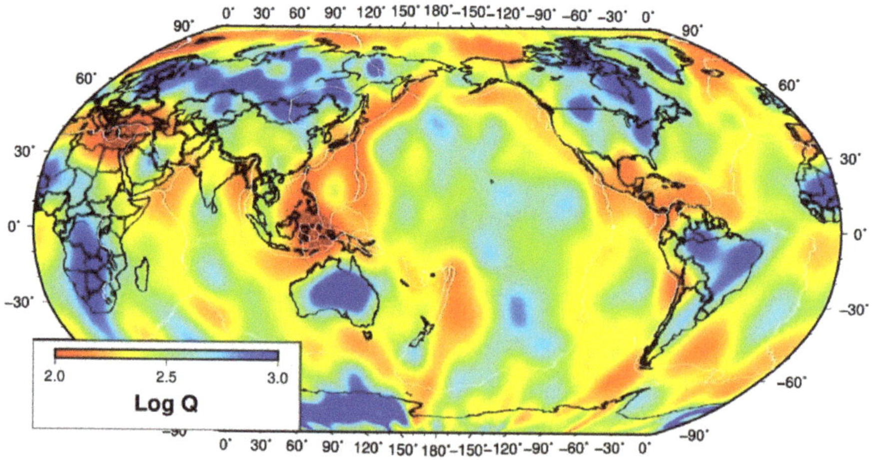

**Figure 5.** Results of attenuation tomography.

## 4. Discussion

### 4.1. Other Studies

A similar study of global Rayleigh wave attenuation at 20 s was carried out by Selby et al. [27], where they imaged corrections to the surface wave magnitude equation. They used an independent data set of amplitudes measured for Ms magnitude determination. Their images show the low-velocity ridges and high-velocity cratons in similar positions as in the ISC image. A study of the global Rayleigh wave group velocity at 20 s was carried out by Ritzwoller [9]. In the oceans, he imagined the ocean depth as the ISC data does. His low velocity features match what was found with the ISC data, and no cratons were imaged. There are also regional studies of Rayleigh waves at 20 s period. Yanovskaya (2003) [10] found low group velocities beneath Tibet. Yang et al. [11] and Zhou et al. [28] were able to image low attenuation beneath the North China and South China Cratons along with the high attenuation Tarim Basin. The two cratons are too small to be resolved with the ISC data, but Mongolia is not. Zhou et al.'s China tomography also imaged Mongolia as a low-attenuation region with Q values approaching 1000, and this does match the ISC tomography. Bao et al. [13] imaged S wave velocities beneath China at 20 s. Most of the features he images are beneath the resolution of the ISC data (Figure S3); however, Tibet has low crustal velocities. Feng et al. [29] and Nascimento et al. [30] imaged South American group velocities at 20 s and found low velocities beneath the Andes but no apparent cratons.

Levshin et al. [8] imaged the Artic group velocity structure at 20 s and found patterns like the ISC tomography. In Asia, Levshin et al. [12] imaged the Rayleigh wave attenuation for Asia at 20 s. They image the low attenuation features in Mongolia, Russia, and the Arabian Plate, as does the ISC data. Their average attenuation value of about 300 agrees with the ISC data, but they have many values over 500. For South America, 20 s waves give an average group velocity of 2.97 km/s with some low velocities beneath the Andes, and this is similar to what the ISC data gives [30]. In comparison to the above results, which generally do not show many Q values greater than 1000, the ISC tomography shows that quite a few of the cratonic regions do have Q values above this. The checkerboard tests show that some of this may be due to some overshoot from the tomography, but that cannot completely explain the high Q values. This may be because other studies isolate the fundamental mode.

At a 25 s period, Babikoff and Dalton [14] imaged the US phase velocities and found a low velocity west and a high velocity east. The ISC tomography only has a resolution to roughly image that. At a 40 s period, Ma et al. [2] have images of both phase velocity and attenuation. They imaged the cratons in both images, but this would be at mantle depths. Studies of global Rayleigh wave attenuation and velocity at longer periods image solely at mantle depths and show low phase velocities [4,5,31–33] and shear wave velocities [15,34,35] and high attenuation beneath midocean ridges [1,4,6,12,36–38]. Group velocity maps at these periods are not available. Many of the same along-ridge variations in attenuation can be found in both the long-period tomography and the ISC tomography. Cratons are also imaged at longer periods in both velocity and attenuation and represent the deeper cratonic roots.

*4.2. Interpretation*

The most visible feature in both images is the contrast between the continents and oceans. Again, this occurs because 20 s waves are sensitive to the crust beneath the continents and sensitive to the mantle beneath the oceans. The contrast is particularly clear for the velocity images where the crust-to-mantle velocity contrast is high. To further explore the sensitivity with depth, sensitivity kernels from a two-layer continental and a three-layer oceanic model were made using the CPS software version 3.3 [39]. The kernels showed the oceanic mantle to have maximum sensitivity over the top 25 km of mantle and the continental crust to have sensitivity over the entire crustal column.

The average oceanic Rayleigh wave group velocity is around 3.6 km/s, and its pattern mimics the bathymetry, with the highest velocities near the ridge crests. Indeed, modeling shows that reducing the ocean depth by two kilometers of ridge height causes a 0.6 km/s difference—about the amount shown in the tomographic image. Any effects of mid-ocean ridge-related temperature are obscured by this. In contrast, modeling shows that ocean depth has little effect on phase velocity and attenuation, and this allows the image to show the effects of mid-ocean ridge cooling in the image with the highest attenuation along the ridge crest. However, this attenuation is not always consistent along the ridge crests, showing that temperature varies along the ridges. There is no obvious connection between ridge spreading rates and temperature, but it is interesting that the Arctic Ridge is both the hottest ridge and, also, one of the slowest spreading subduction zones.

Subduction zones are represented by high-attenuation regions. This occurs because subduction zones disrupt the oceanic wave guide with slabs, volcanism, and backarc spreading. Less attenuation occurs in the simpler subduction zones of South and North America, while the more complicated subduction zones of the South Pacific, Caribbean, West Pacific, and Mediterranean are very attenuative. In contrast, the velocity image does not show most subduction zones. It seems unlikely that attenuation can change by such an extreme amount without changing the velocity if the attenuation is intrinsic. Thus, the subduction zone attenuation must be due to scattering. At 20 s, the wavelength is 66 km, and there are many features at this scale within a subduction zone, including the subducted slab itself.

Continental crustal attenuation shows strong low attenuation features in the crust, with Q values over one thousand. Most of the low-attenuation regions are associated with the Australian, East Antarctica, North American, Amazon, West African, South African, Baltica, Siberian, and Greenland cratons. There are exceptions. The Rio de la Plata Craton in South America is barely visible. Zhou et al. (2020) [28] did image the North China Craton, but it is too small to be resolved in the present study. The India and Madagascar Cratons also seem unresolvable. The Australian low attenuation anomaly lies beneath the central Australian shield but excludes the western and northern cratons, forming an exception to the rule. Furthermore, not all low-attenuation regions are beneath cratons. Mongolia, central Russia, and the eastern USA have low attenuation zones even though they are not cratons.

The key to explaining the above observations is to note that the cratons are not visible in the velocity image. Modeling shows that increasing Q values to 1000 requires the whole crustal column to be involved. This cannot be due to a change in intrinsic attenuation because that would require a large change in temperature or rock type that would be seen in the velocity tomography. The only way to increase Q by that factor is if the intrinsic attenuation of the crust is small and that continental crustal attenuation is mainly due to scattering. The ancient Archean and Paleoproterozoic crust, originally hotter, was annealed, and the internal structure changed as a result. This is supported by receiver function studies, which show few internal boundaries in the Archean crust [40].

## 5. Conclusions

Global Rayleigh wave velocities and attenuation have been imaged for data near the 20 s period, and a new surface wave magnitude equation is derived that is valid from near-source to near-antipode distances. Only geometric spreading and attenuation are required to explain the data, and dispersion-related attenuation is not needed. In the 20 s period range, Rayleigh waves are sensitive to the top of the mantle beneath the oceans and to the crust beneath the continents. Continents and oceans have very different group velocities and attenuations because of this. In the oceans, group velocity is related to water depth, and the tomography images the bathymetry; however, the attenuation is not sensitive to water depth and shows mid-ocean ridges as high attenuation anomalies. These are interpreted as being intrinsic attenuations due to thermal anomalies. Subduction zones disrupt the wave guide and cause attenuation but do not affect the velocity much. This occurs because the attenuation is mostly due to scattering and is not intrinsic. Continental cratons stand out as low-attenuation regions but do not affect velocity tomography. This must occur due to a change in scattering attenuation, with intrinsic attenuation being low. It is concluded that the cratonic crust has been annealed, with the internal structures changed and scatterers removed.

**Supplementary Materials:** The following supporting information can be downloaded at: https://www.mdpi.com/article/10.3390/geosciences14020050/s1. Figure S1: Checkerboard pattern for velocity tomography; Figure S2: Magnitude corrections; Figure S3: Checkerboard pattern for attenuation tomography.

**Funding:** This research received no external funding.

**Data Availability Statement:** ISC Earthquake Bulletin data is from http://www.isc.ac.uk/ (accessed on 2 February 2024) [16]. Plate boundaries on maps are from Bird (2003) [41] at http://peterbird.name/publications/2003_pb2002/2003_pb2002.htm (accessed 2 February 2024).

**Acknowledgments:** Thank you to the International Seismological Centre and the thousands of workers who gathered the data. The CPS software (Herrmann, 2013; https://www.eas.slu.edu/eqc/eqccps.html (accessed 2 February 2024)) was used to model velocities and attenuations along with depth sensitivity plots. Thank you to two reviewers who greatly improved the manuscript.

**Conflicts of Interest:** The author declares no conflicts of interest.

# References


1. Adenis, A.; Debayle, E.; Ricard, Y. Seismic Evidence for Broad Attenuation Anomalies in the Asthenosphere beneath the Pacific Ocean. *Geophys. J. Int.* **2017**, *209*, 1677–1698. [CrossRef]
2. Ma, Z.; Masters, G.; Mancinelli, N. Two-Dimensional Global Rayleigh Wave Attenuation Model by Accounting for Finite-Frequency Focusing and Defocusing Effect. *Geophys. J. Int.* **2016**, *204*, 631–649. [CrossRef]
3. Billien, M.; Lévêque, J.; Trampert, J. Global Maps of Rayleigh Wave Attenuation for Periods between 40 and 150 Seconds. *Geophys. Res. Lett.* **2000**, *27*, 3619–3622. [CrossRef]
4. Dalton, C.A.; Ekström, G. Constraints on Global Maps of Phase Velocity from Surface-Wave Amplitudes. *Geophys. J. Int.* **2006**, *167*, 820–826. [CrossRef]
5. Dalton, C.A.; Ekström, G. Global Models of Surface Wave Attenuation. *J. Geophys. Res.* **2006**, *111*, 1–19. [CrossRef]
6. Dalton, C.A.; Ekström, G.; Dziewoński, A.M. The Global Attenuation Structure of the Upper Mantle. *J. Geophys. Res.* **2008**, *113*, 1–24. [CrossRef]
7. Magrini, F.; Boschi, L.; Gualtieri, L.; Lekić, V.; Cammarano, F. Rayleigh-Wave Attenuation across the Conterminous United States in the Microseism Frequency Band. *Sci. Rep.* **2021**, *11*, 10149. [CrossRef]
8. Levshin, A.L.; Ritzwoller, M.H.; Barmin, M.P.; Villaseñor, A.; Padgett, C.A. New Constraints on the Arctic Crust and Uppermost Mantle: Surface Wave Group Velocities, Pn, and Sn. *Phys. Earth Planet. Inter.* **2001**, *123*, 185–204. [CrossRef]
9. Ritzwoller, M.H. Global Surface Wave Diffraction Tomography. *J. Geophys. Res.* **2002**, *107*, 1–13. [CrossRef]
10. Yanovskaya, T. 3D S-Wave Velocity Pattern in the Upper Mantle beneath the Continent of Asia from Rayleigh Wave Data. *Phys. Earth Planet. Inter.* **2003**, *138*, 263–278. [CrossRef]
11. Yang, X.; Taylor, S.R.; Patton, H.J. The 20-s Rayleigh Wave Attenuation Tomography for Central and Southeastern Asia. *J. Geophys. Res.* **2004**, *109*, 2004JB003193. [CrossRef]
12. Levshin, A.L.; Yang, X.; Barmin, M.P.; Ritzwoller, M.H. Midperiod Rayleigh Wave Attenuation Model for Asia: Rayleigh wave attenuation in Asia. *Geochem. Geophys. Geosyst.* **2010**, *11*. [CrossRef]
13. Bao, X.; Song, X.; Li, J. High-Resolution Lithospheric Structure beneath Mainland China from Ambient Noise and Earthquake Surface-Wave Tomography. *Earth Planet. Sci. Lett.* **2015**, *417*, 132–141. [CrossRef]
14. Babikoff, J.C.; Dalton, C.A. Long-Period Rayleigh Wave Phase Velocity Tomography Using USArray. *Geochem. Geophys. Geosyst.* **2019**, *20*, 1990–2006. [CrossRef]
15. Zhou, Y.; Nolet, G.; Dahlen, F.A.; Laske, G. Global Upper-Mantle Structure from Finite-Frequency Surface-Wave Tomography. *J. Geophys. Res.* **2006**, *111*, B04304. [CrossRef]
16. International Seismological Centre On-Line Bulletin 2023. Available online: https://www.isc.ac.uk (accessed on 2 February 2024).
17. Hearn, T.M. Two-Dimensional Attenuation and Velocity Tomography of Iran. *Geosciences* **2022**, *12*, 397. [CrossRef]
18. Vincenty, T. Direct and Inverse Solutions of Geodesics on the Ellipsoid with Application of Nested Equations. *Surv. Rev.* **1975**, *22*, 88–93. [CrossRef]
19. Paige, C.C.; Saunders, M.A. Algorithm 583, LSQR: Sparse Linear Equations and Least-Squares Problems. *ACM Trans. Math. Softw.* **1982**, *8*, 43–71. [CrossRef]
20. Paige, C.C.; Saunders, M.A. LSQR: An Algorithm for Sparse Linear Equations and Sparse Least Squares, Trans. *ACM Trans. Math. Softw.* **1982**, *8*, 195–209. [CrossRef]
21. Ewing, W.M.; Jardetzky, W.S.; Press, F. *Elastic Waves in Layered Media*; McGraw-Hill Book Company: New York, NY, USA, 1957.
22. Dahlen, F.A.; Tromp, J. *Theoretical Global Seismology*; Princeton University Press: Princeton, NJ, USA, 1998; ISBN 978-0-691-00116-6.
23. Bao, X.; Dalton, C.A.; Ritsema, J. Effects of Elastic Focusing on Global Models of Rayleigh Wave Attenuation. *Geophys. J. Int.* **2016**, *207*, 1062–1079. [CrossRef]
24. Chen, H.; Ni, S.; Chu, R.; Chong, J.; Liu, Z.; Zhu, L. Influence of the Off-Great-Circle Propagation of Rayleigh Waves on Event-Based Surface Wave Tomography in Northeast China. *Geophys. J. Int.* **2018**, *214*, 1105–1124. [CrossRef]

25. Hearn, T.M.; Wang, S.; Ni, J.F.; Xu, Z.; Yu, Y.; Zhang, X. Uppermost Mantle Velocities beneath China and Surrounding Regions. *J. Geophys. Res.* **2004**, *109*, B11301. [CrossRef]

26. Rezapour, M.; Pearce, R.G. Bias in Surface-Wave Magnitude Ms Due to Inadequate Distance Corrections. *Bull. Seismol. Soc. Am.* **1998**, *88*, 43–61. [CrossRef]

27. Selby, N.D.; Bowers, D.; Marshall, P.D.; Douglas, A. Empirical Path and Station Corrections for Surface-Wave Magnitude, Ms, Using a Global Network. *Geophys. J. Int.* **2003**, *155*, 379–390. [CrossRef]

28. Zhou, L.; Song, X.; Yang, X.; Zhao, C. Rayleigh Wave Attenuation Tomography in the Crust of the Chinese Mainland. *Geochem. Geophys. Geosyst.* **2020**, *21*, e2020GC008971. [CrossRef]

29. Feng, M.; Assumpção, M.; Van Der Lee, S. Group-Velocity Tomography and Lithospheric S-Velocity Structure of the South American Continent. *Phys. Earth Planet. Inter.* **2004**, *147*, 315–331. [CrossRef]

30. Nascimento, A.V.D.S.; França, G.S.; Chaves, C.A.M.; Marotta, G.S. Rayleigh Wave Group Velocity Maps at Periods of 10–150 s beneath South America. *Geophys. J. Int.* **2021**, *228*, 958–981. [CrossRef]

31. Ekström, G.; Tromp, J.; Larson, E.W.F. Measurements and Global Models of Surface Wave Propagation. *J. Geophys. Res.* **1997**, *102*, 8137–8157. [CrossRef]

32. Durand, S.; Debayle, E.; Ricard, Y. Rayleigh Wave Phase Velocity and Error Maps up to the Fifth Overtone. *Geophys. Res. Lett.* **2015**, *42*, 3266–3272. [CrossRef]

33. Liu, K.; Zhou, Y. Global Rayleigh Wave Phase-Velocity Maps from Finite-Frequency Tomography. *Geophys. J. Int.* **2016**, *205*, 51–66. [CrossRef]

34. Zhou, Y.; Dahlen, F.A.; Nolet, G.; Laske, G. Finite-Frequency Effects in Global Surface-Wave Tomography. *Geophys. J. Int.* **2005**, *163*, 1087–1111. [CrossRef]

35. Dalton, C.A.; Ekström, G.; Dziewonski, A.M. Global Seismological Shear Velocity and Attenuation: A Comparison with Experimental Observations. *Earth Planet. Sci. Lett.* **2009**, *284*, 65–75. [CrossRef]

36. Selby, N.D.; Woodhouse, J.H. Controls on Rayleigh Wave Amplitudes: Attenuation and Focusing. *Geophys. J. Int.* **2000**, *142*, 933–940. [CrossRef]

37. Selby, N.D. The Q Structure of the Upper Mantle: Constraints from Rayleigh Wave Amplitudes. *J. Geophys. Res.* **2002**, *107*, 2097. [CrossRef]

38. Pilidou, S.; Priestley, K.; Gudmundsson, Ó.; Debayle, E. Upper Mantle S-Wave Speed Heterogeneity and Anisotropy beneath the North Atlantic from Regional Surface Wave Tomography: The Iceland and Azores Plumes. *Geophys. J. Int.* **2004**, *159*, 1057–1076. [CrossRef]

39. Herrmann, R.B. Computer Programs in Seismology: An Evolving Tool for Instruction and Research. *Seism. Res. Lett.* **2013**, *84*, 1081–1088. [CrossRef]

40. Abbott, D.H.; Mooney, W.D.; VanTongeren, J.A. The Character of the Moho and Lower Crust within Archean Cratons and the Tectonic Implications. *Tectonophysics* **2013**, *609*, 690–705. [CrossRef]

41. Bird, P. An Updated Digital Model of Plate Boundaries. *Geochem. Geophys. Geosyst.* **2003**, *4*, 2001GC000252. [CrossRef]

*geosciences*

*Article*

# Subduction as a Smoothing Machine: How Multiscale Dissipation Relates Precursor Signals to Fault Geometry

Patricio Venegas-Aravena [1,*] and Enrique G. Cordaro [2,3]

[1] Department of Structural and Geotechnical Engineering, School of Engineering, Pontificia Universidad Católica de Chile, Vicuña Mackenna 4860, Macul, Santiago 8331150, Chile
[2] Observatorios de Radiación Cósmica y Geomagnetismo, Departamento de Física, FCFM, Universidad de Chile, Casilla 487-3, Santiago 8330015, Chile
[3] Facultad de Ingeniería, Universidad Autónoma de Chile, Pedro de Valdivia 425, Santiago 7500912, Chile
* Correspondence: plvenegas@uc.cl

**Abstract:** Understanding the process of earthquake preparation is of utmost importance in mitigating the potential damage caused by seismic events. That is why the study of seismic precursors is fundamental. However, the community studying non-seismic precursors relies on measurements, methods, and theories that lack a causal relationship with the earthquakes they claim to predict, generating skepticism among classical seismologists. Nonetheless, in recent years, a group has emerged that seeks to bridge the gap between these communities by applying fundamental laws of physics, such as the application of the second law of thermodynamics in multiscale systems. These systems, characterized by describing irreversible processes, are described by a global parameter called thermodynamic fractal dimension, denoted as $D$. A decrease in $D$ indicates that the system starts seeking to release excess energy on a macroscopic scale, increasing entropy. It has been found that the decrease in $D$ prior to major earthquakes is related to the increase in the size of microcracks and the emission of electromagnetic signals in localized zones, as well as the decrease in the ratio of large to small earthquakes known as the b-value. However, it is still necessary to elucidate how $D$, which is also associated with the roughness of surfaces, relates to other rupture parameters such as residual energy, magnitude, or fracture energy. Hence, this work establishes analytical relationships among them. Particularly, it is found that larger magnitude earthquakes with higher residual energy are associated with smoother faults. This indicates that the pre-seismic processes, which give rise to both seismic and non-seismic precursor signals, must also be accompanied by changes in the geometric properties of faults. Therefore, it can be concluded that all types of precursors (seismic or non-seismic), changes in fault smoothness, and the occurrence of earthquakes are different manifestations of the same multiscale dissipative system.

**Keywords:** b-value; electromagnetic signals; multiscale thermodynamics; earthquake precursor

**Citation:** Venegas-Aravena, P.; Cordaro, E.G. Subduction as a Smoothing Machine: How Multiscale Dissipation Relates Precursor Signals to Fault Geometry. *Geosciences* **2023**, *13*, 243. https://doi.org/10.3390/geosciences13080243

Academic Editors: Jesus Martinez-Frias and Masashi Hayakawa

Received: 29 June 2023
Revised: 1 August 2023
Accepted: 7 August 2023
Published: 11 August 2023

## 1. Introduction

The study of pre-earthquake physics holds significant relevance in our efforts to safeguard lives and infrastructure from the destructive impact of seismic events. Extensive research has been conducted, focusing on pre-earthquake measurements, such as groundwater level variations, electromagnetic signals, ionospheric variations, seismic clustering, radon liberation, other gas seeps emissions, or thermal radiation, that offer promising indications of a potential link to impending earthquakes [1–25]. Particularly, these studies highlight the presence of anomalous data during abnormal periods compared to normal background conditions. Nevertheless, it is crucial to recognize that the majority of these studies have primarily focused on establishing spatial and temporal correlations between the observed anomalies and the occurrence of earthquakes.

Although there are studies linking measurements to earthquake magnitude [9,20,26–30], the crucial question of actual causation, which represents the fundamental link between the measured signals and the underlying physics of earthquake rupture, is addressed by only a limited number of researchers within the pre-earthquake signal community [31–36]. This gap in our understanding has generated concerns and skepticism within the seismological community, as the reliability and predictive capabilities of pre-earthquake measurements are called into question [37,38]. This skepticism has made it challenging to overcome the prevailing paradigm that denies the existence of pre-earthquake phenomena [39]. To bridge this gap, considerable attention has been directed toward experiments conducted on rock samples, offering valuable insights into the behavior of pre-failure physics [40–51]. These studies have explored various phenomena, such as multiscale cracking, rock electrification, changes in acoustic emissions, increases in internal damage, or alterations in strain and stress [52–54]. It is thought that the knowledge gained from these rock sample experiments could be extrapolated to understand large-scale lithospheric dynamics.

Significant progress has been achieved in the integration of pre-earthquake signals of the lithosphere with seismic rupture parameters, employing the principles of multiscale thermodynamics and entropy production of rocks [34,35]. A crucial parameter in this framework is the thermodynamic fractal dimension, which accounts for the dissipation of energy across different scales, and specifically characterizes the distribution of multiscale cracking within materials. Notably, the generation of multiscale cracking indicates the dissipation of energy preceding impending earthquakes, marking the culmination of the seismic cycle [36]. This critical stage, which garners significant attention in pre-earthquake signal research, allows for the interpretation of anomalous measurements as manifestations of irreversible processes and impending earthquake occurrence. In this line, Venegas-Aravena et al., 2022 [34] found a relation between the large-scale entropy change to the expected earthquake magnitude. Additionally, Venegas-Aravena and Cordaro 2023 [36] suggested that the multiscale properties of lithospheric dynamics such as the thermodynamic fractal dimension could be linked to fault properties such as the b-value, which indicates the ratio between the larger and smaller earthquakes in a given zone.

In that line, one notable consequence of large-scale entropy production is the emergence of smoother fault surfaces [35,36]. This is relevant because seismological studies describe the fault interface and the seismic source as heterogenous [55–57], implying that friction coefficients depend on the roughness of the surface. For example, rougher surfaces are related to higher friction coefficients as well as smooth surfaces host lower friction coefficients [58]. That is why large slips are more related to smoother faults [59,60]. In that sense, the smoothing of faults indicates the release of accumulated energy and a reduction in resistance to energy storage in multiple seismic cycles [61–63]. To comprehensively understand fault properties, including earthquake magnitude, it becomes essential to establish a connection between fault smoothing and the global parameters of the system. Multiscale thermodynamics provides a suitable framework for analyzing fault behavior and linking it to pre-earthquake signals. In line with these considerations, the present work utilizes a multiscale thermodynamic approach to investigate the relationship between pre-earthquake signals and fault properties. In that line, Section 2 of this study delves into the intricacies of the principles of multiscale thermodynamics and its application to the understanding of the seismic background. Building upon this foundation, Section 3 explores the relationship between two crucial aspects of fault properties: seismic magnitude and fault geometry. Moving forward, Section 4 investigates the connection between multiscale thermodynamics and fracture energy. The discussion section is in Section 5. Here, the focus shifts to the relationship between fault properties, multiscale thermodynamics, and other pre-seismic processes. Finally, Section 6 presents the conclusions drawn from the findings of the study.

## 2. Multiscale Thermodynamics

In the context of multiscale cracking, the study of energy dissipation processes is essential to understand the complex behavior of materials under stress. Cracks in rocks, resulting from external loads, exhibit a multiscale nature as they propagate across different length scales [64,65]. These cracking processes are inherently dissipative, reflecting the irreversible release of accumulated energy within the material [66]. To quantitatively analyze and describe such dynamics, a thermodynamic framework is needed. Recent work on multiscale thermodynamics provides this framework, offering insights into entropy production and the thermodynamic fractal dimension as measures of energy dissipation and complexity. One of the key equations in multiscale thermodynamics work relates the thermodynamic fractal dimension ($D$) to the multiscale entropy production balance [35]:

$$D = -k_V \ln \Omega_V, \tag{1}$$

where $D$ represents the thermodynamic fractal dimension, which characterizes the complexity of the cracking process. The constant $k_V$ is associated with the scaling factor $r$ by the relation $k_V = 1/\ln(r/r_0)$, reflecting the relationship between different length scales. $r_0$ is the size of the smallest components of the system and $\Omega_V$, the multiscale entropy production balance, quantifies the interplay between macroscopic ($dS$) and microscopic ($dS_0$) entropy productions. It captures the relative contribution of entropy production at different scales and provides a measure of the overall energy dissipation in the system. According to Venegas-Aravena et al., 2022 [35], the parameter $\Omega_V$ can be expressed as:

$$\Omega_V = \frac{dS}{dS_0} \times e^{\left(\frac{1-D_E}{k_V}\right)}, \tag{2}$$

where $D_E$ is the Euclidean dimension. By merging Equation (2) into Equation (1), it can be concluded that:

$$\frac{dS}{dS_0} = \frac{1}{\omega_0} e^{-D/k_V}, \tag{3}$$

where $\omega_0 = e^{\left(\frac{1-D_E}{k_V}\right)}$ which corresponds to the exponential term in Equation (2). The equation enables an investigation into how the dominance of macroscopic or microscopic entropy production impacts the thermodynamic fractal dimension. When the macroscopic entropy production dominates (resulting in a larger value of $\Omega_V$), it implies a stronger influence of the dissipation at larger scales, leading to a decrease in the thermodynamic fractal dimension. Conversely, when the microscopic entropy production dominates, the thermodynamic fractal dimension tends to increase, indicating a stronger influence of the smaller scales in the energy dissipation process.

Cracking in materials, such as rocks or brittle solids, involves the propagation and interaction of cracks at various scales. At the macroscopic level, the overall cracking behavior and energy dissipation can be captured by the macroscopic entropy production ($dS$). On the other hand, the microscopic entropy production ($dS_0$) represents the entropy production at smaller scales, capturing the contributions from microcracks, grain boundaries, or other microscopic features. These microscale cracks and defects contribute to the dissipation of energy through processes such as crack propagation, dislocation motion, and local stress concentrations.

## 3. Seismic Moment and Thermodynamic Fractal Dimension

A relationship has been established between the magnitude of an earthquake ($Mw$) and the rate of entropy change ($dS/dt$) [34]. This relationship is given by:

$$M_W \sim \log_{10}\left[\left(\frac{dS}{dt}\right)^p\right], \tag{4}$$

where the exponent is $p = 3/(5 - D)$. This relationship shows a connection between the dissipative processes associated with entropy change and the generation of seismic activity. That is, Equation (4) implies that as the rate of entropy change ($dS/dt$) increases, the magnitude of the earthquake ($M_W$) also tends to increase. Furthermore, the value of $p$ is influenced by the thermodynamic fractal dimension $D$. When $D$ is smaller, closer to 5, $p$

diverges, indicating a stronger relationship between the entropy change and earthquake magnitude. On the other hand, as $D$ increases, $p$ tends to 0, suggesting a weaker coupling between entropy change and earthquake magnitude. It is important to note that the global entropy change, represented by $dS/dt$, provides insights into the overall energy release and dissipation processes occurring within the system. This includes both the cracking generation within the medium and the rupture process during an earthquake. This implies that the entropy production is directly related to the rupture process of faults, including the fault roughness. This can be seen after replacing Equation (3) into Equation (4) after considering that $dS/dt = \left(\frac{dS}{dS_0}\right)\left(\frac{dS_0}{dt}\right)$:

$$M_W \sim \log_{10} e^{-\alpha(D)}, \tag{5}$$

where $\alpha(D) = \frac{pD}{k_V}$. As Equation (5) directly depends on the thermodynamic fractal dimension $D$, which describes the complexity of surfaces, it implies that Equation (5) links the magnitude and the geometrical irregularities of faults. This implies that smoother surfaces, characterized by lower $D$, may be associated with larger magnitude earthquakes. Conversely, more complex, and rough surfaces, represented by higher fractal dimensions, may result in smaller magnitude earthquakes (Figure 1a). In terms of rupture area, Venegas-Aravena et al., 2022 [34] have also shown a relation between entropy change and ruptured area $A$, expressed as follows:

$$A \sim \left(\frac{dS}{dt}\right)^{\frac{2p}{3}}. \tag{6}$$

Just as Equation (5), Equation (6) can be formulated in relation to the thermodynamic fractal dimension as follows:

$$A \sim e^{-\beta(D)}, \tag{7}$$

where $\beta(D) = 2\alpha/3$. Equation (7) highlights the connection between the ruptured area and the fault's irregularities, where the thermodynamic fractal dimension ($D$) serves as a measure of the system, encompassing the fault roughness within this context. Additionally, Equation (7) states that smoother faults, resulting from reductions in microscopic stresses or increases in macroscopic stresses, are associated with larger rupture areas (Figure 1b). This equation implies that larger earthquakes are generated in areas characterized by smoother surfaces. While Equations (4)–(7) emerge from the application of multiscale thermodynamics, further exploration is necessary to provide a more comprehensive seismological description of the rupture process and its relationship to fault surfaces. For instance, Figure 1c offers a visual representation highlighting the relationship between the thermodynamic fractal dimension and fault surface characteristics. In this schematic, the yellow area represents the rupture zone, depicting that larger thermodynamic fractal dimensions are associated with rough fault surfaces, smaller ruptured area, and in consequence, smaller magnitude. In contrast, Figure 1d presents a schematic of a fault with a smaller thermodynamic fractal dimension. The schematic representation of a fault surface shown in this figure appears smoother, without the jagged features present in the schematic representation shown in Figure 1c. A smaller fractal dimension corresponds to smoother fault surfaces. Interestingly, faults with smoother surfaces and a smaller fractal dimension exhibit larger rupture area. Consequently, they also tend to generate greater seismic magnitudes, as indicated by the expanded yellow area in the diagram.

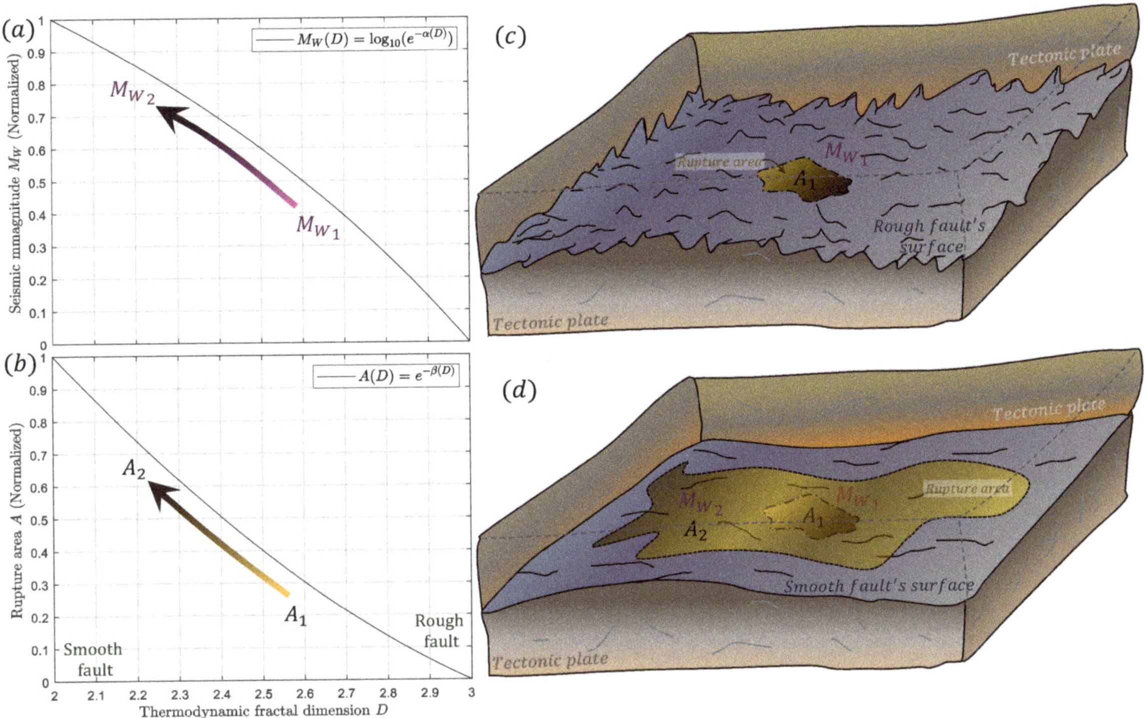



**Figure 1.** Relationship between seismic magnitude, rupture area, and the thermodynamic fractal dimension. (**a**) demonstrates that smaller values of the thermodynamic fractal dimension are associated with larger earthquake magnitudes, while (**b**) shows how smaller fractal dimensions correspond to larger rupture areas. The thermodynamic fractal dimension also influences fault surface characteristics, with larger values indicating rougher surfaces (**c**), whereas smaller fractal dimensions result in smoother fault surfaces (**d**).

## 4. Fracture Energy

The fracture energy, denoted as $G_C$, is a measure of the energy required to propagate an earthquake rupture and extend it further within the medium. The value of $G_C$ depends on various factors, including the material properties and the nature of the fracture process [67]. In terms of material properties, different compositions and regimes, such as brittle or ductile behavior, can significantly affect the fracture energy. Ductile materials are generally more resistant to fracture and require a larger amount of energy to propagate the rupture [68]. In contrast, brittle materials exhibit lower fracture energy, as they are more prone to sudden and catastrophic failure [69–71]. Interestingly, both brittle and ductile regimes are characterized by relatively small fractal dimensions, resulting in smoother surfaces [69,70]. Smoother surfaces indicate a lower degree of complexity or roughness, as described by the fractal dimension [35]. This can be attributed to the nature of the fracture process in these materials, which tends to generate relatively uniform and well-defined fracture surfaces. On the other hand, composite materials, which consist of a combination of different constituents, exhibit rougher surfaces and tend to have larger fractal dimensions [67]. The presence of multiple materials with different properties introduces heterogeneity and increases the complexity of the fracture surfaces. Figure 2a provides a schematic representation that illustrates the variation of the fractal dimension across different material types as shown by Williford (1988) [69]. Specifically, it shows that brittle and ductile materials tend to exhibit smoother crack surfaces. On the other hand, composite materials display a larger fractal dimension, indicating more irregular and complex crack surfaces. Figure 2b serves as a schematic representation that further elucidates the relationship described in Figure 2a.

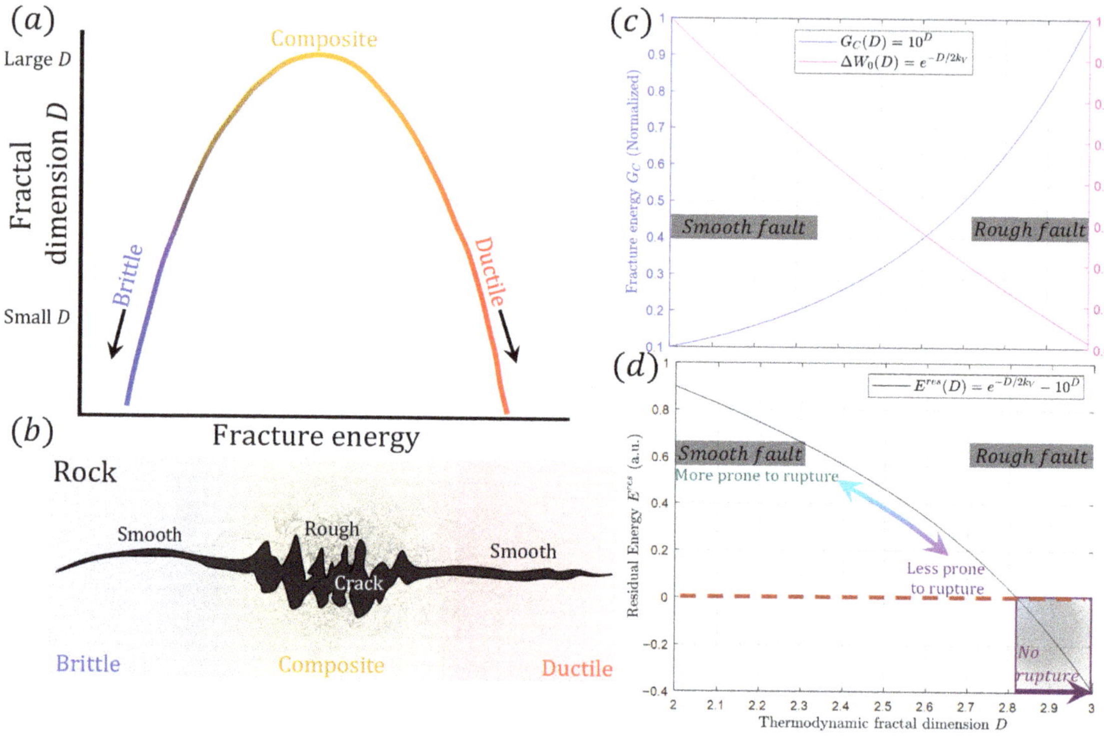

**Figure 2.** The figure presents a comprehensive analysis of the dependence of the fractal dimension on material composition and regime. (**a**) Schematic representation illustrating how the fractal dimension varies according to different material types. Brittle and ductile materials exhibit smoother crack surfaces, while composite materials have a larger fractal dimension; (**b**) Schematic representation of the relationship described in (**a**); (**c**) Demonstrates the interplay between available energy (magenta line) and fracture energy (blue line), both influenced by the thermodynamic fractal dimension; (**d**) Analytic depiction of the residual energy, which is the difference between available energy and fracture energy, as a function of the thermodynamic fractal dimension.

According to Ohnaka (2013) [71], there is a relationship between fracture energy and the geometrical irregularities of fault interfaces. The geometrical irregularities on faults are characterized by a parameter called $\lambda_C$. Ohnaka (2013) [72] suggests that materials with smoother fault interfaces have smaller values of $\lambda_C$ and, therefore, require less fracture energy to propagate the rupture. In contrast, materials with rougher fault interfaces have larger values of $\lambda_C$, resulting in a higher fracture energy requirement to spread the rupture. This can be seen as:

$$G_C = c_0 \lambda_C, \tag{8}$$

where $c_0$ is a proportional factor and represent a material-dependent constant. If $c_0$ is considerably larger for ductile materials compared to brittle materials, it implies that the same amount of geometrical irregularity ($\lambda_C$) or roughness will result in a higher fracture energy ($G_C$) for ductile materials. This is consistent with the observation that ductile materials can absorb more energy due to their ability to accommodate greater plastic deformation and exhibit higher fracture energy, even with similar levels of smoothness on fault interfaces. Thus, Equation (8) implies that the absence of significant roughness reduces the resistance to rupture propagation, resulting in lower energy requirements.

The fracture energy plays an important role in the generation of earthquakes. For instance, according to Noda et al., 2021 [73], earthquakes are more likely to occur in zones where the residual energy ($E^{res}$) is positive. This energy is defined as the difference between the available energy $\Delta W_0$, which is partly produced by stress accumulation, and the fracture energy:

$$E^{res} = \Delta W_0 - G_C. \tag{9}$$

Equation (9) does not directly address the concept of fault smoothing or roughness. However, it can draw a connection based on the underlying mechanisms. For example, the fractal dimension is proportional to the logarithm of the roughness: $D \sim \log \lambda_C$ [74]. Equivalently, $\lambda_C \sim 10^D$. This in Equation (8) leads to $G_C$ being written in function of $D$ as $\sim 10^D$. In that sense, the increase of geometrical roughness implies the increase of fractal dimension and the increase of $G_C$ as shown Figure 2c (blue line). This into Equation (9) leads to Equation:

$$E^{res} = \Delta W_0 - d_0 10^D, \tag{10}$$

where $d_0$ is a constant. Equation (10) means that when a fault surface is smoother, with fewer geometric irregularities or asperities, it requires less energy to propagate the rupture (i.e., lower fracture energy). This means that the energy released during an earthquake is relatively higher compared to the energy needed for fault motion. As a result, the residual energy tends to be positive. In contrast, if the fault surface has more irregularities or roughness, it requires more energy to propagate the rupture (i.e., higher fracture energy). This leads to a lower release of energy during the earthquake relative to the energy needed for fault motion. In such cases, the residual energy may be negative or close to zero and could result in no earthquake generation. Therefore, it can be inferred that smoother fault surfaces, associated with lower fracture energy, are more likely to result in positive residual energy, indicating a higher potential for seismic activity. On the other hand, rougher fault surfaces, associated with higher fracture energy, may lead to lower residual energy and a reduced likelihood of earthquakes.

Therefore, the reduction in fracture energy can lead to an increase in the area characterized by positive residual energy. In other words, more regions become capable of sustaining earthquake propagation due to the lower energy threshold required for rupture. As a result, the areas with reduced fracture energy can increase the areas of potential seismic rupture compared to the pre-smoothing condition. This expansion of the area with positive residual energy increases the overall potential for larger earthquakes to occur.

The available energy $\Delta W_0$ is dependent on a function that describes the initial stress states $S_0(x)$ [73,75], which represents the macroscopic stress states ($\sigma$). By utilizing the relationship between macroscopic ($\dot{\sigma}$) and microscopic ($\dot{\sigma}_0$) stress change balance, expressed as $dS/dS_0 = \omega_{\dot{\sigma}} \, \dot{\sigma}^2/\dot{\sigma}_0^2$, where $\omega_{\dot{\sigma}}$ [36], the macroscopic stress change can be written as $\dot{\sigma} = \dot{\sigma}_0 \gamma_0 e^{-D/2k_V}$, where $\gamma_0 = \left(\omega_{\dot{\sigma}}\omega_0\right)^{-1/2}$. Thus, after temporal integration, the available energy can be described in terms of the thermodynamic fractal dimension as follows:

$$\Delta W_0 \sim e^{-D/2k_V}. \tag{11}$$

This equation shows that the macroscopic available energy decreases as the faults are rougher (magenta line in Figure 2c). Here, it is important to note that the rougher surfaces imply greater degree of irregularity and complexity at the small scale. This implies that the stress concentration phenomena are primarily localized and occur on the microscale, resulting in the increase of small-scale available energy. In that sense, Equation (11) offers a complement perspective such as the decrease of the large-scale available energy. By combining Equations (10) and (11), the residual energy in terms of $D$ is

$$E^{res} \sim e^{-D/2k_V} - 10^D. \tag{12}$$

The relationship between the thermodynamic fractal dimension and residual energy provides a valuable insight into the seismic activity of faults. Specifically, Equation (12) and Figure 2d indicate that smaller values of $D$ are associated with larger residual energy values, while larger values of $D$ correspond to negative values of residual energy. The implication of this relationship is that faults with smoother surfaces and smaller values of $D$ have the potential to host larger amounts of residual energy. Consequently, they may have a higher likelihood of generating future earthquakes. In contrast, rough faults with larger values of $D$ are less likely to accumulate substantial residual energy, resulting in negative values which indicates a fault that is less prone to rupture.

Equations (5), (7), and (12) demonstrate that smaller values of the thermodynamic fractal dimension are correlated with larger areas, magnitudes, and residual energies. Consistent with this, Figure 3a illustrates the relationship between residual energy and the rupture area. Figure 3a confirms that as the residual energy increases, the area prone to rupture also increases. This relationship is captured by the best-fit curve, which correlates residual energy and the area prone to rupture through Equation (13).

$$A(D) \sim (E^{res}(D))^m. \tag{13}$$

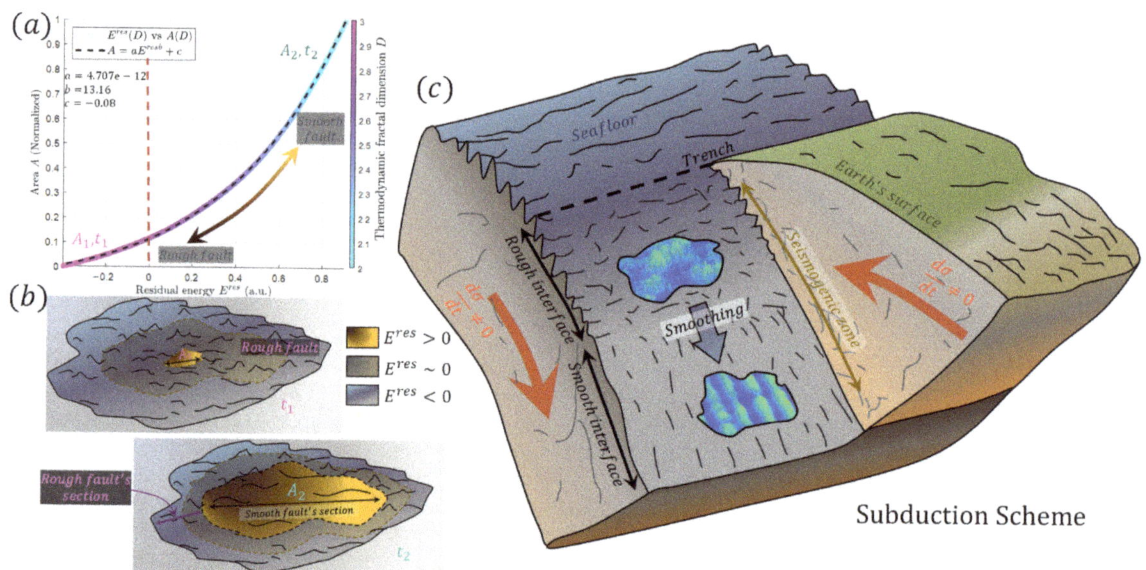



**Figure 3.** (**a**) Relationship between the rupture area $A$ and residual energy $E^{res}$. The segmented line represents the best second-order power fit. The color bar indicates the thermodynamic fractal dimension; (**b**) Schematic representation of the relationship depicted in (**a**) for times $t_1$ and $t_2$. The yellow area highlights regions prone to rupture. Notably, for smoother surfaces, the area prone to rupture is larger at $t_2$ compared to $t_1$; (**c**) Schematic representation of subduction. The shallow sections are characterized by rougher surfaces compared to the deeper sections. Subduction is considered a fault-smoothing mechanism.

Equation (13) and Figure 3a indicate that rough fault surfaces have a lower capacity to store residual energy, resulting in smaller areas prone to rupture. Conversely, smoother fault surfaces allow for a larger portion of the fault to accommodate significant residual energy. In Figure 3b, areas A1 and A2 represent cases for rougher and smoother fault surfaces, respectively. These figures illustrate how smoother surfaces can store more residual energy, leading to larger areas of potential rupture.

## 5. Discussions

During an earthquake, the process of rupture involves the fracturing and sliding of rock layers along the fault surface. This process necessitates overcoming resistance forces and the release of accumulated stress energy. However, few studies manage to link processes inside faults with non-seismic precursors. In recent decades, there have been numerous efforts to explain earthquake precursor phenomena or anomalies [16,76,77]. These efforts involve the deformation of lithospheric material, chemical reactions, or the migration of fluids. In addition to not being able to physically link these effects to the earthquakes they try to predict, there are two major additional challenges. Firstly, experiments demonstrate that rock electrification can occur even in the absence of macroscopic stress changes [78]. Secondly, none of these explanations can be directly associated with the earthquakes they are supposed to precede because they cannot be linked to basic rupture parameters within faults [38]. In order to incorporate seismicity, numerous efforts have been focused on describing pre-earthquake phenomena using more fundamental tools, such as the entropy change of the lithosphere [34–36,79–81]. In that line, the framework proposed by [31,33–36] suggests that fundamental parameters of seismology, such as magnitude, stress drop, fault friction, or changes in b-value, can be linked to precursor measurements when considering the multiscale crack propagation. These small-scale cracks act as pathways for energy dissipation and contribute to the overall change in entropy [34]. The increase of macroscopic entropy, as described by Equations (2) and (3), is associated with a reduction in the thermodynamic fractal dimension (Equation (1)). This reduction in fractal dimension implies smoother fault surfaces or less geometrical irregularities which are associated with lower fracture energy (Equation (8)). As a consequence, the global features of the system, such as the entropy production, the cracking process, and the physical and geometrical faults are linked. Particularly, based on Equations (5), (7), (12) and (13), there exists an analytical relationship among earthquake size, magnitude, residual energy, and the geometric characteristics of faults. This connection suggests that smoother fault surfaces are more likely to produce larger areas of positive residual energy, which, in turn, can give rise to larger earthquakes.

The connection between smooth fault interfaces and large earthquakes finds support in observations of subduction zones. Specifically, studies suggest that significant Chilean earthquakes occurring in subduction zones, like the Valdivia 1960 Mw9.5 earthquake, may be associated with smooth features within the subduction channels [82]. These smooth features result from the extensive accumulation of sediments during the subduction process, which creates fewer resistance barriers [83]. Furthermore, large-scale simulations demonstrate that smoother surfaces have a greater propensity to generate larger ruptures [84]. On the contrary, Equation (8) suggests that rougher faults result in greater fracture energy, which reduces the probability of obtaining positive residual energy. This interpretation of Equations (8) and (12) indicates that rougher faults tend to generate smaller earthquakes, as described by Equation (13). This finding aligns with studies on subduction zones, which have revealed that geometrical irregularities act as barriers to seismic activity [85].

Studies have demonstrated that moderate-to-large earthquakes predominantly occur at deeper zones within subduction areas [86–88]. In contrast, the shallow sections of subduction zones serve as reservoirs for stress accumulation, owing to their higher frictional strength which enables the accumulation of larger stress levels in these shallow regions [89,90]. Hence, deeper zones are more susceptible to earthquake rupture. As illustrated in Figure 3a, this condition aligns with smoother fault surfaces. Consequently, from a multiscale thermodynamic perspective, the shallow sections of the subduction zone exhibit rougher surfaces, while the deeper sections display smoother surfaces. This means that less energy is required to initiate and propagate fractures along these smooth fault interfaces. When the fracture energy is lower, it means that a larger portion of the available energy can be utilized to generate seismic activity (Equation (9)). This can lead to an increase in the area of positive residual energy, as more energy is retained in the system after subtracting the fracture energy. The increase in the area of positive residual energy suggests a greater potential for the occurrence of large earthquakes at deeper zones. This scheme suggests that the subduction of the oceanic crust undergoes a smoothing process as the tectonic plate subducts. Figure 3c provides a schematic representation of this smoothing process, illustrating that the deeper interface sections are smoother compared to the shallower sections. In alignment with this idea, Figure 4a–c indicates the process by which stresses can fracture and smooth out jagged interfaces, resulting in the formation of smoother faults. Figure 4a presents a schematic representation inspired by the experiments conducted by Iquebal et al., 2019 [91], illustrating the polishing of rough surfaces (Figure 6 in Ref. [91]). Figure 4a consists of four surfaces. The first one (1) was created using the code by Chen and Yang [92] to generate a random fractal surface. The other surfaces (2, 3, and 4) were generated by progressively truncating the minimum values. In other words, values smaller than a certain number are set to zero, and this minimum value increases progressively, causing the surfaces to become increasingly gray. These numbered stages resemble the progression of the repetitive sliding contacts shown by reference [91], with higher numbers corresponding to more extensive sliding and consequently smoother surfaces. In this context, Figure 4b,c provide a schematic illustration of how spatial irregularities can store stresses, as demonstrated in Figure 2c (magenta line). In cases where the fractal dimension $D$ is 3, representing a rougher interface (Figure 4b), the storage of stresses is limited due to the lower resistance offered by the geometry, resulting in the smoothing of these irregularities. Conversely, Figure 4c depicts a smoother surface that offers greater resistance. Consequently, smoother surfaces tend to be characterized by larger areas, such as the one-dimensional distance $L_2$ illustrated in this case. As residual energy is dependent on stresses (Equation (9)), it follows that larger residual energy is associated with larger areas, as shown in Figure 3a and described by Equation (13). This analysis suggests that the deeper sections of subduction faults, characterized by multiple stages of slip or earthquakes, may exhibit smoother surfaces. The smoothing process as a function of the slip discussed above has significant implications for fault dynamics. For instance, as the fault roughness decreases, there is a tendency for the fractal dimension of the slip distribution to also decrease [93]. In addition, as noted by Morad et al., 2022 [94], fault surfaces that exhibit exceptionally smooth characteristics experience minimal stress increases and sustained slip. This particular behavior may contribute to the occurrence of slow slip events within the deeper sections of megathrust faults, as reported by Ito et al., 2007 [95]. Consequently, the presence of slow slip events suggests that the smoothing process, influenced by the cyclic macroscopic loads described in Equation (1), has already taken place during the fault's precursor phase. Note that there is evidence supporting the slow slip events as a precursor mechanism [96–99]. This implies that what is commonly referred to as a slow slip is likely the phase in which the fault, aiming to increase entropy and decrease the thermodynamic fractal dimension of the system, starts to slowly be smoothing the fault at the macroscopic scale, thus becoming one of the final mechanisms for releasing the excess energy. Furthermore, the role of the polishing process can be associated with the "Mogi Doughnut" effect, which describes the seismicity surrounding a large rough patch or asperity prior to its eventual rupture or smoothing (representing a major earthquake) [100–102]. In this context, the polishing process reveals the presence of smooth zones surrounding the rough patch, as depicted in Figure 4a. Each rupture event acts as a polishing mechanism that reduces the size of the asperity. Consequently, the immediate surrounding zones of a rough patch are smoother and more prone to generating seismic activity. As the thermodynamic fractal dimension ($D$) decreases, indicating smoother faults, more sections of the fault become susceptible to ruptures in the zones surrounding the large asperity. Thus, the decrease in the thermodynamic fractal dimension provides an explanation for the "Mogi Doughnut" effect through the concept of the polishing process.

According to research conducted by Venegas-Aravena and Cordaro (2023) [36], Equation (1) not only relates to the geometric properties of faults such as the smothering of faults, but also to other global parameters, such as the b-value. For example, it has been observed that when studying systems that span multiple scales, the b-value is proportionate to the fractal dimension [36]. However, in certain cases, a complex positive correlation between the b-value and fractal dimension is observed [36]. This discovery aligns with the positive correlation observed between the b-value and fractal dimension in real natural faults [103]. Therefore, the b-value serves as a measure of the stress states within the lithosphere and can indicate zones that are more prone to seismic activity [104]. Specifically, the b-value has been found to exhibit a negative correlation with stress states [105,106]. This implies that as the load on faults increases, the b-value and thermodynamic fractal dimension decrease [35,36]. Consequently, this phenomenon contributes to the smothering of faults, resulting in the accumulation of residual energy and an expansion of the area prone to rupture. The increase in macroscopic entropy production within the system is also associated with the generation of electromagnetic signals prior to earthquakes or macroscopic failure in rock samples [31,34]. In particular, the propagation of multiscale fractures and the movement of charged particles within the newly formed cracks, as a response or dissipation mechanism to the accumulation of external stress, can give rise to electromagnetic emissions, as demonstrated by experiments conducted on rock samples [78,107–109].

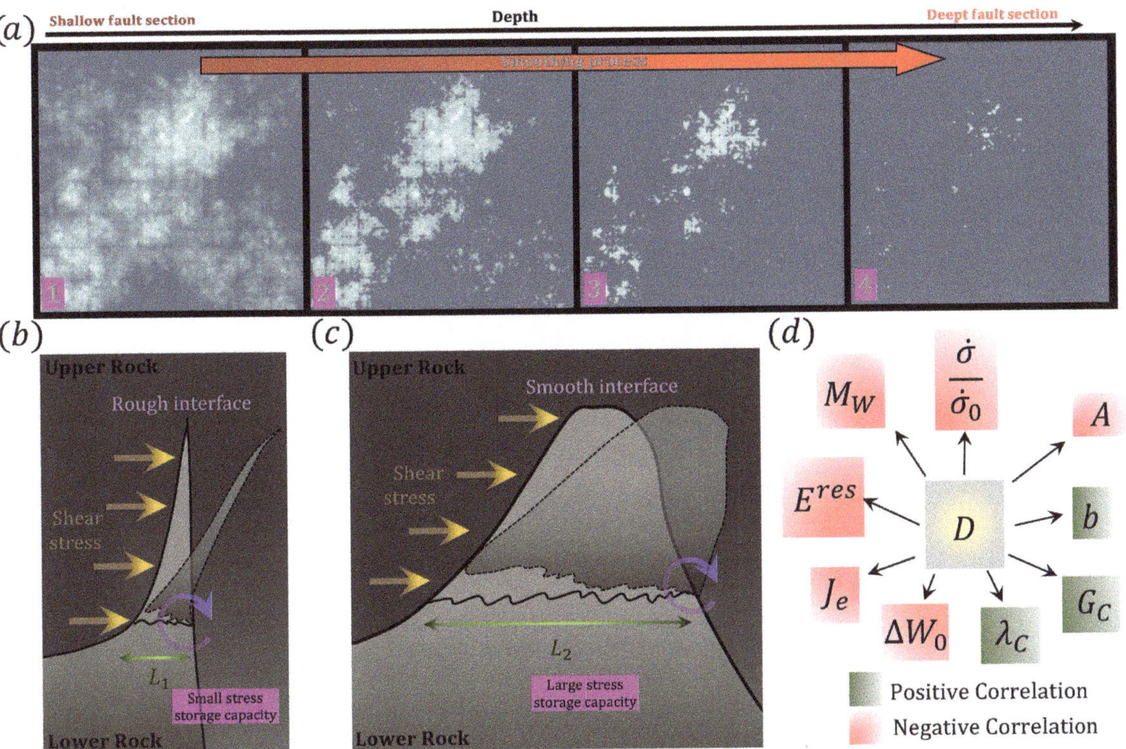


**Figure 4.** (**a**) Smoothing process. The number indicates the number of sliding stages. That is, there are more sliding stages which generate smoother fault at deeper sections of fault. (**b**,**c**) shows schematic representation illustrating the stress storage capacity in two cases. Case (**b**) exhibits a small capacity to hold stresses due to the thin bulge compared to case (**c**); (**c**) Schematic representation highlighting the stress storage capacity. In this case, the bulge is thicker, allowing for a larger capacity to hold stresses; (**d**) Correlation between the thermodynamic fractal dimension and other quantities. Positive correlation is represented by green, while negative correlation is represented by red. Here, the thermodynamic fractal dimension serves as a global parameter controlling various aspects of pre-earthquake physics within the lithosphere.

Furthermore, as the fracture energy decreases, it facilitates the flow of fluids through the fractures, permeating the surrounding rock matrix [110,111]. This migration of fluids can have diverse implications, including the alteration of pore pressure distribution, influencing the stability of the fault zone, and potentially triggering or affecting seismic activity [112]. Consequently, it becomes apparent that the generation of electromagnetic signals, the reduction of fracture energy, fluid migration, fault surface smoothing, increases in the area of positive residual energy, and the occurrence of large earthquakes are interconnected manifestations of the underlying entropy production processes within the Earth's crust. These processes can be analytically described in terms of the thermodynamic fractal dimension, as summarized in Figure 4d, with the green and red colors indicating positive and negative correlations with the thermodynamic fractal dimension.

Finally, adopting a multiscale perspective reveals that the reduction in thermodynamic dimension signifies a diminished capacity of the lithosphere to release excessive energy at a small scale, such as through minor cracks. Consequently, the system strives for release on progressively larger scales. This phenomenon facilitates the development of larger cracks, establishing additional pathways for fluid migration, thereby potentially causing phenomena like heightened surface temperature or the liberation of trapped gases. Furthermore, these enlarged cracks contribute to intensified levels of anomalous electromagnetic signals. Concurrently, a decrease in the b-value and the smoothing of faults can occur, potentially linked to the occurrence of slow slip events, resulting in an expanded area of positive residual energy. When energy dissipation remains inefficient at this level, the predominant mechanism shifts to macroscopic rupture, ultimately culminating in an earthquake on a larger scale.

## 6. Conclusions

The main conclusions are listed below:

- The relationship between the magnitude of earthquakes and thermodynamic fractal dimension was established.
- The increases of large-scale entropy production generate the reduction of geometrical irregularities which leads to larger earthquake magnitudes.
- The large-scale entropy production reduces the fracture energy which increases the probability of generating larger ruptures.
- Smoother surfaces found at the deeper sections of subduction faults are more prone to generating heightened seismic activity.
- Subduction can be seen as a mechanism that contributes to the smoothing of faults because it increases macroscopic entropy production.
- Non-seismic earthquake signals are also a manifestation of this entropy change in the system. This means that the system attempts to release the excess energy through the generation of cracks, which can serve as pathways for fluid migration. This can result in changes in ground temperature or the release of gases trapped underground. Additionally, the increase in entropy causes a decrease in b-value and thermodynamic fractal dimension, while also smoothing the faults, thereby reducing the resistance to earthquake generation. This can lead to precursor seismicity.
- Both the geometry of faults and the stored stresses are heterogeneous. Therefore, future studies should focus on establishing how the smoothing process occurs in faults, both in natural settings and laboratory experiments, while other precursor signals are being produced.

**Author Contributions:** P.V.-A. proposed the core idea, mathematical development, figures, and initial draft of the project. E.G.C. contributed to the scientific discussions of the work. All authors have read and agreed to the published version of the manuscript.

**Funding:** This research received no external funding.

**Data Availability Statement:** Not applicable.

**Acknowledgments:** P.V.-A. would like to thank ANID and PUC for their constant support in the development of science in Chile. P.V.-A. and E.C. would like to express their gratitude to the support staff at the *Observatorios de Radiación Cósmica y Geomagnetismo*.

**Conflicts of Interest:** The authors declare no conflict of interest.

## References

1. Bleier, T.; Dunson, C.; Alvarez, C.; Freund, F.; Dahlgren, R. Correlation of pre-earthquake electromagnetic signals with laboratory and field rock experiments. *Nat. Hazards Earth Syst. Sci.* **2010**, *10*, 1965–1975. [CrossRef]
2. Potirakis, S.M.; Minadakis, G.; Eftaxias, K. Relation between seismicity and pre-earthquake electromagnetic emissions in terms of energy, information and entropy content. *Nat. Hazards Earth Syst. Sci.* **2012**, *12*, 1179–1183. [CrossRef]
3. Petraki, E.; Nikolopoulos, D.; Nomicos, C.; Stonham, J.; Cantzos, D.; Yannakopoulos, P.; Kottou, S. Electromagnetic Pre-earthquake Precursors: Mechanisms, Data and Models—A Review. *J. Earth Sci. Clim. Chang.* **2015**, *6*, 250. [CrossRef]
4. Sanchez-Dulcet, F.; Rodríguez-Bouza, M.; Silva, H.G.; Herraiz, M.; Bezzeghoud, M.; Biagi, P.F. Analysis of observations backing up the existence of VLF and ionospheric TEC anomalies before the Mw6.1 earthquake in Greece, January 26 2014. *Phys. Chem. Earth* **2015**, *85–86*, 150–166. [CrossRef]
5. Schekotov, A.; Hayakawa, M. Seismo-meteo-electromagnetic phenomena observed during a 5-year interval around the 2011 Tohoku earthquake. *Phys. Chem. Earth* **2015**, *85–86*, 167–173. [CrossRef]
6. Scoville, J.; Heraud, J.; Freund, F. Pre-earthquake magnetic pulses. *Nat. Hazards Earth Syst. Sci.* **2015**, *15*, 1873–1880. [CrossRef]
7. Cordaro, E.G.; Venegas, P.; Laroze, D. Latitudinal variation rate of geomagnetic cutoff rigidity in the active Chilean convergent margin. *Ann. Geophys.* **2018**, *36*, 275–285. [CrossRef]
8. Ouzounov, D.; Pulinets, S.; Hattori, K.; Taylor, P. Pre-Earthquake Processes: A Multidisciplinary Approach to Earthquake Prediction Studies. *Am. Geophys. Union Geophys. Monogr. Ser.* **2018**, *12*, 1–365. [CrossRef]
9. De Santis, A.; Marchetti, D.; Pavón-Carrasco, F.J.; Cianchini, G.; Perrone, L.; Abbattista, C.; Alfonsi, L.; Amoruso, L.; Campuzano, S.A.; Carbone, M.; et al. Precursory worldwide signatures of earthquake occurrences on Swarm satellite data. *Sci. Rep.* **2019**, *9*, 20287. [CrossRef]
10. Nakagawa, K.; Yu, Z.-Q.; Berndtsson, R.; Kagabu, M. Analysis of earthquake-induced groundwater level change using self-organizing maps. *Environ. Earth Sci.* **2019**, *78*, 445–455. [CrossRef]
11. Wei, C.; Lu, X.; Zhang, Y.; Guo, Y.; Wang, Y. A time-frequency analysis of the thermal radiation background anomalies caused by large earthquakes: A case study of the wenchuan 8.0 earthquake. *Adv. Space Res.* **2019**, *65*, 435–445. [CrossRef]
12. Sabbarese, C.; Ambrosino, F.; Chiodini, G.; Giudicepietro, F.; Macedonio, G. Continuous radon monitoring during seven years of volcanic unrest at Campi Flegrei caldera (Italy). *Sci. Rep.* **2020**, *10*, 9551–9610. [CrossRef] [PubMed]
13. Warden, S.; MacLean, L.; Lemon, J.; Schneider, D. Statistical Analysis of Pre-earthquake Electromagnetic Anomalies in the ULF Range. *J. Geophys. Res.* **2020**, *125*, e2020JA027955. [CrossRef]
14. Xiong, P.; Long, C.; Zhou, H.; Battiston, R.; Zhang, X.; Shen, X. Identification of Electromagnetic Pre-Earthquake Perturbations from the DEMETER Data by Machine Learning. *Remote Sens.* **2020**, *12*, 3643. [CrossRef]
15. Cordaro, E.G.; Venegas, P.; Laroze, D. Long-term magnetic anomalies and their possible relationship to the latest greater Chilean earthquakes in the context of the seismo-electromagnetic theory. *Nat. Hazards Earth Syst. Sci.* **2021**, *21*, 1785–1806. [CrossRef]
16. Freund, F.; Ouillon, G.; Scoville, J.; Sornette, D. Earthquake precursors in the light of peroxy defects theory: Critical review of systematic observations. *Eur. Phys. J. Spec. Top.* **2021**, *230*, 7–46. [CrossRef]
17. Mehdi, S.; Shah, M.; Naqvi, N.A. Lithosphere atmosphere ionosphere coupling associated with the 2019 Mw 7.1 California earthquake using GNSS and multiple satellites. *Environ. Monit. Assess.* **2021**, *193*, 501. [CrossRef]
18. Vasilev, A.; Tsekov, M.; Petsinski, P.; Gerilowski, K.; Slabakova, V.; Trukhchev, D.; Botev, E.; Dimitrov, O.; Dobrev, N.; Parlichev, D. New Possible Earthquake Precursor and Initial Area for Satellite Monitoring. *Front. Earth Sci.* **2021**, *8*, 586283. [CrossRef]
19. Heavlin, W.D.; Kappler, K.; Yang, Y.; Fan, M.; Hickey, J.; Lemon, J.; MacLean, L.; Bleier, T.; Riley, P.; Schneider, D. Case-Control Study on a Decade of Ground-Based Magnetometers in California Reveals Modest Signal 24–72 hr Prior to Earthquakes. *J. Geophys. Res.* **2022**, *127*, e2022JB024109. [CrossRef]
20. Marchetti, D.; De Santis, A.; Campuzano, S.A.; Zhu, K.; Soldani, M.; D'Arcangelo, S.; Orlando, M.; Wang, T.; Cianchini, G.; Di Mauro, D.; et al. Worldwide Statistical Correlation of Eight Years of Swarm Satellite Data with M5.5+ Earthquakes: New Hints about the Preseismic Phenomena from Space. *Remote Sens.* **2022**, *14*, 2649. [CrossRef]
21. Peleli, S.; Kouli, M.; Vallianatos, F. Satellite-Observed Thermal Anomalies and Deformation Patterns Associated to the 2021, Central Crete Seismic Sequence. *Remote Sens.* **2022**, *14*, 3413. [CrossRef]
22. Anyfadi, E.-A.; Gentili, S.; Brondi, P.; Vallianatos, F. Forecasting Strong Subsequent Earthquakes in Greece with the Machine Learning Algorithm NESTORE. *Entropy* **2023**, *25*, 797. [CrossRef]
23. Bulusu, J.; Arora, K.; Singh, S.; Edara, A. Simultaneous electric, magnetic and ULF anomalies associated with moderate earthquakes in Kumaun Himalaya. *Nat. Hazards* **2023**, *116*, 3925–3955. [CrossRef]
24. Shah, M.; Shahzad, R.; Ehsan, M.; Ghaffar, B.; Ullah, I.; Jamjareegulgarn, P.; Hassan, A.M. Seismo Ionospheric Anomalies around and over the Epicenters of Pakistan Earthquakes. *Atmosphere* **2023**, *14*, 601. [CrossRef]

25. Fidani, C. The Conditional Probability of Correlating East Pacific Earthquakes with NOAA Electron Bursts. *Appl. Sci.* **2022**, *12*, 10528. [CrossRef]
26. Rikitake, T. Earthquake precursors in Japan: Precursor time and detectability. *Tectonophysics* **1987**, *136*, 265–282. [CrossRef]
27. Moriya, T.; Mogi, T.; Takada, M. Anomalous pre-seismic transmission of VHF-band radio waves resulting from large earthquakes, and its statistical relationship to magnitude of impending earthquakes. *Geophys. J. Int.* **2010**, *180*, 858–870. [CrossRef]
28. Han, P.; Zhuang, J.; Hattori, K.; Chen, C.-H.; Febriani, F.; Chen, H.; Yoshino, C.; Yoshida, S. Assessing the Potential Earthquake Precursory Information in ULF Magnetic Data Recorded in Kanto, Japan during 2000–2010: Distance and Magnitude Dependences. *Entropy* **2020**, *22*, 859. [CrossRef]
29. Li, M.; Yang, Z.; Song, J.; Zhang, Y.; Jiang, X.; Shen, X. Statistical Seismo-Ionospheric Influence with the Focal Mechanism under Consideration. *Atmosphere* **2023**, *14*, 455. [CrossRef]
30. Zhang, Y.; Wang, T.; Chen, W.; Zhu, K.; Marchetti, D.; Cheng, Y.; Fan, M.; Wang, S.; Wen, J.; Zhang, D.; et al. Are There One or More Geophysical Coupling Mechanisms before Earthquakes? The Case Study of Lushan (China) 2013. *Remote Sens.* **2023**, *15*, 1521. [CrossRef]
31. Venegas-Aravena, P.; Cordaro, E.G.; Laroze, D. A review and upgrade of the lithospheric dynamics in context of the seismoelectromagnetic theory. *Nat. Hazards Earth Syst. Sci.* **2019**, *19*, 1639–1651. [CrossRef]
32. Martinelli, G.; Plescia, P.; Tempesta, E. "Pre-Earthquake" Micro-Structural Effects Induced by Shear Stress on $\alpha$-Quartz in Laboratory Experiments. *Geosciences* **2020**, *10*, 155. [CrossRef]
33. Venegas-Aravena, P.; Cordaro, E.G.; Laroze, D. The spatial–temporal total friction coefficient of the fault viewed from the perspective of seismo-electromagnetic theory. *Nat. Hazards Earth Syst. Sci.* **2020**, *20*, 1485–1496. [CrossRef]
34. Venegas-Aravena, P.; Cordaro, E.G.; Laroze, D. Natural Fractals as Irreversible Disorder: Entropy Approach from Cracks in the Semi Brittle-Ductile Lithosphere and Generalization. *Entropy* **2022**, *24*, 1337. [CrossRef]
35. Venegas-Aravena, P.; Cordaro, E.; Laroze, D. Fractal Clustering as Spatial Variability of Magnetic Anomalies Measurements for Impending Earthquakes and the Thermodynamic Fractal Dimension. *Fractal Fract.* **2022**, *6*, 624. [CrossRef]
36. Venegas-Aravena, P.; Cordaro, E.G. Analytical Relation between b-Value and Electromagnetic Signals in Pre-Macroscopic Failure of Rocks: Insights into the Microdynamics' Physics Prior to Earthquakes. *Geosciences* **2023**, *13*, 169. [CrossRef]
37. Hough, S.E. *Predicting the Unpredictable: The Tumultuous Science of Earthquake Prediction*; Princeton University Press: Princeton, NJ, USA, 2010.
38. Picozza, P.; Conti, L.; Sotgiu, A. Looking for Earthquake Precursors from Space: A Critical Review. *Front. Earth Sci.* **2021**, *9*, 676775. [CrossRef]
39. Szakács, A. Precursor-Based Earthquake Prediction Research: Proposal for a Paradigm-Shifting Strategy. *Front. Earth Sci.* **2021**, *8*, 548398. [CrossRef]
40. Surkov, V.V.; Molchanov, O.A.; Hayakawa, M. Pre-earthquake ULF electromagnetic perturbations as a result of inductive seismomagnetic phenomena during microfracturing. *J. Atmos. Sol.-Terr. Phys.* **2003**, *65*, 31–46. [CrossRef]
41. Eftaxias, K.; Potirakis, S.M. Current challenges for pre-earthquake electromagnetic emissions: Shedding light from micro-scale plastic flow, granular packings, phase transitions and self-affinity notion of fracture process. *Nonlinear Process. Geophys.* **2013**, *20*, 771–792. [CrossRef]
42. Kamiyama, M.; Sugito, M.; Kuse, M.; Schekotov, A.; Hayakawa, M. On the precursors to the 2011 Tohoku earthquake: Crustal movements and electromagnetic signatures. *Geomat. Nat. Hazards Risk* **2014**, *7*, 471–492. [CrossRef]
43. Matsuyama, K.; Katsuragi, H. Power law statistics of force and acoustic emission from a slowly penetrated granular bed. *Nonlinear Process. Geophys.* **2014**, *21*, 1–8. [CrossRef]
44. Contoyiannis, Y.; Potirakis, S.M.; Eftaxias, K.; Contoyianni, L. Tricritical crossover in earthquake preparation by analyzing preseismic electromagnetic emissions. *J. Geodyn.* **2015**, *84*, 40–54. [CrossRef]
45. Enomoto, Y.; Yamabe, T.; Okumura, N. Causal mechanisms of seismo-EM phenomena during the 1965–1967 Matsushiro earthquake Swarm. *Sci. Rep.* **2017**, *7*, 44774. [CrossRef]
46. Niccolini, G.; Lacidogna, G.; Carpinteri, A. Fracture precursors in a working girder crane: AE natural-time and b-value time series analyses. *Eng. Fract. Mech.* **2019**, *210*, 393–399. [CrossRef]
47. Frid, V.; Rabinovitch, A.; Bahat, D. Earthquake forecast based on its nucleation stages and the ensuing electromagnetic radiations. *Phys. Lett. A* **2020**, *384*, 126102. [CrossRef]
48. Frid, V.; Rabinovitch, A.; Bahat, D. Seismic moment estimation based on fracture induced electromagnetic radiation. *Eng. Geol.* **2020**, *274*, 105692. [CrossRef]
49. Wang, J.-H. Piezoelectricity as a mechanism on generation of electromagnetic precursors before earthquakes. *Geophys. J. Int.* **2020**, *224*, 682–700. [CrossRef]
50. Klyuchkin, V.N.; Novikov, V.A.; Okunev, V.I.; Zeigarnik, V.A. Comparative analysis of acoustic and electromagnetic emissions of rocks. *IOP Conf. Ser. Earth Environ. Sci.* **2021**, *929*, 012013. [CrossRef]
51. Agalianos, G.; Tzagkarakis, D.; Loukidis, A.; Pasiou, E.D.; Triantis, D.; Kourkoulis, S.K.; Stavrakas, I. Correlation of Acoustic Emissions and Pressure Stimulated Currents recorded in Alfas-stone specimens under three-point bending. The role of the specimens' porosity: Preliminary results. *Procedia Struct. Integr.* **2022**, *41*, 452–460. [CrossRef]
52. Triantis, D.; Pasiou, E.D.; Stavrakas, I.; Kourkoulis, S.K. Hidden Affinities Between Electric and Acoustic Activities in Brittle Materials at Near-Fracture Load Levels. *Rock Mech. Rock Eng.* **2022**, *55*, 1325–1342. [CrossRef]
53. Stergiopoulos, C.; Stavrakas, I.; Triantis, D.; Vallianatos, F.; Stonham, J. Predicting fracture of mortar beams under three-point bending using non-extensive statistical modeling of electric emissions. *Phys. A Stat. Mech. Its Appl.* **2015**, *419*, 603–611. [CrossRef]

54. Kourkoulis, S.K.; Pasiou, E.D.; Dakanali, I.; Stavrakas, I.; Triantis, D. Notched marble plates under direct tension: Mechanical response and fracture. *Constr. Build. Mater.* **2018**, *167*, 426–439. [CrossRef]
55. Ruiz, J.; Baumont, D.; Bernard, P.; Berge-Thierry, C. Combining a Kinematic Fractal Source Model with Hybrid Green's Functions to Model Broadband Strong Ground Motion. *Bull. Seismol. Soc. Am.* **2013**, *103*, 3115–3130. [CrossRef]
56. Hinkle, A.R.; Nöhring, W.G.; Leute, R.; Junge, T.; Pastewka, L. The emergence of small-scale self-affine surface roughness from deformation. *Sci. Adv.* **2020**, *6*, eaax0847. [CrossRef] [PubMed]
57. Aghababaei, R.; Brodsky, E.E.; Molinari, J.F.; Chandrasekar, S. How roughness emerges on natural and engineered surfaces. *MRS Bull.* **2022**, *47*, 1229–1236. [CrossRef]
58. Svahn, F.; Kassman-Rudolphi, Å.; Wallén, E. The influence of surface roughness on friction and wear of machine element coatings. *Wear* **2003**, *254*, 1092–1098. [CrossRef]
59. Brodsky, E.E.; Gilchrist, J.J.; Sagy, A.; Collettini, C. Faults smooth gradually as a function of slip. *Earth Planet. Sci. Lett.* **2011**, *302*, 185–193. [CrossRef]
60. Goebel, T.H.W.; Brodsky, E.E.; Dresen, G. Fault Roughness Promotes Earthquake-Like Aftershock Clustering in the Lab. *Geophys. Res. Lett.* **2023**, *50*, e2022GL101241. [CrossRef]
61. Sagy, A.; Brodsky, E.E.; Axen, G.J. Evolution of fault-surface roughness with slip. *Geology* **2007**, *35*, 283–286. [CrossRef]
62. Candela, T.; Renard, F.; Schmittbuhl, J.; Bouchon, M.; Brodsky, E.E. Fault slip distribution and fault roughness. *Geophys. J. Int.* **2011**, *187*, 959–968. [CrossRef]
63. Eijsink, A.M.; Kirkpatrick, J.D.; Renard, F.; Ikari, M.J. Fault surface morphology as an indicator for earthquake nucleation potential. *Geology* **2022**, *50*, 1356–1360. [CrossRef]
64. Shiari, B.; Miller, R.E. Multiscale modeling of crack initiation and propagation at the nanoscale. *J. Mech. Phys. Solids* **2016**, *88*, 35–49. [CrossRef]
65. Hedayati, R.; Hosseini-Toudeshky, H.; Sadighi, M.; Mohammadi-Aghdam, M.; Zadpoor, A.A. Multiscale modeling of fatigue crack propagation in additively manufactured porous biomaterials. *Int. J. Fatigue* **2018**, *113*, 416–427. [CrossRef]
66. Slepyan, S.I. Principle of maximum energy dissipation rate in crack dynamics. *J. Mech. Phys. Solids* **1993**, *41*, 1019–1033. [CrossRef]
67. Xie, H. *Fractals in Rock Mechanics*, 1st ed.; CRC Press: Boca Raton, FL, USA, 1993.
68. Lu, C. Some notes on the study of fractals in fracture. In Proceedings of 5th Australasian Congress on Applied Mechanics, ACAM 2007, Brisbane, Australia, 10–12 December 2007; pp. 234–239.
69. Williford, R.E. Multifractal fracture. *Scr. Metall.* **1988**, *22*, 1749–1754. [CrossRef]
70. Williford, R.E. Scaling similarities between fracture surfaces, energies, and a structure parameter. *Scr. Metall.* **1988**, *22*, 197–200. [CrossRef]
71. Bai, Y.; Lu, C.; Ke, F.; Xia, M. Evolution induced catastrophe. *Phys. Lett. A* **1994**, *185*, 196–200. [CrossRef]
72. Ohnaka, M. *The Physics of Rock Failure and Earthquakes*; Cambridge University Press: Cambridge, UK, 2013. [CrossRef]
73. Noda, A.; Saito, T.; Fukuyama, E.; Urata, Y. Energy-based scenarios for great thrust-type earthquakes in the Nankai trough subduction zone, southwest Japan, using an interseismic slip-deficit model. *J. Geophys. Res. Solid Earth* **2021**, *126*, e2020JB020417. [CrossRef]
74. Underwood, E.E.; Banerji, K. Fractals in fractography. *Mater. Sci. Eng.* **1986**, *80*, 1–14. [CrossRef]
75. Venkataraman, A.; Kanamori, H. Observational constraints on the fracture energy of subduction zone earthquakes. *J. Geophys. Res.* **2004**, *109*, B05302. [CrossRef]
76. Dobrovolsky, I.P.; Gershenzon, N.I.; Gokhberg, M.B. Theory of electrokinetic effects occurring at the final stage in the preparation of a tectonic earthquake. *Phys. Earth Planet. Inter.* **1989**, *57*, 144–156. [CrossRef]
77. Pulinets, S.; Ouzounov, D. Lithosphere–Atmosphere–Ionosphere Coupling (LAIC) model—An unified concept for earthquake precursors validation. *J. Asian Earth Sci.* **2011**, *41*, 371–382. [CrossRef]
78. Loukidis, A.; Tzagkarakis, D.; Kyriazopoulos, A.; Stavrakas, I.; Triantis, D. Correlation of Acoustic Emissions with Electrical Signals in the Vicinity of Fracture in Cement Mortars Subjected to Uniaxial Compressive Loading. *Appl. Sci.* **2023**, *13*, 365. [CrossRef]
79. Ohsawa, Y. Regional Seismic Information Entropy to Detect Earthquake Activation Precursors. *Entropy* **2018**, *20*, 861. [CrossRef]
80. Ramírez-Rojas, A.; Flores-Márquez, E.L.; Sarlis, N.V.; Varotsos, P.A. The Complexity Measures Associated with the Fluctuations of the Entropy in Natural Time before the Deadly México M8.2 Earthquake on 2017. *Entropy* **2018**, *20*, 477. [CrossRef]
81. Posadas, A.; Pasten, D.; Vogel, E.E.; Saravia, G. Earthquake hazard characterization by using entropy: Application to northern Chilean earthquakes. *Nat. Hazards Earth Syst. Sci.* **2023**, *23*, 1911–1920. [CrossRef]
82. Contreras-Reyes, E.; Flueh, E.R.; Grevemeyer, I. Tectonic control on sediment accretion and subduction off south central Chile: Implications for coseismic rupture processes of the 1960 and 2010 megathrust earthquakes. *Tectonics* **2010**, *29*, 1–27. [CrossRef]
83. Contreras-Reyes, E.; Carrizo, D. Control of high oceanic features and subduction channel on earthquake ruptures along the Chile–Peru subduction zone. *Phys. Earth Planet. Inter.* **2011**, *186*, 49–58. [CrossRef]
84. Zielke, O.; Galis, M.; Mai, P.M. Fault roughness and strength heterogeneity control earthquake size and stress drop. *Geophys. Res. Lett.* **2017**, *44*, 777–783. [CrossRef]
85. Menichelli, I.; Corbi, F.; Brizzi, S.; Van Rijsingen, E.; Lallemand, S.; Funiciello, F. Seamount Subduction and Megathrust Seismicity: The Interplay Between Geometry and Friction. *Geophys. Res. Lett.* **2023**, *50*, e2022GL102191. [CrossRef]
86. Lay, T.; Kanamori, H.; Ammon, C.J.; Koper, K.D.; Hutko, A.R.; Ye, L.; Yue, H.; Rushing, T.M. Depth-varying rupture properties of subduction zone megathrust faults. *J. Geophys. Res.* **2012**, *117*, 1–21. [CrossRef]

87. Hayes, G.P.; Herman, M.W.; Barnhart, W.D.; Furlong, K.P.; Riquelme, S.; Benz, H.M.; Bergman, E.; Barrientos, S.; Earle, P.S.; Samsonov, S. Continuing megathrust earthquake potential in Chile after the 2014 Iquique earthquake. *Nature* **2014**, *512*, 295–298. [CrossRef] [PubMed]

88. Melgar, D.; Riquelme, S.; Xu, X.; Baez, J.C.; Geng, J.; Moreno, M. The first since 1960: A large event in the Valdivia segment of the Chilean Subduction Zone, the 2016 M7.6 Melinka earthquake. *Earth Planet. Sci. Lett.* **2017**, *474*, 68–75. [CrossRef]

89. Li, S.; Barnhart, W.D.; Moreno, M. Geometrical and Frictional Effects on Incomplete Rupture and Shallow Slip Deficit in Ramp-Flat Structures. *Geophys. Res. Lett.* **2018**, *45*, 8949–8957. [CrossRef]

90. Moreno, M.; Li, S.; Melnick, D.; Bedford, J.R.; Baez, J.C.; Motagh, M.; Metzger, S.; Vajedian, S.; Sippl, C.; Gutknecht, B.D.; et al. Chilean megathrust earthquake recurrence linked to frictional contrast at depth. *Nature Geosci.* **2018**, *11*, 285–290. [CrossRef]

91. Iquebal, A.S.; Sagapuram, D.; Bukkapatnam, S.T.S. Surface plastic flow in polishing of rough surfaces. *Sci. Rep.* **2019**, *9*, 10617. [CrossRef]

92. Chen, Y.; Yang, H. Numerical simulation and pattern characterization of nonlinear spatiotemporal dynamics on fractal surfaces for the whole-heart modeling applications. *Eur. Phys. J. B* **2016**, *89*, 181. [CrossRef]

93. Bruhat, L.; Klinger, Y.; Vallage, A.; Dunham, E.M. Influence of fault roughness on surface displacement: From numerical simulations to coseismic slip distributions. *Geophys. J. Int.* **2020**, *220*, 1857–1877. [CrossRef]

94. Morad, D.; Sagy, A.; Tal, Y.; Hatzor, Y.H. Fault roughness controls sliding instability. *Earth Planet. Sci. Lett.* **2022**, *579*, 117365. [CrossRef]

95. Ito, Y.; Obara, K.; Shiomi, K.; Sekine, S.; Hirose, H. Slow Earthquakes Coincident with Episodic Tremors and Slow Slip Events. *Science* **2007**, *315*, 503–506. [CrossRef]

96. Ruiz, S.; Metois, M.; Fuenzalida, A.; Ruiz, J.; Leyton, F.; Grandin, R.; Vigny, C.; Madariaga, R.; Campos, J. Intense foreshocks and a slow slip event preceded the 2014 Iquique Mw 8.1 earthquake. *Science* **2014**, *345*, 1165–1169. [CrossRef]

97. Selvadurai, P.A.; Glaser, S.D.; Parker, J.M. On factors controlling precursor slip fronts in the laboratory and their relation to slow slip events in nature. *Geophys. Res. Lett.* **2017**, *44*, 2743–2754. [CrossRef]

98. Voss, N.; Dixon, T.H.; Liu, Z.; Malservisi, R.; Protti, M.; Schwartz, S. Do slow slip events trigger large and great megathrust earthquakes? *Sci. Adv.* **2018**, *4*, eaat8472. [CrossRef]

99. Luo, Y.; Liu, Z. Slow-Slip Recurrent Pattern Changes: Perturbation Responding and Possible Scenarios of Precursor toward a Megathrust Earthquake. *Geochem. Geophys. Geosystems* **2019**, *20*, 852–871. [CrossRef]

100. Mogi, K. Some features of recent seismic activity in and near Japan (2) activity before and after great earthquakes. *Bull. Earthq. Res. Inst.* **1969**, *47*, 395–417.

101. Kanamori, H. The nature of seismicity patterns before large earthquakes. In *Earthquake Prediction: An International Review, Maurice Ewing Series*; Simpson, D.W., Richards, P.G., Eds.; American Geophysical Union: Washington DC, USA, 1981; Volume 4, pp. 1–19. [CrossRef]

102. Schurr, B.; Moreno, M.; Tréhu, A.M.; Bedford, J.; Kummerow, J.; Li, S.; Oncken, O. Forming a Mogi Doughnut in the Years Prior to and Immediately Before the 2014 M8.1 Iquique, Northern Chile, Earthquake. *Geophys. Res. Lett.* **2020**, *47*, e2020GL088351. [CrossRef]

103. Cochran, E.S.; Page, M.T.; van der Elst, N.J.; Ross, Z.E.; Trugman, D.T. Fault Roughness at Seismogenic Depths and Links to Earthquake Behavior. *Seism. Rec.* **2023**, *3*, 37–47. [CrossRef]

104. Morales-Yáñez, C.; Bustamante, L.; Benavente, R.; Sippl, C.; Moreno, M. B-value variations in the Central Chile seismic gap assessed by a Bayesian transdimensional approach. *Sci. Rep.* **2022**, *12*, 21710. [CrossRef] [PubMed]

105. Scholz, C.H. On the stress dependence of the earthquake b value. *Geophys. Res. Lett.* **2015**, *42*, 1399–1402. [CrossRef]

106. Dong, L.; Zhang, L.; Liu, H.; Du, K.; Liu, X. Acoustic Emission b Value Characteristics of Granite under True Triaxial Stress. *Mathematics* **2022**, *10*, 451. [CrossRef]

107. Vallianatos, F.; Tzanis, A. Electric Current Generation Associated with the Deformation Rate of a Solid: Preseismic and Coseismic Signals. *Phys. Chem. Earth* **1998**, *23*, 933–938. [CrossRef]

108. Stavrakas, I.; Triantis, D.; Agioutantis, Z.; Maurigiannakis, S.; Saltas, V.; Vallianatos, F.; Clarke, M. Pressure stimulated currents in rocks and their correlation with mechanical properties. *Nat. Hazards Earth Syst. Sci.* **2004**, *4*, 563–567. [CrossRef]

109. Pasiou, E.D.; Triantis, D. Correlation between the electric and acoustic signals emitted during compression of brittle materials. *Frat. Ed Integrità Strutt.* **2017**, *40*, 41–51. [CrossRef]

110. Nelson, R.A. 1—Evaluating Fractured Reservoirs: Introduction. In *Geologic Analysis of Naturally Fractured Reservoirs*, 2nd ed.; Gulf Professional Publishing: Houston, TX, USA, 2001; pp. 1–100. [CrossRef]

111. Finkbeiner, T.; Zoback, M.; Flemings, P.; Stump, B. Stress, ore pressure, and dynamically constrained hydrocarbon columns in the South Eugene Island 330 field, northern Gulf of Mexico. *AAPG Bull.* **2001**, *85*, 1007–1031.

112. Donzé, F.V.; Tsopela, A.; Guglielmi, Y.; Henry, P.; Gout, C. Fluid migration in faulted shale rocks: Channeling below active faulting threshold. *Eur. J. Environ. Civ. Eng.* **2020**, *27*, 1–15. [CrossRef]

MDPI AG
Grosspeteranlage 5
4052 Basel
Switzerland
Tel.: +41 61 683 77 34

*Geosciences* Editorial Office
E-mail: geosciences@mdpi.com
www.mdpi.com/journal/geosciences

# Salamanders: Distribution, Diversity, and Conservation

# Salamanders: Distribution, Diversity, and Conservation

Editor

**Enrico Lunghi**

Basel • Beijing • Wuhan • Barcelona • Belgrade • Novi Sad • Cluj • Manchester

*Editor*
Enrico Lunghi
University of L'Aquila
L'Aquila
Italy

*Editorial Office*
MDPI
St. Alban-Anlage 66
4052 Basel, Switzerland

This is a reprint of articles from the Special Issue published online in the open access journal *Animals* (ISSN 2076-2615) (available at: https://www.mdpi.com/journal/animals/special_issues/ Salamanders_Distribution_Diversity_Conservation).

For citation purposes, cite each article independently as indicated on the article page online and as indicated below:

Lastname, A.A.; Lastname, B.B. Article Title. *Journal Name* **Year**, *Volume Number*, Page Range.

**ISBN 978-3-7258-0217-3 (Hbk)**
**ISBN 978-3-7258-0218-0 (PDF)**
**doi.org/10.3390/books978-3-7258-0218-0**

Cover image courtesy of Francesco Bacci

# Contents

*Article*

# Evolutionary Insights into the Relationship of Frogs, Salamanders, and Caecilians and Their Adaptive Traits, with an Emphasis on Salamander Regeneration and Longevity

Bin Lu

Chengdu Institute of Biology, Chinese Academy of Sciences, Chengdu 610041, China; lvbin@cib.ac.cn

**Simple Summary:** Amphibians have unique traits, such as regeneration and longevity in salamanders, frequent vocalization in frogs, and degenerative vision in caecilians. The genetic basis of these traits is not well understood. This study aimed to investigate the genetic changes underlying these unique traits, especially salamanders' regeneration and longevity, by comparing the genes of amphibians to other vertebrates. I found that salamander genomes have undergone accelerated adaptive evolution, especially for development-related genes. Several salamanders' genes are under positive selection and/or share mutations with other long-lived and regenerative vertebrates, suggesting that these genes are important for these unique traits. This study could help us to better understand the mechanisms of regeneration and aging, which could lead to the development of new ways to improve human health and well-being.

**Abstract:** The extant amphibians have developed uncanny abilities to adapt to their environment. I compared the genes of amphibians to those of other vertebrates to investigate the genetic changes underlying their unique traits, especially salamanders' regeneration and longevity. Using the well-supported Batrachia tree, I found that salamander genomes have undergone accelerated adaptive evolution, especially for development-related genes. The group-based comparison showed that several genes are under positive selection, rapid evolution, and unexpected parallel evolution with traits shared by distantly related species, such as the tail-regenerative lizard and the longer-lived naked mole rat. The genes, such as *EEF1E1*, *PAFAH1B1*, and *OGFR*, may be involved in salamander regeneration, as they are involved in the apoptotic process, blastema formation, and cell proliferation, respectively. The genes *PCNA* and *SIRT1* may be involved in extending lifespan, as they are involved in DNA repair and histone modification, respectively. Some genes, such as *PCNA* and *OGFR*, have dual roles in regeneration and aging, which suggests that these two processes are interconnected. My experiment validated the time course differential expression pattern of *SERPINI1* and *OGFR*, two genes that have evolved in parallel in salamanders and lizards during the regeneration process of salamander limbs. In addition, I found several candidate genes responsible for frogs' frequent vocalization and caecilians' degenerative vision. This study provides much-needed insights into the processes of regeneration and aging, and the discovery of the critical genes paves the way for further functional analysis, which could open up new avenues for exploiting the genetic potential of humans and improving human well-being.

**Keywords:** extant amphibians; evolution; regeneration; aging

**Citation:** Lu, B. Evolutionary Insights into the Relationship of Frogs, Salamanders, and Caecilians and Their Adaptive Traits, with an Emphasis on Salamander Regeneration and Longevity. *Animals* **2023**, *13*, 3449. https://doi.org/10.3390/ani13223449

Academic Editor: Enrico Lunghi

Received: 25 August 2023
Revised: 1 November 2023
Accepted: 6 November 2023
Published: 8 November 2023



## 1. Introduction

Modern amphibians (Lissamphibia) possess some of the most intriguing features among vertebrates. Salamanders, including both aquatic larval axolotl and metamorphosed newts (*Cynops*; *Notopthalmus*; and *Pleurodeles*, the Iberian newt), have the remarkable ability to regenerate entire limbs, a tail, and even parts of the brain, eye, and heart [1,2]. However, there are some key differences in the regeneration modes of newts and axolotls. Newts regenerate lens and muscle tissues through dedifferentiation, while axolotls regenerate muscle tissues through stem cell activation [3–6]. It remains to be determined whether these differences reflect a higher degree of cell plasticity in newts. In addition, newts can regenerate more body parts than axolotls. Axolotls can only regenerate the eye lens during the first two weeks after hatching [5], while newts can regenerate the eye lens throughout their lifespan, and their ability to regenerate the lens does not decline with age or the number of lens removal/regeneration cycles [6]. Research using salamanders as a model system has gained tremendous insights into the developmental and physiological process of regeneration [7–9]. We now know that the extracellular matrix (ECM) plays a critical role in directing cell growth and migration, and nerves and immune cells are essential for regeneration [8,10–12]. When macrophages were removed, salamanders lost their ability to regenerate and instead formed scar tissue [8]. Other vertebrates, including human, lose their regenerative potential with age, largely because of failure to maintain tissue homeostasis [13]. Furthermore, salamanders have one of the longest life spans for their body size [14], which naturally raises the possibility of genetic interactions underlying regeneration and aging [15,16]. Other traits, such as the diverse body plans of modern amphibians, the vocalization of frogs, and many traits associated with the fossorial lifestyle of caecilians, are just as amazing (Figure 1A). However, the genetic mechanisms behind these traits remain largely elusive. In addition, persisting controversies surrounding the origin of Lissamphibians and the relationships among the three main groups have hindered comparative analysis [17].

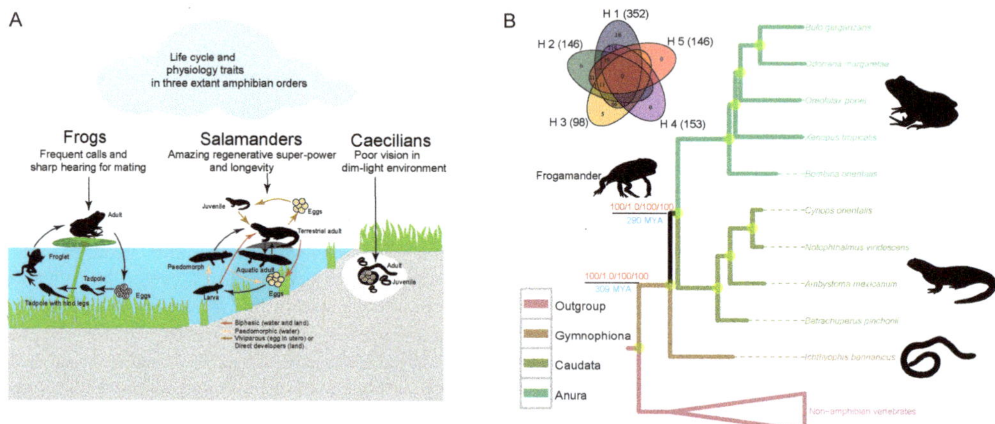



**Figure 1.** Physiology traits and phylogenomic analysis of extant amphibians. (**A**) The most prominent characteristics of frogs, salamanders, and caecilians. (**B**) Phylogenomic tree. The red numbers on the important nodes represent support values from the maximum likelihood (ML), Bayesian (PB), maximum pseudo-likelihood estimates (MPE), and maximum parsimony "Genes as characters" (GAC) methods. The blue numbers below the branch represent the origin date of the extant amphibians (309 MYA) and the living frogs and salamanders (290 MYA). Upper left is a Venn diagram of gene clusters supporting the different phylogenetic hypotheses, as shown in Figure S1. The numbers in brackets represent the quantity of genes that cannot reject a specific hypothesis.

The recent boom in genomic work provides an excellent opportunity to explore the genetic architecture of major evolutionary changes. Amphibians have only one published salamander genome [18], which is largely limited by their extremely large genome sizes. Alternative approaches, such as transcriptome sequencing, are commonly used [19,20]. The enormous amount of available genomic resources enables large-scale and in-depth comparative analyses. Furthermore, several distantly related vertebrates share some phenotypic traits with amphibians; for instance, lizards can regenerate tails, and naked mole rats and Brandt's bats also have long lifespans [21]. This provides opportunities to examine potentially shared genetic mechanisms of the same trait. Although it has been argued that parallel/convergent evolution at the molecular level is often rare because there could be several genomic routes leading to the same phenotype, shared genetic changes would provide strong evidence for adaptive evolution [22].

Using bioinformatics and a comparative analysis of genomic data, I explored the potential genetic mechanisms of major traits in modern amphibians, in particular, the regenerative healing and anti-senescence capabilities of salamanders. My specific objectives were (1) to address the long-debated question of how the three orders of modern amphibians (Anura, Gymnophiona, Caudata) are related to each other as well as to other vertebrate groups; and (2) based on the resulting tree, I explored the genetic architecture of several traits of the three groups of modern amphibians, with a focus on the regeneration capacity and longevity of salamanders, as well as the vocalization and hearing of frogs and the degenerative visual function of caecilians.

## 2. Materials and Methods

### 2.1. Sample Collection, Transcriptome Sequencing, and De Novo Assembly

Transcriptome sequences for five amphibian species were acquired, including Yunnan caecilians (*Ichthyophis bannanicus*; collected in Jinghong, Yunnan Province, China, in 2018), Baoxing tooth toads (*Oreolalax popei*; collected from Baoxing County, Sichuan Province, China, in 2018), oriental fire-bellied toads (*Bombina orientalis*; collected from Qingdao, Shandong Province, China, in 2018), stream salamanders (*Batrachuperus pinchonii*; collected from Baoxing County, Sichuan Province, China, in 2018), and Chinese fire-bellied newts (*Cynops orientalis*; collected from Wuxue, Hubei Province, China, in 2018). All specimens were identified using a morphological method [23]. For evolutionary analyses, I collected samples from two adult individuals of each species, one male and one female. Individuals were euthanized and dissected immediately after death. RNA was separately extracted from the brain, liver, heart, skeletal muscle, and gonad tissues of these individuals using the Trizol protocols (Invitrogen, Carlsbad, CA, USA) and mixed in approximately equal quantities. For the limb regeneration experiment, adult newts were anesthetized and placed in a sterile dish. The right forelimb zeugopod was amputated using sterile scissors. The newts were then returned to their home tanks and monitored for recovery. The proximal healing tissue was harvested at 0 h, 1 day, 5 days, 10 days, and 20 days post-amputation, and RNA was separately extracted using the Trizol protocols (Invitrogen, Carlsbad, CA, USA). Three newts were used for each time window of limb regeneration. The concentration and integrity of total RNA was examined using agarose gel electrophoresis, a NanoPhotometer spectrophotometer (IMPLEN, Westlake Village, CA, USA), and an Agilent Bioanalyzer 2100 system (Agilent Technologies, Santa Clara, CA, USA). cDNA libraries were constructed and subsequently sequenced on an Illumina HiSeq2000 platform, which was carried out by Novogene Inc. (Beijing, China). Approximately 4 G of raw data of paired-end reads was obtained for each transcriptome.

Quality filtration and de novo assembly were performed. The raw reads were first cleaned by filtering out the adapter sequences using Trimmomatic v0.36 [24]. Reads with more than 10% unknown base calls and low-quality reads (<Q20) were also discarded. Preliminary assemblies were produced using Trinity v2.9.0 [25] according to the published protocols [26]. To ensure the quality of the assembly, I also performed a multiple K-mer and multiple coverage cutoff values assembly using ABySS v2.0 [27,28]. Combinations of five K-mer lengths (21, 31, 41, 51, and 61) and six coverage cutoff values (2, 3, 6, 10, 15, and 20) were used, and 30 raw assemblies were produced. Sequence overlaps and redundancies were then eliminated to produce the final assembly using the programs CD-HIT-EST v4.8.1 [29] and CAP3 v10.2011 [30].

### 2.2. Phylogenetic Analyses and Date Estimation

### 2.2.1. Construction of the Raw Phylogenomic Datasets

Genomic data or transcriptome data of five additional amphibian species were obtained from previous studies, including the green odorous frog (*Odorrana margaretae*) [31], Asiatic toads (*Bufo gargarizans*) [32], axolotl (*Ambystoma mexicanum*) [33], eastern newt (*Notophthalmus viridescens*) [34], and western clawed frog (*Xenopus tropicalis*) [35]. A total of ten species were used to represent the three extant orders. Humans and the green anole were used as outgroups.

Putative orthologous genes were identified using the program HaMSTR v1.6.0 (Hidden Markov Model-based Search for Orthologs using Reciprocity) [36] based on its core ortholog database. Amino acid alignments were generated using Clustal Omega v1.2.4 [37] with the default parameters. Codon alignments were generated based on the amino acid alignments. A set of randomly sampled genes was selected, and their alignments were manually checked to ensure the performance of the program. Two raw supermatrix datasets were constructed, one by concatenating the coding regions of the nucleotide sequences (CDS) and the other by concatenating their corresponding amino acid sequences.

### 2.2.2. Data Filtration

To remove or reduce potentially detrimental effects of several confounding factors in phylogenomic reconstruction, I filtered the data using four approaches. First, it is well known that uneven distributions of missing data can affect phylogenetic inferences [38,39]. I used the Alistat v1.7 to quantify the sparseness of the concatenated alignments. Second, the completeness score value ranging from 0 to 1 was estimated. Selecting informative subsets of supermatrices increases the chances of finding the correct trees [40]. I also used mare v0.1.2 [40] to assess the information content of genes in the supermatrices by measuring potential phylogenetic signals and data coverage. Genes with no information content were eliminated. Third, I investigated the influence of base compositional heterogeneity (CH) on the phylogenetic reconstruction using BaCoCa v1.1 [41]. The RCFV (Relative Composition Frequency Variability) values across all taxa were calculated for the complete datasets [42]. The higher the RCFV value, the higher the degree of compositional heterogeneity. Chi-square tests were conducted, and a significance level was set at 0.01, where a *p*-value below 0.01 indicates that the composition significantly deviates from homogeneity, according to the author's suggestion. Heterogeneous genes were excluded from the supermatrices. Fourth, the common phylogenetic assumption assumes that evolutionary processes can be modeled and DNA sequences "evolved under globally stationary, reversible and homogeneous (SRH) conditions" [43–45]. In fact, CH across the sequences is common, and the evolutionary process is more complex than the model assumption. Non-SRH conditions will introduce systematic errors during the tree reconstruction process and even generate false phylogenetic conclusions [46,47]. I used the program SymTest to assess whether the concatenated sequences are consistent with evolution under SRH conditions. In addition, *p*-value heatmaps were generated to visualize which sequence pairs could be assumed to have evolved under SRH conditions and which ones violated this assumption.

After eliminating "noisy" genes and sites, select optimal data subsets were extracted from the raw supermatrices of both the CDS and amino acid sequence and then used in the phylogenomic analyses.

### 2.2.3. Phylogenomic Tree Construction

I analyzed the two filtered supermatrices using both the maximum likelihood (ML) and Bayesian inference (BI) methods. ML trees were inferred using RAxML v8.0.19 [48] with the site-homogeneous JTT-F+G and GTR+G models, and the BI tree was inferred using PhyloBayes v3.3 [49] with the site-heterogeneous CAT-GTR model. The best-fit partitioning schemes and substitution models for the supermatrices were determined by PartitionFinder v2.1.1 [50]. The Bayesian Information Criterion (BIC) was chosen to compare partitioning schemes and models of molecular evolution.

I also applied ML approaches on each filtered gene to obtain all the individual gene trees. The tree sets were used to evaluate alternative phylogenetic hypotheses and to infer species trees.

I evaluated five phylogenetic hypotheses concerning the relationships among the three orders of modern amphibians. The first three hypotheses assume a monophyletic origin for Lissamphibians. (1) The Batrachia hypothesis proposes a frog–salamander sister relationship [51–54]. (2) The Procera hypothesis proposes a salamander–caecilian sister relationship [55–57]. (3) I also examined the frog–caecilian sister relationship, although no previous study has suggested the existence of this relationship. The last two hypotheses assume a paraphyletic origin for Lissamphibians, which were suggested by paleontological data. (4) Caecilians are more closely related to amniotes than to salamanders and frogs [58], while salamanders and frogs form a monophyletic group [59–61]. (5) Caecilians and salamanders are sisters, and together, they cluster with amniotes without frogs [62,63]. For each gene, RAxML was used to compute the per-site log-likelihood values of the five constrained topologies with the GTR+G substitution model. AU tests [64] were then performed with CONSE v 0.20 [65] to calculate $p$-values. A small $p$-value (0.05) for a topology indicates that the topology is significantly worse than the best one and should be rejected. For each hypothesis, I recorded the genes that could not be rejected by the AU tests. The $p$-values of each individual gene for each topology were converted into heatmaps using Phylcon v1.0 [66].

I also used two supertree approaches to construct species trees from individual gene trees. It is well known that gene trees are not always consistent with species trees due to incomplete lineage sorting and other biological reasons [47]. Here, I applied a pseudo-likelihood approach under the multi-species coalescent model [47,67] to overcome potential limitations and to obtain maximum pseudo-likelihood estimates (MPEs) of species trees from a collection of individual gene trees. I also used the parsimony-based "genes as characters" (GAC) approach [68] to infer species trees. This method treats each gene as a single-ordered multi-state character, and individual gene trees are described by step matrices. All the characters were weighted equally because no reasonable a priori information exists. I assumed that these genes evolved largely independently to satisfy the independent character assumption of parsimony analysis. Finally, parsimony analyses were performed in PAUP. A heuristic search was employed with TBR branch swapping and 100 random addition replicates. Bootstrap values were estimated with 1000 replicates.

2.2.4. Divergence Time Estimates

Data were acquired for ten additional vertebrates, including zebrafish (*Danio rerio*), fugu (*Takifugu rubripes*), Amazon molly (*Poecilia formosa*), coelacanth (*Latimeria chalumnae*), Shedao pit-viper (*Gloydius shedaoensis*) [69], alligator (*Alligator mississippiensis*), chicken (*Gallus gallus*), opossum (*Monodelphis domestica*), elephant (*Loxodonta africana*), and mouse (*Mus musculus*). Most data were downloaded from Ensembl [70]. The dataset included 22 vertebrate species representing all major lineages. This dataset provided multiple calibration points, and it also formed the basis for all downstream analyses.

Multiple calibration points provide more realistic divergence time estimates overall [71]; thus, I used four calibration points, including both soft minimum and maximum time constraints: (1) bird–mammal split (min 312.3 MYA, max 330.4 MYA); (2) human–toad split (min 330.4 MYA, max 350.1 MYA); (3) crocodile–lizard split (min 259.7 MYA, max 299.8 MYA); and (4) the origin of crown Osteichthyes (min 416.0 MYA, max 421.75 MYA). A relaxed molecular clock Bayesian method implemented in MCMCTREE v4.9 [72,73] in PAML v4.9 was used to estimate the divergence time among the three modern amphibian orders. I only used the 2nd codon of the CDS and amino acid datasets to estimate the divergence date to avoid potential problems associated with saturation. The results from molecular clock estimates were compared to the ages of Batrachian fossils from the early Permian.

### 2.3. Test for Selections on Lissamphibia

#### 2.3.1. Lineage-Specific dN/dS and dS Estimates

The evolutionary rate ratio of divergence at nonsynonymous and synonymous sites, dN/dS, is a widely used indicator to measure the magnitude of natural selection acting on protein-coding genes [74]. A lower dN/dS ratio indicates a strong purifying selection against protein changes, and an elevated dN/dS ratio suggests a weak purifying selection or a strong positive selection in favor of protein alterations.

The dN/dS calculation was based on the 22-species dataset, which includes all major lineages of vertebrates and allows us to compare Lissamphibia to other vertebrates. I also constructed a reduced dataset with only the ten amphibians. This dataset has fewer taxa but more and longer orthologs, which allows us to perform an in-depth analysis of genes involved in adaptive and parallel evolution.

The dN/dS ratio for each lineage was estimated using a maximum likelihood approach [75], implemented in CODEML of the PAML 4 software [76]. The free-ratio branch model, which allows the dN/dS ratio to vary for different branches, was run on each ortholog and the concatenated supermatrices. Abnormal values (dN/dS > 5) were excluded from the analysis. A mean dN/dS value for each major group was calculated by averaging the ratio of all terminal branches within the group. I also calculated the number of synonymous substitutions (dS) to represent the rate of neutral evolution. Furthermore, I used PHAST v1.5 [77] to estimate the substitution rates for 4-fold degenerate sites in the concatenated supermatrices.

#### 2.3.2. Rapidly Evolving GO Categories

The dN/dS ratio of the Gene Ontology (GO) category can partially reflect the evolutionary rate of a functional module. I identified rapidly evolving GO categories (REGOs) in the five major groups of vertebrates. During the transition from water to land, vertebrates experienced many major changes in their anatomic structure; therefore, I investigated the evolutionary pattern of developmental functions in extant amphibians and compared the proportion of REGOs involved in development between the major groups. I also compared development-related genes within each major group. Salamanders have an extended life span and remarkable regeneration ability during any stage of their life cycle; therefore, I further compared the adaptive evolutionary rate of GOs in modern salamanders with that in their ancestors. If development- and/or aging-related GOs accelerated adaptive evolution in living salamanders relative to their ancestors, this may suggest that salamanders' superpower continually evolved and improved. Conversely, it may imply that these great changes occurred in the salamanders' ancestors, and extant salamanders just inherited this ability. The Wilcoxon rank sum test was used to compare the dN/dS ratios of a particular gene or GO category to that of the other genes or categories as background.

#### 2.3.3. Identification of Genes under Positive Selection

The branch-site model implemented in the program CODEML v4.9 was used to detect positively selected genes (PSGs) along a specific lineage. The model assumes that foreground branches are under positive selection and background branches evolve in a neutral fashion. I compared the selection model (alternative model, dN/dS > 1) and the neutral model (null model, dN/dS = 1) using a likelihood ratio test (LRT). A chi-square test was conducted for each gene to assess statistical significance. Multiple testing was corrected by applying the false discovery rate (FDR) method. For a gene, if the selection model has a significantly higher likelihood than the neutral model (FDR-adjusted $p$-value < 0.05), this indicates that these genes on the foreground branch might have experienced positive selection.

#### 2.3.4. Identification of Fast-Evolving Genes

To identify the fast-evolving genes (FEGs) in Lissamphibia, I ran a one-ratio branch model and a multi-ratio branch model with CODEML in PAML to estimate the global and local dN/dS ratios, respectively. The one-ratio model assumes that all branches have been evolving at the same rate (null hypothesis), and the multi-ratio model allows the foreground branch to evolve at a different rate (alternative hypothesis). Salamanders, frogs, and caecilians were set as the foreground. The LRT was employed to compare the one-ratio and the multi-ratio branch models. The $p$-values of the chi-square test were adjusted using FDR correction for multiple testing. If a gene had an FDR-adjusted $p$-value of <0.05 and a higher dN/dS in the foreground branch than in the background branch, it was considered an FEG in the foreground branch. Functional enrichment analyses for FEGs were carried out using DAVID bioinformatics resources [78].

### 2.4. Test for Parallel Evolution

I tested patterns of parallel evolution between modern amphibians and three distantly related vertebrates, which share the characteristics of interest with amphibians and have relevant data. The green anole (*Anolis carolinensis*) is capable of regenerating its tail; the Brandt's bat (*Myotis brandtii*) has longevity, super hearing for echolocation, and reduced visual capacity [79,80]; and the naked mole rat (*Heterocephalus glaber*) has longevity, cancer resistance, pain insensitivity, degenerated hearing, and poor visual perception [81–85]. In addition, zebrafish, chicken, and humans were also included for comparison. Ancestral sequence reconstruction was carried out using both ANCESTOR v1.0 [86] and CODEML.

To reduce the influence of uncertainty and individuality, I only focused on amino acid changes shared by all members in a group (e.g., all salamanders examined). I first identified amino acid positions where changes only occurred in two lineages: an amphibian group and a distantly related vertebrate species that share the same phenotypic trait of interest. If they share the same amino acid residue and are derived from the same ancestral amino acid residue, these changes were defined as "parallel". If the changes resulted in different amino acid states in the extant species, the changes were classified as "common". Common changes may be a possible indicator of adaptation accomplished via multiple different amino acids at the same position (reference). CONVERG v2.0 [87] was used to compute the probabilities that the observed parallel substitutions are attributable to random chance.

Genes with parallel changes or common changes were then compared to PSGs and FEGs, and overlapping genes were subjected to further analysis. Orthologs from other bats, including three echolocating bats—the little brown bat (*Myotis lucifugus*), David's myotis (*Myotis davidii*), and vampire bat (*Desmodus rotundus*)—and one non-echolocating bat, the black flying fox (*Pteropus alecto*), were downloaded for comparison. The alignment quality of these candidate genes and the degrees of conservation of the region with amino acid changes were manually checked. Furthermore, candidate genes associated with human/mouse diseases were determined through exploring disease databases, such as OMIM (http://omim.org (accessed on 1 November 2022)) and GeneCards (www.genecards. org (accessed on 1 November 2022)) for human and MGI (http://www.informatics.jax.org (accessed on 1 November 2022)) for mouse, which focus on the relationship between disease phenotype and genotype.

### 2.5. Prediction of Functional Impact of Variants

I used SIFT v6.2.1 [88], PROVEAN v1.1.5 [89], and PolyPhen-2 v2.2.2 [90] to predict the possible effect of unique amino acid substitutions and positively selected and parallel mutations on protein structure and function.

I retrieved information on protein domains and important sites and the 3D structure of proteins from the InterPro database [91] and RCSB PDB [92], respectively. Amino acid conservation scores across the sequences of candidate genes were calculated using the Rate4Site algorithm [93,94] on the ConSurf server [95]. The scores were converted by multiplying by $-1$ so that higher scores indicate higher conservation. Local average conservation was calculated by fitting a cubic smoothing spline to the per-site conservation scores using the smooth.spline method in R. Unique amino acid changes were mapped onto conservation domain plots and the protein 3D structure.

### 2.6. Quantitative Real-Time PCR

The proximal healing tissue was harvested at 0 h, 1 day, 5 days, 10 days, and 20 days post-amputation. RNA was extracted using the Trizol protocols (Invitrogen) and then reverse transcribed using a PrimeScript™ RT reagent Kit (Perfect Real Time, TaKaRa) according to the manufacturer's instructions. The primers for OGFR were forward: 5′-CAGCCCAATGGTGTTCCTGAT-3′; and reverse: 5′-GCGGACAAACCTTTCTTTCA-3′. The primers for SERPINI1 were forward: 5′-ATTTAAGGGATCTATCTGAGGCC-3′; and reverse: 5′-CACCCAGCCATTGATGTGTT-3′. The primers for ACTIN were forward: 5′-AGATCTGGCACCACACCTTC-3′; and reverse: 5′-CAGTGGTACGACCAGAAGCA-3′. The qPCR reactions were performed on a BioRad CFX96 Real-Time PCR Detection System using the EvaGreen 2X qPCR MasterMix (Abm). The thermal cycling parameters were 10 min at 95 °C, followed by 40 cycles of 15 s at 95 °C, 30 s at 52 °C, and 30 s at 72 °C. Three replicates were performed at each time point, and failed reactions were excluded for the analysis. The relative expression level of unigenes was normalized to ACTIN as the comparative Ct and calculated using the delta delta Ct ($2^{-\Delta\Delta Ct}$) method.

## 3. Results and Discussion

### 3.1. Robust Support for Frog–Salamander Sister Relationship

I gathered data from ten amphibian species to infer the origin of and relationships between three extant amphibian orders (Figure 1B). After filtering the one-to-one ortholog dataset, I ultimately utilized 369 coding sequences (CDSs) (second codon positions and first + second codon positions) and 772 protein sequences for the phylogenomic analyses. The datasets had relatively full coverage (completeness score: 0.7–1.0; Figure S2A), high information content (information content: 0.6–1.0; Figure S2B), and a low degree of CH (RCFV value < 0.025; Figure S3) and were least affected in terms of violation of the assumption of evolution under global SRH conditions (Figure S4).

Both the partitioned maximum likelihood (ML) and Bayesian inference (BI) analyses of the concatenated data strongly supported the monophyly of Lissamphibia and the frog–salamander sister relationship (the Batrachia hypothesis; Figure 1B). In addition, two independent supertree analyses, including the maximum pseudo-likelihood estimates (MP-EST) from a collection of individual gene trees [96] and the genes as characters approach (GAC) [68], both found the same topology as the concatenated data tree, providing strong support for the Batrachia hypothesis (Figure 1B).

I also examined the conflicts among individual genes and detected the degrees of support of various genes for the alternative hypotheses. The AU test indicated that the Batrachia hypothesis (H1) was supported by the vast majority of the data, with 352 of 369 CDSs and 708 of 772 proteins supporting the hypothesis (Figures 1B and S5), while the Procera hypothesis (H2), which posits that salamanders and caecilians are sister taxa, was supported by only 146 of 369 CDSs and 274 of 772 proteins. Surprisingly, the paraphyletic origin hypothesis (H4), which posits a close relationship between caecilians and amniotes and a sister relationship between salamanders and frogs, was supported by a large number of genes (153 of 369 CDSs and 374 of 772 proteins). The frog–caecilian sister relationship (H3) received the lowest support, with only 98 of 369 CDSs and 194 of 772 proteins supporting it.

Genome-scale data provide opportunities for resolving difficult phylogenetic relationships [97,98]. My data and analysis provide strong support for the monophyly of Lissamphibia and the salamander–frog sister relationship, consistent with the findings of recent studies [99–101].

### 3.2. Fossil-Compatible Divergence Time of Lissamphibia

A total of 22 vertebrate species (Figure 2) were included in this analysis. I used four calibration points from vertebrates (Figure S6) to estimate the divergence time of Lissamphibia. The results indicate that modern amphibians arose in the late Carboniferous, about 309 MYA (Figures 1B and S6). This is consistent with the fossil record; amphibian-like early temnospondyls (e.g., *Amphibiamus*) appeared in the Carboniferous [102]. The origin of

Batrachia was estimated to take place in the early Permian (~290 MYA; Figures 1B and S6). This is extremely close to the estimated age of the "frogamander", *Gerobatrachus hottoni* (~290 MYA), from Texas during the early Permian [59]. *Gerobatrachus hottoni*, who possessed a large frog-like head and a salamander-like tail, was considered the closest relative of Batrachia [59]. Furthermore, the origins of living frogs and salamanders were dated back to the late Triassic (~200 MYA) and middle Jurassic (~170 MYA), respectively (Figure S6), which was close to or matched the fossil study well (~185 MYA for frog origin and ~170 MYA for salamander origin) [103].

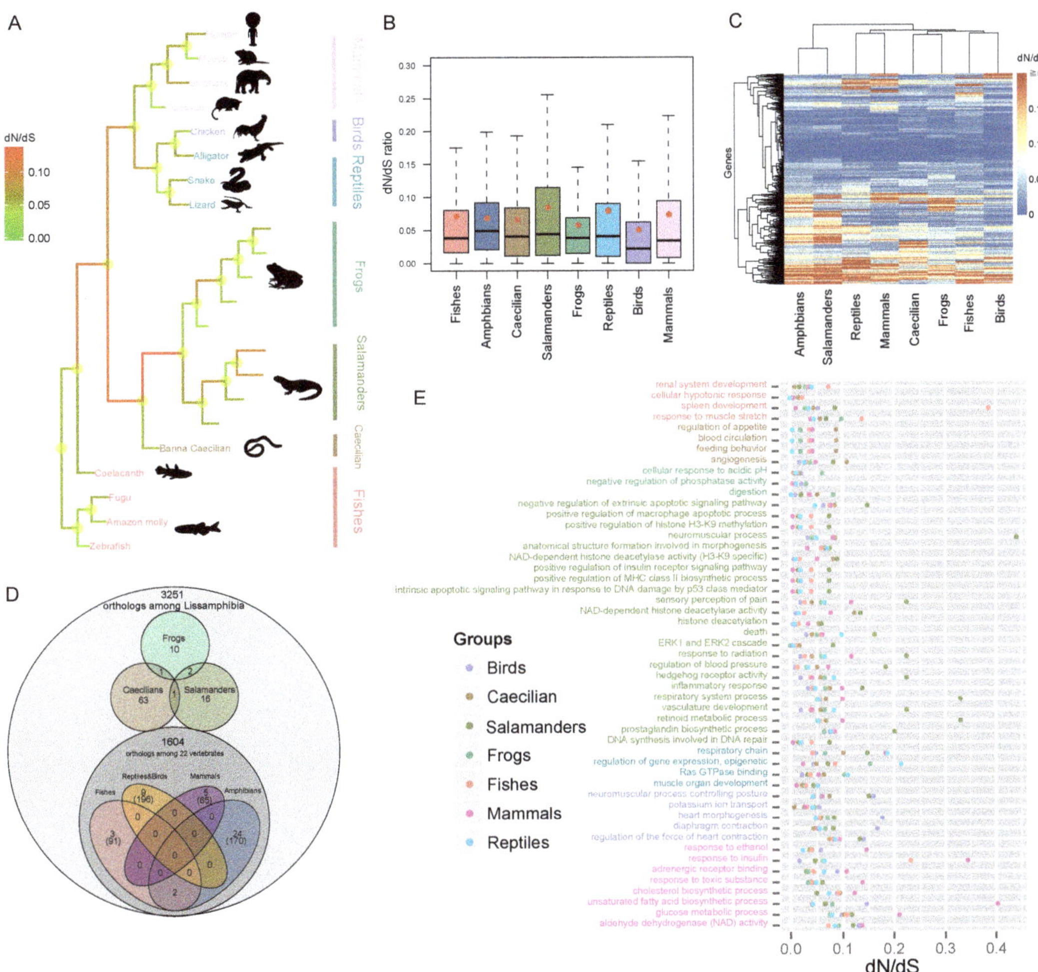


**Figure 2.** Accelerated rate of adaptive evolution in salamanders. (**A**) dN/dS ratio tree of 22 vertebrates shows that salamanders have a higher dN/dS ratio than any other vertebrate group. (**B**) The boxplot of dN/dS ratios for all genes in each major vertebrate group, which shows that salamanders have a much higher mean dN/dS ratio than any other group ($p \leq 0.025$ for all pairwise tests). (**C**) The clustered heatmap of dN/dS ratios of major vertebrate groups also shows that salamanders are distantly related to other vertebrate groups in terms of dN/dS ratio. (**D**) Venn diagram showing number of PSGs and FEGs (in brackets) in Lissamphibia and other vertebrates. (**E**) The rapidly evolving GO categories (REGOs) in Lissamphibia and other vertebrates show that salamanders have a significantly higher number of REGOs than any other vertebrate group.

My age estimates were younger than most previous molecular studies [59] but are mostly consistent with the fossil evidence. For example, the origin of Lissamphibia was estimated to be as young as the early Permian at 294 million years ago (MYA) [54] or as old as the late Devonian period at 369 MYA [104]. Most previous studies used only one or a few genes, which may contribute to the variation and discrepancy.

### 3.3. Rapid Evolution of Salamander Genes

To gain insight into the evolution of Lissamphibia, I compared aspects of their coding gene architecture to those of the other major vertebrate lineages (Figure 2A), including the evolutionary rate and genes or gene clusters that are potentially under positive selection.

The dN/dS ratio has been widely used as a measurement of the rate of adaptive evolution. The colored dN/dS tree shows that salamanders exhibited a higher ratio of dN/dS relative to all the other vertebrate groups (Figure 2A). Furthermore, the distribution of dN/dS ratios among the salamander lineages was significantly larger than that of the other vertebrate groups (Wilcoxon rank sum test (WRST), $p < 0.03$ for all pairwise tests; Figure 2B). Additionally, a cluster analysis for the vertebrate dN/dS ratios found that salamanders were located on a lone branch at the base of the clustering tree and were very different from the other vertebrates (Figure 2C). Evidently, globally elevated evolutionary rates for protein-coding genes likely occurred in salamanders' genomes.

This accelerated evolution can either be due to selection (reduced purifying selection and/or increased positive selection in favor of protein alterations) or a high mutation rate, which can be measured as synonymous mutations [105].

To test whether salamanders have high mutation rates, I compared the mutation rates at synonymous sites (dS) among the major lineages of vertebrates, which has often been used to represent the neutral mutation rate. Interestingly, salamanders exhibited an extremely low dS compared to other vertebrates (Figure S7; WRST $p < 2.9 \times 10^{-40}$ for all pairwise tests). On average, the dS of salamanders was approximately one-third of the dS of caecilians and frogs. I further estimated rates for the four-fold degenerate sites (4D), which evolve the closest to the neutral rate [106]. As expected, the color pattern of the 4D mutation rate tree is very similar to that reconstructed using synonymous sites (Figure S8).

Clearly, salamanders do not have elevated mutation rates, and the high dN/dS ratios are due to selection forces. I further explored aspects of selection forces below. Previous work showed that salamander mitochondrial genomes also display high dN/dS ratios, although their comparisons were restricted to salamanders and frogs [107].

I also noticed that both the common ancestors of Lissamphibia and Batrachia demonstrated higher dN/dS ratios than the extant amphibians (Figure 2A). These are likely a reflection of the rapid genomic changes required during their early adaptation to terrestrial life.

### 3.4. Detection of Selection

I identified a series of fast-evolving genes (FEGs; Table S1) and positively selected genes (PSGs; Table S2) in salamanders, frogs, and caecilians (Figure 2D). The number of candidate genes was conservative because my analyses were conducted on groups of species rather than individual species. This approach helps to reduce the impact of uncertainty and individual variations. Further analysis was conducted on some FEGs and PSGs of interest.

To investigate the lineage-specific evolution of function modules in the genome, I identified rapidly evolving GO categories (REGOs) unique to each major group of vertebrates. For Lissamphibia, many REGOs were associated with developmental processes, including the development of the circulatory (blood circulation, regulation of blood pressure), respiratory, sensory, and immune systems and morphogenesis (Figure 2E). Developmental process-related REGOs accounted for a higher proportion in Lissamphibia (33%) than in other vertebrate groups (13–19%), including birds who have highly specialized anatomic structures (23%) (Figure 3A).

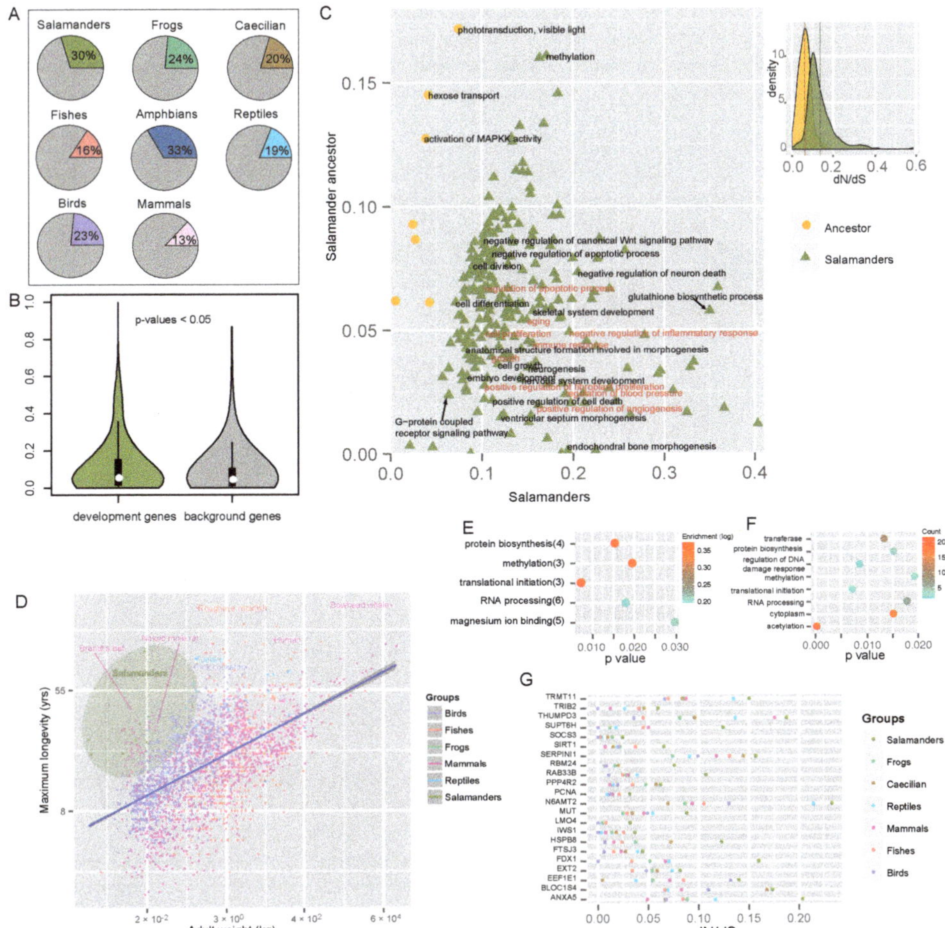

**Figure 3.** Genetic potential underlying salamanders' regenerative capability and longevity. (**A**) Proportion of rapidly evolving GO categories (REGOs) involved in development is highest in salamanders. (**B**) Salamanders' developmental genes evolved significantly faster than other genes. (**C**) Extant salamanders' development-related and aging-related GO categories evolved significantly faster than those of their ancestors. (**D**) Salamanders fall into the category of exception to the rule of "Larger animals live longer", which is shown by the yellow-green color. (**E**) Top five enriched GO terms from salamanders' fast-evolving genes (FEGs). (**F**) Top eight GO terms for salamanders' FEGs. (**G**) Selected salamanders' FEGs may be responsible for regenerative capability and longevity.

## 3.5. Regeneration- and Development-Related Genes in Salamanders

Salamanders have a superior ability to regenerate limbs compared to other vertebrates. I specifically examined genes and GO categories that are linked to development and regeneration in salamanders. I also compared salamanders to another vertebrate, the green anole, which also has a regeneration ability.

The regulation of apoptosis activity plays a key role in local cell dedifferentiation [108,109], which is the first and perhaps the most crucial stage of regeneration. Not surprisingly, I detected several rapidly evolving GO categories associated with the regulation of apoptotic pathways that are unique to salamanders (Figure 2E). Notably, one REGO of regulation of apoptotic cell death was triggered by the tumor suppressor p53 (Figure 2E), which has been proven to be critical for regeneration [110]. In addition, a fast-evolving gene, *EEF1E1*, which is involved in the negative regulation of cell population proliferation and positive regulation of apoptotic processes, shows signals of parallel evolution between salamanders and green anoles (Figure 4A; L108F; $p < 1 \times 10^{-6}$). Furthermore, some REGOs were involved in macrophage regulation and the MHC biosynthetic process (Figure 2E). This is consistent with the essential role of macrophages in successful healing and regeneration [8]. Many FEGs in salamanders, such as *ANXA5*, *SIRT1*, *RAB33B*, *HSPB8*, etc. (Figure 3G), may be involved in the coagulation process, inflammatory processes, and autophagy during the initial stage of regeneration.

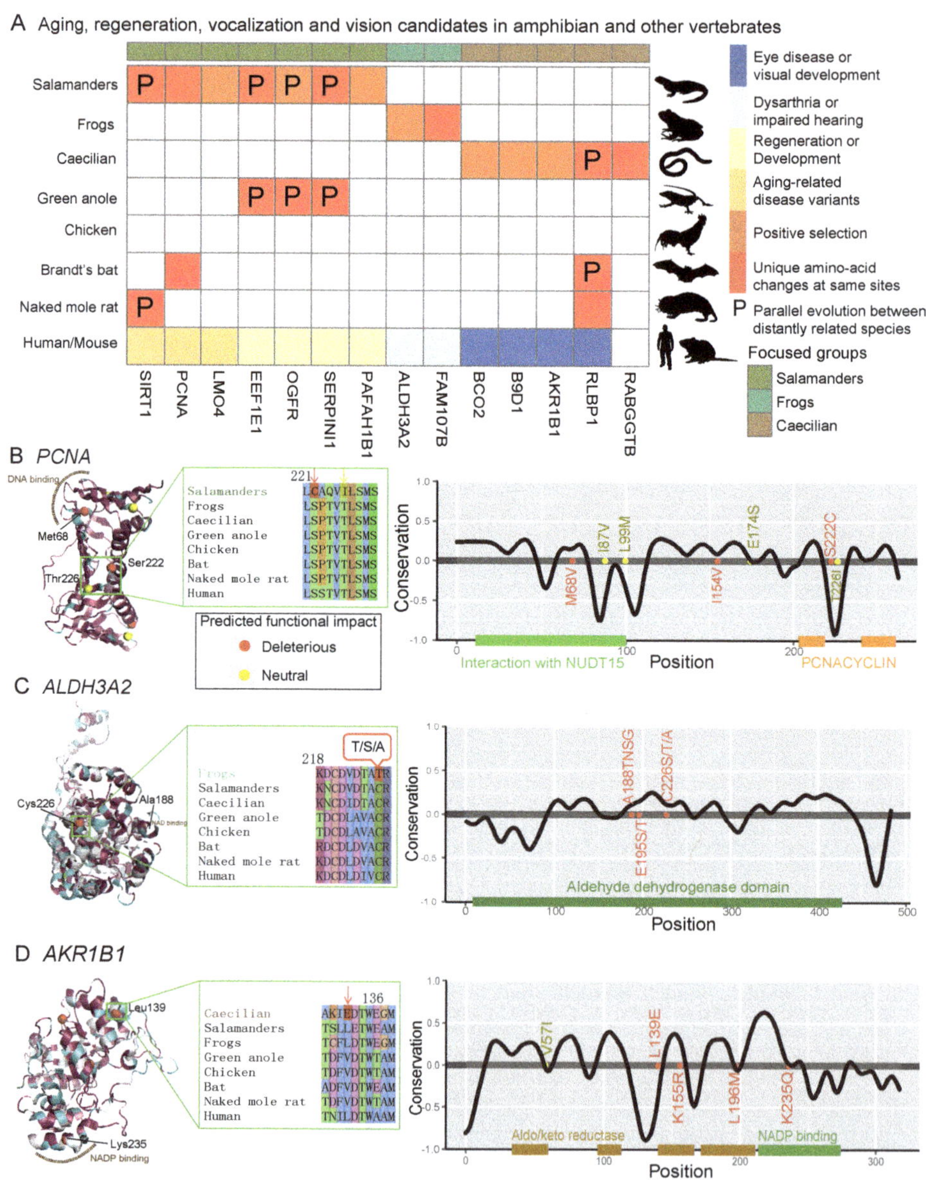


**Figure 4.** Aging, regeneration, vocalization, and vision candidates in extant amphibians and other vertebrates. (**A**) Candidates under positive selection were identified, and those with shared common

or unique position mutations in focused lineages and who have undergone parallel evolution are marked. Associated diseases are also labeled with different colors. (**B–D**) Salamander, frog, and caecilian candidates, respectively. **Left**: Crystal structure of candidates. Purple-red represents conserved regions, and cyan represents variable regions. The important domains and lineage-specific mutations are highlighted in the structure. Residue color represents the strength of the functional impact. Insert: alignment of an example residue with functional impact. **Right**: The conservation scores, domain, and lineage-specific mutation positions. The black curve shows the cubic smoothing spline of the amino acid conservation scores; higher scores indicate higher conservation.

Unlike mammals, salamanders have a scar-free healing capacity, mainly due to their fibroblasts, which form the early blastema rather than scars and control the regeneration process [111]. I detected positive selection acting on the salamanders' *PAFAH1B1* gene, which plays a crucial role in the directed migration of fibroblasts during wound healing and stem cell division [112]. Furthermore, two amino acid changes (L37F and I200V) were specific to salamanders. L37F is located in the LisH domain, which may be required to activate dynein, and I200V is in the WD40 repeat; both mutations were predicted to have strong functional impacts. In addition, the ERK pathway, whose activity is another key difference in cellular reprogramming between salamanders and mammals [9], and the retinoid metabolic process, which may play an important role in the early phase of regeneration [113], were also found to have undergone the most rapid evolution in salamanders (Figure 2E).

The second stage of regeneration is similar to the development process in that it involves proliferation, redifferentiation, and growth. I detected several development-related REGOs that were unique to salamanders, including anatomical structure formation, neuromuscular process, hedgehog receptor activity, and vasculature development, among a few others. (Figure 2E). I detected a strong signal of natural selection on a cell proliferation-related gene, LMO4. One site in its LIM-type zinc finger (Znf) domain (region 23–83), which acts as an interface for protein–protein interactions, was under positive selection (Ala24; Bayes empirical Bayes posterior probabilities (BEBPP) = 0.996). I also detected seven FEGs involved in differentiation and development, including *TRIB2, RBM24, EXT2, SOCS3, BLOC1S4, MUT*, and *SERPINI1* (Figure 3G and Table S1). Furthermore, the gene *SERPINI1*, which is associated with central nervous system development, was detected to have undergone parallel evolution between salamanders and green anoles (Figure 4A; $p < 1 \times 10^{-6}$). A parallel change (F118Y) was predicted to be harmful. The prostaglandin biosynthetic process, a pathway that was recently reported to be closely associated with regenerative capacity in mice (*15-PGDH*) [114] and lizards (*PTGIS* and *PTGS1*) [115], was detected to have evolved most rapidly in salamanders (Figure 2E; *PTGES2* and *PTGS2*). The mutation position of *PTGES2* was shared among salamanders (E183A), echolocating bats (E183D), and naked mole rats (E183D). Unlike *PTGS1*, which is constitutively expressed, *PTGS2* is upregulated during inflammation and contributes to cell proliferation, angiogenesis, apoptosis inhibition, and immune response suppression [116].

Furthermore, I found that the REGOs related to development in salamanders were the most common group in vertebrates (Figure 3A). Did the development-related genes drive the accelerated evolution of salamanders and contribute to regeneration? The answer seems to be yes, as the evolutionary rate of development- and regeneration-related genes were significantly higher than that of the other background genes (Figure 3B; $p = 0.039$). In contrast, a similar pattern was not observed in any other vertebrate group (Figure S9).

How did salamanders obtain such an amazing regenerative ability? To answer this question, I compared the extant salamanders with their most recent common ancestors (MRCAs). Interestingly, many functional modules pertaining to the regeneration process, including the apoptotic process, immune responses, proliferation, angiogenesis, growth, morphogenesis, and aging, have undergone faster adaptive evolution in the extant salamanders relative to their MRCAs (Figure 3C). This is to say, salamanders were constantly evolving their regenerative capacity. Continuous self-improvement may have helped salamanders improve their genetic potential to heal wounds after injury and also delay aging.

For example, the opioid growth factor receptor (*OGFR*) is an important regulator of cell proliferation, tissue growth, cancer, cellular renewal, wound repair, and angiogenesis [117]. In salamanders, *OGFR* likely evolved under positive selection (Figure 4A), and Lys198 was identified as a positively selected site (BEBPP = 0.995). It is intriguing that Val190 and Pro231 of *OGFR* are parallel and common changes in salamanders (V190L, P231K) and lizards that can regenerate their tails (V190L, P231R) (Figure 4A), and all mutations were predicted to probably have a damaging effect on protein function. *OGFR* can interact with *TERF2IP*, a regulator of telomere length that is tightly bound to aging, raising a potential correlation between regeneration ability and aging.

To investigate the potential effect of *OGFR* and *SERPINI1* on regeneration, I used real-time quantitative PCR (qPCR) to measure their expression levels in the healing limb blastemata of Chinese fire-bellied newts (*Cynops orientalis*) at 0 h, 1 day, 5 days, 10 days, and 20 days post-amputation. The expression of both genes varied significantly at almost every time point compared to that at 0 h (Figure 5). Both genes exhibited similar trends along the time course, with gradually decreasing expression over time, reaching a minimum around the fifth day, and then gradually increasing expression. The expression level of one gene, *SERPINI1*, on the 20th day was almost restored to that at 0 h. Previous studies found that activation of *OGFR* prevents cell proliferation [118], and knockdown of *SERPINI1* reduces the outgrowth of neurons [119]. Thus, the observed fluctuations in expression levels over time may point to a dynamic regulation of cell proliferation and axonal growth during limb regeneration.

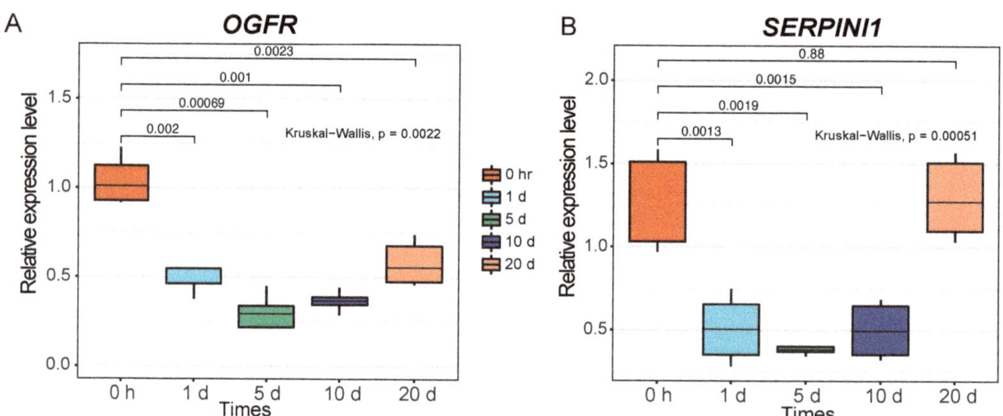


**Figure 5.** Expression patterns measured by qPCR during the time course of limb regeneration. (**A**) *OGFR*. (**B**) *SERPINI1*. Two biological replicates, each with three technical replicates, were measured at each time point.

This study identified several critical genes that are linked to regeneration. Many of these genes were expected and/or consistent with the findings of previous studies [111,120]. Interestingly, three genes (*EEF1E1*, *OGFR*, and *SERPINI1*) displayed significant signals of parallel evolution in salamanders and lizards, both of which possess regenerative capacities, implying a functional convergence for their regenerative mechanisms. Specific changes in these candidates in salamanders may explain, in part, why humans cannot regrow perfectly like salamanders. It is well known that some genes are only activated in specific tissues and/or conditions, such as in healing tissues [121]. Therefore, some genes may have been missed in my screening, which is a limitation of this study. Although I validated two regeneration-related candidates (*OGFR* and *SERPINI1*) by qPCR in healing blastemata and confirmed that their expression levels change significantly during limb regeneration, this was a limited solution.

### 3.6. Longevity-Related Genes in Salamanders

Lifespan commonly correlates with body mass for most animals. This rule is named "Larger animals live longer", although there are several exceptions in mammals, fish, reptiles, and birds (Figure 3D). It is noteworthy that this is not the case for most salamanders (Figure 3D). Olm (*Proteus anguinus*) is possibly the longest-living salamander, which can live over 100 years, even though its body mass is only about 20 g. In contrast, the longest-living mammal (bowhead whale) can live over 200 years but with a 100,000,000 g weight. The average lifespan of salamanders is about 18 years, calculated based on 70 species of salamanders, which is significantly higher than that of frogs (about 12 years; WRST $p = 2.9 \times 10^{-5}$) in the case where there is no significant difference in body mass between the sister groups (WRST $p = 0.219$).

How do salamanders extend their life span? The "rate of living" (ROL) theory of aging (the faster the metabolism, the shorter the lifespan) may partly answer this question (Figure S10; Spearman correlation = $-0.65$, $p = 7.5 \times 10^{-45}$), although there are numerous outliers. Salamanders possess the lowest metabolic rates among tetrapods [122], which can greatly help them reduce the accumulation of damage from reactive oxygen species (ROS) in the body and slow aging. Furthermore, the lowest metabolic rates may also lead to the extremely low mutation rates in salamanders due to the link between the frequency of oxidative damage and the likelihood of DNA change [123]. In other words, longevity seems to be negatively correlated with the mutation rate, which is similar to longer-lived rockfish (although the authors mainly focused on mitochondrial mutations) [124].

It is now clear that the genetic mechanisms underlying aging can be conserved across distantly related species [125]. Increasing evidence indicates that epigenetic factors have critical roles in aging [126]. I examined genes and GO categories that are related to epigenetics and compared them between salamanders and other long-living vertebrates.

A large number of REGOs in salamanders are involved in histone modification, particularly methylation and acetylation, and were significantly enriched (Figure 3E,F). The FEGs included *SIRT1*, *IWS1*, *SUPT6H*, *FTSJ3*, *THUMPD3*, and *TRMT11* (Figure 3G).

I further examined several candidate genes in more detail. *SIRT1*, also known as NAD-dependent deacetylase sirtuin-1, plays a critical role in metabolic regulation. It is an important genetic modulator for aging and longevity in humans, mice, worms, flies, and yeast [127]. *SIRT1* evolved much more rapidly in salamanders (dN/dS = 0.065) than in any other animals (dN/dS ≤ 0. 037). I uncovered two interesting amino acid changes occurring at crucial regions of *SIRT1* in salamanders. One mutation (K375R) located in the catalytic core small domain (365–417) of the sirtuin family domain (deacetylase sirtuin-type), a key component responsible for NAD+ binding and the histone deacetylation reaction (associated with aging), was predicted to be possibly harmful to protein function. In addition, K375R is a parallel change that occurred in both salamanders and the long-lived and cancer-resistant naked mole rat. It is possible that K375R may be directly related to longevity in salamanders. Another salamander-specific mutation (I227L) is in a nuclear localization signal (223–230) and a region that is the site of interaction with *CCAR2* (*DBC1*), a partner that can inhibit *SIRT1* deacetylase activity. I227L may decrease the interaction between *SIRT1* and *CCAR2* and delay aging. In addition, *SIRT1* is involved in the insulin-like signaling (ILS) pathway, which evolved rapidly in salamanders and is a metabolic signaling pathway involved in controlling life span in many species [125]. *SOCS3*, another FEG, is also involved in ILS regulation (Figure 3G). Notably, *IWS1* and *SUPT6H*, both evolving the fastest in salamanders, interact with each other and form a complex to control histone modifications [128]. In addition, *IWS1* has at least eight salamander-specific mutations (K581R, E605D, V612A, S646G, K653R, T700S, R716K, and V763L).

The DNA damage response and repair, which counter the constant assaults by endogenous and environmental agents on DNA, are critical for maintaining genetic stability and are considered to be the key to aging and longevity. I detected that DNA repair-related GOs evolved most rapidly in salamanders (Figure 2E; *PCNA*, *SIRT1*, and *PPP4R2*). Proliferating cell nuclear antigen (*PCNA*), which is essential for DNA replication and damage repair [129], was associated with aging in humans, rats, and long-lived bowhead whales [130,131]. It is highly conserved even between plants and animals, indicating a strong selection pressure for structure conservation in order to interact with its partners. Correspondingly, numerous mutations (Figure 6A) in *PCNA* resulted in impaired DNA replication and damage repair [132]. Notably, salamander *PCNA* evolved at a rate more than two times faster than any other vertebrates (Figure 3G; and Table S1), which results in up to seven unique amino acid mutations present in all salamander species, and of these, three were predicted to be deleterious (Figures 4B and 6A).

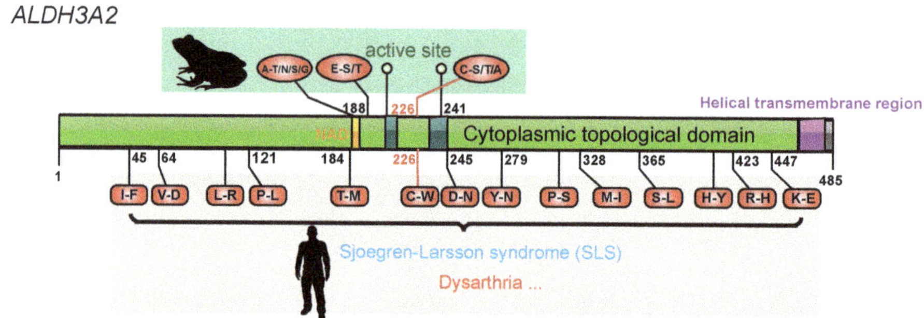


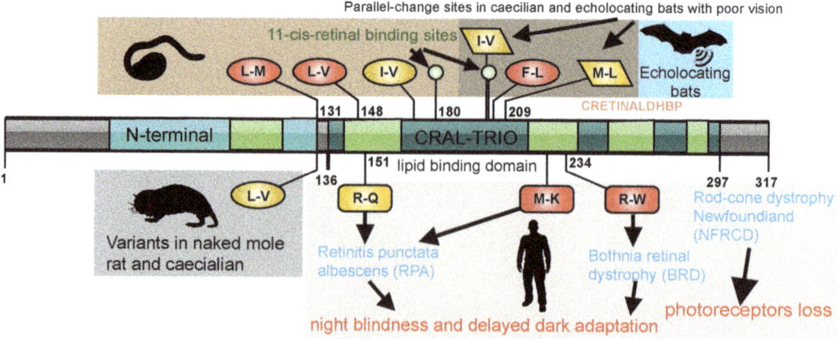


**Figure 6.** Specific and shared variants of candidate genes in extant amphibians and distantly related vertebrates with similar traits. (**A**) PCNA in salamanders. (**B**) ALDH3A2 in frogs. (**C**) RLBP1 in caecilians. Color of residue represents the strength of the functional impact. Red: harmful; yellow: neutral.

It is worth noting that three amino acid substitutions (M68V, I87V, and L99M) are located in the region that interacts with *NUDT15* (Figure 6A), a partner that is important in protecting *PCNA* from degradation [133]. One important mutation (M68V) is in the conserved DNA-binding region (61–79) and was considered to be deleterious (Figures 4B and 6A). These three changes were expected to strengthen the physiological interaction between *PCNA* and *NUDT15* and enhance the stability of *PCNA* in salamanders. It has been shown that the S228I mutation in humans significantly decreases *PCNA* interactions with *FEN1*, *LIG1*, and *ERCC5* and can give rise to an age-related syndrome (ataxia telangiectasia-like disorder 2 (ATLD2)) in which the clinical features include premature aging, a short stature, development delay, neurodegeneration, and hearing loss due to impaired DNA repair (Figure 6A) [134]. Here, I uncovered two salamander-specific amino acid mutations (S222C and T226I) near position 228, and S222C was predicted to have a damaging effect on *PCNA* structure (Figure 4B). Another interesting position in *PCNA* is 174, which is a well-conserved position in vertebrates because common changes were found in salamanders (E174S) and the longer-lived Brandt's bat (E174K) (Figure 6A). In contrast, this is not the case for other vertebrates, even for normal-lived bats (including the little brown bat, vampire bat, David's myotis, and black flying fox). Considering the key role of *PCNA* in the DNA damage response, these amino acid changes may be directly related to minimizing the negative effects of ROS, which aid in slowing the aging of salamanders. Moreover, it is well known that telomeres shorten with age, which is in part because of the end replication problem. *PCNA* may contribute to the maintenance of proper telomere length via semi-conservative replication. Moreover, although salamanders' aging-related module evolved significantly faster than that of their MRCAs (Figure 3C), the DNA repair process did not, raising the possibility that salamanders' MRCAs already possessed a stronger ability to repair DNA damage.

It is intriguing that the sensory perception of pain evolved much faster (greater than two times) in salamanders than in any other vertebrate (Figure 2E). A recent study indicated that loss of pain perception can directly delay aging via knocking out *TRPV1* pain receptors [135]. I studied *PTGS2*, a regenerative candidate that was mentioned previously, which is also responsible for the sensory perception of pain. This study's findings once again suggest that longevity and regeneration are likely to be closely related. Another similar case is *LMO4*. *PCNA*, *EEF1E1*, *OGFR*, and *SIRT1* also play dual roles in life expectancy and regenerative capacity. With this in mind, it is not difficult to imagine how salamanders can perfectly regenerate complex structures, even when aging.

### 3.7. Vocalization- and Hearing-Related Genes in Frogs

Vocalization is the primary form of communication of most frogs during their breeding season. I detected several positively selected genes that are potentially linked to vocalization and hearing.

The gene fatty aldehyde dehydrogenase (*ALDH3A2*) bore the signatures of positive selection. *ALDH3A2* is a critical gene associated with Sjögren–Larsson syndrome (SLS), which is characterized by dysarthria and a few other symptoms in humans. The three sites of *ALDH3A2* (Ala188, Glu195, and Cys226) that were under natural selection in frogs awakened my interest. These sites are located in the highly conserved aldehyde dehydrogenase domain (Figure 4C), and Ala188 is in a small core region (185–190) of the NAD-binding domain (Figure 6B). Importantly, amino acid changes at position 226, which varied in frogs, can cause SLS in humans (Figure 6B). Furthermore, many other variants in *ALDH3A2* can also lead to SLS [136], implying the importance of conserved sites to maintain the function of *ALDH3A2*.

*FAM107B* is a candidate associated with the sensory perception of sound. For frog *FAM107B*, I detected a unique amino acid change (A46S), which probably has a harmful effect on protein structure. I also observed three unique amino acid mutations in *FAM107B* in naked mole rats (H34R, L37F, and Q41L), who have lost much of their ability to localize sounds due to their subterranean lifestyle. In addition, *FAM107B* knockout mice exhibit impaired hearing [137]. Furthermore, TIMM10, which is involved in the sensory perception of sound, evolved the fastest in frogs (Table S1).

*3.8. Vision-Related Genes in Caecilians*

Similar to naked mole rats, most caecilians live a subterranean lifestyle and have degenerated visual functions. I examined the molecular basis for the poor vision of caecilians and identified the top nine candidates. These included six genes that are potentially involved in visual perception/stimulus and/or the retinoic acid metabolic process, including three PSGs (*BCO2*, *JUNB*, and *CALR*), two FEGs (*RABGGTB* and *CLN5*), and one parallel-evolved gene (*RLBP1*).

The retinaldehyde-binding protein 1 (*RLBP1*) gene revealed an interesting pattern of parallel evolution. *RLBP1* is a crucial visual protein expressed in the retinal pigment epithelium and is involved in the retinal "visual cycle" [138]. Mutations in *RLBP1* cause rod–cone dysfunction and severe vision loss, and they are associated with numerous eye diseases such as night blindness, delayed dark adaptation, and loss of color vision (Figure 6C). I observed six unique missense mutations in *RLBP1*, of which five are located in the CRAL-TRIO domain, a hydrophobic binding pocket for 11-cis-retinal binding (Figure 6C). All known mutations in the CRAL-TRIO domain have been proven to cause a series of severe visual diseases due to impaired 11-cis-retinal binding and release triggered by *RLBP1* structural transitions [139]. Furthermore, I detected one common mutation in *RLBP1* between caecilians (L131M) and naked mole rats (L131V; Figure 6C) and two parallel changes between caecilians and echolocating bats (Brandt's bat, little brown bat, and David's myotis; I201V and M209L, $p = 0.000894$), which all live in dim-light environments and display a reduced visual capacity. Interestingly, a non-echolocating bat, the black flying fox, which has excellent eyesight, did not have changes at Ile201 and Met209. Considering the essential role of *RLBP1* in the conversion of photobleached opsin molecules into photosensitive visual pigments, these shared changes may be responsible for the complete or partial color blindness of these species and imply a convergent evolution due to their dark habitats.

A PSG, *AKR1B1*, was detected with the positively selected sites Val57 (BEBPP = 0.975) and Leu139 (BEBPP = 0.982). It is an important gene associated with retinal disease and cataracts in humans. It is noteworthy that all the caecilian-specific deleterious mutations occurred in important regions, especially the NADP-binding motif, which is located in a large, deep, elliptical pocket in the C-terminal end of *AKR1B1* (Figure 4D). A parallel site mutation (T136A, $p = 0.548152$) was only detected in caecilians and echolocating bats but not in non-echolocating bats. Another two PSGs (*B9D1* and *KLF4*) participate in camera-type eye development, whereas positive selection sites (Gln21 for *B9D1*, BEBPP = 0.998; Ser421 for *KLF4*, BEBPP = 0.994) and unique deleterious amino acid changes (T73S for *B9D1* and S421N for *KLF4*) may be associated with the small-sized eyes of caecilians.

## 4. Conclusions

To conclude, this study provides new insights into the origin of Lissamphibia and the genetic basis of adaptive traits of extant amphibians, particularly the regeneration ability and longevity of salamanders. The discovery of these critical genes will set the stage for further functional analyses. With the recent developments in gene editing technology, the importance and function of these candidate genes can be tested, which will provide much-needed clues to understanding the processes of regeneration and aging. All of these will open new avenues to understanding their genetic systems and to exploiting the genetic potential of humans and improving human well-being.

**Supplementary Materials:** The following supporting information can be downloaded at: https://www.mdpi.com/article/10.3390/ani13223449/s1, Figure S1. Phylogenetic hypothesis about origin and relationship of three orders of extant amphibians. (A) Monophyletic origin: Batrachia hypothesis (H1); Procera hypothesis (H2); frog–caecilian sister relationship (H3). (B) Paraphyletic origin: caecilian–amniote and salamander–frog relationships (H4); (salamanders (caecilians, amniotes)) (H5); Figure S2. Completeness score and information content. (A) Completeness score of species pairwise comparison. A high score represents a high degree of shared site coverage (green), while a low score represents a low degree of shared site coverage (red). (B) Information content of potential phylogenetic signal for each gene. Dark blue represents a high information content, light blue represents a low information content, and red (0) represents no information content; Figure S3. Density plot showing degree of compositional heterogeneity for each gene. (A) Amino acid sequences; (B) CDS sequences. The lower the RCFV value, the lower the degree of compositional heterogeneity in that gene; Figure S4. Global stationary, reversible, and homogeneous (SRH) conditions of species pairwise comparison for different datasets (A-I). The higher the $p$-value, the more consistent its evolution with evolution under SRH conditions; Figure S5. Supporting conditions for the five phylogenetic hypotheses concerning the relationship among the three orders of extant amphibians. (A) AU test based on amino acid sequences. Each vertical line represents a gene, and different colors represent different $p$-values from the AU tests. A small $p$-value (dark green) indicates that a gene rejects a topology. (B) AU test based on CDS sequences. (C) Venn diagram based on amino acid sequences. Numbers in brackets represent the quantity of genes that cannot reject a specific hypothesis. (D) Venn diagram based on CDS sequences; Figure S6. Estimation of origin and divergence date of extant amphibians. (A) Clock of protein. (B) Clock of 2nd codon of CDS. Numbers on nodes are age estimations, and blue bars represent 95% confidence intervals. Time unit is ten billion years. Fossil records are in red point on node; Figure S7. Estimation of mutation rates of the synonymous sites (dS). (A) Boxplot of dS values for each major vertebrate group. Red point is the mean value of the dN/dS ratio. Salamanders' dS ratios were significantly smaller than any other vertebrates ($p < 2.9 \times 10^{-40}$ for all pairwise tests). (B) Clustered heatmap of dN/dS ratio of major vertebrate groups. Red color represents a high dN/dS ratio; blue color represents a low dN/dS ratio. (C) dS value tree of 22 vertebrates. Red color represents high dS values, and green represents low dS values; Figure S8. Four-fold degenerate site tree of 22 vertebrates. Color represents mutation rate of 4-fold degenerate sites. Red color represents high rates and green represents low rates; Figure S9. dN/dS ratio comparison of development-related genes and the other genes in three orders of extant amphibians and four other major vertebrates. Wilcoxon rank sum test $p$-values are presented. Only salamanders' dN/dS ratio of development-related genes was significantly greater than the other genes ($p = 0.039$); Figure S10. Scatter diagram of relationship between metabolic rate and max lifespan; Table S1. Fast-evolving genes for salamanders, frogs, and caecilians; Table S2. Positively selected genes for salamanders, frogs, and caecilians.

**Funding:** This research was funded by the National Natural Science Foundation of China (grant number 32170432), Western Lights Young Scholars Plan of Chinese Academy of Sciences (grant number 2021XBZG_XBQNXZ_A_005), and Sichuan Science and Technology Program (grant number 18YYJC0171).

**Institutional Review Board Statement:** All experimental protocols were performed and all animals were handled in strict accordance with the recommendations in the guidelines of the China Council on Animal Care and approved by Chengdu Institute of Biology's Animal Experiments Ethics Committee (approval code: 20170076).

**Informed Consent Statement:** Not applicable.

**Data Availability Statement:** Transcriptome sequencing data were deposited in the Genome Sequence Archive (GSA) of the National Genomics Data Center (NGDC) at https://ngdc.cncb.ac.cn/gsa/ (accessed on 10 August 2023) under accession number PRJCA018958.

**Acknowledgments:** I am grateful to Jinzhong Fu for his helpful suggestions on the manuscript. I would also like to thank Miaozhe Huo and Hong Jin for their help with the expression experiment.

**Conflicts of Interest:** The author declares no conflict of interest.

**Abbreviations**

| | |
|---|---|
| 4D | Four-fold degenerate sites |
| BEBPP | Bayes empirical Bayes posterior probabilities |
| BI | Bayesian inference |
| BIC | Bayesian Information Criterion |
| CDS | Coding sequence |
| CH | Compositional heterogeneity |
| ECM | Extracellular matrix |
| FDR | False discovery rate |
| FEGs | Fast-evolving genes |
| GAC | Genes as characters |
| GO | Gene Ontology |
| HaMSTR | Hidden Markov Model-based Search for Orthologs using Reciprocity |
| LRT | Likelihood ratio test |
| ML | Maximum likelihood |
| MPEs | Maximum pseudo-likelihood estimates |
| PSGs | Positively selected genes |
| RCFV | Relative Composition Frequency Variability |
| REGOs | Rapidly evolving GO categories |
| ROL | Rate of living |
| SRH | Stationary, reversible, and homogeneous conditions |
| WRST | Wilcoxon rank sum test |

## References


1. Brockes, J.P.; Kumar, A. Comparative aspects of animal regeneration. *Annu. Rev. Cell. Dev. Biol.* **2008**, *24*, 525–549. [CrossRef]
2. Tanaka, E.M. The molecular and cellular choreography of appendage regeneration. *Cell* **2016**, *165*, 1598–1608. [CrossRef] [PubMed]
3. Sandoval-Guzmán, T.; Wang, H.; Khattak, S.; Schuez, M.; Roensch, K.; Nacu, E.; Tazaki, A.; Joven, A.; Tanaka, E.M.; Simon, A. Fundamental differences in dedifferentiation and stem cell recruitment during skeletal muscle regeneration in two salamander species. *Cell Stem Cell* **2014**, *14*, 174–187. [CrossRef]
4. Fei, J.-F.; Schuez, M.; Knapp, D.; Taniguchi, Y.; Drechsel, D.N.; Tanaka, E.M. Efficient gene knockin in axolotl and its use to test the role of satellite cells in limb regeneration. *Proc. Natl. Acad. Sci. USA* **2017**, *114*, 12501–12506. [CrossRef] [PubMed]
5. Sousounis, K.; Athippozhy, A.T.; Voss, S.R.; Tsonis, P.A. Plasticity for axolotl lens regeneration is associated with age-related changes in gene expression. *Regeneration* **2014**, *1*, 47–57. [CrossRef] [PubMed]
6. Eguchi, G.; Eguchi, Y.; Nakamura, K.; Yadav, M.C.; Millán, J.L.; Tsonis, P.A. Regenerative capacity in newts is not altered by repeated regeneration and ageing. *Nat. Commun.* **2011**, *2*, 384. [CrossRef]
7. Torres, M. Regeneration: Limb regrowth takes two. *Nature* **2016**, *533*, 328–330. [CrossRef]
8. Godwin, J.W.; Pinto, A.R.; Rosenthal, N.A. Macrophages are required for adult salamander limb regeneration. *Proc. Natl. Acad. Sci. USA* **2013**, *110*, 9415–9420. [CrossRef]
9. Yun, M.H.; Gates, P.B.; Brockes, J.P. Sustained ERK activation underlies reprogramming in regeneration-competent salamander cells and distinguishes them from their mammalian counterparts. *Stem Cell Rep.* **2014**, *3*, 15–23. [CrossRef]
10. Calve, S.; Odelberg, S.J.; Simon, H.-G. A transitional extracellular matrix instructs cell behavior during muscle regeneration. *Dev. Biol.* **2010**, *344*, 259–271. [CrossRef]
11. Kumar, A.; Godwin, J.W.; Gates, P.B.; Garza-Garcia, A.A.; Brockes, J.P. Molecular basis for the nerve dependence of limb regeneration in an adult vertebrate. *Science* **2007**, *318*, 772–777. [CrossRef] [PubMed]
12. Tsai, S.L.; Baselga-Garriga, C.; Melton, D.A. Blastemal progenitors modulate immune signaling during early limb regeneration. *Development* **2019**, *146*, dev169128. [CrossRef]
13. Rando, T.A. Stem cells, ageing and the quest for immortality. *Nature* **2006**, *441*, 1080–1086. [CrossRef] [PubMed]
14. Lunghi, E. Doubling the lifespan of European plethodontid salamanders. *Ecology* **2022**, *103*, e03581. [CrossRef] [PubMed]
15. Yun, M.H. Salamander insights into ageing and rejuvenation. *Front. Cell Dev. Biol.* **2021**, *9*, 689062. [CrossRef]
16. Walters, H.E.; Troyanovskiy, K.E.; Graf, A.M.; Yun, M.H. Senescent cells enhance newt limb regeneration by promoting muscle dedifferentiation. *Aging Cell* **2023**, *22*, e13826. [CrossRef]
17. Zhang, P.; Zhou, H.; Chen, Y.-Q.; Liu, Y.-F.; Qu, L.-H. Mitogenomic perspectives on the origin and phylogeny of living amphibians. *Syst. Biol.* **2005**, *54*, 391–400. [CrossRef]
18. Nowoshilow, S.; Schloissnig, S.; Fei, J.-F.; Dahl, A.; Pang, A.W.; Pippel, M.; Winkler, S.; Hastie, A.R.; Young, G.; Roscito, J.G. The axolotl genome and the evolution of key tissue formation regulators. *Nature* **2018**, *554*, 50–55. [CrossRef]
19. Adams, J. Transcriptome: Connecting the genome to gene function. *Nat. Educ.* **2008**, *1*, 195.

20. Ku, C.-S.; Wu, M.; Cooper, D.N.; Naidoo, N.; Pawitan, Y.; Pang, B.; Iacopetta, B.; Soong, R. Exome versus transcriptome sequencing in identifying coding region variants. *Expert Rev. Mol. Diagn.* **2012**, *12*, 241–251. [CrossRef]

21. Wilkinson, G.S.; Adams, D.M. Recurrent evolution of extreme longevity in bats. *Biol. Lett.* **2019**, *15*, 20180860. [CrossRef] [PubMed]

22. Bailey, S.F.; Blanquart, F.; Bataillon, T.; Kassen, R. What drives parallel evolution? How population size and mutational variation contribute to repeated evolution. *Bioessays* **2017**, *39*, 1–9. [CrossRef] [PubMed]

23. Fei, L.; Ye, C.; Jiang, J. *Colored Atlas of Chinese Amphibians and Their Distributions*; Sichuan Publishing Group: Chengdu, China, 2012.

24. Bolger, A.M.; Lohse, M.; Usadel, B. Trimmomatic: A flexible trimmer for Illumina Sequence Data. *Bioinformatics* **2014**, *30*, 2114–2120. [CrossRef] [PubMed]

25. Grabherr, M.G.; Haas, B.J.; Yassour, M.; Levin, J.Z.; Thompson, D.A.; Amit, I.; Adiconis, X.; Fan, L.; Raychowdhury, R.; Zeng, Q.; et al. Full-length transcriptome assembly from RNA-Seq data without a reference genome. *Nat. Biotechnol.* **2011**, *29*, 644–652. [CrossRef]

26. Haas, B.J.; Papanicolaou, A.; Yassour, M.; Grabherr, M.; Blood, P.D.; Bowden, J.; Couger, M.B.; Eccles, D.; Li, B.; Lieber, M.; et al. De novo transcript sequence reconstruction from RNA-seq using the Trinity platform for reference generation and analysis. *Nat. Protoc.* **2013**, *8*, 1494–1512. [CrossRef]

27. Simpson, J.T.; Wong, K.; Jackman, S.D.; Schein, J.E.; Jones, S.J.M.; Birol, İ. ABySS: A parallel assembler for short read sequence data. *Genome Res.* **2009**, *19*, 1117–1123. [CrossRef]

28. Robertson, G.; Schein, J.; Chiu, R.; Corbett, R.; Field, M.; Jackman, S.D.; Mungall, K.; Lee, S.; Okada, H.M.; Qian, J.Q.; et al. De novo assembly and analysis of RNA-seq data. *Nat. Methods* **2010**, *7*, 909–912. [CrossRef]

29. Li, W.; Godzik, A. Cd-hit: A fast program for clustering and comparing large sets of protein or nucleotide sequences. *Bioinformatics* **2006**, *22*, 1658–1659. [CrossRef]

30. Huang, X.; Madan, A. CAP3: A DNA sequence assembly program. *Genome Res.* **1999**, *9*, 868–877. [CrossRef]

31. Qiao, L.; Yang, W.; Fu, J.; Song, Z. Transcriptome Profile of the Green Odorous Frog (*Odorrana margaretae*). *PLoS ONE* **2013**, *8*, e75211. [CrossRef]

32. Yang, W.; Qi, Y.; Lu, B.; Qiao, L.; Wu, Y.; Fu, J. Gene expression variations in high-altitude adaptation: A case study of the Asiatic toad (*Bufo gargarizans*). *BMC Genet.* **2017**, *18*, 62. [CrossRef] [PubMed]

33. Wu, C.-H.; Tsai, M.-H.; Ho, C.-C.; Chen, C.-Y.; Lee, H.-S. De novo transcriptome sequencing of axolotl blastema for identification of differentially expressed genes during limb regeneration. *BMC Genom.* **2013**, *14*, 434. [CrossRef] [PubMed]

34. Looso, M.; Preussner, J.; Sousounis, K.; Bruckskotten, M.; Michel, C.S.; Lignelli, E.; Reinhardt, R.; Höffner, S.; Krüger, M.; Tsonis, P.A.; et al. A de novo assembly of the newt transcriptome combined with proteomic validation identifies new protein families expressed during tissue regeneration. *Genome Biol.* **2013**, *14*, R16. [CrossRef]

35. Hellsten, U.; Harland, R.M.; Gilchrist, M.J.; Hendrix, D.; Jurka, J.; Kapitonov, V.; Ovcharenko, I.; Putnam, N.H.; Shu, S.; Taher, L.; et al. The genome of the Western clawed frog *Xenopus tropicalis*. *Science* **2010**, *328*, 633–636. [CrossRef] [PubMed]

36. Ebersberger, I.; Strauss, S.; von Haeseler, A. HaMStR: Profile hidden markov model based search for orthologs in ESTs. *BMC Evol. Biol.* **2009**, *9*, 157. [CrossRef]

37. Sievers, F.; Wilm, A.; Dineen, D.; Gibson, T.J.; Karplus, K.; Li, W.; Lopez, R.; McWilliam, H.; Remmert, M.; Soding, J.; et al. Fast, scalable generation of high-quality protein multiple sequence alignments using Clustal Omega. *Mol. Syst. Biol.* **2011**, *7*, 539. [CrossRef]

38. Hartmann, S.; Vision, T.J. Using ESTs for phylogenomics: Can one accurately infer a phylogenetic tree from a gappy alignment? *BMC Evol. Biol.* **2008**, *8*, 95. [CrossRef]

39. Philippe, H.; Snell, E.A.; Bapteste, E.; Lopez, P.; Holland, P.W.H.; Casane, D. Phylogenomics of eukaryotes: Impact of missing data on large alignments. *Mol. Biol. Evol.* **2004**, *21*, 1740–1752. [CrossRef]

40. Misof, B.; Meyer, B.; von Reumont, B.M.; Kück, P.; Misof, K.; Meusemann, K. Selecting informative subsets of sparse supermatrices increases the chance to find correct trees. *BMC Bioinform.* **2013**, *14*, 348. [CrossRef]

41. Kück, P.; Struck, T.H. BaCoCa—A heuristic software tool for the parallel assessment of sequence biases in hundreds of gene and taxon partitions. *Mol. Phylogenetics Evol.* **2014**, *70*, 94–98. [CrossRef]

42. Zhong, M.; Hansen, B.; Nesnidal, M.; Golombek, A.; Halanych, K.M.; Struck, T.H. Detecting the symplesiomorphy trap: A multigene phylogenetic analysis of terebelliform annelids. *BMC Evol. Biol.* **2011**, *11*, 369. [CrossRef] [PubMed]

43. Ho, S.Y.; Jermiin, L. Tracing the decay of the historical signal in biological sequence data. *Syst. Biol.* **2004**, *53*, 623–637. [CrossRef]

44. Jermiin, L.S.; Ho, S.Y.; Ababneh, F.; Robinson, J.; Larkum, A.W. The biasing effect of compositional heterogeneity on phylogenetic estimates may be underestimated. *Syst. Biol.* **2004**, *53*, 638–643. [CrossRef]

45. Jermiin, L.S.; Ho, J.W.K.; Lau, K.W.; Jayaswal, V. SeqVis: A tool for detecting compositional heterogeneity among aligned nucleotide sequences. *Bioinform. DNA Seq. Anal.* **2009**, *537*, 65–91. [CrossRef]

46. Collins, T.M.; Fedrigo, O.; Naylor, G.J. Choosing the best genes for the job: The case for stationary genes in genome-scale phylogenetics. *Syst. Biol.* **2005**, *54*, 493–500. [CrossRef] [PubMed]

47. Rannala, B.; Yang, Z. Phylogenetic inference using whole genomes. *Annu. Rev. Genom. Hum. Genet.* **2008**, *9*, 217–231. [CrossRef]

48. Stamatakis, A. RAxML version 8: A tool for phylogenetic analysis and post-analysis of large phylogenies. *Bioinformatics* **2014**, *30*, 1312–1313. [CrossRef]

49. Lartillot, N.; Lepage, T.; Blanquart, S. PhyloBayes 3: A Bayesian software package for phylogenetic reconstruction and molecular dating. *Bioinformatics* **2009**, *25*, 2286–2288. [CrossRef]
50. Lanfear, R.; Calcott, B.; Ho, S.Y.W.; Guindon, S. PartitionFinder: Combined Selection of Partitioning Schemes and Substitution Models for Phylogenetic Analyses. *Mol. Biol. Evol.* **2012**, *29*, 1695–1701. [CrossRef]
51. Ruta, M.; Coates, M.I.; Quicke, D.L.J. Early tetrapod relationships revisited. *Biol. Rev.* **2003**, *78*, 251–345. [CrossRef]
52. Zardoya, R.; Meyer, A. On the origin of and phylogenetic relationships among living amphibians. *Proc. Natl. Acad. Sci. USA* **2001**, *98*, 7380–7383. [CrossRef] [PubMed]
53. Hugall, A.F.; Foster, R.; Lee, M.S.Y. Calibration Choice, Rate Smoothing, and the Pattern of Tetrapod Diversification According to the Long Nuclear Gene RAG-1. *Syst. Biol.* **2007**, *56*, 543–563. [CrossRef] [PubMed]
54. Zhang, P.; Wake, D.B. Higher-level salamander relationships and divergence dates inferred from complete mitochondrial genomes. *Mol. Phylogen. Evol.* **2009**, *53*, 492–508. [CrossRef] [PubMed]
55. Vallin, G.; Laurin, M. Cranial morphology and affinities of Microbrachis, and a reappraisal of the phylogeny and lifestyle of the first amphibians. *J. Vert. Paleontol.* **2004**, *24*, 56–72. [CrossRef]
56. Feller, A.E.; Hedges, S.B. Molecular Evidence for the Early History of Living Amphibians. *Mol. Phylogen. Evol.* **1998**, *9*, 509–516. [CrossRef]
57. Hedges, S.B.; Maxson, L.R. A Molecular Perspective on Lissamphibian Phylogeny. *Herpetol. Monogr.* **1993**, *7*, 27–42. [CrossRef]
58. Carroll, R.L. The origin and early radiation of terrestrial vertebrates. *J. Paleontol.* **2001**, *75*, 1202–1213. [CrossRef]
59. Anderson, J.S.; Reisz, R.R.; Scott, D.; Frobisch, N.B.; Sumida, S.S. A stem batrachian from the Early Permian of Texas and the origin of frogs and salamanders. *Nature* **2008**, *453*, 515–518. [CrossRef]
60. Anderson, J.S. Focal review: The origin (s) of modern amphibians. *Evol. Biol.* **2008**, *35*, 231–247. [CrossRef]
61. Carroll, R.L. The Palaeozoic Ancestry of Salamanders, Frogs and Caecilians. *Zool. J. Linn. Soc.* **2007**, *150*, 1–140. [CrossRef]
62. Carroll, R.L.; Holmes, R. The skull and jaw musculature as guides to the ancestry of salamanders. *Zool. J. Linn. Soc.* **1980**, *68*, 1–40. [CrossRef]
63. Carroll, R.L.; Boisvert, C.; Bolt, J.; Green, D.M.; Philip, N.; Rolian, C.; Schoch, R.; Tarenko, A. Changing patterns of ontogeny from osteolepiform fish through Permian tetrapods as a guide to the early evolution of land vertebrates. In *Recent Advances in the Origin and Early Radiation of Vertebrates*; Pfeil: Munich, Germany, 2004; pp. 321–343.
64. Shimodaira, H. An approximately unbiased test of phylogenetic tree selection. *Syst. Biol.* **2002**, *51*, 492–508. [CrossRef] [PubMed]
65. Shimodaira, H.; Hasegawa, M. CONSEL: For assessing the confidence of phylogenetic tree selection. *Bioinformatics* **2001**, *17*, 1246–1247. [CrossRef] [PubMed]
66. Susko, E.; Leigh, J.; Doolittle, W.; Bapteste, E. Visualizing and assessing phylogenetic congruence of core gene sets: A case study of the γ-Proteobacteria. *Mol. Biol. Evol.* **2006**, *23*, 1019–1030. [CrossRef]
67. Liu, L.; Yu, L.; Edwards, S.V. A maximum pseudo-likelihood approach for estimating species trees under the coalescent model. *BMC Evol. Biol.* **2010**, *10*, 302. [CrossRef]
68. Lu, B.; Yang, W.; Dai, Q.; Fu, J. Using genes as characters and a parsimony analysis to explore the phylogenetic position of turtles. *PLoS ONE* **2013**, *8*, e79348. [CrossRef]
69. Lu, B.; Wang, X.; Fu, J.; Shi, J.; Wu, Y.; Qi, Y. Genetic adaptations of an island pit-viper to a unique sedentary life with extreme seasonal food availability. *G3 Genes Genomes Genet.* **2020**, *10*, 1639–1646. [CrossRef]
70. Birney, E.; Andrews, T.D.; Bevan, P.; Caccamo, M.; Chen, Y.; Clarke, L.; Coates, G.; Cuff, J.; Curwen, V.; Cutts, T. An overview of Ensembl. *Genome Res.* **2004**, *14*, 925–928. [CrossRef]
71. Benton, M.J.; Donoghue, P.C. Paleontological evidence to date the tree of life. *Mol. Biol. Evol.* **2007**, *24*, 26–53. [CrossRef]
72. Yang, Z.; Rannala, B. Bayesian estimation of species divergence times under a molecular clock using multiple fossil calibrations with soft bounds. *Mol. Biol. Evol.* **2006**, *23*, 212–226. [CrossRef]
73. Dos Reis, M.; Yang, Z. The unbearable uncertainty of Bayesian divergence time estimation. *J. Syst. Evol.* **2013**, *51*, 30–43. [CrossRef]
74. Nielsen, R. Molecular signatures of natural selection. *Annu. Rev. Genet.* **2005**, *39*, 197–218. [CrossRef] [PubMed]
75. Nielsen, R.; Yang, Z. Likelihood models for detecting positively selected amino acid sites and applications to the HIV-1 envelope gene. *Genetics* **1998**, *148*, 929–936. [CrossRef] [PubMed]
76. Yang, Z. PAML 4: Phylogenetic analysis by maximum likelihood. *Mol. Biol. Evol.* **2007**, *24*, 1586–1591. [CrossRef]
77. Hubisz, M.J.; Pollard, K.S.; Siepel, A. PHAST and RPHAST: Phylogenetic analysis with space/time models. *Brief. Bioinform.* **2011**, *12*, 41–51. [CrossRef]
78. Huang, D.W.; Sherman, B.T.; Lempicki, R.A. Systematic and integrative analysis of large gene lists using DAVID bioinformatics resources. *Nat. Protoc.* **2009**, *4*, 44–57. [CrossRef]
79. Seim, I.; Fang, X.; Xiong, Z.; Lobanov, A.V.; Huang, Z.; Ma, S.; Feng, Y.; Turanov, A.A.; Zhu, Y.; Lenz, T.L. Genome analysis reveals insights into physiology and longevity of the Brandt's bat *Myotis brandtii*. *Nat. Commun.* **2013**, *4*, 2212. [CrossRef]
80. Wilkinson, G.S.; South, J.M. Life history, ecology and longevity in bats. *Aging cell* **2002**, *1*, 124–131. [CrossRef]
81. Kim, E.B.; Fang, X.; Fushan, A.A.; Huang, Z.; Lobanov, A.V.; Han, L.; Marino, S.M.; Sun, X.; Turanov, A.A.; Yang, P.; et al. Genome sequencing reveals insights into physiology and longevity of the naked mole rat. *Nature* **2011**, *479*, 223–227. [CrossRef]
82. Larson, J.; Park, T.J. Extreme hypoxia tolerance of naked mole-rat brain. *Neuroreport* **2009**, *20*, 1634–1637. [CrossRef]
83. Park, T.J.; Lu, Y.; Jüttner, R.; Smith, E.S.J.; Hu, J.; Brand, A.; Wetzel, C.; Milenkovic, N.; Erdmann, B.; Heppenstall, P.A. Selective inflammatory pain insensitivity in the African naked mole-rat (*Heterocephalus glaber*). *PLoS Biol.* **2008**, *6*, e13. [CrossRef] [PubMed]

84. Liang, S.; Mele, J.; Wu, Y.; Buffenstein, R.; Hornsby, P.J. Resistance to experimental tumorigenesis in cells of a long-lived mammal, the naked mole-rat (*Heterocephalus glaber*). *Aging cell* **2010**, *9*, 626–635. [CrossRef] [PubMed]

85. Seluanov, A.; Hine, C.; Azpurua, J.; Feigenson, M.; Bozzella, M.; Mao, Z.; Catania, K.C.; Gorbunova, V. Hypersensitivity to contact inhibition provides a clue to cancer resistance of naked mole-rat. *Proc. Natl. Acad. Sci. USA* **2009**, *106*, 19352–19357. [CrossRef] [PubMed]

86. Zhang, J.; Nei, M. Accuracies of ancestral amino acid sequences inferred by the parsimony, likelihood, and distance methods. *J. Mol. Evol.* **1997**, *44*, S139–S146. [CrossRef] [PubMed]

87. Zhang, J.; Kumar, S. Detection of convergent and parallel evolution at the amino acid sequence level. *Mol. Biol. Evol.* **1997**, *14*, 527–536. [CrossRef] [PubMed]

88. Kumar, P.; Henikoff, S.; Ng, P.C. Predicting the effects of coding non-synonymous variants on protein function using the SIFT algorithm. *Nat. Protoc.* **2009**, *4*, 1073–1081. [CrossRef] [PubMed]

89. Choi, Y.; Sims, G.E.; Murphy, S.; Miller, J.R.; Chan, A.P. Predicting the functional effect of amino acid substitutions and indels. *PLoS ONE* **2012**, *7*, e46688. [CrossRef]

90. Adzhubei, I.A.; Schmidt, S.; Peshkin, L.; Ramensky, V.E.; Gerasimova, A.; Bork, P.; Kondrashov, A.S.; Sunyaev, S.R. A method and server for predicting damaging missense mutations. *Nat. Methods* **2010**, *7*, 248–249. [CrossRef]

91. Mitchell, A.; Chang, H.-Y.; Daugherty, L.; Fraser, M.; Hunter, S.; Lopez, R.; McAnulla, C.; McMenamin, C.; Nuka, G.; Pesseat, S. The InterPro protein families database: The classification resource after 15 years. *Nucleic Acids Res.* **2015**, *43*, D213–D221. [CrossRef]

92. Goodsell, D.S.; Dutta, S.; Zardecki, C.; Voigt, M.; Berman, H.M.; Burley, S.K. The RCSB PDB "Molecule of the Month": Inspiring a Molecular View of Biology. *PLoS Biol.* **2015**, *13*, e1002140. [CrossRef]

93. Mayrose, I.; Graur, D.; Ben-Tal, N.; Pupko, T. Comparison of site-specific rate-inference methods for protein sequences: Empirical Bayesian methods are superior. *Mol. Biol. Evol.* **2004**, *21*, 1781–1791. [CrossRef] [PubMed]

94. Pupko, T.; Bell, R.E.; Mayrose, I.; Glaser, F.; Ben-Tal, N. Rate4Site: An algorithmic tool for the identification of functional regions in proteins by surface mapping of evolutionary determinants within their homologues. *Bioinformatics* **2002**, *18*, S71–S77. [CrossRef]

95. Celniker, G.; Nimrod, G.; Ashkenazy, H.; Glaser, F.; Martz, E.; Mayrose, I.; Pupko, T.; Ben-Tal, N. ConSurf: Using evolutionary data to raise testable hypotheses about protein function. *Isr. J. Chem.* **2013**, *53*, 199–206. [CrossRef]

96. Yu, Y.; Nakhleh, L. A maximum pseudo-likelihood approach for phylogenetic networks. *BMC Genomics* **2015**, *16*, S10. [CrossRef] [PubMed]

97. Misof, B.; Liu, S.; Meusemann, K.; Peters, R.S.; Donath, A.; Mayer, C.; Frandsen, P.B.; Ware, J.; Flouri, T.; Beutel, R.G. Phylogenomics resolves the timing and pattern of insect evolution. *Science* **2014**, *346*, 763–767. [CrossRef] [PubMed]

98. Dunn, C.W.; Hejnol, A.; Matus, D.Q.; Pang, K.; Browne, W.E.; Smith, S.A.; Seaver, E.; Rouse, G.W.; Obst, M.; Edgecombe, G.D.; et al. Broad phylogenomic sampling improves resolution of the animal tree of life. *Nature* **2008**, *452*, 745–749. [CrossRef] [PubMed]

99. Hime, P.M.; Lemmon, A.R.; Lemmon, E.C.M.; Prendini, E.; Brown, J.M.; Thomson, R.C.; Kratovil, J.D.; Noonan, B.P.; Pyron, R.A.; Peloso, P.L. Phylogenomics reveals ancient gene tree discordance in the amphibian tree of life. *Syst. Biol.* **2021**, *70*, 49–66. [CrossRef]

100. Chen, M.-Y.; Liang, D.; Zhang, P. Selecting Question-specific Genes to Reduce Incongruence in Phylogenomics: A Case Study of Jawed Vertebrate Backbone Phylogeny. *Syst. Biol.* **2015**, *64*, 1104–1120. [CrossRef]

101. Alexander Pyron, R.; Wiens, J.J. A large-scale phylogeny of Amphibia including over 2800 species, and a revised classification of extant frogs, salamanders, and caecilians. *Mol. Phylogen. Evol.* **2011**, *61*, 543–583. [CrossRef]

102. Werneburg, R.; Witzmann, F.; Schneider, J.W. The oldest known tetrapod (Temnospondyli) from Germany (early Carboniferous, Visean). *PalZ* **2019**, *93*, 679–690. [CrossRef]

103. Marjanović, D.; Laurin, M. Fossils, molecules, divergence times, and the origin of Lissamphibians. *Syst. Biol.* **2007**, *56*, 369–388. [CrossRef] [PubMed]

104. San Mauro, D.; Vences, M.; Alcobendas, M.; Zardoya, R.; Meyer, A. Initial diversification of living amphibians predated the breakup of Pangaea. *Am. Nat.* **2005**, *165*, 590–599. [CrossRef] [PubMed]

105. Wyckoff, G.J.; Malcom, C.M.; Vallender, E.J.; Lahn, B.T. A highly unexpected strong correlation between fixation probability of nonsynonymous mutations and mutation rate. *Trends Genet.* **2005**, *21*, 381–385. [CrossRef] [PubMed]

106. Nei, M.; Kumar, S. *Molecular Evolution and Phylogenetics*; Oxford University Press: Oxford, UK, 2000.

107. Chong, R.A.; Mueller, R.L. Low metabolic rates in salamanders are correlated with weak selective constraints on mitochondrial genes. *Evolution* **2013**, *67*, 894–899. [CrossRef] [PubMed]

108. Rao, N.; Jhamb, D.; Milner, D.J.; Li, B.; Song, F.; Wang, M.; Voss, S.R.; Palakal, M.; King, M.W.; Saranjami, B. Proteomic analysis of blastema formation in regenerating axolotl limbs. *BMC Biol.* **2009**, *7*, 83. [CrossRef]

109. Wang, H.; Lööf, S.; Borg, P.; Nader, G.A.; Blau, H.M.; Simon, A. Turning terminally differentiated skeletal muscle cells into regenerative progenitors. *Nat. Commun.* **2015**, *6*, 7916. [CrossRef]

110. Yun, M.H.; Gates, P.B.; Brockes, J.P. Regulation of p53 is critical for vertebrate limb regeneration. *Proc. Natl. Acad. Sci. USA* **2013**, *110*, 17392–17397. [CrossRef]

111. Hirata, A.; Gardiner, D.M.; Satoh, A. Dermal fibroblasts contribute to multiple tissues in the accessory limb model. *Dev. Growth Differ.* **2010**, *52*, 343–350. [CrossRef]

112. Dujardin, D.L.; Barnhart, L.E.; Stehman, S.A.; Gomes, E.R.; Gundersen, G.G.; Vallee, R.B. A role for cytoplasmic dynein and LIS1 in directed cell movement. *J. Cell Biol.* **2003**, *163*, 1205–1211. [CrossRef]

113. Lee, E.; Ju, B.-G.; Kim, W.-S. Endogenous retinoic acid mediates the early events in salamander limb regeneration. *Anim. Cells Syst.* **2012**, *16*, 462–468. [CrossRef]

114. Zhang, Y.; Desai, A.; Yang, S.Y.; Bae, K.B.; Antczak, M.I.; Fink, S.P.; Tiwari, S.; Willis, J.E.; Williams, N.S.; Dawson, D.M. Inhibition of the prostaglandin-degrading enzyme 15-PGDH potentiates tissue regeneration. *Science* **2015**, *348*, aaa2340. [PubMed]

115. Liu, Y.; Zhou, Q.; Wang, Y.; Luo, L.; Yang, J.; Yang, L.; Liu, M.; Li, Y.; Qian, T.; Zheng, Y. Gekko japonicus genome reveals evolution of adhesive toe pads and tail regeneration. *Nat. Commun.* **2015**, *6*, 10033.

116. Daneau, G.; Boidot, R.; Martinive, P.; Feron, O. Identification of cyclooxygenase-2 as a major actor of the transcriptomic adaptation of endothelial and tumor cells to cyclic hypoxia: Effect on angiogenesis and metastases. *Clin. Cancer. Res.* **2010**, *16*, 410–419. [CrossRef] [PubMed]

117. Zagon, I.S.; Verderame, M.F.; Allen, S.S.; McLaughlin, P.J. Cloning, sequencing, chromosomal location, and function of cDNAs encoding an opioid growth factor receptor (OGFr) in humans. *Brain Res.* **2000**, *856*, 75–83. [PubMed]

118. Zagon, I.S.; Donahue, R.N.; Rogosnitzky, M.; Mclaughlin, P.J. Imiquimod upregulates the opioid growth factor receptor to inhibit cell proliferation independent of immune function. *Exp. Biol. Med.* **2008**, *233*, 968–979.

119. Roet, K.C.; Franssen, E.H.; de Bree, F.M.; Essing, A.H.; Zijlstra, S.-J.J.; Fagoe, N.D.; Eggink, H.M.; Eggers, R.; Smit, A.B.; van Kesteren, R.E. A multilevel screening strategy defines a molecular fingerprint of proregenerative olfactory ensheathing cells and identifies SCARB2, a protein that improves regenerative sprouting of injured sensory spinal axons. *J. Neurosci.* **2013**, *33*, 11116–11135. [CrossRef] [PubMed]

120. Varga, M.; Sass, M.; Papp, D.; Takacs-Vellai, K.; Kobolak, J.; Dinnyes, A.; Klionsky, D.J.; Vellai, T. Autophagy is required for zebrafish caudal fin regeneration. *Cell Death Differ.* **2014**, *21*, 547–556.

121. Gómez, C.M.A.; Echeverri, K. Salamanders: The molecular basis of tissue regeneration and its relevance to human disease. *Curr. Top. Dev. Biol.* **2021**, *145*, 235–275.

122. Gatten, R.; Miller, K.; Full, R. Energetics at rest and during locomotion. *Environ. Physiol. Amphib.* **1992**, 314–377.

123. Martins, S.G.; Zilhão, R.; Thorsteinsdóttir, S.; Carlos, A.R. Linking oxidative stress and DNA damage to changes in the expression of extracellular matrix components. *Front. Genet.* **2021**, *12*, 673002.

124. Hua, X.; Cowman, P.; Warren, D.; Bromham, L. Longevity is linked to mitochondrial mutation rates in rockfish: A test using Poisson regression. *Mol. Biol. Evol.* **2015**, *32*, 2633–2645. [CrossRef] [PubMed]

125. Kenyon, C.J. The genetics of ageing. *Nature* **2010**, *464*, 504–512. [PubMed]

126. Dang, W.; Steffen, K.K.; Perry, R.; Dorsey, J.A.; Johnson, F.B.; Shilatifard, A.; Kaeberlein, M.; Kennedy, B.K.; Berger, S.L. Histone H4 lysine 16 acetylation regulates cellular lifespan. *Nature* **2009**, *459*, 802–807. [CrossRef]

127. Satoh, A.; Brace, C.S.; Rensing, N.; Cliften, P.; Wozniak, D.F.; Herzog, E.D.; Yamada, K.A.; Imai, S.-i. Sirt1 extends life span and delays aging in mice through the regulation of Nk2 homeobox 1 in the DMH and LH. *Cell Metab.* **2013**, *18*, 416–430. [CrossRef]

128. Yoh, S.M.; Lucas, J.S.; Jones, K.A. The Iws1: Spt6: CTD complex controls cotranscriptional mRNA biosynthesis and HYPB/Setd2-mediated histone H3K36 methylation. *Genes Dev.* **2008**, *22*, 3422–3434. [CrossRef] [PubMed]

129. Cazzalini, O.; Sommatis, S.; Tillhon, M.; Dutto, I.; Bachi, A.; Rapp, A.; Nardo, T.; Scovassi, A.I.; Necchi, D.; Cardoso, M.C. CBP and p300 acetylate PCNA to link its degradation with nucleotide excision repair synthesis. *Nucleic Acids Res.* **2014**, *42*, 8433–8448. [CrossRef] [PubMed]

130. Tanno, M.; Ogihara, M.; Taguchi, T. Age-related changes in proliferating cell nuclear antigen levels. *Mech. Ageing Dev.* **1996**, *92*, 53–66. [CrossRef]

131. Keane, M.; Semeiks, J.; Webb, A.E.; Li, Y.I.; Quesada, V.; Craig, T.; Madsen, L.B.; van Dam, S.; Brawand, D.; Marques, P.I. Insights into the evolution of longevity from the bowhead whale genome. *Cell Rep.* **2015**, *10*, 112–122. [CrossRef]

132. Wang, S.-C.; Nakajima, Y.; Yu, Y.-L.; Xia, W.; Chen, C.-T.; Yang, C.-C.; McIntush, E.W.; Li, L.-Y.; Hawke, D.H.; Kobayashi, R. Tyrosine phosphorylation controls PCNA function through protein stability. *Nat. Cell Biol.* **2006**, *8*, 1359–1368. [CrossRef]

133. Yu, Y.; Cai, J.-P.; Tu, B.; Wu, L.; Zhao, Y.; Liu, X.; Li, L.; McNutt, M.A.; Feng, J.; He, Q. Proliferating cell nuclear antigen is protected from degradation by forming a complex with MutT Homolog2. *J. Biol. Chem.* **2009**, *284*, 19310–19320. [CrossRef]

134. Baple, E.L.; Chambers, H.; Cross, H.E.; Fawcett, H.; Nakazawa, Y.; Chioza, B.A.; Harlalka, G.V.; Mansour, S.; Sreekantan-Nair, A.; Patton, M.A. Hypomorphic PCNA mutation underlies a human DNA repair disorder. *J. Clin. Investig.* **2014**, *124*, 3137–3146. [CrossRef] [PubMed]

135. Riera, C.E.; Huising, M.O.; Follett, P.; Leblanc, M.; Halloran, J.; Van Andel, R.; de Magalhaes Filho, C.D.; Merkwirth, C.; Dillin, A. TRPV1 pain receptors regulate longevity and metabolism by neuropeptide signaling. *Cell* **2014**, *157*, 1023–1036. [CrossRef] [PubMed]

136. Sillén, A.; Anton-Lamprecht, I.; Braun-Quentin, C.; Kraus, C.S.; Sayli, B.S.; Ayuso, C.; Jagell, S.; Küster, W.; Wadelius, C. Spectrum of mutations and sequence variants in the FALDH gene in patients with Sjögren-Larsson syndrome. *Hum. Mutat.* **1998**, *12*, 377. [CrossRef]

137. White, J.K.; Gerdin, A.-K.; Karp, N.A.; Ryder, E.; Buljan, M.; Bussell, J.N.; Salisbury, J.; Clare, S.; Ingham, N.J.; Podrini, C. Genome-wide generation and systematic phenotyping of knockout mice reveals new roles for many genes. *Cell* **2013**, *154*, 452–464. [CrossRef]

138. Xue, Y.; Shen, S.Q.; Jui, J.; Rupp, A.C.; Byrne, L.C.; Hattar, S.; Flannery, J.G.; Corbo, J.C.; Kefalov, V.J. CRALBP supports the mammalian retinal visual cycle and cone vision. *J. Clin. Investig.* **2015**, *125*, 727. [CrossRef]
139. He, X.; Lobsiger, J.; Stocker, A. Bothnia dystrophy is caused by domino-like rearrangements in cellular retinaldehyde-binding protein mutant R234W. *Proc. Natl. Acad. Sci. USA* **2009**, *106*, 18545–18550. [CrossRef]

*Article*

# Disentangling Exploitative and Interference Competition on Forest Dwelling Salamanders

Giacomo Rosa, Sebastiano Salvidio and Andrea Costa *

Department of Earth, Environment and Life Sciences (DISTAV), University of Genova, Corso Europa 26, 16126 Genova, Italy
* Correspondence: andrea-costa-@hotmail.it

**Simple Summary:** Exploitative competition and interference competition differ in the way access to resources is modulated by a competitor. Exploitative competition implies resource depletion and usually produces spatial segregation, while interference competition is independent from resource availability and can result in temporal niche partitioning. Here, we inferred the presence of these two patterns of competition on a two-salamander system in Northern Italy. We found evidence supporting interference competition and temporal niche partitioning.

**Abstract:** Exploitative competition and interference competition differ in the way access to resources is modulated by a competitor. Exploitative competition implies resource depletion and usually produces spatial segregation, while interference competition is independent from resource availability and can result in temporal niche partitioning. Our aim is to infer the presence of spatial or temporal niche partitioning on a two-species system of terrestrial salamanders in Northern Italy: *Speleomantes strinatii* and *Salamandrina perspicillata*. We conducted 3 repeated surveys on 26 plots in spring 2018, on a sampling site where both species are present. We modelled count data with N-mixture models accounting for directional interactions on both abundance and detection process. In this way we were able to disentangle the effect of competitive interaction on the spatial scale, i.e., local abundance, and from the temporal scale, i.e., surface activity. We found strong evidence supporting the presence of temporal niche partitioning, consistent with interference competition. At the same time, no evidence of spatial segregation has been observed.

**Keywords:** exploitative competition; interference competition; multi-species N-mixture models; salamanders; temporal niche partitioning

**Citation:** Rosa, G.; Salvidio, S.; Costa, A. Disentangling Exploitative and Interference Competition on Forest Dwelling Salamanders. *Animals* **2023**, *13*, 2003. https://doi.org/10.3390/ani13122003

Academic Editor: Enrico Lunghi

Received: 12 April 2023
Revised: 12 June 2023
Accepted: 13 June 2023
Published: 15 June 2023



## 1. Introduction

Identifying which ecological processes influence the occurrence and local abundance of a species is of fundamental interest among ecologists and conservationists. Local abundance for a given species is expected to reach its maximum when environmental features correspond to the optimum of its ecological requirements [1,2]. However, the local rise in abundance or density of a species as resources or suitability of a given area increase is counterbalanced and limited by biotic interactions with other organisms that share the same ecological niche, usually through density-dependent competitive interactions [3,4].

Even though the study of competition laid its root early in the field of ecology [5], nowadays, the topic is still alive and debated, and the consequences of competitive interactions are still attracting interest among the scientific community. This happens at the theoretical [6–8] and practical level [9,10]. Competitive interactions can occur at the inter- or intra-specific level and can involve various niche axes and mechanisms. For instance, organisms compete directly for mates (e.g., [11,12]) and physical space, such as habitat or shelter use [13], and indirectly for food resources [14] or other niches, such as call space [15]. Regardless of the mechanism or origin of competitive interactions, they usually occur when one species prevents the access of its competitors to a shared resource. This exclusion from use of a resource usually takes place in two different ways: exploitative competition or interference competition [13,16,17]. These two types of competition differ substantially in the way they affect the access of one species to a given resource. In the case of exploitative competition, some organisms make better use of the same limiting resource, directly reducing the availability of the shared resource for other individuals [18]. By contrast, interference competition occurs when one organism actively prevents another organism from accessing a resource, independently from its availability, for instance by means of territorial or aggressive behaviors [13,16].

Both types of competition are assumed to have significant effects on population dynamics, local abundance, community assembly, and species coexistence [5,19,20]. Indeed, exploitative competition can either lead to the exclusion of a species from a given habitat [18], resulting in a density reduction of one competitor, or can trigger resource partitioning among different niche axes [21–23]. Interference competition can trigger competitive exclusion [3], causing behavioral changes, for instance by shifting species activity patterns [13,17]. One example of interference competition is the shift to crepuscular activity of a species that avoids the presence of a more nocturnal competitor [17].

Likewise, different kinds of competitive interactions can result in different layouts of niche and resource partitioning, minimizing niche overlap and competitive exclusion [13]. For instance, when two species or organisms segregate over different microhabitats, we observe spatial niche partitioning; conversely, when organisms exploit the same set of resources but in different time frames, we observe a partitioning along the temporal niche [2,24]. In this respect, exploitative and interference competition are also hypothesized to have different outcomes for what concern spatial or temporal segregation/partitioning. Exploitative competition has been found to have an effect on the local abundance or spatial displacement of one or both competitors (e.g., [21]). This is because of a density-dependent effect on limited resources, in which the under-performing competitor must rely on the leftover resource by the best-performing competitor [25,26]. Conversely, interference competition usually produces niche shift patterns that are consistent with temporal niche partitioning, i.e., when species access the same resource in different timeframes in order to reduce their interactions with the competitor (e.g., [17,27,28]).

In the present study, our aim was to test the occurrence of exploitative or interference competition in a two-species system of forest dwelling salamanders, the Strinati's cave salamander (*Speleomantes strinatii*) and the northern spectacled salamander (*Salamandrina perspicillata*). These two salamanders are found syntopically in NW Italy, where they rely on the same set of trophic resources, particularly soil-dwelling invertebrates [29,30]. We used binomial N-mixture models that account for directional biotic interaction on both the local abundance and the timing of surface activity to alternatively test the occurrence of spatial niche partitioning or temporal niche partitioning, after modelling abundance and activity on environmental covariates. We expected that (i) in the case of exploitative competition, we should observe a negative interaction of one species on the local abundance of the other, this being a negative density-dependent effect; or (ii) in the case of interference competition we should observe a negative interaction of one species on the surface activity of the other, this outcome representing niche partitioning on the temporal axis.

## 2. Materials and Methods

### 2.1. Study Framework

To test for the presence of exploitative or interference competition, we randomly selected 26 permanent plots under a meta-population design [31,32], in one site in Northern Italy where both salamanders are present. We counted salamanders in all plots during three repeated surveys over a short time period. We then used count data to model abundance and activity interaction occurring between the species, using a two-species N-mixture model [31], while incorporating environmental covariates [33–38].

### 2.2. Study Species

The first focal species is the Strinati's cave salamander *Speleomantes strinatii*, a plethodontid salamander with a maximum total length of about 128 mm. The second focal species is the northern spectacled salamander *Salamandrina perspicillata*, that belongs to family Salamandridae; this salamander is smaller, not exceeding 100 mm in total length. Strinati's cave salamanders possess direct development and are fully terrestrial, while northern spectacled salamanders have a biphasic life cycle, lay eggs in the water, and become fully terrestrial after metamorphosis, with only reproductive females entering the water for egg deposition [39,40]. Adults of both species are lungless (*S. strinatii*) or have vestigial lungs (*S. perspicillata*), and gas exchange occurs mainly through their skin. Both species are usually found in the talus and the leaf litter of mixed broadleaved woodlands, and their surface activity is mainly regulated by the amount of rain or soil moisture [29,37,41]. In forest environments, the diet of *S. strinatii* and *S. perspicillata* has been extensively studied, by means of stomach flushing: both species feed on a large number of invertebrate taxa and share many trophic resources [29,30]. However, at the population level *S. strinatii* has been described as a generalist predator [29], while *S. perspicillata* is a specialized predator on springtails and mites [30].

### 2.3. Study Site

The study area is located in North-Western Italy, on the Apennine Mountain range of the Piemonte region, where the two studied species occurs sympatrically and are often found syntopically. The sampling area is located at an altitude of 600 m a.s.l. in the municipality of Mongiardino Ligure (44°38′24″ N; 9°03′00″ E), Alessandria province. The site is crossed by a first order Apennine stream and is characterized by a Supra–Mediterranean mixed deciduous forest, dominated by European Oak (*Quercus pubescens* Willd., 1805; [42]). We selected, individually marked, and GPS-positioned 26 square plots in the proximity of the stream banks, each one measuring 30 m$^2$ (5.5 m side). The minimum distance among plots was about 20 m.

### 2.4. Salamanders' Sampling

In spring 2018 (16 April–25 May), the same observer visited all sites three times during daytime (8–11 a.m.) and in favorable weather conditions for salamander's activity (e.g., during or immediately after light rain). During each survey the observer searched for salamanders within each plot for four minutes [37,43], checking the leaf litter, inspecting rock crevices with a flashlight, and lifting temporary shelters such as superficial rocks and dead wood in search of salamanders. Since both species are known to be active during or immediately after rain and considering that inactive individuals retreat underground and are usually unavailable for sampling, we considered both individuals encountered on the forest floor and those found under temporary shelters as active on the surface [37,38,41]. We assigned each individual to one of the two study species and their abundance in each plot recorded.

### 2.5. Environmental Covariates

Within each marked plot at a depth of 20 cm, we obtained five measurements of soil relative humidity using a digital moisture meter (Extech MO750). All measurements were obtained on the same day (20 April 2018), 4 days following a 50 mm precipitation event. The average of these five measurements was considered as a plot-specific proxy, describing the soil moisture retention potential (MOIST). From a digital elevation model (DEM; 20 m mesh size) of the study area, we calculated two covariates: the duration of direct insolation (INSOL), expressed in hours, and the topographic position index (TPI). This latter index expresses the topographic position of each cell within the landscape, assuming positive values for cells located on ridges or hilltops and negative values for cells located in depressions [44]. Moreover, for each sampling session, we recorded the day of the year (DAY; i.e., the continuous count of the number of days beginning each year from

1 January). We obtained the temperature of the survey (TEMP) and the accumulated rain in the 72 h prior to sampling (RAIN) from local weather stations. We conducted terrain analyses with software SAGA 7 [45].

### 2.6. Data Analysis

We analyzed our repeated count data of *S. strinatii* and *S. perspicillata* using a co-abundance formulation of the static binomial N-mixture model of Royle [31], accounting for directional biotic interactions [32], both on the abundance of salamanders and on their activity pattern. Binomial N-mixture models estimate the latent abundance state ($N$) at site $i$ ($Ni$), assuming $Ni{\sim}Poisson(\lambda)$, where $\lambda$ is the expected abundance over all sites, by using repeated counts $C$ at site $i$ during survey $j$ ($Cij$) to estimate individual detection probability $p$, assuming $Cij \mid Ni{\sim}Binomial(Ni,p)$. Both parameters can be modelled as a function of environmental covariates trough a *log* or *logit* link, respectively. In order to model the possible effect of exploitative competition on local abundance, we stacked two N-mixture models, using the latent abundance of the larger species (i.e., *S. strinatii*) as a covariate in the abundance model of *S. perspicillata* [34–36]. Therefore, we added a co-abundance interaction effect term ($\gamma$) to the model [33] to estimate the overall effect of the abundance of Strinati's cave salamanders on the abundance of the northern spectacled salamander [32]. Likewise, to model the possible presence of interference competition resulting from reduced activity of the smaller species when the larger one is active, we considered the detection probability $p$ of both species to be a rough proxy of species' surface activity. This approach is widely shared and often successfully used in ecological studies assessing species interactions [37,38,46]. Therefore, we added another interaction term ($\varepsilon$) on the detection process, using the detection probability $pij$ of *S. strinatii* as a covariate on the detection of *S. perpicillata*. Prior to building our model, we standardized covariates and checked them for collinearity, considering a cut-off for inclusion of Pearson r < 0.7 [47,48]. We modelled the detection process of two species as follows:

$$[1; \textit{S. strinatii}] \; logit(pij_{Ss}) = \alpha0 + \alpha1{*}DAYij + \alpha2{*}TEMPij + \alpha3{*}RAINij + \delta ij$$

$$[2; \textit{S. perspicillata}] \; logit(pij_{Sp}) = \alpha0 + \alpha1{*}DAYij + \alpha2{*}TEMPij + \alpha3{*}RAINij + \varepsilon{*}pij_{Ss} + \delta ij$$

where the subscripts $Ss$ and $Sp$ stand for *S. strinatii* and *S. perspicillata*, respectively, $\alpha0$ is the intercept, $\alpha1$–$\alpha3$ are covariate effects, $\varepsilon$ is the directional interaction term on the activity between the two species, and $\delta$ is a normally distributed random effect that accounts for possible over dispersion in the detection process [49]. Similarly, we modelled the abundances of the two species as:

$$[3; \textit{S. strinatii}] \; log\,(NiSs) = \beta0 + \beta1{*}MOISTi + \beta2{*}INSOLi + \beta3{*}TPIi + \eta i$$

$$[4; \textit{S. perspicillata}] \; log\,(NiSp) = \beta0 + \beta1{*}MOISTi + \beta2{*}INSOLi + \beta3{*}TPIi + \gamma{*}NiSs + \eta i$$

where the subscripts $Ss$ and $Sp$ stand for *S. strinatii* and *S. perspicillata*, respectively, $\beta0$ is the intercept, $\beta1$–$\beta3$ are covariate effects, $Ni$ is the latent abundance of adults at site $i$, $\gamma$ is the co-abundance effect of adults on juveniles, and $\eta$ is a normally distributed site-level random effect [49]. We estimated model parameters using a Bayesian approach with Markov chain Monte–Carlo methods, using uninformative priors. We ran three chains, each one with an adaptive phase of 10,000 iterations, followed by 350,000 iterations, discarding the first 50,000 as a burn-in and thinning by 100. We considered that chains reached convergence when the Gelman–Rubin statistic (R-hat) was <1.1 [50]. We used the 90% highest density interval of the posterior distribution as a credible interval (CRI), and considered covariates on both abundance, detection, and biotic interaction to have a significant effect when CRI did not cross the zero. In order to assess model fit [51–53], we employed posterior predictive checks based on $\chi^2$ statistics as a measure of the discrepancy between observed and simulated data and calculated a Bayesian $p$-value accordingly [49]. Analyses were conducted calling program JAGS (V4.3.0; [54]) from the R environment with package "JagsUI" (V1.5.1; [55]).

## 3. Results

During our sampling, we counted 96 salamanders: 61 *S. strinatii* and 35 *S. perspicillata*. For both species, the co-abundance N-mixture model was in a good fit (posterior predictive checks: Bayesian *p*-value for *S. strinatii* = 0.46, and for *S. perspicillata* = 0.51). For all monitored parameters, convergence resulted successful (maximum R-hat = 1.059). The complete list of parameters' estimates, along with their respective 90% CRI, for both species, is reported in Table 1.

The two species showed different detection probabilities, with *S. strinatii*, the larger species, being more active on the forest floor (*p* = 0.43; CRI = 0.21–0.66) when compared to *S. perspicillata* (*p* = 0.32; CRI = 0.08–0.62). None of the survey covariates had a clear effect on *p* (CRI crossed zero, in all cases), although there was an 89.7% probability that TEMP had a negative effect on the detection of *S. perspicillata* (Pd, posterior probability of direction; Table 1). The two species also showed different per-site abundances, with *S. strinatii* ($\lambda$ = 0.70; CRI = 0.25–1.15) being less abundant than *S. perspicillata* ($\lambda$ = 2.10; CRI = 0.48–4.02).

Overall, the two species showed differential responses to the same set of environmental predictors, both for the ecological process and their detection/activity. Indeed, the abundance of *S. strinatii* was positively affected by MOIST (Figure 1) and negatively affected by INSOL (Figure 2), while abundance of *S. perspicillata* was only negatively affected by INSOL (Figure 2).

Finally, for what concern the two interaction terms, the one on the surface activity of *S. perspicillata* was negative and significant ($\varepsilon$ = −3.76; CRI = −7.66−−0.09), highlighting that the surface activity of *S. strinatii* has a strong negative effect on the activity of *S. perspicillata* (Figure 3). At the same time, the interaction term on local abundance was also negative ($\gamma$ = −0.18), but its CRI largely crossed the zero (CRI = −0.56–0.20; pd = 79.9%; Figure 3).

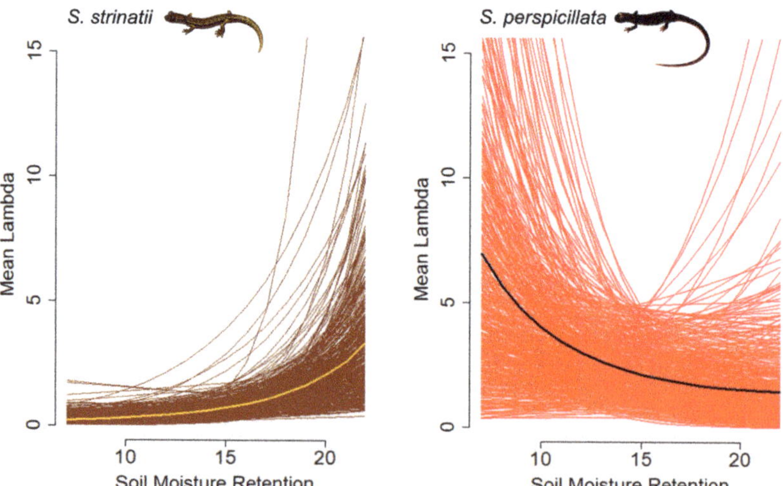


**Figure 1.** Plots showing the effect of the soil moisture retention potential (MOIST) on salamander abundance: *Speleomantes strinatii* on the **left** and *Salamandrina perspicillata* on the **right**. Bold lines represent the posterior mean, while the thin lines represent 500 random draws from the posterior distribution.

**Table 1.** Complete list of the parameters estimated for the detection process, abundance, and interactions for both *S. strinatii* and *S. perspicillata*. Mean = mean obtained from the posterior distribution; 90% CRI = 90% highest density region of the posterior distribution; pd = probability of direction.

| *Speleomantes strinatii* | | | | *Salamandrina perspicillata* | | | |
|---|---|---|---|---|---|---|---|
| **Parameter** | **Mean** | **90% CRI** | **pd** | **Parameter** | **Mean** | **90% CRI** | **pd** |
| **Detection** | | | | **Detection** | | | |
| Mean $p$ | 0.43 | 0.21–0.66 | - | Mean $p$ | 0.32 | 0.08–0.62 | – |
| DAY | 0.04 | −3.9--4.3 | 50.7 | DAY | −0.43 | −4.85–3.61 | 57.0 |
| TEMP | 1.32 | −1.7–4.4 | 76.6 | TEMP | −2.59 | −5.88–0.97 | 89.7 |
| RAIN | 1.33 | −1.56–4.32 | 77.0 | RAIN | −1.81 | −5.2–1.37 | 82.0 |
| **Abundance** | | | | **Abundance** | | | |
| Mean $\lambda$ | 0.70 | 0.25–1.15 | - | Mean $\lambda$ | 2.10 | 0.48–4.02 | - |
| MOIST | 0.88 | 0.25–1.57 | 98.7 | MOIST | −0.48 | −1.30–0.29 | 85.0 |
| INSOL | −0.76 | −1.31--0.20 | 99.2 | INSOL | −1.41 | −2.34--0.37 | 99.5 |
| TPI | −1.11 | −0.48–0.24 | 70.0 | TPI | −0.33 | −0.99–0.36 | 79.6 |
| | | | | **Interactions** | | | |
| | | | | $\varepsilon$ | −3.76 | −7.66--0.09 | 94.8 |
| | | | | $\gamma$ | −0.18 | −0.56–0.20 | 79.9 |

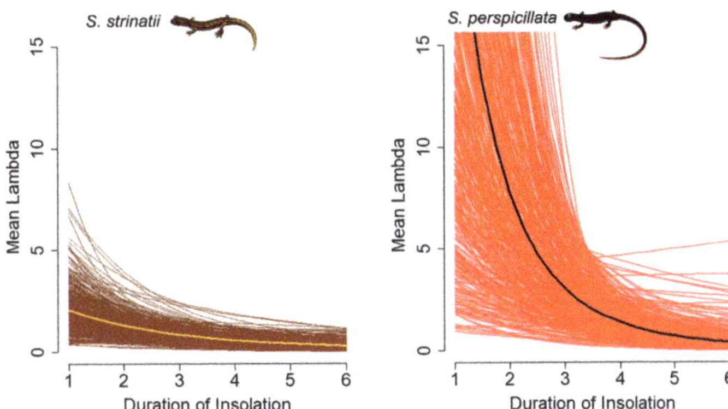


**Figure 2.** Plots showing the effect of duration of direct insolation (INSOL) on salamander abundance: *Speleomantes strinatii* on the **left** and *Salamandrina perspicillata* on the **right**. Bold lines represent the posterior mean, while the thin lines represent 500 random draws from the posterior distribution.

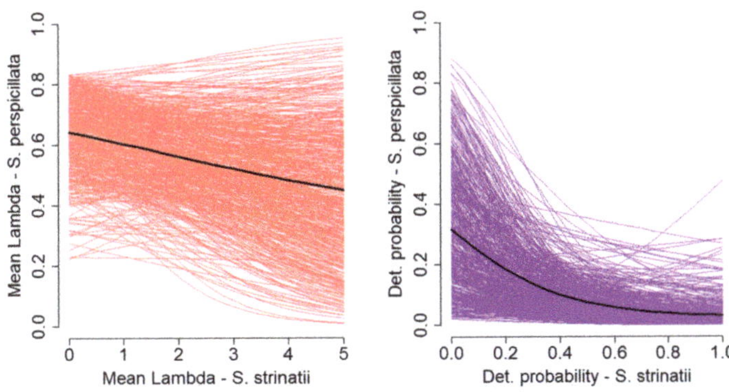


**Figure 3.** Plots showing the effect of surface activity (**left panel**; interaction terms $\varepsilon$) and abundance of *S. strinatii* (**right panel**; interaction term $\gamma$) on the surface activity and abundance of *S. perspicillata*. Bold lines represent the posterior mean, while the thin lines represent 500 random draws from the posterior distribution.

## 4. Discussion

Identifying the ecological processes that impact the presence and population size of a species holds significant importance for ecologists and conservationists, since the maximum local abundance of a species is observed when environmental characteristics align with its ecological requirements. On the other hand, the increase in local abundance or density of a species due to improved resources or suitability of a particular area is constrained by biotic interactions with other organisms that partially overlap the same ecological niche. These interactions typically occur as density-dependent competitive interactions, which can balance and limit local abundance. These competitive interactions can arise within or between species, encompassing a range of niche axes and mechanisms. For instance, organisms engage in direct competition for mates and physical space, such as habitats or shelters, as well as indirect competition for food resources or other niches, such as call space. Irrespective of the mechanism or origin of these competitive interactions, they typically emerge when one species hinders its competitors' access to a shared resource. This exclusion from resource utilization generally occurs through two distinct pathways: exploitative competition or interference competition. These two forms of competition differ significantly in their impact on a species' access to a particular resource. Exploitative competition arises when certain organisms make more efficient use of the same limited resource, thus directly reducing its availability for other individuals. Conversely, interference competition takes place when one organism actively obstructs another organism's access to a resource, regardless of its availability.

In this study, we found evidence that interference competition, rather than exploitative competition, is shaping the co-occurrence of *Speleomantes strinatii* (Strinati's cave salamander) and *S. perspicillata* (northern spectacled salamander), after accounting for differential habitat requirements. That is, after modelling ecological requirements of the two species by means N-mixture models using appropriate covariates on both the abundance and detection processes, we still detected some significant interaction between the two species, which is attributable to a biological interaction [32–36]. Specifically, after accounting for environmental covariates, we found that the activity of *S. strinatii* negatively affected the surface activity of *S. perspicillata*, indicating that the observed difference in surface activity between the two species can be explained by temporal niche partitioning. This suggests that the two species avoided direct interactions on the forest floor by being active at different times. Conversely, we did not find any effect of Strinati's cave salamander's abundance on northern spectacled salamander's abundance at the scale at which the study has been conducted. This indicates that resource depletion (i.e., exploitative competition) probably does not limit the local abundance of either species in the study area. The overall findings of this study align with previous research on interference competition, which has been observed to produce niche shift patterns consistent with temporal niche partitioning [17,27,28].

In contrast, our results did not provide support for the hypothesis of exploitative competition, which influences local abundance or spatial displacement of one or both competitors due to density-dependent effects on limited resources [21,25,26]. The lack of evidence for exploitative competition in this two-species system may imply that resource availability for these salamander species is not a major ecological limiting factor, or that they have developed mechanisms partitioning resources in a way that avoids direct competition [13]. Indeed, *S. perspicillata* and *S. strinatii* share some of the same trophic resources, but (i) display different trophic strategies [29,30], (ii) forage successfully in slightly different weather conditions [56], and (iii) when they both are active on the forest floor, *S. perspicillata* reduces the foraging intensity on the core prey items composing its diet, suggesting some kind of interference with *S. strinatii* [57].

Such niche differentiation likely relaxes competition for food, explaining why we did not detect exploitative effects. Temporal niche partitioning due to interference competition has been observed in several taxa, including amphibians [58], birds [59], and mammals [60]. For other forest salamanders, Jaeger and colleagues [58,61,62] found evidence of temporal niche partitioning at the intraspecific level in the eastern red-backed salamander (*Plethodon cinereus*), which was driven by interference competition for food resources and territories during the breeding season. However, interference competition has also been found to produce negative effects on the fitness and survival of competitors. Hairston et al. [63] found that interference competition between *P. cinereus* and *P. richmondi* resulted in reduced growth rates and survival of *P. richmondi* due to aggressive interactions with *P. cinereus*. Similarly, Petranka and Smith [64] found that interference competition between *P. cinereus* and *P. glutinosus* resulted in decreased survival and growth rates of both species.

*Salamandrina perspicillata* in the study site, despite being the outperformed species (i.e., its overall activity is reduced by the surface activity of *S. strinatii*), is roughly three times more abundant than *S. strinatii*. This is further evidence for the lack of a density-related exploitative competition on this two-species system, otherwise *S. perspicillata* abundance should be reduced. This also suggests the presence of behavioral strategies to minimize the overlap in their realized temporal niches. This may have been facilitated by their ability to successfully exploit similar prey resources at different times, thus minimizing the intensity of their competitive trophic interactions [2,24].

## 5. Conclusions

Our results support the hypothesis that interference competition promotes temporal niche partitioning in terrestrial salamanders. By shifting their surface activity, *S. strinatii* and *S. perspicillata* can co-occur at high densities while minimizing competitive interactions. On the other hand, trophic niche differentiation, divergences in metabolism, prey selection or behavioral traits, together with a high prey abundance may help to prevent the emergence of competition for food resources and may promote species coexistence [65]. Nevertheless, further studies in different ecological situations and adopting different sampling frameworks specifically designed to model surface activity should be conducted to help clarify the role of temporal or spatial niche partitioning in shaping the coexistence of these salamander species.

**Author Contributions:** Conceptualization, A.C., G.R. and S.S.; methodology, A.C., G.R. and S.S; data analysis, A.C.; writing—original draft preparation, A.C. and S.S.; writing—review and editing, A.C., G.R. and S.S.; All authors have read and agreed to the published version of the manuscript.

**Funding:** A.C. is funded by the Italian National Operative Programme "Research and Innovation" (PON—Ricerca e Innovazione, tematica GREEN; CUP N. D31B21008270007); G.R. is fulfilling his PhD at the University of Genoa and S.S. is funded by the University of Genoa (FRA-2018).

**Institutional Review Board Statement:** Capture permits for the present study were issued by the Italian Ministry of Environment (Protocol N#8453/T-A31) and no animals were harmed during the fieldwork.

**Informed Consent Statement:** Not applicable.

**Data Availability Statement:** The data presented in this study are available from the corresponding author upon reasonable request.

**Conflicts of Interest:** The authors declare no conflict of interest.

## References

1. Fretwell, S.D.; Lucas, H.L. On territorial behavior and other factors influencing habitat distribution in birds. *Acta Biotheor.* **1970**, *19*, 16–36. [CrossRef]
2. Chesson, P. Mechanisms of maintenance of species diversity. *Annu. Rev. Ecol. Syst.* **2000**, *31*, 343–366. [CrossRef]
3. Connell, J.H. On the prevalence and relative importance of interspecific competition: Evidence from field experiments. *Am. Nat.* **1983**, *122*, 661–696. [CrossRef]
4. Houston, A.I.; McNamara, J.M. The ideal free distribution when competitive abilities differ: An approach based on statistical mechanics. *Anim. Behav.* **1988**, *36*, 166–174. [CrossRef]
5. Krebs, C.J. *Ecology: The Experimental Analysis of Distribution and Abundance*, 5th ed.; Benjamin Cummings: San Francisco, CA, USA, 2001.

6. Périquet, S.; Fritz, H.; Revilla, E. The Lion King and the Hyaena Queen: Large carnivore interactions and coexistence. *Biol. Rev.* **2015**, *90*, 1197–1214. [CrossRef]

7. Grether, G.F.; Okamoto, K.W. Eco-evolutionary dynamics of interference competition. *Ecol. Lett.* **2022**, *25*, 2167–2176. [CrossRef]

8. Grether, G.F.; Peiman, K.S.; Tobias, A.; Robinson, B.W. Causes and consequences of behavioral interference between species. *Trends Ecol. Evol.* **2017**, *32*, 760–772. [CrossRef]

9. Smallegange, I.M.; Van Der Meer, J.; Kurvers, R.H. Disentangling interference competition from exploitative competition in a crab–bivalve system using a novel experimental approach. *Oikos* **2006**, *113*, 157–167. [CrossRef]

10. Lang, B.; Rall, B.C.; Brose, U. Warming effects on consumption and intraspecific interference competition depend on predator metabolism. *J. Anim. Ecol.* **2012**, *81*, 516–523. [CrossRef]

11. Daly, M. The cost of mating. *Am. Nat.* **1978**, *112*, 771–774. [CrossRef]

12. Parker, G.A. Sperm competition and the evolution of animal mating strategies. In *Sperm Competition and the Evolution of Animal Mating Systems*; Elsevier Academic Press: Cambdridge, MA, USA, 1984; pp. 1–60.

13. Schoener, T.W. Resource Partitioning in Ecological Communities: Research on how similar species divide resources helps reveal the natural regulation of species diversity. *Science* **1974**, *185*, 27–39. [CrossRef]

14. MacArthur, R.; Levins, R. The limiting similarity, convergence, and divergence of coexisting species. *Am. Nat.* **1967**, *101*, 377–385. [CrossRef]

15. Gerhardt, H.C.; Huber, F. *Acoustic Communication in Insects and Anurans: Common Problems and Diverse Solutions*; University of Chicago Press: Chicago, IL, USA, 2002.

16. Case, T.J.; Gilpin, M.E. Interference competition and niche theory. *Proc. Natl. Acad. Sci. USA* **1974**, *71*, 3073–3077. [CrossRef]

17. Carothers, J.H.; Jaksić, F.M. Time as a niche difference: The role of interference competition. *Oikos* **1984**, *42*, 403–406. [CrossRef]

18. Hardin, G. The Competitive Exclusion Principle: An idea that took a century to be born has implications in ecology, economics, and genetics. *Science* **1960**, *131*, 1292–1297. [CrossRef]

19. Adler, F.R.; Mosquera, J. Is space necessary? Interference competition and limits to biodiversity. *Ecology* **2000**, *81*, 3226–3232. [CrossRef]

20. Amarasekare, P. Interference competition and species coexistence. *Proc. R. Soc. Lond. B* **2002**, *269*, 2541–2550. [CrossRef]

21. Petren, K.; Case, T.J. An experimental demonstration of exploitation competition in an ongoing invasion. *Ecology* **1996**, *77*, 118–132. [CrossRef]

22. Schoener, T.W. Competition and the Niche. In *Biology of the Reptilia. Ecology*; Gans, A.C., Tinkle, D.W., Eds.; Academic Press: New York, NY, USA, 1977; Volume 7, pp. 35–136.

23. Schoener, T.W. Field experiments on interspecific competition. *Am. Nat.* **1983**, *122*, 240–285. [CrossRef]

24. Albrecht, M.; Gotelli, N.J. Spatial and temporal niche partitioning in grassland ants. *Oecologia* **2001**, *126*, 134–141. [CrossRef]

25. Andrewartha, H.G.; Birch, L.C. *The Distribution and Abundance of Animals*; University of Chicago Press: Chicago, IL, USA, 1954.

26. Sinclair, A.R.E. Population regulation in animals. In *Ecological Concept*; Cherret, J.M., Ed.; Oxford University Press: Oxford, UK, 1989.

27. Valeix, M.; Chamaillé-Jammes, S.; Fritz, H. Interference competition and temporal niche shifts: Elephants and herbivore communities at waterholes. *Oecologia* **2007**, *153*, 739–748. [CrossRef]

28. Olea, P.P.; Iglesias, N.; Mateo-Tomás, P. Temporal resource partitioning mediates vertebrate coexistence at carcasses: The role of competitive and facilitative interactions. *Basic Appl. Ecol.* **2022**, *60*, 63–75. [CrossRef]

29. Salvidio, S.; Romano, A.; Oneto, F.; Ottonello, D.; Michelon, R. Different season, different strategies: Feeding ecology of two syntopic forest-dwelling salamanders. *Acta Oecol.* **2012**, *43*, 42–50.

30. Costa, A.; Salvidio, S.; Posillico, M.; Matteucci, G.; De Cinti, B.; Romano, A. Generalisation within specialization: Inter-individual diet variation in the only specialized salamander in the world. *Sci. Rep.* **2015**, *5*, 13260. [CrossRef]

31. Royle, J.A. N-mixture models for estimating Population size from spatially replicated counts. *Biometrics* **2004**, *60*, 108–115. [CrossRef]

32. Kéry, M.; Royle, J.A. *Applied Hierarchical Modeling in Ecology: Analysis of Distribution, Abundance and Species Richness in R and BUGS: Volume 2: Dynamic and Advanced Models*; Academic Press: London, UK, 2020.

33. Waddle, J.; Dorazio, R.M.; Walls, S.; Rice, K.; Beauchamp, J.; Schuman, M. A new parameterization for estimating co-occurrence of interacting species. *Ecol. Appl.* **2010**, *20*, 1467–1475. [CrossRef]

34. Clare, J.D.; Linden, D.W.; Anderson, E.M.; MacFarland, D.M. Do the antipredator strategies of shared prey mediate intraguild predation and mesopredator suppression? *Ecol. Evol.* **2016**, *6*, 3884–3897. [CrossRef]

35. Roth, T.; Allan, E.; Pearman, P.B.; Amrhein, V. Functional ecology and imperfect detection of species. *Methods Ecol. Evol.* **2018**, *9*, 917–928. [CrossRef]

36. Brodie, J.F.; Helmy, O.E.; Mohd-Azlan, J.; Granados, A.; Bernard, H.; Giordano, A.J.; Zipkin, E. Models for assessing local-scale co-abundance of animal species while accounting for differential detectability and varied responses to the environment. *Biotropica* **2018**, *50*, 5–15. [CrossRef]

37. Rosa, G.; Salvidio, S.; Costa, A. European plethodontid salamanders on the forest floor: Testing for age-class segregation and habitat selection. *J. Herpetol.* **2022**, *56*, 27–33. [CrossRef]

38. Romano, A.; Rosa, G.; Salvidio, S.; Novaga, R.; Costa, A. How landscape and biotic interactions shape a Mediterranean reptile community. *Landsc. Ecol.* **2022**, *37*, 2915–2927. [CrossRef]

39. Angelini, C.; Vanni, S.; Vignoli, L. Salamandrina terdigitata (Bonnaterre, 1789) Salamandrina perspicillata (Savi, 1821). In *Fauna d'Italia. 42. Amphibia*; Lanza, B., Ed.; Edizioni Calderini: Bologna, Italy, 2007.

40. Lanza, B. Speleomantes strinatii (Aellen, 1958). In *Fauna d'Italia. 42. Amphibia*; Lanza, B., Andreone, F., Bologna, M., Corti, C., Razzetti, E., Eds.; Edizioni Calderini: Bologna, Italy, 2007.

41. Costa, A.; Crovetto, F.; Salvidio, S. European plethodontid salamanders on the forest floor: Local abundance is related to fine-scale environmental factors. *Herpetol. Conserv. Biol.* **2016**, *11*, 344–349.

42. Blondel, J.; Aronson, J. *Biology and Wildlife of the Mediterranean Region*; Oxford University Press: Oxford, UK, 1999.

43. Romano, A.; Costa, A.; Basile, M.; Raimondi, R.; Posillico, M.; Scinti Roger, D.; De Cinti, B. Conservation of salamanders in managed forests: Methods and costs of monitoring abundance and habitat selection. *For. Ecol. Manag.* **2017**, *400*, 12–18. [CrossRef]

44. Guisan, A.; Weiss, S.B.; Weiss, A.D. GLM versus CCA spatial modelling of plant species distribution. *Plant Ecol.* **1999**, *143*, 107–122. [CrossRef]

45. Conrad, O.; Bechtel, B.; Bock, M.; Dietrich, H.; Fischer, E.; Gerlitz, L.; Böhner, J. System for automated geoscientific analyses (SAGA) v. 2.1.4. *Geosci. Model Dev.* **2015**, *8*, 1991–2007. [CrossRef]

46. Amir, Z.; Sovie, A.; Luskin, M.S. Inferring predator–prey interactions from camera traps: A Bayesian co-abundance modeling approach. *Ecol. Evol.* **2022**, *12*, e9627. [CrossRef]

47. Mac Nally, R. Multiple regression and inference in ecology and conservation biology: Further comments on identifying important predictor variables. *Biodiv. Conserv.* **2002**, *11*, 1397–1401. [CrossRef]

48. Dormann, C.F.; Elith, J.; Bacher, S.; Buchmann, C.; Carl, G.; Carré, G.; García Marquéz, R. Collinearity: A review of methods to deal with it and a simulation study evaluating their performance. *Ecography* **2013**, *36*, 27–46. [CrossRef]

49. Kéry, M.; Schaub, M. *Bayesian Population Analysis Using WinBUGS: A Hierarchical Perspective*; Academic Press: London, UK, 2011.

50. Gelman, A.; Rubin, D.B. Inference from iterative simulation using multiple sequences. *Stat. Sci.* **1992**, *7*, 457–472. [CrossRef]

51. Duarte, A.; Adams, M.J.; Peterson, J.T. Fitting N-mixture models to count data with unmodeled heterogeneity: Bias, diagnostics, and alternative approaches. *Ecol. Model.* **2018**, *374*, 51–59. [CrossRef]

52. Knape, J.; Arlt, D.; Barraquand, F.; Berg, Å.; Chevalier, M.; Pärt, T.; Żmihorski, M. Sensitivity of binomial N-mixture models to overdispersion: The importance of assessing model fit. *Met. Ecol. Evol.* **2018**, *9*, 2102–2114. [CrossRef]

53. Costa, A.; Salvidio, S.; Penner, J.; Basile, M. Time-for-space substitution in N-mixture models for estimating population trends: A simulation-based evaluation. *Sci. Rep.* **2021**, *11*, 4581. [CrossRef]

54. Plummer, M. JAGS: A program for analysis of Bayesian graphical models using Gibbs sampling. In Proceedings of the 3rd International Workshop on Distributed Statistical Computing, Vienna, Austria, 20–22 March 2003; Volume 124, pp. 1–10.

55. Kellner, K. jagsUI: A Wrapper around Rjags to Streamline JAGSanalyses. R Package Version 1.5.1. Available online: https://github.com/kenkellner/jagsUI (accessed on 1 March 2023).

56. Rosa, G.; Bosio, M.; Salvidio, S.; Costa, A. Foraging success is differently affected by local climate in two syntopic forest-dwelling salamanders. *Ethol. Ecol. Evol.* **2022**, 1–10. [CrossRef]

57. Costa, A.; Rosa, G.; Salvidio, S. Size-Mediated Trophic Interactions in two Syntopic Forest Salamanders. *Animals* **2023**, *13*, 1281. [CrossRef]

58. Jaeger, R.G.; Gollmann, B.; Anthony, C.D.; Gabor, C.R.; Kohn, N.R. *Behavioral Ecology of the Eastern Red-Backed Salamander: 50 Years of Research*; Oxford University Press: Oxford, UK, 2016.

59. Alatalo, R.V.; Gustafsson, L.; Linden, M.; Lundberg, A. Interspecific competition and niche shifts in tits and the goldcrest: An experiment. *J. Anim. Ecol.* **1985**, *54*, 977–984. [CrossRef]

60. Kronfeld-Schor, N.; Dayan, T. Partitioning of time as an ecological resource. *Annu. Rev. Ecol. Evol. Syst.* **2003**, *34*, 153–181. [CrossRef]

61. Jaeger, R.G.; Kalvarsky, D.; Shimizu, N. Territorial behaviour of the red-backed salamander: Expulsion of intruders. *Anim. Behav.* **1983**, *31*, 490–496. [CrossRef]

62. Jaeger, R.G.; Wicknick, J.A.; Griffis, M.R.; Anthony, C.D. Socioecology of a terrestrial salamander: Juveniles enter adult territories during stressful foraging periods. *Ecology* **1995**, *76*, 533–543. [CrossRef]

63. Hairston, N.G.; Smith, F.E.; Slobodkin, L.B. Community structure, population control, and competition. *Am. Nat.* **1995**, *146*, 796–824. [CrossRef]

64. Petranka, J.W.; Smith, C.K. Interspecific competition between two terrestrial salamanders: Effects on individual growth and survivorship. *Oecologia* **1987**, *73*, 507–511.

65. Lunghi, E.; Corti, C.; Biaggini, M.; Zhao, Y.; Cianferroni, F. The trophic niche of two sympatric species of salamanders (Plethodontidae and Salamandridae) from Italy. *Animals* **2022**, *12*, 2221. [CrossRef] [PubMed]

*Article*

# An Isolated and Deeply Divergent *Hynobius* Species from Fujian, China


Zhenqi Wang [1,†], Siti N. Othman [2,†], Zhixin Qiu [1], Yiqiu Lu [1], Vishal Kumar Prasad [2], Yuran Dong [1], Chang-Hu Lu [1] and Amaël Borzée [2,3,*]

[1] The Co-Innovation Center for Sustainable Forestry in Southern China, College of Biology and the Environment, Nanjing Forestry University, Nanjing 210037, China
[2] Laboratory of Animal Behaviour and Conservation, College of Biology and the Environment, Nanjing Forestry University, Nanjing 210037, China
[3] Jiangsu Agricultural Biodiversity Cultivation and Utilization Research Center, Nanjing 210014, China
* Correspondence: amaelborzee@gmail.com
† These authors contributed equally to this work.



**Simple Summary:** What does not have a name is difficult to understand and protect. Upon the unexpected discovery of an *Hynobius* salamander in Fujian province, China, we worked on understanding its relationship with other species and ultimately describing it. Please welcome the Fujian Bamboo Salamander to science, a segregated species based on genetics and morphology. While it is related to other southern mainland Chinese species, it may have diverged earlier and share some similarities with morphology and behavior with the Anji salamander. The Fujian Bamboo Salamander is special as it produces vocalization when under threat. The species is, however, incredibly rare, fitting the definition of Critically Endangered in the IUCN Red List of Threatened Species.

**Abstract:** It is important to describe lineages before they go extinct, as we can only protect what we know. This is especially important in the case of microendemic species likely to be relict populations, such as *Hynobius* salamanders in southern China. Here, we unexpectedly sampled *Hynobius* individuals in Fujian province, China, and then worked on determining their taxonomic status. We describe *Hynobius bambusicolus* sp. nov. based on molecular and morphological data. The lineage is deeply divergent and clusters with the other southern Chinese *Hynobius* species based on the concatenated mtDNA gene fragments (>1500 bp), being the sister group to *H. amjiensis* based on the *COI* gene fragment, despite their geographic distance. In terms of morphology, the species can be identified through discrete characters enabling identification in the field by eye, an unusual convenience in *Hynobius* species. In addition, we noted some interesting life history traits in the species, such as vocalization and cannibalism. The species is likely to be incredibly rare, over a massively restricted distribution, fitting the definition of Critically Endangered following several lines of criteria and categories of the IUCN Red List of Threatened Species.

**Keywords:** Hynobidae; species description; bamboo forest



Citation: Wang, Z.; Othman, S.N.; Qiu, Z.; Lu, Y.; Prasad, V.K.; Dong, Y.; Lu, C.-H.; Borzée, A. An Isolated and Deeply Divergent *Hynobius* Species from Fujian, China. *Animals* **2023**, *13*, 1661. https://doi.org/10.3390/ani13101661

Academic Editor: Enrico Lunghi

Received: 6 March 2023
Revised: 27 April 2023
Accepted: 1 May 2023
Published: 17 May 2023




## 1. Introduction

Anthropogenisation of landscapes, and other human activities, have brought the world to the sixth great mass extinction [1,2]. The threats to species are not equal, and large-bodied species [3], as well as species with narrow spatial ranges [4], are principally impacted. The toll on species is staggering [5], and between 900 and 130,000 species have become extinct since the 1500s (www.iucnredlist.org/statistics; accessed on 3 May 2023). Similarly, species that have not yet been described are going extinct before being documented [6] and without known impacts [7] (as seen, for instance, in spiders [8]). Biodiversity loss is close to a tipping point [9], and conservation actions are the last tool to maintain evolutionary patterns free of anthropomorphic selection [10]. However, for conservation to be achieved, practitioners need to know what to protect, and thus species need to be described currently, conservation does not correspond to threat status [11], and we can only protect what we know.

With 41% of species listened as threatened by the IUCN Red List of Species (www.iucnredlist.org; accessed on 3 May 2023), amphibians are the most threatened animal class, with habitat loss being one of the principal drivers of species decline [12,13]. For the status of amphibians to improve, a clear taxonomy is first needed, and despite the large number of species described in China every year [14], many species are still in need of formal description or taxonomic revision. This need is especially true for southern China, a hotspot for biodiscovery [15] and conservation needs [16].

All *Hynobius* salamanders species were expected to have been described in China, although the taxonomic resolution of the genus is still an ongoing work, with some recent descriptions in other range countries of the genus, including the Republic of Korea [17] and Japan [18–20]. There are five described *Hynobius* species in mainland China, all part of the Southern Chinese group and all meant to be breeding in lentic water bodies, and four species on Taiwan Island, belonging to a segregated phylogenetic group and breeding in small cool mountain streams [21]. All these species are terrestrial, partially fossorial, and breed through larval development in water bodies. The species in southern China belong to the *Hynobius chinensis* group and include *H. chinensis* Günther, 1889 [22], *H. amjiensis* Gu, 1992 [23], *H. guabangshanensis* Shen, 2004 [24], *H. maoershanensis* Zhou, Jiang and Jiang, 2006 [25] and *H. yiwuensis* Cai, 1985 [26].

A *Hynobius* salamander was reported from Fujian province in 1978 and identified as *H. chinensis* [27]. However, following the multiple species descriptions since then, the distribution of *H. chinensis* is now known to be restricted to the Hubei province, and the morphological differences between the species have been clarified, including embryos [28]. The *Hynobius* salamander collected in 1978 from Fujian is therefore considered an undescribed species, and no additional individual has been found in the area since then, potentially resulting from local extirpations. As a result, upon encountering *Hynobius* salamanders in Fujian, we tested for phylogenetic clustering within the *H. chinensis* clade, then proceeded to determine their taxonomic status with phylogenetic tools, and accordingly described a new species with specific morphological characters.

## 2. Materials and Methods

### 2.1. Field Sampling

We found two individuals of the genus *Hynobius*, one in January and one in August 2022, from Quxi village, Liancheng County, Fujian, China (25.566° N, 116.938° E). We do not provide precise GPS coordinates here to protect the species from harvesting for the pet trade. The two individuals were found in a bamboo forest (*Phyllostachys* cf. *edulis*) about 1500 m above sea level. As ten days (24–26 January and 4–10 August) of fieldwork during adequate sampling seasons and times of day resulted in only two adult individuals, and the population size at this location, and likely for this species, is likely lower than 200 individuals, we followed the IUCN recommendation on ethical sampling [29]. We did not collect the individuals but instead orally swabbed them to obtain genetic materials and measured them (see Section 2.5 Morphometry) before releasing them at the point of capture to avoid a further threat to the species. In addition, we observed two waterbodies with eight pairs of egg sacs in January 2022, and we collected one egg sac deposited by the species for morphological measurement and to study the development of larvae in the species before releasing them at the point of capture. We used the entire body of an embryo from the egg sacks collected that died at an early developmental stage to extract DNA for a third individual (ID: 22HyF007). We consider this individual unlikely to be related to the adults swabbed due to the distance between the egg sacks and the adults in view of the dispersal abilities of the genus [30].

### 2.2. Ethical Approval

All applicable international, national, and/or institutional guidelines for the care and use of animals were strictly followed. All animal sample collection protocols complied with the current laws of the People's Republic of China. All observations and experiments conducted in this study are in agreement with the ethical recommendations of the College of Biology and the Environment at Nanjing Forestry University (IACUC approval number 2022014). We did not collect adult individuals as voucher specimens, in line with the IUCN recommendation on research involving species at risk of extinction [29]. Instead, we relied on one of the individuals raised from the egg mass.

### 2.3. DNA Preparation

We extracted the total genomic DNA for all three samples using the Qiagen DNeasy Blood & Tissue kit (QIAGEN Group, Hilden, Germany) according to the manufacturer's protocol. We then performed standard Polymerase Chain Reaction (PCR) amplifications for all 15 samples in 20 µL of total reaction per tube, containing 35 to 50 ng/µL of template DNA (Table 1). The final concentrations of the other PCR reagents were such as 0.125 µM for each forward and reversed primer, $1\times$ Ex taq Buffer (Takara; Shiga, Japan), 0.2 mM of dNTPs Mix (Takara; Shiga, Japan), 1.875 mM of magnesium chloride ($MgCl_2$), 0.1 unit/µL of Ex taq (HR001A, Takara; Shiga, Japan), and double distilled water added to make up the final volume. The PCR thermal profiles for each primer fragment are described in Table 1.

**Table 1.** PCR information and protocols to amplify DNA fragments. Three mtDNA fragments were selected to study the phylogenetic position of the candidate *Hynobius* species. The primer pair for the Cyt*b* fragment was designed in-house, and the other sequences come from the literature. The length of each PCR amplicon is estimated in base pairs.

| Fragment | Primer | 5′-3′ | Length | Source | PCR Condition |
|---|---|---|---|---|---|
| mtDNA−*COI* | LCO1490 | GGTCAACAAATCATAAAGATATTG | 680 bp | [31] | 95 °C for 3 min, followed by 35 cycles of 95 °C for 30 s, 53 °C for 30 s and 72 °C for 40 s, and final elongation at 72 °C for 7 min |
| | HCO2198 | TAAACTTCAGGGTGACCAAAAAATCA | 680 bp | [31] | |
| 16S rRNA | L02510 | CGCCTGTTTATCAAAAACAT | 500 bp | [32] | 95 °C for 3 min, followed by 35 cycles of 95 °C for 30 s, 53 °C for 30 s and 72 °C for 40 s, and final elongation at 72 °C for 7 min |
| | H03063 | CTCCGGTTTGAACTCAGATC | 500 bp | [33] | |
| mtDNA-Cyt*b* (in-house design) | CytbHy-F1 | TGTAGACCTCCCAACCCCC | 780 bp | This study | 95 °C for 5 min, followed by 35 cycles of 94 °C for 30 s, 55–60 °C for 30 s and 72 °C for 60 s, and final elongation at 72 °C for 10 min |
| | CytbHy-R1 | CGTAGGCGAATAAGAAATACCACT | 780 bp | This study | |

### 2.4. Molecular Analyses

We trimmed all DNA sequences of the three gene fragments, isolated from the three *Hynobius* individuals, and aligned them with their most homologous sequences retrieved from Genbank (see accession numbers and references in Supplementary Table S1). To test the phylogenetic relationship of East Asian Hynobid salamanders, we verified the monophyly of the candidate species and made sure it is not an introduced population of any of the other species; we reconstructed Bayesian Inference (BI) trees for four independent datasets: (i) 16S rRNA gene (*n* sequence = 97, length = 528 bp), (ii) protein-coding Cytochrome b (Cyt*b*) gene (*n* sequence = 117, length = 630 bp), (iii) protein-coding cytochrome c oxidase subunit I (*COI*) gene (*n* sequence = 84, length = 567 bp), and (iv) concatenated 16S rRNA, Cyt*b* and *COI* (*n* sequence = 29, length = 1451 bp; Supplementary Table S1). We searched for the best-fit sequence evolutionary model for each gene fragment using PartitionFinder v.2.1.1 (Canberra, Australia) [34]. We set the model selection to the Bayesian Information Criterion (BIC) and greedy search settings, and we standardized the estimation parameters of the partition model to unlinked branch length and fit to "Mr. Bayes" mode.

Here, the software recovered seven partitions based on a single non-coding and three coding codon's positions (best sequence substitution models in Table 2). Using the information of best substitution models for the gene fragments, we then reconstructed the BI trees based on the four datasets using Mr Bayes v.3.2.7 (Rochester, NY, USA) [35]. For each dataset, we ran the analysis for 10 million iterations of MCMC, with four independent chains, and resampled at each 1000 generation of the run until the trees reached convergences. We defined convergence as split frequency values lower than 0.05 at the end of all analyses and values higher than 200 for the Estimated Sample Size (ESS) for each parameter in the diagnostic reports provided by the software.

**Table 2.** Sequence substitution model. Best sequence substitution model for each gene fragment of the candidate *Hynobius* species are related sequences used for the reconstruction of the Bayesian Inference trees using Mr. Bayes v.3.2.7 [35], predicted and configured in PartitionFinder v. 2.1.1 [34].

| Mitochondrial Gene Fragment | Type | Partition's Strategy | Best Sequence Substitution Model |
|---|---|---|---|
| 16S rRNA | Non-coding ribosomal | 1–528 bp | HKY + I + G |
| Cyt*b* | Protein-coding | Exon by three codons' position (1–630 bp; 2–630 bp; 3–630 bp) | GTR + I + G |
| *COI* | Protein-coding | Exon by three codons' position (1–567 bp; 2–567 bp; 3–567 bp) | GTR + I + G |

Next, we constructed a haplotype network to identify the inheritance pattern between the sequences of our candidate species and its homologous sequences. Due to the strong posterior probability for the candidate species in the *COI* gene tree, and recommendations from the literature [36], we used the same dataset to analyze its haplotype distribution, consisting of 84 Hynobiids individuals (*n* taxa = 18) inferred from the *COI* gene fragment from individuals distributed across continental East Asia. Furthermore, we selected the *COI* alignment dataset due to its adequacy in the number of taxa for robust analysis in comparison to the concatenated dataset. We assigned the sequences to their respective species group and generated the haplotypes using DNAsp v.6.12.03 (Barcelona, Spain) [37]. We then determined the network of haplotypes using a statistical analysis of parsimony with TCS [38], implemented in PopART v.1.7 (Dunedin, New Zealand) [39]. Lastly, we mapped and visualized the distribution of the species involved in the haplotype network using QGIS v.2.18.0 (Chicago, IL, USA) [40].

Finally, to estimate the evolutionary divergence over sequence pairs between species, we analyzed 29 sequences of concatenated 16S rRNA, Cyt*b*, and *COI* using MEGA v.11.0.13 (State College, PA, USA) [41]. We assigned the sequence data to 18 groups based on species identity, and we removed all ambiguous positions for each sequence pair using the pairwise deletion option. There were a total of 1451 positions in the final dataset, and we computed the evolutionary divergence using the Maximum Composite Likelihood model [42].

*2.5. Morphometry*

For the morphological analyses, we measured the characters listed by Shen et al. [24], Lai and Lue [21], Borzée and Min [17], and Chen et al. [43] as they provide the largest dataset available for continental East Asian *Hynobius* species. See Figure 1 in Chen et al. [43] for a graphic representation of the following variables. We used digital calipers (model 1108–150, Insize; Suzhou, China) to the nearest 0.1 mm three times per individual and averaged, following the recommendations of Borzée and Min [17]. The measurements were: total length from the snout to the tail end (TOL); length of body from tip of snout to anterior angle of vent (SVL); tail length from anterior angle of vent to tip of tail (TL); tail height as the height of the tail at its highest point (TH); tail width as the width of the tail at its widest point (TW); head length from tip of snout to gular fold (HL); head width immediately posterior to jaw articulation (HW); snout length from the anterior border of the eye to the tip of the snout (SL); interocular distance, measured between medial margins of eyelids (IOD); laterally measured diameters of eyes (DE); head height as the height of the head at its highest point (HH); internarial distance, measured between the medial margins of the nares (IND); body width at axilla (BW); trunk length from axilla to groin (AG); forelimb length from anterior insertion to tip of second toe (FOL); hindlimb length from anterior insertion to tip of third toe (HIL). We also counted the number of costal grooves (COS) for all individuals, and the number of horny vomerine teeth (VT) for four individuals.

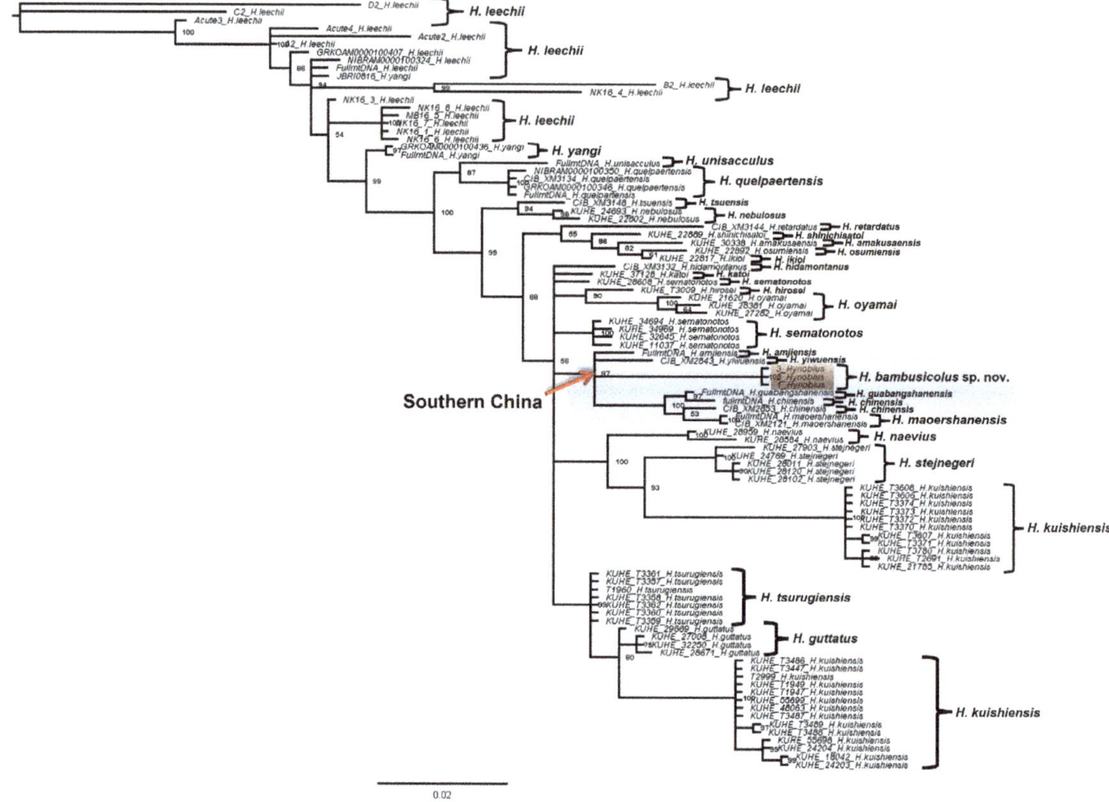


**Figure 1.** Bayesian Inference tree based on a 528 bp-long 16S rRNA fragment for 97 Hynobiidae salamanders in East Asia. Our results recovered *Hynobius bambusicolus* sp. nov. (orange shade) as monophyletic and sharing a sister relationship with the congeneric Hynobius species distributed in southern China (blue shade).

Ideally, a species description should include a morphological comparison between the new species and the most closely related one. Here, it would be a two-by-two comparison between the species described in this paper and *H. amjiensis* (see phylogenetic results). However, due to (1) the rarity of *H. amjiensis*, for instance, Chen et al. [44] found 16 individuals over eight years of surveys, (2) the urgency to describe the species to be able to protect it, (3) the robustness of the phylogenetic analyses (see phylogenetic results), and (4) the presence of great morphological variations; we provide a morphological description for the two adult individuals found, and a morphological key enabling the identification of the species. We do note the presence of morphological data in the literature for *H. maoershanensis* [43] and two Taiwanese species [21]; however, a two-by-two comparison with these species only would not be clarifying the taxonomy as these species are not phylogenetically or geographically close.

The species within the *H. chinensis* species complex, and ranging on the Chinese mainland, are *H. amjiensis*, *H. guabangshanensis*, *H. chinensis*, *H. maoershanensis*, and *H. yiwuensis*. We did not include individuals from northeast China, Taiwan Island, or the Japanese archipelago in the morphological comparisons as they were not closely related in the phylogenetic analyses. We then created a dichotomic key to enable the morphological identification of the species and the determination of discrete characters.

In addition, with the development of "technoecology" [45], 3D printing is becoming widely used, and 3D printed replicas can also benefit taxonomy and systematics [46]. Here, we provide a printable 3D model that can be downloaded online. We photographed an adult individual using a high-resolution camera (Nikon Z9 with a Nikkor MC 50 mm f 2.8 lens) in the wild but on a hard, permeable, and smooth substrate. We took multiple photographs from different angles, with a focus on capturing the characteristics and features of the species. The photos were imported into Autodesk Maya 2019 (Autodesk; San Francisco, CA, USA) as image planes, which were then used as reference images to ensure accuracy and consistency in the details of the model. Using the reference images as a guide, we created a base mesh using a polygon modeling technique. The base mesh was adjusted until it matched the overall shape of the salamander. The sculpting tools were then used to accurately refine and detail the features of the species. We then created a UV map and applied textures to the model, using photos of the specimen's skin as a reference to create a realistic representation. The skeleton of the model was created by using rigging tools. The mesh was then skinned to the skeleton, allowing for a realistic pose of the model. The completed 3D model was exported from Maya in an OBJ format.

### 2.6. Egg and Larval Development

To ascertain that the development of the species does not deviate from that of the *H. chinensis* species complex, we documented the development of the species before the eggs hatched and after hatching at 16, 69, 74, 77, and 87 days. We aimed to document development between stages 29 and 37 when embryos are in the egg sac; tadpoles between stages 40 and 49 when free swimming with balancers; between stages 50 and 59, when the larvae are not yet using their developing hind legs and once after stage 60 when the larvae use their hind legs [47]. The salamanders were kept at 26 °C, in water boiled, and aged for at least 24 h in open-top buckets. Photographs were taken with a Z9 camera (Nikon, Tokyo, Japan) and a Z MC 50 mm f 2.8 lens (Nikkor, Nikon).

### 2.7. Acoustic Signal

When probed, the individual captured in January sometimes emitted a vocalization. We managed to record the individual only once with a linear PCM recorder (Tascam DR-40; Santa Fe Springs, CA, USA) using the built-in microphone. The call was recorded at a sampling rate of 44.1 kHz with a 16-bit resolution. We analyzed the call properties following the method in Prasad et al. [48] using Raven Pro bioacoustics v.1.5 (Center for conservation bioacoustics; Ithaca, NY, USA).

## 3. Results

### 3.1. Molecular Analyses

The rates of evolutionary divergences between sequence pairs provided support for the significant variation between the candidate species and all 18 other species. The highest variation in the average base substitutions per site for sequence pairs was between the candidate species and *Hynobius amjiensis* (average rate = 0.098; marked with * in Table 3). The average divergence rate was comparable to eight other sequence pairs (bolded values in Table 3). Most importantly, the average divergence rate was comparatively higher than the rates of base substitution for the four following pairs: (i) *H. arisanensis* vs. *H. formosanus*, mean = 0.008; (ii) *H. maoershanensis* vs. *H. guabangshanensis*, mean = 0.021; (iii) *H. chinensis* vs. *H. guabangshanensis*, mean = 0.032; and (iv) *H. maoershanensis* vs. *H. chinensis* (mean: 0.035; Table 3).

**Table 3.** Matrix of best substitution rates for evolutionary divergence between the candidate *Hynobius* species and congeneric species. The congeneric *Hynobius* species used in this study are related Hynobiid salamanders distributed in East Asia. We inferred the genetic distance from mitochondrial 16S rRNA, *Cytb*, and *COI* (1451 bp) from 29 taxa representing 18 species. The rate of base substitution provides support to the evolutionary divergence between sequences of *Hynobius bambusicolus* sp. nov. and all 17 species compared. The lowest genetic distance is between the focal species and *H. amjiensis* (0.098; marked with *), but it is higher than the value from four species-pair comparisons (marked with #). All genetic distance values lower than the mean of pairwise difference for *H. bambusicolus* sp. nov. (0.098) are in bold in the table.

| | Species | 1 | 2 | 3 | 4 | 5 | 6 | 7 | 8 | 9 | 10 | 11 | 12 | 13 | 14 | 15 | 16 | 17 | 18 |
|---|---|---|---|---|---|---|---|---|---|---|---|---|---|---|---|---|---|---|---|
| 1 | *H. bambusicolus* sp. nov. | N/A | | | | | | | | | | | | | | | | | |
| 2 | *P. shangchengensis* | 0.166 | | | | | | | | | | | | | | | | | |
| 3 | *H. maoershanensis* | 0.113 | 0.170 | | | | | | | | | | | | | | | | |
| 4 | *H. yiwuensis* | 0.102 | 0.157 | 0.106 | | | | | | | | | | | | | | | |
| 5 | *H. chinensis* | 0.106 | 0.166 | **0.035** # | 0.096 | | | | | | | | | | | | | | |
| 6 | *H. hidamontanus* | 0.101 | 0.146 | **0.093** | **0.090** | 0.091 | | | | | | | | | | | | | |
| 7 | *H. nebulosus* | 0.119 | 0.149 | 0.106 | 0.100 | 0.106 | 0.083 | | | | | | | | | | | | |
| 8 | *H. tokyoensis* | 0.138 | 0.173 | 0.132 | 0.128 | 0.126 | 0.118 | 0.126 | | | | | | | | | | | |
| 9 | *H. tsuensis* | 0.115 | 0.147 | 0.106 | 0.096 | 0.101 | 0.086 | **0.074** | 0.126 | | | | | | | | | | |
| 10 | *H. amjiensis* | **0.098** * | 0.161 | **0.092** | 0.096 | 0.090 | 0.095 | 0.102 | 0.132 | 0.099 | | | | | | | | | |
| 11 | *H. arisanensis* | 0.118 | 0.145 | 0.121 | 0.110 | 0.121 | 0.109 | 0.104 | 0.121 | 0.098 | 0.122 | | | | | | | | |
| 12 | *H. dunni* | 0.122 | 0.156 | 0.109 | 0.107 | 0.106 | 0.091 | **0.060** | 0.121 | **0.077** | 0.099 | 0.105 | | | | | | | |
| 13 | *H. formosanus* | 0.115 | 0.144 | 0.121 | 0.106 | 0.122 | 0.108 | 0.104 | 0.117 | 0.100 | 0.123 | **0.008** # | 0.104 | | | | | | |
| 14 | *H. guabangshanensis* | 0.106 | 0.161 | **0.032** # | **0.093** | **0.021** # | **0.083** | 0.103 | 0.127 | 0.098 | 0.091 | 0.117 | 0.106 | 0.118 | | | | | |
| 15 | *H. quelpaertensis* | 0.114 | 0.146 | 0.105 | 0.099 | 0.100 | **0.085** | 0.092 | 0.123 | **0.088** | 0.095 | 0.111 | **0.094** | 0.107 | 0.095 | | | | |
| 16 | *H. unisacculus* | 0.116 | 0.141 | 0.105 | 0.096 | 0.102 | 0.095 | 0.093 | 0.120 | 0.092 | 0.096 | 0.113 | **0.096** | 0.110 | 0.102 | **0.062** | | | |
| 17 | *H. yangi* | 0.104 | 0.147 | 0.114 | 0.101 | 0.107 | 0.083 | **0.088** | 0.122 | **0.088** | 0.099 | 0.105 | **0.096** | 0.104 | 0.108 | **0.075** | 0.073 | | |
| 18 | *H. leechii* | 0.100 | 0.148 | 0.107 | 0.099 | 0.104 | 0.081 | 0.089 | 0.121 | 0.089 | 0.098 | 0.105 | 0.097 | 0.104 | 0.107 | 0.066 | 0.070 | 0.036 | N/A |

Apart from the significant divergence in DNA sequences, our phylogeny recovered the candidate species as a deeply divergent clade amongst the East Asian Hynobiid (Figures 1–4). All the BI trees, for single genes and concatenated fragments, coherently recovered the candidate species as monophyletic within the Southern Chinese group (posterior probability (PP): between 59% to 99%; Figures 1–4). For the *COI* and concatenated trees, we recovered the candidate species as sister to *H. amjiensis* (PP: 97% and 65%; Figure 1).

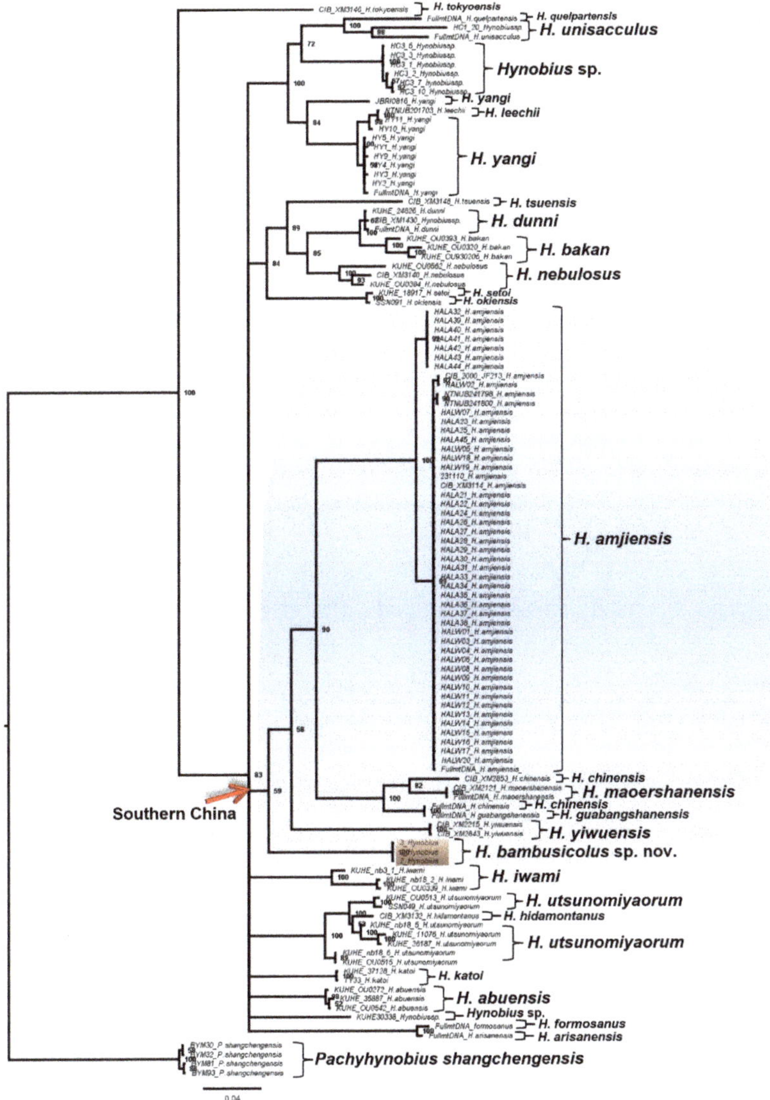


**Figure 2.** Bayesian Inference tree reconstructed from a 630 bp-long *Cytb* gene fragment for 117 Hynobid salamanders distributed across East Asia. Our results recovered *Hynobius bambusicolus* sp. nov. (orange shade) as monophyletic and basal to the congeneric clades of *Hynobius* species distributed in southern China (blue shade).

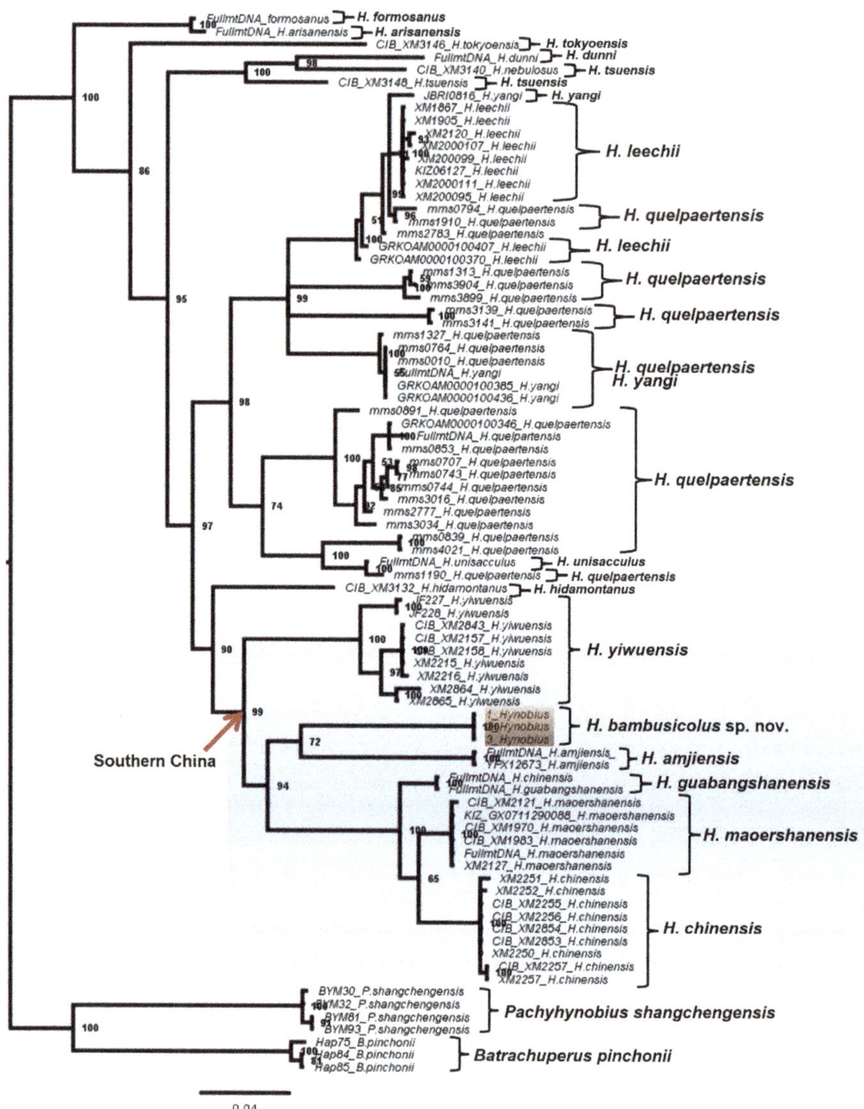



**Figure 3.** Bayesian Inference tree based on a 567 bp-long *COI* gene fragment for 84 Hynobid salamanders distributed across East Asia. Our results recovered *Hynobius bambusicolus* sp. nov. (orange shade) as monophyletic within the southern China group (blue shade) and sharing a sister relationship with *Hynobius amjiensis*.

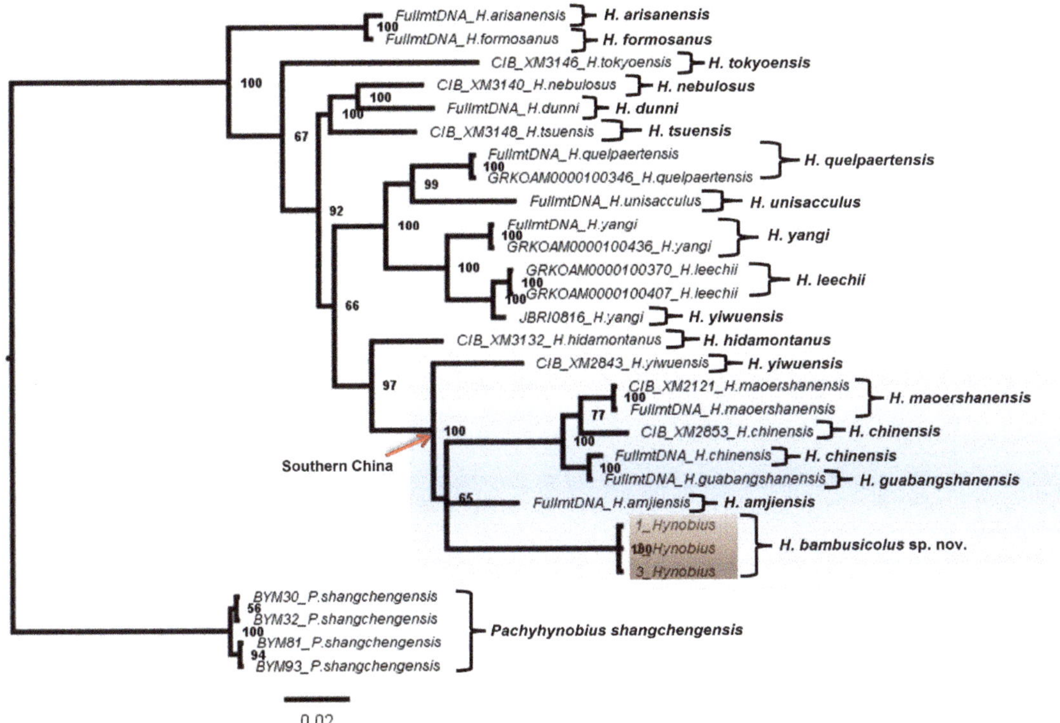



**Figure 4.** Bayesian Inference tree inferred from 1451 bp of 16S rRNA, *Cytb*, and *COI* gene fragments of Hynobiid salamanders distributed across East Asia. Our analyses recovered *Hynobius bambusicolus* sp. nov. (orange shade) as clustered with the congeneric *Hynobius* distributed in southern China (blue shade).

Based on the gene fragment for *COI* (*n* taxa = 84), we obtained 55 haplotypes (h) from 567 sites representing 16 species of Hynobiid salamanders distributed across East Asia, with a haplotype diversity (Hd) of 0.98 (Figure 5). The haplotype distribution further revealed a shared relationship between the haplotype group of the candidate species and the geographically related haplotypes of Southern Chinese *Hynobius*. The haplotype network and distribution also clarified the deep genetic and geographic isolation between the candidate species and the related *H. amjiensis* and *H. maoershanensis*.

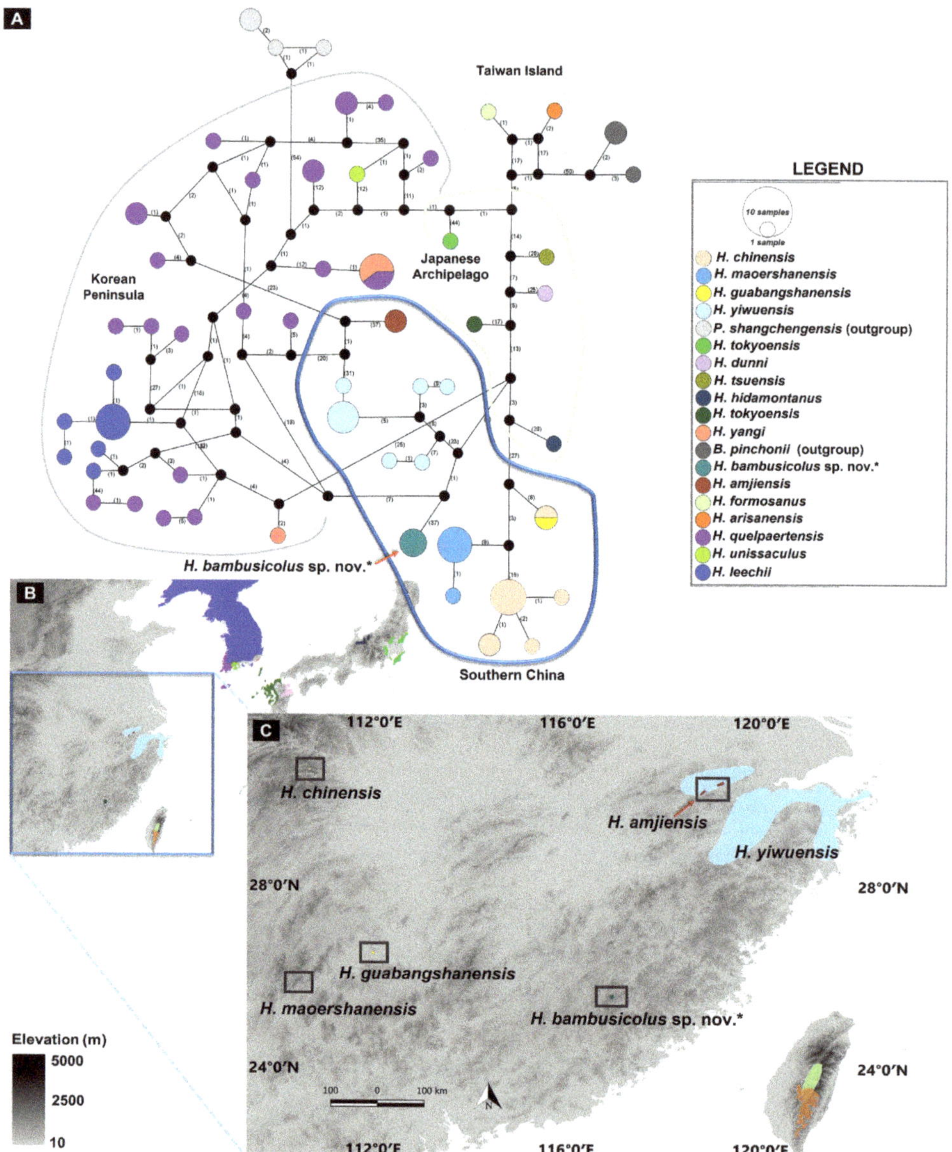


**Figure 5.** Haplotype network and distribution of Hynobiid salamanders distributed in East Asia. The haplotype analysis involved 84 Hynobiids individuals (*n* taxa = 18) inferred from the *COI* gene fragment (Figure 3). (**A**) TSC network and distribution of 55 haplotypes based on their geographic distribution. The number along each branch connecting haplotypes indicates the number of mutations. The asterisk (*) in the figure marks the focal *Hynobius* species described in this study, *Hynobius bambusicolus* sp. nov. (**B**) Distribution of all 18 *Hynobius* species in East Asia, matching with the haplotype network. The blue box highlights the southern Chinese clade of *Hynobius*. (**C**) Distribution of the southern Chinese clade of *Hynobius* species in which *Hynobius bambusicolus* sp. nov. is clustered. The small grey boxes highlight the restricted ranges of four *Hynobius* species within the clade.

*3.2. Morphometric results*

The measurements of the adults highlighted the large size of the species (Table 4), surpassed only by *H. chinensis*, but a low number of coastal grooves, similar to *H. yiwuensis*. Finally, the toe formula was similar to that of three other species in the area (Figure 6). We found a variable number of vomeral teeth (four individuals), likely due to the partial development of the voucher specimen. The holotype was characterized by 12 pairs of vomeral teeth on the right and 13 on the left, the largest individual by 18 and 19 pairs, and the two other individuals by 17 and 16, and 12 and 13 pairs (Figure 7). As this character does not seem fixed yet in our voucher specimen, we do not use it for species identification. For the morphological identification key based on non-invasive identification of the species, we determined the number of costal grooves to be the first character of importance, enabling the assignment of the individual examined into either one of three categories: 10 or fewer grooves, 11 or 12 grooves, or 13 or more grooves (Figure 6). Once assigned to one of these categories, we could develop the following identification key:

- The combination of 10 or fewer grooves with the following:
    - A total length < 151 mm assigns the individual to *H. yiwuensis*;
    - A total length > 180 mm assigns the individual to *H. bambusicolus* sp. nov;
- The combination of 11 or 12 grooves with the following:
    - A total length < 180 mm assigns the individual to *H. maoershanensis*;
    - A total length > 180 mm assigns the individual to *H. chinensis*;
- The combination of 13 or more grooves with the following:
    - A total length < 151 mm assigns the individual to *H. guabangshanensis*;
    - A total length > 152 mm assigns the individual to *H. amjiensis*.

**Table 4.** Morphological measurements. Morphological measurements for two adult and four juvenile *Hynobius bambusicolus* sp. nov. from Fujian, China. The measurements were taken three times per individual and averaged for consistency.

| Age | Adults | | Juveniles | | | |
|---|---|---|---|---|---|---|
| Type | Wild | Wild | Paratype | Paratype | Paratype | Holotype |
| ID | 22HyF001 | 22HyF002 | 22HyF003 | 22HyF004 | 22HyF005 | 22HyF006 |
| TOL | 191.57 | 196.94 | 48.06 | 48.38 | 50.43 | 55.11 |
| SVL | 136.36 | 137.81 | 28.39 | 27.72 | 31.32 | 34.05 |
| TL | 55.21 | 59.13 | 19.57 | 18.74 | 20.32 | 21.73 |
| HL | 22.78 | 21.86 | 7.27 | 8.48 | 8.25 | 10.32 |
| HW | 18.57 | 15.84 | 6.22 | 6.80 | 6.82 | 7.82 |
| IOD | 6.19 | 5.25 | 2.67 | 2.92 | 2.83 | 3.44 |
| IND | 6.29 | 5.77 | 3.16 | 3.29 | 2.72 | 3.16 |
| BW | 17.75 | 16.08 | 5.24 | 6.24 | 6.83 | 7.09 |
| AG | 52.07 | 51.06 | 16.46 | 16.14 | 17.95 | 21.04 |
| FOL | 21.13 | 20.77 | 6.49 | 6.73 | 5.89 | 8.08 |
| HIL | 25.53 | 23.08 | 6.38 | 6.30 | 8.73 | 7.77 |
| COS | 9 | 10 | 9 | 9 | 10 | 9 |
| SL | 19.78 | 20.82 | 4.72 | 5.47 | 4.44 | 5.37 |
| TH | 10.12 | 10.52 | 3.26 | 4.02 | 3.06 | 3.49 |
| TW | 11.03 | 10.90 | 2.21 | 2.29 | 2.83 | 2.48 |
| HH | 10.50 | 10.99 | 3.81 | 3.91 | 3.92 | 3.70 |
| DE | 5.31 | 4.64 | 2.44 | 2.31 | 2.80 | 2.54 |

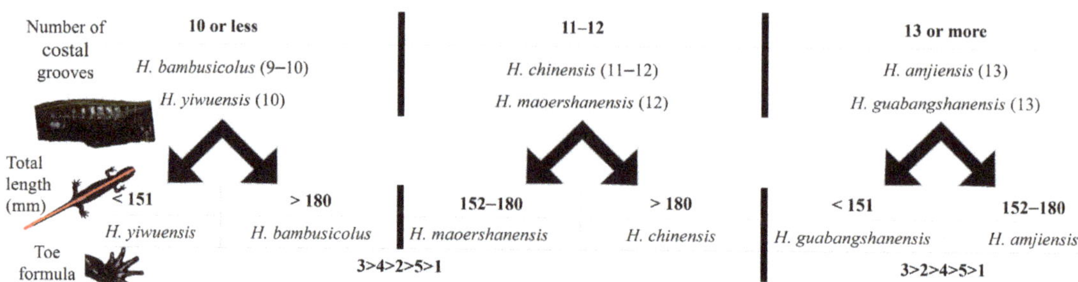



**Figure 6.** Species identification key for southern Chinese *Hynobius* species. With this morphological identification key, it is possible to identify adult *Hynobius* individuals from southern China, based on morphology and without invasive procedures. The data used are from AmphibiaChina.org and our study.

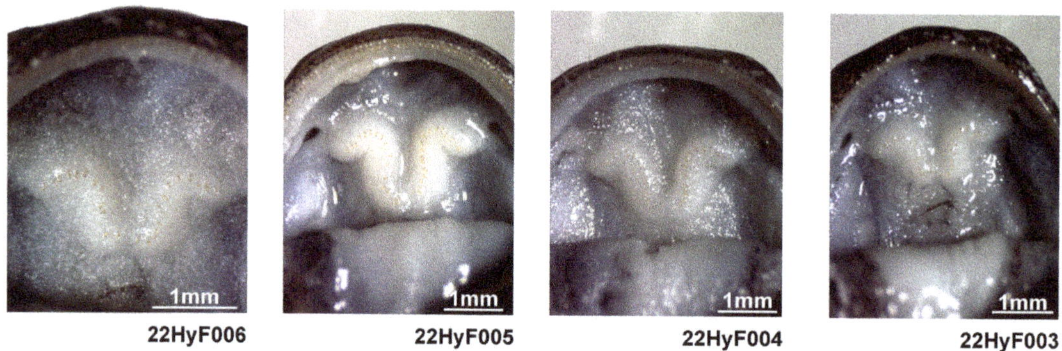



**Figure 7.** Vomeral teeth details for four juvenile *Hynobius bambusicolus* sp. nov. preserved in 70% alcohol. Vouchers were collected in Quxi village, Liancheng county, People's Republic of China (25.566° N, 116.938° E).

### 3.3. Species Description

Following the line of evidence based on molecular analyses for species-level divergence with other southern Chinese *Hynobius* species and a clearly different morphology evidenced by discrete characters, we formally describe the new species. Measurements are summarized in Table 4.

Nomenclature History

Among the species of *Hynobius* currently recognized [49], *Hynobius turkestanicus* Nikolskii is the only enigmatic taxon [50], and it is unlikely to be a member of the *Hynobius* genus [51]. Our focal taxa are also different from *Hynobius yunanicus*, which was invalidated [52] and later re-established [53], but as a synonym of *Pachyhynobius shangchengensis* [49]. The new species belongs to the group of lentic habitat breeders distributed in China and the Korean Peninsula, and morphological comparisons with the closest related species, *H. amjiensis*, and the geographically related species are presented in Figure 5.

*Hynobius bambusicolus* sp. nov. Wang, Othman, Qiu and Borzée

Synonymy:

"*Hynobius chinensis* (partim): [27]" pp. 218–229.

- Holotype

Voucher 22HyF006; sub-adult collected by Zhenqi Wang and Zhixin Qiu on 26 January 2022 in Quxi village, Liancheng county, People's Republic of China (25.566° N, 116.938° E; Figure 5). Measurements and counts are in Table 4 and Figure 8.

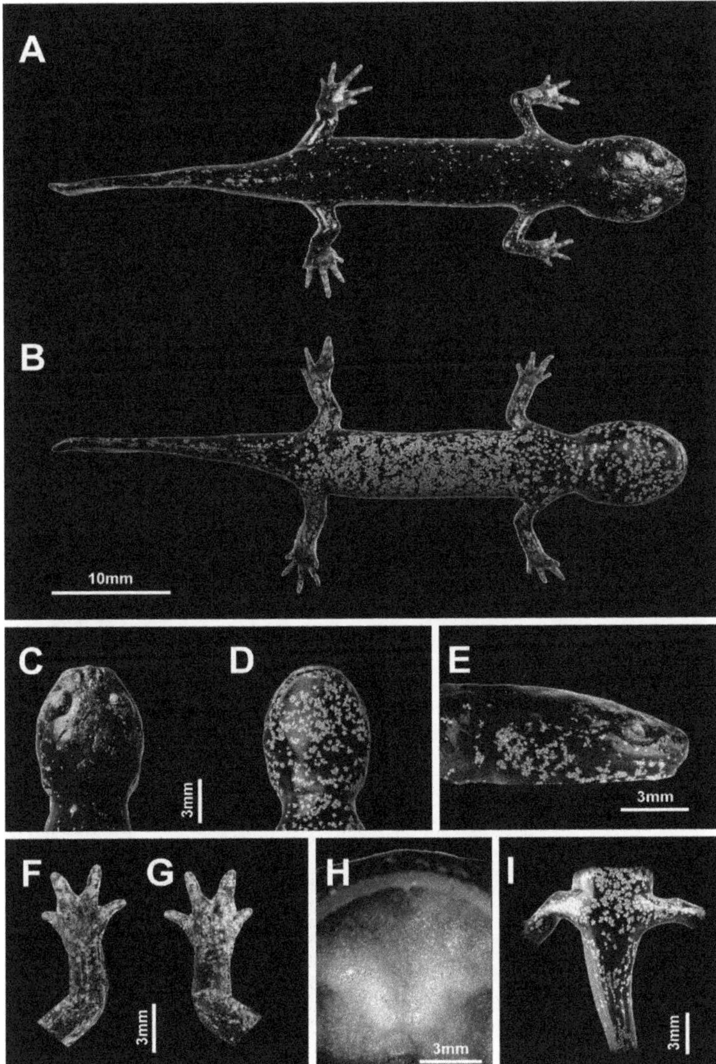

**Figure 8.** Holotype of *Hynobius bambusicolus* sp. nov. preserved in 70% alcohol. Voucher 22HyF006 collected in Quxi village, Liancheng county, People's Republic of China (25.566° N, 116.938° E). Measurements and counts in Table 4. (**A**) Dorsal view. (**B**) Ventral view. (**C**) Head in dorsal view. (**D**) Head in ventral view. (**E**) Head in lateral view. (**F**) Opisthenar view of left hand. (**G**) Volar view of left hand. (**H**) Palatal region showing the vomerine tooth series (pale orange vertical structures on the white background at the center of the palate). (**I**) Ventral view of cloacal area.

- Paratypes

Vouchers 22HyF003, 22HyF004, 22HyF005. Sub-adults collected by Zhenqi Wang and Zhixin Qiu on 26 January 2022 in Quxi village, Liancheng county, People's Republic of China (25.566° N, 116.938° E; Figure 5). Measurements and counts are in Table 4 and Figure 9.

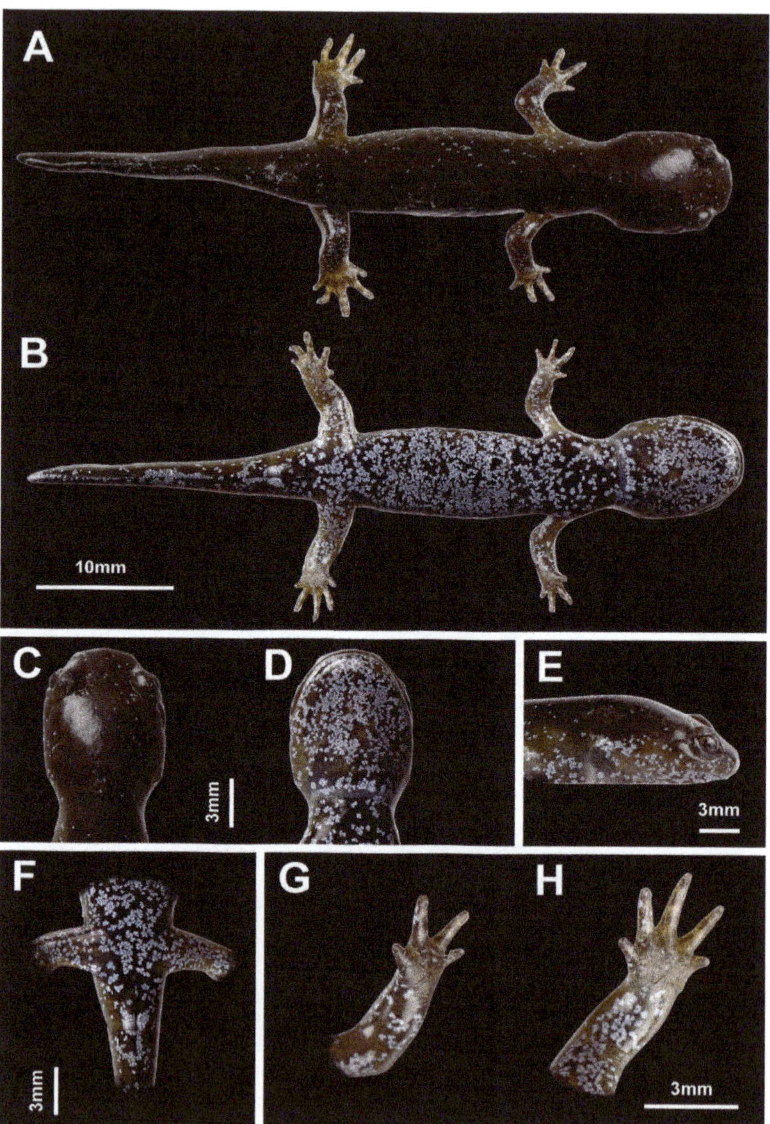

**Figure 9.** Paratype of *Hynobius bambusicolus* sp. nov. preserved in 70% alcohol. Voucher 22HyF003 collected in Quxi village, Liancheng county, People's Republic of China (25.5661° N, 116.9386° E; Figure 5). Measurements and counts in Table 4. (**A**) Dorsal view. (**B**) ventral view. (**C**) Head dorsal view. (**D**) Head ventral view. (**E**) Head lateral view. (**F**) Ventral view of cloacal area. (**G**) Opisthenar view of left hand. (**H**) Volar view of left hand.

- Etymology

The species was first found in Quxi village, Liancheng county, in the west of Fujian province in China. The name *H. bambusicolus* sp. nov. comes from the habitat of the holotype and the only known habitat type for the species: bamboo forests. The vernacular name of the species, Fujian Bamboo Salamander, reflects the scientific name of the species, as does its Chinese name: 虚竹小鲵 (pronounced: Xū Zhú Xiǎo Ní). This salamander is named after a main character from Jin Yong's swordsman fiction "The semi gods and semi devils" [54], with Xuzhu (虚竹) as the main character and where "虚" [xū] means humble, and "竹" [zhú] means bamboo. This character, Xuzhu, was an unknown Shaolin monk, but he inherited the powers of the leader of the Carefree by coincidence and started its legendary journey.

- Identity, diagnosis, and distribution

To date, the species is known from its type locality only, Quxi village, Liancheng county (Figure 5). Larvae are typical of *Hynobius* larvae in shape and color and do not differ from other *Hynobius* species in the region in their development (Figure 10). The embryos develop in egg sacs (Figure 10A), larvae first swimming freely with balancers (Figure 10B; shown at day 16), then develop non-functional hind limbs (Figure 10C; shown at day 69), which slowly become functional (Figure 10D,E; shown at day 74 and 77), and the gills regress before metamorphosis (Figure 10F; shown at day 84). Juveniles are brown, darkening with age, with a large variation in blue speckles on their dorsum, which disappears as they age (Figure 9). Adults of the species are uniform dark chocolate, with light grey and bluish speckles on the venter (Figure 11). Identification is best assessed based on location, although discrete morphological characters include the combination of 10 or fewer costal grooves with a total length > 180 mm (Figure 6). To facilitate the identification and further study, the OBJ file of this model can be downloaded (Supplementary File S1). The visual representation of the model is provided in Appendix A (Figure A1).

**Figure 10.** Representative developmental stages for eggs and larvae of *Hynobius bambusicolus* sp. nov. from Fujian, China. (**A**) pre-hatching; (**B**) 16 days old. (**C**) 69 days old. (**D**) 87 days old. (**E**) 74 days old. (**F**) 77 days old.

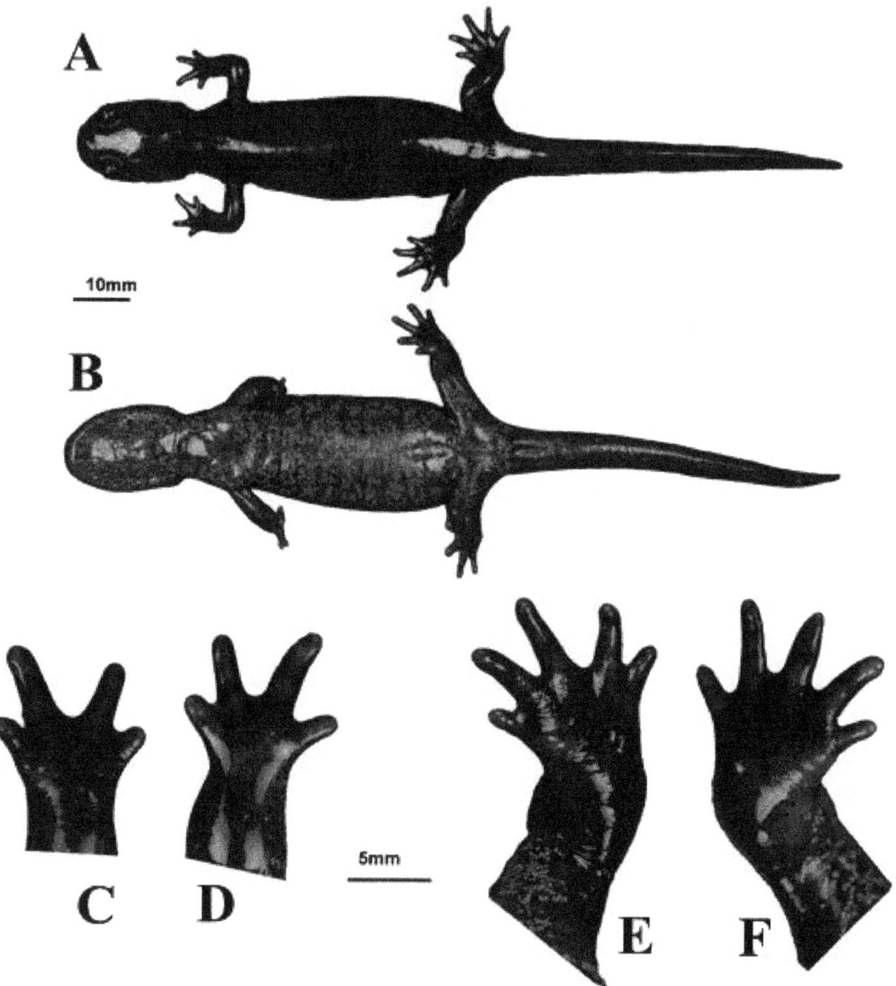

**Figure 11.** Example of *Hynobius bambusicolus* sp. nov. adult in life. Voucher 22HyF001 from Quxi village, Liancheng county, People's Republic of China (25.5661° N, 116.9386° E; Figure 5). Measurements and counts in Table 4. (**A**) Dorsal view. (**B**) ventral view. (**C**) Opisthenar view of left hand. (**D**) Opisthenar view of right hand. (**E**) Opisthenar view of left foot. (**F**) Opisthenar view of right foot.

- ZooBank registration

We hereby state that the present paper has been registered to the Official Register of Zoological Nomenclature (ZooBank) under LSID: urn:lsid:zoobank.org:pub: 9047D736-3394-4B36-AE97-CEEB3359B36D. The new species name *Hynobius bambusicolus* sp. nov., has been registered under LSID: urn:lsid:zoobank.org:act:18CAEC61-DF60-4401-91C5-FEF5614FA08C.

- Nomenclatural acts

The electronic edition of this article conforms to the requirements of the amended International Code of Zoological Nomenclature, and hence the new names contained herein are available under that code from the electronic edition of this article. This published work and the nomenclatural act it contains have been registered in ZooBank, the online registration system for the ICZN. The ZooBank LSIDs (Life Science Identifiers) can be resolved, and the associated information viewed through any standard web browser by appending the LSID to the prefix "http://zoobank.org/". The LSID for this publication is urn:lsid:zoobank.org:pub:9047D736-3394-4B36-AE97-CEEB3359B36D. The electronic edition of this work was published in a journal with an ISSN and has been archived and is available from the following digital repositories: PubMed Central.

- Habitat and behavior

*Hynobius bambusicolus* breeds in shallow pools in bamboo forests (Figure 12) above 1400 m above sea level. All eggs were observed in ruts made from tire tracks. The puddles were 5 to 12 cm deep, and most egg sacs were laid about 10 cm deep without being attached to the substrate, containing between 21 and 27 eggs each. Adult salamanders were found under logs, stones, and dead leaves, in wet soil and humus. These shelters were surrounded by weeds and dry branches in waterlogged areas, and we did not find any individuals on the surface, even at night (Figure 12). Larvae were fed with bloodworms, but some individuals still cannibalized their kin (Figure 13). Once metamorphosed, recently metamorphosed individuals raised their tails when startled, a behavior to divert predation from vital organs [55].

**Figure 12.** Natural habitat and oviposition site for *Hynobius bambusicolus* sp. nov. The site is in Quxi village, Liancheng county, People's Republic of China (25.5661° N, 116.9386° E).

**Figure 13.** Example of cannibalism in *Hynobius bambusicolus* sp. nov. larvae.

The individual recorded produced a single type of call, with a very short duration (153.40 ms) and of low frequency (peak frequency of 129.20 Hz). The vocalization was composed of four strong harmonics (Figure 14), with a peak amplitude of 403 U and a maximum entropy of 3.473 bits. While highly unusual, underwater vocalizations have been reported in salamanders, such as *Siren intermedia* [56]. In this case, the vocalizations were produced by a male being probed, suggesting it could be submissive or alarm calls. In our recording, the individual emitted a similar "alarm call"/"squeak" while half submerged in the water, maybe as an agonistic signal.

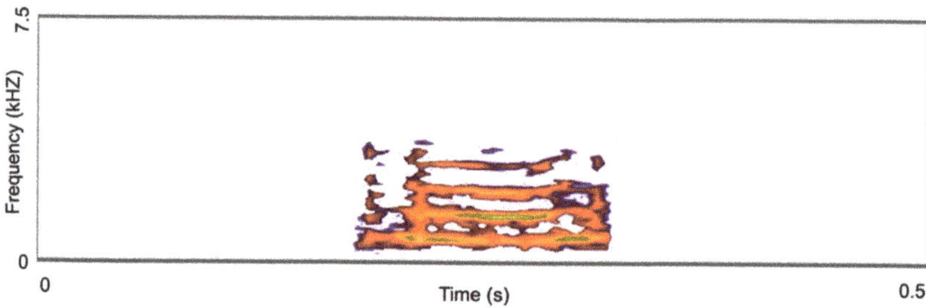

**Figure 14.** Spectrogram of the acoustic signal release by *Hynobius bambusicolus* sp. nov. Dense yellow-orange colour displays the harmonic-like pattern of the acoustic signal. Schematic illustrations based on the data extracted with FFT size = 1024 pts. (Hanning window, 43.1 Hz resolution).

## 4. Discussion and Conclusions

Here we described a Hynobiid salamander species, *Hynobius bambusicolus*, the Fujian Bamboo Salamander, based on deeply divergent molecular analyses, non-overlapping morphological characters, and vastly segregated distribution. Field identification of *Hynobius* individuals based on morphology is not always easy [17], but discrete morphological characters are present to identify *H. bambusicolus* (Figure 6). The easiest character to identify the species perhaps remains the geographic location as the range of the species does not overlap with that of the other *Hynobius* species, with a gap of several hundred kilometers (Figure 5).

Our results, through the phylogenetic trees, haplotype network, and comparative pairwise genetic distance, show that *H. bambusicolus* is alternatively the most divergent and earliest-branching species among Southern Chinese *Hynobius* clades (*Cytb* tree; Figure 2), nested with the other Southern Chinese *Hynobius* clades (16S rRNA and concatenated mtDNA trees; Figures 1 and 4) and only the sister taxon to *H. amjiensis* (*COI* tree; Figure 3). The stem emergence of Southern Chinese *Hynobius* clades (*H. amjiensis* and *H. chinensis*) is dated to c. 20 mya [57], and *H. bambusicolus* is, therefore, an ancient lineage, likely to have seen its distribution regularly shift, expand, and contract with paleogeographic and climatic variations [58]. The area is inhabited by other Caudata, and competition between *H. bambusicolus* and other genera, such as *Pachytriton* and *Paramesotriton*, is not impossible.

The restricted distribution of *H. bambusicolus* is also a negative point to the survival of the species. The species is micro-endemic, presumably composed of a single known relic population with an apparently incredibly small population size, similar to other south Chinese *Hynobius* species (e.g., [44]). The genus is likely to have distributed south broadly since the early Miocene [59], and withdrawn to higher elevations, similarly to other amphibians shifting distribution [60], due to climatic variations as most representatives of the genus are cold-adapted, and the mid-Miocene vegetation in Fujian was tropical [61]. For instance, specific genes in *H. chinensis* are upregulated as a response to cold temperatures [62]. *Hynobius bambusicolus*, however, is adapted to sub-tropical bamboo forests, although cold preference or tolerance for spawning is still a trait in the species, as two egg masses were found in January 2023 at the same site on a snowy day.

The species is known from a single locality, and surveys in 2023 of all water bodies in the area where the species could potentially spawn did not result in more than five extremely small water bodies. As there were eight egg sacks at the known site (therefore four females), even considering this number as a constant for each water body, it would be a maximum of 20 breeding females, and therefore a population size likely to be well below 200 breeding individuals, matching with the criteria B1, B2, and C2 for Critically Endangered following the recommendation of the IUCN Red List of Threatened Species [63]. Due to the threatened status of the species, establishing an ex situ population could help prevent the extinction of the species in the face of growing climate instability and stochastic extinction risks.

Due to the rarity of this new species, we urge all hobbyists to avoid collecting this salamander or divulging local information and resisting any trade. The breeding site is within a planted bamboo forest, surrounded by mixed forest, and the distribution of the species is likely to have been impacted by bamboo plantation and harvest over the last decades or centuries. A site where the species was found about 10 years ago by a ranger on the opposite side of the valley is now entirely dry and unlikely to support a breeding population, making habitat loss the main threat to the species. Climate change is also most likely to impact the species, as seen in other Hynobiids [64], and we can expect the distribution of the species to be contracting. It is, therefore, important to lower other stresses as amphibians can cope with a low number of conservation pressures, but populations crash when there are too many stress factors [65]. As the species is occurring in bamboo forests exploited by humans, we recommend reducing the use of herbicides and restricting water pumping in the area. In addition, restoring the habitat through the creation of additional artificial shallow ponds or rehabilitating old water reservoirs found throughout the bamboo plantation will help boost population growth [66].

**Supplementary Materials:** The following supporting information can be downloaded at: https://www.mdpi.com/article/10.3390/ani13101661/s1, Table S1: GenBank accession numbers; File S1: obj 3D model.

**Author Contributions:** Conceptualization, Z.W., S.N.O. and A.B.; methodology, S.N.O., V.K.P., Y.D. and A.B.; software, S.N.O., V.K.P. and A.B.; validation, Z.W., S.N.O., Z.Q., Y.L., Y.D., C.-H.L. and A.B.; formal analysis, S.N.O., V.K.P. and A.B.; investigation, Z.W., S.N.O., Y.L. and A.B.; resources, A.B.; writing—original draft preparation, Z.W., S.N.O., Y.L., V.K.P. and A.B.; writing—review and editing, Z.W., S.N.O., Z.Q., Y.D. and A.B.; funding acquisition, A.B. All authors have read and agreed to the published version of the manuscript.

**Funding:** This work was supported by the Foreign Youth Talent Program (QN2021014013L) from the Ministry of Science and Technology of the People's Republic of China to A.B.

**Institutional Review Board Statement:** All applicable international, national, and/or institutional guidelines for the care and use of animals were strictly followed. All animal sample collection protocols complied with the current laws of the People's Republic of China. All observations and experiments conducted in this study are in agreement with the ethical recommendations of the College of Biology and the Environment at Nanjing Forestry University (AICUC approval number 2022014). We did not collect adult individuals as voucher specimens, in line with the IUCN recommendation on research involving species at risk of extinction (IUCN 1989). Instead, we relied on one of the individuals raised from the egg mass.

**Informed Consent Statement:** Not applicable.

**Data Availability Statement:** The morphological data are in Table 4, and the genetic sequences are in Supplementary Table S1. Accession numbers and references for all DNA sequences used for the bioinformatic analyses. DNA sequences of the candidate *Hynobius* species generated in this study and homologous sequences are retrieved from Genbank.

**Acknowledgments:** We would like to thank Qing-Xian Lin from the College of The Environment & Ecology, Xiamen University, Jin-Ping Wu, Kang-Peng Luo, Wen-Jie Yu and Jin-Hua Wu from Meihua Mountain National Nature Reserve, Yi Lu from the Geological Survey of Jiangsu province, and Sheng-Ming Lin, for providing the field assistance and important information. We are also grateful to Nikolay Poyarkov for his help and knowledge about the taxonomy of the *Hynobius* genus, and to Ke Jiang, Jinlong Ren, and Chen-Yang Tang from the Chengdu Institute of Biology, Chinese Academy of Science, for their help in accessing the literature.

**Conflicts of Interest:** The authors declare no conflict of interest.

## Appendix A

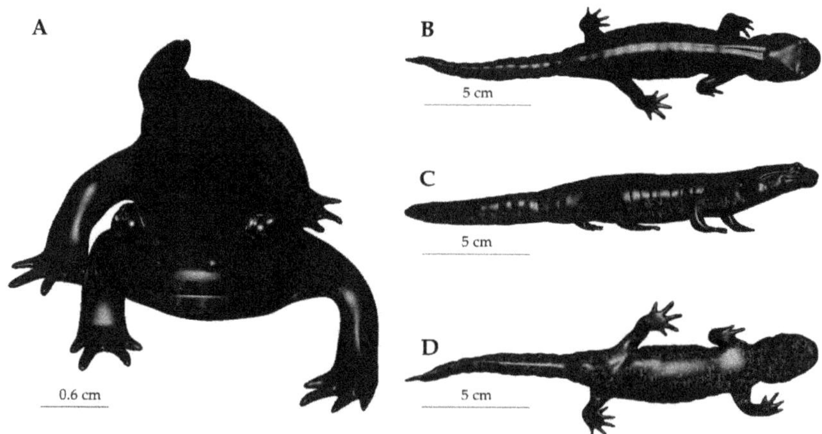



**Figure A1.** Representation of the 3D model. (**A**) Frontal view. (**B**) Dorsal view. (**C**) lateral view. (**D**) Ventral view.

## References

1. Wake, D.B.; Vredenburg, V.T. Are we in the midst of the sixth mass extinction? A view from the world of amphibians. *Proc. Natl. Acad. Sci. USA* **2008**, *105* (Suppl. 1), 11466–11473. [CrossRef]
2. Ceballos, G.; Ehrlich, P.R.; Barnosky, A.D.; García, A.; Pringle, R.M.; Palmer, T.M. Accelerated modern human–induced species losses: Entering the sixth mass extinction. *Sci. Adv.* **2015**, *1*, e1400253. [CrossRef] [PubMed]
3. Smith, F.A.; Elliott Smith, R.E.; Lyons, S.K.; Payne, J.L. Body size downgrading of mammals over the late Quaternary. *Science* **2018**, *360*, 310–313. [CrossRef]
4. Manne, L.L.; Pimm, S.L. Beyond eight forms of rarity: Which species are threatened and which will be next? *Anim. Conserv.* **2001**, *4*, 221–229. [CrossRef]

5. Régnier, C.; Achaz, G.; Lambert, A.; Cowie, R.H.; Bouchet, P.; Fontaine, B. Mass extinction in poorly known taxa. *Proc. Natl. Acad. Sci. USA* **2015**, *112*, 7761–7766. [CrossRef]

6. Tedesco, P.A.; Bigorne, R.; Bogan, A.E.; Giam, X.; Jézéquel, C.; Hugueny, B. Estimating how many undescribed species have gone extinct. *Conserv. Biol.* **2014**, *28*, 1360–1370. [CrossRef]

7. Brodie, J.F.; Aslan, C.E.; Rogers, H.S.; Redford, K.H.; Maron, J.L.; Bronstein, J.L.; Groves, C.R. Secondary extinctions of biodiversity. *Trends Ecol. Evol.* **2014**, *29*, 664–672. [CrossRef] [PubMed]

8. Rix, M.G.; Huey, J.A.; Main, B.Y.; Waldock, J.M.; Harrison, S.E.; Comer, S.; Austin, A.D.; Harvey, M.S. Where have all the spiders gone? The decline of a poorly known invertebrate fauna in the agricultural and arid zones of southern Australia. *Austral Entomol.* **2017**, *56*, 14–22. [CrossRef]

9. Rinawati, F.; Stein, K.; Lindner, A. Climate change impacts on biodiversity—The setting of a lingering global crisis. *Diversity* **2013**, *5*, 114–123. [CrossRef]

10. Otto, S.P. Adaptation, speciation and extinction in the Anthropocene. *Proc. R. Soc. B* **2018**, *285*, 20182047. [CrossRef]

11. Wang, Z.; Zeng, J.; Meng, W.; Lohman, D.J.; Pierce, N.E. Out of sight, out of mind: Public and research interest in insects is negatively correlated with their conservation status. *Insect Conserv. Divers.* **2021**, *14*, 700–708. [CrossRef]

12. Sodhi, N.S.; Brook, B.W.; Bradshaw, C.J. Causes and consequences of species extinctions. In *The Princeton Guide to Ecology*; Levin, S.A., Carpenter, S.R., Godfray, H.C.J., Kinzig, A.P., Loreau, M., Losos, J.B., Walker, B., Wilcove, D.S., Eds.; Princeton University Press: Princeton, NJ, USA, 2009.

13. Lindenmayer, D.B.; Fischer, J. *Habitat Fragmentation and Landscape Change: An Ecological and Conservation Synthesis*; Island Press: Washington, DC, USA, 2013.

14. Wang, K.; Ren, J.; Chen, H.; Lyu, Z.; Guo, X.; Jiang, K.; Chen, J.; Li, J.; Guo, P.; Wang, Y.; et al. The updated checklists of amphibians and reptiles of China. *Biodivers. Sci.* **2020**, *28*, 189–218.

15. Moura, M.R.; Jetzt, W. Shortfalls and opportunities in terrestrial vertebrate species discovery. *Nat. Ecol. Evol.* **2021**, *5*, 631–639. [CrossRef]

16. Button, S.; Borzée, A. An integrative synthesis to global amphibian conservation priorities. *Glob. Chang. Biol.* **2021**, *27*, 4516–4529. [CrossRef] [PubMed]

17. Borzée, A.; Min, M.-S. Disentangling the impact of speciation, sympatry and island effect on the morphology of seven *Hynobius* sp. salamanders. *Animals* **2021**, *11*, 187. [CrossRef] [PubMed]

18. Matsui, M.; Misawa, Y.; Yoshikawa, N.; Nishikawa, K. Taxonomic reappraisal of *Hynobius tokyoensis*, with description of a new species from northeastern Honshu, Japan (Amphibia: Caudata). *Zootaxa* **2022**, *5168*, 207–221. [CrossRef]

19. Sugawara, H.; Naito, J.I.; Iwata, T.; Nagano, M. Molecular phylogenetic and morphological problems of the Aki Salamander *Hynobius akiensis*: Description of two new species from Chugoku, Japan. *Bull. Kanagawa Prefect. Mus.* **2022**, *2022*, 35–46.

20. Sugawara, H.; Fujitani, T.; Seguchi, S.; Sawahata, T.; Nagano, M. Taxonomic re-examination of the Yamato Salamander *Hynobius vandenburghi*: Description of a new species from Central Honshu, Japan. *Bull. Kanagawa Prefect. Mus.* **2022**, *51*, 47–59.

21. Lai, J.S.; Lue, K.Y. Two new *Hynobius* (Caudata: Hynobiidae) salamanders from Taiwan. *Herpetologica* **2008**, *64*, 63–80. [CrossRef]

22. Günther, A.C.L.G. Third contribution to our knowledge of reptiles and fishes from the upper Yangtze-Kiang. *Ann. Mag. Nat. Hist.* **1889**, *6*, 218–229. [CrossRef]

23. Gu, H.-q. A new species of *Hynobius-Hynobius amjiensis*. In *Animal Science Research: A Symposium Issued to Celebrate the 90th Birthday of the Professor Mangven Ly Chang*; Qian, Y., Zhao, E.-M., Zhao, K.-t., Eds.; Chinese Forestry Press: Beijing, China, 1992; pp. 39–43.

24. Shen, Y.-h.; Deng, X.-j.; Wang, B. A new hynobiid species, *Hynobius guabangshanensis*, from Hunan Province, China (Amphiba: Hynobiidae). *Acta Zool. Sin.* **2004**, *50*, 209–215.

25. Zhou, F.; Jiang, A.-W.; Jiang, D.-b. A new species of the genus *Hynobius* from Guangxi Zhuang Autonomous Region, China (Caudata, Hynobiidae). *Acta Zootaxonomica Sin.* **2006**, *31*, 670–674.

26. Cai, C.-m. A survey of tailed amphibians for Zhejiang, with description of a new species of *Hynobius*. *Acta Herpetol. Sin.* **1985**, *4*, 109–114.

27. Hu, S.; Fei, L.; Ye, C. Investigation report of amphibians in Fujian. In *Research Material of Amphibians and Reptiles*; Sichuan Institute of Biology: Chengdu, China, 1978; Volume 4, pp. 22–29.

28. Fei, L.; Hu, S.; Ye, C.; Huang, Y. *Fauna Sinica: Amphibia*; Science Press: Beijing, China, 2006; Volume 1.

29. IUCN. *IUCN Policy Statement on Research Involving Species at Risk of Extinction*; International Union for the Conservation of Nature: Gland, Switzerland, 1989.

30. Kusano, T.; Miyashita, K. Dispersal of the salamander, *Hynobius nebulosus tokyoensis*. *J. Herpetol.* **1984**, *18*, 349–353. [CrossRef]

31. Folmer, O.; Black, M.; Hoeh, W.; Lutz, R.; Vrijenhoek, R. DNA primers for amplification of mitochondrial cytochrome c oxidase subunit I from diverse metazoan invertebrates. *Mol. Mar. Biol. Biotechnol.* **1994**, *3*, 294–299. [PubMed]

32. Palumbi, S. Nucleic acids II: The polymerase chain reaction. In *Molecular Systematics*; Hillis, D., Moritz, C., Mable, B., Eds.; Sinauer Associates: Sunderland, MA, USA, 1996; pp. 205–247.

33. Rassmann, K. Evolutionary age of the Galapagos iguanas predates the age of the present Galapagos Islands. *Mol. Phylogenet. Evol.* **1997**, *7*, 158–172. [CrossRef] [PubMed]

34. Lanfear, R.; Frandsen, P.B.; Wright, A.M.; Senfeld, T.; Calcott, B. PartitionFinder 2: New methods for selecting partitioned models of evolution for molecular and morphological phylogenetic analyses. *Mol. Biol. Evol.* **2017**, *34*, 772–773. [CrossRef]

35. Ronquist, F.; Teslenko, M.; van der Mark, P.; Ayres, D.; Darling, A.; Höhna, S.; Larget, B.; Liu, L.; Suchard, M.; Huelsenbeck, J. MrBayes 3.2: Efficient Bayesian phylogenetic inference and model choice across a large model space. *Syst. Biol.* **2012**, *61*, 539–542. [CrossRef]

36. Xia, Y.; Gu, H.F.; Peng, R.; Chen, Q.; Zheng, Y.C.; Murphy, R.W.; Zeng, X.M. COI is better than 16S rRNA for DNA barcoding Asiatic salamanders (Amphibia: Caudata: Hynobiidae). *Mol. Ecol. Resour.* **2012**, *12*, 48–56. [CrossRef]

37. Rozas, J.; Ferrer-Mata, A.; Sánchez-DelBarrio, J.; Guirao-Rico, S.; Librado, P.; Ramos-Onsins, S.; Sánchez-Gracia, A. DnaSP 6: DNA sequence polymorphism analysis of large data sets. *Mol. Biol. Evol.* **2017**, *34*, 3299–3302. [CrossRef]

38. Templeton, A.R.; Crandall, K.A.; Sing, C.F. A cladistic analysis of phenotypic associations with haplotypes inferred from restriction endonuclease mapping and DNA sequence data. III. Cladogram estimation. *Genetics* **1992**, *132*, 619–633. [CrossRef] [PubMed]

39. Leigh, J.W.; Bryant, D. POPART: Full-feature software for haplotype network construction. *Methods Ecol. Evol.* **2015**, *6*, 1110–1116. [CrossRef]

40. QGIS Development Team. QGIS Geographic Information System.; Open Source Geospatial Foundation Project. Available online: http://qgis.osgeo.org (accessed on 3 May 2023).

41. Tamura, K.; Stecher, G.; Kumar, S. MEGA 11: Molecular Evolutionary Genetics Analysis Version 11. *Mol. Biol. Evol.* **2021**, *38*, 3022–3027. [CrossRef] [PubMed]

42. Tamura, K.; Nei, M.; Kumar, S. Prospects for inferring very large phylogenies by using the neighbor-joining method. *Proc. Natl. Acad. Sci. USA* **2004**, *101*, 11030–11035. [CrossRef]

43. Chen, H.; Bu, R.; Ning, M.; Yang, B.; Wu, Z.; Huang, H. Sexual Dimorphism in the Chinese Endemic Species *Hynobius maoershanensis* (Urodela: Hynobiidae). *Animals* **2022**, *12*, 1712. [CrossRef]

44. Chen, C.; Yang, J.; Wu, Y.; Fan, Z.; Lu, W.; Chen, S.; Yu, L. The breeding ecology of a critically endangered salamander, *Hynobius amjiensis* (Caudata: Hynobiidae), Endemic to Eastern China. *Asian Herpetol. Res.* **2016**, *7*, 54–59. [CrossRef]

45. Allan, B.M.; Nimmo, D.G.; Ierodiaconou, D.; Wal, J.V.D.; Koh, L.P.; Ritchie, E.G. Futurecasting ecological research: The rise of technoecology. *Ecosphere* **2018**, *9*, e02163. [CrossRef]

46. Fernando, M.d.F.L. 3D print so more scholars can access unique specimens. *Nature* **2018**, *563*, 7731.

47. Iwasawa, H.; Yamashita, K. Normal stages of development of a hynobiid salamander, *Hynobius nigrescens* Stejneger. *Jpn. J. Herpetol.* **1991**, *14*, 39–62. [CrossRef]

48. Prasad, V.K.; Chuang, M.F.; Das, A.; Ramesh, K.; Yi, Y.; Dinesh, K.P.; Borzée, A. Coexisting good neighbours: Acoustic and calling microhabitat niche partitioning in two elusive syntopic species of balloon frogs, *Uperodon systoma* and *U. globulosus* (Anura: Microhylidae) and potential of individual vocal signatures. *BMC Zool.* **2022**, *7*, 27. [CrossRef]

49. Frost, D.R. *Amphibian Species of the World: An Online Reference*; Version 6.1; Electronic Database; American Museum of Natural History: New York, NY, USA, 2022.

50. Kuzmin, S.L.; Dunayev, E.A. On the problem of the type territory of the Turkestanian Salamander (*Hynobius turkestanicus* Nikolsky, 1909). *Adv. Amphib. Res. Former Sov. Union* **2000**, *5*, 243–250.

51. Min, M.-S.; Baek, H.; Song, J.-Y.; Chang, M.; Poyarkov Jr, N. A new species of salamander of the genus *Hynobius* (Amphibia, Caudata, Hynobiidae) from South Korea. *Zootaxa* **2016**, *4169*, 475–503. [CrossRef]

52. Xiong, J.L.; Chen, Q.; Zeng, X.M.; Zhao, E.M.; Qing, L.Y. Karyotypic, morphological, and molecular evidence for *Hynobius yunanicus* as a synonym of *Pachyhynobius shangchengensis* (Urodela: Hynobiidae). *J. Herpetol.* **2007**, *41*, 664–671. [CrossRef]

53. Nishikawa, K.; Jiang, J.P.; Matsui, M.; Mo, Y.M.; Chen, X.H.; Kim, J.B.; Tominaga, A.; Yoshikawa, N. Invalidity of *Hynobius yunanicus* and molecular phylogeny of *Hynobius* salamander from continental China (Urodela, Hynobiidae). *Zootaxa* **2010**, *2426*, 65–67. [CrossRef]

54. Jin, Y. *The Demi-Gods and Semi-Devils*; Joint Publishing Press: Beijing, China, 1994.

55. Heo, K.; Shin, Y.; Messenger, K.R. *Hynobius notialis* (Southern Korean Salamander). Behavior. *Herpetol. Rev.* **2022**, *53*, 642.

56. Fauth, J.E.; Resetarits, W.J. Biting in the salamander *Siren intermedia intermedia*: Courtship component or agonistic behavior? *J. Herpetol.* **1999**, *33*, 493–496. [CrossRef]

57. Zhang, P.; Chen, Y.-Q.; Zhou, H.; Liu, Y.-F.; Wang, X.-L.; Papenfuss, T.J.; Wake, D.B.; Qu, L.-H. Phylogeny, evolution, and biogeography of Asiatic salamanders (Hynobiidae). *Proc. Natl. Acad. Sci. USA* **2006**, *103*, 7360–7365. [CrossRef]

58. Lu, B.; Zheng, Y.; Murphy, R.W.; Zeng, X. Coalescence patterns of endemic Tibetan species of stream salamanders (Hynobiidae: *Batrachuperus*). *Mol. Ecol.* **2012**, *21*, 3308–3324. [CrossRef]

59. Li, J.; Fu, C.; Lei, G. Biogeographical consequences of Cenozoic tectonic events within East Asian margins: A case study of *Hynobius* biogeography. *PLoS ONE* **2011**, *6*, e21506. [CrossRef]

60. Zhou, Y.; Wang, S.; Zhu, H.; Li, P.; Yang, B.; Ma, J. Phylogeny and biogeography of South Chinese brown frogs (Ranidae, Anura). *PLoS ONE* **2017**, *12*, e0175113. [CrossRef]

61. Jacques, F.M.; Shi, G.; Su, T.; Zhou, Z. A tropical forest of the middle Miocene of Fujian (SE China) reveals Sino-Indian biogeographic affinities. *Rev. Palaeobot. Palynol.* **2015**, *216*, 76–91. [CrossRef]

62. Che, R.; Sun, Y.; Wang, R.; Xu, T. Transcriptomic analysis of endangered Chinese salamander: Identification of immune, sex and reproduction-related genes and genetic markers. *PLoS ONE* **2014**, *9*, e87940.

63. IUCN Standards and Petitions Committee. *Guidelines for Using the IUCN Red List Categories and Criteria*; Version 14; Prepared by the Standards and Petitions Committee; IUCN: Gland, Switzerland, 2019.

64. Shin, Y.; Min, M.S.; Borzée, A. Driven to the edge: Species distribution modeling of a Clawed Salamander (Hynobiidae: *Onychodactylus koreanus*) predicts range shifts and drastic decrease of suitable habitats in response to climate change. *Ecol. Evol.* **2021**, *11*, 14669–14688. [CrossRef] [PubMed]
65. Green, D.M.; Lannoo, M.J.; Lesbarrères, D.; Muths, E. Amphibian population declines: 30 years of progress in confronting a complex problem. *Herpetologica* **2020**, *76*, 97–100. [CrossRef]
66. Moor, H.; Bergamini, A.; Vorburger, C.; Holderegger, R.; Bühler, C.; Egger, S.; Schmidt, B.R. Bending the curve: Simple but massive conservation action leads to landscape-scale recovery of amphibians. *Proc. Natl. Acad. Sci. USA* **2022**, *119*, e2123070119. [CrossRef]

 *animals*

*Article*

# Survived the Glaciations, Will They Survive the Fish? Allochthonous Ichthyofauna and Alpine Endemic Newts: A Road Map for a Conservation Strategy

Ilaria Bernabò [1,*], Mattia Iannella [2,*], Viviana Cittadino [1,2], Anna Corapi [1], Antonio Romano [3], Franco Andreone [4], Maurizio Biondi [2], Marcellino Gallo Splendore [5] and Sandro Tripepi [1]

[1]  Department of Biology, Ecology and Earth Science, University of Calabria, Via P. Bucci 4/B, I-87036 Rende, Italy
[2]  Department of Life, Health & Environmental Sciences, University of L'Aquila, Via Vetoio—Coppito, I-67100 L'Aquila, Italy
[3]  Consiglio Nazionale delle Ricerche—Istituto per la BioEconomia, Via dei Taurini 19, I-00100 Roma, Italy
[4]  Museo Regionale di Scienze Naturali, Via G. Giolitti 36, I-10123 Torino, Italy
[5]  Museo del Castagno, Via V. Emanuele, I-87013 Fagnano Castello, Italy
*  Correspondence: ilaria.bernabo@unical.it (I.B.); mattia.iannella@univaq.it (M.I.)

**Simple Summary:** In Italy, there is a growing concern over the survival of an endemic subspecies of newt, *Ichthyosaura alpestris inexpectata*, commonly known as the Calabrian Alpine newt, due to the recent fish introduction and acclimatisation into three of the few localised lentic habitats where this glacial relict occurs. Results of a recent survey focused on this rare endemic taxon are presented. We provide updated information on the Calabrian Alpine newt distribution, adding two new localities, documenting local extinction at a few historical sites, providing a rough estimation of population size, and giving a description of breeding habitat features. The results of our pilot study will facilitate future research activities, conservation measures for the amphibian assemblage, and habitat management after fish introduction in the Natura 2000 site. We also pinpoint some actions useful to avoid the extinction of this remarkable taxon.

**Citation:** Bernabò, I.; Iannella, M.; Cittadino, V.; Corapi, A.; Romano, A.; Andreone, F.; Biondi, M.; Gallo Splendore, M.; Tripepi, S. Survived the Glaciations, Will They Survive the Fish? Allochthonous Ichthyofauna and Alpine Endemic Newts: A Road Map for a Conservation Strategy. *Animals* **2023**, *13*, 871. https://doi.org/10.3390/ani13050871

Academic Editor: Enrico Lunghi

Received: 31 January 2023
Revised: 23 February 2023
Accepted: 25 February 2023
Published: 27 February 2023

**Abstract:** The Calabrian Alpine newt (*Ichthyosaura alpestris inexpectata*) is a glacial relict with small and extremely localised populations in the Catena Costiera (Calabria, Southern Italy) and is considered to be "Endangered" by the Italian IUCN assessment. Climate-induced habitat loss and recent fish introductions in three lakes of the Special Area of Conservation (SAC) Laghi di Fagnano threaten the subspecies' survival in the core of its restricted range. Considering these challenges, understanding the distribution and abundance of this newt is crucial. We surveyed the spatially clustered wetlands in the SAC and neighbouring areas. First, we provide the updated distribution of this subspecies, highlighting fish-invaded and fishless sites historically known to host Calabrian Alpine newt populations and two new breeding sites that have been recently colonised. Then, we provide a rough estimate of the abundance, body size and body condition of breeding adults and habitat characteristics in fish-invaded and fishless ponds. We did not detect Calabrian Alpine newts at two historically known sites now invaded by fish. Our results indicate a reduction in occupied sites and small-size populations. These observations highlight the need for future strategies, such as fish removal, the creation of alternative breeding habitats and captive breeding, to preserve this endemic taxon.

**Keywords:** amphibian; Urodela; conservation; *Ichthyosaura alpestris inexpectata*; Calabria; Italy; endemism; invasive fish; morphometrics

## 1. Introduction

The biodiversity crisis is particularly dramatic in freshwater ecosystems [1,2]. This biodiversity crisis requires actions and conservation measures at multiple geographical scales, global, regional, and local, considering both conservation priorities and limited resources [3–5]. The most often used criteria to establish conservation priorities are taxon rarity and endemism (e.g., [6,7]). IUCN criteria consider the limited number of localities of a given taxon as a key factor for assigning high-risk categories [8], while other conservation projects (e.g., the EDGE of Existence programme) involve other evaluation frameworks, such as including endangered species that embody a substantial portion of distinct evolutionary heritage. The insular nature of lentic freshwater habitats has produced a high rate of endemism with small geographic ranges [2] (and reference therein).

Amphibians are nowadays subject to a plethora of possible threats and pressures, ranging from alien species' introduction to pollution, emerging infectious diseases and habitat loss, which sometimes act in synergy and result in detrimental effects at both individual and community levels (e.g., [9–15]). Alien species in freshwater habitats are among the greatest threats to native biodiversity [9]. Indeed, for amphibians in general and newts in particular, fish introduction is considered a determining factor of pond-breeding populations' decline and extinction [12,16–20]. Naturally, the impact of alien species is particularly alarming for native taxa with a restricted range.

*Ichthyosaura alpestris inexpectata* (Dubois & Breuil, 1983), known as the Calabrian Alpine newt, is an endemic taxon of Southern Italy considered to be a post-glacial relict with small and fragmented populations [21,22]. It represents a unicum of high biogeographical and conservation value, with populations exhibiting genetic variation likely originating from the Middle Pleistocene and possessing limited genetic diversity, which supports a long-standing isolation hypothesis [21]. Since the early 1980s, the newt has been known in five localities in the north of the Catena Costiera (a coastal chain representing the first mountain range of the Calabrian Apennines), in ponds and lakes of varying depths localised in clearings within beech forests at an altitude ranging from 850 to 1135 m a.s.l. [22–27]. All these sites currently fall within the Natura 2000 Network established under the EU Habitats Directive (92/43/EEC) (Table 1).

**Table 1.** Summary of the data related to known sites. The last year of observation of *Ichthyosaura alpestris inexpectata* for each site is reported. Maximum number of adults captured or observed is based on information retrieved from the available literature or collected ante 2022 in the DiBEST database. Trifoglietti: TR = large pond Trifoglietti; TRIF = small pond Trifoglietti Inferiore. Fonnente: FO = Lake Fonnente; FA = Fosso Armando.

| Site Code N2000 | Municipality | Locality | Year of Last Observation | Maximum Number Captured/Observed (1983–2018) | References |
|---|---|---|---|---|---|
| IT9310058 Pantano della Giumenta | Malvito | Stagno C/Pantano dorato | 1983 | 4 | [24,25] |
| IT9310060 Laghi di Fagnano | Fagnano Castello | Trifoglietti (pond TR and pond TRIF) | 2018 (TR) and 2022 (TRIF) | 23 | [22–26] |
| | | Due Uomini | 2018 | 21 | |
| | | Fonnente (lake FO and pond FA) | 2022 | 25 | [25,26] |
| IT9310061 Laghicello | San Benedetto Ullano | Laghicello | 2022 | 69 | [22–25,27] |

The Calabrian Alpine newt is assessed as "Endangered (EN)" in the Red List of Italian Vertebrates [28] because of the low number of occurrence localities, the continuous degradation and loss of the few sites due to the impact of human activity, and the recent introduction of allochthonous ichthyofauna in its core range. Since 2017, in fact, three fish species, *Cyprinus carpio*, *Gambusia holbrooki* and *Carassius* sp., have been observed in three newt breeding sites in the Special Area of Conservation (SAC) "IT9310060 Laghi di Fagnano", originally devoid of fish fauna [23,24]. Moreover, other natural pressures, such as the transformation of habitats due to silting phenomena and drought, represent worrying critical issues [23,24], as was also found for another peninsular Italian amphibian species [5] (and references therein).

The aims of this study are threefold. First, we aim to provide an updated framework of the presence/distribution, status and threats of the Calabrian Alpine newt. Second, we aim to supply baseline ecological data for each inhabited site derived from recent systematic monitoring in the SAC Laghi di Fagnano. Third, we aim to provide the first demographic and population parameters. Specifically, we surveyed all spatially clustered wetlands in the protected area where the focal species was previously observed and the two recently discovered sites, thus giving the most updated information about *I. a. inexpectata* breeding habitats (e.g., habitat features and aquatic macroinvertebrate assemblages). Moreover, we collected data regarding abundance, body size and body condition in four sites (two new and two historical), expanding the information on morphometrics and sex ratio to obtain the first perceptions of demography.

## 2. Materials and Methods

### 2.1. Study Taxon

The Alpine newt *Ichthyosaura alpestris* (Laurenti, 1768) is widely distributed in most of western and central Europe and the Balkans [29]. In Italy, this urodele has a discontinuous distribution with three subspecies, two of which have a large distribution area and one of which is extremely localised. *Ichthyosaura a. alpestris* (Laurenti, 1768) is widespread in the Alpine area, *I. a. apuana* (Bonaparte, 1839) occurs in the Langhe hill system, Turin hills, Oltrepo' Pavese (Northern Italy), the Tuscan-Emilian Apennines, the Maritime and Apuan Alps and a restricted and disjointed area within the Laga Mountains (Central Italy); conversely, *I. a. inexpectata* has a very limited range in Southern Italy [22–25,30–33] (Figure 1).

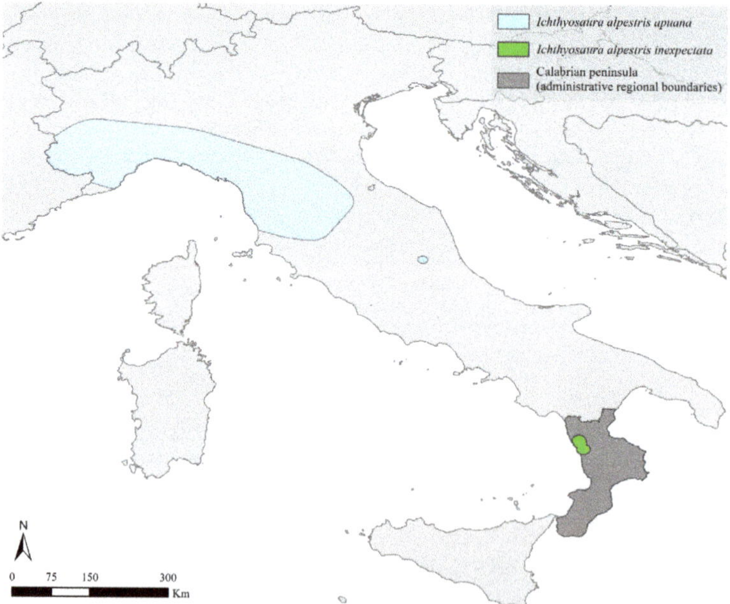


**Figure 1.** Italian ranges (minimum convex polygons based on [30,31]) of the two Italian endemic subspecies of the Alpine newt *Ichthyosaura alpestris*, namely, *I. a. apuana* (light blue) and *I. a. inexpectata* (green); in dark grey, the Calabria region, where our target subspecies (*I. a. inexpectata*) occurs.

Adults of this medium-sized newt present an aquatic lifestyle in spring, enabling them to breed in ponds with aquatic and riparian vegetation, transparent waters and short periods of winter frost. Reproduction occurs mainly in May and June, and no autumn breeding events are known to date [31]. Paedomorphic phenotypes were observed in Lago Due Uomini and Laghicello [31]. Available data on ecological requirements, phenology and life history traits, and demographic information for *I. a. inexpectata* are scarce or outdated. Knowledge of feeding habits only refers to the population of Laghicello [34]. In addition, for this site, the rough estimation of a few hundred individuals of *I. a. inexpectata* (210–300) is the only one reported in the literature [27]. The same authors assumed that the total number of mature individuals in all populations (from Laghicello and Fagnano lakes) could be estimated between 840 and 1200 [27]. Much less is known about populations in Laghi di Fagnano, which have never been assessed.

### 2.2. Study Area and Site Description

We surveyed seven lentic habitats, ranging from temporary ponds to permanent lakes (Figure 2 and Table 2). Five sites fall within the Natura 2000 site "Laghi di Fagnano", which protects a natural system of lakes and ponds with an elevation range from 1000 to 1100 m a.s.l. There are nine other amphibian species within this network of wetlands, in addition to *I. a. inexpectata* (Table 2). Due to this species diversity, the site was recognised as an Area of National Herpetological Relevance by the Societas Herpetologica Italica in 1998 (AREN—ITA022CAL002) [35].

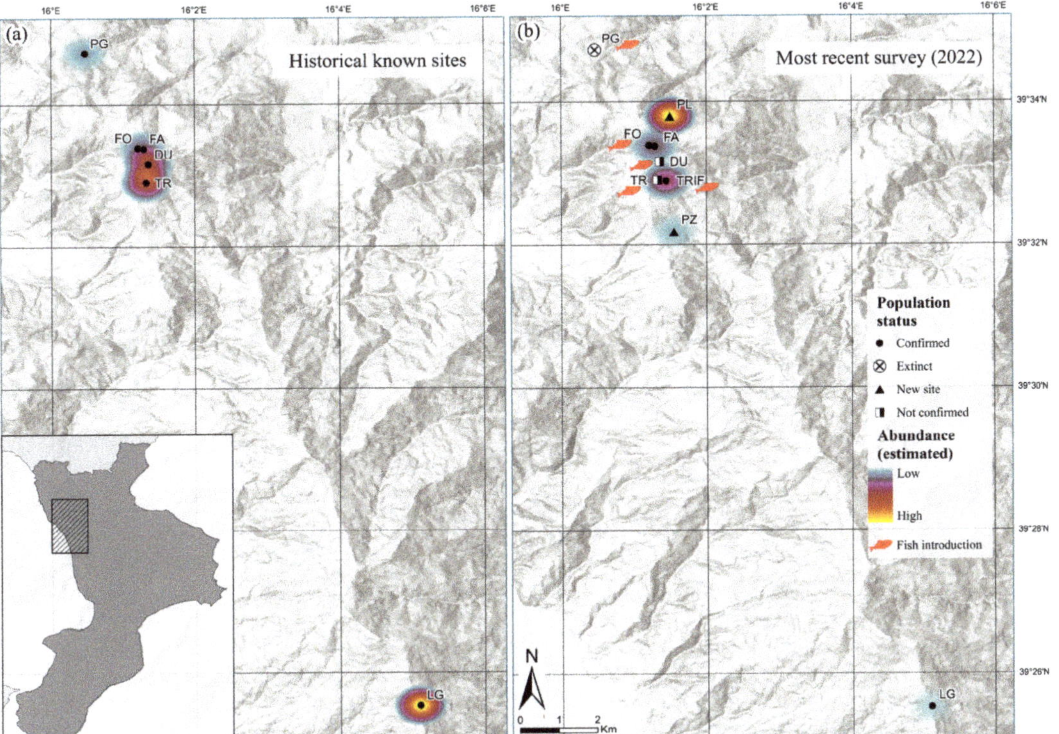


**Figure 2.** (**a**) Distribution and putative population sizes in the known localities of *Ichthyosaura alpestris inexpectata* based on literature data and historical observations; see Table 1 for details; (**b**) Results from recent surveys.

**Table 2.** Geographical locations, morphometric parameters and environmental characteristics measured in the study ponds. Site name: Fonnente (FO), Due Uomini (DU), Trifoglietti (TR), Trifoglietti inferiore (TRIF), Fosso Armando (FA), Piano di Zanche (PZ), Pantano Lungo (PL). The approximate area was calculated using Google Earth Pro satellite images (May 2022). Pond water depth (considering the maximum seasonal depths measured in April–May 2022), pond regime (P: permanent; T: temporary pond subject to drying out at least during a single survey), shading (i.e., the extent of shading of the site during the surveys) and presence or absence of fish are reported. Life stage: adults = A, larvae = L; overwintering larvae = OL; paedomorphs = P. Species abbreviations: Tla: *Typha latifolia*; Pna: *Potamogeton natans*; Pau: *Phragmites australis*; Cve: *Carex vesicaria*; Sau: *Sphagnum auriculatum*; Cro: *Carex rostrata*; Cpa: *C. paniculata*; Jef: *Juncus effusus*; Lvu: *Lysimachia vulgaris*; Eca: *Eupatorium cannabinum*; Epa: *Eleocharis palustris*; Lit: *Lissotriton italicus*; Tca: *Triturus carnifex*; Ssa: *Salamandra salamandra*; Hin: *Hyla intermedia*; Pes: *Pelophylax* sinkl *esculentus*; Rda: *Rana dalmatina*; Rit: *R. italica*; Bbu: *Bufo bufo*.

| | Site Name | | | | | | |
|---|---|---|---|---|---|---|---|
| | **FO** | **DU** | **TR** | **TRIF** | **FA** | **PZ** | **PL** |
| Lat. N<br>Long. E | 39°33′24″<br>16°1′13″ | 39°33′9″<br>16°1′22″ | 39°32′54″<br>16°1′2″ | 39°32′53″<br>16°1′27″ | 39°33′23″<br>16°1′18″ | 39°32′11″<br>16°1′34″ | 39°33′48″<br>16°1′30″ |
| Altitude (m a.s.l.) | 1050 | 1077 | 1048 | 1045 | 1055 | 880 | 1010 |
| Area (m$^2$) | 3561 | 20,679 | 10,800 | 520 | 827 | 742 [1] | 235 [1] |
| Depth (m) | 1 | - | 1.5 [2] | 0.9 | 1.2 | 0.7 | 1.1 |
| Permanence | P/T [3] | P | P | P | T | T | P |
| Shading (%) | 20 | 20 | 30 | 80 | 20 | 90 | 90 |
| Fish presence | *Carassius* sp. | *Cyprinus carpio* | *G. holbrooki* | *G. holbrooki* | - | - | - |
| Aquatic vegetation | Tla, Pna | Pna, Pau, Cve | Sau, Cro, Cpa, Jef, Lvu, Eca, Epa, Pna | Cve, Jef | Jef | absent | absent |
| *I. a. inexpectata* | L | - | - | A, L, P | A, L | A, L | A, L, OL |
| Other breeding species | Tca, Lit, Pes, Rda, Hin | Tca, Bbu, Pes, Rda | Hin, Pes, Rda | Lit, Tca, Ssa, Pes, Rda, Rit, Bbu | Lit, Tca, Ssa, Hin, Pes, Rda, Rit | Lit, Tca, Bbu, Rda | Lit, Ssa, Rda, Rit |

[1] Area calculated by measuring the maximum length and width and assuming an elliptical shape. [2] From [36]. [3] Temporary as referred to the adjacent flooded meadow.

The Special Area of Conservation covers a surface area of around 19 ha, and it extends mainly along the slopes of Monte Caloria (Fagnano Castello) along a sub-plateau belt in which high-grade metamorphic rocks outcrop. The local climate is strongly influenced by the warm and humid air masses from the Tyrrhenian Sea, which give rise to thick fog formations and frequent and abundant rainfall that help mitigate the long, dry summer [37]. The site falls within the lower supratemperate bioclimatic belt with upper subhumid ombrotype [38]. The ponds and lakes network are within beech forest formations (*Fagus sylvatica*), included in habitat 9210*, with a rich shrub and herb layer. The climatic conditions and soil characteristics allowed the formation of peat bogs consisting of a layer of sphagnum and a mosaic of different plants of humid and marshy environments linked with dynamic successions due to variations in water level [37]. A detailed description of the habitats and vegetation in the SAC is available at http://retenatura2000.regione.calabria.it/, (accessed on 12 January 2023).

Briefly, a higher and constant water depth of up to 2 m characterises "Lago Due Uomini" (hereinafter referred to as DU), which hosts submerged vegetation of *Potamogeton natans* and a massive presence of *Phragmites australis*. "Lago Trifoglietti" (TR) is a silting peat bog consisting of a thick moss layer of *Sphagnum auriculatum* with various species of *Carex rostrata, C. paniculata, Juncus effusus, Lysimachia vulgaris, Eupatorium cannabinum* and, in the most flooded areas, *P. natans* and *Eleocharis palustris*. Moreover, the lake holds a unique palynological archive that has yielded important information on the dynamics of Holocene vegetation and climate in southern Italy [36]. A spring flows into the lake from the north, whereas an outflow runs southward. To prevent dry-up during summer and complete infilling, the municipality of Fagnano Castello built a small earthen dam in 2000 that led to the formation of a small underlying pond [36] called "Trifoglietti inferiore" (TRIF). "Lago Fonnente" (FO) refers to an area including the main lake and an adjacent flooded meadow, which dries up in the dry season, and a pond, historically called "Fosso Armando" (FA) [39,40] which is adjacent but not connected with the other wetlands. FO shows well-preserved marsh vegetation with sedges and rushes and is characterised by *Typha latifolia* and *P. natans*. During the study period, the contiguous flooded meadow appeared almost completely dry in June, when numerous dead larvae of all three species of newts and tadpoles of *Hyla intermedia* were found. Therefore, the site was not visited further in subsequent samplings.

The new breeding sites of the Calabrian Alpine newt are constituted by two woodland ponds found in the localities of Piano di Zanche (PZ) and Pantano Lungo (PL) (municipality of Fagnano Castello) during a herpetological survey in 2015 and, after discovering, visited sporadically. The description of these habitats is given in Section 3.

### 2.3. Sampling Procedures

We sampled the five sites in the SAC Laghi di Fagnano and the two new ponds from April to October 2022, during and after the newts' breeding season (May–July), once a month. PZ and PL were visited six times (except in May) due to a weather change (sudden thick fog) preventing sampling. All visits were conducted during daylight hours.

Newt sampling techniques differed across sites due to the intrinsic features of each. Sites TR and DU are the most extensive, and due to the morphology and presence of dense aquatic vegetation, these cannot be easily and completely surveyed. Therefore, the visual encounter survey method (VES) was used for the lakes TR and DU by walking across the entire wadable surface of the wetlands for at least 30 min, so as to cover, in that time, the entire pond perimeter (~14 m/min). While searching for newts, we also reported the detection of fish. For FO, FA, TRIF, PZ and PL sites, survey protocol required at least two observers to enter the wetland and capture animals using dip netting. At least three removal passes lasting 20 min were made during a single visit, removing the individuals captured at each pass to avoid multiple counts and applying the removal sampling scheme to estimate population abundance (see below).

All the captured newts were handled with latex gloves and temporarily kept in plastic containers filled with water from the pond and measured for body mass (to the nearest 0.01 g using digital scales), snout-vent length (SVL) and total length (with a ruler to the nearest 0.1 cm). A randomly selected subset of captured adults (n = 25) for PL in June was measured, so as to make the captured individuals' number homogeneous. The sex was determined by typical secondary sexual characters. Recognition of newt species during larval stages was based on the comparative staging tables of Bernabò and Brunelli [41]. Paedomorphic newts were distinguished by gill slits and three pairs of external gills, and evident secondary sexual characteristics [39]. Once all measurements and data had been collected, newts were released into the capture pond.

At each sampling session, other amphibian species occurring and breeding in the investigated aquatic sites were also recorded.

### 2.4. Habitats' Characteristics and Macroinvertebrates Assemblages

We described the features of the seven sampled aquatic habitats with different hydrological balances and subjected to different pressures (fish-containing and fishless ponds) by recording the size, the water permanence, the distances between each pond and the water's physical–chemical characteristics.

To evaluate the biodiversity levels and trophic availabilities of the four breeding water bodies where adult newts currently occur, we investigated the structure of macroinvertebrate assemblages. Therefore, macroinvertebrate sampling was conducted in June and October, following a time–space-standardised quantitative procedure. In each monitoring site, we sampled three main mesohabitats in fordable zones (emergent and submerged aquatic vegetation, littoral and central sediments) using a Surber-type sampler (square opening: 25 cm × 25 cm) (Scubla s.r.l., Remanzacco UD, Italy) equipped with a 60 cm net of 250 μmesh and a plastic gathering glass. All samples were sieved through a 0.45 μm sieve (Endecotts Ltd., London, UK) and immediately stored in refrigerated containers with 80% ethanol. Once in the laboratory, macroinvertebrates were separated from sediments through a stereomicroscope. All organisms were counted and identified at the family level [42–44] to estimate the following parameters and indices: abundance (N), taxonomic richness (TR), Shannon–Wiener Index (H′) [45] and Margalef Index (D) [46].

Water physical–chemical parameters were assayed twice during the study, in the early summer and autumn, with three measurements along a transect at about 10–20 cm below the surface water. Specifically, temperature, conductivity, pH and total dissolved solids were determined with a multiparameter probe (Hanna Instruments, model HI991300). Dissolved $O_2$ and relative saturation percentage were measured with a portable oxygen meter (Hanna Instruments, model HI98193).

*2.5. Data Analysis*

We analysed the dataset, collecting information on adults of the Calabrian Alpine newts captured in TRIF, FA, PZ and PL (Table 2). The sample size from the four populations in several months was too small since we captured only larvae and <3 individuals; therefore, for each population, we pooled the total number of captured males and females during the sampling period.

Operational sex ratio in each sampling was expressed as the proportion of males over the total adult number (males/(males + females)); deviations from equality were assessed via a two-tailed binomial test when more than ten newts were caught.

To assess the body condition (a proxy for physical condition, overall health and fitness [47]) of the four populations of Calabrian Alpine newts, we calculated the scaled mass index (SMI) [48]. This index provides a comparable measure across individuals of different sizes by rescaling body mass measurements to a standard length (here calculated for SVL) accounting for allometric growth [48]. We calculated SMI separately for males and females within each site, thus avoiding the scaling issue that results when comparing BCIs across groups known to differ in size [49]. SMI is computed as follows (with $Mi$ and $Li$ being the body mass and the linear body length, respectively, $bSMA$ the scaling exponent estimated via the SMA regression of $M$ on $L$, and $L0$ the mean length of the study population):

$$\hat{Mi} = Mi \left[ \frac{L0}{Li} \right]^{bSMA}$$

We quantified the scaling exponent bSMA in R [50] by using 'smatr' package [51] from ln-transformed data.

The normality of morphometric data (body mass, SVL, total length and SMI) was analysed via the Shapiro–Wilk test and the homogeneity of variances was analysed via the Brown–Forsythe test. Data fit normal distributions and resulted homoscedastic; therefore, we used a one-way analysis of variance (ANOVA) followed by Tukey's multiple comparisons test to determine whether there was a difference in body size and body condition between sex within populations and comparing females and males, respectively, among populations.

We also tested the effect of fish presence, macroinvertebrate diversity (in terms of Shannon index), site and month on the SMI using a linear model.

We used OriginPro, Version 2022b (OriginLab Corporation, Northampton, MA, USA) to produce plots.

We estimated newt abundance through the most recent field capture data to obtain updated information about TRIF, FA and PL (sites where sufficient numbers of adult newts were found). Considering that the sampling method we used can be ascribed to the constant probability of capture (each pass does not lower or increase the probability of capture of the subsequent pass, and the chance of sampling each individual is constant, both for their detectability and/or population stability) and the removal of captured individuals was performed at each pass, we used the whole capture dataset, setting the removal method = 'Zippin' [52–54]. The population size was estimated through the 'FSA' package [54] in R. We considered the highest population estimate in the month of greater detectability of adult individuals (present in the water during the breeding season).

## 3. Results

### 3.1. Current Distribution Framework and Habitats' Characteristics

Based on our multi-year observations (1984–1992, 2005, 2015–2018) and during a sampling visit in June 2022, as well as from information gathered from locals and landowners, the occurrence of the Calabrian Alpine newt population reported in the SAC "Pantano della Giumenta" is seriously questionable, since the last observation refers to the early 1980s (Table 1 and Figure 2). In Laghicello, the presence of Alpine newts was verified during a single visit in April 2022; no fish were introduced, but the site appeared subjected to advanced natural silting phenomena (I. Bernabò pers. obs.).

During surveys conducted in 2022 in three study sites in the SAC Laghi di Fagnano (FO, TRIF and FA), Calabrian Alpine newts were observed at least once for each life stage, whereas in DU and TR we did not observe any (Table 2 and Figure 2b). A true absence was not proven, and we could not exclude its presence in the deep central areas of the lakes, which cannot be sampled except with methods other than those used. Individuals of the common carp (Cyprinus carpio) and of the Eastern mosquitofish (Gambusia holbrooki) were highly abundant in the lakes DU and TR, respectively; the presence of cyprinid fish (Carassius sp.) and G. holbrooki was also recorded in FO and TRIF, respectively, but at a low density (Table 2).

The two new records were confirmed to be breeding sites (Figure S1): PZ is a temporary pond, and the water comes almost exclusively from rainfall; conversely, PL is a large woodland puddle that presents a mostly stable hydroperiod thanks to a water flow from a spring. Both sites have significant tree cover and low $O_2$ levels due to a high plant debris load; woods of Fagus sylvatica dominate the vegetation around the two wetlands, and no water vegetation (macrophyte or algae) is present. Water bodies' environmental features (measured during the last field campaigns) are summarised in Table 2 and in Table S1.

Considering the linear distances among clustered wetlands in the SAC of Laghi di Fagnano and the new sites, we found that the nearest was PL (835 m away from FA), whereas PZ occurs at a greater distance (1321 m from TRIF) (Figure 3).

About the amphibian community, among the ten species occurring in the SAC, only *Bombina pachypus* was not found. *Rana dalmatina* was the most common species (in 100% of the sites), followed by *Lissotriton italicus* and *Triturus carnifex*, and then *Pelophylax* sinkl *esculentus* (71%). *Salamandra salamandra*, *Rana italica*, *Hyla intermedia* and *Bufo bufo* were detected in 43% of the aquatic sites (Table 2). Syntopy of *I. a. inexpectata* with *L. italicus* occurred in five sites (FO, TRIF, FA, PL and PZ), and it also lived with *T. carnifex* in four ponds (FO, TRIF, FA and PZ).

Regarding aquatic macroinvertebrate assemblages, a total of 43 families, 15 orders, and 7 classes of invertebrates, cumulatively, were identified in the four study sites (TRIF, FA, PZ and PL). The taxonomic list of all organisms collected at each monitoring site is reported in Table S2. All ponds in both sampling campaigns showed high diversity indices, indicating stable habitats for aquatic invertebrates. In particular, PL showed the highest values of diversity indices (TR: 25, H': 3.056, D: 3.050), followed by FA (TR: 15, H': 2.727,

D: 2.261) and TRIF (TR: 17, H′: 2.555, D: 2.119), whereas PZ had the lowest ones (TR: 12, H′: 2.439, D: 1.591). Moreover, the macroinvertebrate community biodiversity increased in terms of number of taxa from June to October, due to the significant input of trophic resources associated with woody/leaf litter, with an overall autumnal increase of 15.7%.

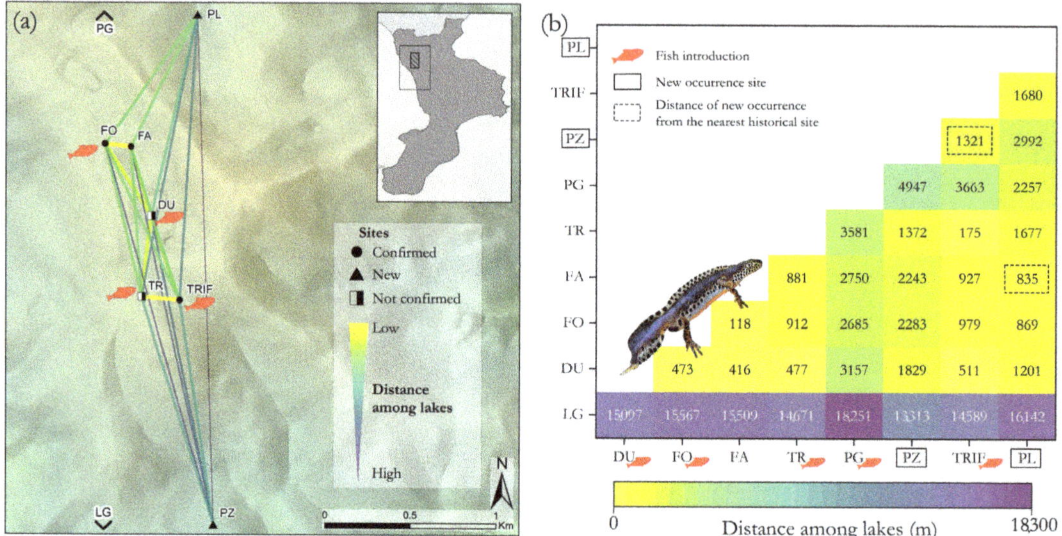


**Figure 3.** (**a**) Water bodies' network within the SAC "Laghi di Fagnano" (Laghicello and Pantano della Giumenta sites are excluded from the view) and (**b**) the distances between each site (including all); the two new sites and the correspondingly shorter distances from which the Calabrian Alpine newt possibly colonised them are highlighted through boxes (full and dashed lines, respectively).

*3.2. Population parameters*

In TRIF, we captured 53 adult Calabrian Alpine newts from May to September, with a maximum of 34 individuals captured in June (20 males and 14 females; estimated abundance 37 ± 3, 95% C.I. = 30–44) (Figure 4), whereas we observed numerous larvae from late summer to autumn, with many of these showing evidence of predation (i.e., tails and legs partially nipped). TRIF was the only site where we detected paedomorphic individuals (four males and three females); paedomorphic males and females had an average SVL (± SE) of 38.8 ± 0.4 mm and 41.8 ± 0.5 mm, respectively. In FA, 30 adult newts were captured from April to June, with a maximum of 15 captured adults in April (10 males and 5 females; estimated abundance 20 ± 8, 95% C.I. = 5–35) when the pond, characterised by a highly variable water level, presented its maximum water level (Figure 4). The pond was almost completely dry (level of about 20 cm) from the end of July until October, and only abundant larvae were observed. In PZ, we collected eight adults (five males and three females) of Calabrian Alpine newts in June, whereas in September, we found one female with a few larvae (Figure 4). The complete drying of the pond was detected in July and August. In PL, we detected adult Calabrian Alpine newts every month except May (see Materials and Methods); larvae were observed from the end of July to October, including overwintering larvae. Altogether, we captured 129 adults; the highest number of individuals was observed in June (33 males and 35 females) (Figure 4). The estimated maximum abundance of newts was 88 ± 14 (estimate ± SE; 95% C.I. = 62–115).

In individual samples with N ≥ 10, our results showed a male-biased sex ratio of 0.59 ($p$ = 0.256) in TRIF. In FA, the proportion significantly differed from 0.5 both in April (0.67) and June (0.73) (all $p$ < 0.05), whereas, in PL, a significant deviation from 50:50 toward males (0.65) occurred only in April ($p$ < 0.05).

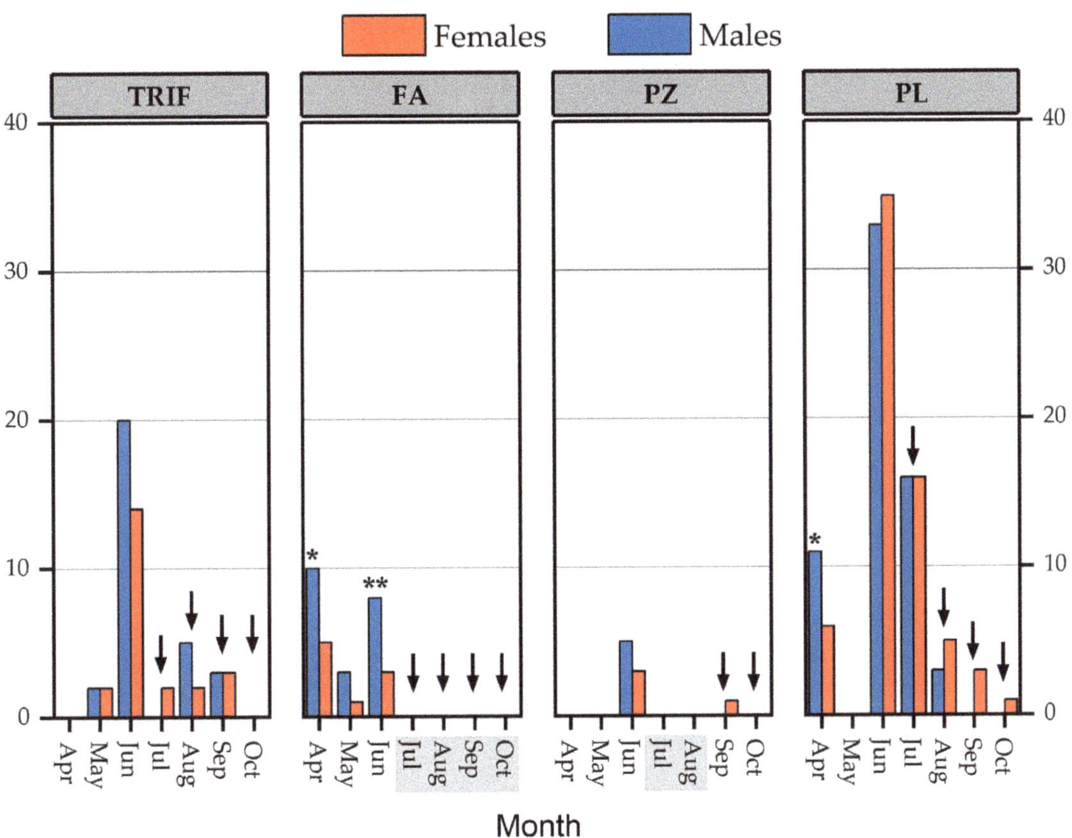



**Figure 4.** Captures of adults of *Ichthyosaura alpestris inexpectata* at four of the study sites during the sampling period. The months with no water are highlighted in grey. Asterisks mark a significant deviation from evenness, * $p < 0.05$, ** $p < 0.01$ (Fisher's exact probability test); arrows indicate when larvae were found.

In all populations except for PZ (due to its small size), females were confirmed to have larger body sizes than males ($F_{7,116} = 10.50$, Tukey's multiple comparisons test, $p < 0.001$) (Figure 5 and Table S3). Overall, the average ($\pm$SD) SVL was 47 $\pm$ 3.5 mm in males and 53 $\pm$ 4.2 mm in females, whereas the average total length was 81 $\pm$ 5.4 mm and 95 $\pm$ 5.5 mm, respectively. Sex differed statistically in body mass within sites ($F_{7,116} = 17.86$, Tukey's multiple comparisons test, $p < 0.001$). The mean body mass was 2.6 $\pm$ 0.5 g in males and 4.4 $\pm$ 1.1 g in females, showing no intra-sexual divergence between populations ($F_{7,116} = 0.55$, Tukey's multiple comparisons test, $p > 0.05$). There were no differences among populations in females' or males' body mass, SVL or total length (Figure 5 and Table S3). SMI differed statistically between sexes in each site, and males exhibited the lowest body condition index (Tukey's multiple comparisons test, all $p < 0.001$); neither male nor female body conditions varied significantly among the four populations (Figure 5 and Table S3). Moreover, we found no significant effect of fish presence, macroinvertebrate diversity, site or month on the SMI ($R^2$ adj. = 0.073, $p > 0.05$ for all of the factors considered).

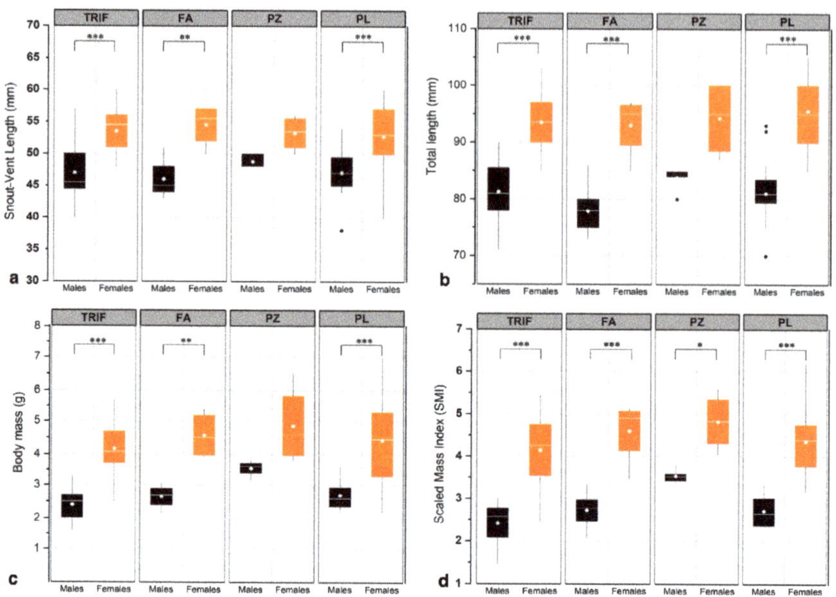

**Figure 5.** Boxplots of: SVL (**a**), total length (**b**), body mass (**c**), and SMI (**d**) for *I. a. inexpectata* males and females in the four sites. The horizontal white bars and circles represent median and mean values, respectively. The box is determined by the 25th and 75th percentiles, and the whiskers refer to the minimum and maximum values observed within data for each population. Outliers are shown as grey circles. One-way ANOVA with post hoc Tukey tests: * $p < 0.05$, ** $p < 0.01$; *** $p < 0.001$.

## 4. Discussion

### 4.1. Updated Distribution, Species Occurrence and Habitat Characteristics

In recent years, no effort has been directed towards surveying and monitoring the Calabrian Alpine newt. Our results update the information on the newt distribution, providing records of local disappearance from three historical sites and the discovery of two new breeding ponds.

The Calabrian Alpine newt was traditionally considered to occur in five localities (see Table 1): three in the SAC Laghi di Fagnano (FO, DU, TR), one in the locality Laghicello and one at the SAC "Pantano della Giumenta". Their occurrence in this protected area, where a goldfish population (*Carassius auratus*) has been stable for almost twenty years, has not been confirmed since 1984, and the population has to be considered extinct. In Laghicello, the endemic newt still occurs. In contrast, in two of the three historically known breeding sites in the SAC Laghi di Fagnano, DU and TR, the newts were not recorded after fish introduction in 2017 and 2019, respectively. However, past surveys (until 2018) applying VES, when the two lakes were fishless, allowed the detection of all newt species. Thus, VES may not be the most efficient way to detect newts in these two studied lakes and could lead to false absences. Furthermore, to avoid predation risk from fish, amphibians can increase shelter use and reduce activities [55,56], resulting in a lower probability of being detected and, likewise, reduced foraging activities and reproduction, two essential fitness components [57]. On the other hand, it is also well known that newts avoid fish-invaded habitats [58,59]. Another Alpine newt subspecies, the Bosnian Alpine newt *I. a. reiseri* (Werner, 1902) from the Prokoško Lake, was dramatically reduced or, more probably, disappeared from the lake following fish introductions [60].

In our case, if the event is not a local extinction, it is a demographic reduction since the detection probability is also a function of the abundance [61].

The progressive modification or loss (early desiccation and silting phenomena) of the few suitable habitats in which *I. a. inexpectata* breeds and the substantial threat posed by the introduction of invasive fish may lead to reproductive failure. This could result in the extinction of the endemic subspecies and those of other newts' populations of community and conservation concern. Fish eradication would also bring benefits to the two other newt species that are syntopic with the Calabrian Alpine newt, i.e., *Triturus carnifex* (Annex II and IV) and *Lissotriton italicus* (Annex IV), which are protected under the Habitats Directive and categorised as NT and LC by the IUCN, respectively.

Knowledge of taxon distribution in a given area underpins most conservation efforts and planning [62]. In the context of the severe threat to the survival of the Calabrian Alpine newt, the discovery of two new occurrence localities (PZ and PL) is crucial for the newts' persistence and highlights the importance of constant field monitoring to assess the species distribution at a local scale [63,64], which is a dynamic trait, determined by local extinction and new colonisation phenomena. These new breeding sites are small woodland ponds formed on natural depressions in well-preserved beech forests, away from the main roads and excessive tourism. Local landscape features include forest continuity and a network of ponds that allow the permeability of matrix habitats for dispersion. The linear distance and the narrow differences in altitude (about 200 m) separating the new ponds from the known ones of Laghi di Fagnano suggest that these sites may have been colonised through the dispersal of newts from the nearby locations. Indeed, the closest connection pathway may have been about 0.8 km from FO and FA toward the northeast for PL and about 1.3 km from TR and TRIF toward the southeast for PZ. Although fidelity to the breeding site is reported in the Alpine newt [65], movement capabilities over long distances, exceeding 1 km and up to 4 km, have also been reported [66,67]. Future studies should map all wetlands in the vicinity of the SAC and consider inter-pond movements, migrations and dispersal to disentangle newt population dynamics and better understand how to plan adequate conservation measures, such as pond creation. In addition, conducting investigations with new approaches, such as eDNA, could be helpful in detecting newts' occurrence in suitable habitats.

At high altitudes (i.e., in the Alps), *I. alpestris* usually lives in oligotrophic sites where the water is very clear. In the Apennines, on the other hand, it inhabits low- and mid-altitude sites where water is frequently turbid, the maximum temperature is high and the environment is generally unpredictable [30]. We provided the first information on water parameters where the subspecies *I. a. inexpectata* breeds. The new ponds can be considered oligotrophic environments based on physical–chemical measurements and total hardness (the total amount of Ca and Mg ranged between 3.29 and 11.4 mg/L) [68]. Dissolved oxygen values are very low, suggesting a strong insaturation, most likely due to the scarce primary productivity and the high oxygen consumption triggered by a large amount of autochthonous and allochthonous plant detritus in the ponds. Here, we describe the first data on the potential feeding resources available for the Calabrian Alpine newt. Overall, the four studied ponds showed high diversity and a well-structured macroinvertebrate assemblage. Our biodiversity indices values are comparable to those calculated for other natural lakes and ponds [69–71] and higher than those measured in artificial, garden and urban ponds [72,73]. In October, the macroinvertebrate assemblages showed, in each pond, an increase in shredding invertebrates feeding on organic detritus derived from the riparian vegetation [74]. With regard to hydrological variation, there was a drop in the water level in October at all sites, particularly in FA and PZ. However, this lower volume of water did not negatively affect the macroinvertebrate assemblages and seemed adequate to sustain the ecological niches occupied by these organisms. Most invertebrates found in breeding ponds are involved in the diet, during the aquatic phase, of adults, paedomorphs and metamorphs of any subspecies of *I. alpestris* [34,75–77]. Typically, the Alpine newt shows generalist dietary habits and exhibits seasonal plasticity; it exploits temporary resources and differing prey taxa in relation to their cycles of availability and local abundance [34,75,76]. However, an individual shift towards a more specialist feeding strategy in artificial sites in response to increasing prey diversity has been reported [78]. In general, the abundance of invertebrate prey in the aquatic breeding sites allows newts not to leave these habitats to forage, a relevant factor for their conservation status [79]. Our preliminary analysis reported no significant correlation between the body condition index, prey species availability and fish presence. The contribution of the newts to the feeding ecology may be elucidated only with dietary studies that may also clarify the trophic strategy of the Calabrian Alpine newt. Newts act as predators in naturally fishless ponds [80,81]; therefore, it is also essential to investigate if the unavoidable trophic competition with introduced fish may become a decisive limiting factor. Indeed, knowledge of the trophic ecology of threatened and protected newts is crucial to evaluating possible threats [82], especially when dealing with the effects of fish introduction, and thus to adopting appropriate conservation policies [83].

### 4.2. Population Parameters

The ecology and demographic parameters of the Calabrian Alpine newt are scanty or outdated. Although long-term monitoring programmes are required for more robust estimations of trends in their distribution and abundance, our survey confirmed that the overall population of the Calabrian Alpine newt in its core range might be much smaller than hitherto assumed. Indeed, the only available study by Dubois and Ohler in 2009 [27] on the population of Laghicello roughly estimated counts of between 210 and 300 individuals. The authors supposed that the total number of mature individuals at the four historically known sites (Table 1) is probably much less than four times the estimated population size of Laghicello. Altogether, 90 adult newts were caught in the ponds within the SAC Laghi di Fagnano, whereas 138 adults were sampled at the two new sites. The highest number was found in permanent ponds. A few adults and larvae were found in PZ; this pond may completely dry out in the warm season and is replenished by late summer rainfall. The detection of larvae of the three newts at the end of September confirmed a late summer reproduction in this water body. However, its interannual hydroperiod is not well-studied, and it remains unclear what reproduction and metamorphosis success occurs at this pond. PL was confirmed to be a suitable breeding habitat of high conservation value for the Calabrian Alpine newt and other amphibians; here, the most abundant population is present.

Population monitoring and demographic estimates in the breeding sites of Laghi di Fagnano are pivotal for the persistence of the Calabrian Alpine newt after fish introduction. To ensure accurate population size estimates, we plan to adopt a monitoring scheme using aquatic funnel traps in lieu of dip-netting and a capture–mark–recapture approach.

Throughout the breeding season, we found statistically significant sex bias in caught newts towards males in FA (every sampling) and PL (in April). In each sampling event, males outnumbered females. The observation of distorted sex ratios in amphibians using counts or captures may reflect an actual ecological trait of the studied populations but may also be an artefact due to different capture probabilities between sexes [84], and references therein]. In other populations of Alpine newts, sex ratios have been reported to be either male- or female-biased [27,85–87]. According to Lanza and coauthors [31], both sexes remain in the water throughout the breeding season. However, males were more numerous than females at the breeding sites early in the year, and females outnumbering males in late spring and early summer represents a weaker trend in many newt species, including the Alpine newt [88]. Given our data, we can speculate that the sex ratio in the studied ponds during the breeding season is quite dynamic and is determined by locally specific factors. Further studies are needed on such dynamics, which may vary between different types of water bodies (e.g., temporary vs. permanent, temperature, food availability and habitat quality), influencing mating behaviour and the duration of the breeding season.

Body condition is a measure of an animal's foraging success and physiological state [47,89,90]. In our estimations of body condition, remarkable intersexual differences were found within each population. Males' and females' body conditions of Calabrian Alpine newts did not vary significantly among the four study ponds, which were both with and without fish and had differences in their hydrological regime. Further long-term monitoring studies are required to determine the seasonal and inter-annual variation in SMI values for each population and the effects of fish presence on newts' body condition.

Our results on body size are comparable with those available in the literature on *inexpectata* and other subspecies. Sexual size dimorphism was detected in our study populations, with females being longer and heavier than males, as previously reported in the literature [21,31].

### 4.3. Threats, Conservation Measures and Active Management

We suppose that within a few years, fish may cause the extirpation of the Calabrian Alpine newt and other native amphibian species from otherwise viable habitats. Meanwhile, the temporary ponds, although protected from the introduction of fish due to their hydroperiod, could likely disappear or become unsuitable habitats for amphibians in a changing climate. Omnivorous fish, such as carp, can cause an increase in water turbidity by damaging plant communities and accelerating eutrophication processes, thus compromising the habitat choice and reproduction of several amphibian species [91,92]. Predatory fish cause numerous adverse effects, such as direct predation on eggs and larvae, injuries due to predation attempts, resource competition, and changes in activity patterns and micro-habitat use [19,55,93–96]. Furthermore, both fish categories may damage native amphibian populations by pathogen transference [97].

We found paedomorphic Calabrian Alpine newts only in TRIF, and to date, paedomorphs have been reported to occur in two localities: Lago Due Uomini and Laghicello [31]. Facultative paedomorphosis is frequent in several Italian populations of *I. alpestris* [22,39,98–100] and is favoured in permanent aquatic habitats and where prey are abundant [77,99]. Evaluating the occurrence and incidence of alternative paedomorphic phenotypes should be of interest, considering the impacts of competition and predation. Indeed, it has been documented that introduced fish have determined the decline of metamorphs and the disappearance of paedomorphs from many newt sites in Europe [100]. A recent study showed several detrimental effects of the invasive *G. holbrooki* on paedomorphs of *Lissotriton graecus*, which exhibited avoidance behaviour and higher metamorphosis rates in the presence of fish [101].

The eradication or control of non-native invasive fish species in the lakes of Fagnano are undoubtedly unique and powerful conservation tools to improve the status of endemic and protected newts as well as the whole amphibian community. Given the different ecological and biological conditions of each water body, a targeted strategy combining several fish management techniques is necessary. Eradication programmes should include physical removal and chemical approaches [13,102–104] after a risk analysis to decide which strategy should be chosen and what the likelihood of success is. Considering the relatively high financial costs of eradication, the prioritisation of fish removal from ponds of higher suitability for newts could buffer the impact of fish presence [19]. Frog and salamander populations are expected to recover within a few years after fish removal from lakes [13,81,104–106]. The proximity of stable newt-breeding populations that can act as both sources and sinks may strongly favour the long-term persistence and newt recolonisation of lakes that have been returned to a fishless condition [12,19,107]. Therefore, in this emergency context, it is mandatory to create a network of small satellite artificial and semi-natural ponds, spatially connected to the historical sites, where to maintain newt populations until programmes to remove fish from lakes can be carried out. The identification of sites where to create suitable habitats should be realised through field surveys and the application of the most recent ecological modelling techniques to indicate the areas with major potential for retaining water, and also predicting changes in temperature and precipitation.

Regarding the newly discovered localities, both sites deserve regular interventions to guarantee water permanence and reduce silting phenomena by enlarging and deepening the ponds and removing excessive biomass accumulation. Furthermore, continuous surveillance is desirable to quickly notice any introduction of fish.

Finally, to deal with the 'emergency' situation, priority conservation action should be taken to preserve the Calabrian Alpine newt by implementing an *ex situ* conservation programme with research and conservation facilities in scientific institutions or zoos in Italy and Europe. The necessary *ex situ* initiatives will complement and support the *in situ* conservation of this highly endangered taxon, also considering that its disjunct distribution and its narrow range make this Alpine newt, by far, the most relevant biogeographic and conservation-related taxon compared to all of the other five *Ichthyosaura alpestris* subspecies [108]. Thus, the conservation of the Calabrian Alpine newt is not only important in terms of protecting taxonomic diversity but, being an evolutionarily significant unit, it assumes relevance for the purpose of better conserving the whole species' genetic diversity [109].

## 5. Conclusions

Updated distribution information and ecological data are mandatory to start a specific management plan for an endemic taxon on the brink of extinction, which is deserving of urgent conservation actions. Our observations facilitate conservation and management activities in the SAC Laghi di Fagnano and the surrounding area. The radical ecological upheaval following fish introductions undermines the survival of the Calabrian Alpine newt and the other two syntopic pond-breeding newts of community and conservation interest. An effective long-term management strategy to restore disturbed habitats by removing fish cannot be further postponed, including pond creation to provide all the amphibian species with alternative habitats. Local biodiversity and ecological functions greatly benefit from the eradication or control of non-native fish. Intensive monitoring and a systematic collection of data to ascertain more in-depth key ecological requirements and population status over a long period are also essential.

**Supplementary Materials:** The following supporting information can be downloaded at: https://www.mdpi.com/article/10.3390/ani13050871/s1, Figure S1: Representative images of the new breeding sites of *Ichthyosaura alpestris inexpectata*; Table S1: Water physical–chemical parameters measured in the study ponds; Table S2: Macroinvertebrates taxonomic list and relative abundance; Table S3: Morphometric parameters of *Ichthyosaura alpestris inexpectata* from the study sites.

**Author Contributions:** Conceptualisation, I.B.; methodology and formal analysis, I.B., V.C., A.R., A.C. and M.I.; investigation, I.B., V.C., A.C., M.G.S. and S.T.; data curation, I.B. and V.C.; writing—original draft preparation, I.B., A.R., V.C. and M.I.; writing—review and editing, I.B., V.C., M.I., F.A., A.R., A.C., M.B. and S.T.; supervision, M.B. and S.T. All authors have read and agreed to the published version of the manuscript.

**Funding:** This study was supported by research grant number 103/2021, funded by Regione Calabria, FESR-POR Calabria 2014–20120-Azione 6.5.A.1-Sub Azione 1.

**Institutional Review Board Statement:** Permits for field work and sampling procedures were obtained by the Italian Ministry of the Environment (prot. number: 0008263—25 January 2022).

**Data Availability Statement:** Not applicable.

**Acknowledgments:** We thank Eleonora Bozzo, Giuseppe De Bonis, Maria Prigoliti and Carlo Terranova for their help in conducting the fieldwork and Antonio Lucadamo for statistical advice. We are grateful to Giovanni Aramini (Environment and Territory Department of the Calabria Region) for his support to the problem of the conservation of the Calabrian Alpine newt. We are also grateful to the three reviewers, whose comments greatly improved the quality of our research.

**Conflicts of Interest:** The authors declare no conflict of interest.

# References


1. Dudgeon, D.; Arthington, A.H.; Gessner, M.O.; Kawabata, Z.I.; Knowler, D.J.; Lévêque, C.; Naiman, R.J.; Prieur-Richard, A.H.; Soto, D.; Stiassny, M.L.; et al. Freshwater biodiversity: Importance, threats, status and conservation challenges. *Biol. Rev.* **2006**, *81*, 163–182. [CrossRef]
2. Strayer, D.L.; Dudgeon, D. Freshwater biodiversity conservation: Recent progress and future challenges. *J. N. Am. Benthol. Soc.* **2010**, *29*, 344–358. [CrossRef]
3. Darwall, W.R.T.; Vie, J.C. Identifying important sites for conservation of freshwater biodiversity: Extending the species-based approach. *Fish. Manag. Ecol.* **2005**, *12*, 287–293. [CrossRef]
4. Groom, M.J.; Meffe, G.K.; Carroll, C.R.; Andelman, S.J. Principles of conservation biology. In *Protected Areas: Goals, Limitations and Design*; Groom, M.J., Meffe, G.K., Carroll, R., Eds.; Sinauer Associates: Sunderland, MA, USA, 2006.
5. Bernabò, I.; Biondi, M.; Cittadino, V.; Sperone, E.; Iannella, M. Addressing conservation measures through fine-tuned species distribution models for an Italian endangered endemic anuran. *Glob. Ecol. Conserv.* **2022**, *39*, e02302. [CrossRef]
6. Reid, W.V. Biodiversity hotspots. *Trends Ecol. Evol.* **1998**, *13*, 275–280. [CrossRef]
7. Schmeller, D.S.; Gruber, B.; Budrys, E.; Framsted, E.; Lengyel, S.; Henle, K. National responsibilities in European species conservation: A methodological review. *Conserv. Biol.* **2008**, *22*, 593–601. [CrossRef]
8. IUCN. *IUCN Red List Categories and Criteria: Version 3.1*, 2nd ed.; IUCN: Gland, Switzerland; Cambridge, UK, 2012; pp. iv + 32.
9. Strayer, D.L. Alien species in fresh waters: Ecological effects, interactions with other stressors, and prospects for the future. *Freshw. Biol.* **2010**, *55*, 152–174. [CrossRef]
10. Bernabò, I.; Guardia, A.; Macirella, R.; Sesti, S.; Tripepi, T.; Brunelli, E. Tissues injury and pathological changes in *Hyla intermedia* juveniles after chronic larval exposure to tebuconazole. *Ecotoxicol. Environ. Saf.* **2020**, *205*, 111367. [CrossRef]
11. Bernabò, I.; Guardia, A.; Macirella, R.; Tripepi, S.; Brunelli, E. Chronic exposures to fungicide pyrimethanil: Multi-organ effects on Italian tree frog (*Hyla intermedia*). *Sci. Rep.* **2017**, *7*, 6869. [CrossRef]
12. Bounas, A.; Keroglidou, M.; Toli, E.A.; Chousidis, I.; Tsaparis, D.; Leonardos, I.; Sotiropoulos, K. Constrained by aliens, shifting landscape, or poor water quality? Factors affecting the persistence of amphibians in an urban pond network. *Aquat. Conserv. Mar. Freshw. Ecosyst.* **2020**, *30*, 1037–1049. [CrossRef]
13. Bosch, J.; Bielby, J.; Martin-Beyer, B.; Rincon, P.; Correa-Araneda, F.; Boyero, L. Eradication of introduced fish allows successful recovery of a stream-dwelling amphibian. *PLoS ONE* **2019**, *14*, e0216204. [CrossRef]
14. Martel, A.; Blooi, M.; Adriaensen, C.; Van Rooij, P.; Beukema, W.; Fisher, M.C.; Farrer, R.A.; Schmidt, B.R.; Tobler, U.; Goka, K.; et al. Recent introduction of a chytrid fungus endangers Western Palearctic salamanders. *Science* **2014**, *346*, 630–631. [CrossRef]
15. Davidson, C.; Knapp, R.A. Multiple Stressors and Amphibian Declines: Dual Impacts of Pesticides and Fish on Yellow-Legged Frogs. *Ecol. App.* **2007**, *17*, 587–597. [CrossRef]
16. Nunes, A.L.; Fill, J.M.; Davies, S.J.; Louw, M.; Rebelo, A.D.; Thorp, C.J.; Vimercati, G.; Measey, J. A global meta-analysis of the ecological impacts of alien species on native amphibians. *Proc. R. Soc. B* **2019**, *286*, 20182528. [CrossRef]
17. Orizaola, G.; Braña, F. Effect of salmonid introduction and other environmental characteristics on amphibian distribution and abundance in mountain lakes of northern Spain. *Anim. Conserv.* **2006**, *9*, 171–178. [CrossRef]
18. Pilliod, S.D.; Peterson, C.R. Local and Landscape Effects of introduced Trout on Amphibians in Historically Fishless Watersheds. *Ecosystems* **2001**, *4*, 322–333. [CrossRef]
19. Tiberti, R. Can satellite ponds buffer the impact of introduced fish on newts in a mountain pond network? *Aquat. Conserv. Mar. Freshw. Ecosyst.* **2017**, *28*, 457–465. [CrossRef]
20. Falaschi, M.; Muraro, M.; Gibertini, C.; Delle Monache, D.; Lo Parrino, E.; Faraci, F.; Belluardo, F.; Di Nicola, M.R.; Manenti, R.; Ficetola, G.F. Explaining declines of newt abundance in northern Italy. *Freshw. Biol.* **2022**, *67*, 1174–1187. [CrossRef]
21. Chiocchio, A.; Bisconti, R.; Zampiglia, M.; Nascetti, G.; Canestrelli, D. Quaternary history; population genetic structure and diversity of the cold-adapted Alpine newt *Ichthyosaura alpestris* in peninsular Italy. *Sci. Rep.* **2017**, *7*, 2955. [CrossRef]
22. Dubois, A.; Breuil, M. Découverte de *Triturus alpestris* (Laurenti, 1768) en Calabre (Sud de l'Italie). *Alytes* **1983**, *2*, 9–18.
23. Dubois, A. Le Triton alpestre de Calabre: Une forme rare et menacée d'extinction. *Alytes* **1983**, *2*, 55–62.
24. Giacoma, C.; Picariello, O.; Puntillo, D.; Rossi, F.; Tripepi, S. The distribution and habitats of the newt (Triturus, Amphibia) in Calabria (southern Italy). *Monit. Zool. Ital.-Ital. J. Zool.* **1988**, *22*, 449–464.
25. Tripepi, S.; Rossi, F.; Serroni, P.; Brunelli, E. Distribuzione altitudinale degli Anfibi in Calabria. *Studi Trent. Sci. Nat. Acta Biol.* **1996**, *71*, 97–101.
26. Sperone, E.; Bonacci, A.; Brunelli, E.; Corapi, B.; Tripepi, S. Ecologia e conservazione dell'erpetofauna della Catena Costiera calabra. *Studi Trent. Sci. Nat. Acta Biol.* **2007**, *83*, 99–104.
27. Dubois, A.; Ohler, A. A quick method for a rough estimate of the size of a small and threatened animal population: The case of the relict *Ichthyosaura alpestris inexpectata* in southern Italy. *Bull. Soc. Nat. Luxemb.* **2009**, *110*, 115–124.
28. Rondinini, C.; Battistoni, A.; Teofili, C. (Eds.) *Lista Rossa IUCN dei Vertebrati Italiani 2022*; Comitato Italiano IUCN e Ministero dell'Ambiente e della Sicurezza Energetica: Roma, Italy, 2022.
29. Sillero, N.; Campos, J.; Bonardi, A.; Corti, C.; Creemers, R.; Crochet, P.A.; Crnobrnja Isailović, J.; Denoël, M.; Ficetola, G.F.; Gonçalves, J.; et al. Updated distribution and biogeography of amphibians and reptiles of Europe. *Amphib. Reptil.* **2014**, *35*, 1–31. [CrossRef]

30. Sindaco, R.; Doria, G.; Razzetti, E.; Bernini, F. *Atlante Degli Anfibi e dei Rettili d'Italia*; Societas Herpetologica Italica, Edizioni Polistampa: Firenze, Italy, 2006.

31. Lanza, B.; Andreone, F.; Bologna, M.A.; Corti, C.; Razzetti, E. *Fauna d'Italia Amphibia*; Edizioni Calderini de Il Sole 24 ORE Editoria Specializzata S.r.l.: Bologna, Italy, 2007; Volume XLII, pp. 254–265.

32. Di Nicola, M.R.; Cavigioli, L.; Luiselli, L.; Andreone, F. *Anfibi & Rettili d'Italia*; Edizione Belvedere: Latina, Italy, 2021; pp. 202–207.

33. Andreone, F. Triturus alpestris alpestris, *Triturus alpestris apuanus*, Tritone alpestre e Tritone appenninico. In *Erpetologia del Piemonte e Della Valle d'Aosta—Atlante Degli Anfibi e dei Rettili*; Andreone, F., Sindaco, R., Eds.; Monografie—Museo Regionale di Scienze Naturali: Torino, Italy, 1999; Volume 26, pp. 162–163.

34. Joly, P.; Giacoma, C. Limitation of similarity and feeding habits in three syntopic species of newts (Triturus, Amphibia). *Ecography* **1992**, *15*, 401–411. [CrossRef]

35. Coppari, L.; Ferri, V.; Marini, D.; Di Nicola, M.; Notomista, T. (Eds.) *Le Aree di Rilevanza Erpetologica in Italia 1995–2021*; Commissione Conservazione della Societas Herpetologica Italica. Available online: http://www-9.unipv.it/webshi/images/files/Volume_ARE_2021.pdf:2021 (accessed on 1 December 2022).

36. de Beaulieu, J.-L.; Brugiapaglia, E.; Joannin, S.; Guiter, F.; Zanchetta, G.; Wulf, S.; Peyron, O.; Bernardo, L.; Didier, J.; Stock, A.; et al. Lateglacial-Holocene abrupt vegetation changes at Lago Trifoglietti in Calabria, Southern Italy: The setting of ecosystems in a refugial zone. *Quat. Sci. Rev.* **2017**, *158*, 44–57. [CrossRef]

37. Aramini, G.; Bernabò, I.; Brandmayr, P.; Brusco, A.; Costa, R.M.S.; Fusillo, R.; Gangale, C.; Infusino, M.; Greco, R.; Musarella, C.M.; et al. *Rete Natura 2000. Biodiversità in Calabria*; Rubbettino Editore: Soveria Mannelli, Italy, 2021; Volume 2, ISBN 978-88-94601-30-5.

38. Pesaresi, S.; Biondi, E.; Casavecchia, S. Bioclimates of Italy. *J. Maps* **2017**, *13*, 955–960. [CrossRef]

39. Andreone, F.; Dore, B. New data on paedomorphism in Italian populations of alpine newts, *Triturus alpestris* (Laurenti, 1768) (Caudata: Salamandridae). *Herpetozoa* **1991**, *4*, 149–156.

40. Andreone, F. *Variabilità Morfologica e Riproduttiva in Popolazioni di Triturus alpestris (Laurenti, 1768) (Amphibia, Urodela, Salamandridae)*; Tesi di Dottorato di Ricerca, Università di Bologna: Bologna, Italy, 1990.

41. Bernabò, I.; Brunelli, E. Comparative morphological analysis during larval development of three syntopic newt species (Urodela: Salamandridae). *Eur. Zool. J.* **2019**, *86*, 38–53. [CrossRef]

42. Campaioli, S.; Ghetti, P.F.; Minelli, A.; Ruffo, S. *Manuale per il Riconoscimento dei Macroinvertebrati Delle Acque Dolci Italiane*; Provincia Autonoma di Trento: Bari, Italy, 1994; Volume I, p. 357.

43. Sansoni, G. *Atlante per il Riconoscimento dei Macroinvertebrati Bentonici dei Corsi D'acqua Italiani*; Provincia Autonoma di Trento: Bari, Italy, 1988; p. 191.

44. Tachet, H.; Richoux, P.; Bournaud, M.; Usseglio-Polatera, P. *Invertébrés d'Eau Douce*; CNRS Editions: Paris, Italy, 2002.

45. Shannon, C.E.; Weaver, W. *The Mathematical Theory of Communication*; The University of Illinois Press: Urbana, IL, USA, 1949; pp. 1–117.

46. Margalef, R. Temporal succession and spatial heterogeneity in phytoplankton. In *Perspectives in Marine Biology*; Buzzati-Traverso, A.A., Ed.; University of California Press: Berkeley, CA, USA, 1958; pp. 323–347.

47. MacCracken, J.G.; Stebbings, J.L. Test of a body condition index with amphibians. *J. Herpetol.* **2012**, *46*, 346–350. [CrossRef]

48. Peig, J.; Green, A.J. New perspectives for estimating body condition from mass/length data: The scaled mass index as an alternative method. *Oikos* **2009**, *118*, 1883–1891. [CrossRef]

49. Peig, J.; Green, A.J. The paradigm of body condition: A critical reappraisal of current methods based on mass and length. *Funct. Ecol.* **2010**, *24*, 1323–1332. [CrossRef]

50. R Core Team. *R: A Language and Environment for Statistical Computing*; R Foundation for Statistical Computing: Vienna, Austria, 2022. Available online: https://www.R-project.org/ (accessed on 15 June 2022).

51. Warton, D.I.; Duursma, R.A.; Falster, D.S.; Taskinen, S. Smatr 3—An R package for estimation and inference about allometric lines. *Methods Ecol. Evol.* **2012**, *3*, 257–259. [CrossRef]

52. Moran, P.A.P. A mathematical theory of animal trapping. *Biometrica* **1951**, *38*, 307–311. [CrossRef]

53. Zippin, C. The removal method of population estimation. *J. Wildl.* **1958**, *22*, 82–90. [CrossRef]

54. Ogle, D.H.; Doll, J.C.; Wheeler, P.; Dinno, A. FSA: Fisheries Stock Analysis. R Package Version 0.9.3. 2022. Available online: https://github.com/fishR-Core-Team/FSA (accessed on 1 December 2022).

55. Orizaola, G.E.; Brana, F.L. Oviposition behaviour and vulnerability of eggs to predation in four newt species (genus Triturus). *Herpetol. J.* **2003**, *13*, 121–124.

56. Teplitsky, C.; Plénet, S.; Joly, P. Tadpoles' responses to risk of fish introduction. *Oecologia* **2003**, *134*, 270–277. [CrossRef]

57. Lima, S.L.; Dill, L.M. Behavioral decisions made under the risk of predation: A review and prospectus. *Can. J. Zool.* **1990**, *68*, 619–640. [CrossRef]

58. Winandy, L.; Darnet, E.; Denoël, M. Amphibians forgo aquatic life in response to alien fish introduction. *Anim. Behav.* **2015**, *109*, 209–216. [CrossRef]

59. Winandy, L.; Legrand, P.; Denoël, M. Habitat selection and reproduction of newts in networks of fish and fishless aquatic patches. *Anim. Behav.* **2017**, *123*, 107–115. [CrossRef]

60. Šunje, E.; Stroil, K.B.; Raffaëlli, J.; Zimić, A.; Marquis, O. A revised phylogeny of Alpine newts unravels the evolutionary distinctiveness of the Bosnian Alpine newt—*Ichthyosaura alpestris reiseri* (Werner, 1902). *Amphib. Reptil.* **2021**, *42*, 481–490. [CrossRef]

61. Tanadini, L.G.; Schmidt, B.R. Population Size Influences Amphibian Detection Probability: Implications for Biodiversity Monitoring Programs. *PLoS ONE* **2011**, *6*, e28244. [CrossRef]
62. Guillera-Arroita, G.; Lahoz-Monfort, J.J.; Elith, J.; Gordon, A.; Kujala, H.; Lentini, P.E.; McCarthy, M.A.; Tingley, R.; Wintle, B.A. Is my species distribution model fit for purpose? Matching data and models to applications. *Glob. Ecol. Biogeogr.* **2015**, *24*, 276–292. [CrossRef]
63. Fusillo, R.; Esse, E.; Marcelli, M.; Mastronardi, D.; Bernabò, I. New record of *Lissotriton vulgaris meridionalis* (Boulenger, 1882) at the southernmost edge of its distribution in Italy. *Herpetol. Notes* **2021**, *14*, 923–926.
64. Bernabò, I.; Cittadino, V.; Tripepi, S.; Marchianò, V.; Piazzini, S.; Biondi, M.; Iannella, M. Updating Distribution, Ecology, and Hotspots for Three Amphibian Species to Set Conservation Priorities in A European Glacial Refugium. *Land* **2022**, *11*, 1292. [CrossRef]
65. Joly, P.; Miaud, C. Fidelity to the breeding site in the alpine newt *Triturus alpestris*. *Behav. Processes* **1989**, *19*, 47–56. [CrossRef]
66. Vilter, A.; Vilter, V. Migration de reproduction chez le triton alpestre des Alpes vaudoises. *C. R. Soc. Biol.* **1962**, *156*, 2005–2006.
67. Jehle, R.; Sinsch, U. Wanderleistung und rientierung von Amphibien: Eine Übersicht. *Z. Feldherpetologie* **2007**, *14*, 137–152.
68. Ghetti, P.L. *I Macroinvertebrati nel Controllo Della Qualità Degli Ambienti di Acque Correnti*; Provincia Autonoma di Trento: Trento, Italy, 1997; p. 222.
69. Thornhill, I.; Batty, L.; Death, R.G.; Friberg, N.R.; Ledger, M.E. Local and landscape scale determinants of macroinvertebrate assemblages and their conservation value in ponds across an urban land-use gradient. *Biodivers. Conserv.* **2017**, *26*, 1065–1086. [CrossRef]
70. Smith, G.R.; Vaala, D.A.; Dingfelder, H.A. Distribution and abundance of macroinvertebrates within two temporary ponds. *Hydrobiologia* **2003**, *497*, 161–167. [CrossRef]
71. Motchié, F.E.; Konan, Y.A.; Koffi, K.B.; Etilé, N.D.R.; Gooré, B.G. Diversity and structure of benthic macroinvertebrates community in relation to environmental variables in Lake Ehuikro, Côte d'Ivoire. *Int. J. Res. Environ. Stud.* **2020**, *7*, 1–13.
72. Hill, M.J.; Mathers, K.L.; Wood, P.J. The aquatic macroinvertebrate biodiversity of urban ponds in a medium-sized European town (Loughborough, UK). *Hydrobiologia* **2015**, *760*, 225–238. [CrossRef]
73. Hill, M.J.; Sayer, C.D.; Wood, P.J. When is the best time to sample aquatic macroinvertebrates in ponds for biodiversity assessment? *Environ. Monit. Assess.* **2016**, *188*, 194. [CrossRef]
74. Merritt, R.W.; Cummins, K.W. Trophic relations of macroinvertebrates. In *Methods in Stream Ecology*; Hauer, F.R., Lamberti, G.A., Eds.; Academic Press: Burlington, NJ, USA, 2006; pp. 585–609.
75. Fasola, M.; Canova, L. Feeding habits of *Triturus vulgaris*, *T. cristatus* and *T. alpestris* (Amphibia, Urodela) in the northern Apennines (Italy). *Ital. J. Zool.* **1992**, *59*, 273–280.
76. Vignoli, L.; Bologna, M.A.; Luiselli, L. Seasonal patterns of activity and community structure in an amphibian assemblage at a pond network with variable hydrology. *Acta Oecol.* **2007**, *31*, 185–192. [CrossRef]
77. Denoël, M.; Andreone, F. Trophic habits and aquatic microhabitat use in gilled immature, paedomorphic and metamorphic Alpine newts (*Triturus alpestris apuanus*) in a pond in central Italy. *Belg. J. Zool.* **2003**, *133*, 95–102.
78. Salvidio, S.; Costa, A.; Crovetto, F. Individual trophic specialisation in the Alpine newt increases with increasing resource diversity. *Ann. Zool. Fenn.* **2019**, *56*, 17–24. [CrossRef]
79. Careddu, G.; Carlini, N.; Romano, A.; Rossi, L.; Calizza, E.; Sporta Caputi, S.; Costantini, M.L. Diet composition of the Italian crested newt (*Triturus carnifex*) in structurally different artificial ponds based on stomach contents and stable isotope analyses. *Aquat. Conserv. Mar. Freshw. Ecosyst.* **2020**, *30*, 1505–1520. [CrossRef]
80. Schabetsberger, R.; Jersabek, C.D. Alpine newts (*Triturus alpestris*) as top predators in a high-altitude karst lake: Daily food consumption and impact on the copepod *Arctodiaptomus alpinus*. *Freshw. Biol.* **1995**, *33*, 47–61. [CrossRef]
81. Aronsson, S.; Stenson, J.A. Newt–fish interactions in a small forest lake. *Amphib. Reptil.* **1995**, *16*, 177–184.
82. Iannella, M.; Console, G.; D'Alessandro, P.; Cerasoli, F.; Mantoni, C.; Ruggieri, F.; Di Donato, F.; Biondi, M. Preliminary Analysis of the Diet of *Triturus carnifex* and Pollution in Mountain Karst Ponds in Central Apennines. *Water* **2020**, *12*, 44. [CrossRef]
83. Solé, M.; Rödder, D. Dietary assessments of adult amphibians. In *Amphibian Ecology and Conservation: A Handbook of Techniques*; Dodd, J., Ed.; Oxford University Press: Oxford, UK, 2010; pp. 167–184.
84. Romano, A.; Costa, A.; Basile, M. Skewed sex ratio in a forest salamander: Artefact of the different capture probabilities between sexes or actual ecological trait? *Amphib. Reptil.* **2018**, *39*, 79–86. [CrossRef]
85. Kalezić, M.L.; Džukić, G.; Popadić, A. Paedomorphosis in Yugoslav Alpine newt (*Triturus alpestris*) populations: Morphometric variability and sex ratio. *Arhiv. Bioloških. Nauka.* **1989**, *41*, 67–79.
86. Schabetsberger, R.; Goldschmid, A. Age structure and survival rate in alpine newts (*Triturus alpestris*). *Alytes* **1994**, *12*, 41–47.
87. Naumov, B.Y.; Popgeorgiev, G.S.; Kornilev, Y.V.; Plachiyski, D.G.; Stojanov, A.J.; Tzankov, N.D. Distribution and Ecology of the Alpine Newt *Ichthyosaura alpestris* (Laurenti, 1768) (Amphibia: Salamandridae) in Bulgaria. *Acta Zool. Bulg.* **2020**, *72*, 83–102.
88. Arntzen, J.W. Seasonal variation in sex ratio and asynchronous presence at ponds of male and female Triturus newts. *J. Herpetol.* **2002**, *36*, 30–35. [CrossRef]
89. Sztatecsny, M.; Schabetsberger, R. Into thin air: Vertical migration, body condition, and quality of terrestrial habitats of alpine common toads, *Bufo bufo*. *Can. J. Zool.* **2005**, *83*, 788–796. [CrossRef]
90. Bancila, R.I.; Hartel, T.; Plaiasu, R.; Smets, J.; Cogalniceanu, D. Comparing three body condition indices in amphibians: A case study of yellow-bellied toad *Bombina variegata*. *Amphib. Reptil.* **2010**, *31*, 558–562.

91. Vilizzi, L.; Tarkan, A.S.; Copp, G.H. Experimental Evidence from Causal Criteria Analysis for the Effects of Common Carp *Cyprinus carpio* on Freshwater Ecosystems: A Global Perspective. *Rev. Fish. Sci. Aquacult.* **2015**, *23*, 253–290. [CrossRef]
92. Kloskowski, J.; Nieoczym, M. Strong behavioral effects of omnivorous fish on Amphibian oviposition habitat selection: Potential consequences for ecosystem shifts. *Front. Ecol. Evol.* **2022**, *10*, 323. [CrossRef]
93. Segev, O.; Mangel, M.; Blaustein, L. Deleterious effects by mosquitofish (*Gambusia affinis*) on the endangered fire salamander (*Salamandra infraimmaculata*). *Anim. Conserv.* **2009**, *12*, 29–37. [CrossRef]
94. Vannini, A.; Bruni, G.; Ricciardi, G.; Platania, L.; Mori, E.; Tricarico, E. *Gambusia holbrooki*, the "tadpole fish": The impact of its predatory behaviour on four protected species of European amphibians. *Aquat. Conserv. Mar. Freshwat. Ecosyst.* **2018**, *28*, 476–484. [CrossRef]
95. Gamradt, S.C.; Kats, L.B. Effect of introduced crayfish and mosquitofish on California newts. *Conserv. Biol.* **1996**, *10*, 1155–1162. [CrossRef]
96. Monello, R.J.; Wright, R.G. Predation by goldfish (*Carassius auratus*) on eggs and larvae of the eastern long-toed salamander (*Ambystoma macrodactylum columbianum*). *J. Herpetol.* **2001**, *35*, 350–353. [CrossRef]
97. Kiesecker, J.M.; Blaustein, A.R.; Miller, C.L. Transfer of a pathogen from fish to amphibians. *Conserv. Biol.* **2001**, *15*, 1064–1070. [CrossRef]
98. Fasola, M. Resource partitioning by three species of newts during their aquatic phase. *Ecography* **1993**, *16*, 73–81. [CrossRef]
99. Denoël, M.; Duguet, R.; Džukić, G.; Kalezić, M.L.; Mazzotti, S. Biogeography and ecology of paedomorphosis in *Triturus alpestris* (Amphibia, Caudata). *J. Biogeogr.* **2001**, *28*, 1271–1280. [CrossRef]
100. Denoël, M.; Džukić, G.; Kalezić, M.L. Effect of widespread fish introductions on paedomorphic newts in Europe. *Conserv. Biol.* **2005**, *19*, 162–170. [CrossRef]
101. Toli, E.A.; Chavas, C.; Denoël, M.; Bounas, A.; Sotiropoulos, K. A subtle threat: Behavioral and phenotypic consequences of invasive mosquitofish on a native paedomorphic newt. *Biol. Invasions* **2020**, *22*, 1299–1308. [CrossRef]
102. Fried, L.M.; Boyer, M.C.; Brooks, M.J. Amphibian response to rotenone treatment of ten alpine lakes in Northwest Montana. *N. Am. J. Fish. Manag.* **2018**, *38*, 237–246. [CrossRef]
103. Schnee, M.E.; Clancy, N.G.; Boyer, M.C.; Bourret, S.L. Recovery of freshwater invertebrates in alpine lakes and streams following eradication of non-native trout with rotenone. *J. Fish Wildl. Manag.* **2021**, *12*, 475–484. [CrossRef]
104. Tiberti, R.; Bogliani, G.; Brighenti, S.; Iacobuzio, R.; Liautaud, K.; Rolla, M.; von Hardenberg, A.; Bassano, B. Recovery of high mountain Alpine lakes after the eradication of introduced brook trout *Salvelinus fontinalis* using non-chemical methods. *Biol. Invasions* **2019**, *21*, 875–894. [CrossRef]
105. Denoël, M.; Winandy, L. The importance of phenotypic diversity in conservation: Resilience of palmate newt morphotypes after fish removal in Larzac ponds (France). *Biol. Conserv.* **2015**, *192*, 402–408. [CrossRef]
106. Miró, A.; O'Brien, D.; Tomàs, J.; Buchaca, T.; Sabás, I.; Osorio, V.; Lucati, F.; Pou-Rovira, Q.; Ventura, M. Rapid amphibian community recovery following removal of non-native fish from high mountain lakes. *Biol. Conserv.* **2020**, *251*, 108783. [CrossRef]
107. Denoël, M.; Scimè, P.; Zambelli, N. Newt life after fish introduction: Extirpation of paedomorphosis in a mountain fish lake and newt use of satellite pools. *Curr. Zool.* **2016**, *62*, 61–69. [CrossRef]
108. Recuero, E.; Buckley, D.; García-París, M.; Arntzen, J.W.; Cogălniceanu, D.; Martínez-Solano, I. Evolutionary history of *Ichthyosaura alpestris* (Caudata, Salamandridae) inferred from the combined analysis of nuclear and mitochondrial markers. *Mol. Phylogenet. Evol.* **2014**, *81*, 207–220. [CrossRef]
109. Casacci, L.P.; Barbero, F.; Balletto, E. The "Evolutionarily Significant Unit" concept and its applicability in biological conservation. *Ital. J. Zool.* **2014**, *81*, 182–193. [CrossRef]

*Article*

# Monitoring of the Endangered Cave Salamander *Speleomantes sarrabusensis*

**Roberto Cogoni** [1]**, Milos Di Gregorio** [2,*]**, Fabio Cianferoni** [3,4] **and Enrico Lunghi** [5,6,7]

[1]  Unione Speleologica Cagliaritana, Quartu Sant'Elena, 09045 Cagliari, Italy
[2]  Dipartimento di Biologia, Università degli Studi di Pisa, 56126 Pisa, Italy
[3]  Istituto di Ricerca sugli Ecosistemi Terrestri (IRET), Consiglio Nazionale delle Ricerche (CNR), Sesto Fiorentino, 50019 Firenze, Italy
[4]  Museo di Storia Naturale dell'Università degli Studi di Firenze, "La Specola", 50125 Firenze, Italy
[5]  Dipartimento di Medicina Clinica, Sanità Pubblica, Scienze Della Vita e dell'ambiente (MeSVA), Universià degli Studi dell'Aquila, 67100 L'Aquila, Italy
[6]  Associazione Natural Oasis, 59100 Prato, Italy
[7]  Unione Speleologica Calenzano, 50041 Calenzano, Italy
*  Correspondence: m.digregorio4@studenti.unipi.it

**Simple Summary:** Here, we provide the information derived from the first monitoring activities performed on the endangered Sette Fratelli cave salamander, *Speleomantes sarrabusensis*. Adopting two different monitoring schemes, we estimated the abundance of four populations of *S. sarrabusensis*, providing important data for future status assessments of the species.

**Abstract:** In this study, we performed the first monitoring activities on one of the most endangered amphibians in Europe, the Sette Fratelli cave salamander *Speleomantes sarrabusensis*. The data presented here are derived from two monitoring activities aiming to assess the status and abundance of four populations of *S. sarrabusensis*. With the first monitoring, we surveyed the well-known population occurring within artificial springs during the period 2015–2018, providing monthly data on the number of active individuals. With the second monitoring performed during spring to early summer of 2022, we surveyed four populations at three time points (the one from artificial springs and three from forested areas) and we provided the first estimation of the populations' abundance. Furthermore, we analyzed for the first time the stomach contents from a population of *S. sarrabusensis* only occurring in forested environments. With our study, we provided the first information on the abundance of different populations of *S. sarrabusensis*, representing the starting point for future status assessments for this endangered species.

**Keywords:** *Hydromantes*; Plethodontidae; abundance; wildlife; amphibian; conservation

**Citation:** Cogoni, R.; Di Gregorio, M.; Cianferoni, F.; Lunghi, E. Monitoring of the Endangered Cave Salamander *Speleomantes sarrabusensis*. *Animals* **2023**, *13*, 391. https://doi.org/10.3390/ani13030391

Academic Editor: José Martín

Received: 20 December 2022
Revised: 11 January 2023
Accepted: 20 January 2023
Published: 24 January 2023

## 1. Introduction

The Sette Fratelli cave salamander, *Speleomantes sarrabusensis* Lanza et al., 2001 (Figure 1), is one of the European amphibian species with the smallest distribution ($\leq 70$ km$^2$); it can be found only in the most south-eastern area of Sardinia (Italy) [1].

Before 2001, *S. sarrabusensis* was considered a sub-species of *S. imperialis* [1]. Although called "cave salamanders", all the European plethodontid salamanders of the genus *Speleomantes* are epigean species that occur in caves and other subterranean environments to avoid unsuitable climatic conditions [2,3]. However, *S. sarrabusensis* is distributed over a granitic area where no caves exist, and thus it can be found only in surface environments or within anthropic structures characterised by a suitable inner microclimate [1,4]. The very small distribution and the lack of potential subterranean refuges are within the characteristics that make *S. sarrabusensis* highly sensitive to extinction risk [5].

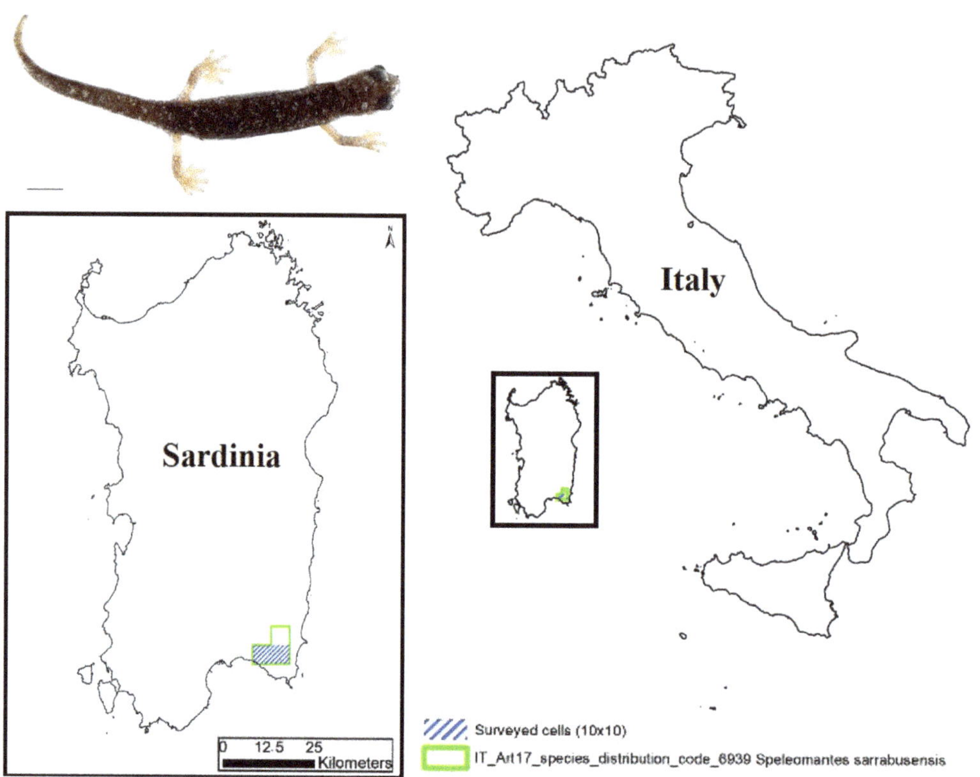

**Figure 1.** Map indicating the area of interest in our monitoring activity. In the upper-left corner, an adult of *Speleomantes sarrabusensis* (scale bar = 10 mm).

The Sette Fratelli cave salamander is the *Speleomantes* species for which the available knowledge is the poorest and the most anecdotal. For example, observing captive individuals, Lanza and Leo [6] hypothesised that the species might be viviparous; however, recent studies confirmed its oviparity as for the rest of the genus [7]. Most of our lack of knowledge is probably due to the difficulties in observing the species when it hides in its refuges. Subterranean *Speleomantes* usually show a high detection probability, as individuals can be easily observed clinging onto cave walls [8–10]. On the contrary, detecting *Speleomantes* in a forested area is more challenging as individuals have plenty of places to hide, and they can be confused with the background [11]. The only well-known population of *S. sarrabusensis* is the one that occurs inside the artificial springs located on the Sette Fratelli Mountain (squared structures made of concrete with water tanks that cover about 80% of the inner surface; Figure 2) [1], where salamanders cling onto the flat walls and experience microclimatic conditions that are comparable with those found in subterranean environments [4]. Consequently, our knowledge of *S. sarrabusensis* is only based on studies performed on the population from artificial springs [7,12–14].

**Figure 2.** (**A**,**B**) show the two artificial springs in which *Speleomantes sarrabusensis* occurs. In (**C**), it is possible to see part of the inner environment of one spring; arrows are pointing to a few individuals that are clinging on the concrete walls.

The present study provides new information on the distribution and life history of the strictly protected *S. sarrabusensis*. Specifically, we provide data that originates from two different monitoring schemes. The first was long-term monitoring which was interested in the well-known artificial springs aiming to estimate the population trend over a period of four years [15]. Although in this part of the study we provide the number of individuals observed within each spring, we should remark that the two structures are not far enough from each other to assume that they hold two independent populations [16]. In the second part, we performed repeated surveys in both the artificial springs (considered as a single population) and in three different forested areas to provide the first estimation of their abundance (i.e., the abundance of active individuals at that time). The occurrence of these three independent epigean populations was previously assessed from literature and due to novel discoveries. The need for improving the information on this poorly known species was recently boosted by a large death toll observed in the summer of 2021, where 14 individuals were found dead. We still do not know the cause of such a death toll, but this dramatic event was one of the reasons behind the change in the species conservation status to critically endangered [5].

## 2. Materials and Methods

The first long-term monitoring of the artificial springs (Figure 2A,B) involved monthly surveys from March 2015 until June 2018. During each survey, one operator counted the number of active individuals of *S. sarrabusensis* which are clearly visible clinging onto the flat concrete walls (Figure 2C).

At the end of the survey, at about 5 m from the entrance, air temperature and humidity were recorded at the ground level using a thermohygrometer (accuracy $\pm 0.5$ °C, $\pm 2.5\%$). We used Linear Models (LM) [17] to evaluate the potential variation of the number of observed individuals during the period of monitoring. The number of observed individuals was used as the dependent variable, the month and the year as fixed factors, and the spring identity as a random factor. The likelihood ratio test was used to test the significance of variables.

The second monitoring occurred in 2022. Besides the well-known population from the artificial springs, we surveyed three additional epigean populations aiming to perform the first estimation of their abundance and to provide much-improved information on species status. Two sites were characterized by forested areas mainly composed of evergreen oak *Quercus ilex* L., 1753 and the strawberry tree *Arbutus unedo* L., 1753, while one was characterized by Mediterranean scrub. The exact coordinates of the populations are not provided due to conservation concerns [18]. The three epigean populations were surveyed in March, while the one occurring inside the artificial springs was surveyed in June. These periods were chosen based on the general knowledge of *Speleomantes* phenology, where individuals are more active in surface environments during periods characterized by higher precipitation and relatively low temperatures, while during hot and dry periods the subterranean abundance should be the highest [9,11,19]; see also Table 1. Each population was surveyed three times within a period of 30 days, to guarantee population closure and to avoid strong microclimatic fluctuations [9,20]. During each survey, individuals were searched within the defined area and counted without providing any further disturbance. In each site, we selected a defined area to monitor according to the local characteristics of the territory. For the first site (forest #1) we selected a transect of 962 m, for the second site (forest #2) we selected a transect of 219 m; both transects were defined using GPS tracks. In the third site (scrub #1) we defined a plot of 320 m$^2$ in which we adopted a serpentine (bustrophedic) search of salamanders on the ground and under the stones, avoiding inaccessible dense bushes. The overall surveyed area of the artificial springs is 70 m$^2$ of building footprint (about 4 × 7.5 m the first and 4 × 10 m the second). Inside the artificial springs, a single operator searched for *Speleomantes* on the flat walls, with a constant sampling effort of 7.5 min per 3 linear meters [19]. For the epigean populations, two operators simultaneously searched for individuals on the ground (also lifting logs/stones) and on the trees [21] within the defined area. The two forested areas were surveyed with an average sampling effort of 1 h/1270 m$^2$ per person (1 h/1227 m$^2$ for the first and 1 h/1314 m$^2$ for the second site), while for the scrub we dedicated 1 h for the whole plot (320 m$^2$). During the third survey on each site, we measured temperature and humidity at the ground with a thermohygrometer (accuracy ±0.5 °C, ±2.5%). Estimations of the populations' abundance were performed using $N$-mixture models [22].

Considering the large number of individuals observed from site forest #1 (see Results), we selected this population to perform an analysis on the stomach contents of individuals. During the last survey, individuals were captured and then we proceeded as follows: each individual was weighed using a digital scale (0.01 g), photographed for post hoc estimation of both their snout-vent length (SVL) and total length (TL) both SVL and TL; [23,24], and stomach flushed to collect the residuals of their last foraging activity [25]. Adult salamanders showing male's sexual characteristics (mental chin, conical shape of the head) were considered males [1]. The SVL was then used to distinguish between adults females (≥55 mm) and juveniles (<55 mm) [7]. Individuals were then stabled in situ inside fauna boxes; this allowed us to perform a second round of capture–photograph–stomach-flushing the following night, without the risk of recapturing the same individuals. The recognised prey items were counted and divided following Lunghi, et al. [26]. In brief, consumed prey were divided according to their taxonomic order and possibly also according to their life stage (larva vs. adult). In two circumstances, the family Staphylinidae (Coleoptera) and the family Formicidae (Hymenoptera) were considered separately. Considering the limited number of individuals and the single-season sampling, we simply describe the seasonal diet of this population, comparing this information with those available for the population from the artificial spring [26], avoiding performing weak statistical analyses. The full data related to salamanders' stomach contents is provided as supplementary material (Table S1).

## 3. Results

During the first long-term monitoring, we performed a total of 929 observations of *Speleomantes sarrabusensis* within the two artificial springs (737 in the first and 102 in the second). A detailed summary of the observed individuals is shown in Table 1.

**Table 1.** Monthly data from the long-term monitoring of *Speleomantes sarrabusensis*. In the upper part of the table, we show the monthly count performed within the two artificial springs coded following Lunghi, et al. [27]. The symbol (-) indicates that the survey has not been performed. In the lower part, we show the average monthly microclimatic conditions recorded inside the two artificial springs during the monitoring period. For each month, we show the average (±SD) temperature (°C) and relative humidity (%) recorded during the period 2015–2018. When the standard deviation is not shown, it means that the relative microclimatic variable has been recorded only once.

| Number of Observed Individuals | | | | | | | | | | | | | |
| --- | --- | --- | --- | --- | --- | --- | --- | --- | --- | --- | --- | --- | --- |
| Site | Year | January | February | March | April | May | June | July | August | September | October | November | December |
| Spring 1 | 2015 | - | - | 4 | 40 | 72 | 79 | 66 | 50 | 83 | 16 | 1 | - |
| | 2016 | 3 | - | 10 | 70 | 84 | 90 | 36 | 39 | 42 | 22 | 0 | 1 |
| | 2017 | 1 | 1 | 5 | 29 | 18 | 23 | 36 | 33 | 20 | 35 | 2 | 1 |
| | 2018 | 2 | 6 | 9 | 21 | 39 | 59 | - | - | - | - | - | - |
| Spring 2 | 2015 | - | - | 0 | 6 | 10 | 2 | 5 | 5 | 7 | 18 | 3 | - |
| | 2016 | 3 | - | 8 | 8 | 5 | 3 | 3 | 1 | 1 | 4 | 5 | 3 |
| | 2017 | 2 | 2 | 8 | 12 | 6 | 5 | 6 | 5 | 3 | 7 | 5 | 3 |
| | 2018 | 3 | 5 | 6 | 0 | 8 | 6 | - | - | - | - | - | - |

| Data on the inner microclimate | | | | | | | | | | | | | |
| --- | --- | --- | --- | --- | --- | --- | --- | --- | --- | --- | --- | --- | --- |
| Site | | January | February | March | April | May | June | July | August | September | October | November | December |
| Spring 1 | Temperature | 10.2 (±0.8) | 10.3 (±0.1) | 10.3 (±0.3) | 12.2 (±1.3) | 13.2 (±1) | 14.2 | 16.4 (±0.8) | 17.3 (±0.8) | 16.5 | 15.2 (±0.3) | 13.0 (±0.7) | 9.8 |
| | Humidity | 81.2 (±2.2) | 81.8 (±0.3) | 82.7 (±0.9) | 83.4 (±2.1) | 84.1 (±3) | 81.0 | 78.8 (±1.4) | 79.7 (±2.6) | 79.5 | 82.2 (±1) | 82.4 (±1.5) | 81.6 |
| Spring 2 | Temperature | 10.0 (±1.3) | 9.9 (±0.3) | 10.9 (±0.1) | 11.9 (±0.8) | 12.6 (±0.7) | 13.9 | 15.4 (±0.6) | 16.0 (±0.9) | 15.7 | 14.2 (±0.1) | 12.8 (±1.3) | 9.3 |
| | Humidity | 82.7 (±1) | 82.6 (±1.1) | 82.6 (±0.9) | 84.1 (±2.8) | 84.5 (±3.4) | 82.6 | 82.2 (±0.4) | 81.8 (±1) | 81.6 | 82.9 (±1.2) | 80.0 (±2.8) | 82.4 |

The number of observed individuals was significantly affected by both the month ($F_{11,61}$ = 3.17, $p$ = 0.002) and the year of the survey ($F_{1,61}$ = 4.72, $p$ = 0.034). Overall, in June, the number of observed individuals was the highest, while in November, it was the lowest. The number of individuals significantly declined over the monitoring period. In Table 1 are also shown data on the monthly average ($\pm$SD) of the air temperature and humidity recorded within the two springs.

With the second part of our study, we were able to provide the first estimation of the abundance for four different populations of *S. sarrabusensis* (Table 2).

**Table 2.** Data related to the monitored populations of *Speleomantes sarrabusensis*. Along with general information on the location of the site, the surveyed areas, and the local microclimate, for each population we provide the number of observed individuals during each survey and the estimation of the population abundance. The elevation represents the average elevation of the monitored area. The abundance (mode) is accompanied by the confidence interval (2.5–97.5%).

| Population | Latitude | Longitude | Elevation (m a.s.l.) | Surveyed Area (m²) | Microclimate | Survey #1 | Survey #2 | Survey #3 | Estimated Abundance | 2.5% | 97.5% |
|---|---|---|---|---|---|---|---|---|---|---|---|
| Springs | 39°29′ | 9°44′ | 783 | 70 | 15.4 °C—79.2% | 21 | 14 | 3 | 34 | 28 | 41 |
| Forest #1 | 39°12′ | 9°28′ | 444 | 2454 | 10.3 °C—80% | 10 | 12 | 28 | 41 | 35 | 48 |
| Forest #2 | 39°14′ | 9°28′ | 569 | 2628 | 10.3 °C—80.4% | 2 | 0 | 1 | 9 | 5 | 16 |
| Scrub #1 | 39°15′ | 9°23′ | 620 | 320 | 8.7 °C—79.9% | 0 | 2 | 0 | 9 | 4 | 15 |

The species detection probability was relatively low (0.328 $\pm$ 0.081 SE). For the population inhabiting the two artificial springs on Sette Fratelli mountain, we estimated 34 individuals, while the estimated abundance for the other three epigean populations was 41, 9, and 9, respectively (Table 2).

In site forest #1 we captured a total of 37 individuals (24 + 15) of which 11 were males, seven females, and 21 juveniles (Table S1). Six juveniles were too small and we did not perform stomach flushing; among the other individuals, none had an empty stomach. We recognised a total of 356 prey items belonging to 31 different prey categories; three categories accounted for more than 43% of the consumed prey (Entomobryomorpha, 25%; Araneae and Diptera both >9%) (Table S1).

## 4. Discussion

With this study we provided the results of the first monitoring performed on *Speleomantes sarrabusensis*. Our results provided quantitative information on the monitored populations, data that represents the starting point for future conservation assessments. During the systematic monitoring performed in this study (2015–2018), we noticed a significant decline in the number of active individuals inside the springs. This can be considered just a red flag that should stimulate further monitoring because we have no additional information that helps us in understanding the potential causes. According to the data on the microclimate recorded inside the springs throughout the monitoring period (Table 1), we can exclude a drastic change in the inner microclimatic conditions as a potential cause for the reduction of the number of active individuals [8]. Indeed, from our data we can notice a yearly natural fluctuation of the inner microclimate correlated to the external climatic conditions [9]. Furthermore, when averaging the monthly data of the inner microclimate recorded throughout the survey period, we obtain a standard deviation whose maximum reaches 1.3 °C and 3.4% for temperature and humidity, respectively, a condition that maintains the inner microclimate within the preferred range for the species [4]. It is possible that some human-induced factors negatively affected the populations causing a sensible decline, but we also cannot exclude the possibility that we simply recorded a natural contraction of the population abundance; further monitoring activities are therefore necessary to better comprehend the causes of such decline.

In the second part of this study we provided the first quantitative information on the abundance of different populations of *S. sarrabusensis*. However, we must say that the observed individuals and the related estimation of abundance may be lower than expected. According to the studies performed on different subterranean populations of *Speleomantes*, during the period April–May, the number of active individuals was the highest [8,9,19], a condition that allows to produce estimations that are the closest to the real population size [22]. Considering the fully epigean habits of *S. sarrabusensis*, we chose to perform our surveys within the forested area in March, a month in which we expected suitable environmental conditions [8]. Unfortunately, the microclimatic conditions registered at the sites during our surveys were sub-optimal for the species [4] (Table 2), a circumstance that strongly affects the activity of individuals [8,9]. Therefore, our data should be interpreted as a snapshot taken in a sub-optimal period when the number of active individuals was very low, producing a highly underestimated estimation. Similarly, the surveys within artificial springs were performed in June, a period in which the subterranean activity of *Speleomantes* is the highest [4]. However, the extremely harsh climatic conditions occurred in that period likely induced an earlier aestivation of the population [1], allowing us to observe only a small number of active individuals, and therefore produced unexpected underestimated abundances.

The trophic spectrum of this population was larger than that of the population from artificial springs (31 vs. 19 recognized prey categories) [26], suggesting that resource availability may play an important role in defining the diet and the foraging behaviour of these generalist salamanders [13,28,29]. According to the optimal forage theory, individuals should adopt a foraging strategy that maximises their food intake with the least effort [30]. Aggregative behaviour is often adopted by prey to dilute predation risk but, at the same time, it may allow predators to easily locate a very fruitful target. Considering the > 3000 prey items recognised from the *S. sarrabusensis* population occurring in the artificial spring [26], it can be noted that about 94% of the consumed prey belongs to just two categories: Diptera (76,45%) and Coleoptera Staphylinidae (17.42%). These relatively small prey are often observed in very dense clumps inside those artificial springs, giving insights on the reasons promoting the poorly diverse diet of salamanders occurring therein [13]. Indeed, if we only consider individuals that consumed these two types of prey, we observe that the average number (±SD) of consumed prey is $20.36 \pm 25.29$ for Diptera and $10.62 \pm 14.3$ for Coleoptera Staphylinidae. On the other hand, the individuals from the epigean population studied here showed more variability in their diet (Table S1), and the most consumed prey was Entomobryomorpha, which represents the most diverse group of Collembola [31]. Although these Collembola can reach high densities [up to 1800 individuals per $dm^3$; 31], in our case, each individual did not consume a very high number of Entomobryomorpha ($3.42 \pm 2.96$), meaning that these prey may be locally common but do not show particularly gregarious behaviour.

## 5. Conclusions

Our study provided new information on the poorly known *Speleomantes sarrabusensis* through standardized methodologies that should be systematically adopted in the future to monitor the conservation status of this endangered amphibian species. Monitoring of epigean populations of *Speleomantes* is more challenging compared to subterranean ones [9,11,19], but they also require more attention as they have fewer opportunities to oppose human-induced alterations of the environment [32,33]. Furthermore, monitoring both epigean and hypogean populations may allow us to discover potential divergences in life traits between conspecific populations [34]. In conclusion, *S. sarrabusensis* needs particular attention and we are willing to continue with our data collection to provide important information useful for the conservation of this highly endangered species.

**Supplementary Materials:** The following supporting information can be downloaded at: https://www.mdpi.com/article/10.3390/ani13030391/s1, Table S1: Data on the recognized stomach contents in the epigean population of *Speleomantes sarrabusensis*. The columns represent: (A) the site code; (B–D) the coordinates and elevation of the site; (E,F) Region and Province; (G,H) the month and year of the survey; (I) the sex of individuals (male, female, and juvenile); (J,K) weight and total length of individuals; (L) stomach condition (1 = empty; 0 = full); (M) recognizable contents (yes = 0, no = 1); (N-AR) the number of recognized prey items for each prey category. NA = not available data.

**Author Contributions:** R.C. and E.L. conceived the study; R.C., M.D.G. and E.L. performed data collection; M.D.G. and F.C. recognized prey from stomach contents; E.L. analyzed the data and drafted the manuscript. All authors reviewed the manuscript and accepted its final form. All authors have read and agreed to the published version of the manuscript.

**Funding:** EL was supported by Societas Europaea Herpetologica through the Conservation Grant in Herpetology. FC was partially supported by the Ministry of University and Research of Italy (MUR), project FOE 2020-Capitale naturale e risorse per il futuro dell'Italia-Task Biodiversità. The study was conducted under the authorization of the Italian Ministry of Environment (prot. 9384 of 12/5/15 and further integrations), of the Autonomous Region of Sardegna (RAS) (prot. 3607 of 2/9/16 and further integrations), and the Agenzia Forestas (prot. 8947 of 15/12/17).

**Institutional Review Board Statement:** None.

**Informed Consent Statement:** Not applicable.

**Data Availability Statement:** Data for *Speleomantes sarrabusensis* stomach contents are provided as supplementary material (Table S1).

**Conflicts of Interest:** The authors declare no conflict of interest.

# References


1. Lanza, B.; Pastorelli, C.; Laghi, P.; Cimmaruta, R. A review of systematics, taxonomy, genetics, biogeography and natural history of the genus *Speleomantes* Dubois, 1984 (Amphibia Caudata Plethodontidae). *Atti. Mus. Civ. Stor. Nat. Trieste* **2006**, *52*, 5–135.
2. Ficetola, G.F.; Pennati, R.; Manenti, R. Do cave salamanders occur randomly in cavities? An analysis with *Hydromantes strinatii*. *Amphib.-Reptil.* **2012**, *33*, 251–259. [CrossRef]
3. Lunghi, E.; Manenti, R.; Ficetola, G.F. Do cave features affect underground habitat exploitation by non-troglobite species? *Acta. Oecol.* **2014**, *55*, 29–35. [CrossRef]
4. Ficetola, G.F.; Lunghi, E.; Canedoli, C.; Padoa-Schioppa, E.; Pennati, R.; Manenti, R. Differences between microhabitat and broad-scale patterns of niche evolution in terrestrial salamanders. *Sci. Rep.* **2018**, *8*, 10575. [CrossRef]
5. IUCN SSC Amphibian Specialist Group. Speleomantes Sarrabusensis. The IUCN Red List of Threatened Species 2022. 2022. Available online: https://www.iucnredlist.org/species/135825/89699335 (accessed on 3 December 2022).
6. Lanza, B.; Leo, P. Prima osservazione sicura di riproduzione vivipara nel genere *Speleomantes* (Amphibia: Caudata: Plethodontidae). In Proceedings of the Atti 3° Congresso Nazionale Societas Herpetologica Italica (Pavia), Pianura, Cremona, 14–16 September 2000; pp. 317–319.
7. Lunghi, E.; Corti, C.; Manenti, R.; Barzaghi, B.; Buschettu, S.; Canedoli, C.; Cogoni, R.; De Falco, G.; Fais, F.; Manca, A.; et al. Comparative reproductive biology of European cave salamanders (genus *Hydromantes*): Nesting selection and multiple annual breeding. *Salamandra* **2018**, *54*, 101–108.
8. Lunghi, E.; Manenti, R.; Mulargia, M.; Veith, M.; Corti, C.; Ficetola, G.F. Environmental suitability models predict population density, performance and body condition for microendemic salamanders. *Sci. Rep.* **2018**, *8*, 7527. [CrossRef] [PubMed]
9. Lunghi, E.; Manenti, R.; Ficetola, G.F. Seasonal variation in microhabitat of salamanders: Environmental variation or shift of habitat selection? *PeerJ* **2015**, *3*, e1122. [CrossRef]
10. O'Donnell, M.K.; Deban, S.M. Cling performance and surface area of attachment in plethodontid salamanders. *J. Exp. Biol.* **2020**, *233*, jeb211706. [CrossRef]
11. Costa, A.; Crovetto, F.; Salvidio, S. European plethodontid salamanders on the forest floor: Local abundance is related to fine-scale environmental factors. *Herpetol. Conserv. Biol.* **2016**, *11*, 344–349.
12. Lunghi, E.; Romeo, D.; Mulargia, M.; Cogoni, R.; Manenti, R.; Corti, C.; Ficetola, G.F.; Veith, M. On the stability of the dorsal pattern of European cave salamanders (genus *Hydromantes*). *Herpetozoa* **2019**, *32*, 249–253. [CrossRef]
13. Lunghi, E.; Cianferoni, F.; Ceccolini, F.; Veith, M.; Manenti, R.; Mancinelli, G.; Corti, C.; Ficetola, G.F. What shapes the trophic niche of European plethodontid salamanders? *PLoS ONE* **2018**, *13*, e0205672. [CrossRef] [PubMed]
14. Lunghi, E.; Ficetola, G.F.; Mulargia, M.; Cogoni, R.; Veith, M.; Corti, C.; Manenti, R. *Batracobdella* leeches, environmental features and *Hydromantes* salamanders. *Int. J. Parasitol. Parasites Wildl.* **2018**, *7*, 48–53. [CrossRef] [PubMed]
15. Ficetola, G.F.; Romano, A.; Salvidio, S.; Sindaco, R. Optimizing monitoring schemes to detect trends in abundance over broad scales. *Anim. Conserv.* **2017**, *21*, 221–231. [CrossRef]
16. Lunghi, E.; Bruni, G. Long-term reliability of Visual Implant Elastomers in the Italian cave salamander (*Hydromantes italicus*). *Salamandra* **2018**, *54*, 283–286.
17. Pinheiro, J.; Bates, D.; DebRoy, S.; Sarkar, D.; Team, R.C. nlme: Linear and Nonlinear Mixed Effects Models. *R Package Version* **2016**, *3*, 1–128.

18. Lunghi, E.; Corti, C.; Manenti, R.; Ficetola, G.F. Consider species specialism when publishing datasets. *Nat. Ecol. Evol.* **2019**, *3*, 319. [CrossRef]

19. Lunghi, E.; Corti, C.; Mulargia, M.; Zhao, Y.; Manenti, R.; Ficetola, G.F.; Veith, M. Cave morphology, microclimate and abundance of five cave predators from the Monte Albo (Sardinia, Italy). *Biodivers Data. J.* **2020**, *8*, e48623. [CrossRef]

20. MacKenzie, D.I.; Nichols, J.D.; Royle, J.A.; Pollock, K.H.; Bailey, L.L.; Hines, J.E. *Occupancy Estimation and Modeling. Inferring Patterns and Dynamics of Species Occurrence*; Academic Press: San Diego, CA, USA, 2006.

21. Cogoni, R.; Mulargia, M.; Manca, S.; Croubu, V.; Giachello, S.; Lunghi, E. New observations on the tree-dwelling behaviour of European cave salamanders (genus *Speleomantes*). In Proceedings of the XIV Congresso Nazionale della Societas Herpetologica Italica, Torino, Italia, 13–17 September 2022.

22. Ficetola, G.F.; Barzaghi, B.; Melotto, A.; Muraro, M.; Lunghi, E.; Canedoli, C.; Lo Parrino, E.; Nanni, V.; Silva-Rocha, I.; Urso, A.; et al. N-mixture models reliably estimate the abundance of small vertebrates. *Sci. Rep.* **2018**, *8*, 10357. [CrossRef] [PubMed]

23. Lunghi, E.; Bacci, F.; Zhao, Y. How can we record reliable information on animal colouration in the wild? *Diversity* **2021**, *13*, 356. [CrossRef]

24. Lunghi, E.; Giachello, S.; Manenti, R.; Zhao, Y.; Corti, C.; Ficetola, G.F.; Bradley, J.G. The post hoc measurement as a safe and reliable method to age and size plethodontid salamanders. *Ecol. Evol.* **2020**, *10*, 11111–11116. [CrossRef] [PubMed]

25. Costa, A.; Salvidio, S.; Posillico, M.; Altea, T.; Matteucci, G.; Romano, A. What goes in does not come out: Different non-lethal dietary methods give contradictory interpretation of prey selectivity in amphibians. *Amphib.-Reptil.* **2014**, *35*, 255–262. [CrossRef]

26. Lunghi, E.; Cianferoni, F.; Ceccolini, F.; Mulargia, M.; Cogoni, R.; Barzaghi, B.; Cornago, L.; Avitabile, D.; Veith, M.; Manenti, R.; et al. Field-recorded data on the diet of six species of European *Hydromantes* cave salamanders. *Sci. Data.* **2018**, *5*, 180083. [CrossRef] [PubMed]

27. Lunghi, E.; Giachello, S.; Zhao, Y.; Corti, C.; Ficetola, G.F.; Manenti, R. Photographic database of the European cave salamanders, genus *Hydromantes*. *Sci. Data.* **2020**, *7*, 171. [CrossRef]

28. Lunghi, E.; Manenti, R.; Cianferoni, F.; Ceccolini, F.; Veith, M.; Corti, C.; Ficetola, G.F.; Mancinelli, G. Interspecific and inter-population variation in individual diet specialization: Do environmental factors have a role? *Ecology* **2020**, *101*, e03088. [CrossRef]

29. Araújo, M.S.; Bolnick, D.L.; Layman, C.A. The ecological causes of individual specialisation. *Ecol. Lett.* **2011**, *14*, 948–958. [CrossRef]

30. Roughgarden, J. Evolution of niche width. *Am. Nat.* **1972**, *106*, 683–718. [CrossRef]

31. Bellinger, P.F.; Christiansen, K.A.; Janssens, F. Checklist of the Collembola of the World. Available online: https://www.collembola.org/ (accessed on 3 December 2022).

32. Araújo, M.B.; Pearson, R.G.; Thuiller, W.; Erhard, M. Validation of species–climate impact models under climate change. *Glob. Chang. Biol.* **2005**, *11*, 1504–1513. [CrossRef]

33. Halliday, T.R. Declining amphibians in Europe, with particular emphasis on the situation in Britain. *Environ. Rev.* **1993**, *1*, 21–25. [CrossRef]

34. Taylor, S.J.; Krejca, J.K.; Niemiller, M.L.; Dreslik, M.J.; Phillips, C.A. Life history and demographic differences between cave and surface populations of the Western slimy salamander, *Plethodon albagula* (Caudata: Plethodontidae), in central Texas *Herpetol. Conserv. Biol.* **2015**, *10*, 740–752.

*Article*

# Different Traits, Different Evolutionary Pathways: Insights from *Salamandrina* (Amphibia, Caudata)

**Claudio Angelini** [1,*], **Francesca Antonucci** [2], **Jacopo Aguzzi** [3] **and Corrado Costa** [2]

1    Salamandrina Sezzese Search Society, Via G. Marconi 30, 04018 Sezze, Italy
2    Consiglio per la ricerca in Agricoltura e l'Analisi dell'Economia Agraria (CREA), Centro di Ricerca Ingegneria e Trasformazioni Agroalimentari, Via della Pascolare 16, 00015 Monterotondo, Italy
3    Instituto de Ciencias del Mar (ICM-CSIC), Paseo Marítimo de la Barceloneta, 37–49, 08003 Barcelona, Spain
*    Correspondence: oppela@gmail.com

**Simple Summary:** In this paper we studied the body and colour features of the endemic amphibian *Salamandrina* across its range. Using a multimodel approach and machine learning clustering, we found much difference relative to body size and shape, as well as colour, among populations and between the habitats in which the individuals dwell, while differences between the two species of *Salamandrina* genus are lower. We suppose that selective pressures are more similar across species' ranges than within them. Taking into account also previous knowledge on this genus, here we point out that different traits of an organism could result from different evolutionary routes, something not unexpected but often neglected.

**Abstract:** Species delimitation is often based on a single or very few genetic or phenetic traits, something which leads to misinterpretations and often does not provide information about evolutionary processes. Here, we investigated the diversity pattern of multiple phenetic traits of the two extant species of *Salamandrina*, a genus split only after molecular traits had been studied but the two species of which are phenetically very similar. The phenetic traits we studied are size, external body shape and head colour pattern, in a model comparison framework using non-linear mixed models and unsupervised and supervised clustering. Overall, we found high levels of intra-specific variability for body size and shape, depending on population belonging and habitat, while differences between species were generally lower. The habitat the salamanders dwell in also seems important for colour pattern. Basing on our findings, from the methodological point of view, we suggest (i) to take into account the variability at population level when testing for higher level variability, and (ii) a semi-supervised learning approach to high dimensional data. We also showed that different phenotypic traits of the same organism could result from different evolutionary routes. Local adaptation is likely responsible for body size and shape variability, with selective pressures more similar across species than within them. Head colour pattern also depends on habitat, differently from ventral colour pattern (not studied in this paper) which likely evolved under genetic drift.

**Keywords:** phenetic characters; multivariate clustering; image analysis; local adaptation; evolutionary processes

**Citation:** Angelini, C.; Antonucci, F.; Aguzzi, J.; Costa, C. Different Traits, Different Evolutionary Pathways: Insights from *Salamandrina* (Amphibia, Caudata). *Animals* **2022**, *12*, 3326. https://doi.org/10.3390/ani12233326

Academic Editor: Enrico Lunghi

Received: 30 October 2022
Accepted: 23 November 2022
Published: 28 November 2022

**Publisher's Note:** MDPI stays neutral with regard to jurisdictional claims in published maps and institutional affiliations.

## 1. Introduction

Whichever species concept is used, the taxonomic assignment of individuals to a species relies on the (combined) analysis of phenotypic and genetic characters [1,2]. Thanks to spectacular technological advancements in molecular biology, genetics has become the backbone of evolutionary studies more and more during the previous decades. However, researchers have become aware that "gene trees and species trees are not the same" [3], and multilocus approaches are advisable. Furthermore, genetic surveys require well-equipped and funding-sustained laboratories, a combination of conditions not always available. In practice, the phenotype is largely the way by which species are still recognised, at least as the first step, and genetic studies are usually carried out for deeper and further investigation. Therefore, most field ecologists primarily rely on the phenetic species concept.

On the other hand, genetic evidence of speciation stimulates the study of differences among species, and differences or resemblances in turn stimulate evolutionary hypotheses, i.e., speciation pathways. Usually, searching for differences among species leads one to think in terms of "there is no difference" (null hypothesis, $H_0$) or "there is a difference" (a single alternative hypothesis, $H_1$), which is null hypothesis testing. However, while the difference (or resemblance) detected by hypothesis testing can be true, it is not necessarily the stronger signal the dataset bears, it could even be the weaker. In fact, any dataset brings several signals, and while it is very difficult to detect the true information (i.e., the strongest signal), it is possible to search for and likely find various signals (i.e., patterns). On the other hand, not all signals have the same strength. Instead, multimodel inference and model selection [4] allow to compare the strength of different signals, and thus each of multiple hypotheses ($H_i$). Thus, model selection is a natural companion for finding discrete clusters within an apparently homogenous group of organisms or for comparing the partitions within a group based on different biological rationales (species, populations, sexes, ecotypes, etc.).

In this paper, multimodel inference was applied to investigate if any phenetic trait differs between the two only extant species of the genus *Salamandrina*, the Northern Spectacled Salamander *S. perspicillata* (Savi, 1821), occurring from Liguria to northern Campania, and the Southern Spectacled *S. terdigitata* (Bonnaterre, 1789), from Campania to Calabria, in the Italian peninsula. The species are clearly separated taxa based on allozymes [5] and mitochondrial DNA [6,7]. However, nuclear DNA reveals incomplete lineage sorting and a hybrid zone [7–9]. The split between the two species happened between 11 and 2.2 mya [6,8], likely as the result of a climatic cooling trend and geographical segregation [8]. It is known that differences in colouration and morphology allow for a percentage of correct species assignment larger than 90% when using discriminant statistics [10,11]. Furthermore, it is known that salamanders' body size is affected by the type of breeding site, and it shows high intra-specific variability, at least for *S. perspicillata* [12,13].

The main goal of this study is to analyse multiple phenetic traits to evaluate if differences between the two species exist, and if differences due to the habitat are stronger than differences between species. In doing this, we also took into account the contribution to phenetic variability attributable to intra-specific differences at a population level, a factor often underestimated or even neglected when investigating differences between species, which instead turned out to be important.

## 2. Materials and Methods

Three populations of the Northern spectacled salamander *S. perspicillata* and two populations of the Southern spectacled salamander *S. terdigitata* were compared (Table 1). All populations inhabited mixed deciduous forests. The northern species populations were sampled in the Lepini mountains (about 800 km$^2$, Lazio region) as this area harbours the largest body size individuals and populations discovered in this species [12,13]. By doing so, we maximized body size differences between species, a caution which allowed us to evaluate the role of species belonging versus habitat and population belonging in shaping salamanders' bodies (in size). We are aware the study design is incomplete, because a southern pond population is missing (as a matter of fact, very few *terdigitata* populations oviposit in lentic water), however, this is not a limitation for the study, as will result from our findings. Data for both species were collected during oviposition seasons. Only data collected from females were analysed.

**Table 1.** Summary information about the sites of *Salamandrina perspicillata* and *S. terdigitata* used in this study; site codes are the same as in supplementary Table S1.

| Species | Site (Code) | Location | Habitat Features | Sampling Year and Size |
|---|---|---|---|---|
| *S. perspicillata* | Acqua della chiesa (AC) | Lepini Mountains, Latium | Trough, 900 m a.s.l. | 2007 $n = 28$ |
| *S. perspicillata* | Ciccopano nuove (CN) | Lepini Mountains, Latium | Rocky spring ponds, 700 m a.s.l. | 2006, 2007 $n = 19$ |
| *S. perspicillata* | Sant'Angelo (SA) | Lepini Mountains, Latium | Brook, 940 m a.s.l. | 2007 $n = 27$ |
| *S. terdigitata* | Torrente Cerasuolo (TC) | Picentini Mountains, Campania | Brook, 750 m a.s.l. | 2007 $n = 41$ |
| *S. terdigitata* | Torrente Rosa (TR) | Pollino, Calabria | Brook, 550 m a.s.l. | 2007 $n = 41$ |

*2.1. Biometries*

Animals were captured on sight. We measured the following body features: AG: distance between axilla and groin, CL: cloaca length, HL: head length, HW: head width, No: distance between nostrils, RU: radius-ulna (forearm) length, SVL: snout-vent length, TF: tibia-fibula (leg) length, TL: total length, tL: tail length and tW: tail width (Figure 1, Supplementary Table S1). They were measured directly on the individuals after anaesthesia (by using a 0.01% solution of MS-222-Sandoz) with a calliper (to the nearest 0.1 mm), but HL, HW and No were measured from photographs using the software tpsDig2 [14]. SVL and TL showed high collinearity with AG ($r_{Pearson}$ = 0.94 and 0.9 respectively, $p < 0.001$ in both cases) and thus were not used for the main analyses. All salamanders were released at their site after they recovered from the anaesthesia.

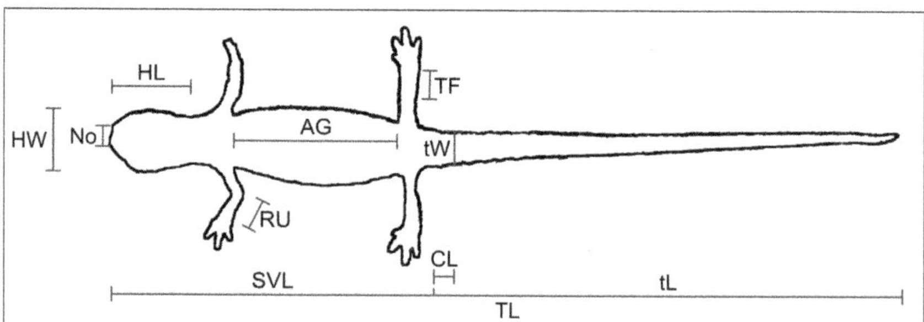


**Figure 1.** Biometric features of *Salamandrina perspicillata* and *S. terdigitata* used for the study: AG: distance between axilla and groin, CL: cloaca length, HL: head length, HW: head width, No: distance between nostrils, RU: radius-ulna (forearm) length, SVL: snout-vent length, TF: tibia-fibula (leg) length, TL: total length, tL: tail length, tW: tail width.

We investigated how biometric variation was related to species (*S. perspicillata* or *S. terdigitata*), type of oviposition site (brooks versus lentic water) and population belonging, by using model selection based on an information theoretic approach [15]. Namely, linear mixed-effects models were fitted to evaluate how the three factors explain size (AG) and multivariate morphological variation. Both "species" and "site typology" predictors were used as fixed factors separately and never crossed each other; "population" was always used as a random factor, either nested within "species" and "site typology" or alone. For AG we compared five models (Table 2) (since SVL and TL are often used as proxies for body size, we also show the results of the same analysis for these features in Supplementary Tables S2 and S3). For the analysis of the multivariate morphological variation, we considered all (eight) body features, except AG, and then compared five models built as described above, plus the same five models also including AG as a covariate, as well as a further model with only AG as a continuous predictor, thus obtaining 11 candidate models in total (Table 3). Model comparisons were done using Akaike's information criterion for small sample size (AICc) and AICc weight, the likelihood that a given model is actually the best among the candidate ones [15]. Analyses were performed in the R environment [16], using the packages lme4 (version 1.1-31, [17]) for fitting the models: lmerTest (version 3.1-3, [18]) was used to obtain the (multivariate) analysis of variance [(M)ANOVA] output from the linear mixed model fitted with lme4, and MuMIn (version 1.9.13, [19]) for the computation of model weight.

**Table 2.** Model ranking of candidate models used to analyse the dependence of axilla-groin distance of salamanders based on the species (*Salamandrina perspicillata* or *S. terdigitata*), the type of breeding site (brook or lentic water) and their population (five populations). Species and site type were used as fixed factors, population as a random factor (when together, the random factor is nested in the fixed one). AICc is the Akaike's information criterion for small sample size, w is the model weight. *p*-value refers to the fixed factor for the given ANOVA model.

| Model | AICc | W | *p*-Value |
|---|---|---|---|
| site typology:population | 609.25 | 0.74 | 0.018 |
| species:population | 612.38 | 0.16 | 0.1 |
| population | 613.27 | 0.1 | |
| site typology | 644.94 | 0 | <0.001 |
| species | 668.81 | 0 | <0.001 |

**Table 3.** Model ranking of candidate models used to analyse the multivariate dependence of body features (see text for details) of salamanders based on the species (*Salamandrina perspicillata* or *S. terdigitata*), the type of breeding site (brook or lentic water), their population (five populations) and the individual axilla-groin distance (AG; "*" denote the use of AG as covariate). Species, site type and axilla-groin distance were been used as fixed factors, population as random factor (when together, the random factor is nested in the fixed one). AICc is the Akaike's information criterion for the small sample size, w is the model weight. *p*-value refers to the categorical fixed factor species or site typology for the given MAN(C)OVA model.

| Model | AIC | W | *p*-Values |
|---|---|---|---|
| population*AG | 4260.98 | 0.986 | |
| species*AG:population | 4270.87 | 0.008 | =0.04 |
| site typology*AG:population | 4271.35 | 0.006 | =0.04 |
| species*AG | 4448.34 | 0 | <0.001 |
| site typology*AG | 4470.28 | 0 | <0.001 |
| AG | 4722.57 | 0 | |
| population | 4851.38 | 0 | |
| site typology:population | 4862.03 | 0 | =0.03 |
| species:population | 4863.34 | 0 | =0.14 |
| site typology | 5509.29 | 0 | <0.001 |
| species | 5718.67 | 0 | <0.001 |

*2.2. Head Colour Topographical Analysis*

*Salamandrina* is characterized by a yellow V-shaped patch on the head. Analyses were based on digital photographs (1200 × 1600 pixels; 72 dpi) of the dorsal view of salamander heads. Animals were placed horizontally, as much as possible following the main body axis, the camera was parallel to the animals and a metric reference was placed aside. Then, a geometric morphometric analysis was performed on each picture [20–22]. Nine landmarks and 24 equidistant points along the outline (semi-landmarks) [21] from the head of each salamander were digitized using the software TpsDig2 (Ver. 2.17 [14]). All landmark and semi-landmark configurations for each specimen were aligned, translated, rotated and scaled to a unit centroid size by the Generalized Procrustes Analysis (GPA [23]), using the consensus configuration of all specimens as the starting form. Residuals from the fitting were modelled with the thin-plate spline interpolating function [21–23]. The shape and head colour pattern of each individual were morphologically adjusted to a standard view by means of the consensus configuration. Before the morphometric standardization, the head colour pattern was manually segmented into pure black and white. The final grey-scale images were 2288 × 1712 pixels at 96 ppi, for an amount of 26,748 pixels constituting the head's area.

We used the information on position and colour (black or white) of every single pixel of each individual to cluster salamanders on the basis of their head colour pattern by applying both unsupervised and supervised clustering. For unsupervised clustering we compared all the models with a candidate number of clusters from two to six. Candidate supervised clustering was based on: partition from the best model from unsupervised clustering (two clusters), species (two clusters), population belonging (five clusters) and habitat (two clusters). This has been done in the R environment [16] by using the package MixAll (version 1.5.1 [24]). MixAll allows for model-based clustering and classification. Notably, we used multivariate categorical mixture models. For model selection we used the Integrated Completed Likelihood (ICL) criterion [25], a model selection criterion specifically introduced for model-based clustering. Since MixAll does not allow for reducing dimensionality and does not show the main features underlining differences, we applied the Partial Least Square Discriminant Analysis for this purpose. Details and results from this analysis are reported in the Supplementary Material S4.

## 3. Results

### 3.1. Biometries

Model selection shows that the effect of both the type of oviposition site and species on salamanders' body sizes (AG) appear evident only when accounting for intra-specific differences at the population level (Table 2). Based on models' weight, population belonging is the main determinant for body size, then it depended on site typology (salamanders from brooks are smaller than ones from lentic waters; Supplementary Table S1) and finally, there is poorly if any effect from salamander species (second best model, *S. perspicillata* are larger; supplementary Table S1). Nevertheless, the fixed factors always obtained significant *p*-values (marginally so in one case). We want also to stress that our findings are not invalidated by the absence of a *S. terdigitata* pond population from the dataset: since salamanders from ponds tended to be larger, a *terdigitata* pond population would have further made the size of *S. terdigitata* similar to *perspicillata*.

Model selection for multivariate morphological variation clearly indicated that it depended on intra-specific population variability and body size (all body features are positively correlated with AG, with $r_P$ ranging from 0.37 to 0.84, $p < 0.001$), and the role of both species and oviposition site is poor and only detectable when accounting for body size and population variability (Table 3). At both the univariate (size) and multivariate (morphology) level, species is the predictor that performed the worst. However, fixed factors again achieved significant *p*-values, except in one case.

### 3.2. Head Colour Topographical Analysis

Unsupervised clustering indicates that individuals group in two clusters (ICL = 2,382,622), since the ICL from three clusters (2,393,146) onward increasingly augment. There were associations between these two clusters and habitat (chi-square $p < 0.01$) and population ($p < 0.001$), but not with species ($p = 0.28$). Supervised clustering ranked as best the partition obtained from unsupervised clustering (ICL = 5,441,850), then the partition based on site typology (Table 4).

**Table 4.** Model ranking of candidate models for supervised clustering. Unsupervised partition refers to the best clustering obtained from unsupervised clustering. Accuracy is the classification accuracy of the model.

| Model | ΔICL | Accuracy |
|---|---|---|
| unsupervised partition | 0 | 0.97 |
| site typology | 32,919 | 0.72 |
| species | 44,959 | 0.73 |
| population | 116,985 | 0.52 |

## 4. Discussion

Both kinds of analyses show that, even though *Salamandrina* species differ, interspecific difference is not the stronger signal the data bear. A one-way (M)ANOVA approach would have detected a significant effect of the fixed factors species and site typology on morphometry for (almost) all alternative models (Tables 2 and 3), which is hardly useful. On the other hand, testing only one hypothesis (e.g., difference between species) would have led us to neglect other, more likely, hypothesis (e.g., difference between site typology taking into account population, see Table 2). Model selection allowed us to figure out: (i) the importance of population belonging for both univariate and multivariate variability; (ii) the partial importance of the site typology for body size variation (when accounting for population belonging); (iii) the dependence of multivariate variability on population belonging and on body size (two factors that being related reinforce their signal in explaining the multivariate information).

The unsupervised analysis of the head colour pattern found two clusters. These clusters are not randomly associated with site typology and population, but reveal a further, more important, even though not interpretable, pattern. Apart from these clusters, supervised clustering ranked as the best the partition between site typologies, showing that local factors are determinants also for the shape of head spots. Despite the very high difference among the information criterion values, the accuracies of the classifications are rather good for site typology, species and population (taking into account that the probability of random assignment among populations is 20%). Again, also for the head colour pattern, as for body shape and size, testing only one model could have led us to a spurious conclusion, as well as testing more models without an estimator of their relative quality would have been uninformative (see Supplementary Material S4). However, it depends on the objective of the investigation: if it is finding a way to distinguish groups, supervised learning based on only one hypothesis can be useful. For example, when applied to *Salamandrina*, it actually allows for a percentage of correct species assignment larger than 90% [10,11]. However, if the issue is to find the most relevant biological pattern, model selection among unsupervised learning and/or supervised models should be the choice.

The problem of statistically approaching unknown classification patterns is emerging with the evolution of machine learning approaches; consequently, clear protocols on whether using supervised or unsupervised techniques are not yet clearly established [26]. Unsupervised learning methods significantly outperform the supervised ones when the number of clusters [27] or the observation attributions [28] to classes are unknown or unclear. On the other hand, supervised methods could produce a more informative output, e.g., finding out the latent dimensions responsible for a given clustering or identifying the main features underlining differences. The combination of supervised and unsupervised approaches is called semi-supervised learning and offers a promising direction of future research [26,29]. We suggest as an optimal workflow, especially when dealing with high multidimensional data, such as data from image analysis: first to use unsupervised modelling, then compare hypotheses by supervising learning and finally, apply to the best clustering a statistical tool which reduces dimensionality and shows the features responsible for differences among clusters. As an example, in the supplementary material (S4) we show the implementation of Partial Least Square Discriminant Analysis on our models.

Apart from methodological hints, our analyses yielded also some biological findings. It was already known that *S. terdigitata* is smaller than the northern species [10] as well as that the typology of the oviposition site is even more important in determining the body size [13]. We add to this information the key role played by population belonging in determining both size and shape of salamanders. Despite the fact we opportunistically introduced a bias in the analysis by sampling the largest body size populations of *S. perspicillata*, we found a high level of biometric and morphometric diversity within both *Salamandrina* species compared to diversity between them. This confirms that the selective pressures acting on these traits are more similar across ranges than at a local scale. The split between the two species happened in an allopatric speciation scenario [8], which is expected to cause slow diversification [30–32]. In fact, if the physical barrier does not produce any ecological difference between the new-species' ranges, the species continue to share the same kind of environment, they will still be under the same selective pressures, and traits under such pressures will differ little between species. In this framework, local pressures could become the main ones. Since the *perspicillata* populations we studied are closely related (same mitochondrial haplogroup and microsatellite cluster [8]) and are no more than 4 km away from each other, we are prone to attribute population differences to very local factors (i.e., excluding climate). Based on our results, we suppose that local topography is a candidate to take into consideration. This is in accordance with what was previously suggested, that salamanders from brook populations experience a lower survival rate due to floods, thus leading to a smaller body size [12]. However, lower survival could cause smaller body size just as a demographic by-product (younger individuals are smaller) and/or by adaptative reaction leading to earlier sexual maturity [33] and thus smaller adult size ([34] and reference therein). It is known that *Salamandrina* shows high phenotypic plasticity, with close populations having different oviposition periods [35,36], however, only knowledge on individual ages can disentangle the role of demography and reaction norms on body size and shape variability, as well as the validation of our hypothesis requires studies involving (local) environmental factors.

The V-shaped patch on the head depends mostly on the type of oviposition site, little if any on species and even less on population. It is difficult to explain this finding, because no hypothesis exists about the role of the yellow head patch (we could just speculate it is a disruptive or distractive colouration). Differently than the head yellow patch, the extension of red colouration on the tail and the ventral colour pattern proved to efficiently discriminate between *Salamandrina* species [10,11]. The discriminatory performances of such colour features suggest that they are not subjected to the same selective pressures as the morphometric and colouration traits we analysed in the current study. Likely, the anterior part of the ventral colouration is involved in intra-specific communication [11,37–39], as well as the tail [40]. The speciation scenarios suggested for *Salamandrina* [8,9] involve the reduction of population size in separated refugial areas, from which both species colonized the extant ranges. In particular, both studies pointed out that the *perspicillata* lineage survived in a single refugial area. In the small, ancestral population, the importance of genetic drift was likely higher than natural selection. For any colour pattern involved in intraspecific communication, the random process would have been positively reinforced by the efficacy of signalling [41], thus leading to fixation of colour differences between the two *Salamandrina* species. Something that did not happen for the head V-patch, or that has been blurred later by any selective pressures.

## 5. Conclusions

Our findings confirm that within-species variability is not negligible when studying phenetic and functional differences between species, and that such variability can even challenge the discrimination between species [42,43]; thus, taking into account differentiation at population level is an important caution. We also showed how different phenotypic traits of the same organism could follow different evolutionary routes. On one hand, genetic drift probably led to strong differentiation of some, but not all, colour patterns between *Salamandrina* species. On the other hand, natural selection seems to act on size and shape of individuals in similar ways across the genus' range, but depending on local factors, it produces ecotypes that differ at least as much as the species and are shared between them. Focussing the analysis only on species discrimination would have hidden more important differences within a genus, as well as intriguing insights.

**Supplementary Materials:** The following supporting information can be downloaded at: https://www.mdpi.com/article/10.3390/ani12233326/s1, Table S1: Data summary for the measured body features; Table S2: Snout-vent length model comparison; Table S3: Total length model comparison; Supplementary material S4: Example of second-step further analysis. References [44–51] are cited in the supplementary materials.

**Author Contributions:** Conceptualization, C.A. and C.C.; methodology, C.A. and C.C.; software, C.A. and C.C.; validation, C.A., F.A., J.A. and C.C.; formal analysis, C.A. and C.C.; investigation, C.A.; data curation, C.A. and C.C.; writing—original draft preparation, C.A.; writing—review and editing, C.A., F.A., J.A. and C.C.; visualization, C.A. All authors have read and agreed to the published version of the manuscript.

**Funding:** This research received no external funding.

**Institutional Review Board Statement:** The animal study protocol was approved by Ministero dell'Ambiente e della Tutela del Territorio (protocol code DCN/2D/2003/15281).

**Informed Consent Statement:** Not applicable.

**Data Availability Statement:** The data presented in this study are partly available in the supplementary material, and are available on request from the corresponding author.

**Conflicts of Interest:** The authors declare no conflict of interest.

# References


1. Hey, J. On the failure of modern species concepts. *Trends Ecol. Evol.* **2006**, *21*, 447–450. [CrossRef] [PubMed]
2. De Queiroz, K. Species concepts and delimitation. *Systematic. Biol.* **2007**, *56*, 879–886. [CrossRef] [PubMed]
3. Nichols, R. Gene trees and species trees are not the same. *Trends Ecol. Evol.* **2001**, *16*, 358–364. [CrossRef] [PubMed]
4. Anderson, D.R. *Model Based Inference in the Life Science: A primer on Evidence*; Springer: Berlin/Heidelberg, Germany, 2008.
5. Canestrelli, D.; Zangari, F.; Nascetti, G. Genetic evidence for two distinct species within the Italian endemic *Salamandrina terdigitata* (Bonnaterre, 1789) (Amphibia: Urodela: Salamandridae). *Herpetol. J.* **2006**, *16*, 221–227.
6. Mattoccia, M.; Romano, A.; Sbordoni, V. Mitochondrial DNA sequence analysis of the spectacled salamander, *Salamandrina terdigitata* (Urodela: Salamandridae), supports the existence of two distinct species. *Zootaxa* **2005**, *995*, 1–19. [CrossRef]
7. Hauswaldt, J.S.; Angelini, C.; Pollok, A.; Steinfartz, S. Hybridization of two ancient salamander lineages: Molecular evidence for endemic spectacled salamanders (genus *Salamandrina*) on the Apennine peninsula. *J. Zool.* **2011**, *4*, 248–256. [CrossRef]
8. Hauswaldt, J.S.; Angelini, C.; Gehara, M.; Benavides, E.; Polok, A.; Steinfartz, S. From species divergence to population structure: A multimarker approach on the most basal lineage of Salamandridae, the spectacled salamanders (genus *Salamandrina*) from Italy. *Mol. Phylogenet. Evol.* **2014**, *70*, 1–12. [CrossRef]
9. Mattoccia, M.; Marta, S.; Romano, A.; Sbordoni, V. Phylogeography of an Italian endemic salamander (genus *Salamandrina*): Glacial refugia, postglacial expansions, and secondary contact. *Biol. J. Linn. Soc.* **2011**, *104*, 903–922. [CrossRef]
10. Romano, A.; Mattoccia, M.; Marta, S.; Bogaerts, S.; Pasmans, F.; Sbordoni, V. Distribution and morphological characterization of the endemic Italian salamanders *Salamandrina perspicillata* (Savi, 1821) and *S. terdigitata* (Bonnaterre, 1789) (Caudata: Salamandridae). *Ital. J. Zool.* **2009**, *76*, 422–432. [CrossRef]
11. Angelini, C.; Costa, C.; Raimondi, S.; Menesatti, P.; Utzeri, C. Image analysis of the ventral colour pattern discriminates between Spectacled salamanders, *Salamandrina perspicillata* and *S. terdigitata* (Amphibia, Salamandridae). *Amphibia-Reptilia* **2010**, *31*, 273–282. [CrossRef]
12. Angelini, C.; Antonelli, D.; Utzeri, C. A multi-year and multi-site population study on the life history of *Salamandrina perspicillata* (Savi, 1821) (Amphibia, Urodela). *Amphibia-Reptilia* **2008**, *29*, 161–170.
13. Romano, A.; Ficetola, G.F. Ecogeographic variation of body size in the spectacled salamanders (*Salamandrina*): Influence of genetic structure and local factors. *J. Biogeogr.* **2010**, *37*, 2358–2370. [CrossRef]
14. Rohlf, F.J. *Digitalized Landmarks and Outlines, TpsDig Ver. 2.17*; Department of Ecology and Evolution, State University of New York at Stony Brook: Stony Brook, NY, USA, 2013.
15. Burnham, K.P.; Anderson, D.R. *Model Selection and Multimodel Inference: A Practical Information-Theoretic Approach*; Springer: New York, NY, USA, 2002.
16. R Core Team. *R: A Language and Environment for Statistical Computing*; R Foundation for Statistical Computing: Vienna, Austria, 2021; Available online: https://www.r-project.org/ (accessed on 29 October 2022).

17. Bates, D.; Mächler, M.; Bolker, B.; Walker, S. Fitting Linear Mixed-Effects Models Using lme4. *J. Stat. Softw.* **2015**, *67*, 1–48. [CrossRef]
18. Kuznetsova, A.; Brockhoff, P.B.; Christensen, R.H.B. lmerTest Package: Tests in Linear Mixed Effects Models. *J. Stat. Softw.* **2017**, *82*, 1–26. [CrossRef]
19. Bartoń, K. MuMIn: Multi-Model Inference, R package version 1.9.13. 2013. Available online: https://CRAN.R-project.org/package=MuMIn (accessed on 29 October 2022).
20. Antonucci, F.; Boglione, C.; Cerasari, V.; Caccia, E.; Costa, C. External shape analyses in *Atherina boyeri* (Risso, 1810) from different environments. *Ital. J. Zool.* **2012**, *79*, 60–68. [CrossRef]
21. Bookstein, F.L. *Morphometric Tools for Landmark Data: Geometry and Biology*; Cambridge University Press: New York, NY, USA, 1991.
22. Zelditch, M.L.; Swiderski, D.L.; Sheets, H.D.; Fink, W.L. *Geometric Morphometrics for Biologists: A Primer*; Elsevier: San Diego, CA, USA, 2004.
23. Rohlf, F.J.; Slice, D.E. Extensions of the Procrustes method for the optimal superimposition of landmarks. *Syst. Zoo.* **1990**, *39*, 40–59. [CrossRef]
24. Iovleff, S. MixAll: Clustering and Classification Using Model-Based Mixture Models, R package version 1.5.1. 2019. Available online: https://CRAN.R-project.org/package=MixAll (accessed on 29 October 2022).
25. Biernacki, C.; Celeux, G.; Govaert, G. Assessing a mixture model for clustering with the integrated completed likelihood. *IEEE Trans. Pattern. Anal. Mach. Intell.* **2000**, *22*, 719–725. [CrossRef]
26. Laskov, P.; Düssel, P.; Schäfer, C.; Rieck, K. Learning intrusion detection: Supervised or unsupervised? In *International Conference on Image Analysis and Processing*; Springer: Berlin/Heidelberg, Germany, 2005; pp. 50–57.
27. Antonucci, F.; Costa, C.; Pallottino, F.; Paglia, G.; Rimatori, V.; De Giorgio, D.; Menesatti, P. Quantitative method for shape description of almond cultivars (*Prunus amygdalus* Batsch). *Food. Bioproc. Tech.* **2012**, *5*, 768–785. [CrossRef]
28. Kwon, B.C.; Eysenbach, B.; Verma, J.; Ng, K.; De Filippi, C.; Stewart, W.F.; Perer, A. Clustervision: Visual supervision of unsupervised clustering. *IEEE Trans. Vis. Comput. Graph.* **2017**, *24*, 142–151. [CrossRef]
29. Albalate, A.; Minker, W. *Semi-Supervised and Unsupervised Machine Learning: Novel Strategies*; John Wiley & Sons: Hoboken, NJ, USA, 2013.
30. McCune, A.R.; Lovejoy, N.R. The relative rate of sympatric and allopatric speciation in fishes: Tests using DNA sequence divergence between sister species and among clades. In *Endless Forms: Species and Speciation*; Howard, D.J., Berlocher, S.H., Eds.; Oxford University Press: Oxford, UK, 1998; pp. 172–185.
31. Orr, M.R.; Smith, T.B. Ecology and speciation. *Trends Ecol. Evol.* **1998**, *13*, 502–506. [CrossRef] [PubMed]
32. Schluter, D. Ecology and the origin of species. *Trends. Ecol. Evol.* **2001**, *16*, 372–380. [CrossRef] [PubMed]
33. Bruce, R.C.; Hairston, N.G. Life-history correlates of body-size differences between two population of the salamander, *Desmognathus monticola*. *J. Herpetol.* **1990**, *24*, 124–134. [CrossRef]
34. Angelini, C.; Sotgiu, G.; Tessa, G.; Bielby, J.; Doglio, S.; Favelli, M.; Garner, T.W.J.; Gazzaniga, E.; Giacoma, C.; Bovero, S. Environmentally determined juvenile growth rates dictate the degree of sexual size dimorphism in the Sardinian brook newt. *Evol. Ecol.* **2015**, *29*, 169–184. [CrossRef]
35. Corsetti, L. Osservazioni sulla ecologia e biologia riproduttiva di *Salamandrina terdigitata* nei Monti Lepini (Lazio) (Amphibia Salamandridae). *Quad. Mus. Stor. Nat. Patrica* **1994**, *4*, 111–130.
36. Angelini, C.; Vanni, S.; Vignoli, L. *Salamandrina terdigitata* (Bonnaterre, 1789) *Salamandrina perspicillata* (Savi, 1821). In *Fauna d'Italia; Amphibia*; Lanza, B., Andreone, F., Bologna, M.A., Corti, C., Razzetti, E., Eds.; Calderini: Bologna, Italy, 2007; Volume 42, pp. 228–237.
37. Utzeri, C.; Antonelli, D.; Angelini, C. Notes on the behavior of the Spectacled Salamander *Salamandrina terdigitata* (Lacépède, 1788). *Herpetozoa* **2005**, *18*, 182–185.
38. Costa, C.; Angelini, C.; Scardi, M.; Menesatti, P.; Utzeri, C. Using image analysis on the ventral colour pattern in *Salamandrina perspicillata* (Amphibia: Salamandridae) to discriminate among populations. *Biol. J. Linn. Soc.* **2009**, *96*, 35–43. [CrossRef]
39. Ancillotto, L.; Vignoli, L.; Martino, J.; Paoletti, C.; Romano, A.; Bruni, G. Sexual dichromatism and throat display in spectacled salamanders: A role in visual communication? *J. Zool.* **2002**, *318*, 75–83. [CrossRef]
40. Bruni, G.; Romano, A. Courtship behaviour, mating season and male sexual interference in *Salamandrina perspicillata* (Savi, 1821). *Amphibia-Reptilia* **2011**, *32*, 63–76. [CrossRef]
41. Hoskin, C.; Higgie, M.; McDonald, K.; Moritz, C. Reinforcement drives rapid allopatric speciation. *Nature* **2005**, *437*, 1353–1356. [CrossRef]
42. Albert, C.H.; Thuiller, W.; Yoccoz, N.G.; Douzet, R.; Aubert, S.; Lavorel, S. A multi-trait approach reveals the structure and the relative importance of intra- vs. interspecific variability in plant traits. *Funct. Ecol.* **2010**, *24*, 1192–1201.
43. Albert, C.H.; Grassein, F.; Schurr, F.M.; Vieilledent, G.; Violle, C. When and how should intraspecific variability be considered in trait-based plant ecology? *Perspecti. Plant. Ecolo. Evol. Syst.* **2011**, *13*, 217–225. [CrossRef]
44. Costa, C.; Antonucci, F.; Boglione, C.; Menesatti, P.; Vandeputte, M.; Chatain, B. Automated sorting for size, sex and skeletal anomalies of cultured seabass using external shape analysis. *Aquacult. Eng.* **2013**, *52*, 58–64. [CrossRef]
45. De Jong, S. SIMPLS: An alternative approach to partial least squares regression. *Chemometr. Intell. Lab.* **1993**, *18*, 251–263. [CrossRef]

46. Forina, M.; Oliveri, P.; Casale, M.; Lanteri, S. Multivariate range modeling, a new technique for multivariate class modeling: The uncertainty of the estimates of sensitivity and specificity. *Anal. Chim. Acta* **2008**, *622*, 85–93. [CrossRef]
47. Forina, M.; Oliveri, P.; Lanteri, S.; Casale, M. Class-modeling techniques, classic and new, for old and new problems. *Chemometr. Intell. Lab.* **2008**, *93*, 132–148. [CrossRef]
48. Infantino, A.; Aureli, G.; Costa, C.; Taiti, C.; Antonucci, F.; Menesatti, P.; Pallottino, F.; De Felice, S.; D'Egidio, M.G.; Mancuso, S. Potential application of PTR-TOFMS for the detection of deoxynivalenol (DON) in durum wheat. *Food Control* **2015**, *57*, 96–104. [CrossRef]
49. Kennard, R.W.; Stone, L.A. Computer aided design of experiments. *Technometrics* **1969**, *11*, 137–148. [CrossRef]
50. Sabatier, R.; Vivein, M.; Amenta, P. Two approaches for discriminant partial least square. In *Between Data Science and Applied Data Analysis*; Schader, M., Gaul, W., Vichi, M., Eds.; Springer: Berlin/Heidelberg, Germany, 2003; pp. 100–108.
51. Sjöström, M.; Wold, S.; Söderström, B. PLS discrimination plots. In *Pattern Recognition in Practice II*; Gelsema, E.S., Kanals, L.N., Eds.; Elsevier: Amsterdam, The Netherlands, 1986; pp. 461–470.

 *animals*

*Article*

# The Trophic Niche of Two Sympatric Species of Salamanders (Plethodontidae and Salamandridae) from Italy

Enrico Lunghi [1,2,3,4,*], Claudia Corti [5], Marta Biaggini [5], Yahui Zhao [1,*] and Fabio Cianferoni [5,6]

1    Key Laboratory of the Zoological Systematics and Evolution, Institute of Zoology, Chinese Academy of Sciences, Beijing 100101, China
2    Division of Molecular Biology, Institut Ruđer Bošković, 10000 Zagreb, Croatia
3    Natural Oasis, 59100 Prato, Italy
4    Unione Speleologica Calenzano, 50041 Florence, Italy
5    Museo di Storia Naturale dell'Università degli Studi di Firenze, Museo "La Specola", 50125 Florence, Italy
6    Istituto di Ricerca sugli Ecosistemi Terrestri (IRET), Consiglio Nazionale delle Ricerche (CNR), 50019 Florence, Italy
*    Correspondence: enrico.arti@gmail.com (E.L.); zhaoyh@ioz.ac.cn (Y.Z.)

**Simple Summary:** Studies on species' trophic niches are essential to understand the characteristics of species' ecology and life traits, as well as to improve conservation strategies. In the absence of competitors, species realize their trophic niche including in their diet the most profitable food resources. In the presence of competitors, species modify their preferences to reduce competition and maintain the highest benefits at the same time. In this study, we assessed the trophic niche of two species of salamanders coexisting in a forested area of Italy and evaluated which might be the mechanisms that these two species adopted to reduce competition. We found that the Italian cave salamander (*Speleomantes italicus*) mostly consumed flying prey with a hard cuticle, while the fire salamander (*Salamandra salamandra*) preferred worm-like and soft-bodied prey. In conclusion, we hypothesize that in our case, the two species of salamanders did not have to change their prey preference in order to avoid competition, but divergences in metabolism and behavioral traits likely worked as natural deterrent.

**Abstract:** The trophic niche of a species is one of the fundamental traits of species biology. The ideal trophic niche of a species is realized in the absence of interspecific competition, targeting the most profitable and easy-to-handle food resources. However, when a competitor is present, species adopt different strategies to reduce competition and promote coexistence. In this study, we assessed the potential mechanisms that allow the coexistence of two generalist salamanders: the Italian cave salamander (*Speleomantes italicus*) and the fire salamander (*Salamandra salamandra*). We surveyed, in April 2021, a forested area of Emilia-Romagna (Italy) during rainy nights. Analyzing the stomach contents of the captured individuals, we obtained information on the trophic niche of these two sympatric populations. Comparing our results with those of previous studies, we found that the two species did not modify their trophic niche, but that alternative mechanisms allowed their coexistence. Specifically, different prey preferences and predator metabolisms were likely the major factors allowing reduced competition between these two generalist predators.

**Keywords:** *Speleomantes*; *Hydromantes*; *Salamandra*; diet; forest; competition; prey selection

**Citation:** Lunghi, E.; Corti, C.; Biaggini, M.; Zhao, Y.; Cianferoni, F. The Trophic Niche of Two Sympatric Species of Salamanders (Plethodontidae and Salamandridae) from Italy. *Animals* **2022**, *12*, 2221. https://doi.org/10.3390/ani12172221

Academic Editor: José Martín

Received: 28 July 2022
Accepted: 21 August 2022
Published: 29 August 2022

**Publisher's Note:** MDPI stays neutral with regard to jurisdictional claims in published maps and institutional affiliations.

## 1. Introduction

The trophic niche is one of the fundamental traits of species biology [1]. The study of the trophic niche provides important information on multiple species traits such as behavior (e.g., foraging strategy and prey selection), physiology (e.g., specific nutritional requirements and metabolism), and its trophic position in the local community [2–5]. Nonetheless, this type of study may be pivotal to implement conservation strategies for species that need particular protection [6]. In the absence of heterospecific competitors, a species tends to develop its trophic niche by targeting the most profitable resources in terms of nutritional intake and handling ability [4,7]. Under this condition, the overall trophic spectrum of the population can mirror the shared preference of individuals (when they exploit the same typologies of food resources), or it can result from the combination of different preferences when intraspecific competition forces individuals to target only a subset of the food resources consumed by the entire population [8,9]. When co-occurring species compete for the same food resources, their realized trophic niche usually differs (at least for the weaker competitor) from the ideal one, as individuals have to switch to alternative resources to reduce the competition and be able to coexist [10,11].

We here assessed for the first time the trophic niche of two sympatric generalist salamanders: the Italian cave salamander *Speleomantes italicus* and the fire salamander *Salamandra salamandra*. The trophic niche of these two species has only been assessed in the absence of potential competitors e.g. [12–14], leaving unknown the mechanisms and the extent to which these species are able to modify their trophic niche to coexist with competitors. *Speleomantes italicus* is one of the eight species of Plethodontidae occurring in Europe, and its distribution encompasses the Apennine chain, from northern Tuscany to Abruzzo (Italy), where it occurs in forested areas and subterranean environments as well [15,16]. *Speleomantes* are lungless salamanders that require specific microhabitat conditions (i.e., high humidity and relatively low temperature) to maintain the high efficiency of cutaneous respiration [17,18]. *Speleomantes* are fully terrestrial amphibians that live and reproduce exclusively in subaerial environments [15]. Courtship can occur all year round, while gravid females lay their eggs twice per year (beginning of spring or autumn) in hidden places where they provide prolonged parental care ($\geq 4$ months) until the hatchlings are ready to leave the nest [19–22]. The narrow microhabitat requirements and the *k*-selected reproductive strategy of these species make them deserving of special protection [18,23,24]. Although being epigean species, *Speleomantes* gained the vernacular name of "cave salamanders" because they can be easily observed in natural and artificial subterranean environments [15,25]. These species are therefore able to prey in both surface and subterranean environments [26–29]. The trophic niche of *S. italicus* was only studied in subterranean populations [12]. Researchers have observed high variability in the consumed prey among populations, with a clear predominance of Diptera, one of the most abundant prey in subterranean environments [30,31].

*Salamandra salamandra* is the most widespread species of Salamandridae in Europe, ranging from the Iberian Peninsula to the western part of Ukraine, including the Balkans and all central European countries [32]. Despite being a widely spread Urodela, the fire salamander recently faced a huge decline in some parts of its distribution due to infection with the fungus *Batrachochytrium salamandrivorans* [33]. The fire salamander is a typical biphasic amphibian that has aquatic larvae and terrestrial adults [34]. Courtship occurs from spring to autumn, and females usually maintain the fertilized eggs in their bodies until the following spring. When eggs are ready, the female enters into a body of water to release the newly hatched larvae, which need at least 6 months to metamorphose into the adult form [34]. Although *S. salamandra* usually reproduces in surface water, several cases of reproduction in subterranean environments are also known [35,36]. The larvae feed upon multiple aquatic invertebrate species, but they can also adopt cannibalism when prey are scarce [14,37,38]. Adults forage in terrestrial environments and they often prey upon "worm-like" prey such as annelids, diplopods (among arthropods), and "slugs", defined as apparently shell-less terrestrial gastropods (among mollusks) [13,39,40].

The present study aimed to assess whether *S. italicus* and *S. salamandra* may be potential competitors when they occur in sympatry and to determine the mechanisms preventing the competition. In order to coexist, species tend to reduce the competition by targeting different types of prey [41,42]; therefore, we expected a little overlap in the realized trophic niche of these populations when occurring in sympatry. In addition, we evaluated potential divergences in foraging behavior among conspecific individuals.

## 2. Materials and Methods

We carried out surveys in a forested area of the northern Apennines in the province of Bologna (Emilia-Romagna, Italy); precise information on the site is omitted to ensure species protection [43]. The surveys (six in total) were carried out in April on rainy nights (7 p.m.–2 a.m.) with random frequency, but with at least a one-day interval between two surveys. We surveyed an area of about 4000 m$^2$ in search of individuals of *Speleomantes italicus* and *Salamandra salamandra* on the ground, on trees, and on dry stone walls [44]. The captured salamanders were photographed using a portable photo-studio [45]. We estimated the snout-vent length (SVL, in mm) of salamanders from the images using ImageJ software [46–48]. We focused only on SVL as this measure provides more accurate information on both the age and size of the salamanders [23], as the tail can be lost (and possibly regenerate) due to predation events [49]. The images were also used to individually recognize the salamanders of both species by observing their dorsal pattern [50,51]; for *S. italicus*, an additional marking method was also employed (i.e., visual implant elastomers) [52]. Individuals of both species were sexed based on their size and the presence of distinctive sexual characters. For *S. italicus*, we used the size of 50 mm (SVL) as a threshold to distinguish adults ($\geq$) and juveniles ($<$) [23]. Among adults, male recognition was based on the presence of their typical secondary sexual characters (i.e., mental gland, prominent pre-maxillary teeth, and conical shape of the head) [15]; all adult individuals lacking these traits were considered females. In *S. salamandra*, we used the size of 90 mm (SVL) to distinguish between adults ($\geq$) and juveniles ($<$). Among adults, those with swelling at the base of the cloaca were considered to be males [49]. We weighed all the salamanders using a digital scale (accuracy 0.01 g) and then inspected their stomach residues by stomach flushing, a harmless technique that allows to obtain information on the individual's latest foraging activity [26,53]. Prey residues were recognized at the order level, and each order represents a single prey category [26]. In some cases, further distinctions were also made at the family level, or between larval stages, i.e., when these groups are morphologically distinct and characterized by different ecology (e.g., aquatic vs. subaerial) (see Table 1). Each of these groups was considered an independent category. For additional information on the method, see [26,44].

We used the analysis of similarity (ANOSIM with 10,000 permutations) to assess whether the similarity in the trophic niche between the two populations was higher than that occurring within each population [54,55]. Furthermore, we tested whether the two populations diverged in terms of multivariate dispersion of their diet (*betadispr* function, 999 permutations) [56]. We used PERMANOVA to assess potential interspecific differences in diet composition [57,58]. The nonmetric multidimensional scaling (NMDS) plot with Euclidean distance was used to show the trophic niche differences between the two populations of salamanders. Although the dataset contained data on individuals captured multiple times (see Results), we decided to maintain all the data in this analysis to have a more complete information on the species' overall trophic niche.

Generalized linear mixed models (GLMMs) [59,60] were used to assess whether differences between sex and ontogenetic stage in terms of the number and diversity of prey consumed existed. In the first GLMM, the log-transformed number of consumed prey by each individual was used as the dependent variable, while individual SVL and sex/stage (male, female, or juvenile; hereafter only referred as "sex" for the sake of brevity) were used as independent variables; the day of the survey was used as the random factor. In the second GLMM, we used the Shannon index of the prey consumed by each individual as the dependent variable; all the other variables remained the same. The analysis of the two GLMMs were separately performed for *S. italicus* and *S. salamandra*. To avoid bias due to pseudoreplication, we purged the datasets used in GLMM analyses removing the data on recaptured individuals and maintaining only the event in which the number of recognized prey was the largest. We decided not to analyze individuals with empty stomachs or with unrecognized prey, as they represented a very small percentage of the overall pool of individuals (about 6% for *S. italicus* and 3% for *S. salamandra*; see Results).

**Table 1.** List of the prey items found in the stomach contents of *Speleomantes italicus* and *Salamandra salamandra*. To each group of prey, we assigned a code (first column), which we used in the NMDS plot to increase its clarity. In brackets is the relative importance (%) of each group of prey within the trophic niche of each species.

| Prey Code | Prey Order | Number Recognized in *Speleomantes italicus* and Relative Importance (%) | Number Recognized in *Salamandra salamandra* and Relative Importance (%) |
|---|---|---|---|
| A | Pulmonata | 54 (1.86) | 59 (35.12) |
| B | Sarcoptiformes | 78 (2.69) | 0 |
| C | Mesostigmata | 15 (0.52) | 0 |
| S | Trombidiformes | 7 (0.24) | 0 |
| E | Araneae | 359 (12.38) | 9 (5.36) |
| F | Pseudoscorpiones | 125 (4.31) | 0 |
| G | Opiliones | 30 (1.03) | 8 (4.76) |
| H | Lithobiomorpha | 22 (0.76) | 2 (1.19) |
| I | Geophilomorpha | 14 (0.48) | 0 |
| J | Scolopendromorpha | 5 (0.17) | 0 |
| K | Julida | 16 (0.55) | 7 (4.17) |
| L | Polydesmida | 92 (3.17) | 13 (7.74) |
| M | Isopoda | 81 (2.79) | 2 (1.19) |
| N | Symphypleona | 11 (0.38) | 0 |
| O | Poduromorpha | 35 (1.21) | 0 |
| P | Entomobryomorpha | 288 (9.93) | 0 |
| Q | Blattodea | 4 (0.14) | 0 |
| R | Hemiptera | 186 (6.41) | 0 |
| S | Hymenoptera | 22 (0.76) | 0 |
| T | Hymenoptera-Formicidae | 121 (4.17) | 1 (0.6) |
| U | Coleoptera | 275 (9.48) | 2 (1.19) |
| V | Coleoptera-Staphylinidae | 82 (2.83) | 0 |
| W | Coleoptera-larvae | 35 (1.21) | 4 (2.38) |
| X | Trichoptera-larvae | 3 (0.10) | 0 |
| Y | Plecoptera | 179 (6.17) | 3 (1.79) |
| Z | Lepidoptera | 1 (0.03) | 1 (0.6) |
| AA | Lepidoptera-larvae | 24 (0.83) | 0 |
| AB | Diptera | 582 (20.07) | 10 (5.95) |
| AC | Diptera-larvae | 109 (3.76) | 23 (13.69) |
| AD | Archaeognatha | 10 (0.34) | 0 |
| AE | Speleomantes-skin | 5 (0.17) | 0 |
| AF | Haplotaxida | 22 (0.76) | 24 (14.29) |
| AG | Siphonaptera | 3 (0.10) | 0 |
| AH | Dermaptera | 4 (0.14) | 0 |
| AI | Ixodida | 1 (0.03) | 0 |

Finally, we estimated the degree of individual diet specialization that occurred in these two sympatric populations [61]. For each species, we calculated the index of individual specialization (IS) as shown in [62,63]. We focused only on IS, as this index is significantly correlated to the other diet specialization niche metrics [64]. To improve the clarity, we used the index V =1—IS proposed in Bolnick, et al. [65], where values tending to 1 indicate a high degree of individual diet specialization, while values tending to 0 indicate that the population is mostly made up of generalists [62]. Bootstrapping (repeated 9999 times) was used to test whether the observed index of individual diet specialization significantly diverged from the simulated one, i.e., a scenario in which all individuals randomly choose their prey. In this analysis, we further purged the dataset used in GLMM, removing individuals from which we recognized <3 prey items; this was a precaution to not overinflate the individual diet specialization index.

The data on *S. italicus* used here were retrieved from [44], while the data on *S. salamandra* are provided as Supplementary Materials (Table S1).

## 3. Results

We captured and inspected the stomach contents from 315 *Speleomantes italicus* (128 females, 146 males, and 41 juveniles) (average captured individuals $\pm$ SD per night; $52 \pm 20.5$) and 32 *Salamandra salamandra* (22 females, 9 males, and 1 juvenile) ($5 \pm 6$). Individuals of *S. italicus* were captured during all six surveys, while individuals of *S. salamandra* were only captured in three. Some of them (30 *S. italicus* and 4 *S. salamandra*) were recaptured several times. In the stomachs of most of the captured salamanders, we found residuals of consumed prey; only 18 *S. italicus* had an empty stomach. In two *S. italicus* and one *S. salamandra*, the stomach contents were in an advanced state of digestion, so we were unable to recognize the prey consumed at the established taxonomic level. We recognized 2,900 prey items from *S. italicus* (average $\pm$ SD per individual; $9.83 \pm 6.56$) and 168 from *S. salamandra* ($5.42 \pm 3.08$), belonging to 35 groups of prey (Table 1).

The trophic niche of *S. italicus* included all the prey categories described in Table 1, where just four (Araneae, Entomobryomorpha, Coleoptera, and Diptera) accounted for 51.86% of the consumed prey. The trophic niche of *S. salamandra* included only 15 of the prey categories, with 3 of them (Gastropoda, Diptera-larvae, and Haplotaxida) accounting for 63.1% of the consumed prey. The analysis of similarity identified a significant divergence between the trophic niche of *S. italicus* and that of *S. salamandra* (R = 0.476, $p$ = 0.001) (Figure 1A); the diet of the two populations showed a significant heterogeneity of multivariate dispersion (permutation test: $p$ = 0.001). The analysis of PERMANOVA confirmed the divergence of the trophic niche between the two populations ($R^2$ = 0.07, $p$ = 0.001). The trophic niche of *S. italicus* was much larger than that of *S. salamandra* (Figure 1B).

In *S. italicus*, neither the SVL ($F_{1,262.13}$ = 1.13, $p$ = 0.289) nor the sex ($F_{2,260.75}$ = 1.35, $p$ = 0.261) significantly affected the number of prey consumed. Similar results were obtained for the diversity of prey consumed (SVL: $F_{1,261.43}$ = 0.1, $p$ = 0.748; sex: $F_{2,260.19}$ = 0.95, $p$ = 0.387). In *S. salamandra*, none of the variables affected the number (SVL: $F_{1,22.52}$ = 0.03, $p$ = 0.863; sex: $F_{2,22.99}$ = 0.6, $p$ = 0.558) or diversity of the prey consumed (SVL: $F_{1,23}$ = 0.26, $p$ = 0.612; sex: $F_{2,23}$ = 0.07, $p$ = 0.933). In *S. italicus*, we found a significantly high proportion of specialized individuals (V = 0.621, $p$ < 0.001), while in *S. salamandra*, we observed a similar proportion of both generalist and specialist individuals (V = 0.506, $p$ = 0.019).

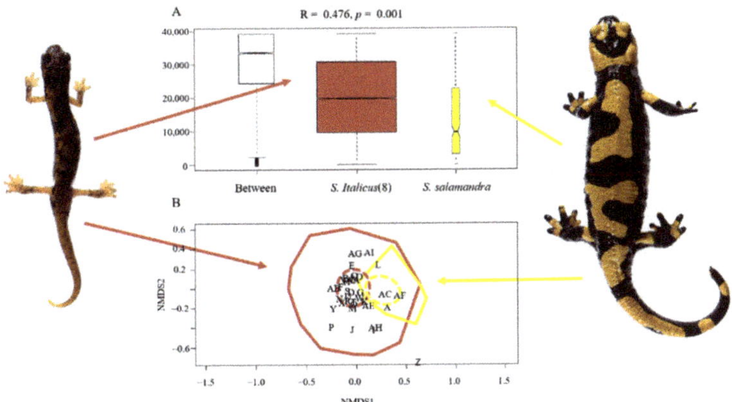



**Figure 1.** (**A**) Box whisker plot of ANOSIM analysis comparing the diets of *Speleomantes italicus* and *Salamandra salamandra*. Boxes indicate values from 25th (bottom) to 75th (top) percentile; horizontal black line indicates the median; box width is proportional to sample size. (**B**) Cumulative NMDS and dashed 95% confidence ellipses with relative position of each species. NMDS plot needs to be carefully interpreted due to the unbalanced datasets (stress = 0.26). The number (8) indicates the population code for *S. italicus*, which is aligned with that in [44,66]. The two sample animals (*S. italicus* on the left and *S. salamandra* on the right) are not to scale.

## 4. Discussion

The trophic niche of the plethodontid salamander, *Speleomantes italicus*, was the widest, including in its diet all the prey categories recognized in this study (Table 1, Figure 1B). These results are in line with what has been observed in other populations of *S. italicus* and for the entire genus, as *Speleomantes* generally show a wide variability in the prey consumed regardless of sex and age [12,26–28,44,67]. In this epigean population, most of the prey consumed are flying prey (about 43%), similar to what has been observed in subterranean populations of *Speleomantes* [4,35,68]. It may be possible that *Speleomantes'* protrusible tongue increases their ability to capture flying prey, a skill that has proven to be extremely useful when individuals cling to vertical surfaces [69–71]. Conversely, *Speleomantes* are not very attracted by slow wormlike prey [71,72], so the poor representation of these types of prey in their diet may be the result of individual choice [8,64]. Indeed, the overall worm-like prey consumed by this population of *S. italicus* did not even cover 14% of its diet (Table 1). Soft-bodied prey was usually underrepresented in previous studies on *Speleomantes'* diet, as this type of prey is likely rapidly digested without a trace [12,26,27,66]. In the studied epigean population, this type of prey (i.e., Pulmonata and Haplotaxida) represented approximately 2.5% of the prey consumed (Table 1). Notably, not only snails were consumed in this population, but also slugs, a type of prey never reported for *Speleomantes* before. It could be possible that, in our case, individuals underwent stomach flushing shortly after ingesting the prey, without having completed their digestion. The stomach contents of the subterranean *Speleomantes* populations were usually obtained during the day (9 a.m.–6 p.m.) [26,44]. *Speleomantes* come out of their subterranean shelter to feed on the surface, especially on cold and humid nights [15,73]; therefore, checking the stomach contents the day after foraging may be too late to obtain information on prey being quickly digested. No significant effect on the number of consumed prey was found in this population of *S. italicus*. In a previous study, it was observed that juvenile *Speleomantes* from subterranean populations consumed significantly less prey than adults [4]. In this study, we surveyed a fully epigean population, meaning that individuals only forage in the surface environment, where the diversity and availability of prey are remarkably different from that found in subterranean environments [74]. A large number of prey consumed in this study were Collembola (Table 1), animals of millimetric size that can be eaten in large number also by juveniles. These prey were widely underrepresented in the stomach contents from subterranean populations [4], despite their steady presence in subterranean environments [75]. This divergence could be due to a variability in the abundance of Collembola between subterranean and surface environments rather than different prey preferences between juveniles inhabiting different environments. This hypothesis remains to be tested.

The trophic niche of *Salamandra salamandra* was narrower, as it included less than half of the overall prey consumed by *S. italicus* (Table 1, Figure 1B). *Salamandra salamandra* behaved exactly the opposite from *Speleomantes*, preferring mostly slow-moving worm-like prey [40,76]. This type of prey consumed by *S. salamandra* accounted for 79% of its diet (Table 1), corroborating similar results obtained from different populations throughout its distribution [13,39,53,77,78]. Among the stomach contents recognized from *S. salamandra*, we observed a very low frequency of prey with hard cuticles, and generally they were <1 cm in size (Table 1). In a recent study, it was shown that soft-bodied prey can be underrepresented when analyzing the species' diets through stomach flushing [13]. In our study, we observed that most (>63%) of the prey consumed by *S. salamandra* exclusively consisted of soft-bodied prey (Table 1), which means that these taxa are not necessarily underrepresented in this type of study. As we discussed for *S. italicus* (see above), this inconsistency may have been due to the delay between the foraging event and the inspection of salamanders' stomachs. In their study, Marques, Mata and Velo-Antón [13] investigated the trophic niche of *S. salamandra*, analyzing their feces. This means that the prey had undergone the entire digestive process and that the salamanders had assimilated most of the wet mass by defecating only the solid residues that they had not been able to digest. In this circumstance, it is understandable how soft-bodied prey can only be detected through DNA analysis.

Our hypothesis in which we predicted limited overlap of the trophic niche between the two syntopic species was supported by our results, as only a small portion of the 95% CI of the species trophic spectrum overlaps (Figure 1B). The divergence in prey preference, as well as their size, may play a fundamental role in reducing the competition between these two species, and those preferences may be related to different energy requirements. Lungless and relatively small salamanders such as *S. italicus* do not need a large energy supply because they have a low-energy and efficient metabolism, while larger salamanders with lungs require much more energy to maintain their relatively higher metabolism [79,80]. Soft-bodied prey is mostly composed of moist mass that can be fully digested, providing much more energy (kilojoules) than those that have hard parts that cannot be digested [79]. This may be one of the reasons why *S. salamandra* more frequently consumes soft-bodied prey. This also supports the hypothesis that *S. italicus* rather chooses to not prey on this type of prey, as it does not seem that the surveyed area was characterized by a particularly low abundance of worm-like prey. Although we lack specific data to show, Pulmonata and Haplotaxida consumed by *S. salamandra* had an impressive size, certainly unsuitable for *S. italicus*. A study focusing on six species of *Speleomantes* showed that the size of prey consumed can vary between species and between individuals of different sizes [4]. Therefore, although the overall categories of prey consumed by *S. salamandra* can also be consumed by *S. italicus* (Figure 1B), prey size can be a powerful mechanism to reduce competition. This is a hypothesis we would like to test in the near future.

In our study the analyzed dataset was quite unbalanced; the data for *Speleomantes italicus* was almost 10 times higher than that of *Salamandra salamandra*. This asymmetry may have affected the robustness of some analyses, so we recommend a careful interpretation of our results.

The two species also diverged in terms of proportion of specialized individuals: the population of *S. italicus* had a higher proportion of specialized individuals, while in *S. salamandra* the proportion of both generalists and specialist individuals were similar. Considering the co-occurrence of these two populations, we can exclude the potential effects of the environment on the frequency of specialized individuals [64]. Alternatively, intraspecific competition may play a major role here [8]. In our study, the observed density of *S. italicus* (i.e., the overall captured individuals) was about 0.07 individuals/m$^2$, while for *S. salamandra*, it was ten folds lower (0.007 individuals/m$^2$). Previous studies on *Speleomantes* did not provide straightforward information about the potential occurrence of intraspecific competition or the related effects on the diet specialization of individuals [64,81–83]. To the best of our knowledge, we are not aware of similar studies performed on adult individuals of *S. salamandra*. Therefore, this remains an open hypothesis that deserves to be tested.

## 5. Conclusions

Our study provides indications on the potential mechanisms used by two co-occurring generalist salamanders to avoid competition. In our case, the two generalist salamanders, although able to target similar prey, mostly chose different prey types, and when they consumed similar prey, they did target prey of different size. *Speleomantes italicus* mostly consumed flying prey with hard cuticle, while *Salamandra salamandra* did the opposite, more frequently consuming worm-like soft-bodied prey. The prey consumed by both species mostly differed in size, making morphological constraints a potential tool useful for reducing competition. In conclusion, morphological constraints, together with other characteristics such as species metabolism (i.e., digestion ability) and prey preference, might play an important role in reducing the competition between sympatric generalist predators.

**Supplementary Materials:** The following supporting information can be downloaded at: https://www.mdpi.com/article/10.3390/ani12172221/s1, Table S1: Dataset used for the analysis on *Salamandra salamandra* trophic niche. The dataset contains: the site in which the study was performed together with low-resolution coordinates and elevation (m a.s.l); the region and province in which the population was located; the population code; the date of the survey; the unique individual code (Tag), together with sex and snout-vent length (SVL, in mm); the stomach condition (1 = empty; 0 = full); Not_identifiable indicates whether the prey were recognized (0) or not (1) at order level; the abundance of each prey type recognized from stomach contents.

**Author Contributions:** E.L. conceived the study, analyzed the data, prepared the figure and drafted the manuscript. E.L., C.C. and M.B. performed data collection. C.C. contributed in data interpretation. E.L. and F.C. recognized prey from stomach contents. E.L., C.C., M.B., Y.Z. and F.C. reviewed the manuscript. All authors have read and agreed to the published version of the manuscript.

**Funding:** This study was supported by the National Natural Science Foundation of China (NSFC-31972868). Enrico Lunghi was supported by the Chinese Academy of Sciences President's International Fellowship Initiative for postdoctoral researchers. Fabio Cianferoni was partially supported by the Ministry of University and Research of Italy (MUR), project FOE 2020-Capitale naturale e risorse per il futuro dell'Italia-Task Biodiversità.

**Institutional Review Board Statement:** The study was authorized by the Italian Ministry of Environment (DPR 357/97–PNM 25526, of 23 November 2017) and (Prot. 67681 DEL 27/11/2018 T-A31 2019-2021).

**Data Availability Statement:** Data for *Speleomantes italicus* were retrieved from [44], while data for *Salamandra salamandra* are provided as Supplementary Materials (Table S1).

**Conflicts of Interest:** The authors declare no conflict of interest.

## References

1. Hutchinson, G.E. Concluding remarks. *Cold Spring Harb. Symp. Quant. Biol.* **1957**, *22*, 415–427. [CrossRef]
2. Campbell, C.J.; Nelson, D.M.; Ogawa, N.O.; Chikaraishi, Y.; Ohkouchi, N. Trophic position and dietary breadth of bats revealed by nitrogen isotopic composition of amino acids. *Sci. Rep.* **2017**, *7*, 15932. [CrossRef] [PubMed]
3. Manenti, R.; Denoël, M.; Ficetola, G.F. Foraging plasticity favours adaptation to new habitats in fire salamanders. *Anim. Behav.* **2013**, *86*, 375–382. [CrossRef]
4. Lunghi, E.; Cianferoni, F.; Ceccolini, F.; Veith, M.; Manenti, R.; Mancinelli, G.; Corti, C.; Ficetola, G.F. What shapes the trophic niche of European plethodontid salamanders? *PLoS ONE* **2018**, *13*, e0205672. [CrossRef]

5.	Fraenkel, G.; Blewett, M. The basic food requirements of several insects. *J. Exp. Biol.* **1943**, *20*, 28–34. [CrossRef]

6.	Stuart, S.N.; Chanson, J.S.; Cox, N.A.; Young, B.E.; Rodrigues, A.S.L.; Fischman, D.L.; Waller, R.W. Status and trends of amphibian declines and extinctions worldwide. *Science* **2004**, *306*, 1783–1786. [CrossRef]

7.	Persson, L. Optimal foraging: The difficulty of exploiting different feeding strategies simultaneously. *Oecologia* **1985**, *67*, 338–341. [CrossRef]

8.	Araújo, M.S.; Bolnick, D.L.; Layman, C.A. The ecological causes of individual specialisation. *Ecol. Lett.* **2011**, *14*, 948–958. [CrossRef]

9.	Bolnick, D.I.; Ingram, T.; Stutz, W.E.; Snowberg, L.K.; Lee Lau, O.; Paull, J.S. Ecological release from interspecific competition leads to decoupled changes in population and individual niche width. *Proc. R. Soc. B* **2010**, *277*, 1789–1797. [CrossRef]

10.	Costa-Pereira, R.; Araújo, M.S.; Souza, F.L.; Ingram, T. Competition and resource breadth shape niche variation and overlap in multiple trophic dimensions. *Proc. R. Soc. B* **2019**, *286*, 20190369. [CrossRef]

11.	Costa-Pereira, R.; Rudolf, V.H.W.; Souza, F.L.; Araújo, M.S. Drivers of individual niche variation in coexisting species. *J. Anim. Ecol.* **2018**, *87*, 1452–1464. [CrossRef] [PubMed]

12.	Vignoli, L.; Caldera, F.; Bologna, M.A. Trophic niche of cave populations of *Speleomantes italicus*. *J. Nat. Hist.* **2006**, *40*, 1841–1850. [CrossRef]

13.	Marques, A.J.D.; Mata, V.A.; Velo-Antón, G. COI metabarcoding provides insights into the highly diverse diet of a generalist salamander, Salamandra salamandra (Caudata: Salamandridae). *Diversity* **2022**, *14*, 89. [CrossRef]

14.	Costa, A.; Baroni, D.; Romano, A.; Salvidio, S. Individual diet variation in *Salamandra salamandra* larvae in a Mediterranean stream (Amphibia: Caudata). *Salamandra* **2017**, *53*, 148–152.

15.	Lanza, B.; Pastorelli, C.; Laghi, P.; Cimmaruta, R. A review of systematics, taxonomy, genetics, biogeography and natural history of the genus *Speleomantes* Dubois, 1984 (Amphibia Caudata Plethodontidae). *Atti Mus Civ Stor Nat Trieste* **2006**, *52*, 5–135.

16.	Ficetola, G.F.; Lunghi, E.; Manenti, R. Microhabitat analyses support relationships between niche breadth and range size when spatial autocorrelation is strong. *Ecography* **2020**, *43*, 724–734. [CrossRef]

17.	Spotila, J.R. Role of temperature and water in the ecology of lungless salamanders. *Ecol. Monogr.* **1972**, *42*, 95–125. [CrossRef]

18.	Ficetola, G.F.; Lunghi, E.; Canedoli, C.; Padoa-Schioppa, E.; Pennati, R.; Manenti, R. Differences between microhabitat and broad-scale patterns of niche evolution in terrestrial salamanders. *Sci. Rep.* **2018**, *8*, 10575. [CrossRef]

19.	Lunghi, E.; Corti, C.; Manenti, R.; Barzaghi, B.; Buschettu, S.; Canedoli, C.; Cogoni, R.; De Falco, G.; Fais, F.; Manca, A.; et al. Comparative reproductive biology of European cave salamanders (genus *Hydromantes*): Nesting selection and multiple annual breeding. *Salamandra* **2018**, *54*, 101–108.

20.	Lunghi, E.; Manenti, R.; Manca, S.; Mulargia, M.; Pennati, R.; Ficetola, G.F. Nesting of cave salamanders (*Hydromantes flavus* and *H. italicus*) in natural environments. *Salamandra* **2014**, *50*, 105–109.

21.	Bruni, G. Tail-straddling walk and spermatophore transfer in *Hydromantes italicus*: First observations for the genus and insights about courtship behavior in plethodontid salamanders. *Herpetol. Rev.* **2020**, *51*, 673–680.

22.	Oneto, F.; Ottonello, D.; Pastorino, M.V.; Salvidio, S. Posthatching parental care in salamanders revealed by infrared video surveillance. *J. Herpetol.* **2010**, *44*, 649–653. [CrossRef]

23.	Lunghi, E. Doubling the lifespan of European plethodontid salamanders. *Ecology* **2022**, *103*, e03581. [CrossRef] [PubMed]

24.	Rondinini, C.; Battistoni, A.; Peronace, V.; Teofili, C. *Lista Rossa IUCN dei Vertebrati Italiani*; Comitato Italiano IUCN e Ministero dell'Ambiente e della Tutela del Territorio e del Mare: Roma, Italy, 2013; p. 54.

25.	Lunghi, E.; Manenti, R.; Ficetola, G.F. Seasonal variation in microhabitat of salamanders: Environmental variation or shift of habitat selection? *PeerJ* **2015**, *3*, e1122. [CrossRef]

26.	Lunghi, E.; Cianferoni, F.; Ceccolini, F.; Mulargia, M.; Cogoni, R.; Barzaghi, B.; Cornago, L.; Avitabile, D.; Veith, M.; Manenti, R.; et al. Field-recorded data on the diet of six species of European *Hydromantes* cave salamanders. *Sci. Data* **2018**, *5*, 180083. [CrossRef]

27.	Salvidio, S.; Pasmans, F.; Bogaerts, S.; Martel, A.; van de Loo, M.; Romano, A. Consistency in trophic strategies between populations of the Sardinian endemic salamander *Speleomantes imperialis*. *Anim. Biol.* **2017**, *67*, 1–16. [CrossRef]

28.	Salvidio, S. Diet and food utilization in the European plethodontid *Speleomantes ambrosii*. *Vie et Milieu* **1992**, *42*, 35–39.

29.	Salvidio, S.; Romano, A.; Oneto, F.; Ottonello, D.; Michelon, R. Different season, different strategies: Feeding ecology of two syntopic forest-dwelling salamanders. *Acta Oecol.* **2012**, *43*, 42–50.

30.	Lunghi, E.; Ficetola, G.F.; Zhao, Y.; Manenti, R. Are the neglected Tipuloidea crane flies (Diptera) an important component for subterranean environments? *Diversity* **2020**, *12*, 333. [CrossRef]

31.	Manenti, R.; Lunghi, E.; Ficetola, G.F. Distribution of spiders in cave twilight zone depends on microclimatic features and trophic supply. *Invertebr. Biol.* **2015**, *134*, 242–251. [CrossRef]

32.	Speybroeck, J.; Beukema, W.; Bok, B.; Van Der Voort, J.; Velikov, I. *Field Guide to the Amphibians & Reptiles of Britain and Europe*; Bloomsbury: London, UK, 2016.

33.	Martel, A.; Spitzen-van der Sluijs, A.; Blooi, M.; Bert, W.; Ducatelle, R.; Fisher, M.C.; Woeltjes, A.; Bosman, W.; Chiers, K.; Bossuyt, F.; et al. *Batrachochytrium salamandrivorans* sp. nov. causes lethal chytridiomycosis in amphibians. *Proc. Natl. Acad. Sci. USA* **2012**, *110*, 15325–15329. [CrossRef]

34. Caldonazzi, M.; Nistri, A.; Tripepi, S. Salamandra salamandra. In *Fauna d'Italia. Vol. XLII. Amphibia*; Lanza, B., Andreone, F., Bologna, M.A., Corti, C., Razzetti, E., Eds.; Edizioni Calderini de Il Sole 24 ORE Editoria Specializzata S.r.l.: Bologna, Italy, 2007; pp. 221–227.

35. Lunghi, E.; Guillaume, O.; Blaimont, P.; Manenti, R. The first ecological study on the oldest allochthonous population of European cave salamanders (*Hydromantes* sp.). *Amphib.-Reptil.* **2018**, *39*, 113–119. [CrossRef]

36. Manenti, R.; Lunghi, E.; Ficetola, G.F. Cave exploitation by an usual epigean species: A review on the current knowledge on fire salamander breeding in cave. *Biogeographia* **2017**, *32*, 31–46. [CrossRef]

37. Bressi, N.; Dolce, S.; Stoch, F. Observations on the feeding of amphibians: IV. Larval diet of the fire salamander, *Salamandra salamandra salamandra* (Linnaeus, 1758), near Trieste (North-eastern Italy) (Amphibia, Caudata, Salamandridae). *Atti Mus Civ Stor Nat Trieste* **1996**, *47*, 275–283.

38. Melotto, A.; Ficetola, G.F.; Manenti, R. Safe as a cave? Intraspecific aggressiveness rises in predator-devoid and resource-depleted environments. *Behav. Ecol. Sociobiol.* **2019**, *73*, 68. [CrossRef]

39. Balogová, M.; Miková, E.; Orendáš, P.; Uhrin, M. Trophic spectrum of adult *Salamandra salamandra* in the Carpathians with the first note on food intake by the species during winter. *Herpeto. Notes* **2015**, *8*, 371–377.

40. Maier, A.-R.-M.; Ferenți, S.; Dumbravă, A.-R.; Cadar, A.-M. A late autumn feeding: Diet of *Salamandra salamandra* (Linnaeus, 1758) (Amphibia: Salamandridae) in October and November in the Iron Gates natural park, Romania. *Acta Zoologica Bulgarica* **2022**, *74*, 195–201.

41. Cloyed, C.S.; Eason, P.K. Niche partitioning and the role of intraspecific niche variation in structuring a guild of generalist anurans. *R. Soc. Open Sci.* **2017**, *4*, 170060. [CrossRef]

42. Novak, T.; Tkavc, T.; Kuntner, M.; Arnett, A.E.; Lipovšek Delakorda, S.; Perc, M.; Janžekovič, F. Niche partitioning in orbweaving spiders *Meta menardi* and *Metellina merianae* (Tetragnathidae). *Acta Oecol.* **2010**, *36*, 522–529. [CrossRef]

43. Lunghi, E.; Corti, C.; Manenti, R.; Ficetola, G.F. Consider species specialism when publishing datasets. *Nat. Ecol. Evol.* **2019**, *3*, 319. [CrossRef]

44. Lunghi, E.; Corti, C.; Biaggini, M.; Merilli, S.; Manenti, R.; Zhao, Y.; Ficetola, G.F.; Cianferoni, F. Capture-mark-recapture data on the strictly protected *Speleomantes italicus*. *Ecology* **2022**, *103*, e3641. [CrossRef]

45. Lunghi, E.; Bacci, F.; Zhao, Y. How can we record reliable information on animal colouration in the wild? *Diversity* **2021**, *13*, 356. [CrossRef]

46. Lunghi, E.; Giachello, S.; Manenti, R.; Zhao, Y.; Corti, C.; Ficetola, G.F.; Bradley, J.G. The post hoc measurement as a safe and reliable method to age and size plethodontid salamanders. *Ecol. Evol.* **2020**, *10*, 11111–11116. [CrossRef]

47. Lunghi, E.; Biaggini, M.; Corti, C. Reliability of the post-hoc measurement on *Salamandra salamandra*. *Il Nat. Sicil.* 2021; in press.

48. Schneider, C.A.; Rasband, W.S.; Eliceiri, K.W. NIH Image to ImageJ: 25 years of image analysis. *Nature Methods* **2012**, *9*, 671–675. [CrossRef]

49. Lanza, B.; Nistri, A.; Vanni, S. Classe Amphibia Gray, 1825 (Morfologia e Biologia). In *Fauna d'Italia. Vol. XLII. Amphibia*; Lanza, B., Andreone, F., Bologna, M.A., Corti, C., Razzetti, E., Eds.; Edizioni Calderini de Il Sole 24 ORE Editoria Specializzata S.r.l.: Bologna, Italy, 2007; p. 2.

50. Lunghi, E.; Romeo, D.; Mulargia, M.; Cogoni, R.; Manenti, R.; Corti, C.; Ficetola, G.F.; Veith, M. On the stability of the dorsal pattern of European cave salamanders (genus *Hydromantes*). *Herpetozoa* **2019**, *32*, 249–253. [CrossRef]

51. Speybroeck, J.; Steenhoudt, K. A pattern-based tool for long-term, large-sample capture-markrecapture studies of fire salamanders *Salamandra* species (Amphibia: Urodela: Salamandridae). *Acta Herpetol.* **2017**, *12*, 55–63.

52. Lunghi, E.; Bruni, G. Long-term reliability of Visual Implant Elastomers in the Italian cave salamander (*Hydromantes italicus*). *Salamandra* **2018**, *54*, 283–286.

53. Ferenti, S.; David, A.; Nagy, D. Feeding-behaviour responses to anthropogenic factors on *Salamandra salamandra* (Amphibia, Caudata). *Biharean Biol.* **2010**, *4*, 139–143.

54. R Development Core Team. *R: A Language and Environment for Statistical Computing*; R Foundation for Statistical Computing: Vienna, Austria.

55. Kohl, M. *MKmisc: Miscellaneous Functions from M. Kohl.* R Package Version 0.993. 2016. Available online: http://www.stamats.de (accessed on 13 August 2022).

56. Oksanen, J.; Blanchet, F.G.; Friendly, M.; Kindt, R.; Legendre, P.; McGlinn, D.; Minchin, P.R.; O'Hara, R.B.; Simpson, G.L.; Solymos, P.; et al. Vegan: Community Ecology Package. R Package Version 2.5-7. Available online: https://github.com/vegandevs/vegan (accessed on 13 August 2022).

57. Anderson, M.J.; Walsh, D.C.I. PERMANOVA, ANOSIM, and the Mantel test in the face of heterogeneous dispersions: What null hypothesis are you testing? *Ecol. Monogr.* **2013**, *83*, 557–574. [CrossRef]

58. Anderson, M.J. Distance-based tests for homogeneity of multivariate dispersions. *Biometrics* **2006**, *62*, 245–253. [CrossRef]

59. Kuznetsova, A.; Brockhoff, B.; Christensen, H.B. lmerTest: Tests in Linear Mixed Effects Models, R package version 2.0-2.9; 2016. Available online: www.r-project.org (accessed on 13 August 2022).

60. Douglas, B.; Maechler, M.; Bolker, B.; Walker, S. Fitting Linear Mixed-Effects Models using lme4. *J. Stat. Softw.* **2015**, *67*, 1–48. [CrossRef]

61. Bolnick, D.I.; Svanbäck, R.; Fordyce, J.A.; Yang, L.H.; Davis, J.M.; Hulsey, C.D.; Forister, M.L. The ecology of individuals: Incidence and implications of individual specialization. *Am. Nat.* **2003**, *161*, 1–28. [CrossRef]

62. Bolnick, D.I.; Yang, L.H.; Fordyce, J.A.; Davis, J.M.; Svanbäck, R. Measuring individual-level resource specialization. *Ecology* **2002**, *83*, 2936–2941. [CrossRef]
63. Zaccarelli, N.; Bolnick, D.I.; Mancinelli, G. RInSp: An R package for the analysis of individual specialization in resource use. *Methods Ecol. Evol.* **2013**, *4*, 1018–1023. [CrossRef]
64. Lunghi, E.; Manenti, R.; Cianferoni, F.; Ceccolini, F.; Veith, M.; Corti, C.; Ficetola, G.F.; Mancinelli, G. Interspecific and inter-population variation in individual diet specialization: Do environmental factors have a role? *Ecology* **2020**, *101*, e03088. [CrossRef]
65. Bolnick, D.I.; Svanbäck, R.; Araújo, M.S.; Persson, L. Comparative support for the niche variation hypothesis that more generalized populations also are more heterogeneous. *Proc. Natl. Acad. Sci. USA* **2007**, *104*, 10075–10079. [CrossRef]
66. Lunghi, E.; Giachello, S.; Zhao, Y.; Corti, C.; Ficetola, G.F.; Manenti, R. Photographic database of the European cave salamanders, genus *Hydromantes*. *Sci. Data* **2020**, *7*, 171. [CrossRef]
67. Lunghi, E.; Cianferoni, F.; Giachello, S.; Zhao, Y.; Manenti, R.; Corti, C.; Ficetola, G.F. Updating salamander datasets with phenotypic and stomach content information for two mainland *Speleomantes*. *Sci. Data* **2021**, *8*, 150. [CrossRef]
68. Lunghi, E.; Cianferoni, F.; Merilli, S.; Zhao, Y.; Manenti, R.; Ficetola, G.F.; Corti, C. Ecological observations on hybrid populations of European plethodontid salamanders, genus *Speleomantes*. *Diversity* **2021**, *13*, 285. [CrossRef]
69. Deban, S.M.; Marks, S.B. Metamorphosis and evolution of feeding behaviour in salamanders of the family Plethodontidae. *Zool. J. Linn. Soc.* **2002**, *134*, 375–400. [CrossRef]
70. O'Donnell, M.K.; Deban, S.M. Cling performance and surface area of attachment in plethodontid salamanders. *J. Exp. Biol.* **2020**, *233*, jeb211706. [CrossRef]
71. Roth, G. Experimental analysis of the prey catching behavior of *Hydromantes italicus* Dunn (Amphibia, Plethodontidae). *J. Comp. Physiol. A* **1976**, *109*, 47–58. [CrossRef]
72. Roth, G. Responses in the optic tectum of the salamander *Hydromantes italicus* to moving prey stimuli. *Exp. Brain Res.* **1982**, *45*, 386–392. [CrossRef]
73. Lunghi, E.; Manenti, R.; Mulargia, M.; Veith, M.; Corti, C.; Ficetola, G.F. Environmental suitability models predict population density, performance and body condition for microendemic salamanders. *Sci. Rep.* **2018**, *8*, 7527. [CrossRef]
74. Culver, D.C.; Pipan, T. *The Biology of Caves and Other Subterranean Habitats*, 2nd ed.; Oxford University Press: New York, NY, USA, 2019; p. 336.
75. Lunghi, E.; Valle, B.; Guerrieri, A.; Bonin, A.; Cianferoni, F.; Manenti, R.; Ficetola, G.F. Complex patterns of environmental DNA transfers from surface to subterranean soils revealed by analyses of cave insects and springtails. *Sci. Total Environ.* **2022**, *826*, 154022. [CrossRef]
76. Luthardt, G.; Roth, G. The relationship between stimulus orientation and stimulus movement pattern in the prey catching behavior of *Salamandra salamandra*. *Copeia* **1979**, *1979*, 442–447. [CrossRef]
77. Bas Lopez, S.; Gutian Rivera, J.; Castro Lorenzo, A.D.; Sanchez Canals, J.L. Datos sobre l'alimentacion de la salamandra (*Salamandra salamandra* L.) in Galicia. *Bol. De La Estac. Cent. De Ecol.* **1979**, *8*, 73–78.
78. Wang, Y.; Smith, H.K.; Goossens, E.; Hertzog, L.; Bletz, M.C.; Bonte, D.; Verheyen, K.; Lens, L.; Vences, M.; Pasmans, F.; et al. Diet diversity and environment determine the intestinal microbiome and bacterial pathogen load of fire salamanders. *Sci. Rep.* **2021**, *11*, 20493. [CrossRef]
79. Feder, M.E. Integrating the ecology and physiology of plethodontid salamanders. *Herpetologica* **1983**, *39*, 291–310.
80. Feder, M.E. Oxygen consumption and activity in salamanders: Effect of body size and lunglessness. *J. Exp. Zool.* **1977**, *202*, 403–414. [CrossRef]
81. Lunghi, E.; Cianferoni, F.; Ceccolini, F.; Zhao, Y.; Manenti, R.; Corti, C.; Ficetola, G.F.; Mancinelli, G. Same diet, different strategies: Variability of individual feeding habits across three populations of Ambrosi's cave salamander (*Hydromantes ambrosii*). *Diversity* **2020**, *12*, 180. [CrossRef]
82. Salvidio, S.; Pastorino, M.V. Spatial segregation in the European plethodontid *Speleomantes strinatii* in relation to age and sex. *Amphib.-Reptil.* **2002**, *23*, 505–510.
83. Ficetola, G.F.; Pennati, R.; Manenti, R. Spatial segregation among age classes in cave salamanders: Habitat selection or social interactions? *Popul. Ecol.* **2013**, *55*, 217–226. [CrossRef]

*Article*

# A New Disease Caused by an Unidentified Etiological Agent Affects European Salamanders

**Raoul Manenti** [1,*], **Silvia Mercurio** [1,*], **Andrea Melotto** [2], **Benedetta Barzaghi** [1], **Sara Epis** [3], **Marco Tecilla** [3], **Roberta Pennati** [1], **Giorgio Ulisse Scarì** [3] and **Gentile Francesco Ficetola** [1,4,5]

[1]  Department of Environmental Science and Policy, University of Milano, 20133 Milano, Italy; benedetta.barzaghi@unimi.it (B.B.); roberta.pennati@unimi.it (R.P.); francesco.ficetola@gmail.com (G.F.F.)
[2]  Centre of Excellence for Invasion Biology, Department of Botany and Zoology, Stellenbosch University, Stellenbosch 7602, South Africa; mel8@hotmail.it
[3]  Department of Biosciences, University of Milano, 20133 Milano, Italy; sara.epis@unimi.it (S.E.); marco.tecilla@gmail.com (M.T.); giorgio.scari@unimi.it (G.U.S.)
[4]  Laboratoire d'Ecologie Alpine (LECA), University Grenoble Alpes, CNRS, 38400 Grenoble, France
[5]  Laboratoire d'Ecologie Alpine (LECA), University Savoie Mont Blanc, CNRS, 38400 Grenoble, France
*  Correspondence: raoul.manenti@unimi.it (R.M.); silivia.mercurio@unimi.it (S.M.); Tel.: +39-3490733107 (R.M.)

**Simple Summary:** In the last few years, multiple new infectious diseases have affected amphibians, causing unprecedented declines and extinctions at the global scale. Many of these diseases are caused by pathogens that have been described during recent decades. In this study we report a novel disease, affecting European amphibians. In a protected area of Northern Italy, since the autumn of 2013, we started to find some adult salamanders with cysts at the throat level (subgular region). Following this observation, we ran a regular monitoring of the salamander population and performed multiple morphological and molecular investigations to identify the cause of this undescribed disease. The cysts surround peculiar cells, probably protists, of about 30 μm covered by numerous motile cilia/undulipodia. Despite multiple attempts with a broad spectrum of techniques, the detailed identification remains challenging as we have been unable to match its features with previously described organisms. We provide the results achieved till now to promote a rapid dissemination on this new enigmatic wildlife pathogen and to create a basis for further and deeper studies.

**Abstract:** New pathologies are causing dramatic declines and extinctions of multiple amphibian species. In 2013, in one fire salamander population of Northern Italy, we found individuals with undescribed cysts at the throat level, a malady whose existence has not previously been reported in amphibians. With the aim of describing this novel disease, we performed repeated field surveys to assess the frequency of affected salamanders from 2014 to 2020, and integrated morphological, histological, and molecular analyses to identify the pathogen. The novel disease affected up to 22% of salamanders of the study population and started spreading to nearby populations. Cysts are formed by mucus surrounding protist-like cells about 30 μm long, characterized by numerous cilia/undulipodia. Morphological and genetic analyses did not yield a clear match with described organisms. The existence of this pathogen calls for the implementation of biosecurity protocols and more studies on the dynamics of transmission and the impact on wild populations.

**Keywords:** protist; Mesomycetozoea; pathogenic; amphibians; fire salamander; *Salamandra salamandra*

**Citation:** Manenti, R.; Mercurio, S.; Melotto, A.; Barzaghi, B.; Epis, S.; Tecilla, M.; Pennati, R.; Scarì, G.U.; Ficetola, G.F. A New Disease Caused by an Unidentified Etiological Agent Affects European Salamanders. *Animals* **2022**, *12*, 696. https://doi.org/10.3390/ani12060696

Academic Editor: José Martín

Received: 10 February 2022
Accepted: 4 March 2022
Published: 10 March 2022

**Publisher's Note:** MDPI stays neutral with regard to jurisdictional claims in published maps and institutional affiliations.



## 1. Introduction

### Epigraph

"Но как же это так? Ведь это же чудовищно! Это чудовищно, господа", -повторил он, обращаясь к жабам в террарии, но жабы спали и ничего ему не ответили." ("But how can it be? It's monstrous! Quite monstrous, gentlemen," he repeated, addressing the toads in the terrarium, who were asleep and made no reply). M. Bulgakov, Роковые яйца

The emergence and spreading of novel diseases in wildlife is one of the most challenging threats to both biodiversity and human well-being [1–4]. Emerging wildlife diseases can quickly have profound broad-scale impacts, including ecological disturbance, economic and agricultural impacts, and even human losses [4,5]. The rapid detection of novel wildlife diseases is thus crucial, and often requires an interdisciplinary approach, including for instance morphological and genetic investigations, ecological assessments, and management consultations [5–7]. However, when diseases emerge in wildlife organisms, their detection is not easy and some pathogens may remain unnoticed till their spreading becomes difficult to control [8,9]. This could be particularly true for elusive or poorly studied animals, as 'non-mammals' vertebrates, and when pathologies are strongly different from the already known diseases.

Amphibians provide a clear example of the damage that the increasing spread of novel diseases may play over short periods to biodiversity at a global scale. In the last decades, recently discovered pathogens have caused dramatic declines and extinctions of populations and even species of amphibians [10,11]. The most studied pathologies threatening amphibians are chytridiomycoses caused by the fungi *Batrachochytrium salamandrivorans* and *Batrachochytrium dendrobatidis* [12–14], which are involved in the decline of >500 amphibian species and determine the greatest loss of wildlife biodiversity linked to a pathogen [13]. Amphibians are also susceptible to viruses such as *Ranavirus* sp. [15], which is the main infectious pathogen of multiple aquatic ectothermic vertebrates and is often implicated in mass die-offs of amphibians [16,17], even causing 100% mortality in tadpoles of some anuran species [18,19]. Moreover, amphibians also host understudied pathogens such as the fungal-like protists Mesomycetozoea [20–23]. Their biological features are typical of pathogens able of determining emerging infectious diseases in different amphibians as in the case of the genus *Amphibiocistydium* [24,25], but information on these organisms and their impacts remains scanty.

With this paper, we want to bring to the attention of the world's scientific community a novel disease affecting a widespread European amphibian, the fire salamander (*Salamandra salamandra* Linnaeus, 1758). In 2013, two individuals with an odd cyst at the level of the throat (Figure 1) were discovered in a small, protected area of Northern Italy. During following surveys, we observed similar cysts in multiple individuals. Therefore, we directed great effort to identify the cause of the cysts and to understand the extent of their spreading. In particular, we monitored the occurrence of the cysts during the following years and performed detailed morphological, ultrastructural and molecular analyses in order to characterize their content. We believe this information must be disseminated across scientists and managers before this phenomenon strikes other amphibian populations.

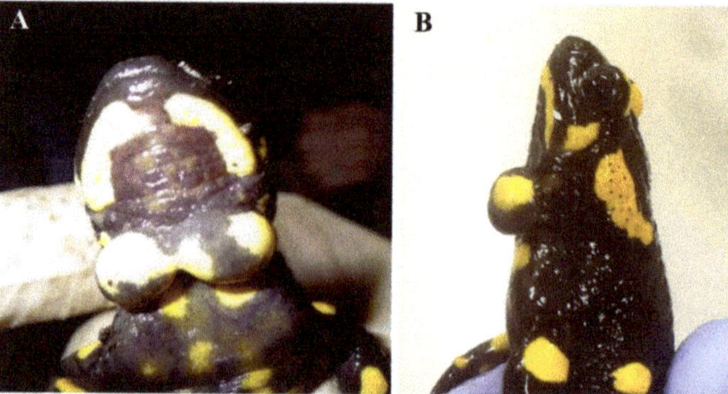


**Figure 1.** Salamander cysts. Ventral (**A**) and lateral (**B**) view of a male *Salamandra salamandra* with two turgid cysts occurring in a median position at the throat level.

## 2. Materials and Methods

### 2.1. The Fire Salamander: Study Area and Monitoring

The fire salamander is ovoviviparous and widespread in Europe; this species mainly inhabits hilly landscapes covered by broadleaf forests and breeds in streams and other freshwater sites [26]. When adult, fire salamanders show strong terrestrial habits, with dispersal abilities range from 200 to 1300 m (but generally up to 500 m), and nocturnal behaviour [27]. Although widespread, the species has recently suffered catastrophic declines in the northern part of its range because of the chytrid fungus *Batrachochytrium salamandrivorans* [14]. Our study focuses on the regional protected reserve named "Fontana del Guercio" located around 30 km North of the city of Milan (NW Italy) and that is recorded as a European Site of Community Interest. The area is characterized by an extended broadleaf wood crossed by a slow flowing stream and numerous springs where the fire salamander breeds. This area has undergone annual monitoring since 2010. Since 2014, after the observation of the first two individuals bearing the subgular cysts, till 2020, we established yearly surveys with three transects covering the main area of the reserve (Figure 2). From 2016, to monitor the spreading of the pathogen occurrence in the surroundings, we also sampled three additional transects in a nearby (1.6 km distant) wooded area (named "Olgelasca") occurring in a parallel valley and connected only through some wooded stretches Transects are on average $\pm$ SE 462.7 $\pm$ 95.7 m long and 18.7 $\pm$ 2.6 m wide; they follow the main paths and cover all the terrestrial habitats surrounding multiple breeding sites of the fire salamander. We performed at least three-night surveys in each autumn season (October–December) for each transect. On each survey, we collected all the salamanders we encountered; after recording the position with a GPS, we checked each individual for the occurrence of the pathogen, and then we photographed on millimetre paper and weighed it, before releasing it in the same place of collection.

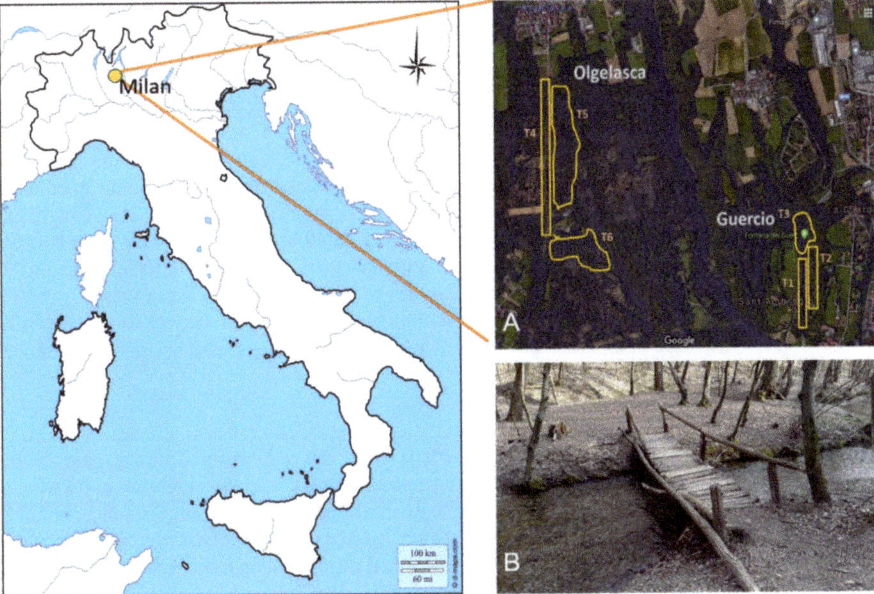


**Figure 2.** Location of the study area. (**A**) location and area of the transects monitored, identified by the letter T; "Guercio" identifies the main locality on the regional protected area named "Riserva del Guercio" in which the pathogen was first detected. "Olgelasca" identifies the adjacent locality in which the pathogen appeared in 2016. (**B**) picture of the protected area between transect 1 and transect 2.

The occurrence of the cyst was detected by examining the throat of the individuals. If at least one cyst was visible, we considered the salamander as affected; we considered swellings larger than 2 mm of thickness, paying attention to the fact that the throat of the fire salamanders often shows small plicas that can resemble small bumps. In cases where a second swelling occurred, we recorded it. To prevent spreading the disease during surveys we handled the salamanders with nitrile gloves that we changed for each individual; each salamander was placed for weighing and picturing on a disposable bag that was then immediately disposed of. At the end of each survey, all the material used, including shoes, was disinfected with 10% bleach.

### 2.2. Cyst Extraction

To understand the origin of the cyst under the skin, a few infected animals were collected during sampling and we extracted cyst samples in vivo from salamanders' throat using a non-invasive surgical protocol. Prior to surgery, the animals were anesthetized by a subcutaneous injection of tricaine/carbocaine 4 mg/mL (0.32 mL/g of weight). Loss of superficial reflexes was tested prior to the incision. The samples were removed by using a 0.3 cm, sterile, disposable biopsy punch with plugger (Bioseb lab, Vitrolles, France) and immediately processed for subsequent analyses (see below). After complete remission, usually occurring within 24 h in ozonized water, the salamanders were reintroduced in their collection site. In total we extracted samples of 15 cysts from 8 individuals from 2014 to 2018.

### 2.3. Cyst Cell Isolation

Under aseptic conditions, two cysts were surgically removed and immediately processed for in vitro analysis. Samples were dissected into pieces using sterile fine-tipped tweezers and disaggregated in sterile amphibian Ringer's solution. Then, cells were centrifuged at low speed (1200 rpm/9 g) and resuspended in Medium 199 with 10% foetal bovine serum and 50 mg/L gentamicin (Sigma, Igea Marina RN, Italy). Cells were cultured in petri dishes at 12 °C for 20 days. All cultures were observed daily using an inverted phase contrast microscope. Replacement of 50% of the culture medium was carried out every day.

### 2.4. Histology

Three cysts were processed for light microscopy analysis. Samples were immediately fixed with Bouin's solution (picric acid, formaldehyde, acetic acid, 75:25:5) for at least 24 h and rinsed several times in tap water until all the fixative solution was completely removed. Samples were then dehydrated with an ethanol series (70%, 90%, 95% and 100%), cleared in xylene and left overnight in a solution of xylene and paraffin wax 56–58 °C (1:1). Samples were then immersed in three changes of paraffin wax and finally embedded. Sections 5–7 µm thick were cut with a standard microtome and stained with Haematoxylin and Eosin (HE).

### 2.5. Electron Microscopy

To better describe cyst structure and the cells trapped inside, electron microscopy was also performed. Samples were pre-fixed with 2% glutaraldehyde in 0.1 M cacodylate buffer for two hours and, after overnight washing in the same buffer, post-fixed with 1% solution of OsO4 in 0.1 M cacodylate buffer. After standard dehydration in ethanol series (25%, 70%, 90%, and 100%), samples were washed in propylene oxide and embedded in Epon-Araldite 812 resin (Bio Optica, Milan, Italy). Semi-thin (about 0.9 µm) and ultra-thin (70 nm) sections were cut with a Reichert–Jung ULTRACUT E using glass knives. Semi-thin sections were stained with crystal violet and basic fuchsine, mounted with Eukitt (Bio Optica, Milan, Italy) and observed under a Leica light microscope. Ultrathin sections were mounted on copper grids and stained with uranyl acetate and lead citrate for electron microscopy, then observed and photographed in a Talos L120C TEM microscope.

### 2.6. Scanning Electron Microscopy

Cyst samples were immediately fixed in 1% glutaraldehyde and 0.4% formaldehyde in 0.1 M sodium cacodylate buffer (pH 7.2, 300–400 mOsm) for 30 min at 4 °C. Then, samples were washed 3 times in 0.1 M sodium cacodylate buffer over 5 min and post fixed with 1% OsO4 in 0.1 M sodium cacodylate buffer for 30 min. Cysts were washed again 3 times, dehydrated in ethanol series (50%, 70%, 80%, 90%, 96%, and absolute alcohol), and critical point dried [CPD] (Balzers CPD 030 Critical Point Dryer; Bal-Tec AG, Balzers, Liechtenstein) in carbon dioxide. Dried samples were mounted on stabs with carbon adhesive discs, then Au sputtered using a Bal-Tec SCD 050 Sputter Coater (Bal-Tec AG, Balzers, Liechtenstein), and examined with a scanning electron microscope (LEO-1430).

### 2.7. Genetic Analyses

Standard protocol for genomic DNA extraction with proteinase K from whole cyst samples was performed with some modifications. Briefly, overnight proteinase K digestion was preceded with triple liquid nitrogen/water bath in 100 °C for 2 min to guarantee effective genomic extraction [28]. Based on literature we selected different primers for DNA amplification. A pair of universal non-metazoan primers, 18S-EUK581-F (5′-GTGCCAGCAGCCGCG-3′) and 18S-EUK1134-R (5′-TTTAARKTTCAGCGCTTGSG-3′), were chosen to amplify a 544 bp fragment of 18S rDNA from protists without getting animal DNA [29]. Two protozoa-specific forward primers (P-SSU-342f and PR900f) in combination with a protozoa-specific reverse primer (PR900r) or an eukarya-specific reverse primer were used for targeting 18S rDNA gene. Primers sequences were: PR900f (5′-TTTCGATGGTAGATTGGAC-3′), PR900r (5′-CTTGTTACGACTTCTCCTTCC-3′) amplifying a 900 bp fragment [30,31]); P-SSU-342f (5′-CTTTCGATGGTAGTGTATTGGACTAC-3′) and Medlin B (5′-TGATCCTTCTGCAGGTTCACCTAC-3′) amplifying a 1360 bp fragment [32].

A second attempt to identify the host trapped in the salamander's cyst consisted of extracting DNA directly from the cells isolated for in vitro analysis. We selected the ciliated cells occurring in the cyst using 20 µL of TE Buffer (Tris-EDTA, 100× Solution, pH 8.0) plus 3.5 µL of liquid cystic (about 10–20 cells). We performed DNA extraction using a Proteinase K protocol. All the reagents were sterile. 2 µL of PK 20 mg/mL were added to the sample mixing the solution very well. Samples were incubated in the Thermomixer at 56 °C for two hours, then at 95 °C for 5 min and finally centrifuged at 22 °C for 10 min at 20,000× *g*. The small subunit ribosomal RNA (SSrRNA) genes were amplified with the universal eukaryotic primers forward 18S F9 [59-CTG GTT GATCCT GCC AG-39] [33] and reverse 18SR1513 Hypo [59-TGA TCC TTC (CT)GC AGG TTC-39] [34]. PCR products were purified and directly sequenced in both directions. To minimize the possibility of amplification errors, high-fidelity Taq was used. The internal primers used for sequencing were 18S R536 [5′-CTGGAATTACCGCGGCTG-3′] and 18S R1052 [5′-AACTAAGAACGGCCATGCA-3′].

Finally, we used a metabarcoding approach, trying to target the largest number of eukaryotes. The DNA extracted with proteinase K was amplified using the Euka02 primers (Forward: TTTGTCTGSTTAATTSCG; Reverse: CACAGACCTGTTATTGC) which are highly generalist primers that amplify most of eukaryotes with limited bias [35,36]. These primers amplify a short (~120 bp) region of the 18S rDNA (V7), and are thus suitable for the analysis of degraded or poor quality DNA [36,37]. DNA was amplified following the same protocol of [37] and sequenced using an Illumina Hiseq 2000 platform (see [37] for complete details on sequencing and bioinformatics processing). The detected sequences were taxonomically assigned using the Ecotag program of the OBITOOLS package, on the basis of NCBI database [38]. We analysed two DNA aliquots, each with four PCR replicates. We also ran 9 PCR controls, each replicated four times, including PCR mix but no template DNA to identify environmental contaminants [39]. To avoid the risk of false positives we only considered taxa detected in >50% of PCR replicates with >10 reads [40].

*2.8. Ethics*

The study design, the samplings and the surgery on the fire salamanders was approved by the ethical committee of the Lombardy Region Authority and was authorized as complying with the regional law 10/2008, permission number: T1.2016.0052349. After the removal of the cysts and recovery, each individual was released in the exact place of its collection.

## 3. Results

*3.1. Cysts Occurrence in Fire Salamander Populations*

The cyst's occurrence was first detected in 2013 in two males out of 66 salamanders observed. One of the males showed two turgid swellings occurring in a median position at the throat level, one left and one right separated by less than a millimetre (Figure 1); the cyst diameter was 10 and 8 mm respectively. The cyst size varied among individuals, while its position was always similar. We did not detect signs of external damage or injuries at the throat level, but on the cysts some small points (2 and 1 respectively) characterized by very thin skin stratus and appearing of a light black colour, were visible. The number of individuals detected per sampling in Fontana del Guercio locality, varied from five in a sampling time in 2017 to 202 during a sampling period in 2018 (Figure S1). From 2014 to 2020 the proportion of affected individuals was on average ($\pm$SE) 18.5% $\pm$ 8.1% per year. Both the proportion of individuals displaying the pathogen and the total number of adult salamanders detected along the transects varied across the years of monitoring (Figure S2); on average, in the Guercio protected area, considering the different samplings and transects, the largest proportion of affected salamanders was observed during 2015. In most cases, salamanders showed one single cyst, but 12.09 $\pm$ 6.08% of the affected individuals per year showed two of them.

In 2016, we observed for the first time, an affected adult in the transect performed in the second locality. Here, since 2016 the percentage of affected salamanders has been on average 1.18 $\pm$ 0.4%.

*3.2. Histological Analysis*

Macroscopically, cyst masses were spherical, located on the ventral surface of the jugular region, frequently bilateral, and, once removed, ranged between 0.2 cm and 0.4 cm in size (Figure 1A,B). They were histologically composed of a thin capsule encircling a central area of mucus material admixed with numerous granulocytes (morphologically compatible with heterophils), and a lesser number of plasma cells and lymphocytes (Figure 3D,E). Embedded in the mucus, numerous ciliated protists were detected (Figure 3F). A final diagnosis of a heterophilic cyst with intralesional microorganisms was made.

*3.3. In Vitro Isolation*

In cyst cell cultures, different cellular phenotypes were observed. Most of them were recognized as salamander leukocytes, mainly represented by granulocytes, among which peculiar cells were always present (Figure 3A). These latter appeared as spherical protist-like cells of about 30 μm in total length. Their cytoplast was generally heterogeneous and the cell membrane was covered by numerous motile cilia/undulipodia (Figure 3B,C).

*3.4. Electron Microscopy*

Electron microscopy analyses provided a detailed description of the protist-like cells in the salamander cysts. They appeared as unicellular organisms with a diameter of 10 μm, characterized by numerous undulipodia (Figure 4A,F,G). These structures protruded from the cell body (Figure 4B–D) and completely covered the cellular membrane (Figure 4H). Undulipodia were as long as the cell body, about 10 μm; together, the cell body and the undulipodia constituted what appeared as an unicellular organism of 30 μm in total length (Figure 4A,H). The cytoplasm was rich in electron dense and transparent inclusions (Figure 4A). At higher magnification, it was possible to appreciate the exogenous materials contained in the vesicles (Figure 4B). Mitochondria with peculiar cristae were also observed (Figure 4E).

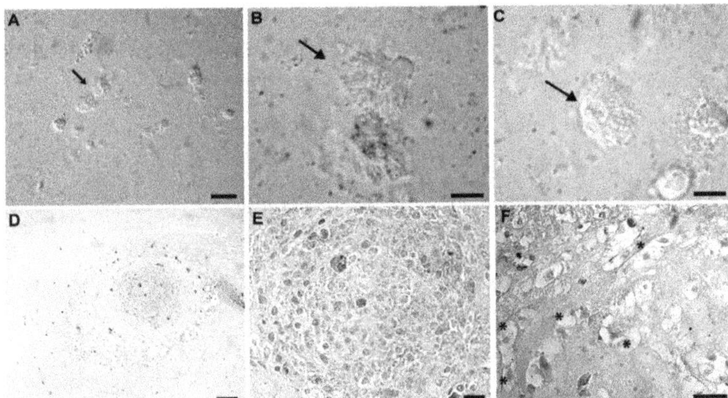

**Figure 3.** Morphological analyses. (**A–C**) Cyst cell culture in which peculiar ciliated cells are observable (arrow). These protist-like cells displayed heterogeneous cytoplast and their cell membrane was covered by numerous motile cilium-like structures (**B,C**). (**D–F**) Histological analysis revealed that the cyst masses consisted of a thin capsule of connective tissue encircling a central area of mucus material in which numerous granulocytes, plasma cells, and lymphocytes were detected (**D,E**). Among these, ciliated protist-like cells were present trapped in the mucus (**F**); asterisks identify trapped cells. Scale bars: (**A**) 20 μm; (**B,C**) 10 μm; (**D**) 250 μm; (**E,F**) 50 μm.

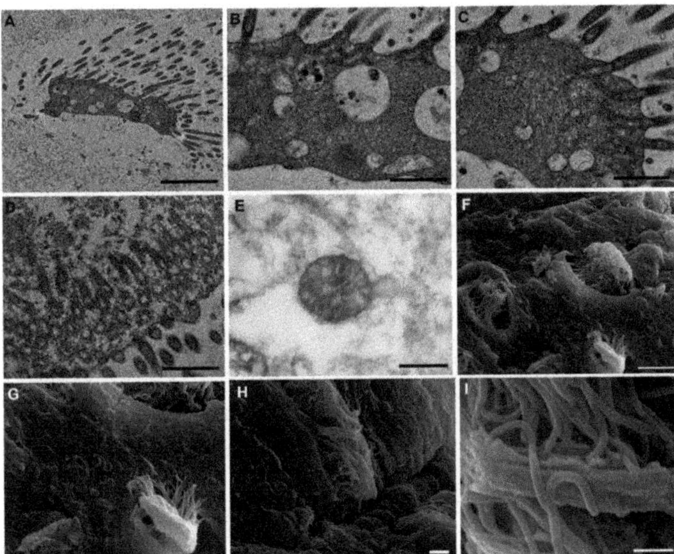

**Figure 4.** Electron microscopy analyses. (**A–E**) Transmission electron microscopy of the pathogen. In cyst mucus, numerous protist-like cells (**A**) with long undulipodia were found. At higher magnification (**B**), electron dense and transparent vacuoles were observed as well as undulipodium rootlets regularly distributed along cytoplasm periphery (**C,D**). Mitochondria with peculiar cristae were detected (**E**). (**F–I**) Scanning electron microscopy. Trapped inside mucus niches, protist-like cells were observed (**F,G**). Their main feature was the numerous and long cilia/undulipodia which covered the cell membrane (**H,I**). Scale bars: (**A**) 2 μm; (**B–D**) 1 μm; (**E**) 500 nm; (**F**) 10 μm; (**G**) 5 μm; (**H**) 2 μm; (**I**) 1 μm.

### 3.5. DNA Analyses

None of the several DNA amplifications and sequencing attempts clarified the identity of the organisms. Using genomic DNA extracted from whole cyst samples as a template, no rDNA fragment was amplified employing either protists primers [29] or protozoan-specific ones [30–32]. DNA amplification from in vitro cell isolates was also not successful. Sequencing analysis of the only amplified DNA fragment revealed the presence of fungi belonging to the genus *Cladosporium*, which was probably due to contamination. Using metabarcoding, we detected only one molecular taxonomic unit in >50% of PCR replicates (sequence: ctcaaacttccatctactaaacgtagatagtccctctaagaagccaaaaaagccaac-caaagtcgacccggctatttagcaggttaaggtctcgttcgttat). The Ecotag program assigned this sequence to the fungal genus *Meira* (Brachybasidiaceae). However, this fungus was also detected in 25% of controls, suggesting that it is a contaminant.

## 4. Discussion

The combination of extensive field surveys with molecular and morphological analyses allowed us to detail a new disease affecting one of the most widespread amphibians in Europe. Morphological analyses showed that the turgescence was composed mainly by mucus encapsulating numerous protist-like cells, which appeared to be embedded in small niches among the tissue. The main feature of these cells was the presence of long peculiar cilia similar to undulipodia covering the cell membrane. In the cyst, numerous leukocytes of the salamander were also recognizable, suggesting an active immune response. In vitro studies confirmed the morphology of the cells and provided more information about their cilia/undulipodia which were motile. Electron microscopy further characterized the cells. Their heterogeneous cytoplasm rich in digestive vacuoles strongly suggested phagocytosis activity. However, a definitive diagnosis remained elusive because morphological and ultrastructural data did not allow identification of the cells encapsulated by the cysts, as we were unable to match the morphology of these cells with a known pathogen.

Amphibians are affected by a large spectrum of parasites and pathologies [11,41,42] and are parasitized by multiple protists [41]. Initially we supposed a similarity between the encapsulated microorganisms and some mesomycetozoan stages [23] that parasitize fish; however, they had a different position and shape of their flagella [23]. This similarity led us to perform molecular investigations following [29], which reported for the first time the occurrence of *Amphibiocystidium* sp. in the red-spotted newts. Despite some similarities with *Amphibiocystidium*, we failed to detect its DNA and, even though we attempted multiple PCR primers and techniques, we only amplified the DNA of contaminants. The mucus of cysts could inhibit DNA extraction or amplification by reducing their efficiency [43,44], still a modification of the DNA extraction protocol according to [28] did not provide significant improvements. The absence of DNA amplification could be because, even if the cysts may attain conspicuous sizes, the encapsulated cells were quite sparse, possibly limiting the amount of DNA. Furthermore, the amplification success of primers can be highly heterogeneous across taxa [37]. Even if we employed both specific and universal primer combinations, it is possible that the potential pathogen has mismatches in the priming region that hamper amplification. In principle, the location of the cysts could also be suggestive of internal disorders linked to production of salamander cells with abnormal morphologies. Nevertheless, we can consider neoplastic formations as unlikely because the encapsulated cells are completely different from described amphibian cells and the encapsulation reaction clearly seems to underline an immune reaction toward an external agent. Ciliated amphibian cells have been recorded only in the gills of some urodeles, but they show very different shapes and features compared to the detected ones [45,46].

## 5. Conclusions

Despite the numerous morphological, ultrastructural, and molecular analyses, we failed to unambiguously identify the cause of the disease. The significant frequency of salamanders showing cysts and the features of the cells encapsulated by those cysts, suggest that this disease could be caused by a slow growing and enigmatic wildlife pathogen.

Our study suggests that this disease could be caused by a pathogen that might be an unidentified protist; if dispersal by infected salamanders is likely to become a threat for the surrounding populations, the occurrence of other possible vectors should not be excluded. Waterways are considered highly suitable for the survival and spread of several pathogens [47]: a stream flows down from the Guercio protected area and connects with a complex hydrographic network in the basin of the river Lambro. Moreover, the Guercio protected area has numerous visitors each year, represented mainly by local people but also including schools and even people from abroad. Efforts will thus be necessary to make all the stakeholders of the area, from wildlife managers to visitors, aware of the presence of the phenomenon and of the caution measures to adopt to avoid its spreading. Moreover, we hope the findings of our study stimulate further studies, both on the biology of the new discovered cysts and on the occurrence of other potential animals affected, from fishes to other amphibian species.

**Supplementary Materials:** The following supporting information can be downloaded at: https://www.mdpi.com/article/10.3390/ani12060696/s1, Figure S1: Box-plot of the total number of adult fire salamanders recorded each year in the protected area "Fontana del Guercio". In some years, only a single survey was performed. Figure S2: Frequency of salamanders affected by cysts in the study populations through time. Error bars are Jeffrey's Bayesian 95% confidence intervals, computed using the binom package in R.

**Author Contributions:** R.M., G.F.F., G.U.S. and R.P. conceived the study; A.M., B.B., G.F.F., G.U.S., R.P. and R.M. performed field samplings; G.U.S. performed surgery on salamanders; G.U.S., S.M. and M.T. performed in vitro isolation and histology; G.U.S. and S.M. performed electron microscopy; G.F.F. and S.E. performed genetic analyses; A.M. and G.F.F. performed statistical analyses; R.M. wrote the first draft. All authors have read and agreed to the published version of the manuscript.

**Funding:** This research received no external funding.

**Institutional Review Board Statement:** The study was conducted in accordance with the Declaration of Helsinki; the study design, the samplings and the surgery of fire salamanders were approved by the ethical committee of the LOMBARDY REGION AUTHORITY and was authorized as complying with the regional law 10/2008, permission number: T1.2016.0052349. Each salamander after the removal of the cyst was released in the exact place of finding.

**Informed Consent Statement:** Not applicable.

**Data Availability Statement:** Not applicable.

**Acknowledgments:** Part of this study was carried out at NOLIMITS, an advanced imaging facility established by the Università degli Studi di Milano. We are grateful to B. Lombardi, M. Sala, I. Caccamo and the numerous students who participated in monitoring fire salamanders across the years. S. Tripepi, E. Brunelli and F. De Bernardi provided useful comments to understand the features of the encapsulated cells. Authors acknowledge support from the University of Milan through the APC initiative.

**Conflicts of Interest:** The authors declare no conflict of interest.

## References

1. Cleaveland, S.; Haydon, D.T.; Taylor, L. Overviews of pathogen emergence: Which pathogens emerge, when and why? *Curr. Top. Microbiol. Immunol.* **2007**, *315*, 85–111. [CrossRef] [PubMed]
2. Lu, M.M.; Wang, X.D.; Ye, H.; Wang, H.M.; Qiu, S.; Zhang, H.M.; Liu, Y.; Luo, J.H.; Feng, J. Does public fear that bats spread COVID-19 jeopardize bat conservation? *Biol. Conserv.* **2021**, *254*, 9. [CrossRef] [PubMed]
3. Maurin, M.; Fenollar, F.; Mediannikov, O.; Davoust, B.; Devaux, C.; Raoult, D. Current Status of Putative Animal Sources of SARS-CoV-2 Infection in Humans: Wildlife, Domestic Animals and Pets. *Microorganisms* **2021**, *9*, 868. [CrossRef] [PubMed]
4. Schrag, S.J.; Wiener, P. Emerging infectious disease: What are the relative roles of ecology and evolution? *Trends Ecol. Evol.* **1995**, *10*, 319–324. [CrossRef]

5.  Murray, A.G.; Smith, R.J.; Stagg, R.M. Shipping and the spread of infectious salmon anemia in Scottish aquaculture. *Emerg. Infect. Dis.* **2002**, *8*, 1–5. [CrossRef] [PubMed]
6.  Gupta, S.K.; Minhas, V. Wildlife population management: Are contraceptive vaccines a feasible proposition? *Front. Biosci. (Sch. Ed.)* **2017**, *9*, 357–374. [CrossRef]
7.  Saldanha, I.F.; Lawson, B.; Goharriz, H.; Rodriguez-Ramos Fernandez, J.; John, S.K.; Fooks, A.R.; Cunningham, A.A.; Johnson, N.; Horton, D.L. Extension of the known distribution of a novel clade C betacoronavirus in a wildlife host. *Epidemiol. Infect.* **2019**, *147*, e169. [CrossRef]
8.  Morner, T.; Obendorf, D.L.; Artois, M.; Woodford, M.H. Surveillance and monitoring of wildlife diseases. *Rev. Sci. Tech.* **2002**, *21*, 67–76. [CrossRef]
9.  Verner-Carlsson, J.; Lohmus, M.; Sundstrom, K.; Strand, T.M.; Verkerk, M.; Reusken, C.; Yoshimatsu, K.; Arikawa, J.; van de Goot, F.; Lundkvist, A. First evidence of Seoul hantavirus in the wild rat population in the Netherlands. *Infect. Ecol. Epidemiol.* **2015**, *5*, 27215. [CrossRef]
10.  Niederle, M.V.; Bosch, J.; Ale, C.E.; Nader-Macias, M.E.; Aristimuno Ficoseco, C.; Toledo, L.F.; Valenzuela-Sanchez, A.; Soto-Azat, C.; Pasteris, S.E. Skin-associated lactic acid bacteria from North American bullfrogs as potential control agents of *Batrachochytrium dendrobatidis*. *PLoS ONE* **2019**, *14*, e0223020. [CrossRef]
11.  Rollins-Smith, L.A. Amphibian immunity-stress, disease, and climate change. *Dev. Comp. Immunol.* **2017**, *66*, 111–119. [CrossRef] [PubMed]
12.  Berger, L.; Speare, R.; Daszak, P.; Green, D.E.; Cunningham, A.A.; Goggin, C.L.; Slocombe, R.; Ragan, M.A.; Hyatt, A.D.; McDonald, K.R.; et al. Chytridiomycosis causes amphibian mortality associated with population declines in the rain forests of Australia and Central America. *Proc. Natl. Acad. Sci. USA* **1998**, *95*, 9031–9036. [CrossRef] [PubMed]
13.  Daszak, P.; Berger, L.; Cunningham, A.A.; Hyatt, A.D.; Green, D.E.; Speare, R. Emerging infectious diseases and amphibian population declines. *Emerg. Infect. Dis.* **1999**, *5*, 735–748. [CrossRef] [PubMed]
14.  Martel, A.; Spitzen-van der Sluijs, A.; Blooi, M.; Bert, W.; Ducatelle, R.; Fisher, M.C.; Woeltjes, A.; Bosman, W.; Chiers, K.; Bossuyt, F.; et al. *Batrachochytrium salamandrivorans* sp nov causes lethal chytridiomycosis in amphibians. *Proc. Natl. Acad. Sci. USA* **2013**, *110*, 15325–15329. [CrossRef] [PubMed]
15.  Campbell, L.J.; Garner, T.W.J.; Tessa, G.; Scheele, B.C.; Griffiths, A.G.F.; Wilfert, L.; Harrison, X.A. An emerging viral pathogen truncates population age structure in a European amphibian and may reduce population viability. *PeerJ* **2018**, *6*, e5949. [CrossRef] [PubMed]
16.  Duffus, A.L.; Andrews, A.M. Phylogenetic analysis of a frog virus 3-like ranavirus found at a site with recurrent mortality and morbidity events in southeastern Ontario, Canada: Partial major capsid protein sequence alone is not sufficient for fine-scale differentiation. *J. Wildl. Dis.* **2013**, *49*, 464–467. [CrossRef]
17.  Duffus, A.L.J.; Garner, T.W.J.; Nichols, R.A.; Standridge, J.P.; Earl, J.E. Modelling Ranavirus Transmission in Populations of Common Frogs (*Rana temporaria*) in the United Kingdom. *Viruses* **2019**, *11*, 556. [CrossRef] [PubMed]
18.  Hoverman, J.T.; Gray, M.J.; Haislip, N.A.; Miller, D.L. Phylogeny, life history, and ecology contribute to differences in amphibian susceptibility to ranaviruses. *Ecohealth* **2011**, *8*, 301–319. [CrossRef]
19.  Mihaljevic, J.R.; Hoverman, J.T.; Johnson, P.T.J. Co-exposure to multiple ranavirus types enhances viral infectivity and replication in a larval amphibian system. *Dis. Aquat. Organ.* **2018**, *132*, 23–35. [CrossRef]
20.  Ayres, C.; Acevedo, I.; Monsalve-Carcano, C.; Thumsova, B.; Bosch, J. Triple dermocystid-chytrid fungus-ranavirus co-infection in a Lissotriton helveticus. *Eur. J. Wildl. Res.* **2020**, *66*, 1–3. [CrossRef]
21.  Gonzalez-Hernandez, M.; Denoel, M.; Duffus, A.J.L.; Garner, T.W.J.; Cunningham, A.A.; Acevedo-Whitehouse, K. Dermocystid infection and associated skin lesions in free-living palmate newts (Lissotriton helveticus) from Southern France. *Parasitol. Int.* **2010**, *59*, 344–350. [CrossRef]
22.  Martínez-Silvestre, A.; Fernandez-Guiberteau, D.; Pérez-Sorribes, L.; Velarde, R. Infección por dermocistidios en *Lissotriton helveticus* en Cataluña: Nuevos datos y apuntes sobre su diagnóstico. *Boletín Asoc. ÓN Herpetológica Española* **2017**, *28*, 66–69.
23.  Mendoza, L.; Taylor, J.W.; Ajello, L. The class mesomycetozoea: A heterogeneous group of microorganisms at the animal-fungal boundary. *Annu. Rev. Microbiol.* **2002**, *56*, 315–344. [CrossRef] [PubMed]
24.  Adamovicz, L.; Woodburn, D.B.; Virrueta Herrera, S.; Low, K.; Phillips, C.A.; Kuhns, A.R.; Crawford, J.A.; Allender, M.C. Characterization of *Dermotheca* sp. Infection in a midwestern state-endangered salamander (Ambystoma platineum) and a co-occurring common species (Ambystoma texanum). *Parasitology* **2020**, *147*, 360–370. [CrossRef] [PubMed]
25.  Fagotti, A.; Rossi, R.; Canestrelli, D.; La Porta, G.; Paracucchi, R.; Lucentini, L.; Simoncelli, F.; Di Rosa, I. Longitudinal study of Amphibiocystidium sp. infection in a natural population of the Italian stream frog (*Rana italica*). *Parasitology* **2019**, *146*, 903–910. [CrossRef] [PubMed]
26.  Manenti, R.; Ficetola, G.F.; De Bernardi, F. Water, stream morphology and landscape: Complex habitat determinants for the fire salamander *Salamandra salamandra*. *Amphib.-Reptil.* **2009**, *30*, 7–15. [CrossRef]
27.  Manenti, R.; Conti, A.; Pennati, R. Fire salamander (*Salamandra salamandra*) males' activity during breeding season: Effects of microhabitat features and body size. *Acta Herpetol.* **2017**, *12*, 29–36.
28.  Adamska, M.; Leonska-Duniec, A.; Maciejewska, A.; Sawczuk, M.; Skotarczak, B. Comparison of efficiency of variuous DNA extraction methods from cysts of *Giardia intestinalis* measured by PCR and Taqman Real Time PCR. *Parasite* **2010**, *17*, 299–305. [CrossRef]

29. Raffel, T.R.; Bommarito, T.; Barry, D.S.; Witiak, S.M.; Shackelton, L.A. Widespread infection of the Eastern red-spotted newt (*Notophthalmus viridescens*) by Amphibiocystidium, a genus of a new species of fungus-like mesomycetozoan parasites not previously reported in North America. *Parasitology* **2008**, *135*, 203–215. [CrossRef]

30. Karnati, S.K.R.; Yu, Z.; Sylvester, J.T.; Dehority, B.A.; Morrison, M.; Firkins, J.L. Technical note: Specific PCR amplification of protozoal 18S rDNA sequences from DNA extracted from ruminal samples of cows. *J. Anim. Sci.* **2003**, *81*, 812–815. [CrossRef]

31. Shin, E.C.; Cho, K.M.; Lim, W.J.; Hong, S.Y.; An, C.L.; Kim, E.J.; Kim, Y.K.; Choi, B.R.; An, J.M.; Kang, J.M.; et al. Phylogenetic analysis of protozoa in the rumen contents of cow based on the 18S rDNA sequences. *J. Appl. Microbiol.* **2004**, *97*, 378–383. [CrossRef] [PubMed]

32. Shanan, S.; Abd, H.; Bayoumi, M.; Saeed, A.; Sandstrom, G. Prevalence of Protozoa Species in Drinking and Environmental Water Sources in Sudan. *Biomed. Res. Int.* **2015**, *2015*, 345619. [CrossRef] [PubMed]

33. Medlin, L.; Elwood, H.J.; Stickel, S.; Sogin, M.L. the characterization of enzymatically amplified eukaryotic 16S-Like RRNA-coding regions. *Gene* **1988**, *71*, 491–499. [CrossRef]

34. Petroni, G.; Dini, F.; Verni, F.; Rosati, G. A molecular approach to the tangled intrageneric relationships underlying phylogeny in Euplotes (Ciliophora, Spirotrichea). *Mol. Phylogenetics Evol.* **2002**, *22*, 118–130. [CrossRef] [PubMed]

35. Guardiola, M.; Uriz, M.J.; Taberlet, P.; Coissac, E.; Wangensteen, O.S.; Turon, X. Deep-Sea, Deep-Sequencing: Metabarcoding Extracellular DNA from Sediments of Marine Canyons. *PLoS ONE* **2016**, *11*, e0139633. [CrossRef] [PubMed]

36. Taberlet, P.; Bonin, A.; Zinger, L.; Coissac, E. *Environmental DNA for Biodiversity Research and Monitoring*; Oxford University Press: Oxford, UK, 2018.

37. Ficetola, G.F.; Boyer, F.; Valentini, A.; Bonin, A.; Meyer, A.; Dejean, T.; Gaboriaud, C.; Usseglio-Polatera, P.; Taberlet, P. Comparison of markers for the monitoring of freshwater benthic biodiversity through DNA metabarcoding. *Mol. Ecol.* **2021**, *30*, 3189–3202. [CrossRef] [PubMed]

38. Boyer, F.; Mercier, C.; Bonin, A.; Le Bras, Y.; Taberlet, P.; Coissac, E. OBITOOLS: A UNIX-inspired software package for DNA metabarcoding. *Mol. Ecol. Resour.* **2016**, *16*, 176–182. [CrossRef] [PubMed]

39. Zinger, L.; Bonin, A.; Alsos, I.G.; Balint, M.; Bik, H.; Boyer, F.; Chariton, A.A.; Creer, S.; Coissac, E.; Deagle, B.E.; et al. DNA metabarcoding-Need for robust experimental designs to draw sound ecological conclusions. *Mol. Ecol.* **2019**, *28*, 1857–1862. [CrossRef]

40. Ficetola, G.F.; Pansu, J.; Bonin, A.; Coissac, E.; Giguet-Covex, C.; De Barba, M.; Gielly, L.; Lopes, C.M.; Boyer, F.; Pompanon, F.; et al. Replication levels, false presences and the estimation of the presence/absence from eDNA metabarcoding data. *Mol. Ecol. Resour.* **2015**, *15*, 543–556. [CrossRef] [PubMed]

41. Densmore, C.L.; Green, D.E. Diseases of amphibians. *ILAR J.* **2007**, *48*, 235–254. [CrossRef]

42. Lunghi, E.; Ficetola, G.F.; Mulargia, M.; Cogoni, R.; Veith, M.; Corti, C.; Manenti, R. Batracobdella leeches, environmental features and Hydromantes salamanders. *Int. J. Parasitol. Parasites Wildl.* **2018**, *7*, 48–53. [CrossRef]

43. Divar, M.R.; Sharifiyazdi, H.; Kafi, M. Application of polymerase chain reaction for fetal gender determination using cervical mucous secretions in the cow. *Vet. Res. Commun.* **2012**, *36*, 215–220. [CrossRef] [PubMed]

44. Rumpho, M.E.; Mujer, C.V.; Andrews, D.L.; Manhart, J.R.; Pierce, S.K. Extraction of DNA from mucilaginous tissues of a sea slug (Elysia chlorotica). *Biotechniques* **1994**, *17*, 1097–1101.

45. Brunelli, E.; Sperone, E.; Maisano, M.; Tripepi, S. Morphology and ultrastructure of the gills in two Urodela species: Salamandrina terdigitata and Triturus carnifex. *Ital. J. Zool.* **2009**, *76*, 158–164. [CrossRef]

46. Brunelli, E.; Tripepi, S. Effects of low pH acute exposure on survival and gill morphology in Triturus italicus larvae. *J. Exp. Zool. Part A Comp. Exp. Biol.* **2005**, *303*, 946–957. [CrossRef] [PubMed]

47. Spitzen-van der Sluijs, A.; Stegen, G.; Bogaerts, S.; Canessa, S.; Steinfartz, S.; Janssen, N.; Bosman, W.; Pasmans, F.; Martel, A. Post-epizootic salamander persistence in a disease-free refugium suggests poor dispersal ability of *Batrachochytrium salamandrivorans*. *Sci. Rep.* **2018**, *8*, 3800. [CrossRef] [PubMed]

*Article*

# Temperature and Diet Acclimation Modify the Acute Thermal Performance of the Largest Extant Amphibian

Chun-Lin Zhao [1,2,†], Tian Zhao [2,†], Jian-Yi Feng [2], Li-Ming Chang [2], Pu-Yang Zheng [2], Shi-Jian Fu [3], Xiu-Ming Li [3], Bi-Song Yue [1,*], Jian-Ping Jiang [2] and Wei Zhu [2,*]

[1] Key Laboratory of Bioresources and Ecoenvironment (Ministry of Education), College of Life Sciences, Sichuan University, Chengdu 610064, China; zhaocl_1@163.com

[2] CAS Key Laboratory of Mountain Ecological Restoration and Bioresource Utilization & Ecological Restoration Biodiversity Conservation Key Laboratory of Sichuan Province, Chengdu Institute of Biology, Chengdu 610041, China; zhaotian@cib.ac.cn (T.Z.); fengjy@cib.ac.cn (J.-Y.F.); changliming16@mails.ucas.ac.cn (L.-M.C.); zhengpy@cib.ac.cn (P.-Y.Z.); jiangjp@cib.ac.cn (J.-P.J.)

[3] Laboratory of Evolutionary Physiology and Behavior, Chongqing Key Laboratory of Animal Biology, Chongqing Normal University, Chongqing 400047, China; shijianfu9@hotmail.com (S.-J.F.); xiumingli418@hotmail.com (X.-M.L.)

[*] Correspondence: bsyue@scu.edu.cn (B.-S.Y.); zhuwei@cib.ac.cn (W.Z.); Tel.: +86-028-82890935 (B.-S.Y.)

[†] These authors contributed equally to this work.

**Citation:** Zhao, C.-L.; Zhao, T.; Feng, J.-Y.; Chang, L.-M.; Zheng, P.-Y.; Fu, S.-J.; Li, X.-M.; Yue, B.-S.; Jiang, J.-P.; Zhu, W. Temperature and Diet Acclimation Modify the Acute Thermal Performance of the Largest Extant Amphibian. *Animals* 2022, 12, 531. https://doi.org/10.3390/ani12040531

Academic Editor: Enrico Lunghi

Received: 10 January 2022
Accepted: 18 February 2022
Published: 21 February 2022

**Publisher's Note:** MDPI stays neutral with regard to jurisdictional claims in published maps and institutional affiliations.

**Simple Summary:** The Chinese giant salamander (*Andrias davidianus*) is one of the largest extant amphibian species, and it is considered critically endangered by the IUCN Red List. Previous studies have demonstrated that future climate change could strongly affect this species. However, how to conduct the related conservation activities are still unclear. Understanding the thermal physiology of *A. davidianus* is meaningful in guiding its conservation, e.g., habitat selection and preadaptation before population translocation. In this study, the influences of temperature and diet on the metabolic capacity and thermal limits were studied for *A. davidianus* larvae based on laboratory experiments. Our results indicated prominent physiological plasticity in the thermal tolerance of *A. davidianus* in response to temperature and diet changes. This thermal plasticity likely, to some extent, buffers the effects of climate change on the Chinese giant salamander. In addition, the potential mechanisms underlying this plasticity were discussed. Our results provide insights for the formulation of conservation strategies for this species.

**Abstract:** The Chinese giant salamander (*Andrias davidianus*), one of the largest extant amphibian species, has dramatically declined in the wild. As an ectotherm, it may be further threatened by climate change. Therefore, understanding the thermal physiology of this species should be the priority to formulate related conservation strategies. In this study, the plasticity in metabolic rate and thermal tolerance limits of *A. davidianus* larvae were studied. Specifically, the larvae were acclimated to three temperature levels (7 °C, cold stress; 15 °C, optimum; and 25 °C, heat stress) and two diet items (red worm or fish fray) for 20 days. Our results indicated that cold-acclimated larvae showed increased metabolic capacity, while warm-acclimated larvae showed a decrease in metabolic capacity. These results suggested the existence of thermal compensation. Moreover, the thermal tolerance windows of cold-acclimated and warm-acclimated larvae shifted to cooler and hotter ranges, respectively. Metabolic capacity is not affected by diet but fish-fed larvae showed superiority in both cold and heat tolerance, potentially due to the input of greater nutrient loads. Overall, our results suggested a plastic thermal tolerance of *A. davidianus* in response to temperature and diet variations. These results are meaningful in guiding the conservation of this species.

**Keywords:** *Andrias davidianus*; animal conservation; metabolic compensation; physiological plasticity; respiration rate; thermal limits

## 1. Introduction

Temperature is one of the most important climatic factors for ectotherms. Variations of environmental temperatures can not only impact the fitness of ectotherms directly by reducing the behavior and physiological performance [1], but they also disrupt the homeostasis of ecosystems and cause the spread of pathogens or shrink of habitat [2,3]. The critical thermal limit ($T_c$, including upper $T_c$ and lower $T_c$) is the most widely used index representing the thermal tolerance of animals [4,5]. It is defined as the thermal point at which locomotory activity becomes disorganized and the animal loses its ability to escape from conditions in gradually heating or cooling thermal conditions [6]. Many studies suggested that the $T_c$ are linked to species geographical distributions by matching with the local climate [7–10], making it an important index in predicting the population dynamics of animals under climate change [11,12]. However, the biological processes that determine the thermal tolerances of animals are still controversial. One of the best-known hypotheses is the "Oxygen Limitation of Thermal Tolerance" [13], which claims that a mismatch between the demand for oxygen and the capacity of oxygen supply to tissues is the critical mechanism restricting the whole animal's tolerance to thermal extremes [14]. The association between aerobic metabolism and upper $T_c$ has been supported by a large body of evidence from fish [15] and arthropods [16]. Moreover, a higher metabolic rate is always associated with better locomotory performance under cold stress [17,18]. These results suggest that the metabolic architecture of animals can provide mechanistic insight into their thermal tolerance.

The thermal tolerances of many ectotherms are plastic with the change of the environment. These animals can remodel their cellular processes and structure to reduce the variation degree of physiological activities in response to environmental changes [19]. For example, ant foragers in late summer had higher average upper $T_c$ compared to those in March and December [20]. More evidence is provided from laboratory studies [21,22]. This physiological plasticity is called acclimation or acclimatization capacity. From a metabolic perspective, thermal acclimation induces metabolic compensation in individuals to offset the thermodynamic variations in metabolic reactions [23]. Plasticity enabled by thermal acclimation is expected to broaden the range of temperatures at which animals are active [24] and is often highlighted as a powerful mechanism to buffer the impact of climate change [25,26]. In addition to environmental temperature, variations in other factors (e.g., oxygen level, water level, and nutrient conditions) can also affect the thermal tolerance (called the cross-talk effect) [27,28]. Therefore, it is necessary to consider the plasticity in thermal tolerance at multiple conditions to better understand species' thermal physiology.

Global decline in the amphibian populations is an urgent environmental and ecological problem [29]. The Chinese giant salamander (*Andrias davidianus*) is one of the largest extant amphibian species, and it is often referred to as a living fossil [30]. This species was once widely distributed in central and southern China [31,32]. In the past 60 years, the wild populations of *A. davidianus* have declined dramatically due to habitat degradation, pollution, and overexploitation [33,34]. Recently, this species was evaluated as critically endangered by the International Union for Conservation of Nature Red list [33] and Chinese specialists [35]. Its evolutionary and ecological significance makes it a flagship species for biodiversity conservation in China. In particular, the effect of climate change (e.g., rise of temperature) has been considered a severe challenge to its survival in the wild [34,36]. In addition to the climatic factors, empirical evidence from farmers and our laboratory indicate that the diet types (e.g., fish, red worm, and pork liver in the farming industry) and temperature can jointly affect the growth rate, a primary fitness index, of the *A. davidianus* larvae. It is interesting to know whether the diet type can shape the thermal physiology of these animals. This knowledge can be implicative in their conservation in the context of climate change.

In this study, the influences of acclimation temperature and diet on the thermal performance (i.e., metabolic capacity and thermal limits) were studied for the Chinese giant salamander larvae. Our main target was to study the plasticity in the thermal performance of this species. This knowledge may extend our understanding of the physiology of this important species and provide useful information for the conservation of the wild *A. davidianus* population under future climate change.

## 2. Materials and Methods

### 2.1. Animals and Acclimation

Larvae of *A. davidianus* from the same clutch were purchased. They were cultured under the same environmental conditions in a farm located at Hongya, Sichuan Province, China (103°10′05″ E, 29°52′36″ N). Specifically, the average water temperature was 15 ($\pm$1.1 SD) °C, and their main food throughout the first year after hatching was red worm. Once collected from the farm, the larvae were cultured in artificial rearing tanks (length $\times$ width $\times$ height = 29 cm $\times$ 20 cm $\times$ 9.7 cm; with 3000 mL water) in laboratory conditions (light: dark = 12: 12; 15 $\pm$ 0.5 °C; dissolved oxygen level >90%) for two weeks before treatments. The larvae were fed with sufficient red worm twice a day at 09:00 a.m. and 18:00 p.m., respectively. Water was replaced daily. Animal procedures were approved by the Animal Care and Use Committee of the Chengdu Institute of Biology, Chinese Academy of Sciences.

According to laboratory studies, *A. davidianus* larvae are sensitive to temperature variations [37,38]. The optimal growth occurs in water with a temperature range of 15–21 °C [39]. Their feeding behavior was inhibited by water temperature higher than 25 °C or lower than 8 °C [40,41]. Therefore, three discrete temperature levels (7 °C/cold stress, 15 °C/optimum, and 25 °C/heat stress) were selected for thermal acclimation. The red worm (*Limnodrilus* sp.) was an empirical diet for captive giant salamanders at the first year after hatching, while wild individuals likely have more diverse prey, including fish fray, frogs, and arthropods. In this study, the red-worm and fish fray were selected to feed the experimental larvae.

Two independent acclimation programs were conducted successively to measure the respiration rate and thermal limits, respectively. For the measurement of respiration rate, 210 larvae were collected on their 60th day after hatching. After two weeks of laboratory acclimation, these larvae (1.65 $\pm$ 0.3 g, mean $\pm$ SD) were randomly divided into six groups: worm diet at 7 °C, worm diet at 15 °C, worm diet at 25 °C, fish diet at 7 °C, fish diet at 15 °C, and fish diet at 25 °C (Figure 1A). Each group included two tanks, and each tank kept 15–20 individuals. The nutrient composition of red worm and fish fray is presented in Table 1. The whole acclimation duration lasted for 20 days. The daily culture followed the conditions described above. The body weight of the larvae was measured at the nearest 0.01 g at the end of treatment, as well as before the measurement of respirate rate. To measure thermal limits, 150 larvae were collected from the farm on their 120th day after hatching. After two weeks of laboratory acclimation, these larvae (3.26 $\pm$ 0.5 g, mean $\pm$ SD) were randomly divided into five groups (30 individuals for each group): worm diet at 7 °C, worm diet at 15 °C, worm diet at 25 °C, worm diet at room temperature (15–25 °C), and fish diet at room temperature (15–25 °C) (Figure 2A). The whole acclimation duration lasted 30 days. For worm diet at 7 °C, worm diet at 15 °C, worm diet at room temperature (15–25 °C), and fish diet at room temperature (15–25 °C) acclimation groups, 15 individuals were tested for either upper or lower thermal limits. However, only 10 individuals were tested at the lower limit for worm diet in the 25 °C-acclimated group as 5 individuals died during the acclimation.

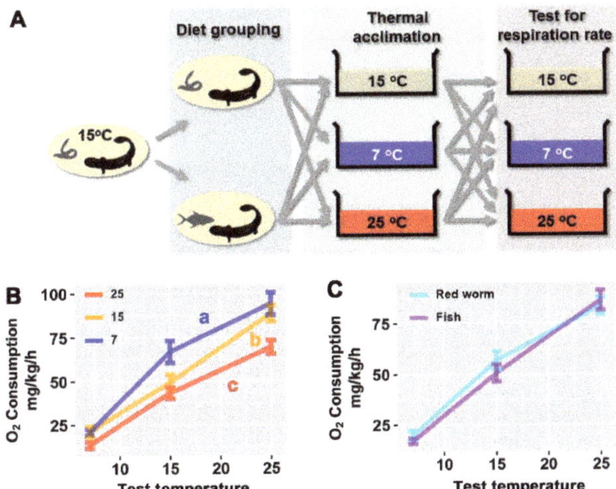



**Figure 1.** Plasticity in metabolic capacity after thermal and diet acclimation. (**A**) Experimental design. (**B**) Variations between thermal-acclimated groups. (**C**) Variations between diet-acclimated groups. The data were analyzed by ANCOVA (test temperature as covariant) and LSD post hoc test; the results are detailed in Table 2. Different letters denote significant differences between groups.

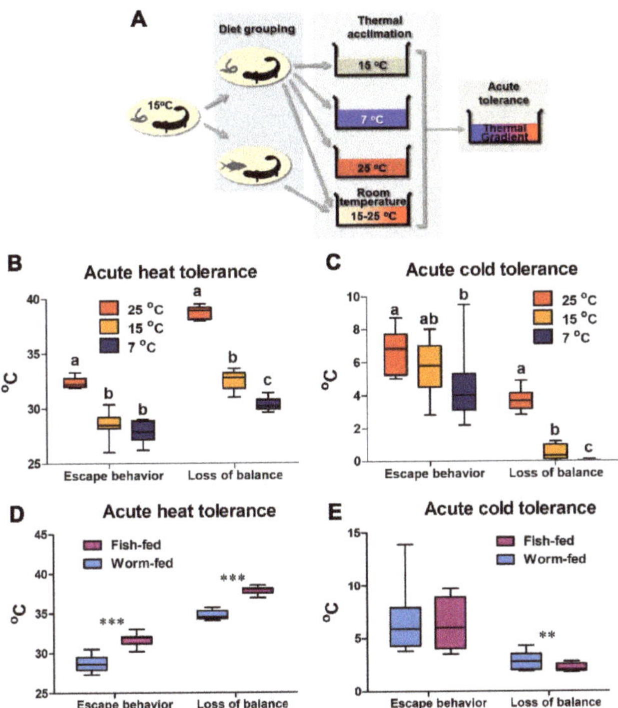



**Figure 2.** Plasticity in thermal limits after thermal and diet acclimation. (**A**) Experimental design. (**B,C**) Variations between thermal-acclimated groups. Different letters denote significant differences ($p < 0.05$) between groups; Kruskal–Wallis test. (**D,E**) Variations between diet-acclimated groups. The differences between worm and fish diets were analyzed by Mann–Whitney U test; **, $p < 0.01$; ***, $p < 0.001$.

**Table 1.** Nutrient compositions of worm and fish diet. Values are presented as mean $\pm$ SE.

| Nutrients | Worm Diet (g/100 g) | Fish Diet (g/100 g) |
|---|---|---|
| Water | $90.01 \pm 0.3$ | $69.61 \pm 0.6$ |
| Lipids | $1.21 \pm 0.1$ | $8.66 \pm 0.8$ |
| Proteins | $5.11 \pm 0.2$ | $17.70 \pm 0.3$ |
| Carbohydrates | $1.65 \pm 0.2$ | $0.39 \pm 0.0$ |

**Table 2.** Influences of temperature and diet acclimation on metabolic rate. The differences between groups were analyzed with ANCOVA, with test temperature as covariant.

| Factors | Type III Sum of Square | df | Mean Square | F Value | Sig. |
|---|---|---|---|---|---|
| Acclimation temperature (AT) | 1023.872 | 2 | 511.936 | 10.987 | 0.002 |
| Test temperature (TT) | 12,955.612 | 1 | 12,955.612 | 278.041 | <0.001 |
| Diet | 21.827 | 1 | 21.827 | 0.468 | 0.508 |
| AT $\times$ Diet | 199.824 | 2 | 99.912 | 2.144 | 0.164 |

*2.2. Respiration Rate*

Oxygen consumption of *A. davidianus* individuals was measured by intermittent respirometer. For each round of tests, the respiration rate of one larva was measured. All individuals were fasted for 24 h before measurement, and the animal could swim freely in the chamber. For any acclimation groups, 11–18 individuals were randomly selected and measured at each test temperature (i.e., 7, 15, or 25 °C). For each individual, the respiration rates at 7, 15, and 25 °C were not measured continuously. Instead, these measurements were separated by recovery intervals (24 h) at the acclimation temperatures to make sure the larvae returned to their initial status. During the recovery intervals, the larvae were placed back in their respective tanks. It should be noted that the larvae from the same tank (more than one individual in each tank) could not be distinguished from each other, as there was no suitable method to label them. This meant that the same larva might be selected to measure the respiration rate at two or three different test temperatures. Thus, to avoid pseudo-replication in the statistical models, the average value of the respiration measurements, which shared the same test temperature, acclimation temperature, and diet type (n = 11–18 individuals or measurements for each condition), was treated as the only replication for each condition. See the raw data in Supplementary files.

The measurement began when the larvae became quiet. During measurement, their movements, if any, were transient and limited in small ranges as the chamber is small, and most of the time, these larvae were stationary. The respirometer consisted of one chamber (0.15 L) and one dissolved oxygen analyzer (HQ30d, HACH company, New York, USA). In addition, the intermittent respirometer was equipped with an internal circulation pump to fully mix the dissolved oxygen in the internal water. One chamber in each group without *A. davidianus* individual was used as a blank control chamber to calculate the background oxygen consumption. Then, oxygen consumption rate ($VO_2$) was measured one time at 1 min intervals for 30 min. The following formula was used to calculate $VO_2$ ($mg\ O_2 kg^{-1}h^{-1}$) [42]:

$$VO_2 = \Delta O_2 \times v/m$$

where $\Delta O_2$ is the difference in the oxygen concentration level ($mg\ O_2\ L^{-1}$) between the experimental and blank control chamber, v is the water flow rate in the chamber ($Lh^{-1}$), and m is the body weight of the *A. davidianus* individual (kg).

### 2.3. Measurement of Thermal Limits

The thermal limits of the larvae were measured after acclimation. The larvae were placed in a digital water bath, and a square space (height $\times$ width $\times$ depth = 23 $\times$ 11 $\times$ 18 cm) was delimited by a plastic net to avoid the direct touch between animals and the bath. The water temperature was initially set at 20 °C, and it increased or decreased by ~1 °C/min. The larvae exhibited free swimming at the beginning and then exhibited behavioral agitation (struggled to escape the confinement). The water temperature for the onset of escaping behavior was recorded, and it was defined as the escaping temperature ($T_{esp}$). Once the larvae exhibited escaping behavior, they were turned over by a glass hook every 1 min. The water temperature was considered to have reached the critical temperature ($T_c$) once the larvae exhibited a loss of righting response (five seconds). The loss of righting response was characterized by the animal's inability to correct their orientation after being flipped upside down using a glass rod, while submerged in water [43]. Note that the absolute Tc values might be overestimated, as there was likely a lag in the variation of the body temperature following the variation of the water temperature. Each individual was tested for either lower thermal limits or upper thermal limits.

### 2.4. Statistical Analyses

Statistical analyses were conducted on SPSS v25.0 (SPSS Inc., Chicago, IL, USA). The differences in growth rate between groups were analyzed by ANOVA, with acclimation temperature and diet as two independent factors. The variations in respiration rates were analyzed by ANCOVA and LSD post hoc test, with test temperature as a covariant. The interactive effects between the independent factors were considered in the ANOVA and ANCOVA models. The differences in upper or lower $T_c$ and $T_{esp}$ were analyzed by Kruskal–Wallis or Mann–Whitney U tests. Graphs were generated by Graphpad Prism 5 or ggplot2, an R package [44].

## 3. Results

The interaction between environmental temperature and diet affects the larvae growth rate ($F_{2,87}$ = 20.356, $p < 0.01$, two-way ANOVA; Table S1). In worm-fed groups, 7 °C larvae had decreased somatic growth compared to their 15 °C counterparts. However, no significant acceleration in growth was detected between larvae from the 7 °C group and those from the 25 °C group (simple effect analysis; Figure S1). In fish-fed groups, the growth rate tended to increase with the rise of environmental temperature, but the inter-group variations were not significant (simple effect analysis; Figure S1).

### 3.1. Influences of Temperature and Diet Acclimation on Larvae Metabolism

Acclimation temperature affect the larvae's metabolic rate (presented as oxygen consumption rate) independently ($F_{2,11}$ = 10.987, $p$ = 0.002; Table 2). Cold acclimation enhanced the metabolic capacity of larvae, while warm acclimation reduced their metabolic capacity ($p < 0.05$, LSD post hoc test; Figure 1B). The larvae metabolism was not significantly affected by diet ($F_{1,11}$ = 0.468, $p$ = 0.508; Figure 1C).

### 3.2. Influences of Temperature and Diet Acclimation on the Acute Thermal Tolerance Window

The influences of acclimation temperature and diet on acute thermal tolerance were studied (Figure 2A). Temperature acclimation caused a shift of the acute thermal tolerance windows of giant salamanders (Figure 2B,C). The upper $T_{esp}$ and $T_c$ of warm-acclimated larvae increased from 28.57 $\pm$ 1.2 °C and 32.52 $\pm$ 0.8 °C to 32.39 $\pm$ 0.5 °C and 38.72 $\pm$ 0.6 °C, respectively (mean $\pm$ SD, and similarly hereinafter). This was accompanied by an increment in their lower $T_{esp}$ and $T_c$ from 5.51 $\pm$ 1.5 °C and 0.48 $\pm$ 0.4 °C to 6.63 $\pm$ 1.3 °C and 3.72 $\pm$ 0.7 °C, respectively. Cold-acclimation caused opposite changes, with their lower $T_{esp}$ and $T_c$ decreasing to 4.64 $\pm$ 2.1 °C and 0 $\pm$ 0.02 °C, respectively, which was accompanied by a decrease in the upper $T_{esp}$ and $T_c$. Fish-fed larvae exhibited a wider thermal tolerance window than worm-fed individuals, characterized by increased upper $T_{esp}$ and $T_c$ and decreased lower $T_c$ (Figure 2D,E).

## 4. Discussion

### 4.1. Temperature-Induced Shift of Thermal Limits

Our results indicated that acclimating *A. davidianus* larvae to higher and lower environmental temperatures decreased and increased their metabolic capacity, respectively, suggesting metabolic compensation and adaptive plasticity in response to thermal fluctuation. Thermal compensation enables the maintenance of physiological rates across environmental conditions. This was frequently observed in animals facing mildly cold environments [23,45,46]. This biological phenomenon is believed to partly offset the reduced physiological performance under cold and thus allows better exploitation of the environment by maintaining physiological activities such as locomotion, feeding, and development at low temperatures [47]. For animals in warming conditions, their increased metabolic activity due to thermodynamic effects resulted in accelerated resource consumption. Downregulation of metabolism can benefit their long-term survival from the perspective of resource-saving [48–50]. Given that amphibians' larval stage is devoted to somatic growth and energy storage [51], either an increased metabolic rate at a lower temperature or decreased metabolic rate at a higher temperature are aligned with their life strategy.

This study measured the thermal limits of *A. davidianus* larvae in the early developmental stage. It demonstrated their plasticity in thermal tolerance. The ability to acclimatize to changing thermal conditions is expected to be a primary factor that dictates the vulnerability of taxa to climate change. Generally speaking, the plasticity in thermal tolerance means better outcomes for *A. davidianus* larvae in response to global warming or extreme weather than expected. Currently, the existence and distribution of wild *A. davidianus* in the future has become a topic that is in great need of research and discernment. The ecological niche model raised by Zhao, et al. [34] suggested that climate change can potentially affect the distribution of these animals. To have a more reliable and refined prediction of the fate of the wild populations, physiological models considering their thermal limits and plasticity should be considered. Our results may provide a foundation for these studies. However, two factors should be still be considered, the variation of thermal tolerance with life stages [4] and the potential differences in thermal traits between different phylogenic clades of *A. davidianus* [52]. Further studies should focus on these questions, particularly in clarifying which life stage or phylogenic clades exhibit the lowest thermal tolerance and plasticity. This knowledge could be meaningful in guiding the conservation of wild *A. davidianus*, e.g., making more reliable predictions of the individual survival from different geographical populations and optimizing the trophic structure of their habitats to enhance their tolerance to extreme weather conditions.

Making clear the mechanisms underlying the thermal tolerance may guide the conservation of Chinese giant salamander in the context of climate change. Despite the complexity in the mechanisms of acute thermal intolerance, metabolic capacity is always an important contributor to the thermal tolerance of aquatic ectotherms [14,53]. This is particularly true for cold conditions, where the ectotherms have difficulty maintaining their metabolic activities. For example, the *atu* mutant *Drosophila melanogaster* has increased metabolic capacity, which improves its cold tolerance [54]. Accordingly, variations in metabolic capacity explained the improved and weakened cold tolerance in cold- and warm-acclimated individuals, respectively. For heat tolerance, as the activities of most biological reactions increase with ambient temperatures within certain thermal scopes, the metabolic capacity may no longer be a major limiting factor for heat tolerance. Instead, the overactive metabolic capacity can limit heat tolerance by overburdening the oxygen supply [13,14]. Additionally, it was reported that downregulation of metabolism could improve the acute heat tolerance of animals by reducing their requirement for oxygen [55]. Accordingly, decreased metabolic capacity after heat acclimation is likely associated with higher upper limits and vice versa. Taken together, the metabolic rate may be an important factor for predicting the tolerance of *A. davidianus* to extreme temperature. In conservation practice, to improve the survival rate of reintroduced populations, we might screen individuals whose thermal tolerance matches the climatic characteristics of the translocation habitats, e.g., introducing cold-tolerant individuals to regions with severe winters or introducing warm-tolerant individuals to hot environments. As direct measurement of the thermal tolerance is detrimental or even lethal to animals, the metabolic rate can be an applicable indicator for their thermal-tolerance properties.

*4.2. Diet-Induced Shift of Thermal Limits*

Unlike the temperature acclimation, which induced a unidirectional shift of thermal limits, diet acclimation broadened the thermal-tolerance window of *A. davidianus* larvae in both directions. Diet did not change the metabolic capacity of giant salamander. It implied different mechanisms between diet and temperature acclimations in affecting thermal tolerance. Nutrition was reported to modify the critical thermal limits of ectotherms [56,57]. Despite the effects of nutrition depending on the animal size or thermal conditions [58], most studies supported that nutrient supplements (e.g., carbohydrates and amino acids) can improve thermal tolerance in animals [56,59–64]. This is reasonable, as rich nutrient storage can improve cellular metabolic maintenance and benefit the synthesis of protectants, which are necessary for survival in stressful conditions. In our study, the most prominent difference between fish and red-worm diets was that the former was richer in total energy, protein, and lipid levels (Table 1). This might be a reason for the superior of fish-fed larvae in thermal tolerance. This finding suggests that the productivity or prey abundance of the natural habitats could be a potential determinator for the thermal tolerance of wild *A. davidianus*. Introducing suitable prey with great nutrient loads to the natural or artificial habitats of *A. davidianus* may be an alternative approach to reduce the impact of climate change and extreme weather on these animals.

Some mechanisms by which the nutrients modify the thermal physiology of animals were revealed. For example, it was reported that the storage of carbohydrates, the anaerobic substrate, is associated with the tolerance of animals to extreme heat stress [16,65] when the oxygen supply cannot match the metabolic requirement [14]. Although the fish diet contains less sugar than the worm diet, the protein level in the former is much higher. Amino acids derived from proteolysis can be easily converted into carbohydrates. For cold tolerance, the abundance of lipids should be important, as these compounds can be major metabolic substrates in cold conditions [47,66]. More importantly, lipids are required for cellular membrane remodeling, which is an important mechanism underlying cold tolerance [67]. Accordingly, the rich lipid in the fish diet might contribute to the cold tolerance of fish-fed larvae. Overall, these findings shed light on the new thoughts referring to the conservation of this species under climate change. Currently, the reintroduction of captive-bred individuals to the historical natural habitats has been an important measure for the recovery of wild giant salamander populations [68]. Since the diet is a determinant of thermal performance, the prey abundance and diversity in the habitat should be considered before reintroduction. Moreover, to improve the survival rate of the released giant salamander, a preadaptation procedure should be conducted before reintroduction.

## 5. Conclusions

In this study, we demonstrated the plasticity in thermal physiology of *A. davidianus* larvae in response to environmental temperature and diet changes. These larvae exhibited apparent thermal compensation in metabolic capacity after thermal acclimation. Specifically, cold- or heat-acclimation improved their tolerance to more extreme thermal stress but compromised their tolerance to the opposite thermal extremes. Diet did not affect the metabolic rate; however, the fish diet, which was richer in protein, lipid, and total energy, broadened the thermal-tolerance window of *A. davidianus* larvae. This knowledge provides some implications in the conservation of this endangered animal.

**Supplementary Materials:** The following supporting information can be downloaded at: https://www.mdpi.com/article/10.3390/ani12040531/s1, Table S1: Influences of temperature and diet acclimation on growth rate, Figure S1: Body weight at the end of the acclimation program.

**Author Contributions:** Conceptualization, T.Z., W.Z., J.-P.J. and B.-S.Y.; methodology, C.-L.Z., T.Z., S.-J.F. and X.-M.L.; formal analysis, W.Z. and C.-L.Z.; investigation, C.-L.Z., J.-Y.F., L.-M.C. and P.-Y.Z.; resources, C.-L.Z. and J.-Y.F.; writing—original draft preparation, W.Z.; writing—review and editing, W.Z. and C.-L.Z.; visualization, W.Z. and C.-L.Z.; supervision, T.Z., W.Z., J.-P.J. and B.-S.Y.; project administration, T.Z., W.Z. and J.-P.J.; funding acquisition, T.Z. and J.-P.J. All authors have read and agreed to the published version of the manuscript.

**Funding:** This work was supported by the National Key Research and Development Program of China (2016YFC0503200), the Biodiversity Survey and Assessment Project of the Ministry of Ecology and Environment, China (2019HJ2096001006), the Construction of Basic Conditions Platform of Sichuan Science and Technology Department (2019JDPT0020), and the China Biodiversity Observation Networks (Sino BON).

**Institutional Review Board Statement:** The animal study protocol was approved by the Institutional Ethics Committee of Animal Ethical and Welfare Committee of the Chengdu Institute of Biology, the Chinese Academy of Sciences (CIB20191105).

**Data Availability Statement:** The datasets presented in this study can be found in online repositories (https://figshare.com/articles/dataset/Respiration_rate_and_thermal_limits_of_acclimated_Andrias_davidianus/19188443 accessed on 17 February 2022).

**Acknowledgments:** The authors thank Zi-Jian Sun and Wen-Bo Fan for their help in sample collection.

**Conflicts of Interest:** The authors declare no conflict of interest. The funders had no role in the design of the study; in the collection, analyses, or interpretation of data; in the writing of the manuscript; or in the decision to publish the results.

## References


1. Galloy, V.; Denoël, M. Detrimental effect of temperature increase on the fitness of an amphibian (*Lissotriton helveticus*). *Acta Oecol.* **2010**, *36*, 179–183. [CrossRef]
2. Rohr, J.R.; Raffel, T.R. Linking global climate and temperature variability to widespread amphibian declines putatively caused by disease. *Proc. Natl. Acad. Sci. USA* **2010**, *107*, 8269–8274. [CrossRef] [PubMed]
3. Alan Pounds, J.; Bustamante, M.R.; Coloma, L.A.; Consuegra, J.A.; Fogden, M.P.L.; Foster, P.N.; La Marca, E.; Masters, K.L.; Merino-Viteri, A.; Puschendorf, R.; et al. Widespread amphibian extinctions from epidemic disease driven by global warming. *Nature* **2006**, *439*, 161–167. [CrossRef]
4. Dahlke, F.T.; Wohlrab, S.; Butzin, M.; Pörtner, H.O. Thermal bottlenecks in the life cycle define climate vulnerability of fish. *Science* **2020**, *369*, 65. [CrossRef] [PubMed]
5. Stillman, J.H. Acclimation capacity underlies susceptibility to climate change. *Science* **2003**, *301*, 65. [CrossRef] [PubMed]
6. Lutterschmidt, W.I.; Hutchison, V.H. The critical thermal maximum: History and critique. *Can. J. Zool.* **1997**, *75*, 1561–1574. [CrossRef]
7. Calosi, P.; Bilton, D.T.; Spicer, J.I.; Votier, S.C.; Atfield, A. What determines a species' geographical range? Thermal biology and latitudinal range size relationships in european diving beetles (coleoptera: Dytiscidae). *J. Anim. Ecol.* **2010**, *79*, 194–204. [CrossRef]
8. Kellermann, V.; Overgaard, J.; Hoffmann, A.A.; Fløjgaard, C.; Svenning, J.C.; Loeschcke, V. Upper thermal limits of *Drosophila* are linked to species distributions and strongly constrained phylogenetically. *Proc. Natl. Acad. Sci. USA* **2012**, *109*, 16228–16233. [CrossRef]
9. Andersen, J.L.; Manenti, T.; Sørensen, J.G.; MacMillan, H.A.; Loeschcke, V.; Overgaard, J. How to assess drosophila cold tolerance: Chill coma temperature and lower lethal temperature are the best predictors of cold distribution limits. *Funct. Ecol.* **2015**, *29*, 55–65. [CrossRef]
10. Eliason, E.J.; Clark, T.D.; Hague, M.J.; Hanson, L.M.; Gallagher, Z.S.; Jeffries, K.M.; Gale, M.K.; Patterson, D.A.; Hinch, S.G.; Farrell, A.P. Differences in thermal tolerance among sockeye salmon populations. *Science* **2011**, *332*, 109–112. [CrossRef]
11. Pinsky, M.L.; Eikeset, A.M.; McCauley, D.J.; Payne, J.L.; Sunday, J.M. Greater vulnerability to warming of marine versus terrestrial ectotherms. *Nature* **2019**, *569*, 108–111. [CrossRef] [PubMed]
12. Sunday, J.; Bennett, J.M.; Calosi, P.; Clusella-Trullas, S.; Gravel, S.; Hargreaves, A.L.; Leiva, F.P.; Verberk, W.C.E.P.; Olalla-Tárraga, M.Á.; Morales-Castilla, I. Thermal tolerance patterns across latitude and elevation. *Philos. Trans. R. Soc. London. Ser. B Biol. Sci.* **2019**, *374*, 20190036. [CrossRef] [PubMed]
13. Pörtner, H.O.; Knust, R. Climate change affects marine fishes through the oxygen limitation of thermal tolerance. *Science* **2007**, *315*, 95–97. [CrossRef] [PubMed]

14. Pörtner, H.O. Climate variations and the physiological basis of temperature dependent biogeography: Systemic to molecular hierarchy of thermal tolerance in animals. *Comp. Biochem. Physiol. A* **2002**, *132*, 739–761. [CrossRef]
15. Martin, B.T.; Dudley, P.N.; Kashef, N.S.; Stafford, D.M.; Reeder, W.J.; Tonina, D.; Del Rio, A.M.; Scott Foott, J.; Danner, E.M. The biophysical basis of thermal tolerance in fish eggs. *Proc. Biol. Sci.* **2020**, *287*, 20201550. [CrossRef]
16. Zhu, W.; Meng, Q.; Zhang, H.; Wang, M.L.; Li, X.; Wang, H.T.; Zhou, G.L.; Miao, L.; Qin, Q.L.; Zhang, J.H. Metabolomics reveals the key role of oxygen metabolism in heat susceptibility of an alpine-dwelling ghost moth, *Thitarodes xiaojinensis* (lepidoptera: Hepialidae). *Insect Sci.* **2019**, *26*, 695–710. [CrossRef]
17. Javal, M.; Roques, A.; Roux, G.; Laparie, M. Respiration-based monitoring of metabolic rate following cold-exposure in two invasive *Anoplophora* species depending on acclimation regime. *Comp. Biochem. Physiol. Part A Mol. Integr. Physiol.* **2018**, *216*, 20–27. [CrossRef]
18. Coughlin, D.J.; Shiels, L.P.; Nuthakki, S.; Shuman, J.L. Thermal acclimation to cold alters myosin content and contractile properties of rainbow smelt, *Osmerus mordax*, red muscle. *Comp. Biochem. Physiol. Part A Mol. Integr. Physiol.* **2016**, *196*, 46–53. [CrossRef]
19. Hoffmann, A.A.; Chown, S.L.; Clusella-Trullas, S.; Fox, C. Upper thermal limits in terrestrial ectotherms: How constrained are they? *Funct. Ecol.* **2013**, *27*, 934–949. [CrossRef]
20. Bujan, J.; Roeder, K.A.; Yanoviak, S.P.; Kaspari, M. Seasonal plasticity of thermal tolerance in ants. *Ecology* **2020**, *101*, e03051. [CrossRef]
21. Yu, X.; Yu, K.; Huang, W.; Liang, J.; Qin, Z.; Chen, B.; Yao, Q.; Liao, Z. Thermal acclimation increases heat tolerance of the scleractinian coral *Acropora pruinosa*. *Sci. Total Environ.* **2020**, *733*, 139319. [CrossRef] [PubMed]
22. Enriquez, T.; Colinet, H. Cold acclimation triggers major transcriptional changes in *Drosophila suzukii*. *BMC Genom.* **2019**, *20*, 413. [CrossRef] [PubMed]
23. Guderley, H. Functional significance of metabolic responses to thermal acclimation in fish muscle. *Am. J. Physiol. Regul. Integr. Comp. Physiol.* **1990**, *259*, R245–R252. [CrossRef] [PubMed]
24. Seebacher, F.; White, C.R.; Franklin, C.E. Physiological plasticity increases resilience of ectothermic animals to climate change. *Nat. Clim. Change* **2015**, *5*, 61–66. [CrossRef]
25. Rohr, J.R.; Civitello, D.J.; Cohen, J.M.; Roznik, E.A.; Sinervo, B.; Dell, A.I. The complex drivers of thermal acclimation and breadth in ectotherms. *Ecol. Lett.* **2018**, *21*, 1425–1439. [CrossRef]
26. Chevin, L.M.; Lande, R.; Mace, G.M.; Kingsolver, J.G. Adaptation, plasticity, and extinction in a changing environment: Towards a predictive theory. *PLoS Biol.* **2010**, *8*, e1000357. [CrossRef]
27. Verberk, W.C.E.P.; Overgaard, J.; Ern, R.; Bayley, M.; Wang, T.; Boardman, L.; Terblanche, J.S. Does oxygen limit thermal tolerance in arthropods? A critical review of current evidence. *Comp. Biochem. Phys. A* **2016**, *192*, 64–78. [CrossRef]
28. Mitchell, K.A.; Boardman, L.; Clusella-Trullas, S.; Terblanche, J.S. Effects of nutrient and water restriction on thermal tolerance: A test of mechanisms and hypotheses. *Compa. Biochem. Phys. A* **2017**, *212*, 15–23. [CrossRef]
29. Stuart, S.N.; Chanson, J.S.; Cox, N.A.; Young, B.E.; Rodrigues, A.S.; Fischman, D.L.; Waller, R.W. Status and trends of amphibian declines and extinctions worldwide. *Science* **2004**, *306*, 1783–1786. [CrossRef]
30. Zhang, L.; Jiang, W.; Wang, Q.J.; Zhao, H.; Zhang, H.X.; Marcec, R.M.; Willard, S.T.; Kouba, A.J. Reintroduction and post-release survival of a living fossil: The chinese giant salamander. *PLoS ONE* **2016**, *11*, e0156715. [CrossRef]
31. Todd, W.P.; Fang, Y.; WANG, Y.; Theodore, P. A survey for the chinese giant salamander (*Andrias davidianus*; blanchard, 1871) in the qinghai province. *Amphib. Reptile Conse.* **2014**, *8*, 1–6.
32. Wang, J.; Zhang, H.; Xie, F.; Wei, G.; Jiang, J.P. Genetic bottlenecks of the wild chinese giant salamander in karst caves. *Asian Herpetol. Res.* **2017**, *8*, 174–183.
33. Liang, G.; Geng, B.R.; Zhao, E.M. Andrias davidianus. In *The IUCN Red List of Threatened Species*; IUCN: Gland, Switzerland, 2004; e.T1272A3375181.
34. Zhao, T.; Zhang, W.Y.; Zhou, J.; Zhao, C.L.; Liu, X.K.; Liu, Z.D.; Shu, G.S.; Wang, S.S.; Li, C.; Xie, F.; et al. Niche divergence of evolutionarily significant units with implications for repopulation programs of the world's largest amphibians. *Sci. Total Environ.* **2020**, *738*, 140269. [CrossRef] [PubMed]
35. Jiang, J.P.; Xie, F.; Zang, C.X.; Cai, L.; Li, C.; Wang, B.; Li, J.T.; Wang, J.; Hu, J.H.; Wang, Y.; et al. Assessing the threat status of amphibians in china. *Biodivers. Sci.* **2016**, *24*, 588–597. [CrossRef]
36. Zhang, Z.; Mammola, S.; Liang, Z.; Capinha, C.; Wei, Q.; Wu, Y.; Zhou, J.; Wang, C. Future climate change will severely reduce habitat suitability of the critically endangered chinese giant salamander. *Freshwater Biol.* **2020**, *65*, 971–980. [CrossRef]
37. Hu, Q.; Tian, H.; Xiao, H. Effects of temperature and sex steroids on sex ratio, growth, and growth-related gene expression in the chinese giant salamander andrias davidianus. *Aquat. Biol.* **2019**, *28*, 79–90. [CrossRef]
38. Zhang, L.; Kouba, A.; Wang, Q.J.; Zhao, H.; Jiang, W.; Willard, S.; Zhang, H.X. The effect of water temperature on the growth of captive chinese giant salamanders (*Andrias davidianus*) reared for reintroduction: A comparison with wild salamander body condition. *Herpetologica* **2014**, *70*, 369–377. [CrossRef]
39. Chen, X.; Shen, J. The effect of water temperature on intake of *Andrias davidianus* (in chinese). *Fish. Sci.* **1999**, *1*, 20–22.
40. Mu, H.m.; Li, Y.; Yao, J.j.; Ma, S. A review: Current research on biology of chinese giant salamander. *Fisheries Sci.* **2011**, *30*, 513–516.
41. Wang, H.W. The artificial culture of chinese giant salamander. *J. Aquacult.* **2004**, *25*, 41–44.

42. Fu, S.J.; Xie, X.J.; Cao, Z.D. Effect of dietary composition on specific dynamic action in southern catfish *Silurus meridionalis* chen. *Aquac. Res.* **2005**, *36*, 1384–1390. [CrossRef]

43. Layne, J.R.; Claussen, D.L. Seasonal variation in the thermal acclimation of critical thermal maxima (ctmax) and minima (ctmin) in the salamander eurycea bislineata. *J. Therm. Biol.* **1982**, *7*, 29–33. [CrossRef]

44. Wickham, H. *Ggplot2: Elegant Graphics for Data Analysis*; Springer Publishing Company, Incorporated: New York, NY, USA, 2009.

45. Guderley, H. Metabolic responses to low temperature in fish muscle. *Biol. Rev.* **2004**, *79*, 409–427. [CrossRef] [PubMed]

46. Abe, T.; Kitagawa, T.; Makiguchi, Y.; Sato, K. Chum salmon migrating upriver adjust to environmental temperatures through metabolic compensation. *J. Exp. Biol.* **2019**, *222*, jeb.186189.

47. Zhu, W.; Zhang, H.; Li, X.; Meng, Q.; Shu, R.; Wang, M.; Zhou, G.; Wang, H.; Miao, L.; Zhang, J.; et al. Cold adaptation mechanisms in the ghost moth *Hepialus xiaojinensis*: Metabolic regulation and thermal compensation. *J. Insect Physiol.* **2016**, *85*, 76–85. [CrossRef]

48. Storey, K.B.; Storey, J.M. Metabolic regulation and gene expression during aestivation. In *Aestivation: Molecular and Physiological Aspects*; Arturo Navas, C., Carvalho, J.E., Eds.; Springer: Berlin/Heidelberg, Germany, 2010; pp. 25–45.

49. Staples, J.F. Metabolic flexibility: Hibernation, torpor, and estivation. *Compr. Physiol.* **2016**, *6*, 737–771.

50. Sandblom, E.; Grans, A.; Axelsson, M.; Seth, H. Temperature acclimation rate of aerobic scope and feeding metabolism in fishes: Implications in a thermally extreme future. *Proceedings. Biol. Sci.* **2014**, *281*, 20141490. [CrossRef]

51. Chang, L.; Wang, B.; Zhang, M.; Liu, J.; Zhao, T.; Zhu, W.; Jiang, J. The effects of corticosterone and background colour on tadpole physiological plasticity. *Comp. Biochem. Phys. D* **2021**, *39*, 100872. [CrossRef]

52. Yan, F.; Lü, J.C.; Zhang, B.L.; Yuan, Z.Y.; Zhao, H.P.; Huang, S.; Wei, G.; Mi, X.; Zou, D.H.; Xu, W.; et al. The chinese giant salamander exemplifies the hidden extinction of cryptic species. *Curr. Biol.* **2018**, *28*, R590–R592. [CrossRef]

53. O'Brien, K.M.; Rix, A.S.; Egginton, S.; Farrell, A.P.; Crockett, E.L.; Schlauch, K.; Woolsey, R.; Hoffman, M.; Merriman, S. Cardiac mitochondrial metabolism may contribute to differences in thermal tolerance of red- and white-blooded antarctic notothenioid fishes. *J. Exp. Biol.* **2018**, *221*, jeb177816. [CrossRef]

54. Takeuchi, K.; Nakano, Y.; Kato, U.; Kaneda, M.; Aizu, M.; Awano, W.; Yonemura, S.; Kiyonaka, S.; Mori, Y.; Yamamoto, D.; et al. Changes in temperature preferences and energy homeostasis in dystroglycan mutants. *Science* **2009**, *323*, 1740–1743. [CrossRef] [PubMed]

55. Semsar-kazerouni, M.; Boerrigter, J.G.J.; Verberk, W.C.E.P. Changes in heat stress tolerance in a freshwater amphipod following starvation: The role of oxygen availability, metabolic rate, heat shock proteins and energy reserves. *Comp. Biochem. Phys. A* **2020**, *245*, 110697. [CrossRef] [PubMed]

56. Bujan, J.; Kaspari, M. Nutrition modifies critical thermal maximum of a dominant canopy ant. *J. Insect Physiol.* **2017**, *102*, 1–6. [CrossRef] [PubMed]

57. Gomez Isaza, D.F.; Cramp, R.L.; Smullen, R.; Glencross, B.D.; Franklin, C.E. Coping with climatic extremes: Dietary fat content decreased the thermal resilience of barramundi (*Latescalcarifer*). *Comp. Biochem. Phys. A* **2019**, *230*, 64–70. [CrossRef] [PubMed]

58. Turko, A.J.; Nolan, C.B.; Balshine, S.; Scott, G.R.; Pitcher, T.E. Thermal tolerance depends on season, age and body condition in imperilled redside dace clinostomus elongatus. *Conserv. Physiol.* **2020**, *8*, coaa062. [CrossRef] [PubMed]

59. Wolfe, G.R.; Hendrix, D.L.; Salvucci, M.E. A thermoprotective role for sorbitol in the silverleaf whitefly, *Bemisia argentifolii*. *J. Insect Physiol.* **1998**, *44*, 597–603. [CrossRef]

60. Kumar, N.; Minhas, P.S.; Ambasankar, K.; Krishnani, K.K.; Rana, R.S. Dietary lecithin potentiates thermal tolerance and cellular stress protection of milk fish (*Chanos chanos*) reared under low dose endosulfan-induced stress. *J. Therm. Biol.* **2014**, *46*, 40–46. [CrossRef]

61. Kumar, N.; Ambasankar, K.; Krishnani, K.K.; Kumar, P.; Akhtar, M.S.; Bhushan, S.; Minhas, P.S. Dietary pyridoxine potentiates thermal tolerance, heat shock protein and protect against cellular stress of milkfish (*Chanos chanos*) under endosulfan-induced stress. *Fish Shellfish Immun.* **2016**, *55*, 407–414. [CrossRef]

62. Tejpal, C.S.; Sumitha, E.B.; Pal, A.K.; Shivananda Murthy, H.; Sahu, N.P.; Siddaiah, G.M. Effect of dietary supplementation of l-tryptophan on thermal tolerance and oxygen consumption rate in cirrhinus mrigala fingerlings under varied stocking density. *J. Therm. Biol.* **2014**, *41*, 59–64. [CrossRef]

63. Gupta, S.K.; Pal, A.K.; Sahu, N.P.; Dalvi, R.S.; Akhtar, M.S.; Jha, A.K.; Baruah, K. Dietary microbial levan enhances tolerance of *labeo rohita* (Hamilton) juveniles to thermal stress. *Aquaculture* **2010**, *306*, 398–402. [CrossRef]

64. Koštál, V.; Šimek, P.; Zahradníčková, H.; Cimlová, J.; Štětina, T. Conversion of the chill susceptible fruit fly larva (*Drosophila melanogaster*) to a freeze tolerant organism. *P. Nati. Acad. Sci. USA* **2012**, *109*, 3270–3274. [CrossRef]

65. Verberk, W.C.E.P.; Sommer, U.; Davidson, R.L.; Viant, M.R. Anaerobic metabolism at thermal extremes: A metabolomic test of the oxygen limitation hypothesis in an aquatic insect. *Integr. Comp. Biol.* **2013**, *53*, 609–619. [CrossRef] [PubMed]

66. Lu, D.L.; Ma, Q.; Wang, J.; Li, L.Y.; Han, S.L.; Limbu, S.M.; Li, D.L.; Chen, L.Q.; Zhang, M.L.; Du, Z.Y. Fasting enhances cold resistance in fish through stimulating lipid catabolism and autophagy. *J. Physiol.* **2019**, *597*, 1585–1603. [CrossRef] [PubMed]

67. Teets, N.M.; Gantz, J.D.; Kawarasaki, Y. Rapid cold hardening: Ecological relevance, physiological mechanisms and new perspectives. *J. Exp. Biol.* **2020**, *223*, jeb203448. [CrossRef] [PubMed]

68. Shu, G.C.; Liu, P.; Zhao, T.; Li, C.; Hou, Y.M.; Zhao, C.L.; Wang, J.; Shu, X.X.; Chang, J.; Jiang, J.P.; et al. Disordered translocation is hastening local extinction of the chinese giant salamander. *Asian Herpetol. Res.* **2021**, *12*, 271–279.

*animals*

MDPI

*Article*

# The Reproductive Success of *Triturus ivanbureschi* × *T. macedonicus* F$_1$ Hybrid Females (Amphibia: Salamandridae)

Tijana Vučić [1,2,*], Ana Ivanović [1], Maja Ajduković [2], Nikola Bajler [1,2] and Milena Cvijanović [2,*]

[1]   Faculty of Biology, Institute of Zoology, University of Belgrade, Studentski trg 16, 11000 Belgrade, Serbia; ana@bio.bg.ac.rs (A.I.); nikola.bajler@outlook.com (N.B.)

[2]   Department of Evolutionary Biology, Institute for Biological Research "Siniša Stanković", National Institute of the Republic of Serbia, University of Belgrade, Bulevar Despota Stefana 142, 11000 Belgrade, Serbia; maja.ajdukovic@ibiss.bg.ac.rs

*   Correspondence: tijana.vucic@bio.bg.ac.rs (T.V.); milena.cvijanovic@ibiss.bg.ac.rs (M.C.)

**Simple Summary:** Two moderately related large-bodied newt species endemic to the Balkan Peninsula, the Balkan crested newt (*Triturus ivanbureschi*) and the Macedonian crested newt (*T. macedonicus*), coexist and hybridize in central Serbia. Many generations of mutual hybrid crossings and backcrossings with parental species shaped the genetic composition of hybrid populations. Natural populations have admixed nuclear DNA (nuDNA) of parental species and *T. ivanbureschi* mitochondrial DNA (mtDNA), which is usually maternally inherited. The mechanisms that direct gene flow and shape the first generations of hybrids could explain the formation of hybrid zones and their maintenance in nature. We followed and compared life history traits related to reproduction of the first generation of reciprocal hybrids obtained by experimental crossing. Our results suggested that possible incompatibilities between mitochondrial and nuclear genomes, which could lead to the exclusion of *T. macedonicus* mtDNA in natural populations, most likely act at later stages of development or subsequent hybrid generations. Results from this study add to the growing knowledge of *Triturus* hybrid biology and ecology, which is the baseline for conservation programs necessary to protect these highly endangered amphibians.

**Citation:** Vučić, T.; Ivanović, A.; Ajduković, M.; Bajler, N.; Cvijanović, M. The Reproductive Success of *Triturus ivanbureschi* × *T. macedonicus* F$_1$ Hybrid Females (Amphibia: Salamandridae). *Animals* **2022**, *12*, 443. https://doi.org/10.3390/ani12040443

Academic Editor: Enrico Lunghi

Received: 24 December 2021
Accepted: 10 February 2022
Published: 12 February 2022

**Publisher's Note:** MDPI stays neutral with regard to jurisdictional claims in published maps and institutional affiliations.



**Abstract:** Two large-bodied newt species, *Triturus ivanbureschi* and *T. macedonicus*, hybridize in nature across the Balkan Peninsula. Consequences of hybridization upon secondary contact of two species include species displacement and asymmetrical introgression of *T. ivanbureschi* mtDNA. We set an experimental reciprocal cross of parental species and obtained two genotypes of F$_1$ hybrids (with *T. ivanbureschi* or *T. macedonicus* mtDNA). When hybrids attained sexual maturity, they were engaged in mutual crossings and backcrossing with parental species. We followed reproductive traits over two successive years. Our main aim was to explore the reproductive success of F$_1$ females carrying different parental mtDNA. Additionally, we tested for differences in reproductive success within female genotypes depending on the crossing with various male genotypes (hybrids or parental species). Both female genotypes had similar oviposition periods, number of laid eggs and hatched larvae but different body and egg sizes. Overall reproductive success (percentage of egg-laying females and viability of embryos) was similar for both genotypes. The type of crossing led to some differences in reproductive success within female genotypes. The obtained results suggest that processes that led to exclusion of *T. macedonicus* mtDNA in natural populations may be related to the survival at postembryonic stages of F$_2$ generation or reproductive barriers that emerged in subsequent hybrid generations.

**Keywords:** egg size; hybrid breakdown; life-history traits; newts; mtDNA introgression

## 1. Introduction

The reproduction of genetically divergent taxa is a frequent phenomenon in natural populations (e.g., [1,2]). Hybrids could be sterile or produce further generations by mutual crossings and/or backcrossing with parental genotypes. The novel combination of genotypes in hybrids could be adaptive and beneficial for fitness, leading to hybrid speciation or these combinations could cause fitness loss [2–8]. The aforementioned outcomes of hybridization are largely dependent on the phylogenetic relatedness of parental species and/or the time of divergence from the most common ancestor. More closely related taxa, i.e., phylogenetically recent lineages, are expected to have viable hybrid offspring that could have adaptive advantages. The hybrid breakdown could be expected in the second or even in subsequent generations as a result of an accumulation of incompatibilities that produce reproductive barriers. On the other hand, hybridization between phylogenetically more divergent taxa could result in largely decreased viability or sterility of one or both sexes of $F_1$ hybrids [6,9–17]. One of the hybridization consequences is asymmetrical gene flow resulting in introgression of mtDNA from one species to another, which could disturb mito-nuclear compatibility and potentially lead to substantial loss of hybrid fitness (e.g., [18–20]). There are several models that describe and explain the accumulation of genetic incompatibilities over time. For mito-nuclear incompatibilities, the Dobzhansky–Muller model was proposed as the most probable model of the evolution of incompatibilities (see [18] and references therein).

Phylogenetic relations within large-bodied newts (*Triturus*, Salamandridae) and a lack of complete reproductive barriers among species, with a wide range of hybridization outcomes, provide a substantial base for evolutionary studies of hybridization consequences. This monophyletic genus consists of two major groups, marbled and crested newts, which diverged from each other around 24 mya [21]. The marbled newts are *T. marmoratus* and *T. pygmeus*, while within crested newts, four groups are recognized based on genetic data and ecomorphological characteristics: (1) *T. ivanbureschi, T. anatolicus, T. karelinii*; (2) *T. carnifex, T. macedonicus*; (3) *T. cristatus*; and (4) *T. dobrogicus* [22]. The species have mostly parapatric distribution throughout Europe and parts of adjacent Asia. In the zones of species contact, interspecific hybridization occurs [23], resulting in dynamic hybrid zones in space and time. The movement of a hybrid zone implies advantages of one species at the expense of the other, which eventually leads to species spatial displacement. Most often, this scenario includes asymmetrical introgression of mtDNA from outcompeted to colonizing species [24–31]. Hybridization between *Triturus* species was also confirmed between autochthonous species and anthropogenically introduced species in areas well outside their natural range [32–34].

Natural hybrid populations could consist mostly of $F_1$ hybrid generation, as in the case of two genetically well-separated species with different ecology and morphology, *T. cristatus* and *T. marmoratus*, which hybridize in western France (see [35] and references therein). These hybrids express typical heterosis; they are larger than the parental species and mostly sterile [36]. The opposite example is that of *T. ivanbureschi* × *T. macedonicus* hybrid populations on the Balkan Peninsula [27,37], which consist of an unknown generation of hybrid individuals derived from a long history of mutual hybrid crossings and backcrossing with both parental species [27,38,39]. *Triturus ivanbureschi* × *T. macedonicus* hybrids have intermediate body lengths related to parental species, and they are morphologically largely similar to both parental species [39]. In central Serbia, all populations of *T. macedonicus*, as well as *T. ivanbureschi* × *T. macedonicus* hybrids, have *T. ivanbureschi* mtDNA [25–27]. Asymmetrical mtDNA introgression is hypothesized to be a consequence of *T. macedonicus* range expansion over the range of *T. ivanbureschi* [25]. *Triturus ivanbureschi* and *T. macedonicus* are two moderately related, morphologically divergent species [22]. They diverge in body shape during postembryonic, larval development [40,41] as well as in the postmetamorphic period [42,43]. These species reproduce in similar aquatic habitats [44] but differ in reproductive traits, where *T. macedonicus* lay more eggs and have greater embryonic survival [43,45].

The mito-nuclear mismatch in *Triturus* populations in the central Balkans caused confusion in interpretations of species distribution and taxonomy (see [37] and references therein). In Serbia, many populations were considered to be *T. ivanbureschi* (syn. *T. karelinii*) according to mtDNA analysis, or *T. macedonicus* (syn. *T. carnifex*), according to morphology, were later confirmed as hybrid populations. Therefore, in previous studies, the reproductive data of some species are actually data for hybrid populations (Pirot [46,47]; Đurđevac [47]). These hybrid populations consist of *T. ivanbureschi* × *T. macedonicus* individuals of unknown generation ($F_n$) with *T. ivanbureschi* mtDNA [27]. Body and egg sizes of hybrid females [46,47] were similar to those reported for parental species [43,46,48].

The first generation of *T. ivanbureschi* × *T. macedonicus* hybrids has not yet been confirmed in nature [27,38,39]. Since processes occurring in the $F_1$ generation are of special interest for an explanation of forming and maintenance of hybrid zones [49], we set up experimental reciprocal crossings to obtain two types of hybrid genotypes: *T. ivanbureschi*-mothered hybrids (with *T. ivanbureschi* mtDNA, confirmed in natural populations) and *T. macedonicus*-mothered hybrids (with *T. macedonicus* mtDNA, not confirmed in natural populations). Our main goal was to explore the eventual differences in reproductive success of reciprocal hybrids carrying different parental mtDNA. Since growth and reproduction in newts are negatively interconnected due to resource allocation [50,51] and the period between the first and second reproductions is characterized by considerable growth [52], we followed life history traits in the first two consecutive years of reproduction. We recorded the life history traits directly related to reproductive output: body size (length and mass), the number of egg-laying females, duration of oviposition, the number and size of eggs and the number of hatchlings. As an estimation of reproductive success, we calculated the percentage of egg-laying females and the viability of their embryos. We also tested for eventual differences in reproductive success when hybrid females were bred with the hybrid males (with the same or reciprocal mtDNA) and with the males of parental species *T. ivanbureschi* and *T. macedonicus* (backcross combinations). Overall, we expected that hybrids with *T. macedonicus* mtDNA would have at least lower reproductive success, considering that *T. macedonicus* mtDNA is not present in natural populations. Failure in their reproductive success would indicate early exclusion of *T. macedonicus* mtDNA in the first generations of crossings upon *T. ivanbureschi* and *T. macedonicus* secondary contact.

## 2. Materials and Methods

Individuals of *T. ivanbureschi* and *T. macedonicus* were collected from natural populations away from their contact zones: *T. ivanbureschi* from Zli Dol, Serbia (42°25 N; 22°27 E) with permission obtained from the Serbian Ministry of Energy, Development and Environmental Protection (permit No. 353-01-75/2014-08) and *T. macedonicus* in Ceklin, Montenegro (42°21 N; 18°59 E) with permission obtained from the Agency for Environmental Protection, Montenegro (permit No. UPI-328/4). Genetic data confirm that these were *T. ivanbureschi* and *T. macedonicus* populations [27]. A series of crossbreeding experiments were carried out at the Institute for Biological Research "Siniša Stanković" and approved by the Ethics Committee of the Institute (decision Nos. 03-03/16 and 01-1949). By crossing parental species, two genotypes of hybrids (HI—with *T. ivanbureschi* mtDNA and HM—*T. macedonicus* mtDNA) were obtained in two different years:

- HI in 2016: *T. ivanbureschi* ♀ ($n = 8$) × *T. macedonicus* ♂ ($n = 5$);
- HM in 2017: *T. macedonicus* ♀ ($n = 4$) × *T. ivanbureschi* ♂ ($n = 4$).

Since growth and survival, as well as reproductive traits, can be affected by external factors (e.g., temperature, density, availability of food) and their mutual interactions (e.g., [53]), we kept the experimental settings and conditions the same throughout different years of the experiment. Crossing of parental species was done outdoors in 500 L plastic vats (separate vat for each breeding crossing). Plastic strips were provided for egg deposition. When females started laying eggs, they were transferred to the laboratory in separate 10 L aquariums half-filled with dechlorinated tap water. Eggs, embryos and larvae were raised in the same controlled experimental conditions in both years. The temperature was kept constant (18–19 °C) with a natural day/night regime. Eggs and embryos were raised in Petri dishes (up to 10 eggs/embryos per dish). Dechlorinated tap water was changed every other day. All Petri dishes were checked daily to remove non-developing eggs and arrested embryos. Larvae were raised in 2 L plastic containers (single larva per container). Containers were half-filled with dechlorinated tap water, which was changed every other day. Larvae were fed *ad libitum* with *Artemia* sp. and *Tubifex* sp. After metamorphosis, animals hibernated during winter in a cold chamber at a constant temperature (4 °C). During the first postmetamorphic year, between first and second hibernation, juveniles were kept in 200 L plastic vats placed outdoors with a similar number of animals per vat. Plastic vats were enriched with perforated bricks, providing shelter and plastic lids, which were used as floating platforms. Newts were fed *Tubifex* sp. and *Lumbricus* sp. twice a week.

When hybrids had developed secondary sexual characters during the second postmetamorphic year, they were engaged in breeding series of mutual crossings and backcrossings with parental species (Figure 1). Animals hibernated before each breeding. HI females mated in 2018 (first reproduction) and 2019 (second reproduction). HM females mated in 2019 (first reproduction) and 2020 (second reproduction). The numbers of females involved in the breeding series were 23 HI and 20 HM in the first reproductive year and 25 HI and 20 HM in the second year of reproduction. To exclude the possibility of spermatozoid retention [54], the same females were crossed with the same males in two consecutive years. The only exceptions are two HI females, which were included in the crossing with HM males. Each crossing (see Figure 1) was conducted in a separate 200 L plastic container. Containers were set in the backyard of the Institute in semi-natural conditions. After females started laying eggs, they were transferred to separate 10 L aquariums, half-filled with dechlorinated water. Eggs and embryos were raised in the same controlled experimental conditions (see above). For more details of animals' housing and experimental conditions, see [40,41,52].

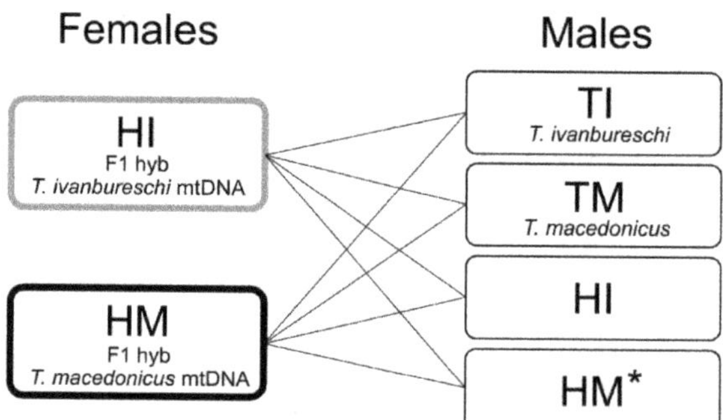



**Figure 1.** Schematic presentation of the experimental design. The $F_1$ hybrid females with *T. ivanbureschi* mtDNA (HI) were crossed and backcrossed in 2018 and 2019. The $F_1$ hybrid females with *T. macedonicus* mtDNA (HM) were crossed and backcrossed in 2019 and 2020. * Sexually mature HM males were available only in 2019, and therefore not involved in crossings with HI females in their first reproduction in 2018.

Data for this study were collected during two consecutive years, in 2018 and 2019 for HI and in 2019 and 2020 for HM. To obtain measures of body size, females were photographed and weighed each year right after hibernation. Imaging was performed using a camera (Nikon D7100 or Sony DSC-F828) on a fixed stand with millimeter paper as a background. Body length (snout to vent length—SVL) was measured as the distance between the tip of the snout and the level of the posterior edge of the hind legs from the dorsal view using ImageJ software v. 1.50i [55]. Body mass (BM) was weighted to the nearest 0.01g using an electronic scale (MP300, Chyo Balance Corporation, Kyoto, Japan).

The eggs of newts consist of vitellus and a protective jelly layer. Females deposit individual eggs and wrap them in submerged vegetation [56], thus providing transparent plastic strips for egg deposition enabled easy observation of newly laid eggs. To estimate the number of laid eggs and egg size, we collected eggs on a daily basis and photographed them with a Nikon Digital Sight Fi2 camera attached to a Nikon SMZ800 stereo zoom microscope immediately after removal from the plastic strips for measurements. The number of eggs within each genotype was calculated from the first laid egg in common containers to the last laid egg in separate aquaria. Egg dimensions (maximum width of vitellus; maximum length and width of jelly) were taken using ImageJ software. Based on these measures, we calculated the volume of the vitellus and the volume of the entire egg. We calculated vitellus volume as the volume of a sphere: $Vv = 4/3 \times r^3\pi$, where r is the vitellus width/2. For the volume of egg, we measured ellipsoid volume: $Ve = 4/3 \times r_1 r_2^2 \pi$, where $r_1$ and $r_2$ are the radii length/2 and width of egg/2, respectively. The volume of the jelly was calculated as a difference between the volume of egg and the volume of vitellus: $Vj = Ve - Vv$. For statistical analysis of egg size, we used 100 eggs per genotype per year, which were taken randomly (RAND function in Microsoft Excel).

We also observed changes in other reproductive traits between the two reproductions, such as the number of egg-laying females, duration of oviposition, the total number of laid eggs and the total number of hatched larvae. Duration of oviposition was estimated as the number of days between the first and last laid egg within each genotype. To estimate reproductive success, we calculated the percentage of egg-laying females and the viability of their embryos. The percentage of egg-laying females was calculated as the ratio between the number of egg-laying females and the total number of females involved in breeding per hybrid genotype multiplied by 100. The viability of embryos was calculated as the ratio between the numbers of hatched larvae and the deposited eggs per hybrid genotype multiplied by 100.

Hybrid females of both genotypes were subdivided into four breeding groups exposed to different males (see Figure 1). As the exposure of females to different males can affect reproductive success in amphibians (e.g., [57]), we compared the percentages of egg-laying females and the viability of embryos separately within HI and HM females.

We used separated repeated measures ANOVAs to test for differences in the average body size (SVL and BM) and egg size (vitellus and jelly volumes), with female genotype as the categorical factor and year of reproduction as the within-subject (repeated measure) factor on the measured variables. Tukey's HSD post hoc test was used to determine the statistical significance of between-group differences. Other reproductive traits (duration of oviposition, total number of eggs and total number of hatched larvae) were compared by the Kruskal–Wallis test since their values did not meet the criteria for ANOVA (assumptions of normality and sphericity). Differences in reproductive success were tested using differences between two proportions. Statistical analyses were done using Statistica10 software (StatSoft Inc., Tulsa, OK, USA, 2011).

## 3. Results

The repeated measures ANOVA showed that reproductive year (SVL: $F_{1,38} = 317.53$, $p < 0.0001$; BM: $F_{1,38} = 109.94$, $p < 0.0001$), genotype (SVL: $F_{1,38} = 15.34$, $p = 0.004$; BM: $F_{1,38} = 10.95$, $p = 0.002$) and their interaction (SVL: $F_{1,38} = 36.99$, $p < 0.0001$; BM: $F_{1,38} = 14.57$, $p = 0.0005$) had a significant effect on the increase of body size between two reproductive years. Post-hoc tests showed a significant ontogenetic increase in body size (SVL and BM) within both genotypes. Differences in body size between genotypes were evident only in the second year of reproduction (Figure 2, Table 1).

For the volume of vitellus, repeated measures ANOVA (Figure 2) showed a significant effect of reproductive year ($F_{1,198} = 62.10$, $p < 0.0001$) and year $\times$ genotype interaction ($F_{1,198} = 27.17$, $p < 0.0001$), but non-significant effect of genotype alone ($F_{1,198} = 0.08$, $p = 0.7714$). Volume of jelly was significantly affected by both factors, reproductive year ($F_{1,198} = 34.92$, $p < 0.0001$) and genotype ($F_{1,198} = 21.41$, $p < 0.0001$), as well as their interaction ($F_{1,198} = 7.28$, $p = 0.0076$). Post-hoc tests revealed that there was no ontogenetic difference in the volume of vitellus and jelly within HI, while HM eggs were significantly larger in the second reproductive year. HI had larger eggs (volume of vitellus and jelly) than HM in the first reproductive year. In the second reproductive year, HM had a larger volume of vitellus but a similar jelly volume to HI (Figure 2, Table 1). Kruskal–Wallis analyses showed that there were no differences in other reproductive traits (duration of oviposition, total number of laid eggs and total number of hatched larvae) between and within hybrid genotypes ($p > 0.05$ in all comparisons, Figure 2).

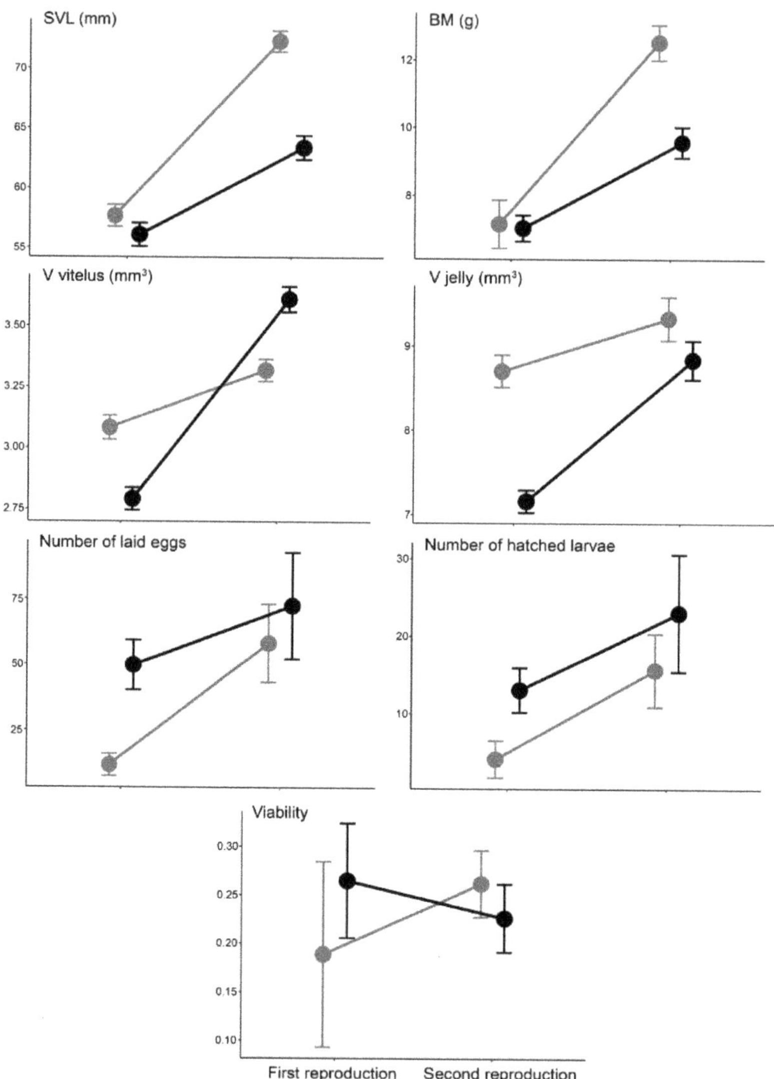


**Figure 2.** Ontogenetic differences in body size (length—SVL and mass—BM) and reproductive traits in the first two reproductive years of F1 hybrid females with *T. ivanbureschi* mtDNA, HI (gray) and with *T. macedonicus* mtDNA, HM (black). Values of each trait are represented as mean ± standard error per each genotype and year of reproduction.

**Table 1.** Differences in body size and egg volume between F1 hybrid females with *T. ivanbureschi* mtDNA (HI) and *T. macedonicus* mtDNA (HM) during the first two reproductive years. Analyzed traits: SVL—snout to vent length, BM—body mass, Vv—vitellus volume, Vj—jelly volume. Level of significances of pairwise comparisons: ns = not significant.

| Comparisons | Females | | Eggs | |
|---|---|---|---|---|
| | SVL | BM | Vv | Vj |
| within genotypes (I vs. II year) | | | | |
| HI | 0.0002 | 0.0002 | ns | ns |
| HM | 0.0002 | 0.0004 | <0.0001 | <0.0001 |
| between genotypes (HI vs. HM) | | | | |
| I year | ns | ns | 0.0011 | <0.0001 |
| II year | 0.0002 | 0.0002 | 0.0051 | ns |

As the percentage of egg-laying females and viability of embryos also did not differ between HI and HM females in both reproductive years ($p > 0.05$ in all comparisons, Figure 3), we tested for differences in reproductive success between types of crossings (male effect) within each hybrid genotype. The percentage of egg-laying females did not differ within HI in both reproductive years regardless of different types of crossings, i.e., exposure to males of various genotypes. Within the HM females, the crossing of HM ♀ × ♂ had a relatively low percentage of egg-laying females, which was significantly lower from backcrossing of HM females with males of both parental species (*T. ivanbureschi* and *T. macedonicus*) in the first year, but this difference was lost in the second reproductive year (Tables 2 and 3).

Offspring were obtained from all available breeding crossings in both reproductive years (see Figure 1). The viability of embryos obtained by different hybrid crossings and backcrossing was similar in the first reproductive year for both HI and HM females. For HI females, embryos obtained from the backcrossing of HI ♀ × *T. macedonicus* ♂ had greater viability than from the backcrossing of HI ♀ × *T. ivanbureschi* ♂. Embryos obtained from the reciprocal hybrid crossing (HI ♀ × HM ♂) had greater viability than those obtained from all other crossings of HI females and different male genotypes (HI, *T. ivanbureschi* and *T. macedonicus*) (Tables 2 and 3). For HM females, embryos obtained from the crossing of HM ♀ × ♂ in the second reproductive year had the largest viability, significantly different from the crossing of reciprocal hybrids (HM ♀ × HI ♂), as well as from the backcrossing of HM females with *T. ivanbureschi* males (Tables 2 and 3).

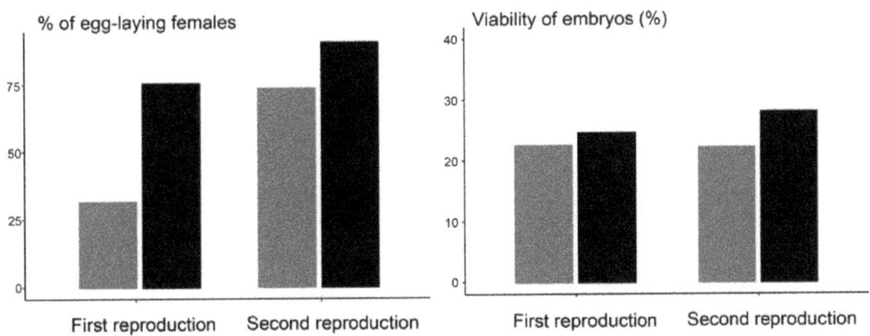


**Figure 3.** Differences in the percentage of egg-laying females and viability of embryos, as a measure of reproductive success during the first two reproductive years of $F_1$ hybrid females with *T. ivanbureschi* mtDNA, HI (gray) and with *T. macedonicus* mtDNA, HM (black).

**Table 2.** Reproductive success (percentages of egg-laying females and viability of embryos) of $F_1$ hybrid females with *T. ivanburechi* mtDNA (HI) and *T. macedonicus* mtDNA (HM) exposed to different males (see Figure 1 for various breeding crossings and abbreviations). Sexually mature HM males were not available in 2018 and therefore not involved in crossings with HI females in their first reproductive year.

| Breeding Crossings (♀ × ♂) | Egg-Laying Females (%) | | Viability (%) | |
|---|---|---|---|---|
| | I Year | II Year | I Year | II Year |
| HI × HI | 22 | 43 | 26 | 20 |
| HI × HM | / | 50 | / | 35 |
| HI × TI | 14 | 100 | 8 | 18 |
| HI × TM | 57 | 100 | 33 | 25 |
| HM × HM | 20 | 100 | 19 | 35 |
| HM × HI | 80 | 100 | 20 | 23 |
| HM × TI | 100 | 100 | 26 | 26 |
| HM × TM | 100 | 60 | 23 | 27 |

**Table 3.** Pairwise comparison of reproductive success (percentages of egg-laying females and viability of embryos) of $F_1$ hybrid females with *T. ivanbureschi* mtDNA (HI) and *T. macedonicus* mtDNA (HM) exposed to different males (see Figure 1 for various breeding crossings and abbreviations). The comparisons were done within HI and HM females separately. Sexually mature HM males were not available in 2018 and therefore not involved in crossings with HI females in their first reproductive year. Level of significances of pairwise comparisons: ns = not significant.

| Compared Breeding Crossings (♀ × ♂) | Egg-Laying Females (%) | | Viability (%) | |
|---|---|---|---|---|
| | I Year | II Year | I Year | II Year |
| HI × HI vs. HI × TI | ns | ns | ns | ns |
| HI × HI vs. HI × TM | ns | ns | ns | ns |
| HI × TI vs. HI × TM | ns | ns | ns | 0.0310 |
| HI × HI vs. HI × HM | / | ns | / | 0.0002 |
| HI × TI vs. HI × HM | / | ns | / | <0.0001 |
| HI × TM vs. HI × HM | / | ns | / | 0.0140 |
| HM × HM vs. HM × TI | 0.0320 | ns | ns | <0.0001 |
| HM × HM vs. HM × TM | 0.0320 | ns | ns | ns |
| HM × TI vs. HM × TM | ns | ns | ns | ns |
| HM × HM vs. HM × HI | ns | ns | ns | 0.0160 |
| HM × TI vs. HM × HM | ns | ns | ns | ns |
| HI × HI vs. HM × HM | ns | ns | ns | ns |

## 4. Discussion

In natural populations, overall reproductive output and success of $F_1$ hybrid generation, particularly divergence of reciprocal hybrids, have an important impact on the long-term consequence of hybridization, as they can alter the direction of gene flow [58,59]. Therefore, the mito-nuclear incompatibility could shape patterns of genetic variation in hybrid zones and impact their dynamics. *Triturus* newts are a suitable model system for studies of mechanisms that lead to asymmetrical direction of mtDNA and introgression. Hybridization of the phylogenetically distant *T. cristatus* and *T. marmoratus* resulted in mtDNA introgression asymmetry due to divergence in the reproductive success of reciprocal crossings. More than 90% of individuals in natural populations have *T. cristatus* mtDNA [60], while $F_2$ and further hybrid generations are rare [35,60]. Hybridization of *T. ivanbureschi* and *T. macedonicus*, two moderately related species, also resulted in an asymmetrical mtDNA introgression. *Triturus ivanbureschi* mtDNA is found across populations of *T. ivanbureschi*, *T. macedonicus* and *T. ivanbureschi* × *T. macedonicus*, in the Balkan Peninsula with the loss of *T. macedonicus* mtDNA in central Serbia [25–27].

Studies of *T. ivanbureschi* × *T. macedonicus* hybridization showed that hybrids have intermediate morphology compared to parental genotypes [39–41]. Hybridization affected physiological response resulting in raised oxidative stress parameters in hybrid larvae, which might have a negative impact on their survival [61–63]. However, fitness consequences of hybridization comprising slower growth rate, sex ratio disturbance and lower survival were not evident in $F_1$ hybrids compared to maternal species during the early postmetamorphic period [52]. The morphological divergences between reciprocal hybrids were not recorded in previous morphological studies of larvae and recently metamorphosed juveniles. They have similar developmental patterns and growth rates [40,41].

We found that, at the beginning of the second postmetamorphic year, both $F_1$ hybrids showed secondary sexual characters but did not diverge in body size. The divergence in size between reciprocal hybrids was recorded after the first reproduction, which coincides with the timing of sexual size dimorphism and divergence in size between hybrids and maternal species [52].

The other significant difference between reciprocal hybrids is related to egg size, which can have a profound impact on embryonic development and survival, as well as on larval growth and developmental rates [64]. Hybrids with *T. ivanbureschi* mtDNA laid larger eggs in the first reproduction (no differences between reciprocal hybrids in female body size), while eggs of hybrids with *T. macedonicus* mtDNA had substantially larger vitellus in the second reproduction, despite significantly smaller body size compared to hybrid females with *T. ivanbureschi* mtDNA. Additionally, reciprocal hybrids differed in the pattern of change in egg size between the first and second reproductions, where hybrids with *T. macedonicus* mtDNA showed a greater increase in vitellus size (see Figure 2). It is considered that the size of an amphibian egg is positively correlated with female body size. Thus, it is expected that older and larger females produce larger eggs, i.e., that ontogenetic changes of female body size should affect egg size [46,65,66]. Our results do not support previous notations. We suggest three mutually non-exclusive hypotheses that can be applied to explain the observed pattern of ontogenetic changes and divergences in vitellus size between reciprocal hybrids. First, the observed divergences may indicate that hybrids with *T. ivanbureschi* mtDNA invest more in somatic growth than in reproduction, while the opposite trend could be true for hybrids with *T. macedonicus* mtDNA, which could invest more in reproduction by enlarging vitellus volume. Second, our results could indicate that egg size is a heritable trait that is inherited by maternal investments through egg cytoplasm. In newts, egg size is a species-specific trait [47]. In a comparison of parental species, *T. macedonicus* has larger eggs than *T. ivanbureschi* [43]. The cytoplasm of the oocyte is full of maternal cytoplasmic components subsequently present in each cell of the new embryo [67], having an important role in the determination of egg size and possibly leading to significant divergences in egg size between two reciprocal hybrids. The third possible explanation for the divergence in egg size and female growth could be different sensitivity to external conditions of reciprocal hybrids. Previously, it was proposed that *T. macedonicus* is a thermophilic species [43,68]. Vitellogenesis occurs way before the breeding season, usually before hibernation [64], and it is dependent on different external and internal factors (e.g., [56]). It could be possible that external conditions were less suitable for $F_1$ hybrids with *T. macedonicus* mtDNA at the time of their preparation for the first reproduction and vitellogenesis, i.e., hybrids with *T. macedonicus* mtDNA could be more susceptible to environmental variation during these periods. Studies on the plasticity of egg size and early life-history traits under different environmental conditions could give insight into the sensitivity of different genotypes to environmental conditions, with emphasis on different mtDNA groups. Hybrid females with both *T. ivanbureschi* and *T. macedonicus* mtDNA deposited a similar number of eggs in two reproductive years during a similar oviposition period. The number of eggs that hybrid females laid was in the range reported for both parental species in the same experimental settings [43], as well as for females from *T. ivanburechi* × *T. macedonicus* natural population [46], suggesting similar reproductive potential to that of the parental species. The observed pattern is opposite to *T. cristatus* and

*T. marmoratus* hybridization, where hybrids have higher reproductive potential compared to parental species in experimental conditions, but low viability of eggs [69].

In our study, hybrid females were bred with four male genotypes (see Figure 1). In some amphibian taxa, females tend to choose males with whom they will engage in mating [56,57,70,71], which could affect overall reproductive success by decreasing or increasing the percentage of females that are involved in active breeding. Moreover, in hybridizing species, females' choice could be a prezygotic barrier causing unidirectional mtDNA exchange if females of one species mate with males of the other and not vice versa [71]. Considering hybrid mutual mating and mating with parental species, it has been proposed that intermediate phenotypes can be disadvantageous for hybrid males, as they could be less attractive to females [7]. This could be true for newts, as it was shown that females do prefer larger males with a prominent crest [72]. In our experimental settings, the absence of differences between reciprocal hybrids in the percentage of egg-laying females, i.e., females engaged in active reproduction, suggests that females did not refuse to breed with either male genotype and their pickiness did not represent a barrier for either mtDNA to be passed on to the next generation. In comparisons within each female genotype, slight differences in percentages of egg-laying females in the first reproduction were not recorded in the second year of reproduction. Possibly, in their first reproduction, females preferred older (i.e., larger) males with prominent crests [72], as their conspecific males were of the same age or hybrid males were less successful in their first year of reproduction.

One characteristic of *Triturus* newts is that half of embryos die during development at the tail bud stage due to balanced lethal syndrome [73]. Viable eggs that had developed to hatchlings were obtained from all types of crossings (Figure 1) with no difference in embryo mortality between females with *T. ivanbureschi* and *T. macedonicus* mtDNA. However, viability differed in the second reproduction within reciprocal hybrid females when a different type of crossing (i.e., different male genotype) was included. Differences in the viability of embryos from different crossings and backcrossings indicate a possible decrease in survival of the $F_2$ generation. For example, the backcrossing of hybrid females with *T. ivanbureschi* mtDNA and *T. macedonicus* was more successful than the one including *T. ivanbureschi* males. The obtained result could back up the hypothesis of postglacial species displacement and *T. macedonicus* nuclear DNA spreading [25]. This hypothesis suggests that, throughout generations, subsequent backcrossing of hybrid females with *T. ivanbuerschi* mtDNA with males of *T. macedonicus* diminished *T. ivanbureschi* nuclear DNA, rebuilt *T. macedonicus* nuclear DNA and retained *T. ivanbureschi* mtDNA [25]. The difference in the viability of $F_2$ generations should be confirmed and further investigated in a larger study that would include postembryonic developmental stages, as it was shown that asymmetry in the frequency of reciprocal hybrids emerged after embryonic development in *T. cristatus* × *T. marmoratus* hybridization [59].

## 5. Conclusions

We found that the reproductive traits and success of reciprocal *T. ivanbureschi* × *T. macedonicus* $F_1$ hybrids (with *T. ivanbureschi* or *T. macedonicus* mtDNA) is largely similar, expressing some differences depending on trait, year of reproduction and male genotype involved in mating. Overall, the obtained results suggest that the loss of *T. macedonicus* mtDNA emerged in $F_2$ or subsequent hybrid generations. Major cito-nuclear incompatibilities are often masked in the first generation and manifested in $F_2$ or subsequent hybrid generations due to their accumulation, the phenomenon known as hybrid breakdown (e.g., [10,16]). The viability of $F_2$ embryos, obtained from mutual hybrid crossings and backcrossing with both parental species, points out a possible decrease of fitness in the $F_2$ generation. However, these results are only preliminary and should be tested further.

Females were monitored in the first two years of reproduction, so possibly results considering success in crossings with different male genotypes could change with the age of females. The postembryonic viability of $F_2$ generation should be analyzed, as possible differences in survival can occur during larval development, metamorphosis or even during the postmetamorphic period. An experiment, which would include females and males of various ages of *T. ivanbureschi*, *T. macedonicus* and hybrids with both types of mtDNA, would enable females to choose between different male genotypes. The suggested experimental design could give another insight into processes that took place a long time ago, upon the initial secondary contact between two moderately related species of *T. ivanbureschi* and *T. macedonicus*.

**Author Contributions:** Conceptualization, T.V., A.I. and M.C.; methodology, T.V., A.I. and M.C.; software, M.C.; validation, T.V.; formal analysis, M.C. and T.V.; investigation, T.V., M.A., N.B. and M.C.; resources, T.V., A.I., M.A. and M.C.; data curation, T.V. and M.C.; writing—original draft preparation, T.V.; writing—review and editing, T.V., A.I., M.A., N.B. and M.C.; visualization, T.V. and A.I.; supervision, A.I. All authors have read and agreed to the published version of the manuscript.

**Funding:** This research was funded by the Serbian Ministry of Education, Science and Technological Development, grant numbers 451-03-9/2021-14/ 200007 and 451-03-9/2021-14/ 200178. The APC was funded by the Institute for Biological Research "Siniša Stanković", National Institute of the Republic of Serbia.

**Institutional Review Board Statement:** The animal study protocol was approved by the Ethics Committee of the Institute for Biological Research "Siniša Stanković", National Institute of the Republic of Serbia decision Nos. 03-03/16 (date of approval: 1 April 2016) and 01-1949 (date of approval: 1 October 2018).

**Data Availability Statement:** Data is available upon request from the corresponding author.

**Acknowledgments:** We would like to thank S. Nikolić and J. Jovanović for their help in 2018. We thank M. Lakušić for her assistance in 2019. We also thank many undergraduate students of University of Belgrade—Faculty of biology for their technical assistance during the experiments. We would like to thank three anonymous reviewers for their constructive comments and suggestions and Enrico Lunghi for his editorial advices.

**Conflicts of Interest:** The authors declare no conflict of interest. The funders had no role in the design of the study; in the collection, analyses, or interpretation of data; in the writing of the manuscript, or in the decision to publish the results.

## References


1. Barton, N.H.; Hewitt, G.M. Analysis of hybrid zones. *Annu. Rev. Ecol. Syst.* **1985**, *16*, 113–148. [CrossRef]
2. Abbott, R.; Albach, D.; Ansell, S.; Arntzen, J.W.; Baird, S.J.; Bierne, N.; Boughman, J.; Brelsford, A.; Buerkle, C.A.; Buggs, R.; et al. Hybridization and speciation. *J. Evol. Biol.* **2013**, *26*, 229–246. [CrossRef]
3. Arnold, M.L.; Hodges, S.A. Are natural hybrids fit or unfit relative to their parents? *Trends Ecol. Evol.* **1995**, *10*, 67–71. [CrossRef]
4. Dowling, T.E.; Secor, C.L. The role of hybridization and introgression in the diversification of animals. *Annu. Rev. Ecol. Syst.* **1997**, *28*, 593–619. [CrossRef]
5. Barton, N.H. The role of hybridization in evolution. *Mol. Ecol.* **2001**, *10*, 551–568. [CrossRef] [PubMed]
6. Orr, H.A.; Turelli, M. The evolution of postzygotic isolation: Accumulating Dobzhansky-Muller incompatibilities. *Evolution* **2001**, *55*, 1085–1094. [CrossRef] [PubMed]
7. Coyne, J.A.; Orr, H.A. *Speciation*; Sinauer: Sunderland, MA, USA, 2004.
8. Seehausen, O. Hybridization and adaptive radiation. *Trends Ecol. Evol.* **2004**, *19*, 198–207. [CrossRef]
9. Endler, J.A. *Geographic Variation, Speciation, and Clines*; Princeton University Press: Princeton, NJ, USA, 1977.
10. Burton, R.S. Hybrid breakdown in physiological response: A mechanistic approach. *Evolution* **1990**, *44*, 1806–1813. [CrossRef]
11. Orr, H.A. The population genetics of speciation—The evolution of hybrid incompatibilities. *Genetics* **1995**, *139*, 1805–1813. [CrossRef]
12. Burke, J.M.; Arnold, M.L. Genetics and the fitness of hybrids. *Annu. Rev. Gen.* **2001**, *35*, 31–52. [CrossRef]
13. Burton, R.S.; Ellison, C.K.; Harrison, J.S. The sorry state of F₂ hybrids: Consequences of rapid mitochondrial DNA evolution in allopatric populations. *Am. Nat.* **2006**, *168*, S14–S24. [CrossRef]
14. Stelkens, R.; Seehausen, O. Genetic distance between species predicts novel trait expression in their hybrids. *Evolution.* **2009**, *63*, 884–897. [CrossRef]
15. Arnold, M.L.; Martin, N.H. Hybrid fitness across time and habitats. *Trends Ecol. Evol.* **2010**, *25*, 530–536. [CrossRef] [PubMed]
16. Stelkens, R.B.; Schmid, C.; Seehausen, O. Hybrid breakdown in cichlid fish. *PLoS ONE* **2015**, *10*, e0127207. [CrossRef] [PubMed]
17. Rometsch, S.J.; Torres-Dowdall, J.; Meyer, A. Evolutionary dynamics of pre- and postzygotic reproductive isolation in cichlid fishes. *Philos. T. Roy. Soc. B.* **2020**, *375*, 20190535. [CrossRef] [PubMed]

18. Burton, R.S.; Barreto, F.S. A disproportionate role for mtDNA in Dobzhansky–Muller incompatibilities? *Mol. Ecol.* **2012**, *21*, 4942–4957. [CrossRef] [PubMed]

19. Healy, T.M.; Burton, R.S. Strong selective effects of mitochondrial DNA on the nuclear genome. *Proc. Natl. Acad. Sci. USA* **2020**, *117*, 6616–6621. [CrossRef] [PubMed]

20. Coughlan, J.M.; Matute, D.R. The importance of intrinsic postzygotic barriers throughout the speciation process: Intrinsic barriers throughout speciation. *Philos. T. Roy. Soc. B.* **2020**, *375*, 20190533. [CrossRef]

21. Steinfartz, S.; Vicario, S.; Arntzen, J.W.; Caccone, A. A Bayesian approach on molecules and behavior: Reconsidering phylogenetic and evolutionary patterns of the Salamandridae with emphasis on *Triturus* newts. *J. Exp. Zool. Part B.* **2007**, *308*, 139–162. [CrossRef] [PubMed]

22. Wielstra, B.; McCartney–Melstad, E.; Arntzen, J.W.; Butlin, R.K.; Shaffer, H.B. Phylogenomics of the adaptive radiation of *Triturus* newts supports gradual ecological niche expansion towards an incrementally aquatic lifestyle. *Mol. Phylogenet. Evol.* **2019**, *133*, 120–127. [CrossRef] [PubMed]

23. Arntzen, J.W.; Wielstra, B.; Wallis, G.P. The modality of nine *Triturus* newt hybrid zones assessed with nuclear, mitochondrial and morphological data. *Biol. J. Linn. Soc.* **2014**, *113*, 604–622. [CrossRef]

24. Arntzen, J.W.; Wallis, G.P. Restricted gene flow in a moving hybrid zone of the newts *Triturus cristatus* and *T. marmoratus* in western France. *Evolution.* **1991**, *45*, 805–826. [CrossRef]

25. Wielstra, B.; Arntzen, J.W. Postglacial species displacement in *Triturus* newts deduced from asymmetrically introgressed mitochondrial DNA and ecological niche models. *BMC Evol. Biol.* **2012**, *12*, 161. [CrossRef] [PubMed]

26. Wielstra, B.; Arntzen, J.W. Extensive cytonuclear discordance in a crested newt from the Balkan Peninsula glacial refugium. *Biol. J. Linn. Soc.* **2020**, *130*, 578–585. [CrossRef]

27. Wielstra, B.; Burke, T.; Butlin, R.K.; Arntzen, J.W. A signature of dynamic biogeography: Enclaves indicate past species replacement. *Proc. R. Soc. B Biol. Sci.* **2017**, *284*, 20172014. [CrossRef]

28. Wielstra, B.; Burke, T.; Butlin, R.K.; Avcı, A.; Üzüm, N.; Bozkurt, E.; Kurtuluş, O.; Arntzen, J.W. A genomic footprint of hybrid zone movement in crested newts. *Evol. Lett.* **2017**, *1*, 93–101. [CrossRef] [PubMed]

29. Wielstra, B. Historical hybrid zone movement: More pervasive than appreciated. *J. Biogeogr.* **2019**, *46*, 1300–1305. [CrossRef]

30. Arntzen, J.W.; López-Delgado, J.; van Riemsdijk, I.; Wielstra, B. A genomic footprint of a moving hybrid zone in marbled newts. *J. Zool. Syst. Evol. Res.* **2021**, *59*, 459–465. [CrossRef]

31. López-Delgado, J.; van Riemsdijk, I.; Arntzen, J.W. Tracing species replacement in Iberian marbled newts. *Ecol. Evol.* **2021**, *11*, 402–414. [CrossRef]

32. Brede, E.G.; Thorpe, R.S.; Arntzen, J.W.; Langton, T.E.S. A morphometric study of a hybrid newt population (*Triturus cristatus/T. carnifex*): *Beam Brook Nurseries, Surrey, UK. Biol. J. Linn. Soc.* **2000**, *70*, 685–695. [CrossRef]

33. Meilink, W.R.; Arntzen, J.W.; van Delft, J.J.; Wielstra, B. Genetic pollution of a threatened native crested newt species through hybridization with an invasive congener in the Netherlands. *Biol. Conserv.* **2015**, *184*, 145–153. [CrossRef]

34. Wielstra, B.; Burke, T.; Butlin, R.K.; Schaap, O.; Shaffer, H.B.; Vrieling, K.; Arntzen, J.W. Efficient screening for 'genetic pollution' in an anthropogenic crested newt hybrid zone. *Conserv. Genet. Resour.* **2016**, *8*, 553–560. [CrossRef]

35. Arntzen, J.W.; Jehle, R.; Wielstra, B. Genetic and morphological data demonstrate hybridization and backcrossing in a pair of salamanders at the far end of the speciation continuum. *Evol. Appl.* **2021**, *14*, 2784–2793. [CrossRef] [PubMed]

36. Cogălniceanu, D.; Stănescu, F.; Arntzen, J.W. Testing the hybrid superiority hypothesis in crested and marbled newts. *J. Zool. Syst. Evol. Res.* **2020**, *58*, 275–283. [CrossRef]

37. Vučić, T.; Tomović, L.; Ivanović, A. The distribution of crested newts in Serbia: An overview and update. *Bull. Nat. Hist. Mus.* **2020**, *13*, 237–252. [CrossRef]

38. Wielstra, B.; Arntzen, J.W. Kicking *Triturus arntzeni* when it's down: Large–scale nuclear genetic data confirm that newts from the type locality are genetically admixed. *Zootaxa* **2014**, *3802*, 381–388. [CrossRef]

39. Arntzen, J.W.; Üzüm, N.; Ajduković, M.D.; Ivanović, A.; Wielstra, B. Absence of heterosis in hybrid crested newts. *PeerJ.* **2018**, *6*, e5317. [CrossRef]

40. Vučić, T.; Vukov, T.D.; Tomašević Kolarov, N.; Cvijanović, M.; Ivanović, A. The study of larval tail morphology reveals differentiation between two *Triturus* species and their hybrids. *Amphib–Reptilia* **2018**, *39*, 87–97. [CrossRef]

41. Vučić, T.; Sibinović, M.; Vukov, T.D.; Tomašević Kolarov, N.; Cvijanović, M.; Ivanović, A. Testing the evolutionary constraints of metamorphosis: The ontogeny of head shape in *Triturus* newts. *Evolution.* **2019**, *73*, 1253–1264. [CrossRef]

42. Cvijanović, M.; Ivanović, A.; Kalezić, M.L.; Zelditch, M.L. The ontogenetic origins of skull shape disparity in the *Triturus cristatus* group. *Evol. Dev.* **2014**, *16*, 306–317. [CrossRef]

43. Vučić, T.; Ivanović, A.; Nikolić, S.; Jovanović, J.; Cvijanović, M. Reproductive characteristics of two *Triturus* species (Amphibia: Caudata). *Arch. Biol. Sci.* **2020**, *72*, 321–328. [CrossRef]

44. Džukić, G.; Vukov, T.D.; Kalezić, M.L. *The Tailed Amphibians of Serbia*; Serbian Academy of Science and Arts: Belgrade, Serbia, 2016.

45. Cvijanović, M.; Ivanović, A.; Tomašević Kolarov, N.; Džukić, G.; Kalezić, M.L. Early ontogeny shows the same interspecific variation as natural history parameters in the crested newt (*Triturus cristatus* superspecies)(Caudata, Salamandridae). *Contrib. Zool.* **2009**, *78*, 43–50. [CrossRef]

46. Furtula, M.; Ivanović, A.; Džukić, G.; Kalezić, M.L. Egg size variation in crested newts from the western Balkans (Caudata, Salamandridae, *Triturus cristatus* superspecies). *Zool. Stud.* **2008**, *47*, 585–590.

47. Furtula, M.; Todorović, B.; Simić, V.; Ivanović, A. Interspecific differences in early life-history traits in crested newts (*Triturus cristatus* superspecies, Caudata, Salamandridae) from the Balkan Peninsula. *J. Nat. Hist.* **2009**, *43*, 469–477. [CrossRef]
48. Lukanov, S.; Tzankov, N. Life history, age and normal development of the Balkan-Anatolian crested newt (*Triturus ivanbureschi* Arntzen and Wielstra, 2013) from Sofia district. *North-West. J. Zool.* **2016**, *12*, 22–32.
49. Nolte, A.W.; Tautz, D. Understanding the onset of hybrid speciation. *Trends Genet.* **2010**, *26*, 54–58. [CrossRef]
50. Halliday, T.R.; Verrell, P.A. Body size and age in amphibians and reptiles. *J. Herpetol.* **1988**, *22*, 253–265. [CrossRef]
51. Arntzen, J.W. A growth curve for the newt *Triturus cristatus*. *J. Herpetol.* **2000**, *34*, 227–232. [CrossRef]
52. Bugarčić, M.; Ivanović, A.; Cvijanović, M.; Vučić, T. Hybridization and early postmetamorphic growth in *Triturus macedonicus*. *Amphib–Reptilia* **2022**. [CrossRef]
53. Semlitsch, R.D. Critical elements for biologically based recovery plans of aquatic-breeding amphibians. *Conserv. Biol.* **2002**, *16*, 619–629. [CrossRef]
54. Eddy, S.L.; Vaccaro, E.A.; Baggett, C.L.; Kiemnec-Tyburczy, K.M.; Houck, L.D. Sperm mass longevity and sperm storage in the female reproductive tract of *Plethodon shermani* (Amphibia: Plethodontidae). *Herpetologica.* **2015**, *71*, 177–183. [CrossRef]
55. Schneider, C.A.; Rasband, W.S.; Eliceiri, K.W. NIH Image to ImageJ: 25 years of image analysis. *Nat. Methods.* **2012**, *9*, 671–675. [CrossRef]
56. Duellman, W.E.; Trueb, L. *Biology of Amphibians*; JHU Press: Baltimore, MD, USA, 1994.
57. Reyer, H.U.; Frei, G.; Som, C. Cryptic female choice: Frogs reduce clutch size when amplexed by undesired males. *Proc. R. Soc. B.* **1999**, *266*, 2101–2107. [CrossRef] [PubMed]
58. Arnold, M.L.; Ballerini, E.S.; Brothers, A.N. Hybrid fitness, adaptation and evolutionary diversification: Lessons learned from Louisiana irises. *Heredity.* **2012**, *108*, 159–166. [CrossRef] [PubMed]
59. Dong, C.M.; Rankin, K.J.; McLean, C.A.; Stuart-Fox, D. Maternal reproductive output and $F_1$ hybrid fitness may influence contact zone dynamics. *J. Evol. Biol.* **2021**, *34*, 680–694. [CrossRef]
60. Arntzen, J.W.; Jehle, R.; Bardakci, F.; Burke, T.; Wallis, G.P. Asymmetric viability of reciprocal-cross hybrids between crested and marbled newts (*Triturus cristatus* and *T. marmoratus*). *Evolution* **2009**, *63*, 1191–1202. [CrossRef]
61. Prokić, M.D.; Despotović, S.G.; Vučić, T.Z.; Petrović, T.G.; Gavrić, J.P.; Gavrilović, B.R.; Radovanović, T.B.; Saičić, Z.S. Oxidative cost of interspecific hybridization: A case study of two *Triturus* species and their hybrids. *J. Exp. Biol.* **2018**, *221*, jeb182055. [CrossRef]
62. Petrović, T.G.; Vučić, T.Z.; Nikolić, S.Z.; Gavrić, J.P.; Despotović, S.G.; Gavrilović, B.R.; Radovanović, T.B.; Faggio, C.; Prokić, M.D. The effect of shelter on oxidative stress and aggressive behavior in crested newt larvae (*Triturus* spp.). *Animals* **2020**, *10*, 603. [CrossRef]
63. Prokić, M.D.; Petrović, T.G.; Despotović, S.G.; Vučić, T.; Gavrić, J.P.; Radovanović, T.B.; Gavrilović, B.R. The effect of short-term fasting on the oxidative status of larvae of crested newt species and their hybrids. *Comp. Biochem. Physiol. Part A Mol. Integr. Physiol.* **2021**, *251*, 110819. [CrossRef]
64. Kaplan, R.H.; Cooper, W.S. The evolution of developmental plasticity in reproductive characteristics: An application of the" adaptive coin-flipping" principle. *Am. Nat.* **1984**, *123*, 393–410. [CrossRef]
65. Kaplan, R.H.; Salthe, S.N. The allometry of reproduction: An empirical view in salamanders. *Am. Nat.* **1979**, *113*, 671–689. [CrossRef]
66. Morrison, C.; Hero, J.M. Geographic variation in life-history characteristics of amphibians: A review. *J. Anim. Ecol.* **2003**, *72*, 270–279. [CrossRef]
67. Maurel, M.C.; Kanellopoulos-Langevin, C. Heredity—Venturing beyond genetics. *Biol. Reprod.* **2008**, *79*, 2–8. [CrossRef]
68. Litvinchuk, S.N.; Rosanov, J.M.; Borkin, L.J. Correlations of geographic distribution and temperature of embryonic development with the nuclear DNA content in the Salamandridae (Urodela, Amphibia). *Genome* **2007**, *50*, 333–342. [CrossRef]
69. Arntzen, J.W.; Hedlund, L. Fecundity of the newts *Triturus cristatus*, *T. marmoratus* and their natural hybrids in relation to species coexistence. *Ecography.* **1990**, *13*, 325–332. [CrossRef]
70. Wilbur, H.M.; Rubenstein, D.I.; Fairchild, L. Sexual selection in toads: The roles of female choice and male body size. *Evolution* **1978**, *32*, 264–270. [CrossRef]
71. Wirtz, P. Mother species–father species: Unidirectional hybridization in animals with female choice. *Anim. Behav.* **1999**, *58*, 1–12. [CrossRef]
72. Halliday, T.R. The courtship of European newts: An evolutionary perspective. In *The Reproductive Biology of Amphibians*; Taylor, D.H., Guttman, S.I., Eds.; Plenum Press: New York, NY, USA, 1977; pp. 185–232.
73. Wielstra, B. Balanced lethal systems. *Curr. Biol.* **2020**, *30*, R742–R743. [CrossRef]

MDPI
St. Alban-Anlage 66
4052 Basel
Switzerland
www.mdpi.com

*Animals* Editorial Office
E-mail: animals@mdpi.com
www.mdpi.com/journal/animals